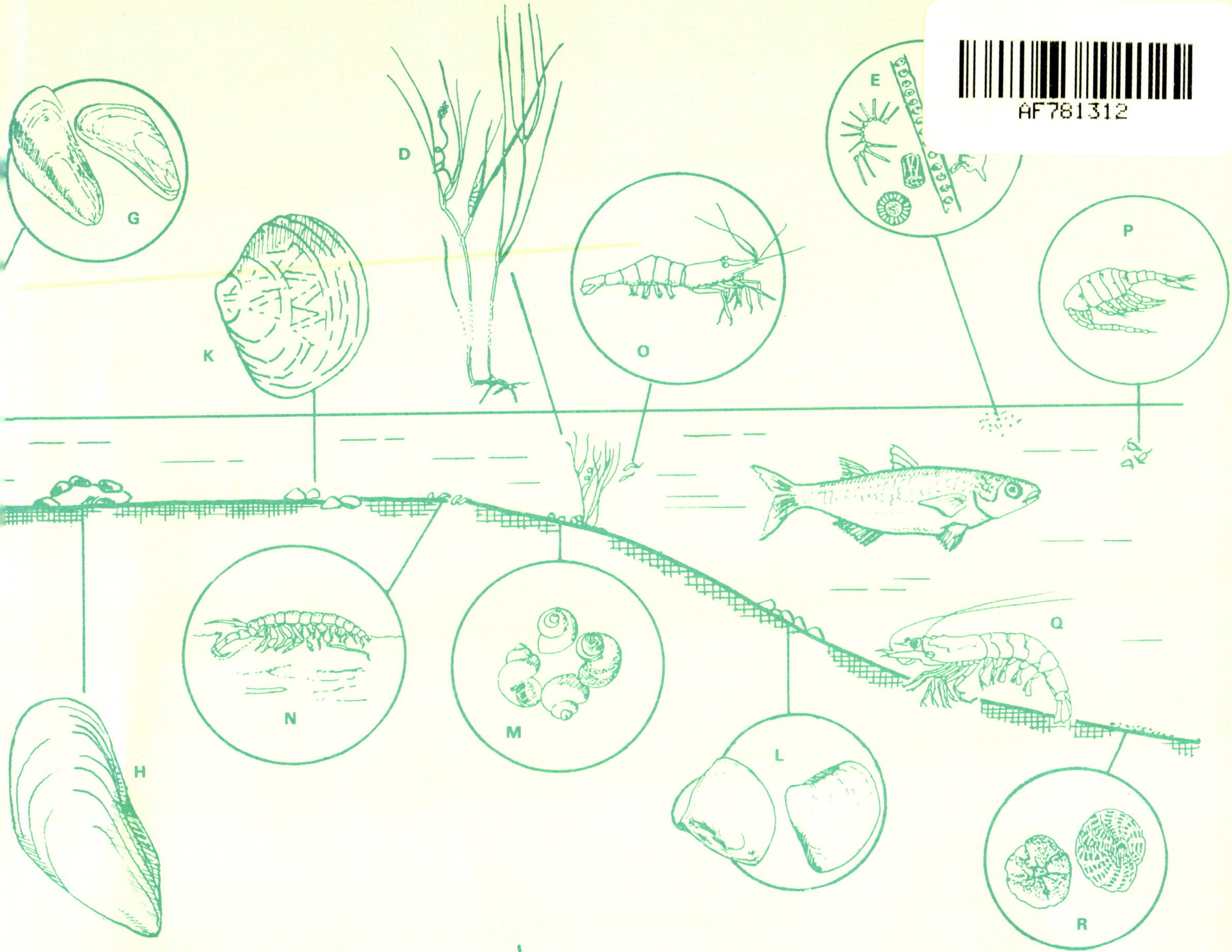

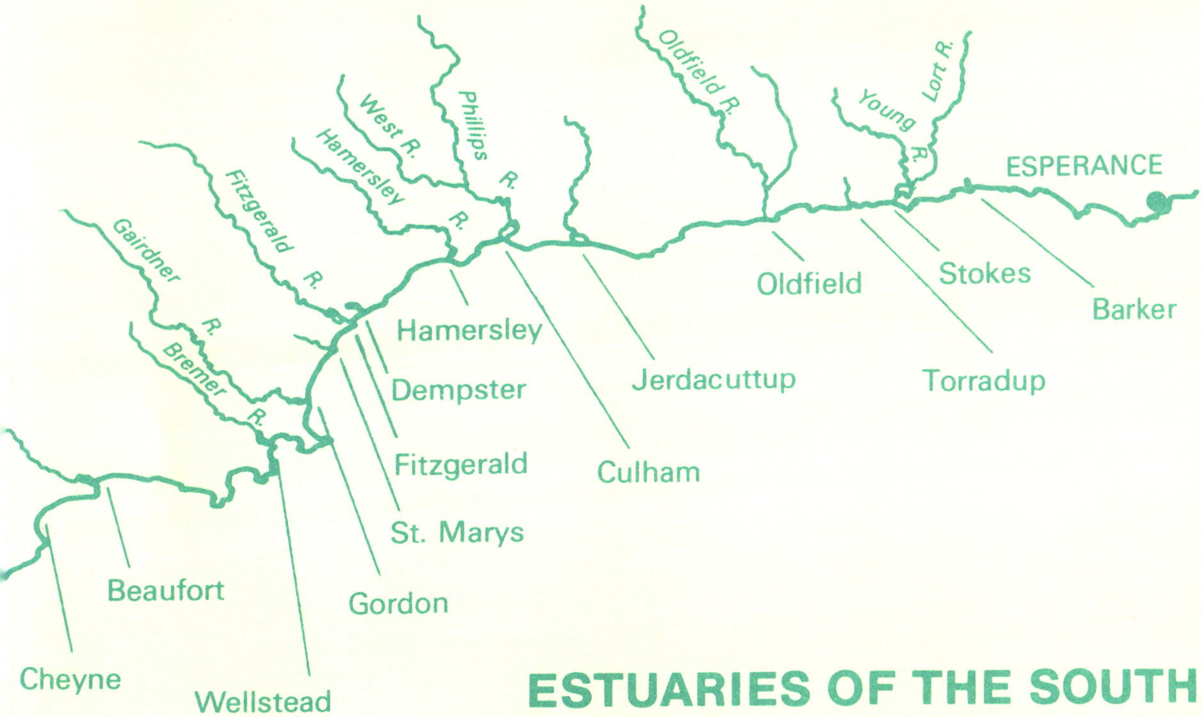

ESTUARIES OF THE SOUTH COAST OF WESTERN AUSTRALIA

Illustration from the *Estuarine Study Series*, the Blue Books, Hodgkin and Clark 1987–1999.

0 100

Kilometres

Ernest Hodgkin's

SWANLAND

Ernest Hodgkin's

SWANLAND

Estuaries and Coastal Lagoons of South-western Australia

Anne Brearley

Anne Brearley

University of Western Australia Press

First published in 2005 by
University of Western Australia Press
Crawley, Western Australia 6009
www.uwapress.uwa.edu.au

for the Ernest Hodgkin Trust for Estuary Education and Research
and National Trust of Australia (WA)

Ernest Hodgkin Trust
for Estuary Education and Research

National Library of Australia
Cataloguing-in-Publication entry:

Brearley, Anne, 1949– .
Ernest Hodgkin's Swanland: estuaries and coastal lagoons of Southwestern Australia.

Bibliography.
Includes index.
ISBN 1 920694 38 2.

1. Estuarine ecology—Western Australia—Swan River Estuary. 2. Estuarine biology—Western Australia—Swan River Estuary. 3. Swan River Estuary (W.A.). I. Hodgkin, E. P. (Ernest Pease). II. Title.

577.786

Cover image: Hardy Inlet, estuary of the Blackwood River at Augusta, February 2005. Jamie Scott, Margaret River (jamiescottimages.com).

Consultant editor: Brenda Hamilton
Designed by Sandra Nobes, Tou-Can Design Pty Ltd, Melbourne
Typeset in 10 pt Janson by Lasertype, Perth
Prepress by Hell Colour Australia
Printed by BPA Print Group

Foreword

Ernest Hodgkin has affected the lives of many Western Australians. Generations of students heard him lecture in biology at The University of Western Australia—students in science, medicine, agriculture and education—or heard him speak about his love of coasts and estuaries.

Perhaps you are one of the privileged few who were his research students, and whose careers benefited from his influence and support. Or you may have been a colleague, working with him through a university, government department or the museum. Possibly you were in a group drawn from the public, who heard him on a field trip explaining how estuaries are at the mercy of their catchments. If you catch fish, prawns and crabs in one of the many estuaries along our coast, the management of your favourite estuary is sure to have been strongly influenced by Ernest Hodgkin. I will not sketch his career in this foreword, as it is outlined in greater detail within this excellent book. Suffice to say that *Swanland* addresses the estuaries of our coast from Ernest Hodgkin's perspective, drawing on his many scientific and general publications, and his extensive files.

My own interaction with Ernest began when, early in my university career, we both contributed to biology teaching, Ernest from zoology, me from botany. Our collaboration increased when, because of my background in plant physiology, he inveigled me into working on the primary producers of the Blackwood River Estuary. We began working with others from scientific institutions and the general public he had drawn into the program on the Blackwood. Soon I was hooked into ongoing, collaborative work on estuaries and other coastal waters, a collaboration that came into its own with work on the Peel–Harvey Estuarine System, which so dominated the later phase of Ernest's long career.

You can see from my own precedent how much Ernest Hodgkin has influenced others, and many such examples could be given. *Swanland* encapsulates the breadth of Ernest's interests, and his interactions with others in the scientific and management arenas. I hope you dip into and savour this book, pausing to look at a photograph or map, or read some detail that catches your attention. Enjoy!

Emeritus Professor A. J. McComb FAA
School of Environmental Science
Murdoch University

Publication of *Ernest Hodgkin's Swanland* was sponsored by the following organisations:

University of Western Australia
Government of Western Australia Department of Fisheries
Fish for the future
Fremantle Ports
Swan River Trust
Water Corporation
Government of Western Australia Department of Environment
Government of Western Australia Department of Conservation and Land Management
Conserving the nature of WA

Contents

Author's note and acknowledgements

The often-quoted description of Australia as a 'land of sweeping plains, of droughts and flooding rains' encapsulates the great forces of the natural events affecting our rivers and estuaries. In years of drought the flow through many estuaries is limited; they become very salty, large areas may dry out completely, the ocean bar may not open and conditions are very stressful. Rains and floods bring sediments and nutrients that initiate new cycles in the estuaries, resetting the biological clock. Today, however, the volume of river flow may be quite different as many rivers have been dammed, and with rivers flowing through modified landscapes to estuaries surrounded by urban developments, the waters carry more nutrients and sediment so effects downstream are dramatic. Blooms of algae are more common and sustained, and toxic blooms more frequent. The vision for this book was to increase appreciation for the land we live in; to present a story of how the estuaries we treasure were formed and are evolving, how they function and how we are influencing these processes. In the past, responses to our environment led directly to many problems we have to deal with today. Although we still have much to learn about how these systems operate, we do know enough about key processes to make a difference: in some cases to prevent the problems getting worse or repair damage, and perhaps to prevent some things happening again if we learn from our past mistakes. To do so, however, requires community concern and political willingness behind which lies a love and knowledge of our environment, both of which I hope the book will serve to enhance.

During the writing of *Swanland* I had the great privilege of following in some of the footsteps taken by Ernest Hodgkin—'Hodge' as he was fondly known by students, friends and colleagues. Some of these were visits to places, and conversations with his many associates. On trips to familiarise myself with some of the localities and processes described here, I noted features or feelings that I have later found expressed in notes Hodge made during his long voyage of discovery. At times it has been difficult to determine if they were mine or if I had read some of his thoughts and buried them in my subconscious. Compiling this book has been difficult but a privilege and at times great fun, and I in turn am indebted to a large number of people.

I followed the south coast footsteps with Geoff Bastyan on a water sampling run for the Water and Rivers Commission. Together we ventured westwards and eastwards covering fourteen estuaries in six days. I thank Geoff for his good company and his willingness to share his great store of observations and knowledge. It is truly amazing to sit in a tiny boat with only one other person in the middle of an estuary such as the Beaufort in the wind and rain taking water samples, discussing the salinity observed, bar openings and closures, the extent of macrophytes growing on the bottom, and still be able to marvel at the rich colour of the ancient siltstones with their cover of towering eucalypts as a flock of hundreds of Black Swans pass over head. These are unique experiences that have made writing such a privilege; yet they also highlight why these places are special. Some, I hope, will be shared by others who set out to explore our country to look, listen and wonder.

I make no claim that this work is mine or that it stems from my own research. To acknowledge the work of others, I chose to use many older published illustrations rather than redrafting or modifying to a particular style. I have not referenced every statement, and have compiled information from numerous sources detailed in the bibliography and here in the formal acknowledgements. Quotations and sources of tables, maps, figures and graphs are referenced in either the bibliography 'General themes and Chapter 1' if pertaining to that chapter or to more than one chapter, or if specific to a regional area in the bibliography for that chapter. I thank the numerous authors, the researchers and writers of the papers whose carefully considered works and tomes I have honed down to simple phrases. The bibliography is a testament to them. I hope this book is a conduit for encouraging a wider readership of those articles.

I have endeavoured to update many details of natural systems that vary so greatly, and even as I write some information has been superseded. Generally I have remained true to the original authors' intent and conclusions. In other areas I have perhaps extended or generalised their findings to amplify a particular principal. A number of experts reviewed sections of the book and I have followed the advice given. I apologise for any errors made, and there will be some in a volume of this size and scope—they are however mine alone. I ask that where they occur you look beyond to the larger picture I have tried to create.

Throughout Ernest Hodgkin's career, or perhaps careers, a number of people worked closely with him, and their organisation of his papers and photographs left me a great legacy. Rosemary Jasper spent many hours with him in his last months collecting references and preparing the papers for the person who would write this book. Similarly, Ruth Clark, co-author of the *Estuarine Study Series*, contributed to the archives.

The archives also contain a large collection of photographs used throughout this book. Most were taken by Ernest, however when the provenance was in doubt they are referred to as from the E. P. Hodgkin Collection. The collected papers, notes and photographs will be archived at the Scholars' Centre, The University of Western Australia, and some of his collecting equipment is housed in the museum in the School of Animal Biology. Sources of other illustrations are detailed in the captions. Abbreviations of organisations and institutions have been used in most captions, so a full listing of these and acronyms used in the text is included following these acknowledgements.

I would like to acknowledge Rod Lenanton, Supervising Scientist for Finfish Research at the Department of Fisheries, Government of Western Australia (DoF) for discussions about fish biology, estuaries and the environment of the south-west coast. Many of the ideas about how fish in the South-West use estuaries were originally developed by Rod while he was a student and later working in collaboration with Ernest Hodgkin. His numerous scientific papers and reports on fish biology and the development of the fishing industry in Western Australia proved invaluable to me. In 2002 this wide knowledge was recognised by his peers at the Department of Fisheries with the awarding of the Outstanding Service Award. The award also epitomises the debt that I, and the readers of this book, have to him.

In addition, a number of Rod's colleagues also assisted in checking text about commercial and recreational use of estuaries, and providing assistance with illustrations. I thank Suzy Ayvazian, Gabrielle Nowara, Rod Pern, Neil Sumner, Jenny Shaw, Cathy Anderson, Shaun Sims and Steve Emery. Life cycles of the blue-swimmer crabs, prawns and mulloway, and graphs of commercial catch and effort were provided by the DoF.

Ian Potter of Murdoch University (Murdoch) devoted considerable and much appreciated time to discuss fish biology with me, and checked my interpretation of his research along with that of his colleagues and students. In addition, members of his research group, Margaret Platell and Simon de Lestang were also helpful, as was David Morgan with advice and photographs of freshwater fishes. Greg Jenkins, WA Maritime Training Centre, Fremantle Challenger TAFE, provided information about the estuarine black bream. Glenn Hyndes of Edith Cowan University (ECU) advised and assisted with the preparation of illustrations of fish use of estuaries. Alan Kendrick of the Department of Conservation and Land Management (CALM) provided advice on the Leschenault Inlet and fish communities in south coast estuaries.

I received valuable advice about birds from John Dell, Department of the Environment (DoE). Jim Lane, CALM, Busselton gave me access to information on birds of the Vasse–Wonnerup and Peel–Harvey, and assisted with illustrations. I also thank Kylie Ryan, John Hunter, Ron Kawalilak, Kevin Bancroft and Nick D'Adamo for assistance. John assisted with access to photographs in the Babs and Bert Wells Collection, and others were provided by Cliff Winfield and Graeme Chapman. Mike Bamford of Bamford Consulting Ecologists also provided information on birds of the Swan Estuary.

Over the years, Ernest Hodgkin developed a great interest in estuarine evolution and his final paper was written with Patrick Hesp of Massey University, Palmerston North, New Zealand. Patrick's research and the field trips during the 1980s in which I was occasionally included were of particular use to me in writing this book. George Kendrick's work at the Western Australian Museum (WAM) on the fossils shells hidden below the estuaries also contributed greatly to Ernest Hodgkin's ideas about estuarine evolution, and for more than 30 years George has also enriched my understanding of molluscan communities. Cleve Hassell assisted with information about sediments and evolution of the south coast estuaries, and reviewed some text. I also acknowledge the many years of estuarine research undertaken by Vic Semeniuk of V. & C. Semeniuk Research Group, and thank him for illustrations of the Leschenault Inlet Estuary.

A number of colleagues at the Western Australian Museum (WAM) provided advice and assistance to me as they had to Ernest in previous years. I thank Shirley Slack-Smith, a long-time friend and associate of Ernest, who guided my early work on molluscs and with whom I still have the most enthralling conversations. For support in my first forays into work at the Western Australian Museum, I thank Barry Wilson and Ray George. For assistance with photographs I thank Clay Bryce, Barry Hutchins, Sue Morrison and Di Jones. Ken McNamara provided advice and illustrations of fossils in the Pallinup Siltstone and Tumblagooda. Fred Wells advised on the Peel–Harvey and Albany Harbours and, through the Marine Biological Workshop program, provided me with many opportunities to participate in fieldwork along the coastline of Western Australia from Dampier to Esperance. The workshops also provided me with the opportunity to meet many international marine scientists. In particular I would like to acknowledge Carole Hickman, The University of California at Berkeley, for her inspiration and encouragement over many years including those predating the workshops. Similarly, Emeritus Professor Alan Kohn and Dr Marian Kohn University of Washington supported my endeavours and made a generous donation to the Ernest Hodgkin Trust. Charles Dortch and Moya Smith (WAM) assisted with advice and photographs of the structures built by Aboriginal people around the estuaries. Irene Stainton, past Chairperson Aboriginal Cultural Material Committee, provided information on the fish traps in Oyster Harbour and the Barragup Weir. I thank Harley Coyne of Albany for assistance with my interpretation of aspects of Aboriginal culture and use of estuaries. Robert Reynolds, Department of Indigenous Affairs liaised on my behalf with Vernice Gillies and the Albany Heritage Reference Group, Wayne Webb, a custodian of the south coast estuaries and Joe Walley, Lorna Little and other members of the Bilya Noongar Organisation.

Staff at the Department of Environment, Government of Western Australia (DoE), previously the Water and Rivers Commission (WRC) provided advice and prepared illustrations; in particular I thank Malcolm Robb,

Verity Klemm, Beth Hughes, Tom Rose, Donna Weston, Richard Bretnall, Peter Muirden, Phil Commander and Rod Nowrajee, also Wendy Yorke of the Swan River Trust (SRT). Dianne Abbott and Nicholas Flett brought the maps to reality. Dieter Tracey, Ben Boardman, Mischa Cousins and Helen Astill sorted the archives for water quality graphs and compiled these into suitable formats. Vas Hosja and Ashrafi Begum of the Phytoplankton Ecology Unit also provided expert help and photographs. At Albany, Naomi Arrowsmith, Steve Janicke, Kaylene Parker and Julie Pech provided advice. At Bunbury, Kathryn Hardcastle responded to my numerous requests for information about Leschenault Inlet, Vasse–Wonnerup and the Hardy Inlet. At Geraldton, Jessica Scott, Bronte Grant and Kim Griffith provided some recent reports and checked details. The Department of Environment, Government of Western Australia, is acknowledged for permission to use illustrations (photographs, maps and graphs) attributed to the Water and Rivers Commission, and those from publications produced by agencies in existence before the current organisational structure.

Kevin McAlpine and Ray Masini (DoE) provided illustrations, checked details and advised on historic research on the Swan–Canning, Albany Harbours and Leschenault Inlet. Steve Appleyard assisted with access to information about ground water on the Swan Coastal Plain.

Geological and tectonic maps of Western Australia are reproduced with permission from the Geological Survey of Western Australia (GSWA), Department of Industry and Resources (DoIR). Philip Playford's writings on geology and history have been a valuable resource and inspiration, and his permission to adapt maps of changes in sea level and Vlamingh's explorations along the west coast are appreciated. At UWA David Haig provided advice on the geology of Swanland and assisted with illustrations.

I received advice on terrestrial vegetation from Steve Hopper at The University of Western Australia, previously of the Botanic Garden and Parks Authority (BGPA). Information on soils, landscapes and nutrient flows was given by Noel Schoknecht and David Weaver of the Department of Agriculture, Government of Western Australia (DAWA). Angas Hopkins (CALM) and Damian Shepherd (DAWA) provided maps of pre-European and remnant native vegetation. Noel Schoknecht, David Weaver and Rob Summers (DAWA) provided information and illustrations of soils, landscapes and nutrient flows.

Alan Pearse, CSIRO Floreat, provided advice on the Leeuwin Current. I obtained advice from Chari Pattiarachi (UWA) on currents, tides and water levels. John Cramb, Bureau of Meteorology (BOM, WA) assisted with the sections on weather and climate, and isobaric charts are reproduced courtesy of the Commonwealth Bureau of Meteorology. Christine Stack, Danielle Crake and Danielle Stefani at the Department of Land Information (DLI) helped with information on the origin and correct spelling of place names, and Sue Burrows facilitated use of aerial and Landsat photographs. I also acknowledge the assistance of Peter Sanders, Satellite Remote Sensing Services (SRRS) Leeuwin Centre, Department of Land Information (DLI), Perth, Western Australia for Landsat imagery from the Australian Centre of Remote Sensing (ACRES), and Rodney Hoath, Department of Planning and Infrastructure (DPI) for aerial photographs.

A number of people in regional areas were also of great assistance. I thank David Allen-Williams of Augusta who showed me Ernest Hodgkin's view of the Hardy Inlet bar. Rob Rippingale of Curtin University (Curtin) checked sections of my text. Don Bradshaw (UWA) discussed current issues of the Blackwood and Scott Rivers and Hardy Inlet. Brian Clay and Neil McDonald of the Toby Inlet Catchment Group Inc provided permission for the use of images supplied by M. P. Rogers and Associates Pty Ltd. Martina Scheltema, ATA Environmental, liased with a number of people for the use of figures about the northern rivers and estuaries. Lesley Borowitzka of Cognis

Australia Pty Ltd discussed the growth of *Dunaliella* for β-carotene at Hutt Lagoon. Graham Grundy, Irwin Historical Society, provided photographs of the Irwin at Dongara as did Sarah Grimes.

Karen Hillman and Bruce Hegge of D. A. Lord Science and Engineering (DALSE now Oceanica) read sections of my manuscript and assisted with photographs of Bandy Creek, Murchison River, Peel–Harvey and the Fremantle coastline. I thank Celeste Wilson and Eric Paling of the Marine and Freshwater Research Group (MAFRL/MAFRA) Murdoch University (Murdoch) for the opportunity to participate in a water sampling trip to the Peel–Harvey that allowed me a view of the estuary and ocean in the Dawesville Channel, and for illustrations of the current state of the macroalgae in that system. I thank Halina Kobryn for assistance with illustrations of the Peel–Harvey. Keith Bradby's story *Peel–Harvey: The Decline and Rescue of an Ecosystem* guided me into the massive array of publications about that system, while the foreword by Ernest Hodgkin gave me insight into his aim for *Swanland*. Similarly *A Sense of Place* one of George Seddon's books, has been an ongoing source of pleasure and an influence to how I view the landscape, as it was to Ernest Hodgkin.

Gillian Craig of Ravensthorpe, Julia Phillips of CSIRO Floreat and Paul Lavery of ECU checked my interpretation of material about the eastern south coast estuaries. I also acknowledge Paul's research on the algae in the Peel–Harvey and Swan–Canning that are a major influence on our understanding of estuarine and marine communities in south-western Australian.

Janda Gooding and Robyn Walton of the Art Gallery of Western Australia provided assistance with selecting historical material from the collection. Similarly, staff at the Battye Library, Library Information Service of Western Australia and the National Library Canberra assisted with access to images.

The Department of Premier and Cabinet, Government of Western Australia provided the quotation and permission to include a passage from the speech by the then Premier Hon. Richard Court at the opening of the Dawesville Channel. The Department also provided permission to use the State Crest that includes a Black Swan in the section discussing the demise of an iconic species. Additional crests using the Black Swans were supplied by local government authorities, and institutions and industries are thanked for their help with Figure 2.40. A quotation from Baudin (1974), translation by C. Cornell, is reproduced with permission of the South Australian Libraries Board. Figures from Hillman, K., McComb, A. J. & Walker, D. I. (1995) 'The distribution, biomass and primary production of the seagrass *Halophila ovalis* in the Swan–Canning Estuary, Western Australia', *Aquatic Botany*, vol. 51, pp. 1–54, are reproduced with permission from Elsevier. Permission to use a quotation from Bunbury and Morrell (1930) was granted by Oxford University Press. An excerpt from an interview with Phil Smeeton by Rod Moran is reproduced courtesy of The West Australian Newspaper. A quotation and figures (6A and 10) from Hodgkin, E. P. & Hesp, P. (1998) is reproduced with permission from Marine and Freshwater Research, CSIRO Publishing, Victoria, Australia. Hon. David and Marie Louise Wordsworth kindly allowed me to use photographs of their paintings.

Although this book celebrates the ideas of Dr Ernest Hodgkin and his passion for understanding the complexities of the estuaries and the greater environment of Western Australia, much of the research on estuaries was done with the vision and guidance of Emeritus Professor Arthur McComb of Murdoch University. I first studied with Arthur McComb at The University of Western Australia where through the 1970s and 1980s he led a dynamic team working on the estuaries and wetlands. Many of the young researchers went on to lead work on environmental issues in industry, government, universities, schools and the wider community. After his move to Murdoch, Arthur McComb continued to guide and enthuse me. I was fortunate to have been advised by him through my PhD research on seagrass communities, and most recently during the writing of *Swanland*. My story of Arthur's help and guidance is echoed by many, and to him we all owe a great debt—thank you Arthur McComb.

At The University of Western Australia, I thank Di Walker without whose encouragement I would never have undertaken or completed my PhD and this book. I also thank Gary Kendrick for his insightful comments on my marine research as well as many other topics, and for including me in his fun times. Carol Wilkins, Tim Carruthers and Michelle Waycott, fellow students of botany and the class of 1989, as well as past and present members of the Marine Group of the Botany Department, now School of Plant Biology, provided ongoing encouragement. In particular I thank Kieryn Kilmister for discussions on chemistry; Justin McDonald, Grey Coupland and Vanessa Forbes for discussions on invertebrates, mangroves and estuaries. Helen Penrose helped with collation of the first draft, and Emma Kitson organised scans and line drawings. Marion Cambridge, one of Hodge's great friends who journeyed with him on the first south coast estuary foray in the 1970s, listened with patience to my deliberations of how to organise a large volume of disparate information, and interceded in my times of confusion with considered words. I also acknowledge the support of Professor Hans Lambers, Head of School of Plant Biology, Professor Alistar Robertson, Dean Faculty of Natural and Agricultural Sciences and the Vice Chancellor Professor Alan Robson.

At University of Western Australia Press, I thank Janine Drakeford, Publishing Manager, and Sam Wilson, and lately Olivier Breton, as intermediaries for the production team. I acknowledge the editor Brenda Hamilton for patience with my long delays and for dealing with the complex tome I have created, and Keith Feltham, Lasertype for advice on graphics formatting. I also thank Jenny Gregory, Director of UWA Press, and other staff including Maureen de la Harpe for assistance.

The Trustees of the Ernest Hodgkin Trust, Graham Hodgkin, Bruce Hamilton, Professor Di Walker (UWA), Karen Majer (WA Maritime Museum), Malcolm Robb (DoE) and Tom Perrigo (National Trust, WA), guided this work and I thank them for their support and faith in me. To the Hodgkin family, I hope you find this product of your forbear's vision and passion to your taste and satisfaction, and thank you for your trust in me to undertake this work. Additional financial support for the book's publication and the ongoing education and research component of the Trust was received from The University of Western Australia, the Department of Environment, Department of Conservation and Land Management, Department of Fisheries, Swan River Trust, Water Corporation, Ecoscape Australia Pty Ltd and Fremantle Ports.

On the other side of life I would like to honour personal friends who never thought I would finish, but never dared to say so. I thank my dear friends Elizabeth and John Rippey for expression of their pleasure at the natural world, for what I was trying to do, and for their advice and encouragement. My father Colin Paterson fuelled my curiosity in the natural world, and allowed me to pursue my dreams. My mother Isabel Paterson worried about my long hours. My brother Owen Paterson and his family supported my endeavours from afar. My husband Reg, my companion in life and on visits to many estuaries, was party to my anxious nights as I struggled with the chapters and the overrun schedules. My sons Winston and Charles gave love and faith, and represent the generation I hope will find this book of relevance.

Dr Anne Brearley
School of Plant Biology
Faculty of Agricultural and Natural Sciences
The University of Western Australia
2005

Abbreviations

ACRES	Australian Centre For Remote Sensing, Geoscience Australia, Canberra (*see also* SRSS)
AGPS	Australian Government Publishing Service
AGSO	Australian Geological Survey Organisation, Canberra
AHD	Australian Height Datum (*see also* Glossary)
AMSA	Australian Marine Sciences Association
ANCA	Australian Nature Conservation Agency, Commonwealth of Australia (now a division of the Australian Government Department of the Environment and Heritage (DEH))
ANZECC	Australian and New Zealand Environment and Conservation Council
ARMCANZ	Agricultural and Resource Management Council of Australia and New Zealand
BGPA	Botanic Garden and Parks Authority, Western Australia
BOM, WA	Bureau of Meteorology, Western Australia
CALM	Department of Conservation and Land Management, Government of Western Australia
CAMBA	China–Australia Migratory Bird Agreement (*see also* JAMBA & RAMSAR)
CSIRO	Commonwealth Scientific and Industrial Research Organisation, Australia
CURTIN	Curtin University of Technology, Western Australia
DALSE	DAL Science & Engineering Pty Ltd, previously D. A. Lord & Associates Pty Ltd (DAL), August 2004 name change to Oceanica
DAWA	Department of Agriculture, Government of Western Australia
DEH	Australian Government Department of the Environment and Heritage (previously referred to as Environment Australia (EA))
DEP	Department of Environmental Protection, previously Department of Conservation and Environment (DCE); name changed to Department of Environment (DoE) in July 2004 and amalgamated with the Water and Rivers Commission (*see also* EPA)
DLI	Department of Land Information, Government of Western Australia, previously Department of Land Administration (DOLA also referred to as DLA)
DoE	Department of Environment, Government of Western Australia, previously Department of Environmental Protection, and amalgamated with the Water and Rivers Commission (WRC)

DoF	Department of Fisheries, Government of Western Australia, previously Fisheries Western Australia, Fisheries Department of Western Australia and Department of Fisheries and Wildlife.
DoIR	Department of Industry and Resources, Government of Western Australia
DOLA	Department of Land Administration, Western Australia (*see also* DLI & DLA)
DPI	Department for Planning and Infrastructure, Government of Western Australia
ECU	Edith Cowan University
EPA	Environmental Protection Authority of Western Australia (*see also* DEP)
ERMP	Environmental Review and Management Programme
ESS	*Estuarine Study Series* (*see* Hodgkin & Clark references)
FRDC	Fisheries Research Development Corporation
FRNP	Fitzgerald River National Park
GIS	Geographic Information System
GSWA	Geological Survey of Western Australia, Department of Industry and Resources Western Australia (DoIR)
IBRA	Interim Biogeographic Regionalisation for Australia
JAMBA	Japan–Australia Migratory Bird Agreement (*see also* CAMBA & RAMSAR)
MAFRA	Marine and Freshwater Research Laboratory, Environmental Science, Murdoch University (also MAFRL)
Murdoch	Murdoch University
NEMP	National Eutrophication Management Program
NHT	Natural Heritage Trust (*see also* Australian Government Natural Resource Management (NRM))
NP	National Park
NRM	Australian Government Natural Resource Mangement Programmes Division within the Departments of Agriculture, Fisheries and Forestry, and Environment and Heritage in cooperation with State and Territory agencies and regional bodies responsible for coordinating the National Action Plan for Salinity and Water Quality, and the Natural Heritage Trust
ppm	parts per million
ppt (‰)	parts per thousand
RAMSAR	Ramsar (Convention), Iran 1971 (*see also* CAMBA & JAMBA)
RAOU	Royal Australasian Ornithological Union
SCCP	Swan–Canning Cleanup Program
SRSS	Satellite Remote Sensing Service, Department of Land Information (DLI), Perth, Western Australia (*see also* ACRES)
SRT	Swan River Trust
UWA	The University of Western Australia
UWA Press	University of Western Australia Press, Crawley, Western Australia
WAERF	Western Australian Estuarine Research Foundation
WAM	Western Australian Museum
WC	Water Corporation
WIMA	Wilson Inlet Management Authority
WRC	Water and Rivers Commission (Western Australia), now Department of Environment (DoE)

Prelude to *Swanland*

Dr Ernest Hodgkin

This book is based on, and inspired by, the life and work of Dr Ernest Hodgkin, a man intrigued by the natural world who pursued a lifetime of enquiry related to it. In his latter years Ernest Hodgkin specialised in understanding the form, function and history of estuaries in the south-west of Australia. His study of the estuaries was firmly grounded in the discipline of science but he married this with the best traditions of the naturalist and the historian. He was also a dedicated educator, and it was his ambition to be able to present his understanding of estuaries to a wider audience.

The book has been funded through a Trust established by him before his death in September 1998 and guided by trustees, family members and colleagues from The University of Western Australia, Department of Environment, Water and Rivers Commission, Western Australian Maritime Museum and The National Trust of Australia (WA).

The book is intended as a guide for people interested in estuaries or simply inquisitive about their natural surroundings, those who enjoy a beautiful landscape or waterscape but who would also like to understand some of its intricacies. Hodge wanted to tell the story of the estuaries in an interesting and informative way to increase people's enjoyment of the environment. By its nature the book calls for a respect for estuaries, a respect that grows from understanding.

> To make sound strategic decisions about how to manage estuaries to conserve their environmental and cultural values, we need to understand the importance of how they have changed in response both to natural environmental factors and to human activities. Otherwise, we may well expend resources managing one factor while another unmanaged could lead to the demise of estuaries. It is vital to undertake the research necessary to provide a broad strategic understanding for estuary management beyond such immediate problems as eutrophication.
>
> Hodgkin & Hesp, 1998.

Where is Swanland?

> I am inclined to resuscitate the name 'Swanland' instead of the cumbersome 'South-western Australia'. Swanland is a convenient and appropriate term which was applied early this century to that portion of Western Australia which is bounded on the west and south by the ocean, the eastern boundary being determined by the catchments of all river systems south of the Murchison which drain into the sea.
>
> Hodgkin, 1998.

In 1915 the physiographer Griffith Taylor in his work titled *The Climatic Control of Australian Production* (an attempt to gauge the potential wealth of the Commonwealth) gave the following reasons for the term:

> The south-west of the State ('Swanland') is different in character from any other portion of Australia. Its rains fall chiefly in winter and reach 40 inches per annum near Cape Leeuwin, which differentiates it from the rest of the State; while its geology and flora are notably different from that of the winter regions in Eastern Australia. For this geographic entity—well-watered region in the south-west of Western Australia—a comprehensive name akin to Gippsland, New England, &c is badly needed. A euphonious name connoting the chief river, the most characteristic animal, and the national emblem would be 'Swanland'.

Although the boundaries and names used to define the area varied, the distinctive nature of the South-West was also recognised by botanists such as Ferdinand von Mueller in 1867, Ludwig Diels in 1906, Charles Gardner in 1942 and more recently by John Beard in 1980 and 1990, as well as geologists such as Edward de Courcy Clarke in 1926. None of the terms, however, are quite as distinctive or as appropriate for a book about estuaries as Swanland.

The map of Swanland (Map 1) and the description of its boundaries indicate the scope of this book. In times of floods, the river catchments extend into the Salt Lake Division, 'Salinaland'; another old term with a new relevance as problems of salinity increase each year in the eastern agricultural lands (Maps 2, 1.10 and 1.11). The estuaries span almost 2,000 kilometres of coastline where the mediterranean climate is characterised by cool wet winters and hot dry summers that dictate a highly seasonal pattern of river flow. However like all natural patterns there are exceptions. Summer rain often brought by tropical cyclones can also bring floods and high river flows.

The area of Swanland covers the most heavily populated and intensively farmed areas of south-western Australia, where the landscape since the 1830s has been extensively modified. The nature of estuaries at the end of rivers means they receive the products of the catchments. Nutrients and sediments arriving from the catchments make estuaries some of the most productive ecosystems on earth. This also means however, that nutrients along with

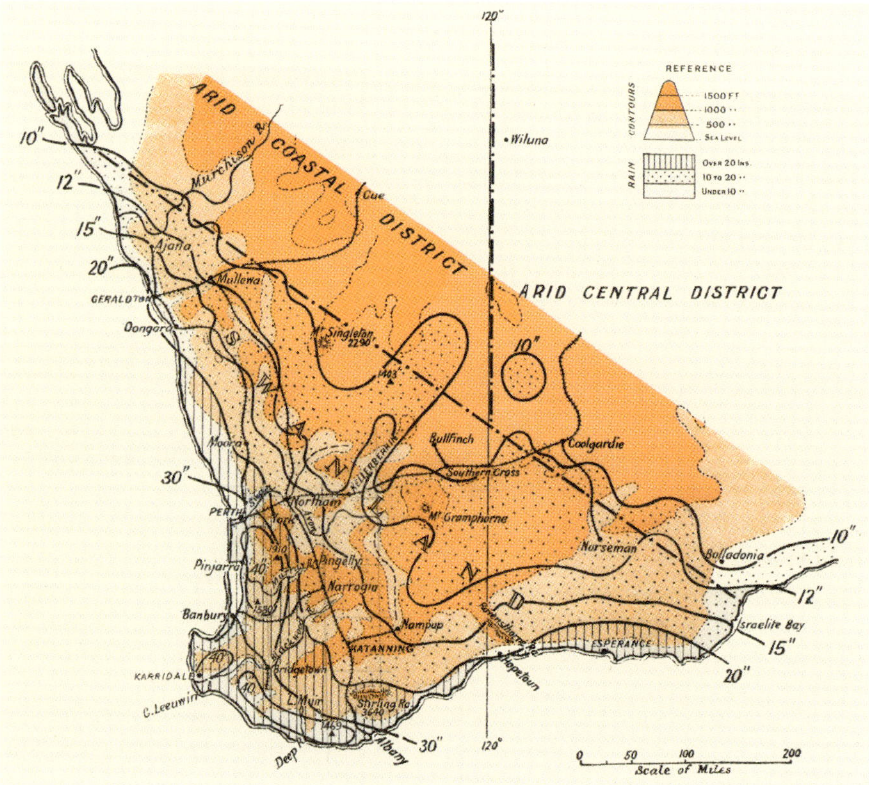

MAP 1 *Swanland: The term 'Swanland', a shortened form of 'Swan River Settlement', was used by Griffith Taylor in 1915 'for the temperate region in Western Australia with a rainfall over 10 inches per annum'. Taylor was of the opinion that the phrase 'South-west Division of Western Australia' 'is too great a mouthful. Moreover, it does not indicate the fact that this region differs materially from other portions of the State, not only in the quantity and time of rainfall and in the climate generally, but in its flora, its topography, and especially in its capabilities as regards future development.'* Taylor (1918), Commonwealth of Australia.

other pollutants and products of erosion are often in excess. The nutrients lead to excessive algal growth that changes the ecology, or noxious blooms unfavourable for many life forms including humans.

The life behind the book

Ernest Hodgkin loved to spend time in the environment, wading out into the estuaries with collecting equipment, photographing the scenery, or observing the tiny details that would escape most people. He loved to gather stories and share his enjoyment and vast knowledge with other people. The text that follows has been prepared from the extensive collection of published papers and notes assembled over his long and varied career.

His career continued far beyond the traditional years of employment. World War II and his four-year internment in Singapore interrupted his early work in Malaya as a medical entomologist working principally on malarial mosquitoes (Wood, 2003). Peace brought him to Perth to join his wife Mary and children, where an appointment to The University of Western Australia brought him into contact with all the students of zoology. Early work on the estuaries, much of which was done with students, included studies of water levels, tides, as well as the organisms living in them, have since been written and some are detailed in Table 2.7. Retirement in 1973 was short-lived as he was asked to coordinate an environmental study of the Blackwood River Estuary, the results of which are described in Chapter 6 'Hardy Inlet and estuaries of the Naturaliste–Leeuwin Ridge'. These were exciting times for people interested in the environment and estuaries. Environmental problems were emerging in the Peel–Harvey, and the first Environmental Protection Act was proclaimed in 1971. Following the Blackwood River Study, Ernest was asked to lead studies of the Peel–Harvey for the Environmental Protection Authority.

These wide-ranging studies revealed how little was known about estuarine processes, and the need for management based on knowledge. In response, Ernest Hodgkin was asked to compile the information of the south coast estuaries—'An inventory of information on the estuaries and coastal lagoons of south Western Australia'—that became the *Estuarine Studies Series* of eight volumes (1987–1990). These, the 'Blue Books', are still in use today. Although many scientific papers and reports have been written about the estuaries, many of the studies reported in them were initiated in response to the books that remain the only synthesis and overview of those estuaries. Along with his involvement with committees, Ernest also wrote and co-authored numerous scientific papers. I hope *Swanland* reflects the wide interests and overview of the different environments that so fascinated Ernest Hodgkin, an unusual trait in a world where expert knowledge is now finely focused.

Ernest Hodgkin's library included files and papers for each estuary; observations made on numerous visits; photographs taken over the decades from estuary banks and from the air; cuttings from local newsletters; copies of scientific papers covering all aspects of geology, hydrology, zoology, botany and agriculture as well as history; notes on all manner of things connected with each estuary. Many of the notes document observations made by local residents, and there are long lists of contacts in each area of coast—Hodge had a wide network of friends from all walks of life.

The chapters that follow cover the range of topics that Hodge wanted to include and, in keeping with his interest in the 'latest news', new information only available since his death in 1998. All the estuaries of Swanland receive a mention, following Hodge's own philosophy and that of the Landcare movement—recognising that people identify with what they know or are familiar with. I have attempted to make each estuary *special* while identifying themes and processes common to all so that a greater appreciation of one area can be transferred to other areas, and even other environments (see Maps 3 and 4).

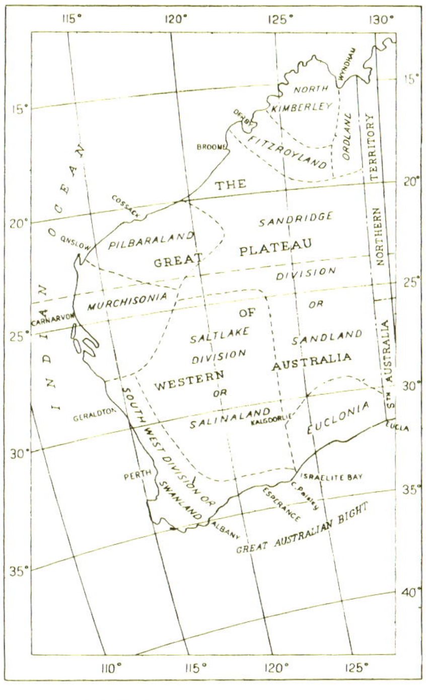

MAP 2 *Outline map of Western Australia showing the nine physiographic divisions of Jutson in 1934. The term Swanland was used by J. T. Jutson in various editions of 'The Physiography (Geomorphology) of Western Australia' in the* Geological Survey Bulletins, *but was not used in the first editions. The term later fell into disuse.* Reproduced with permission from GSWA.

The following quotation used by Hodge in discussions with student researchers deliberating of questions to ask, and the relevance of their work, may underpin Ernest Hodgkin's philosophy and the influence of his role in Australian environmental studies.

> In order to penetrate further into their subjects, the hosts of specialists narrow their field and dig down even deeper till they can't see each other from hole to hole. But the treasures their toil brings to light they place on the ground above. A different kind of specialist should be sitting there, the only one still missing. He would not go down any hole, but would stay on top and piece all the different facts together.
>
> Heyerdahl (1958).

How the book is organised

Chapter 1 'Understanding Swanland' introduces general themes and estuarine processes. More detailed chapters describe the estuaries large and small by regions: the northern and southern parts of the west coast; the Leeuwin–Naturaliste Ridge; the western and eastern parts of the south coast and the Albany region (Maps 3 and 4). It follows that there is less known about some of the smaller estuaries and those more distant from the largest population centres, such as those north of the Swan River. The four largest estuaries, the most modified and studied, the Swan–Canning, Peel–Harvey, Wilson, and Albany Harbours are the subject of individual chapters.

'The Swan–Canning Estuary', the first of the detailed chapters, is used as a model for understanding other estuaries. As the site of the first European settlement and now surrounded by a major city, the estuary is familiar to many people who value it greatly but understand it poorly. Environmental changes around the Swan–Canning are extensive. Nutrients and other pollutants from urban and industrial areas, both current and historic, cause periodic algal blooms or affect organisms. Issues of management for the Swan–Canning are particularly complex and range from the distant to immediate catchment. Around the shoreline, development has to be balanced against maintenance of equitable access and strategies for maintaining estuarine function for a healthy and productive system. Despite the changes it is worth noting that many fish can still be caught in its waters and generally they are still safe to eat.

The Swan–Canning is followed by the story of 'The Peel–Harvey Estuarine System' (Chapter 3), where nutrients from the catchment fuelled algal blooms of plague proportions and ignited years of public debate and scientific studies. The solution, to construct a channel at Dawesville (opened 1994) to flush nutrients to the sea and increase the marine exchange, created a new marine phase in the estuary's life. The water is now saltier and clearer due to the larger tidal influence, but the higher water levels are causing erosion of river banks and changes to the vegetation. Nutrients in the catchment soils and river sediments are still high however, and algal blooms now occur in the river reaches. The Peel–Harvey still supports productive commercial and recreational fisheries.

The southern west coast estuaries are described in Chapter 4 'Leschenault Inlet and Estuary and smaller estuaries of the Geographe Bay area'. Although smaller, they are all interesting environments subject to similar processes and problems as the larger estuaries, and often the foci for wildlife or humans. Commercial fishing in the Leschenault has now ceased but recreational fishers reap a rich bounty from its waters. The estuaries from the Murchison River to the Moore River are detailed in Chapter 5 'Estuaries north of the Swan'. Like the smaller estuaries further south, water bodies flowing to the sea are always intriguing, but many of these are small in area, very shallow and river flow is more sporadic so that some dry out over summer.

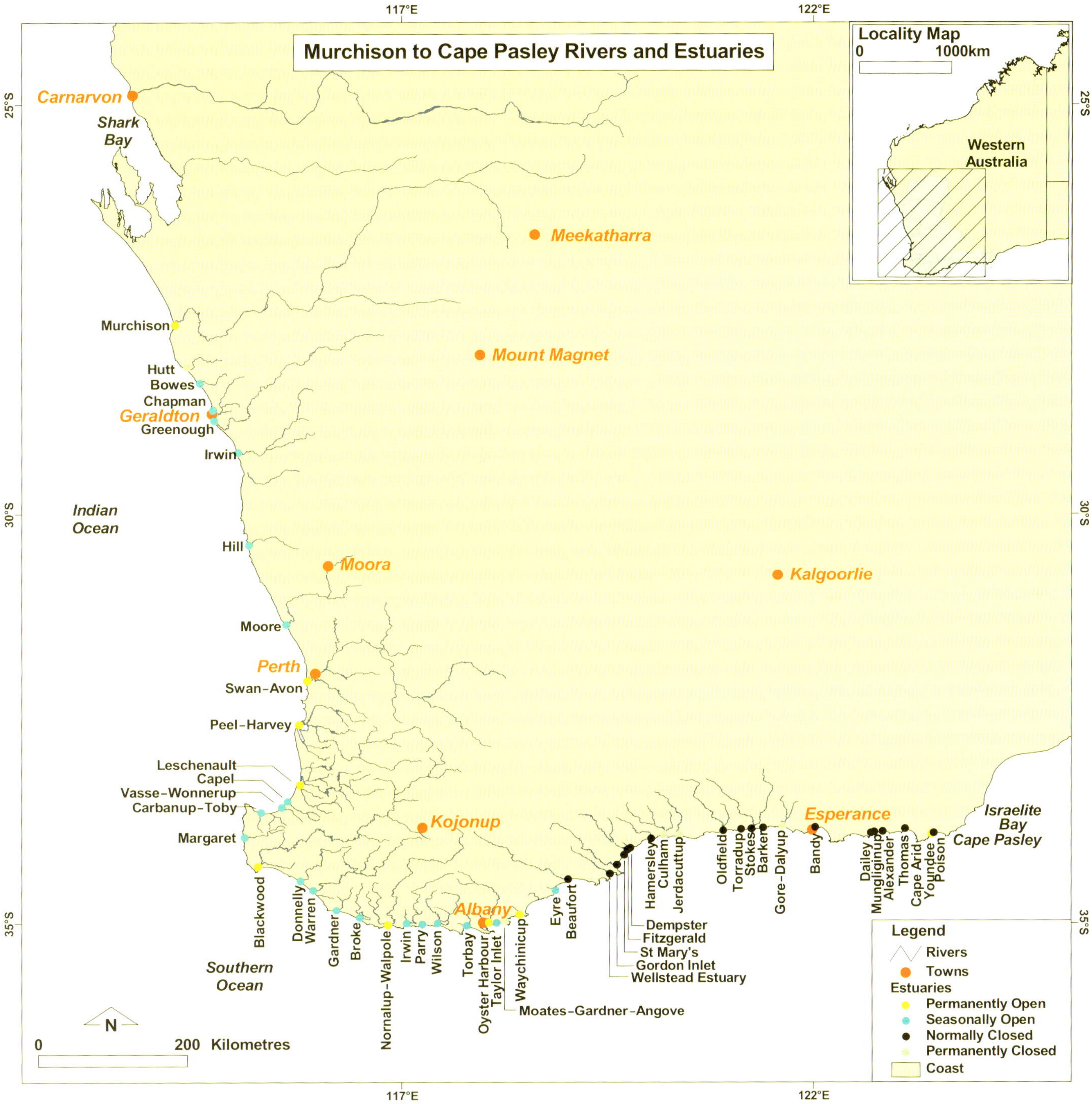

MAP 3 *The rivers and estuaries of Swanland: between the Murchison River at Kalbarri and Cape Arid area near Esperance there are 38 rivers, 50 major creeks and 180 minor streams that discharge into the ocean or to a coastal inlet or lake.* Based on Hodgkin & Hesp, 1998 and Pen, 1999. Courtesy DoE.

The Blackwood River Estuary, the 'Hardy Inlet and estuaries of the Naturaliste–Leeuwin Ridge' (Chapter 6), follows the west coast chapters and begins the journey across the south coast, where rounded granite rocks dominate the landscape and rainfall is higher. Fronting the Southern Ocean at the end of the mighty Blackwood, the Hardy Inlet receives water from the now salt-affected wheatbelt and from forest country. In response to proposals to mine mineral sands in the early 1970s, the Hardy became the site of the first comprehensive estuarine study. Organised by Ernest Hodgkin, the research documented the hydrology, water chemistry, plants and animals of the estuary and set the scene for future studies on algal blooms in the Peel–Harvey.

Along the south coast the estuaries are divided into regions from west to east following the rainfall patterns. The first region, in Chapter 7, is the 'Estuaries of the wet Western South Coast', dealing with estuaries that lie in forest areas where river flow is larger and more reliable. 'Wilson Inlet', although very similar, is dealt with separately in Chapter 8 as management issues are more complex. Surrounded by cleared land as well as forest and the expanding township of Denmark, the estuary supports commercial and recreational fisheries. Luxuriant growth of the seagrass *Ruppia* or swan grass however, indicates that nutrients are abundant and need to be monitored.

Chapter 9 'Albany Harbours and the smaller estuaries of the Albany region' includes a number of small estuaries: Torbay, Normans, Taylors and Cheynes Inlets that lie behind the coastal dunes, and Waychinicup Inlet, which lies in a rocky chasm, as well as the larger Oyster Harbour and Princess Royal Harbour. Although conditions in Oyster Harbour are quite marine, large algal growths due to nutrients from the catchment were similar to early problems in the Peel–Harvey. Inclusion of Princess Royal Harbour as an estuary is perhaps extending the definition, as freshwater inflows come from drains and ground water rather than riverine sources. The problems in both water bodies, however, are similar, although those in Princess Royal Harbour have been more widely documented as the severity of mercury contamination in the 1980s resulted in closure of the harbour to fishing.

To the east of Albany the rainfall gradually decreases and Chapter 10 'Estuaries to the east of Albany' includes estuaries where the bar closes for many months. Some become hypersaline and may even dry out. Fossil deposits indicate that they were open to the sea more regularly only a few thousand years ago. The bars open today only after particularly strong river flow and, while the bars are open, marine organisms enter and flourish until conditions become too harsh. These estuaries illustrate the final step in estuarine evolution, where sediments have filled the estuary and for much of the time they resemble coastal lakes.

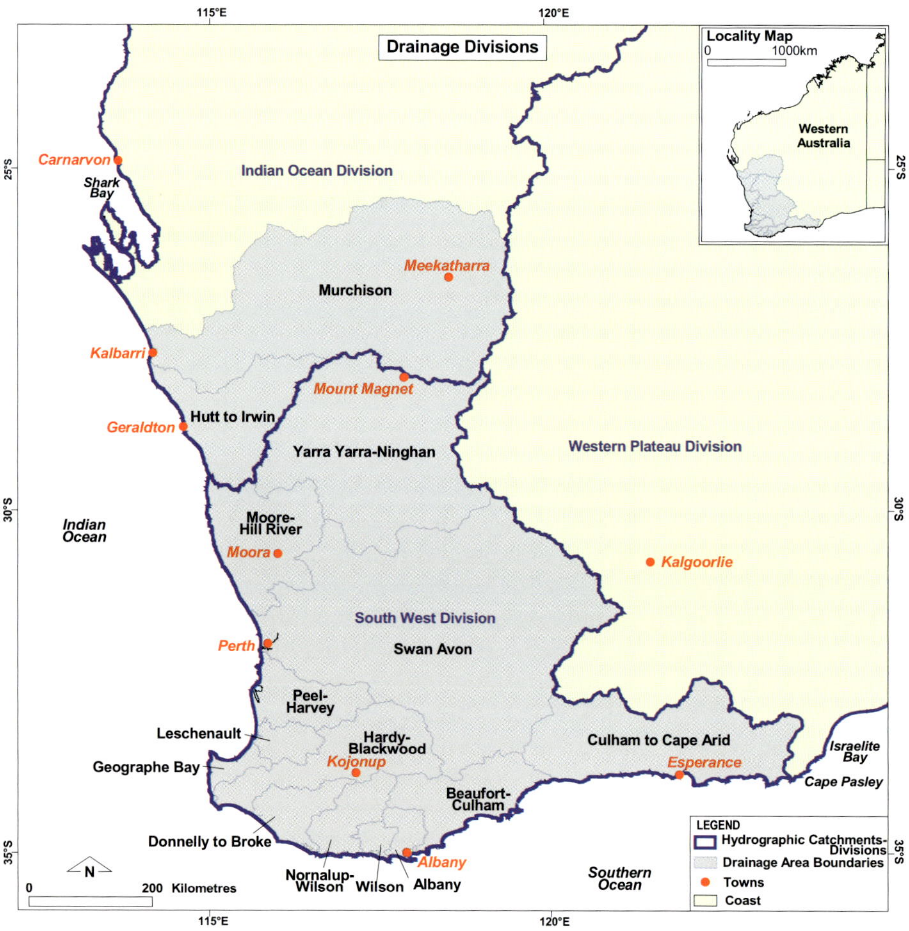

MAP 4 *Divisions used in chapters of* Swanland*: these are based on the Drainage Divisions and river catchments.* Courtesy DoE.

Each chapter is slightly different and unique, as the estuaries themselves are in nature. These estuaries have been used and modified differently, and knowledge about them varies although general patterns are also apparent. The format follows a pattern throughout covering the place in the landscape and general features of the area, together with an overview of the history, of European discovery and how the areas have been used. Features of the individual estuaries are then described from the catchment to the mouth, including the biological communities and processes, using information known about particular systems. Themes of particular interest are developed further in 'Boxes'. The boxes are often relevant to other chapters and are cross-referenced. The format allows the text to be used in a number of ways, either 'geographically', reading about one area or estuary, or by 'topics' of interest.

Captions accompanying the figures, maps, tables and graphs are designed to lead readers into the text. Many 'bird's-eye view' aerial photographs taken by Ernest Hodgkin have been used although they were taken some years ago. Their inclusion is appropriate as they reflect a great understanding of the environment, showing the important characteristics; and many are of exceptional quality being taken at low altitude at a good angle without sun-glint. More recent photographs taken by other people have been used to illustrate features, particularly entrance bars, which have changed over time, sometimes with profound effects on the estuarine ecology.

Sources of information are vast and varied, and I have not referenced statements as they are often from multiple or composite sources. Sources used in boxes, tables and captions of maps, figures and graphs are listed in the general list if pertinent to more than one chapter, and by chapter if regionally specific.

Swanland is a natural history book. The topics range from the tiny algae of the water column to the birds that fly over the top using the estuary perhaps for only short but critical times of the year. I have tried to include natural history stories that intrigue me as they did Ernest Hodgkin, bringing together snippets sourced from scattered reports and books that are not generally available. As a result, this book encompasses the work of a great many people to whom I—as was Hodge and the community of Western Australia—am greatly indebted.

In many ways this is also a history of south-western Australia. A history of how estuaries formed, how they evolved and are still evolving in response to natural and human pressures. It is a history of our interactions with important features of the landscape and of changing attitudes. Aboriginal use of the estuaries and coastal wetlands in many cases is little known, and the traditional stories remain to be told by their descendants and custodians in each particular area. More importantly for this Swanland story, the use by Aboriginal people did not greatly change the estuaries, with the greatest modifications following European settlement in 1829. Historic notes also include impressions and responses to the environment through European eyes, the use of the estuaries for transport, as a source of food and for recreation by the original inhabitants and the people who followed.

What is an estuary—what's in a name?

Generally estuaries can be defined as 'the mouth of a river where tidal effects are evident and where fresh water and sea water mix' (Moore, 1988). Generally the salinity can be expected to change progressively from fresh at the river end to marine near the entrance, varying slightly as the tides flow and ebb. With the progression from fresh through brackish to sea water, the plant and animal communities also change from organisms tolerating only fresh water, through species whose physiology or behaviour enables them to cope with varying salinities (euryhaline species) to organisms requiring stable marine conditions (stenohaline species). Some mobile organisms actively move in and out with the tides making use of the sheltered productive waters while conditions are favourable.

In Western Australia and in some other areas of Australia however, as well as in South Africa, this definition falls short. In Swanland river flow is seasonal, often non-existent in summer, and tidal flows are small. Salinity does not change with tidal cycles but with seasons, and ranges from fresh to marine or even hypersaline. Additionally, all estuaries are ephemeral features of the landscape. In many areas they have filled with sediment brought by the rivers over the last 10,000 years, in the Holocene period, and since the last ice age. This demise has accelerated as entrance channels have become narrow and shallow with the natural and human-influenced movement of sand along the coast. Isolated from the sea for months or years, some estuaries may become hypersaline or dry out completely. Progression from a tidally dominated estuary where a diverse fauna once flourished to a limited or depauperate fauna is recorded by fossils, particularly shells of molluscs, foraminifera and crustaceans, below the estuary floor. These extremes of state have led to the definition of an estuary as used in *Swanland* to be: 'an estuary is a place which was the mouth of a river where tidal effects were evident and where river water and sea water mixed when first flooded by the Holocene marine transgression' (Hodgkin & Hesp, 1998).

Confusing names

The names of the estuaries of Swanland add to the confusion. In the lower reaches the Swan River is not a river; it is the estuary of the Avon River (from All Saint's Church, Henley Brook at the junction of Ellen Brook) to the sea, a distance of about 60 kilometres. In addition, the estuary has two main sources of flow from the Swan and Canning Rivers, hence the use of the term Swan–Canning. Many of the south-western estuaries are called 'inlets', while in Victoria and New South Wales they are commonly called 'lakes' (for example, Gippsland Lakes, Lake Macquarie, and Myall Lakes). The entrance to Leschenault Inlet has now been so extensively modified that the isolated area around the old channel is no longer estuarine and is termed the Inlet, while the larger original inlet now open to the sea via The Cut is termed the Estuary. More recently, to commemorate its origins and its ecology, the term Leschenault Inlet Estuary has been suggested for the larger water body. In dictionaries an 'inlet' is described as 'a small opening into the coastline or into the shore of a lake' (Moore, 1988).

A general theme of this book is to describe the variety and diversity of estuarine habitats and processes within a common framework to encourage people to go out and look. I hope to show that the estuarine areas we live around or visit are not only fascinating but are indicative of how our larger environment functions, and vital to our continued wellbeing.

Chapter 1 Understanding Swanland

Estuaries located at the mouth of a river, where tidal effects are evident and where river and sea water mix, are ephemeral features of the landscape. There have been estuaries in the past that no longer exist, and the estuaries of today will cease to exist as they fill with sediments and become isolated from the sea due to coastal processes of dune building associated with sea level changes on short or long time scales.

The estuaries of Swanland, like estuaries the world over, lie on layers of sediments deposited by rivers and seas and sculptured by winds in quite recent time—the last 10,000 years since the most recent ice age when the sea level was low. The rising sea levels that followed the melting of the ice caps flooded old river valleys forming the estuaries of today. These 'proto' estuaries would have been open to the sea, influenced by tides and inhabited by marine life. Fluctuating sea levels gradually led to isolation from the sea that continues today. The estuaries of today are therefore influenced by the landscapes of the past, the geology, landforms, rivers, climate past as well as present, the material brought by rivers, the eroding flow, the sediments and nutrients, the terrestrial and marine flora and fauna, and, more recently, human influences that have altered water flows and added nutrients.

THE SWANLAND LANDSCAPE—CATCHMENT TO THE SEA

Along the west coast the sediments lie in the Perth and Southern Carnarvon Sedimentary Basins, while those along the south coast form the Bremer Basin. The basins surround the Precambrian core of the continent, including the Yilgarn Craton (Block), a mass of granite and gneiss formed 2,900 to 2,500 million years ago in the Archaean. In the South-West this ancient core is often called the Darling Plateau or Range, and the Yilgarn Plateau in the south-east. The landscape of the Yilgarn Craton reflects its long history with periods of sedimentation, volcanic activity, intrusion of granites, folding, metamorphism due to pressure and heat, and millions of years of erosion. During the latest Carboniferous and earliest Permian (290 Ma), ice sheets ground down the landscape to the low, generally less than 400 metres, undulating terrain of today. Later periods of higher rainfall brought the development of river systems that are marked today by chains of salty lakes. The headwaters of the largest river systems such as the Murchison, Moore, Swan and Blackwood lie in this land of salt lakes hundreds of kilometres inland. Heavy rains occasionally reactivate these old drainage lines, and flood waters flow to the coast where the effect on the estuaries is dramatic.

Along the south coast, slightly younger deformed rocks in the Stirling and Barren Ranges form the Albany–Fraser Orogen dating back to the

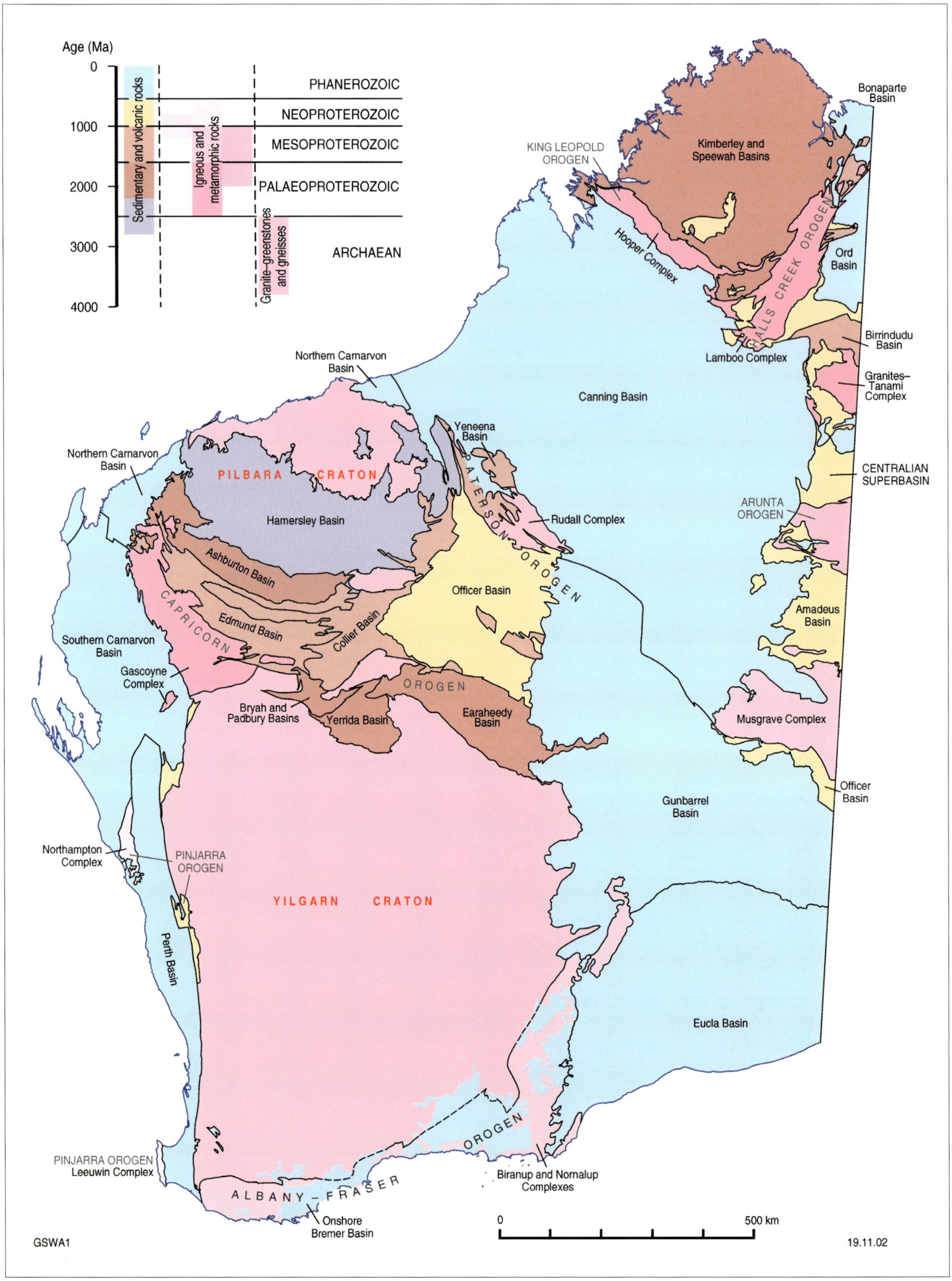

MAP 1.1 *Main tectonic units of Western Australia.* Courtesy GSWA & DoIR.

Proterozoic Era about 1,300 to 1,100 million years ago (Maps 1.1 and 1.2 and Fig. 1.1).

Carnarvon Basin

The Carnarvon Basin extends about 1,000 kilometres south from the Dampier Archipelago to the lower Murchison area, and contains marine sediments deposited in shallow seas and on coastal plains during the Ordovician through to the Cretaceous and Quaternary periods. In many areas, however, the full sequence of deposits is absent or limited in extent. The oldest rock, the Tumblagooda Sandstone of Ordovician age, outcrops along the Murchison River where the type section is found, and in headlands and cliffs further south. (See **Box 5.2 Tumblagooda Sandstone and Giant Sea Scorpions**.) The Tumblagooda Sandstone also extends into the Perth Basin where generally it is overlain by younger sands of Pleistocene and Holocene age, although it outcrops along the Bowes and Hutt Rivers. It is over 1,000 metres thick near the Murchison although it may be 3,000 metres thick in other areas. The Carnarvon Basin is bounded to the east by the Precambrian rocks of the Gascoyne Province, the south-east by the Northampton Block (Complex), and the south by the Perth Basin, extending westwards onto the continental shelf.

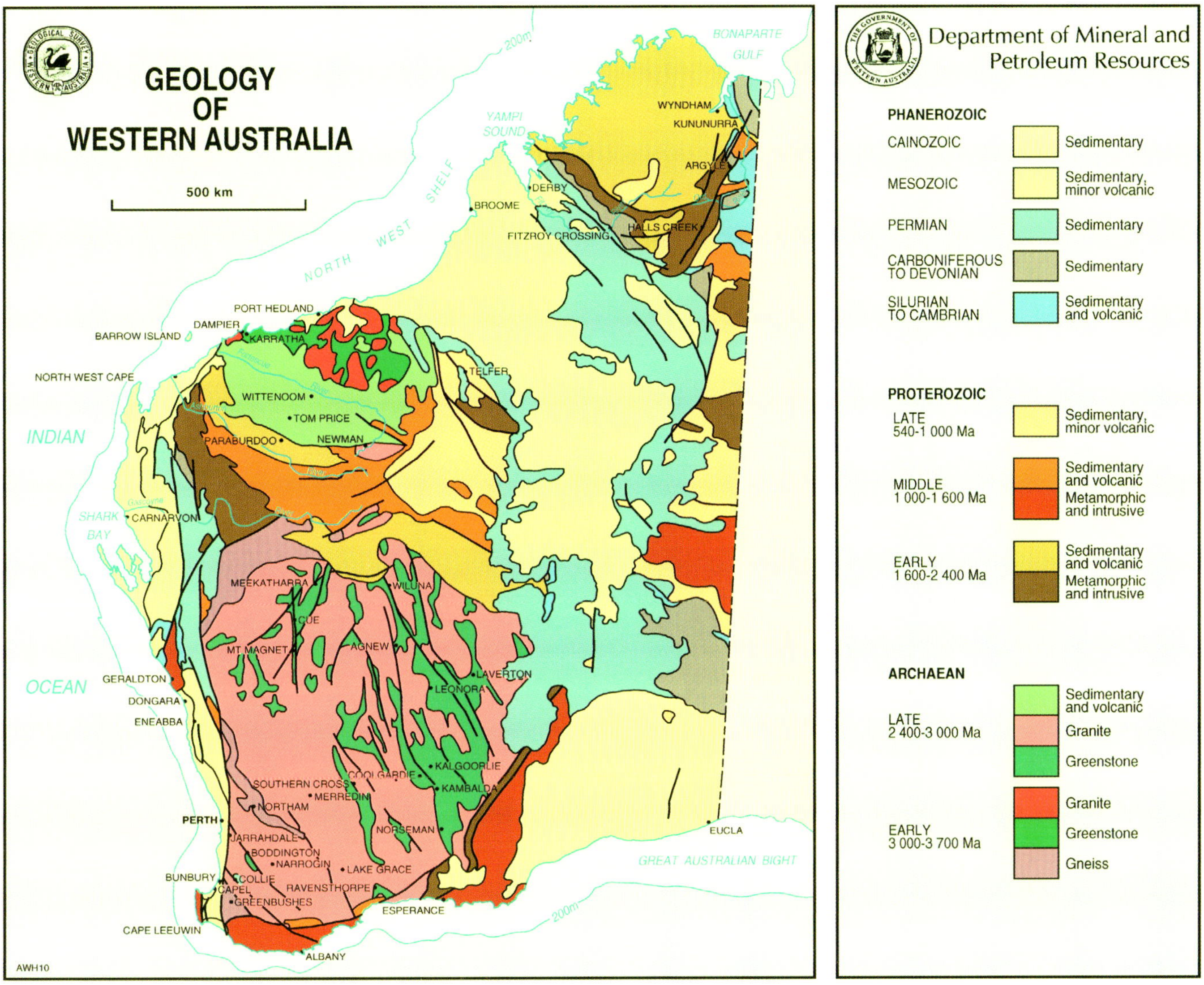

MAP 1.2 *Geology of Western Australia. Although the west coast estuaries of Swanland lie on the younger surface sediments there are older rocks below the surface; these outcrop north of Perth and on the south coast, while the catchments extend into the ancient core of the continent.* Courtesy GSWA & DoIR.

Perth Basin and Swan Coastal Plain

The Perth Basin is a sediment-filled narrow trough extending from the Murchison River to the south coast, a distance of over 1,000 kilometres. To the east the basin is bounded by the Darling Scarp, and it extends to the west under the continental slope. The Darling Scarp marks a fault line active throughout the Phanerozoic (Cambrian to the Quaternary), including the break up of the ancient super continent of Gondwana and separation of southern Australia from Antarctica.

The Perth Basin contains 15 kilometres of sediments laid down in ancient seas before the formation of the Darling Fault. Following the faulting and formation of a rift valley, sediments eroded from the ancient landmass continued to accumulate. The thickest of the sediments were formed during the Permian glaciation that ground down the landscape of adjacent high areas. After continental break up (136 million years ago in the Early Cretaceous), the Perth Basin became a continental shelf. A progressively deeper sea flooded this shelf from about 110 million years ago to about 80 million years ago. The continent started to rise at the

FIG. 1.1 *Simplified time scale of the earth's history,* modified from the International Stratigraphic Chart, Internal Commission for Stratigraphy, 2003 following the colour code of the US Geological Survey http://micropress.org/stratigraphy (Stratigraphic Information System pages). *Note: The term Quaternary is used in older publications for the most recent part of Tertiary; the terms Paleogene and Neogene now replace the term Tertiary. Block size does not indicate relative time. This table indicates only the period in which the organisms first appear in the fossil record.*

	EON	ERA	PERIOD	EPOCH	Million of years ago MA	Appearance of the first Plants	Appearance of the first Animals
	PHANEROZOIC	CENOZOIC	Quaternary	Holocene	0.01		
				Pleistocene	1.81		
			Neogene	Pliocene	5.32		
				Miocene	23.8		
			Paleogene	Oligocene	33.7		
				Eocene	55.0		
				Paleocene	65.5		
		MESOZOIC	Cretaceous		142		
			Jurassic		205	flowering plants	birds
			Triassic		250		Mammals dinosaurs
		PALAEOZOIC	Permian		292		
			Carboniferous		354		reptiles
			Devonian		417	conifers	amphibians ammonoids
			Silurian		440	ferns and 'seedless' land plants	jawed fish
			Ordovician		495		Fishes corals bryozoans crinoids
			Cambrian		545	marine eukaryotic algae	trilobites brachiopods bivalves gastropods
PRECAMBRIAN	PROTEROZOIC				2,500		
	ARCHAEAN				3,600	procaryotes	

end of the Cretaceous 65 million years ago when marine sediments were deposited in some areas. The Kings Park Formation that formed during this time represents a shallow estuarine deposit laid down in the ancestral Swan River.

The older sediments are generally lying hidden below much younger marine, aeolian (wind) and fluvial (river) sediments of Quaternary age. These older sediments only reach the surface around the Murchison, Geraldton, Mingenew, Hill River and Gingin on the Victoria, Darling and Dandaragan Plateaus.

West of the Darling Scarp are red gravels rich in iron and aluminium that gradually merge with alluvial soils of the Pinjarra Plain, that in turn abut a series of coastal dunes on the Swan Coastal Plain. Generally the deposits decrease in height and age from east to west and reflect sea level changes. Sea level was generally associated with the growth and retreat of the ice caps (eustatic changes), although tectonic forces that warped or elevated the earth's crust may have also played a role. The Ridge Hill Shelf is probably of late Neogene, once referred to as 'Tertiary' or early Pleistocene age; the Bassendean Dune System of early to middle Pleistocene; the Spearwood Dune System of late Pleistocene; and the Quindalup Dune System of Holocene age.

On the south coast, the Scott Coastal Plain is of similar formation to the Swan Coastal Plain. There are two divisions: a coastal dune belt backed by Tamala Limestone, and the Warren Dunes of scattered hills and ridges of leached sand interspersed with swamps.

Bremer Basin

Like the Perth Basin, much older sediments that occasionally outcrop along the coast underlie the Bremer Basin. Of Eocene age (40–35 million years ago), the sediments were deposited when the sea flooded about 270 kilometres inland leaving only the highest points in the landscape, such as the Stirlings, Porongurups and Barren Ranges, uncovered and forming islands and headlands as they still do today. Scattered outcrops of marine and terrestrial origin deposited during this time occur in a 60-kilometre wide belt from Northcliffe eastwards to Cape Arid. Inland near Norseman these are referred to as the Eundynie Group and nearer to the coast as the Plantagenet Group. The Plantagenet Group consists of the Werillup Formation and the Pallinup Siltstone. Coals (lignites) of the Werillup outcrop in the Fitzgerald River area and around Nornalup Inlet. Of late Eocene age these contain algae, foraminiferans and bryozoans as well as echinoids, brachiopods and molluscs.

The Pallinup Siltstone or white, brown or red siltstone and spongolite is exposed along the north side of Beaufort Inlet and Pallinup River, the Gordon Inlet, in various parts of the Fitzgerald River National Park and in the Stirling Ranges. (See **Box 10.1 Sponge garden to rock formation—the Pallinup Siltstone or spongolite**.)

Later the sediments were raised and tilted along the Jarrahwood Axis with further tilting leading to the formation of the Ravensthorpe Ramp that slopes southwards to the sea. This upward tilting probably resulted in the short deeply cut rivers that currently flow to the Southern Ocean. In the west, rivers like the Blackwood were 'rejuvenated' as the valleys deepened and widened, and new southward flowing rivers such as the Frankland formed by cutting back into the plateau. East of Albany, rivers like the Beaufort and Gordon cut deep into the spongolite forming spectacular gorges.

The older sediments of the Bremer Basin, like those of the Perth Basin, are covered by younger marine and shore sediments deposited since the estuaries first formed in the Pleistocene. At that time the estuaries were much larger and the fauna living in them more marine.

FLAT COASTAL LANDSCAPE OF WATER AND WIND

'Flatness' and sand dunes are the dominant features of the coastal landscape. On the Swan Coastal Plain, the background hills of the Darling Range generally reach less than 400 metres in height and are hardly worthy of the term 'ranges'. People of Western Australia generally become accustomed to the flatness, but for new arrivals this is very characteristic. This land of dunes has no doubt led to the nickname for Western Australians of sandgropers (Fig. 1.2). (See **Box 1.1 Sandgropers—the people of Western Australia**.)

The coastal plains of Swanland were created by invading and retreating seas due to global changes in sea level in response to environmental conditions dictated by global warming and cooling. The processes that created the landscape also created the estuaries of the past, the estuaries of today and the estuaries of the future due to both natural processes and human activities. For our story of the estuaries we need therefore to think about geological time scales, millions to thousands of years, and human influences most particularly in the last 170 years, which are marked in decadal, yearly and daily scales.

The flat landscape of Swanland also affects processes within the estuaries. Water movement is slow and drainage is poor. Nutrients leach through the sand, so the soils are infertile and fertilisers are needed to boost plant growth. The fertilisers however also leach to the ground water and through to the lakes and estuaries, where they enrich the waters fuelling algal and plant growth. Sediments and nutrients accumulate over time, as estuaries are low points in the landscape where flushing rates can be relatively limited.

Lines of dunes in Swanland

Although there were local and regional differences in the rise and fall of sea level during the last glaciation, and uplift of the land due to tectonism, the dunes on the Swan Coastal Plain are probably typical of similar formations along the whole Swanland coast.

Originally the dunes would have contained silica and other minerals eroded from the ancient inland granitic rocks, together with pieces of calcium carbonate from shells, coralline algae and single-celled organisms, the foraminifera that once lived in the surrounding sea. The youngest dunes along the coast retain the carbonate material, but it has since dissolved away in the older dunes under the influence of rain made slightly acid by carbon dioxide in the air. It has also been suggested that the eastern dunes and the underlying Ascot Formation had a more complex history influenced by rivers and marine events, and that they may have contained more silica than carbonate.

About 130,000 years ago, the coastline was about 10 kilometres further inland and the oldest of the three dune lines, the Bassendean Dunes that abut the Pinjarra Plain, were deposited. Although except for the western margin most of the dune topography disappeared long ago, remnants of the dune origin remain in higher areas of grey to white sands with a core of yellow sand. In lower areas, the white sands are underlain by a clay or 'coffee rock' layer and the watertable is only a few metres below the surface. In winter the small depressions link up to form an almost continuous wetland. The Bassendean sands are siliceous, acidic, infertile, and do not hold nutrients. These old leached sands and low topography hold the key to the 'problems' in the Peel–Harvey and other estuaries (Map 1.3). The dunes

Box 1.1

SANDGROPERS—THE PEOPLE OF WESTERN AUSTRALIA

The flat coastal plain of sand leads to the title given to Western Australians of 'sandgropers', derived from the name of the mole cricket *Cylindracheta* that tunnels through the sand just below the surface (Fig. 1.2).

FIG. 1.2 *Mole cricket or sandgroper* Cylindracheta, *and its tracks on a sand dune. The tracks are quite distinctive showing where the cricket has tunnelled beneath the sand.*
Courtesy Dr Terry Houston, WAM.

and the underlying Ascot Formation are also important features affecting shallow ground water flows and contamination.

Sea level fell with the Würm Ice Age and the coastline moved westwards reaching a low point 17,000–18,000 years ago (Map 1.5). At this time sea level was about 130 metres lower than present and the coast about 40 kilometres further west. The Spearwood Dunes were formed as sea level fell. In the Perth area a number of subdivisions of the Spearwood Dune

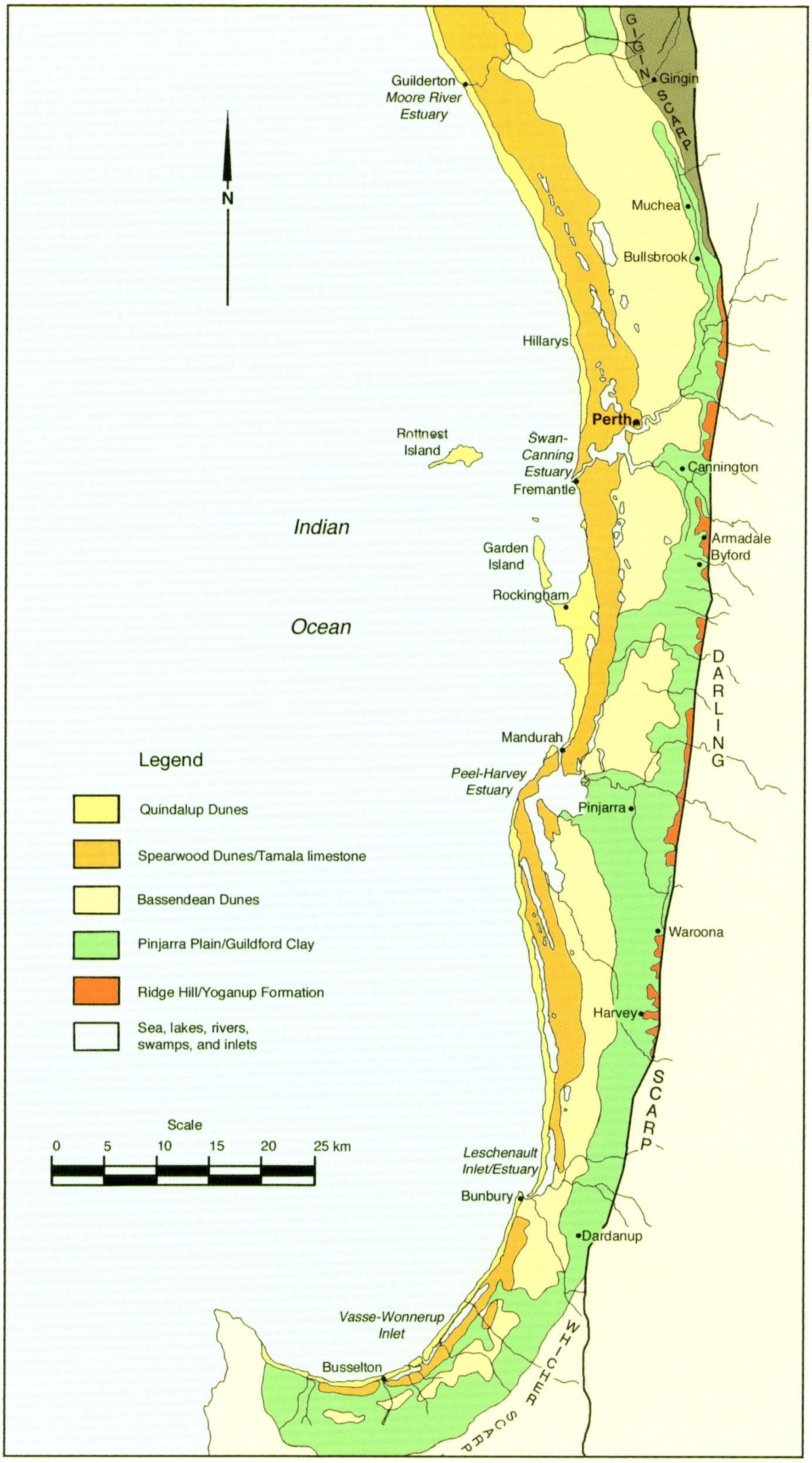

MAP 1.3 *Generalised surface geology and landforms of the lower Perth Basin.*
Adapted from Seddon, 1972; & Davidson, 1995.

System are now recognised: the Trigg, Karrinyup, Gwelup, Balcatta and Yokine Dunes, and the western margin of the Bassendean Dunes termed the Gnangara dunes (Map 1.4). In some areas the dune base contains sediment deposited in shallow marine areas and beaches. These younger dunes are much higher, steeper and dune-shaped. They often have a limestone core where the shelly carbonate material leached from the top sands has been redeposited. Limestone from this period, also called the Coastal, Aeolian or Tamala Limestone, is found along the coast through to Shark Bay and in the offshore islands such as Penguin, Garden, Carnac and Rottnest and their linking reefs. Soils associated with the Spearwood Dunes are subdivided according to colour and depth to limestone. Near the coast, the thin layer over limestone is referred to as Cottesloe Soil Association while those further inland as Karrakatta sand or soil. These yellow-brown soils coloured by iron oxides are generally infertile although some nutrients are retained bound to the iron. The limestone holds a thin layer of fresh water that floats on top of the saltier water. Rich in carbonate, the fresh water flows westwards and in places seeps to the surface along the shore of Lake Clifton near Mandurah, where it contributes to the growth of stromatolites or thrombolites (Fig. 1.3). (See **Box 1.2 Stromatolites and blue-green algae—early forms of life, Box 2.2 Seashells at Peppermint Grove** and **Box 3.3 Blue-greens—ancient forms of life a modern day nuisance**.)

The end of the last ice age about 10,000 years ago brought another rise in sea level, flooding the western dunes that today form the shallow reefs and islands of Garden and Rottnest Islands. Sea level reached its present height about 6,000 years ago, and brought the formation of the Quindalup Dunes along the coast. The dune-forming processes are still active under the influence of wind-blown sand from the beaches. The white sands of the Quindalup Dunes (Safety Bay Sands) are largely composed of shell material and gradually stabilise as vegetation adapted to the salty, dry and hot conditions becomes established. Near the Peel–Harvey at Mandurah the dunes are narrow and higher, but further north near Rockingham and Jurien Bay these dunes are lower and form an extensive area of beach ridges.

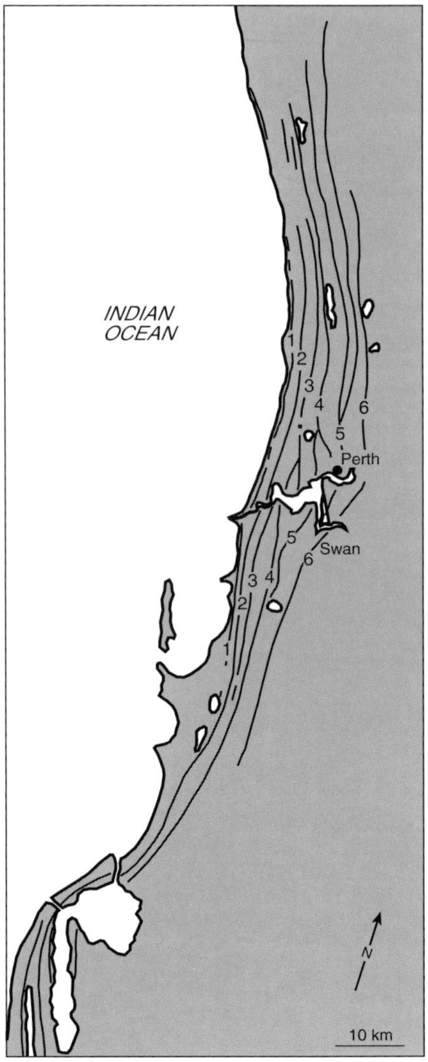

MAP 1.4 *Major dune units in the Perth Region: 1 Trigg, 2 Karrinyup, 3 Gwelup, 4 Balcatta, 5 Yokine, and 6 Gnangara Dunes.* Adapted from Bastian, 1996; Courtesy D. Haig, UWA.

Lakes, inlets and estuaries—following the dune lines

The landscape of the coastal plain is not only flat, but is only slightly higher than sea level and waterlogged in winter. Water accumulates in low points of the landscape, often at the boundaries between the dune systems. Wetlands of swamps, lakes, floodplains, rivers and estuaries on the Swan Coastal Plain cover over 360,000 hectares or 25 per cent of the land area. Many of these reflect the shallow underlying watertable. Some are 'lakes' and hold permanent water, some are 'sump-lands' only containing water during winter and spring, while others, 'damp-lands', are waterlogged in wet seasons. Creeks and rivers about 3,869 kilometres in length but only about 1,074 hectares in area drain to about 29,249 hectares of estuaries.

This land of winter waterlogging posed great problems for the first European settlers who attempted to drain the flat landscape. The problem of winter waterlogging continues today even though over 80 per cent of the wetlands on the Swan Coastal Plain have been filled, drained or cleared. Many low-lying areas are still inundated when rainfall is heavy, and water cannot flow to the ocean as much of the coastal plain lies just above sea level.

The three dune systems, all characteristically infertile, were leached of nutrients long ago, yet because of their proximity to population nodes and the ease of cultivation, these areas are some of most intensively developed areas of horticulture and agriculture. The secret to the success of cultivation lies in the use of fertilisers. The downside for cultivation is the poor nutrient retention properties of the soils that require continuous applications. Fertilisers gradually enter the ground water and move through to the rivers, eventually reaching the estuaries. Fertilisers are also a major contaminant to ground water in urban areas, where along with other diffuse sources from septic tanks and point sources from landfill and industry, they pose a threat to potable ground water supplies as well as the lakes and estuaries. (See **Chapter 2 The Swan–Canning Estuary: With the water flow—nutrients; Box 3.1 Ingredients of a green soup**.)

Digging the estuaries of Swanland—rising and falling seas

Evidence of the formation and history of the estuaries is revealed by the submarine canyons opposite the Swan River and the Peel–Harvey Inlets on the west coast, and many estuaries along the south coast including some that have been closed for long periods of time. The estuaries of the Swan Coastal Plain cut down into Pleistocene sediments or overlie even older sediments, while those on the south coast are in bays or valleys in rocks of Eocene or older age. Although the timing of sea level rise and fall is still debated, land plants and peat beds below the present Swan River indicate that sea level was about 20–30 metres lower about 10,000 years ago (Table 1.1, Maps 1.5 and 1.6). The sediments also reveal that the area was flooded during the high sea levels of the last interglacial 8,000–4,000 years ago (the Holocene marine transgression). When flooded, the entrance channels were deep (Swan 30 metres, Peel 11 metres, Hardy more than 10 metres and Culham about 25 metres) and there was free tidal exchange with the sea. The fossil deposits indicate that there was a rich and diverse marine–estuarine fauna. (See **Box 2.1 An older estuary hidden under Guildford; Box 2.2 Seashells at Peppermint Grove; Chapter 3 The Peel–Harvey Estuarine System: Part of the sea—the past, and now?; Chapter 6 Hardy Inlet and estuaries of the Naturaliste—Leeuwin Ridge: Hidden history;** and **Box 8.1 Sediments revealing times past**.)

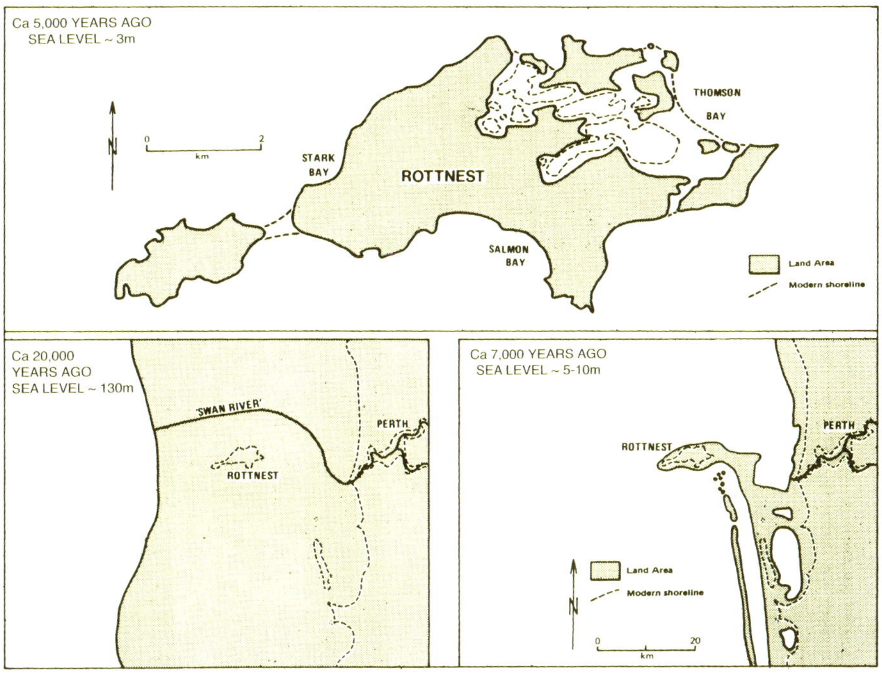

MAP 1.5 *Alterations in the landscape with sea level changes in the vicinity of Perth and Rottnest Island. About 20,000 years ago the sea level was lower and the Swan River crossed the wide coastal plain discharging to the north of Rottnest Island. The sea level then rose gradually flooding the area between the dunes: Rottnest and Garden Islands were probably cut off from the mainland about 7,000 years ago while Cockburn Sound was probably flooded 5,000 to 6,000 years ago.* Adapted from Playford & Leech, 1977.

Box 1.2 **STROMATOLITES AND BLUE-GREEN ALGAE—EARLY FORMS OF LIFE**

Stromatolites are sedimentary rocks formed by blue-green algae or cyanobacteria that bind sediment together, so they could be thought of as 'living rocks'. Stromatolites resemble the oldest form of organised life. They have a rich fossil record dating back over 3,500 million years, and the oldest forms are found near Marble Bar in the Pilbara of Western Australia. They were most diverse about 1,000 million years ago.

Living forms are only found in a few areas in the Bahamas, Bermuda and Australia. In Western Australia they occur at Hamelin Bay in Shark Bay, in Hutt Lagoon, on the Swan Coastal Plain in Lake Thetis at Cervantes, Lake Richmond at Rockingham, Lake Clifton near Mandurah, as well as the lakes on Rottnest Island (Fig. 1.3). The discovery of living forms has been important as their biology helps in understanding Precambrian environments and the evolution of life.

FIG. 1.3 *Stromatolites along the north-east of Lake Clifton exposed during low water levels in summer, March 2001.* A. Brearley.

When stromatolites were abundant, the atmosphere consisted mostly of nitrogen and carbon dioxide with little oxygen. Life consisted of single cells of bacteria or prokaryotes. About 2,500 million years ago the cyanobacteria started to release free oxygen in the chemical reaction of photosynthesis. This extra oxygen first caused the iron dissolved (ferrous iron) in the ancient oceans to settle out forming the iron oxide deposits of the Pilbara, but as oxygen concentrations rose it 'leaked' to the atmosphere until it formed about 21 per cent of the volume as it does today.

Between 2,000 and 1,700 million years ago, two different cell forms merged and formed one cell or one cell within another, giving an internal structure to the eukaryotic cell. This basic cell structure became the building block for the development of more complex forms of life. Then 600 to 650 million years ago, oxygen concentrations were sufficiently high for the first oxygen-using organism to evolve.

In the ancient seas before other life evolved, stromatolites formed great reefs. Stromatolites actually grow very slowly—about 3.5 centimetres in 100 years; coral reefs in comparison grow over ten times faster. As the cells grow upwards, chemical interactions between the cells and water rich in dissolved calcium and carbonate favour the settling out of these compounds over the cells, and the cells resume growth towards the light creating a layered structure. The layers, like tree rings, show daily and seasonal growth and resting times that can be used to work out how day length and yearly cycles have altered. Sun angles can also be calculated as they grow towards the sun. Using information revealed by stromatolites and other organisms, we now know that the rate of rotation of the earth has decreased over time.

Stromatolites can take a columnar, dome or club-shaped form, and those found along the Swan Coastal Plain are called microbialites or thrombolites signifying their less organised, unlayered structure in which calcium carbonate accumulates in clots or clumps.

At Lake Clifton, stromatolites are exposed to seasonal changes in salinity. The water is fresher during winter and more saline during summer, but always less saline than sea water. Stromatolites are mostly found along the eastern margin associated with low salinity, highly alkaline ground water seeps, unlike the situation in Shark Bay where stromatolites grow in hypersaline water. Other more saline coastal lakes do not generally contain stromatolites, but mats of algae grow across the bottom.

Only a limited number of animal species are found around the stromatolites but they can be quite abundant, forming an important food source for the fish and birds that use the wetlands. Some of the fish such as the gobies and atherinids as well as amphipods, isopods, and the shrimp *Palaemonetes australis* are related to marine species that have been isolated since the lake was formed, while the snails *Potamopyrgus* sp., the bivalve *Arthritica semen*, beetle and caddis fly larvae are more common in freshwater lakes and estuaries. The salt lake snail *Coxiella* is an indication of the high salinities experienced over summer.

Since they grow slowly, today stromatolites live only where conditions are harsh and other organisms do not out-compete them. In this changing world the competitors are increasing, and stromatolite responses can be sentinels of a changing environment.

Recently at Lake Clifton, some stromatolite formations have been overgrown by the green alga *Cladophora vagabunda*. The growth appears to be fuelled by nutrients in the ground water, probably in response to the further clearing for farming and housing, and greater use of fertilisers. To ensure that these ancient life forms survive the modern era, we will need to improve land use practices and reduce nutrient flows. Buffers of reeds and rushes have been shown to reduce nutrient flows in other areas. Around Lake Clifton maintenance or restoration of buffers 100–150 metres wide would not only decrease nutrient flow but maintain the aesthetics and values of the waterbird habitats.

Source: Chivas et al., 1990; Davies & Lane, 1996; McNamara, 1992; Moore, 1987; Moore et al., 1983.

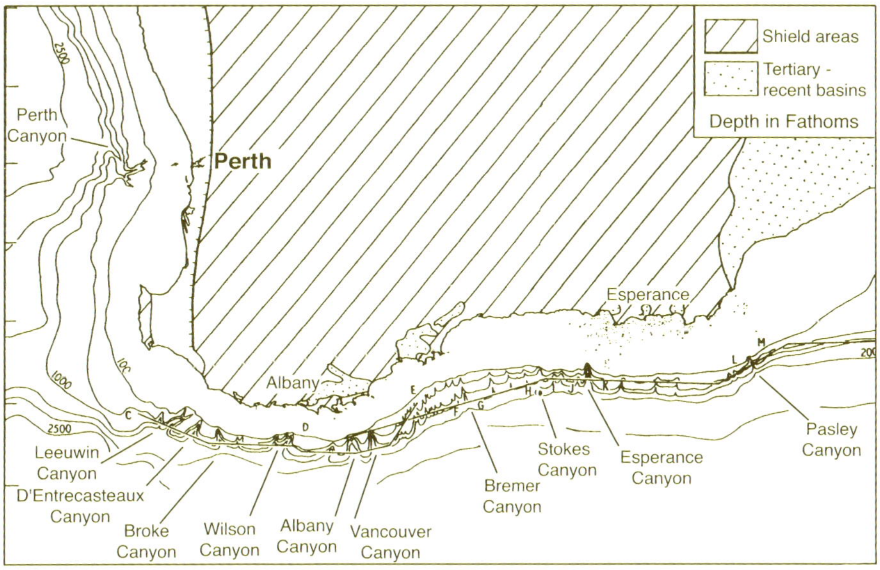

MAP 1.6 *Location of canyons carved by ancient river systems when the sea level was lower. Note the northern arm of the Perth Canyon aligns with the Swan River, while the southern arm may have been formed by outflow from the Murray and Serpentine Rivers through what is now the Peel–Harvey. Although many canyons around the coast align with the present drainage systems others do not. The timing (Pleistocene or much earlier in the Paleocene) and mechanism of their formation is still debated. Some scientists believe that they may have been formed by submarine mudslides off the continental shelf.* Adapted from Conolly & Von der Borch, 1967, Von der Borch, 1968.

TABLE 1.1 *Age of fossil deposits from estuaries in Swanland. There are similar shell deposits in other estuaries and coastal lagoons in the north to Thomas River in the south-east.* Adapted from Hodgkin & Hesp, 1999.

	Location	Height (m)	Years before present
Swan	Ascot Formation		
	40 km from mouth	− 4	6,660–4,500
	Shore	+ 1	6,000–4,000
	Fremantle	+ 3	Mid to late Pleistocene
	Minim Cove	+ 7	Pleistocene
		+ 2	
	Peppermint Grove	+ 7	130,000–170,000
Peel	Basin		5,250–4,500
	Shore	+ 1–2	6.910–4,240
	North of lagoon	+ 1.8	4,910
Leschenault	Lagoon		8,055–5,555
Hardy		− 5	4,475–3,880
Wellstead	Lagoon terrace	+ 1.5	4,900–4,430
	Sandbank 5 km from entrance		3,930
Culham	Lagoon shore		3,660
Stokes	Lagoon shore		4,380

Source: Swan: Kendrick, 1977; Yassini & Kendrick, 1988; Peel: Brown et al., 1980; Semeniuk & Semeniuk, 1991; Kendrick, unpubl.; Leschenault: Semeniuk, 1985; Hardy: Hodgkin, 1978; Wellstead: Kendrick, unpubl.; Hodgkin & Clark, 1987; Culham: Kendrick, unpubl.; Hodgkin & Clark, 1990; Stokes: Kendrick, unpubl.; Hodgkin & Clark, 1989.

CLIMATE AND WEATHER—PAST, PRESENT AND FUTURE?

Over its long geological history Australia has been through major changes of climate that helped to shape the landscape of hills, valleys and soils. The landscape we see today, shaped by past climates, together with current precipitation determine the flow of water to the estuaries. Records of the last few hundred years, however, are a reminder that climate varies over shorter time scales, and coastal marine and estuarine processes need to be viewed with an eye on the past.

Climate past—shaping the landscape

During the Carboniferous or Permian periods when the southern continents were one great landmass of Gondwana, ice sheets flattened and ground down the landscape. Later during the Cretaceous (142 million years ago) as the continents split apart, Australia had a more polar position, however the temperature range across the globe was not as extreme as it is today. The climate of Australia would have been similar to that of today, although there was a cooling trend and then a warming trend at the beginning of the Tertiary, now referred to as the Paleogene and Neogene.

Faulting and rifting associated with the break-up of Gondwana also created the Darling Fault, and the Perth Basin along the western margin of Australia, where sediments accumulated. Separation of Australia and India was probably completed in the Early Cretaceous although the fault lines remained active and caused subsidence of the continental margin. Later in the Cretaceous when the sea level reached the highest point during the Phanerozoic, about 250 metres higher than present, marine sediments such as Molecap Greensand, Gingin Chalk and Poison Hill Greensand were laid down along the edge of the continent (Table 5.9 and Figure 5.34).

About 65 million years ago Australia and Antarctica were still partly connected in the region near Tasmania, but the temperatures were higher and there was no ice cap. The Eocene Epoch was characterised by high sea

Box 1.3 ISOBARS, HIGH AND LOW PRESSURE SYSTEMS, AND COLD FRONTS

Isobars are lines drawn on a map joining places of equal air pressure measured at sea level. The direction and speed of the wind are related to the mean sea level pressure pattern, with close spacing of isobars meaning stronger winds. In the Southern Hemisphere wind flows in a clockwise direction around low pressure systems and anticlockwise around high pressure systems.

Cold fronts form when a mass of cold dense air wedges under a mass of warm lighter air. On a weather chart, cold fronts are shown as a 'barbed-line', the barbs indicating the direction of movement. There are many factors that influence how cold fronts affect the weather. For example, the speed and shape of the advancing cold wedge, and dry or moist state of the warm air determines how the clouds form and the extent of rain.

When a cold front advances, the winds usually from the north and north-west gradually increase in strength, the clouds become darker and lower, and rain falls. As the front passes the wind shifts to the west and south, the temperature drops, the air pressure rises and clouds begin to clear.

A **trough** is an elongated area of low air pressure, often an extension to a depression, that is generally an approximately circular or oval-shaped low pressure area. A front is always marked by a trough, though many other troughs are not associated with fronts at all. The 'west coast trough', a dominant influence on the meteorology of Western Australia during warmer months, forms as a result of the temperature contrast between land and ocean.

levels and the deposition of the Kings Park Formation in the Perth Basin and the Pallinup Siltstone on the south coast. As Australia separated still further from Antarctica (40 million years ago), the development of a sea way and the circumpolar ocean currents brought greater temperature differences between the poles and the equator, and the development of polar ice sheets. Later Australia's northwards migration brought a decrease in rainfall, and the land became drier as the mediterranean climate established.

The last part of the Neogene, the Pliocene, was characterised by waxing and waning of the polar ice sheets, the glacial and interglacial periods that exerted the major control on global sea level. The rising and falling sea levels formed the dune systems of the coastal plains where the estuaries of Swanland lie.

There has been great variability in global temperatures over the last thousand years and from the Medieval Warm Epoch 1500 to 1800s, the time of the 'little ice age'. Differences from year to year overlain by decadal and longer-term variations confuse projections of climate change induced by human activity. Computer modelling, where the different influences can be incorporated, and better separation of the 'noise' of natural variability that makes detection of real changes difficult, now allows a greater interpretation of observations.

Climate present

The overriding influence on water flow to the estuaries of Swanland comes from the mediterranean climate, a pattern of cool wet winters and long dry summers dictated by the movement of high pressure systems or anticyclones from west to east across the continent (Box 1.3). In the Southern Hemisphere the winds blow in an anticlockwise direction around a high pressure system. In winter these anticyclones usually pass over the southern continent (26° S to 34° S). As winter gives way to summer the anticyclones moves further south and they pass below the south coast (35° S to 45° S) (Map 1.7).

In winter the winds come from the west off the ocean bringing colder, moister air and rain. The winds are strengthened by a series of cold fronts, depressions or low-pressure areas that move across the south-west. Storms generally arrive from the north-west with the wind direction moving to the west, south-west and south with the passage of the front and the low-pressure system. Between the depressions, clear sunny but cool conditions are features of the environment. With the periods of rain the soil moisture builds up, and the creeks and small rivers start to flow to the estuaries and sea, although the amount of water reaching the estuaries varies from year to year. In years when the fresh water run-off is low, the lighter fresh water flows seawards along the surface. In other years when the river flow is stronger or

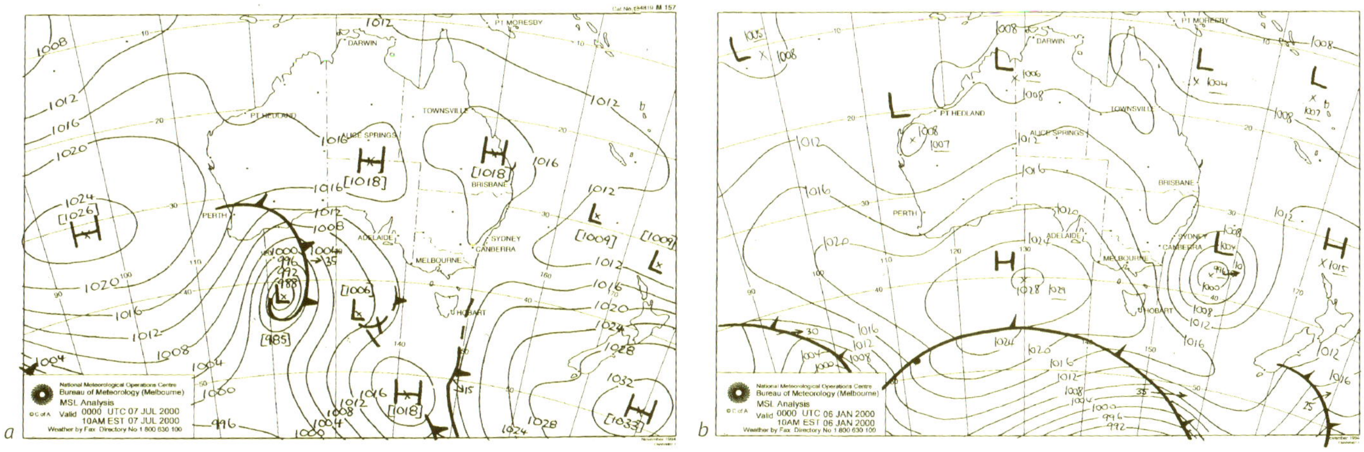

MAP 1.7 *Weather charts showing the typical (a) winter and (b) summer positions of the pressure systems. In the Southern Hemisphere, winds blow anticlockwise around a high pressure system and clockwise around a low pressure systems. Winds in Swanland tend to be from the west off the ocean in winter and from the east off the dry continent in summer.* Courtesy BOM, WA.

more persistent, the estuaries may be flushed completely and a freshwater environment persists until river flow ceases and tides again bring marine water upstream. (See Graph 2.3.)

In summer the main winds blow from the east off the warm dry heartland, particularly in the morning. Temperatures are high but sea breezes usually blow from the cooler ocean (where pressure is slightly higher) in the late afternoon, sometimes reaching speeds in excess of 50 kilometres per hour or Force 7 on the Beaufort scale, a near gale. Westerly winds are generally too far south to bring summer rain to the west coast although some showers still fall along the south coast. In the north of the state, the high temperatures also favour the development of monsoonal depressions or cyclones that bring cloudy conditions and some heavy rainfall. Most cyclones cross the coast north of Exmouth Gulf, but heavy rainfall associated with dissipating cyclones brings floods to the interior. As a result cyclonic events, although infrequent and unreliable, can exert a major influence on the rivers and estuaries. Rain deposited on the parched landscape can be very erosive, particularly in agricultural areas where there is no binding vegetation during summer. Flood waters fill the shallow river courses rejuvenating the often forgotten old drainage systems, mobilising sediment and salt as they spread out across the land turning the upper catchments into a vast red-brown lake system that may extend for thousands of kilometres. Notable events of this type were cyclones Bruno and Errol in January 1982 that brought 100–200 millimetres of rain to a wide area; the resulting floods breaking the bars of many south coast estuaries. The freshwater flush from this event also sparked a second bloom of *Nodularia* in the Peel–Harvey. Similarly, summer floods following heavy rain in the Avon catchment initiated a blue-green algal bloom in the Swan in February 2000 that resulted in closure of the estuary to all recreational activity. (See **Box 2.4 Tiny cells on the rampage take a city's playground off limits**.)

Rainfall to river flow

Rainfall is highest (1,400 millimetres per year) and relatively reliable in the extreme south-west of Swanland, but decreases and becomes more variable inland and to the north and east (Map 1.8). Evaporation shows a reverse pattern with higher rates away from the coast, and is generally greater than rainfall except in the extreme south-west. Most of the rainfall in Swanland occurs from May to September, and the south coast also receives rain in summer and autumn. Infrequently heavy rainfall occurs in summer with dissipating tropical cyclones.

As a result, the size of a catchment may have little relevance to the volume and pattern of river flow to the estuaries. For example, much of the largest river catchment, the Blackwood, lies in areas of low rainfall and generally flow to Hardy Inlet comes in winter from the middle and lower catchment. There are, however, occasional summer floods. Similarly, although the two rivers the Frankland and the Deep that flow to Nornalup Inlet contribute similar amounts of water to the inlet, the catchment of the Frankland River is six times the area of the Deep River. Much of the Deep catchment lies in drier inland areas, and most flow comes from a relatively small area near the coast.

Remembering the past—is there such as thing as a normal year?

Rainfall may have decreased but in many areas run-off has increased due to rising ground water following clearing. This creates a problem for low-lying areas where structures have been built. Although levee banks and diversion drains have been constructed, many only offer partial or temporary

MAP 1.8 *Rainfall isohyets across Swanland, south-western Australia.*

Map drafted by DoE.

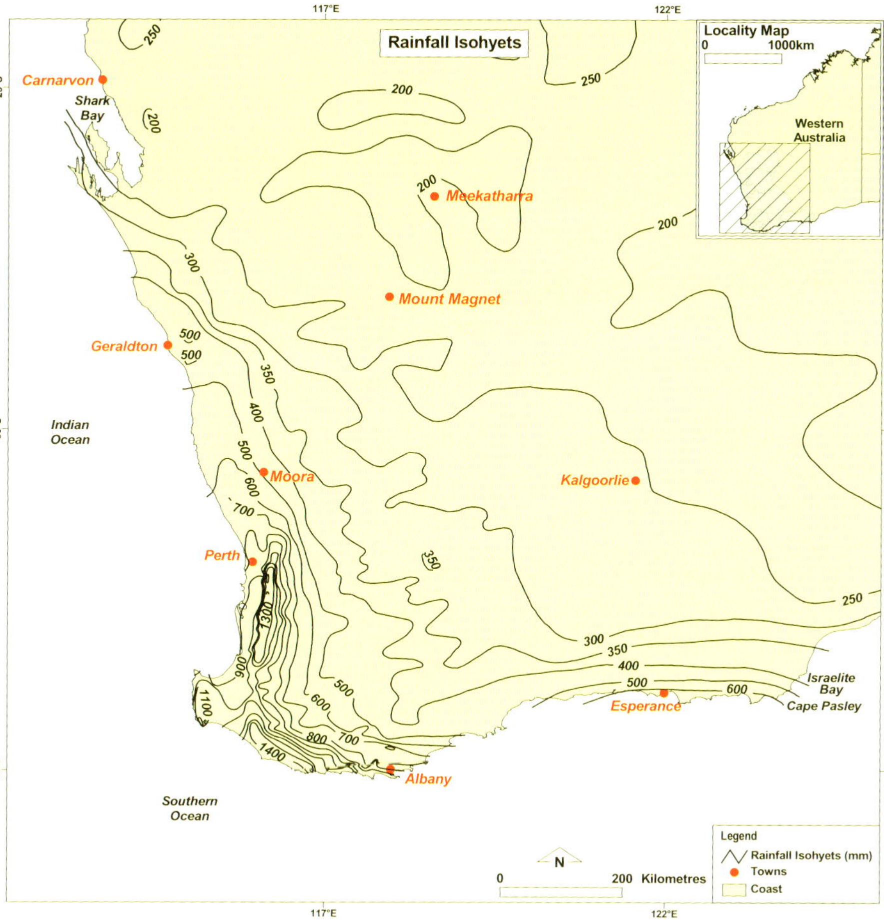

protection during flood events as gradients are low, and there are few options for diverting floodwaters without creating other problems downstream.

For example, the Avon flooded sixteen times between 1900 and 1964, and the 1955 flood was so devastating that it led to the Avon Training Scheme of the 1960s. Similarly, the Harvey, Preston, Collie and Vasse Rivers were also straightened and deepened. The drier 1970s, however, reduced the need for such measures and land use was modified with less regard for past problems. More recent floods around the Greenough River, Moora and Busselton in the late 1990s are a reminder that rainfall can be heavy over very short periods and the landscape is flat and drains slowly.

Climate future—less rainfall?

In addition to variation in rainfall from year to year throughout Swanland, there has also been a marked decrease in rainfall since the 1930s. Of particular importance has been a marked downward step in the 1970s to a lower rainfall regime that has persisted. (See Graph 2.1 and Table 7.1.)

Where the decrease has been most pronounced, inland of the lower west coast and to the south near Walpole, the decline has been 3–4 per cent per decade (1910–1989). It remains to be established, however, to what extent the decline in rainfall may be part of a natural climatic cycle so that wetter times will return, and to what extent it may be part of longer-term global change caused by human activities. In addition to the decline in rainfall there appears to be a delay in the onset of winter rains, the largest decreases being evident in late autumn and early winter months. Projections by CSIRO of rainfall changes in the South-West due to enhanced greenhouse effect tend towards a decrease but range from –20 per cent to +5 per cent by 2003, and

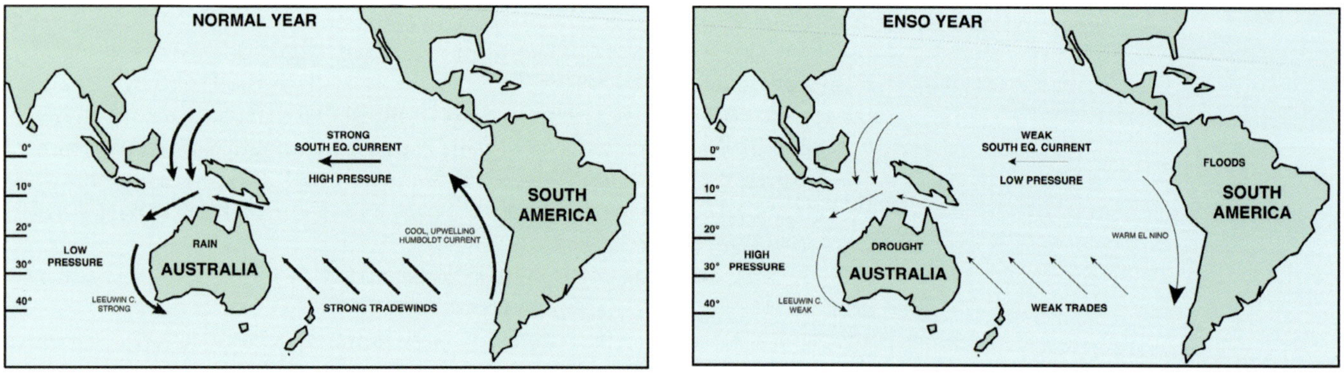

MAP 1.9 *Normal and El Niño–Southern Oscillation (ENSO) conditions.* Adapted from Pearce et al., 1997.

–60 per cent to +10 per cent by 2070. Evaporation rates across southern Australia also appear to be increasing, and this may compound the effects of declining rainfall.

In Swanland, the decline in rainfall appears to be associated with a general warming of the Indian Ocean over the last few decades, possibly caused by variation in the large-scale atmospheric circulation. The warming is also linked to El Niño–Southern Oscillation (ENSO) events that influence tropical flow through the Indonesian Archipelago, and consequently the strength of the Leeuwin Current that brings warm water southwards along the Western Australian coast (Map 1.9).

Changes in rainfall lead of course to changes in river flow, and impact on the estuarine areas. Lower river flows reduce flushing of the estuaries allowing sediments and nutrients to accumulate, and increase the chance of estuaries drying out in summer. Reduced flushing, particularly in estuaries in drier areas or in estuaries where the bars close more often, also hastens progression to a closed lagoon state that brings major changes to the communities of organisms living in them.

Water movements— the tides and currents of Swanland

The astronomical tides due to solar and lunar attraction along the south-western Australian coastline are termed 'microtidal' as they have only a small range. This is generally about 0.4 metre with a maximum range of 1.1 metres at Fremantle, and slightly greater at Albany and Esperance. The tides are also mainly diurnal, ranging from high to high or low to low once only in a 24-hour cycle.

Within estuaries astronomical tides are 'damped', the rise and fall less than in the nearby ocean. The amount of damping depends on the length of the entrance channel, the depth of the channel and the extent of the bar. When the entrance channels through the bars are open and as river flow decreases in spring, tidal movements bring marine water into the estuaries in what is termed a salt wedge along the estuary bottom. With progression of the salt wedge upstream, the waters become salty. Saline conditions persist over the usually dry summer months until the autumn rains initiate another fresher phase. In the Swan–Canning the tides are now much greater than they were at the time of European settlement due to the removal of the rocky bar at the entrance and deepening of the estuary channels. Similarly, tides in the Peel–Harvey have increased in amplitude since the Dawesville Channel was constructed.

Water level along the coast and in the open estuaries is also affected by meteorological changes due to atmospheric pressure and wind. Locally, pressure changes modify astronomical tides with high pressure lowering low tides, or low pressure allowing water levels to rise to greater heights.

Water level may also rise with onshore winds, while winds blowing offshore can lower water level along the coast. It is of interest that these non-tidal changes in water level were first described and ascribed to pressure changes by Hodgkin and Di Lollo in 1958. Events such as tropical cyclones in the north can also affect estuaries through the propagation of shelf waves that travel southwards over a period of days raising water. (See **Chapter 2 The Swan–Canning Estuary: Water movements—tides, currents and winds**.)

The Leeuwin Current that flows most strongly in winter from the tropics along the edge of the continental shelf at a rate of 0.2–0.5 metre per second, about one knot, also raises water levels in winter by 10–20 centimetres more than the height of predicted astronomical tides. This body of less saline, low nutrient, warm water brings tropical life southwards accounting for the unusual catches of fish and mud crabs in the estuaries of the Moore and Swan as well as Hardy and Wilson Inlets on the south coast. Similarly, tropical molluscs occasionally occur along the coast particularly at Rottnest Island where the higher than expected winter water temperatures due to the Leeuwin Current allow some coral growth (Fig. 1.4).

In summer, the strong southerly winds prevailing along the Swanland coast weaken the Leeuwin Current and it moves offshore. The Capes Current then flows northwards up the inner shelf from the vicinity of Cape Leeuwin past Cape Naturaliste to the north of Rottnest Island. Satellite imagery reveals that the southerly winds may also cause an upwelling of cold water between the capes. (See **Box 1.4 Leeuwin Current, El Niño and ENSO years; Chapter 4 Leschenault Estuary and the smaller estuaries of the Geographe Bay area: Toby Inlet**.)

The rivers of Swanland—back to front rivers

The rivers in south-western Australia are quite different to those of many parts of the world; the river forms dictated by a long geological history. In the early 19th century rivers were considered to run from mountains where they cut steep banks, through hilly areas that were once mountainous, cutting moderate slopes, and then in the oldest stage through flat featureless plains where sediments were deposited or the land had eroded away. Our rivers in some ways have features in reverse, reflecting their history: the mature parts lie in a flat landscape east of the Darling Range, while the youngest parts are found nearer to the coast cutting through the Darling Scarp and Ravensthorpe Ramp (Maps 1.10 and 1.11). The parts of the rivers beyond the ranges are ancient, the remnants of river systems that existed before the super continent of Gondwana split apart and Australia was formed (Map 1.6).

During the rifting that created the Australian continent, the rivers that once flowed to and from Antarctica were truncated and areas of inland drainage formed as the edges of the Yilgarn Block rose. New rivers later formed along the rift valleys cutting back into the plateau, redistributing the sediments. Some of the new rivers such as the Bowes, and the Swan and Frankland 'captured' older rivers. Sediments accumulated, up to 60 metres deep, where the gradients were low, a process that accelerated as the climate became drier. Inland these broad sediment-filled valleys of the old rivers *palaeochannels* (fossil channels) in the upper catchments of rivers such as the Moore (Monger) and Avon (Lochart, Yilgarn and Ninghan) are occupied by salt lakes. Occasionally after heavy rain the lakes link up and overflow into the younger drainage zones, the flood waters eventually flowing through to the sea. The effect on the landscape, particularly in cleared agricultural areas, is of course devastating, even downstream in the estuaries (Fig. 5.33). Sediments and nutrients are mobilised leading to algal blooms like those that closed the Swan in January 2000.

Box 1.4 LEEUWIN CURRENT, EL NIÑO AND ENSO YEARS

The Leeuwin Current starts as a broad fan of warm (>24°C), nutrient poor, less saline water off the north-west coast and moves southwards at a rate of 0.5–1 knot, maximum of 3 knots, to Cape Leeuwin and then eastwards to the Great Australian Bight. As it moves along the coast the current forms eddies, meanders and offshoots, and the water properties change due to mixing, air–sea heat exchange and evaporation (Fig. 1.4).

The Leeuwin Current is ultimately driven by flow from the Pacific Ocean through the Indonesian Archipelago into the Indian Ocean. The strength of the current varies from year to year depending on global weather patterns. In 'normal years' (La Niña conditions) the South Equatorial Current in the Pacific Ocean raises sea levels in the western tropical Pacific. This warm water in turn flows westwards through the Indonesian Archipelago, raising sea levels in the equatorial Indian Ocean up to 55 centimetres higher than along the southern coast, and the tropical water flows southwards down the Western Australian coast. Although the current flows all year, it is weaker in summer as it flows against the summer winds from the south. In El Niño–Southern Oscillation (ENSO) years, flow across the Pacific from South America is minimal and the Leeuwin Current is weaker.

El Niño was originally used to describe unusual warming of the ocean in the eastern Pacific occurring at Christmas time in some years, hence the name derived from 'Christ-child'. The term is now more generally used to describe a sustained warming over the central and eastern tropical Pacific. It represents a major change in the global atmospheric and oceanic conditions caused by a shift of atmospheric pressure (Southern Oscillation) between the Pacific Ocean and the Indian Ocean. El Niño events typically occur every four to seven years and last for twelve to eighteen months, although there is much variability in this pattern. Warm water from the western Pacific is transported eastwards to Peru and Ecuador where it replaces the more usual cold, nutrient-rich upwelling water from the Humboldt Current that accounts for the area's highly productive fisheries (Map 1.9).

In Australia, ENSO years are usually associated with drought conditions in eastern Australia, however during some events rainfall may also be above average. The dry conditions of the El Niño in 1982–1983 contributed to the Ash Wednesday bushfires in South Australia and Victoria. Other El Niño events occurred in 1986–1987, 1990–1994 and 1997–1998 with heavy rains falling during some months of the last event. Another El Niño in 2002 was associated with severe drought over wide areas of Australia.

El Niño events are a natural part of the climate system, however given the impact of rainfall particularly to the rural economy there is great interest in the development of computer models using ocean as well as atmospheric (weather) data to predict the likelihood of new events.

The strength and timing of the Leeuwin Current influence many features of the Western Australian environment and the commercial fisheries. During years when the Leeuwin Current is strong, coastal sea level is relatively high and settlement of rock lobster larvae is higher as larvae are brought shorewards to coastal reefs. In addition, higher temperatures may affect spawning. In Shark Bay, the prawn catch is higher after a strong Leeuwin Current flow but catches of scallops and recruitment of juveniles are higher in the year following a weak Leeuwin Current, probably due to differences in flushing, lower temperatures or higher nutrients in the absence of nutrient poor tropical water.

Variability in finfish catches also appears to be related to changes in the strength of the current. Salmon on the south coast may move into deeper water where they are more difficult to catch in years of strong Leeuwin current, and larvae may be transported further east. Catches of Australian herring follow a similar pattern. Low nutrient tropical waters also reduce plankton growth and food for pilchards stocks.

Source: Pearce & Walker, 1991.

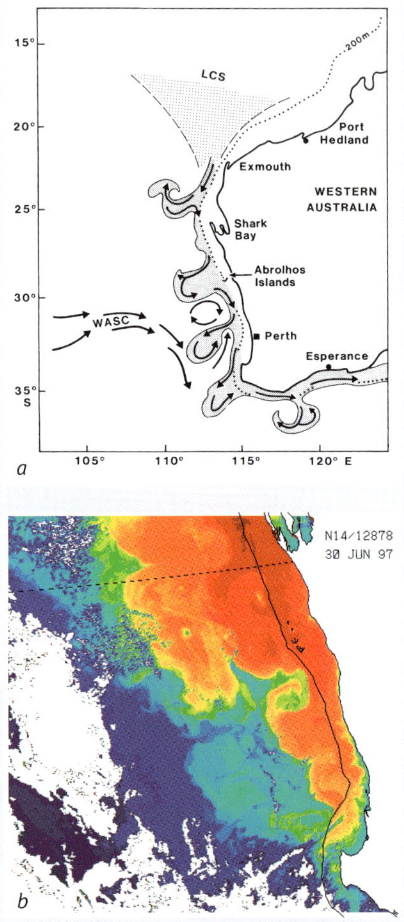

FIG. 1.4 *The Leeuwin Current. (a) Shaded areas show the main current and arrows the direction, typical eddies and jets. The strength of the current depends on flow of tropical water through the Indonesian Archipelago from the Pacific Ocean.* Source: Pearce, 1991 modified from Andrews, 1977. *(b) Satellite image of the current (shown in red) meandering southwards along the Western Australian coast in June 1997; the black line marks the approximate edge of the continental shelf.* Image by Alan Pearce, CSIRO Marine Research, Courtesy of WASTAC.

MAP 1.10 *Surface drainage of the south-west of Western Australia.* Adapted from van de Graff et al., 1977; Beard, 1999. Courtesy Phil Commander, WRC.

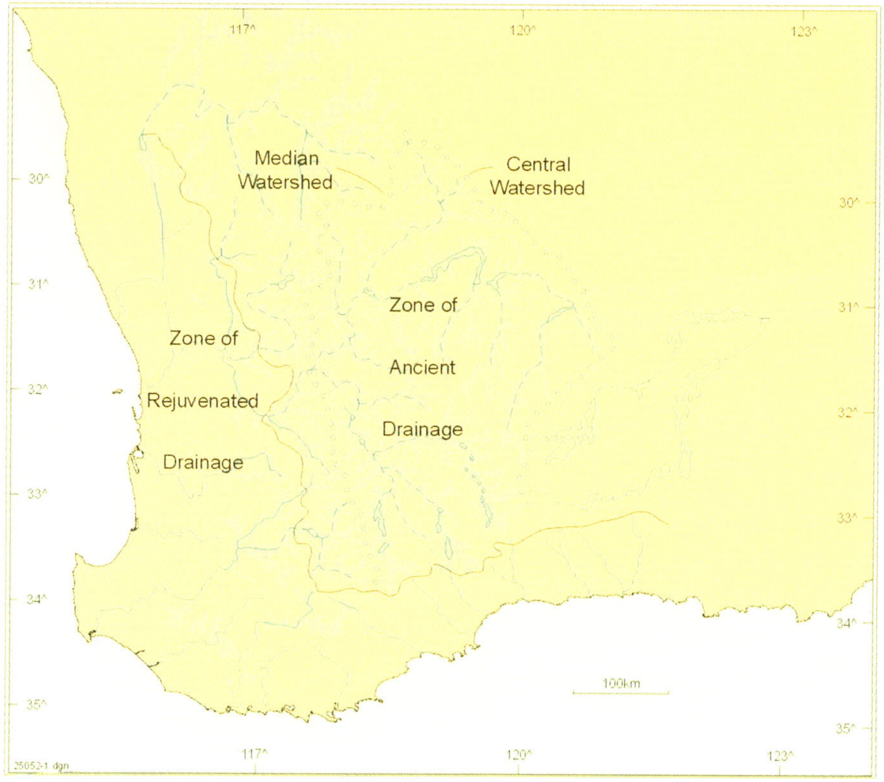

The link between the obvious river channels and the old drainage channels creates a problem for determining catchment size. The diffuse and vast extent of many of the river systems, together with a tendency for localised rainfall, makes estimations of river influence downstream difficult. The volumes and rate of discharge, and sediment and nutrient loads reaching the estuaries may depend on how much rain falls, on the extent of the rainfall, and on which particular part of a catchment the rain falls. Similarly, calculations of estuary basin size also vary depending on the defining boundaries (topography and water levels), the mapping techniques used, or conditions (salinity and turbidity) when photographs used for mapping were taken (Fig. 1.5).

Drainage zones

The river systems and estuaries can be grouped according to the types of drainage that reflect geological history, and the types of rocks and soils. In many cases these zones are associated with distinctive vegetation.

Ancient Drainage Zone. These are remnants of rivers that probably once traversed the super continent of Gondwana originally draining to the south and following the NNW trend in the basement rocks. This was a landscape of low relief probably first eroded in the Proterozoic (1,800 million years) and glaciated in the Carboniferous–Permian (280 million years). These rivers were cut off by the uplift along the south of the old Plateau in the Eocene (43 million years) with separation from Antarctica when another cycle of erosion was initiated. There are three sub areas: northern, eastern and south-western reflecting differences in formation. Generally however the landscape is an undulating plain on weathered granite, dominated by salt lakes in the main valleys with sluggish drainage due to low gradients. Although rainfall is generally low, occasional heavy rains bring floods. Some of the longest rivers such as the Murchison,

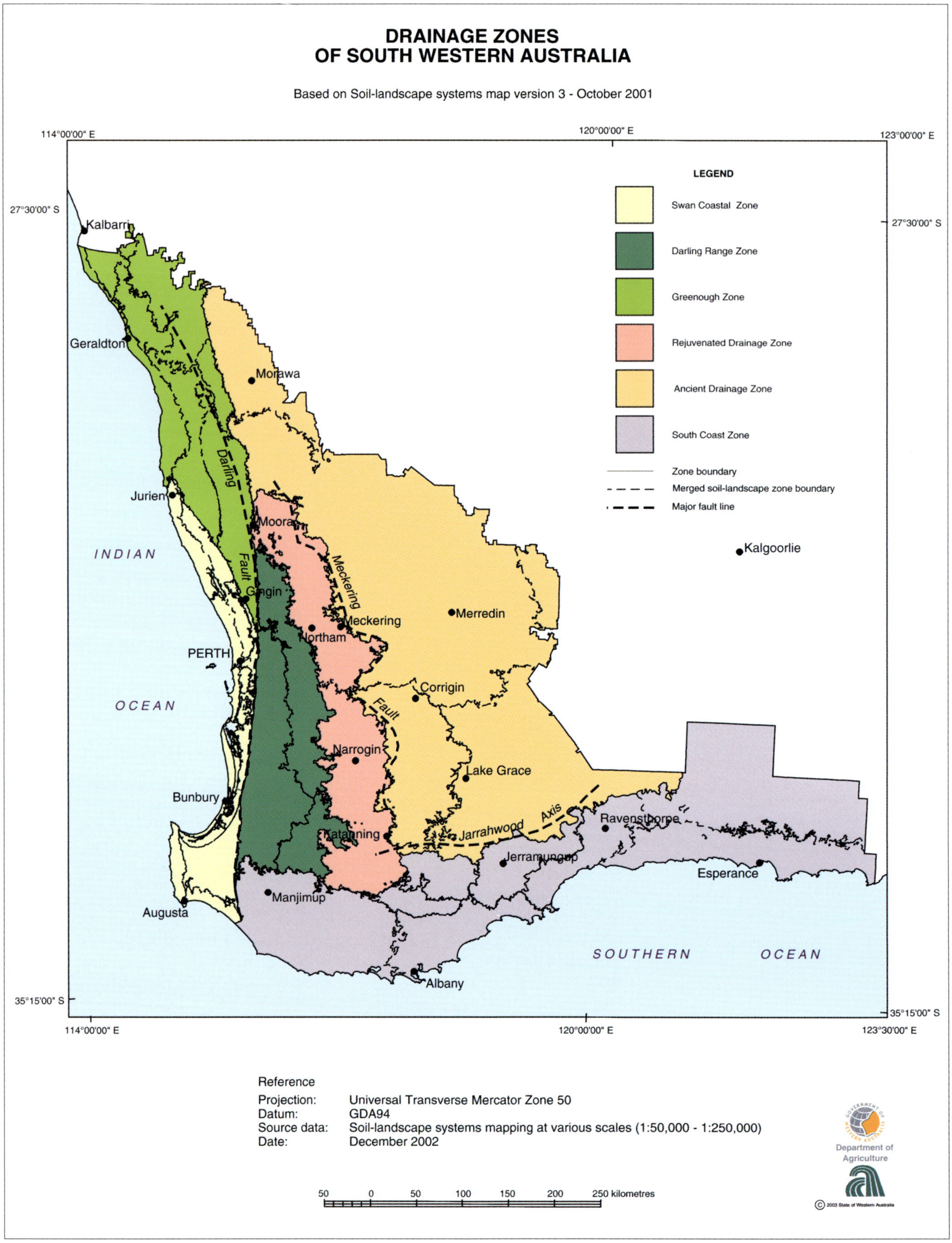

MAP 1.11 *Broad drainage zones based on groupings of soil–landscape zones, which in turn are derived from more detailed soil–landscape mapping covering all of south-western Australia.* Courtesy N. Schoknecht, DAWA.

Moore, Avon and Blackwood drain from this zone as do some of the shorter rivers to the south-east of Albany where it nears the coast, but in many years there may be little water flow through them to the coast. There is mallee vegetation on duplex soils, with proteaceous vegetation on gravels and sands.

South Coast Drainage Zone. The uplift of the edge of the old Plateau in the region of the Jarrahwood Axis along with down warping further north probably occurred during separation from Antarctica. This resulted in the truncation and reversal of some of older southward flowing rivers and development of short, deep, south-flowing rivers.

Rejuvenated Drainage Zone. The older Eocene drainage system was probably modified 'rejuvenated' through the Miocene–early Pliocene by uplift along the Darling Range. This area is characterised by more incised river channels than those further east of the Meckering Line, which coincides with the South West Seismic Zone. Here the streams flow in most years. Soil is of colluvium or in-situ weathered mainly metamorphic rock.

Darling Range Drainage Zone. On the Darling Range streams flow in most years during winter through dissected lateritic soils lying above a granite basement. Soils are composed of laterite, lateritic colluvium, and weathered in-situ granite and gneiss.

Coastal Drainage Zone. West of the Darling Fault and the Whicher Range in the Geographe Bay area, and the Scott River Plain on the south coast, the rivers flow across flat coastal dunes formed by rising and falling seas. North of Perth where the band of coastal dunes is narrow, the

FIG. 1.5 *A high resolution Digital Elevation Model (DEM), now available for south-western Australia, enables detailed analysis of landforms and drainage. This example of a DEM derivative is a 3D sun-shaded image (sun angle 45°, sun direction north-east) of a salt lake system of lakes and lunettes among rises west of Wubin (approximately 90 km north-west of Moora) in the Ancient Drainage Zone. Ground dimensions of image approximately 8.5 km by 7 km.*
Courtesy N. Schoknecht, DAWA.

short rivers have cut through hills of older rocks and the rivers hold alluvial sediments. Included in this grouping are the short rivers flowing seawards from the Leeuwin Ridge, and the steep-sided river valleys west of Nannup and the Darling Fault on the Blackwood Plateau. This latter group cut through alluvial sediments in the area sometimes referred to as the Bunbury Trough and Donnybrook Sunklands.

VEGETATION AND CLEARING LEADING TO CHANGES

The natural vegetation of south-western Australia is unique and diverse, and varies with the soil and rainfall (Map 1.12, see also Prelude Maps 1 and 2). Many of the plants possess unusual adaptations for acquiring nutrients from old leached soils or for reducing water loss during the summer drought.

Generally forests of large eucalypt trees are found where rainfall is highest and most reliable and where soils are deep, with the trees gradually decreasing in size and growing in a mallee habitat where rainfall is lower and less reliable. Members of the Proteaceae and Myrtaceae dominate shrub lands, and heaths are a feature in areas of lower rainfall and sandy soils. Much of the natural deep-rooted vegetation has now been cleared for agriculture leading to rising watertables, increased run-off, salinisation and soil erosion, all of which impact downstream in the estuaries (Map 1.13). (See **Box 5.3 Kwongan**.)

MAP 1.12 *Broad vegetation types of Swanland at the time of European settlement. The green line defines the South–West Botanical Province and follows in part the boundary of Swanland as defined by Taylor in 1915. The Province is modified from Beard 1980 and is now incorporated into the Interim Biogeographic Regionalisation for Australia (IBRA) system (Environment Australia, 2001), a system used for Conservation Planning and Natural Resource Management.* Courtesy A. Hopkins, CALM & D. Shepherd, DAWA. See also Beard, 1990.

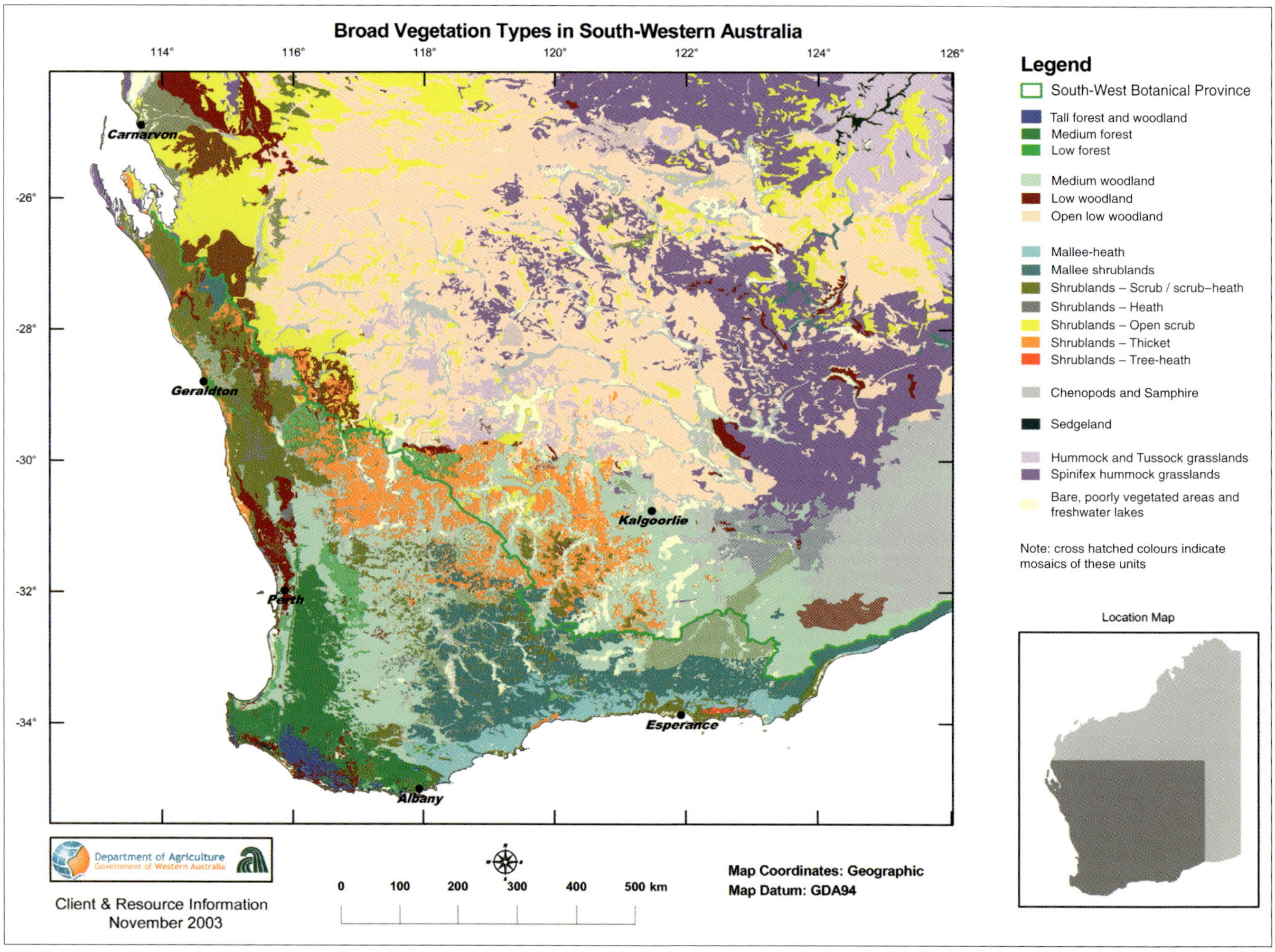

The plants of the coastal plains where the estuaries are found show distinct patterns related to the major dune systems that lie parallel to the coast. Banksia woodlands dominate the higher ground and deep sandy soils while paperbarks, sedges and rushes occupy the low-lying areas and rivers, and border the estuaries. The species change with soil depth, water inundation and salinity. In salty areas around the estuary margins samphires and rushes are dominant, the species varying with water depth and time of immersion. Mangroves are only found in one Swanland estuary, the Leschenault. (See also individual chapters for descriptions of fringing vegetation and **Box 4.1 Mangroves**.)

On the youngest dunes near the estuary mouths, dune plants, both native and recently introduced species, stabilise the sand reducing dune 'blow-outs' that smother other vegetation and even spill over into the estuaries. (See **Chapter 5 Estuaries north of the Swan: Greenough River, Features of the wind**.) The dune plants are often grey in colour to reduce heat, covered with hairs or possess other features to reduce water loss. Many of the introduced species are annuals or bulbs of African or Mediterranean origin: they grow in winter and dominate disturbed areas.

Little of the original peripheral or fringing vegetation remains along the river and estuary banks in the highly populated areas, and weeds are gradually replacing much of the fragmented remnant vegetation. Changes in the fringing vegetation have severe consequences as these are the plants that stabilise the sandy estuary banks, trap nutrients, form a habitat for aquatic life when water levels are high or provide rich feeding grounds for terrestrial life, including wading birds, when water levels are low.

MAP 1.13 *Remnant native vegetation of Swanland 2001.* Courtesy A. Hopkins, CALM & D. Shepherd, DAWA.

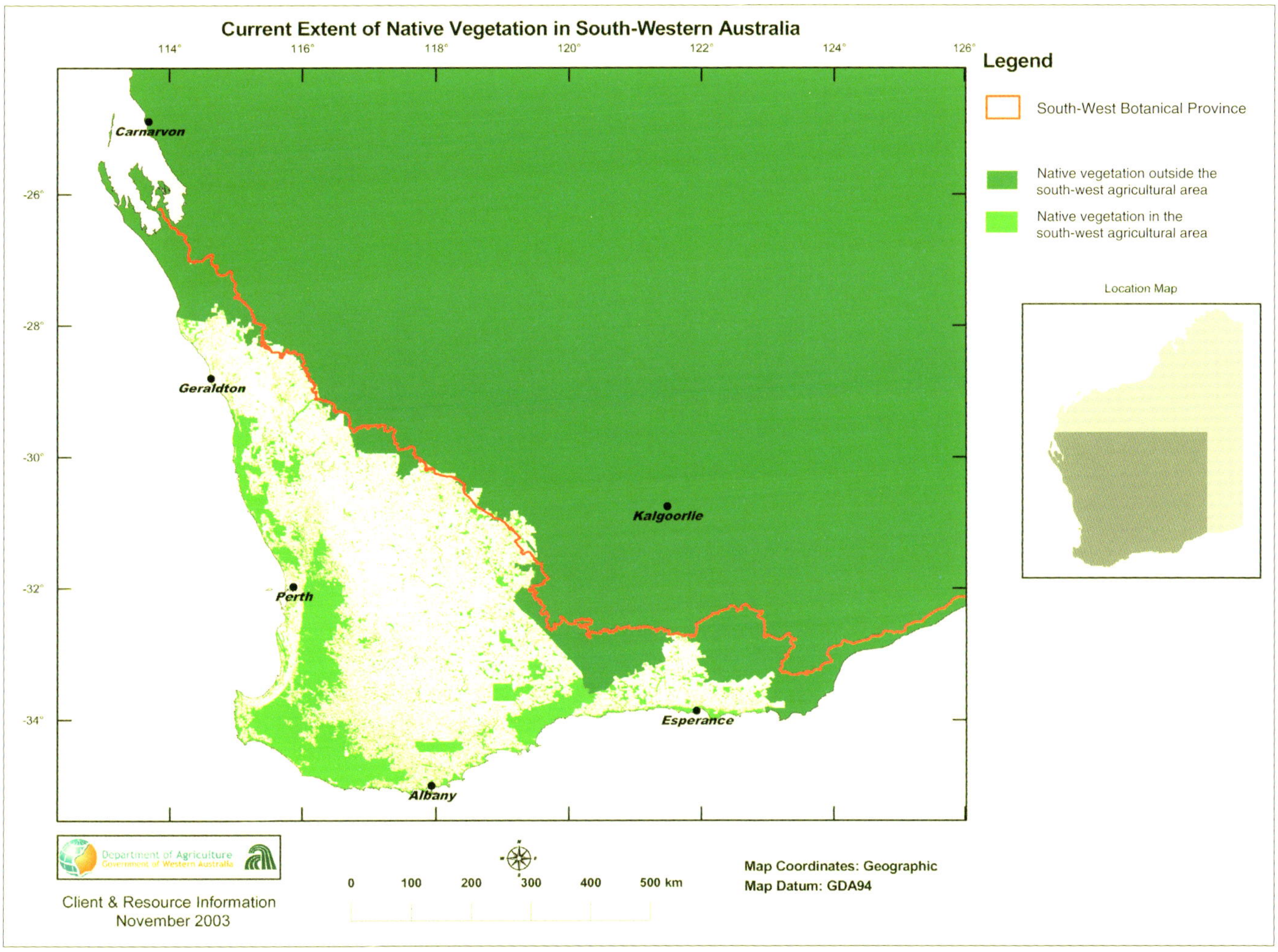

Arriving in the estuaries

The rivers of Swanland bring the products of the catchment to the estuaries, the water and all that it contains. The river flows vary from year to year due to interactions between the rainfall, run-off, state of the catchment, and type of vegetation. In some areas removal of long-lived vegetation and compaction of soils has increased erosion, exposing the underlying hard pan. Water no longer infiltrates the soil, but instead flows across the surface moving sediment that accumulates in low points of the landscape filling watercourses and estuaries, or is lost to sea.

Sediment

The flowing rivers bring sediment to the estuaries, but how much is difficult to gauge. Heavier material is moved along the bottom at a rate determined by the strength and duration of river flow and the type of river channel, settling where water movement slows. Where river water meets the salt water of estuaries, fine suspended material flocculates and falls to the bottom forming deltas and sandbars.

In the last 10,000 years large amounts of sediment have accumulated in the estuaries. There are about 30 metres in the Swan, 3 metres in the Peel–Harvey and about 5 metres in the Hardy. Although measurements of sediment movements are complicated, it would appear that in the 1940s an estimated 93,000 tonnes was carried by floodwaters of the Collie after heavy rain, 77,000 tonnes was deposited in the Leschenault and a further 16,000 tonnes was carried out to sea. In 1982, 100,000 cubic metres of sediment reached the Beaufort Inlet in floodwaters dumped by dissipating cyclones in the Pallinup River catchment, and Stokes Inlet received 50–60 centimetres of sediment in a 30-year period after the catchment was cleared in the 1970s. In 1986 floods in the Chapman River passed about 22,000 tonnes to sea, while the 1971 flood of the Irwin River appears to have been a particularly significant erosive event, bringing large amounts of material to the estuary.

Over longer and shorter time scales, due to natural and human changes, sediments have transformed the estuaries and will continue to do so. Today river reaches are narrower and shallower, filled with sediment that is gradually moving downstream to the estuaries during floods, where the turbid waters probably affect seagrass growth and animals living there. River deltas are more extensive, the lagoons shallower and the bars larger. The poor scouring effect of the small tides and low river flushing leads to extensive flood tide deltas amplifying the size of entrance bars, which are also affected by the longshore transport of sand along the coast.

Hidden in the flowing waters—salt and nutrients

As water passes through the soil on the way to the rivers and estuaries, the salt and nutrient content gradually rises. The diaries kept by the early explorers and settlers indicate that many of the rivers of Swanland, particularly the upper reaches that extended into Salinaland, and the eastern estuaries, were naturally brackish. Ground water flows were often saline, although flood and surface flows were fresh. Other rivers were fresh in the 1800s but became saline as the land was cleared of the deep-rooted native vegetation (Fig. 1.6). The increase in salinity followed an increase in ground water flows that mobilised the salt moving it into the streams, while higher watertables brought salt to the surface. Water supplies gradually turned too salt for domestic use, stock and the boilers of steam trains (Tables 1.2, 1.3 and 1.4, Prelude Map 2). (See **Box 1.5 Landscape of salt—above and below the surface**.)

Box 1.5

LANDSCAPE OF SALT—ABOVE AND BELOW THE SURFACE

Salt lakes are major features of the ancient Australian landscape, a surface indication of a vast hidden salt store. The salt store (connate salt) developed over millions of years as spray-laden winds blowing off the ocean created a salty atmosphere that the rains brought downwards. Over the millennia this 'fossil salt' accumulated, particularly where rainfall was too low to create strong ground water flow and flush the salt away.

In coastal regions of the South-West salt is accumulating at a rate of 100–250 kilograms per hectare per year while further inland the rate is a less, about 20–50 kilograms per hectare per year. Salt also originates from the natural weathering of underlying rock and from 'fossil' salt left behind when sea that once covered the land retreated. The process of transpiration by plants also concentrates salt.

Estimates suggest that there are 100–10,000 tonnes of salt stored below each hectare of land in Swanland. The salt store is highest in the lower rainfall areas. In the natural state, salt tolerant plants grow where salt is near the surface, around salt lakes, estuaries and along the coast. In other areas, deep-rooted plants that pump water in the transpiration stream help to keep the salty watertable depressed, allowing growth of less salt tolerant plants. The removal of deep-rooted native vegetation that tapped and trapped the ground water, however, has allowed the watertable to rise bringing the hidden salt to the surface.

Source: Stokes et al., 1980; George et al., 1996.

The changing salinity has also brought changes to aquatic life. A shift from freshwater to brackish water species has been documented in the Avon and the Blackwood. The freshwater mussel *Westralunio carteri* and freshwater cobbler *Tandanus bostocki* have disappeared from the Avon, and in the Blackwood River freshwater species, most notably the marron, are now only found in the forested areas where the waters are still fresh. Rising salinity will continue to impact on the human population as land becomes unproductive, water supplies are threatened, and the need increases for expensive and complicated remedial work, particularly as the success of many restorative actions is as yet unknown. The impact on estuaries where the biota is accustomed to extremes of salinity, however, may be less catastrophic. Nevertheless, with a decrease in freshwater influences changes in conditions for aquatic life can be expected.

Rising watertables and changed flows

Under natural vegetation little of the rainfall flowed to the creeks and rivers. Most was evaporated, held by the vegetation or transpired. The amount of rainfall flowing to rivers was only 20–30 per cent of rainfall in the wetter regions, and only about 1 per cent in drier parts of the wheatbelt. Widespread clearing, however, reduced the interception of rainfall leading to an increase in a recharge of the ground water.

It has been estimated 30 per cent of agricultural land will be affected by salt or rising watertables by the year 2050. Even if the higher rainfall conditions of the 1930s and 1940s do not return, stream flow and the frequency of flooding are still likely to increase.

TABLE 1.2 *Water salinity: *human drinking water unpalatable 500–1,500 mg/L and unsafe at 2,500 mg/L; beef cattle can withstand water of 15,500 mg/L.*

Adapted from George et al., 1996.

Water quality	Concentration				Conductivity		Limit of use
	(ppm)	(ppt or ‰)	(gr/gal)	(mg/L)	(mS/m)	(mS/cm)	(mS/m)
Fresh							Dam water clear
							Productive pastures (50); hot water systems (160); some loss of less tolerant riparian vegetation (100–150); blue gum (100)
	1,100	1	77	<500	200	2	Plant stress and loss of less tolerant sp. (180)
							Human drinking water (250)
Marginal	1,640	1.6	116	500–1,500*	300	3	
Brackish	2,740	2.7	193	1,500–5,000	500	5	Poultry (470)
	3,280	3.3	230		600	6	Dairy cattle milking (550)
							Decline in some gum trees and shift to salt tolerant sp. (≈600)
	4,280	4.3	270		700	7	Pigs (700)
	5,460	5.5	385		1,000	10	Horses (1,000)
Saline	6,160	6.2	425	>5,000	1,100	11	Young & breeding sheep, dry dairy cattle (1,100)
							Yabbie growth slows
	9,360	9.4	616	15,500*	1,600	16	Beef cattle, septic tanks (1,550)
	13,360	13.4	850		2,200	22	Adult sheep (1,650–2,200)
Marine	35,600	35	2,040		5,300	53	Sea water
Hyper-saline							

Parts per million (ppm); parts per thousand (ppt or ‰); grains per gallon (gr/gal); milligrams per litre (mg/L); millisiemens per metre (mS/m); millisiemens per centimetre (mS/cm).

Salinity	Conversion factor	Salinity
Electrical conductivity (mS/m)	× 5.5	= total soluble salts (mg/L TSS)
Electrical conductivity (mS/m)	× 0.385	= total soluble salts (gr/gal TSS)
Total soluble salts (gr/gal)	× 14.25	= total soluble salts (mg/L)
Milligrams per litre (1 mg/L)	÷ 1	= one part per million (ppm)
Milligrams per litre (1 mg/L)	÷ 1,000	= one part per thousand (ppt or ‰)
Grams per litre (1 g/L)	÷ 1	= one part per thousand (ppt or ‰)

TABLE 1.3 *Standard conversions between different units used to measure salinity. Note care needs to be taken with conversions as electrical conductivity is affected by temperature. Measurements are usually corrected to a standard temperature (25 °C). In very saline waters (> 10,000 mg/L), the multiplication factor is much greater.* Adapted from George et al., 1996.

Water quality	Ecologists	Engineers and agricultural scientists
Fresh	< 3,000	< 500
Marginally fresh		500–1,500
Brackish	3,000–10,000	1,500–5,000
Saline	10,00–35,000	> 5,000

TABLE 1.4 *Categories and units of salinity (total dissolved solids (mg/L) is a term in common use).* Adapted from Pen, 1999.

Changes to the ground water are largely unknown, however it has been estimated that there could be a 6–10 millimetre increase in recharge in the 350-millimetre rainfall area. If this figure is added to the Blackwood catchment of 17,900 square kilometres, most of which lies in areas receiving more than 350 millimetres per year, this amounts to about 25 per cent of the current annual flow of the Blackwood River. The amount could be even higher around its tributaries, the Beaufort and Coblinine Rivers, where this would amount to increases of 40 and 250 per cent of current run-off. Although the magnitude of evaporation of this increased run-off may also increase, it is worth noting that the average flow of the Coblinine River was about 900 megalitres prior to clearing and 11,000 megalitres after clearing in the 1990s. With an estimated 40 per cent of the upper Blackwood catchment now waterlogged, floods and flows could be two to four times the size of those brought by cyclone Errol in 1982.

The effects of waterlogging and salinisation are likely to be extensive and cumulative. The remaining vegetation is likely to die due to waterlogging and salt. This will then expose the soil to erosive forces of wind when dry and floodwater when wet. Higher ground water will generate higher flows and raise the salinity in downstream areas previously receiving fresh water. Inland catchments such as the Yarra Lakes in the upper Moore catchment, the Yenenning Lakes on the Avon and Lake Dumbleyung on the Blackwood are likely to flood more often and flow to the coast. As a result, effects of flooding downstream in the estuaries are likely to become more common. Low-lying areas around the estuarine margins will be covered more often, and there will be larger inflows of nutrients and organic materials accumulated in the river channels.

In addition to rising watertables, mobilisation of salt and the increase in surface flow, there are also problems with reduced and altered water flows to other rivers and estuaries due to the impoundment for waste supplies and the overuse of ground water. Dams have reduced water flow through the rivers to the estuaries and to the sea. Reduced flows have altered salinity and flushing regimes of sediments and nutrients. Increased use of ground water in addition to changing the watertable near the surface also has implications for deep-rooted vegetation, and water seepage to rivers and estuaries. In many instances little is known about the origin or extent of underground aquifers nor the rates of recharge, and increasing use of resources such as the Yarragadee may have profound long-term ecological effects throughout Swanland.

FIG. 1.6 *The role of native vegetation in the control of watertables and salinity: (a) the deep-rooted plants tap the deep soil profile; (b) the pumping action of plants (the process of transpiration) mediated by evaporation (evapotranspiration) reduces the watertable; the foliage and organic matter in the soil also hold water reducing run-off; (c) shallow rooted annual vegetation only removes shallow water during the growing season, allowing the watertables to rise, bringing salt higher into the soil profile; and (d) rainwater contributes to erosion or compaction.* Modified from Pen, 1999.

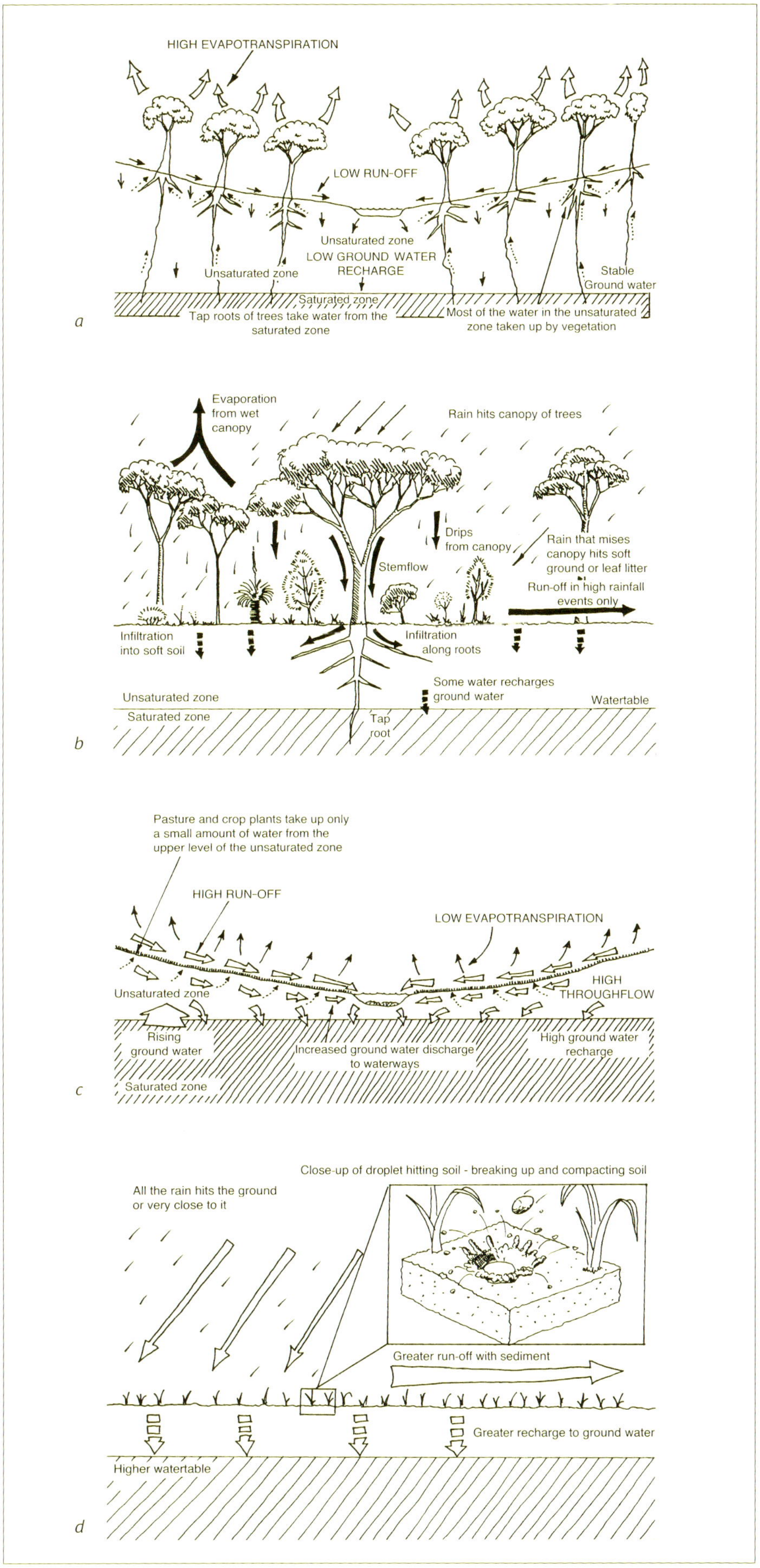

Additional nutrients and the onset of eutrophication

Australian soils are naturally low in nutrients due to weathering and leaching over millions of years. Development of agriculture has depended on a range of nutrient supplements including phosphates, nitrogen and trace elements. This is an Australia-wide problem, although the critical issues vary depending primarily on the geological history, climate, vegetation and current land use (Graph 1.1). In particular, the sandy soils around estuaries such as the Peel–Harvey with poor binding capacity for phosphate have required applications of superphosphate. (See **Box 3.1 Ingredients of a green soup, Box 3.2 Understanding the system—test tubes, cages and computers** and **Box 9.2 The P problem—can it be controlled?**) Over time phosphate leached into the estuaries making them eutrophic, and algal blooms became common. The increase in some algal blooms has been directly related to the increase in superphosphate use. (See **Box 1.6 Terms used for water quality**.) The nutrient increase and shift in ratio of nitrogen and phosphorous favoured the growth of blue-green algae or cyanobacteria such as *Nodularia spumigera* in the Peel–Harvey and *Microcystis aeruginosa* in the fresher parts of the Swan–Canning. Both of these can be independent of dissolved nitrogen as they can also use atmospheric nitrogen for growth. Strains of these species

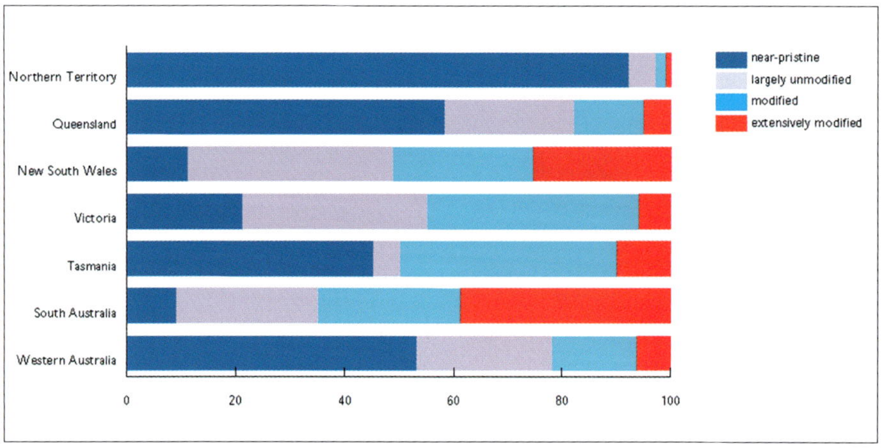

GRAPH 1.1 *Comparison of the condition of estuaries in Western Australia to those in other Australian states. Throughout Australia 50% are near pristine, 22% largely unmodified, 19% modified and 9% extensively modified. Most of the near-pristine estuaries are located away from population centres, although some are in more developed areas they are usually near or within areas such as national parks.* Courtesy National Land and Water Resources Audit 2002, Ozestuaries (See www.ozestuaries.org)

Box 1.6 TERMS USED FOR WATER QUALITY

The term **eutrophic** is derived from a Greek word *eutrophus* that means 'well-fed'. Eutrophication is the process by which aquatic systems are enriched with nutrients that are usually various forms of nitrogen and phosphate. This can occur naturally over geological time scales or may be accelerated by human activities (cultural eutrophication) such as fertilising soils and through sewage disposal. Rapid nutrient changes caused by humans result in extreme responses by plants and animals, such as accumulations of macroalgae and large phytoplankton blooms including blue-green algae, toxic to humans. As algae die they release nutrients for further algal growth. The process of decay also reduces oxygen concentrations creating unfavourable conditions for aquatic life and favourable conditions for release of nutrients in the sediment. Cycles of algal growth, death, decay and growth are typical of eutrophic waters like the Peel–Harvey and Vasse–Wonnerup.

The term **oligotrophic** is derived from the Greek word *oligos*, meaning small or few. Oligotrophic waters have low concentrations of nutrients and generally low production of organic material (primary production).

The term **mesotrophic** is derived from the Greek word *mesos*, meaning middle. Mesotrophic waters have moderate amount of nutrients and are moderately productive.

Estuaries in south-western Australia are particularly vulnerable to nutrient enrichment, eutrophication, because of the evolutionary history of low, oligotrophic conditions.

also have the ability to produce toxins. (See **Chapter 2 The Swan–Canning Estuary: The fuels for algal growth—nitrogen and phosphorus** and **Box 2.4 Tiny cells on the rampage take a city's playground off limits**.)

The nitrogen content of soils has risen with fertiliser applications but also the cultivation of plants such as clover, lupins and lucerne. The roots of these leguminous plants form nodules that are colonised by nitrogen-fixing bacteria. As with phosphates, excess nitrogen or nitrogen released as plants decay leaches into the ground water and through to the rivers and estuaries. Large amounts of nitrogen also enter in surface flows with the first freshwater flush after a dry period. In the Swan–Canning, the rise in nitrogen appears to be the critical factor for initiating phytoplankton blooms.

The amount of carbon in the rivers and estuaries increases with inflow of organic material, which increases the Biological Oxygen Demand (BOD) of the system as bacteria use oxygen in the decay process. In turn, this reduces the amount of oxygen in the sediment and water column. Freshwater systems usually need about 5 milligrams per litre of oxygen to remain healthy although this varies with temperature and water depth. When water bodies are healthy, oxygen enters water from the atmosphere and through the process of photosynthesis of plants and algae. During blooms, however, waters may be supersaturated during daylight when algae and plants are photosynthesising but at night, when respiration rates are higher, oxygen concentrations decrease creating unfavourable conditions for animals such as fish. (See **Chapter 4 Leschenault Estuary and the smaller estuaries of the Geographe Bay area: Dead fish—symptoms of a distressed environment**.)

Deoxygenation of sediments occurs in estuaries when the water column is stratified, with fresher lighter water at the surface and more saline denser water at the bottom. This may be due to the intrusion of salty marine water or fresher river flows covering more saline estuary water. The salinity–density stratification isolates the bottom waters and oxygen concentrations decline. In the estuaries of Swanland marine water intrudes when the bars are open and river flow slows. Deoxygenation is a problem in deeper areas of the estuaries and the riverine reaches of Swan–Canning and Peel–Harvey. In

TABLE 1.5 *Comparison of nutrients in water available or not available for plant and algal growth.*

Nutrient	Available for plants and algae	Not available for plants and algae
Phosphorus	Orthophosphate = Dissolved Inorganic Phosphorus (DIP) or Filterable Reactive Phosphorus (FRP)	Dissolved Organic Phosphorus (DOP) Particulate Phosphorus (PP)
Nitrogen	Ammonium (NH_4^+) and Nitrate (NO_3^-) together also termed Dissolved Inorganic Nitrogen (DIN)	Dissolved Organic Nitrogen (DON) Particulate Nitrogen (PN) Nitrogen gas (N_2)

TABLE 1.6 *ANZECC Guidelines 1992 and 2000 nutrient trigger values for assessing water quality in rivers and estuaries; ANZECC values are marked on many graphs and tables in this publication. Water quality may be classed as low (close to natural), moderate (mildly enriched), high (enriched), very high (highly enriched) or extreme (extremely enriched).*

	NH_3N (mg/L)	NOx (mg/L)	TN (mg/L)	FRP (mg/L)	TP (mg/L)	CHl *a* (mg/L)
ANZECC 1992 estuarine	< 0.005	0.1	NA	0.015	NA	0.01
ANZECC 1992 riverine	NA	NA	0.75	NA	0.1	NA
ANZECC 2000 estuarine	0.04	0.045	0.75	0.005	0.03	0.003
ANZECC 2000 lowland rivers	0.08	0.15	1.2	0.04	0.065	0.005

shallow estuarine areas the stratification is usually broken down by wind generated waves.

Low oxygen (anoxic) concentrations facilitate the release of nutrients stored in the sediment (Fig. 1.7). In shallow well-oxygenated areas, nitrogen in forms used by plants is converted to nitrogen gas that is released to the atmosphere in the process known as denitrification (Table 1.5). (See **Box 1.7 Nutrient processes—nitrification and denitrification**.) The amount of oxygen in the sediment also influences the various forms of iron, which in turn influence the binding of phosphate in the sediments. Phosphate bound to iron is not available for plant growth. More recently, iron released from acid sulfate soils during clearing has been implicated in the growth of *Lyngbya* 'mermaids hair', also called 'mermaids curse', in the large estuary of Moreton Bay, Queensland. *Lyngbya*, a blue-green alga, is toxic in Moreton Bay. Where it comes in contact with skin it causes severe dermatitis with the skin peeling off leaving red scaly patches. The toxins can also cause asthma and eye irritations. Another environmental problem is the effect on crab and fish catches as *Lyngbya* smothers seagrasses where these animals live.

Should we be concerned about *Lyngbya* in the estuaries of Swanland? The answer is yes. The species is present in the Peel–Harvey near the Murray and Serpentine Rivers, although we do not as yet know if this form is toxic or if toxicity can be induced. The lesson from Moreton Bay is that the

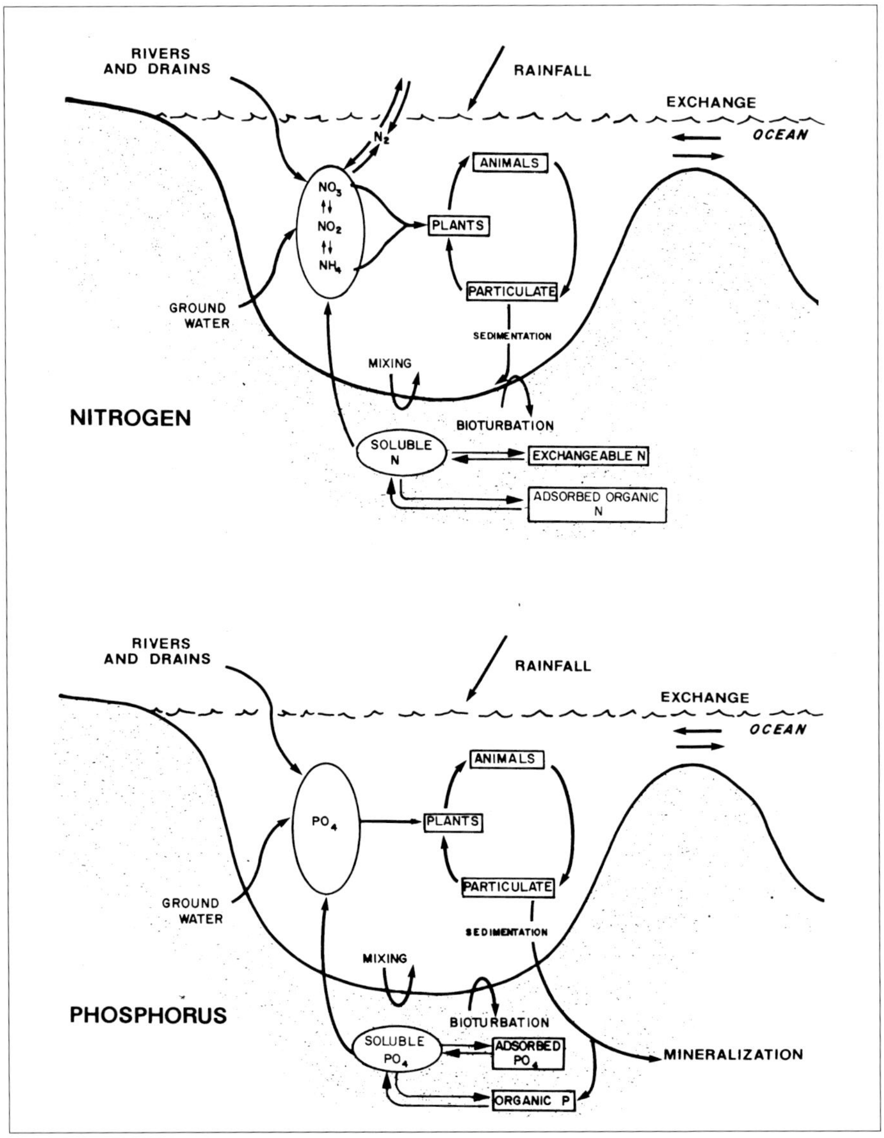

FIG. 1.7 *Nitrogen and phosphorus processes. The double arrows on N_2 indicate that the total nitrogen in the system is determined by two dynamic processes: nitrogen fixation by blue-green algae and denitrification by bacteria. Similarly, the total phosphorus is determined by dynamic processes.* Courtesy D.E. Adapted from Hodgkin et al., 1980.

Lyngbya problem has only developed in the last decade since bridge works to Bribie Island disturbed the sediments and changed the water circulation.

Phosphate also binds to carbonate forming inert apatite phosphorus when oxygen concentrations are high and calcium carbonate is abundant. This binding capacity was one of the rationales for increasing seawater exchange in the Peel–Harvey and why increasing the flow of phosphate into the nearby ocean (which is high in carbonate) was not considered a major problem. (See also **Box 3.1 Ingredients of a green soup**.)

Understanding water health—components we measure

Assessing the health of a water body is a complex task, requiring standard measurements of physical and chemical conditions that are used to build up a picture of the sources of the water, what it contains, and the physical, chemical and biological processes occurring. Water quality objectives in south-western Australia are based on the guidelines of the Australian and New Zealand Environment and Conservation Council (ANZECC) for fresh and marine water quality (Table 1.6). Physical parameters can include pH, salinity (fresh to marine), conductivity, dissolved oxygen, temperature, turbidity, water clarity, light penetration and nutrients.

The pH, a measure of acidity or alkalinity, varies in different water bodies. Typically, inshore marine waters have a higher pH of 8.0–8.5 than estuarine waters (7.5–8.5) and lowland rivers (6.5–8.0). High pHs can occur during algal blooms, however salinity and pH are closely linked. Higher salinities usually 'buffer' pH (resist change to pH in the 7–8 range), so the most extreme values are found in low salinity environments. The pH of ground water flows also has an influence, depending on the surrounding soils and rocks. For example, water flowing through limestone has a pH greater than 8.0. pH also varies with biological activity and usually increases during an algal bloom. Low pH can also indicate that oxygen concentrations are low, which has implications for ammonia release.

Salinity can vary from fresh at the riverine end to marine (35 parts per thousand, 35 ppt or 35‰) near the ocean. Hypersaline conditions occur when water evaporates and the salinity increases past the marine condition of 35‰. Ground water inputs may also be a major factor affecting salinity in an estuary. Salinity is usually calculated from measurements of electrical conductivity and temperature. Conductivity measures the ability of a water body to conduct electricity, which increases with the concentration of all dissolved salts in the water (Tables 1.2 and 1.3).

Box 1.7 NUTRIENT PROCESSES—NITRIFICATION AND DENITRIFICATION

Nitrification: the oxidation of ammonium

$$2NH_4^+ + 3O_2 \rightleftarrows 2NO_2^- + 2H_2O + 4H^+$$

ammonium + oxygen ⇄ nitrite + water + hydrogen ions

$$2NO_2^- + O_2 \rightleftarrows 2NO_3^-$$

nitrite + oxygen ⇄ nitrate

Denitrification: the reduction of nitrate to dinitrogen

$$5CH_2O + 4NO_3^- + 4H^+ \rightarrow 2N_2 + 5CO_2 + 7H_2O$$

carbohydrate unit + nitrate + hydrogen ions → nitrogen gas + carbon dioxide + water

The dissolved oxygen (DO) in a water body is probably the single most important measure of habitat quality, as most living organisms do not survive without oxygen. Dissolved oxygen is measured as a concentration (milligrams per litre or mg/L) but sometimes as the percentage saturation. The concentration of dissolved oxygen is controlled by many factors, such as temperature, and usually decreases with rising temperatures. Oxygen concentrations may also increase during the day with photosynthetic activity of algae and plants. However it can decrease during the night when they respire. During large blooms oxygen levels may be very low and conditions are termed anoxic: this can cause fish kills.

Total suspended solids (TSS) indicates the amount of sediment and organic matter suspended in the water column, and is generally related to flow conditions. When TSS concentrations are high the system is turbid, usually as a result of erosion from the catchment of riverbanks and/or re-suspension of the bottom sediment. High turbidity can reduce light penetration through the water column and therefore affect plant growth and productivity. Suspended solids may also disrupt filter-feeding activity and clog fish gills.

The most common nutrients are total nitrogen (TN), which as the name suggests includes all forms of nitrogen, particulate nitrogen (PN), dissolved inorganic nitrogen (DIN), nitrogen gas (N_2), and dissolved organic nitrogen (DON). DIN includes ammonia (NH_3), ammonium (NH_4^+), nitrate (NO_3^-) and nitrite (NO_2^-), the total amount of readily available nitrogen for plant growth. Total oxidised nitrogen (NOx) is the sum of nitrate and nitrite. Laboratories often report total ammonia–N, which includes aqueous ammonia (NH_3) and ammonium (NH_4^+). The relative amount of NH_3 or NH_4^+ depends on pH, salinity and temperature with NH_3 increasing with higher pH. Ammonia, a breakdown product of organic material (plants, animals and their wastes), can be toxic to aquatic animals. Ammonium is slightly less toxic; the toxicity depending on the pH and temperature usually when these are higher. Generally DIN consists predominantly of NOx and the ammonia component is small, however under low oxygen conditions the proportion of ammonia increases as it is less easily converted to nitrate. DIN is often presented in graphs as a percentage of total nitrogen as it is useful for examining where nutrients originate and enter a water body. For example, if most of the nitrogen is in soluble form then it is likely to have come from surface run-off, ground water, or decaying plants and animals. In contrast, most organic nitrogen not readily used by plants is more likely to come from phytoplankton and undecomposed plants and animal waste.

Nitrification is the process by which bacteria in the presence of oxygen convert ammonium NH_4^+ to nitrite (NO_2^-) and nitrate (NO_3^-). Denitrification can occur with or without oxygen and is the process by which other bacteria convert nitrate to nitrogen gas (N_2), a form no longer available for algal growth. However before denitrification can proceed, nitrification must first occur and this does require oxygen. When oxygen concentrations are low and the nitrification process is blocked, ammonium concentrations rise. Shallow well-aerated areas are particularly important for the denitrification process.

Total phosphorus (TP) includes the sum of particulate phosphorus, dissolved organic phosphorus and dissolved inorganic phosphorus. Dissolved inorganic phosphorus (DIP), also termed orthophosphate (PO_4^{3-}), filterable reactive phosphorus (FRP), and soluble reactive phosphorus (SRP) are usually expressed as micrograms per litre (µg/L), milligrams per litre (mg/L), parts per million (ppm) and as a percentage of total phosphorus. When expressed as a percentage of TP it is not, however, always clear if this is based on weight or atomic weight. DIP, the form readily used by plants, enters in run-off and ground water but increases in low oxygen conditions when released from the sediment. Soluble forms of phosphorus are also released when oxygen concentrations are low.

In addition to the individual concentrations of various nutrients, the atomic ratio of N, C and P can also be useful for assessing which nutrients are important for plant growth, and if nutrient additions have occurred to throw a system out of balance. The ratio C:N:P = 106:16:1 was originally defined by A. C. Redfield in 1934 for the open ocean, thus termed the Redfield Ratio. The ratio is very important as limiting the supply of one nutrient automatically influences growth no matter how much of the second is available—hence if the system is in balance, limiting the supply of nitrogen will limit growth even when phosphorus is abundant. As a result, reducing nitrogen inputs can have an immediate effect on water quality in comparison to reducing phosphorus, which would have to be reduced by a far greater amount.

In fresh and estuarine waters the ratio varies but the same concept applies. In the estuarine reaches of the Swan the midsummer ambient N:P ratios can be near unity in the water column, so there is not enough nitrogen and phytoplankton growth is limited. In addition, the nuisance blue-green alga may proliferate when the ratio of N:P is low as these organisms are able to fix atmospheric nitrogen. Experiments in the Swan indicate that nitrogen given as nitrate increases phytoplankton growth (measured by chlorophyll *a*) suggesting that the availability of nitrogen is critical for controlling phytoplankton. Furthermore, the positive relationship between nitrate concentrations and rainfall suggests that river flow following rain is the mechanism for supplying nitrate to the surface waters.

Ratios of C:N and C:P are also used, but again it is not always clear if these are based on weight or atomic weight. In sea water these ratios are much greater than those of phytoplankton, and this suggests that the supply of nitrogen and phosphorus is more likely to restrict plant growth in the sea than is the supply of inorganic carbon.

The amount of other elements such as silica is also very important in assessing water quality. For example, diatoms that are generally non-toxic require silica, and if concentrations are high they tend to dominate a water body outcompeting less desirable toxic microalgae. Silica is also a component of the outer skeleton of crustaceans, and higher silica concentrations favour the zooplankton that feed on microalgae potentially reducing bloom formation.

Chlorophyll *a*, one of the most common photosynthetic pigments in a plant cell, is used to indicate the amount of plant growth in the water column. As chlorophyll *a* varies with different types and sizes of phytoplankton cells however, this needs to be viewed in context with other information, such as the abundance of different types of phytoplankton, to allow assessment of how the system is functioning. The term 'bloom' is usually used when the abundance of a single species exceeds 20,000 cells of phytoplankton per millilitre. Generally the nuisance types are dinoflagellates and blue-green algae. Similarly, diversity of zooplankton is indicative of a healthy system.

Water colour or Gilvin caused by dissolved organic matter that makes the water look tea coloured due to light scattering and absorption of light is another useful measure in assessment of water quality. It affects light levels that influence photosynthesis and plant growth. Gilvin is generally higher when water flows are high in winter.

To ensure validity of measurements and allow comparison of data collected either at different times or places requires a number of samples. The values then need to be presented in a consistent manner. Commonly the average or mean is used and the spread of the values from high to low is shown by an error term. Box plots are generally used in graphs of water quality in this publication to show the spread of 50 per cent of the measurements, and the median the 'midpoint' of the values. The 'whiskers' at either end are used to show the more extreme values. However some older graphs from other publications may use a different form, and the spread of the data may be indicated by a standard deviation or standard error. In a healthy water body most values are low with occasional high values, generally due to high inflow of water during storms or unusual conditions creating situations where nutrients in sediments are released.

How healthy are the estuaries of Swanland?

The way estuaries respond to nutrient enrichment depends on their geological, physical and chemical characteristics. These include the depth of the water column, the time and rate of flushing, the types of sediment, and how nutrients are stored or recycled through the sediments and into the water column. For example, although deep estuaries such as the Swan may receive large quantities of nutrients these may be flushed out if tides or river flow are strong. In addition, the deep areas may be below the photic zone where plants flourish. In contrast, shallow estuaries such as the Old Peel–Harvey with less tidal flow receive and store nutrients, and light levels are conducive to algal growth.

The estuaries of Swanland with limited summer river flows, restricted bar openings and small tides favour the accumulation of nutrients in the sediments. In turn, the estuarine environment of calm, warm, shallow water with good light is also very favourable for aquatic plant growth. The plant growth too retains nutrients within the system bound up in the living tissues or in the dead material (detritus), and with nutrients still arriving from the catchment the problem grows.

There is little export of nitrogen and phosphorus from the relatively undisturbed catchment (only 4 per cent cleared) of the Shannon River that flows into Broke Inlet, the only estuary in the south-west of Western Australia classified as 'pristine'. In contrast, export of nutrients from cleared catchments particularly with the first flush of run-off with winter rain is high (Graph 1.2). In the Swan–Canning Estuary phosphorus concentrations have increased two to threefold since the 1940s. Similarly, the amount of inorganic phosphorus has doubled in the sediments of the Peel Inlet, and increased seven times in Harvey Inlet since the 1950s.

Efforts to control nutrients entering the estuaries have focused on protecting remaining vegetation, rehabilitating degraded waterways, reducing or modifying nutrient additions to agricultural land (diffuse sources), and reducing point source additions from urban, industrial and horticultural areas. Recently investigations have included the management of nutrients stored in sediments of rivers and estuaries through maintenance of high oxygen concentrations, and control of stratification and anoxic conditions that favour release of biologically active nutrients. A modified clay Phoslock™ that binds to phosphorus in the sediment reducing the amount released into the water has also been trialled. (See **Chapter 2 The Swan–Canning Estuary: Sediment remediation**.)

It is relatively easy to describe the status or trophic state of an estuary or river as oligotrophic, mesotrophic or eutrophic, but because of natural variability in nutrients it is far more difficult to detect a trend towards another state. (See **Box 1.6 Terms used for water quality.**) It is even more difficult to be predictive about how and when changes are likely to be manifest. Estuarine systems in Swanland, like those throughout the world, are generally resilient to changes such as the natural variations of river flow, changing salinity, cycles of wetting and drying due to tides, and low oxygen. Communities living in estuarine environments are also well adapted to changing conditions either through the organisms' physiology, reproductive strategies or migration out of the system or to less stressful areas within the system. Although large episodic events decimate populations they do, however, recover quickly from such disturbances by either recolonising from outside or from unaffected individuals reproducing rapidly when favourable conditions return.

Estuaries in south-western Australia are, however, particularly vulnerable to nutrient enrichment because of the evolutionary history of low (oligotrophic) conditions. Following nutrient enrichment, species' abundances change as opportunistic species, able to utilise the additional nutrients, proliferate (Graph 1.3).

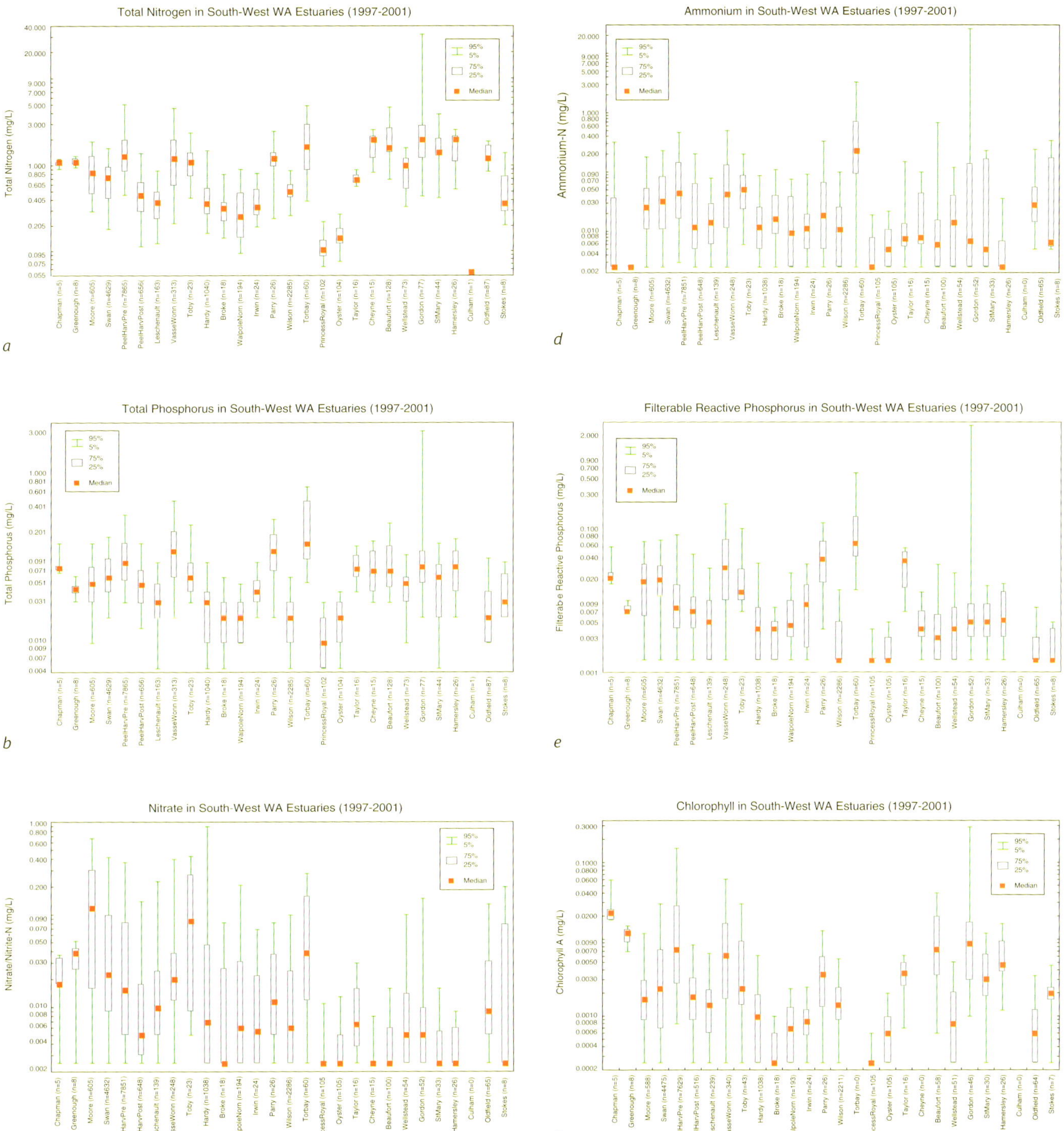

GRAPH 1.2 *Nutrient state of the estuaries from Chapman River on the west coast to Stokes Inlet on the south coast: (a) total nitrogen, (b) total phosphorus, (c) nitrate, (d) ammonium, (e) filterable reactive phosphorus, and (f) chlorophyll* a, *1997–2001. Note the great variability in nutrient concentrations in part due to variations in river flow and the flush of nutrients after dry periods. Although some of the estuaries are surrounded by native vegetation, the catchments of many lie in cleared land. Broke Inlet is the only pristine estuary on the wet western south coast. Further east the estuaries of the Fitzgerald River National Park, Gordon, Hamersley, Fitzgerald and Dempster are undisturbed but have evolved to a normally closed state where sediments and nutrients accumulate.*
Courtesy WRC.

Overview of issues, status and health of the estuaries in Swanland

As we have seen, rainfall and river discharge in Swanland vary greatly throughout the year due to the mediterranean climate, and also from year to year particularly with the influence of summer rain as weakening tropical cyclones move south. Discharge to rivers is also increasing with rising ground water following removal of the native vegetation. Variability in river flow affects the estuaries downstream, particularly influencing the opening and closing of the bars. Changes in water flow through damming, extraction of ground water or interference with bar openings and drainage also affect water levels or salinity. These in turn influence the fringing vegetation (trees, rushes and samphires), a major part of the estuarine habitat that would normally buffer nutrient additions and prevent erosion.

The Swan–Canning estuary, surrounded by the city and suburbs of Perth, is now highly modified. Removal of the rocky bar across the entrance channel and dredging of channels, along with construction of breakwaters to create Fremantle Port, increased the marine influence and the proliferation of a 'salt wedge'. The typical tides of this region, although less than a metre in amplitude, are now more obvious and only slightly less than on the open coast. Extensive deepening of channels, removal of the wide sandy banks along with most of the fringing vegetation and dumping of the spoil along the margins have changed circulation and altered chemical and biological processes. The

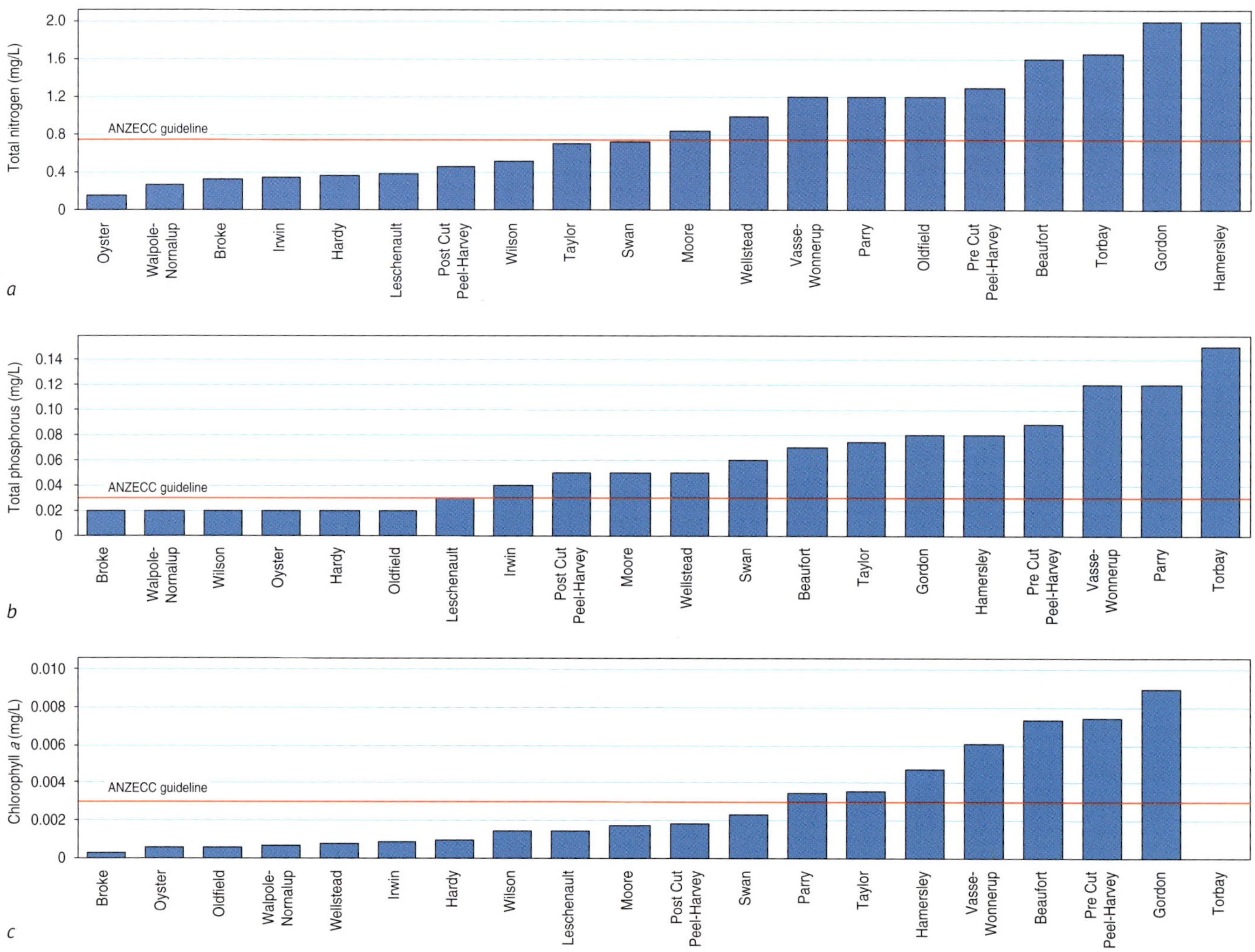

GRAPH 1.3 *(a) Total nitrogen, (b) total phosphorus and (c) chlorophyll* a *in the estuaries of Swanland ranged according nutrient status.* Courtesy WRC.

long history of nutrient additions from agriculture in the catchments and around the estuary, from industry, landfill and rubbish disposal, septic tanks and suburban gardens has resulted in a huge nutrient store in soils and in the river sediments, every ready fuel for algal blooms. These include blue-green algae blooms of toxic *Microcystis* that were particularly spectacular in 2000 following summer floods in the catchment. Similarly, in autumn 2003, fresh water brought a flush of nitrogen that fuelled growth of the dinoflagellate *Karlodinium micrum*. *Karlodinium* cells multiplied rapidly in the warm sunny conditions, reducing oxygen in the water and damaging fish gills resulting in spectacular fish kills. The latest blooms are a reminder that there are many other algal species ever present in low numbers in the sumps, drains and river reaches ready to be flushed to the estuary and proliferate when conditions are ideal.

Similarly, the plagues of macroalgae and toxic blue-green algal blooms in the Peel–Harvey in the 1970s and 1980s resulted from a long history of catchment changes including modifications to the drainage, changes to the bar, development of agriculture, and fertiliser additions. Although algal blooms in the estuary have decreased with the greater marine influence since the opening of the Dawesville Channel, nutrient concentrations in water flowing from the catchment are still high and blooms now occur in the river reaches. The saltier tidal river reaches now favour the release of nutrients deposited years ago which, together with nutrients still arriving in run-off, fuel blooms of algae including toxic *Microcystis*.

Like the New Peel–Harvey, the Leschenault has a direct channel to sea via The Cut that allows nutrients in fresh water to pass from the main basin. The marine influence in the Leschenault is also stronger, and algal blooms plague the deeper river reaches. The Leschenault is under pressure from land use in the past, substantial changes to drainage, and removal of the buffering fringing vegetation. Landscape changes are ongoing with new problems of nutrients from urban development and from agricultural land in the relatively isolated northern part of the estuary (Fig. 1.8).

FIG. 1.8 *Landsat view of plumes of fresh water laden with nutrients flowing from the Swan–Canning and Peel–Harvey, 14 August 1991, 0930 hours Western Standard Time. The plume from the Peel–Harvey extended about 100 km northwards influencing the coastal areas of Comet Bay, Sepia Depression and Warnbro Sound. Water from the Swan extended about 60 km northwards, but also moved southwards into Owen Anchorage and Cockburn Sound under the influence of northerly winter winds.* Courtesy DLI.

The Vasse–Wonnerup, a vast wetland draining to Geographe Bay extending from the Leschenault to Cape Naturaliste, is the most highly modified system in Swanland. Most of the original small diffuse rivers and swamps are now fast flowing drains to sea, the low-lying land is used for intensive grazing with the whole area a large flood compensation basin for Busselton. Although it has the dubious reputation as the most eutrophic body in Swanland where algal blooms are common, the Vasse–Wonnerup is still a most important refuge for birds.

Further south, fresh water flows to the tiny Margaret River and other tiny rivers, seasonal 'microestuaries', are diminishing year by year as ground water is applied to vineyards or used by the escalating population. The changing land use is contributing more nutrients each year. Around Cape Leeuwin, the Hardy at the mouth of the mighty Blackwood is influenced by changes in the catchment and reduction in river flows, both climatic and anthropogenic, manifest as increasing nutrients and sedimentation. These also appear to have affected the black bream. Like all the estuaries the Hardy is experiencing people pressure; with more people living in the area there are more nutrients from gardens and sewerage, and more people fishing. The eastward growth of the bar, a repeat performance of the 1940s scenario, although largely a natural phenomenon due to lower river flows and variation in longshore currents along the coast, has affected boating access and probably the marine exchange.

Many of the estuaries along the western section of the south coast, although lying in the area of highest rainfall where water flow is more reliable, also have extensive bar formations at the mouth with Walpole–Nornalup the only permanently open system. Marine exchange is limited to times when the bar is open, and many are opened artificially when water levels are high, but close naturally due to a combination of reduced outward flow and onshore movement of sand brought by swell waves and longshore currents.

Although some of the smaller rivers that briefly exhibit estuarine features and the larger estuary basins lie in uncleared catchments, the larger estuaries have catchments that extend into highly modified salt-affected agricultural land. Broke Inlet with a wide shallow basin and long narrow channel to the sea is, however, the only estuary in Swanland classified as pristine, lying entirely within the D'Entrecasteaux National Park. Thus processes in Broke can be viewed as a 'benchmark' for some of the other estuaries. While the Walpole–Nornalup is surrounded by giant trees and partially protected in state forests used for timber logging and national parks, the inlets are classified as slightly modified due to nutrients from Walpole into the smaller enclosed basin. Although some of the estuaries are surrounded by cleared land and conditions are generally reasonable, most, including Wilson Inlet, have increased growth of algae, *Ruppia* and microalgae, and are highly productive supporting large fish and bird populations. Low-lying, naturally waterlogged areas from Wilson Inlet through to Torbay and Princess Royal Harbour are now drained and the water levels controlled, and Lake Powell with a long history of modifications experiences blue-green algal blooms in summer.

The Albany Harbours, surrounded by a city, receiving water from a modified catchment were the scene of great seagrass losses in the 1970s and 1980s due to overgrowth of algae, along with contamination of water and fish by bacteria and industrial pollutants. Water quality has improved due to a reduction in point sources in Princess Royal Harbour, however nutrients from the diffuse sources originating from urban areas are still high as are those from the extensively cleared catchments of the Kalgan and King Rivers that flow to Oyster Harbour. Although efforts to control algal growth by harvesting were not successful on a large scale, the amounts of algae have decreased and seagrass is expanding in some areas.

To the east of Albany with the decrease in rainfall and more erratic flows occurring in either winter or summer, the bars are closed for long periods and the estuaries show a progression towards greater isolation from the sea. Culham Inlet, isolated from the sea for decades, has opened several times in the last ten years with increased flow from the cleared catchment. The shallow sediment-filled estuaries such as the Beaufort, Gordon and Wellstead receive flows from extensively cleared catchments and are exhibiting signs of stress manifest as algal blooms. The Oldfield, in contrast, is regarded as relatively pristine despite extensive clearing in the catchment, the dark tannin-stained waters perhaps less favourable for algal growths that have plagued other estuaries. The Fitzgerald and St Marys in the Fitzgerald National Park and International Biosphere Reserve, surrounded by native vegetation, are little influenced by changes to catchments although they are progressing naturally towards an isolated or lagoon state. Historically, clearing of the catchments has been a concern and community groups have largely focused on this issue. However recognition of the area's unique geology and flora is leading to an increase in development pressure from tourism and mining.

Estuaries in the north of Swanland on the mid-west coast are similar in some ways to eastern estuaries of the south coast. These estuaries are also characterised by erratic flows, and many are small, shallow and filled with sediment. Although they lie in a landscape similar to the southern coastal plain, they are outcrops of much older formations nearby that affect the surrounding soils and estuary sediments. The largest, the Murchison River estuary at Kalbarri, receives water from a huge catchment that extends far into the pastoral country from which floodwaters rush seawards through the deep valleys of Tumblagooda Sandstone. The estuary entrance although permanently open is shallow and the sediment filling the estuary is contaminated by lead eroded from mine tailings along the river banks.

The Chapman, surrounded by the city and suburbs of Geraldton, is highly modified with nutrient-rich waters. While urban developments around the Greenough are not as extensive, the surrounding land is heavily cultivated with the result that nutrients and algal blooms in the estuary are increasing. The Irwin is filled with sediment and usually contains less water than the other estuaries of the area, and the infilling process appears to be accelerating.

The estuary of the Moore River just to the north of metropolitan Perth is increasingly under pressure from development—the sprawl of Perth. Here the remaining vegetation offers an insight into the original nature of the Swan–Canning. The Moore catchment extends into saline affected flood-prone cleared land. The bar opens and closes through the year in response to river flows, but also with fresh ground water inflows in the lower reaches. Nutrient concentrations in the estuary are high with phosphorus originating principally from Gingin Brook, and nitrogen from winter flows from both rivers and from groundwater. Algal blooms would probably be more frequent if it were not for the dark tannins in the water that limit light penetration.

Viewing and dealing with the problems of estuary health

Problems of algal blooms and fish kills that plague estuarine environments could be compared to those of human health where illness resulting from lifestyle may not manifest for many years, and where the most effective and sustainable cures require large effort, changes in attitude, money and time before improvements are obvious.

Management of estuaries is the responsibility of the Water and Rivers Commission under the *Waterways Conservation Act 1976* (WA). The Commission has been combined with the Department of Environmental Protection as the Department of Environment (DoE) to include powers of the *Environmental Protection Act 1986* (WA). Particular issues in estuaries are dealt with by a number of organisations. Information on water quality is collected by the DoE, while the Department of Fisheries collects information on the fisheries. Management is principally involved with eutrophication, the presence and effects of algal blooms along with issues specific to a particular estuary such as the artificial opening of the bar, dredging, and building of breakwaters and marinas. The major estuaries at risk, Peel–Harvey, Leschenault Inlet, Wilson Inlet and Albany Harbours were until recently managed by local community-based waterways management authorities. The Swan–Canning is managed by the Swan River Trust, a statutory authority with unique powers over planning and development. (See **Chapter 2 The Swan–Canning Estuary**.) The July 2003 release of 'River Plan' An Environmental Management Framework for the Swan and Canning Rivers with an Implementation Strategy for the 1998 Environmental Protection (Swan and Canning Rivers) Policy with the key function of coordinating and drawing together existing initiatives, is timely given the most recent algal blooms and fish kills.

Currently estuaries are managed along with the catchments by the government agencies working in partnership with community-based Regional Natural Resource Management bodies with the assistance of community advisory committees. Regional Strategies for all natural resources are being prepared, which will be accredited by the State and Commonwealth Governments. Investment Plans will be developed for the accredited strategies that will set out how funds will be used to improve the condition of natural resources according to agreed priorities.

Institutional and management arrangements are still evolving, but they are a positive move because for the first time the estuaries will be managed in the context of all land uses in their catchments and all natural resources. In a broader context an understanding of how the estuaries of Swanland function, and how they have been and are currently being modified—the purpose of this book—is a major issue for future management as some of our values and actions need reassessment if the estuaries are to remain healthy.

CLASSIFYING ESTUARIES

A classification of the estuaries of Swanland is most useful for understanding similarities and differences in the processes operating in different systems. The processes are necessary for keeping the areas beautiful, productive and healthy. There are many ways of classifying estuaries: according to their formation, to water flow, or to the biological processes. Like any classification there are exceptions and overlaps, and in some respects each estuary is unique or has unique features that make it special or notable.

Classifying estuary formation—the dominant processes

In the first years of the 21st century a classification based on dominant processes was developed through The Australian Catchment, River and Estuary Assessment, the National Land and Water Resources Audit, a program conducted under the Natural Heritage Trust as a basis for cost-effective management strategies. The classification grouped the estuaries throughout Australia into six classes defined by the influences of waves, tides and rivers. Throughout Australia 17 per cent are wave dominated, 11 per cent tide dominated, 10 per cent wave dominated deltas, 9 per cent tide dominated deltas, 5 per cent strand plains, coastal lakes and lagoons and 35 per cent tidal creeks and flats.

Tidal-dominated systems exist mainly in tropical northern Australia. In tidal-dominated systems a major management issue is the maintenance of tidal flows and minimisation of restrictions to flow due to structures such as barrages, causeways and bridges. In Swanland, tidal amplitudes are small and the inward flow of sea water is restricted as many estuary entrances are barred. River flows also vary throughout the year, and winds are a major influence generating waves that affect circulation and stir sediments.

Wave-dominated systems occur mainly in southern temperate regions including Swanland. In wave-dominated systems the major issues are minimisation of incoming nutrients and sediments as well as the management of the ocean entrance.

Classification by formation—geomorphology—combining geological history and water movement

The following classification according to geomorphology was part of a major review of estuarine evolution developed by Ernest Hodgkin with Patrick Hesp and published shortly before Hodgkin's death in 1998. Based on notes in the Hodgkin Library, I have extended the classification to include all the estuaries in Swanland, as was his intent for this book. The major distinction was between riverine estuaries and the three lagoonal-type estuaries (Fig. 1.9).

Riverine estuaries. These are confined to narrow riverine channels throughout their length, and lie roughly perpendicular to the coast. In the high-rainfall area some are relatively long, such as the Donnelly (12 kilometres) that flows through swamps and coastal dunes. In the low-rainfall areas of the west (Murchison, Chapman, Oakabella, Oakajee, Buller, Irwin, Hill, Moore) and south coast (Saint Marys and the small estuaries before the desert) all are less than 2 kilometres long and retain little water. Little is known about their evolution.

On the south coast there are also two small, narrow estuaries Waychinicup and Quoin, ria-types that are particularly unusual. In form they are river valleys carved into the ancient hard Precambrian rocks and flooded as the sea level rose.

Inter-barrier estuaries. These elongate lagoons are parallel to the coast behind the barrier dune and on-shore reefs of the west coast; the Greenough and Hutt are in the low-rainfall area north of the Swan while the Harvey and Leschenault are in the high-rainfall area further south. While very similar in form, the Leschenault and Harvey lie between different dune systems. The Harvey lies to the east of the Spearwood Dunes and connects to the sea through the Peel Lagoon, while the Leschenault lies east of the Quindalup Dunes. These most recently formed Holocene dunes are still mobile, and sand from the dune barrier is still migrating eastwards into the Leschenault. The dunes also extend along Geographe Bay where they enclose the Vasse–Wonnerup Estuary.

Valley estuaries. These are in drowned valleys roughly perpendicular to the coast. Most receive flow from a single river (Beaufort) but some have a second tributary river (Swan–Canning and Stokes). The Swan is in coastal limestone and Pleistocene sediments in the Swan Coastal Plain. South coast valley estuaries of the low-rainfall area are carved in Precambrian and Eocene siltstone rocks and the coastal dunes.

Basin estuaries. In the high rainfall area, Peel is in a depression in the Swan Coastal Plain across which tributary rivers meander from their sources on the plateau. On the south coast the estuaries are in segments of narrow coastal plain between the drowned edge of the Ravensthorpe Ramp and the headlands and coastal dune barrier. Nornalup and Wilson

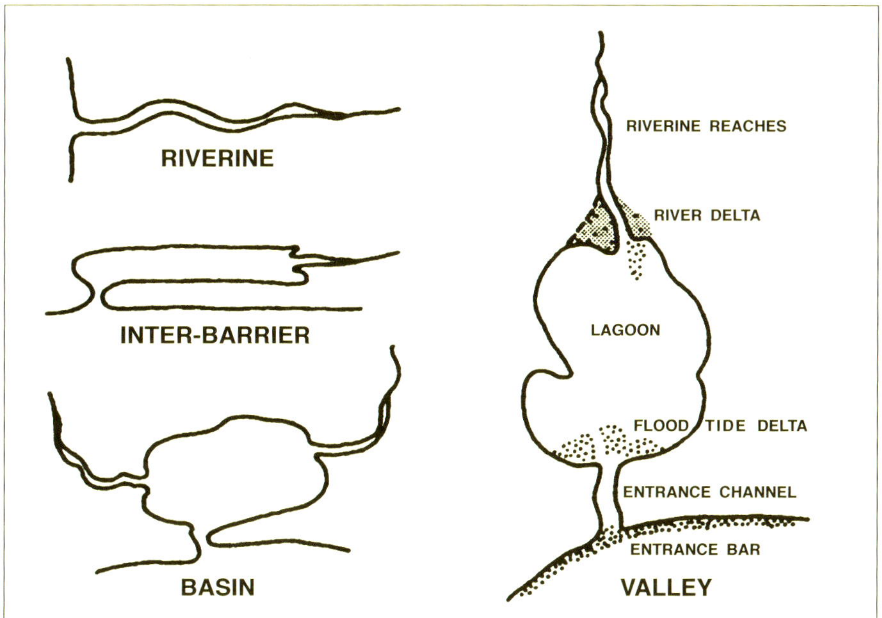

FIG. 1.9 *Geomorphic types and features of estuaries of south-western Australia.* Hodgkin & Hesp, 1998. Courtesy Marine and Freshwater Research, CSIRO Publishing, Victoria, Australia.

were probably already sheltered bays during the last interglacial (when sea level was high). In the low rainfall area further east, Culham is a bay now enclosed by a dune-bar, but Jerdacuttup lagoon basin is elongate between the Precambrian and a Pleistocene dune and Holocene bar.

Classification by hydrodynamic condition—an ecosystem approach

Birth, life and death of the estuaries of Swanland

Over recent time the character of the estuaries throughout Swanland has changed as sediment filled the upper river channels, deltas formed and the lagoons shallowed. Along the coast longshore movement of sand, predominantly northwards on the west coast and eastwards on the south coast, formed bars and flood-tide deltas in the entrances. Over time these consolidated, undisturbed by the small tides and long periods of insignificant river flow, processes later enhanced by falling sea level. Today the ocean bars and deltas hamper tidal exchange and river flow dominates in winter, although tides bring marine water when flow is low in summer. As a result salinity is highly seasonal, fresh in winter and salty in summer. These long periods of salinity extremes are only tolerated by the most euryhaline biota, so that in the last 3,500 years or so the fauna of the seasonally extreme estuaries has been dominated by only a few estuarine species.

Continued isolation from the sea from permanently open (PO) to permanently closed (PC) appears to result from interactions of many factors. These include volumes and patterns of river flow, large and reliable or small and episodic; size of the lagoon or the extent to which flood water can spread out before the entrance bar breaks; the degree of shelter given to an entrance by headlands and/or reef lagoons; the supply of sand along the coast to build bars and flood-tide deltas; and the amount of sediment brought to the estuary by rivers (Table 1.7).

The principal features leading to isolation of the estuaries from a permanently open to a permanently closed state are:

- volume and pattern of river flow: large and reliable versus small and episodic
- volume of river flow relative to available volume of receiving lagoon
- extent to which entrances are sheltered from the south-west winds and swell
- volume of available sand.

TABLE 1.7 *Summary of the main characteristics of estuaries in Swanland.*
Source: Hodgkin & Hesp, 1998.

	River flow	River flow relative to lagoon area	Entrance	Salinity regime	Community
PO	Large and reliable	Large flow relative to available volume of lagoon	Sheltered by headlands or reefs	Changes extreme	Recruitment continuous, diverse marine–estuarine fauna, but euryhaline animals dominate
SO	Lower but reliable	Less flow relative to lagoon volume	Less sheltered	Changes extreme	Recruitment limited, fewer species
NC	Small mean flow but large episodic flows	High episodic flows greater than lagoon volume; bars break	Relatively sheltered by headlands or reefs	More extreme to hypersaline	Limited fauna, periodic marine recruitment, massive mortality when hypersaline
PC	Small mean flow but large episodic flows	High episodic flow but lagoons large and high bars break only rarely	Relatively sheltered by headlands or reefs	Saline tributaries hypersaline to dry	Few estuarine species, massive mortalities when hypersaline

The estuaries of south-western Australia have also been classified according to the connection to the sea, how often and for how long the bars are open (Table 1.8). This classification considers the hydrodynamics and effects on biology. Open bars allow the entry (recruitment) of animals and the persistence of favourable conditions for communities to develop. Conversely, bars that close quickly and may remain closed for long periods limit the entry of animals and lead to more stressful estuarine environments.

The classification illustrates how the estuaries become isolated from the sea, a progression from 'textbook' estuaries when they were formed about 6,000 years ago when tidal influences were strong, to those that close over for months to years.

Stages of isolation

The classification like all such groupings is not rigid. For example, the Peel permanently open bar at Mandurah sometimes closed before training walls were built; seasonally closed Wilson and Broke would probably stay closed after dry winters if not opened artificially; Wellstead and Gordon although normally closed sometimes stay open for several years; and Culham regarded in some reports as permanently closed has broken through to the sea in 1993, 1997 and again in 2000.

Permanently open (PO). The bars and tidal-deltas obstruct tidal exchange but rarely close the entrances. Hydrologically the lagoons are nearly fresh in winter, and brackish to marine salinity or slightly hypersaline in summer. An exception is Oyster Harbour where exchange with the sea is always good. The Swan and Leschenault, although permanently open today, were once more isolated from marine influences before the entrance channels were modified.

Seasonally open–closed (SO). The bars close for several months in summer–autumn and open annually following major river flow in winter. Hydrologically they are similar to PO estuaries but there is no tidal exchange or access for marine fauna while the bars are closed.

Normally closed (NC). The bars stay closed for several years and open only in response to major river flow, however most remain open only briefly. Salinity varies from about 10‰ following river flow to hypersaline while the bars are closed, and those with shallow lagoons may dry up altogether. Normally closed estuaries are sometimes referred to as intermittently open.

Permanently closed (PC). The bars never or rarely break. These former estuaries are now salt lakes. Saline rivers flow to them but there is no replenishment with sea water or access for marine animals. The salinity range is similar to that in normally closed estuaries, and they sometimes dry up altogether.

Comparing the estuarine environments

The individual estuaries are described more fully in the following chapters, grouped according to location along the coast, but here we consider the range of different types according to geomorphology, hydrology and how they function. A comparison of the environmental factors is particularly useful as this dictates the biological processes as well as the types of plants and animals that live in them (Table 1.9).

TABLE 1.8 *Morphological type, stage of isolation and size of the larger lagoonal estuaries of south-western Australia* (Hodgkin & Hesp, 1998). *The riverine estuaries are not included although they fall most commonly into the categories of normally closed–intermittently open (such as the Hutt, Oakajee, Oakabella, Buller, Chapman, Greenough, Hill, Moore, Vasse–Wonnerup, Tobys, Margaret, Donnelly, Gardner, Warren, and the estuaries east of Esperance). Note that the ocean entrance may be manually opened.*

	Type and area					
Stage	**Inter-barrier**	**Area (km^2)**	**Valley**	**Area (km^2)**	**Basin**	**Area (km^2)**
Permanently open	Murchison	4	Swan	50	Peel–Harvey	131
	Harvey	56	Hardy	9	Nornalup	13
	Leschenault	20 or 27			Oyster	16
Seasonally open			Cheyne	1	Broke	48
					Irwin	11
					Wilson	48
					Parry	1
Normally closed	Greenough	2	Beaufort	5	Barker	2
			Wellstead	2		
			Gordon	3		
			Fitzgerald	7		
			Dempster	2		
			Hamersley	2		
			Oldfield	1		
			Stokes	14		
Permanently closed			Lake Nameless	2	Culham	11
					Jerdacuttup	30

Permanently open estuaries

Oyster Harbour and Princess Royal Harbour are the most marine of the estuaries. In Oyster Harbour, the deep permanently open channels allow flows of fresh water to move through relatively quickly and marine waters to flow in and out with the tides. Similarly, in Princess Royal Harbour, marine waters brought by tides are more dominant than the fresh ground water inflows.

In contrast other permanently open estuaries, except the Murchison, receive relatively large and reliable river flows; the entrances are sheltered by headlands; and tidal exchange is obstructed by flood-tide deltas and/or shallow bars. As a result the salinity in the estuary changes seasonally from fresh to marine.

The Swan–Canning, another permanently open estuary, has undergone a number of changes affecting marine exchange. During the low sea levels of the last ice age the entrance to the Swan was through a rocky channel bordered by coastal limestone and a coral limestone reef. Later as the seas rose again, sand accumulated in the channel. At the time of European discovery in the late 18th century, a reef at about sea level and sandbars restricted the Swan entrance. Settlement brought engineering works to allow access and create a harbour; the reef and channels were deepened and the tidal influence increased from about 10 per cent of the ocean tides to 80 per cent. Today the main basin contains fresh water in most winters at least to the depth of the remaining sill (about 5 metres) and marine waters in summer.

The Peel–Harvey has also undergone change. Until 1994 the Peel–Harvey opened northwards in the lee of a limestone headland. Tidal exchange was limited by an offshore ebb-tide bar, flood-tide delta and a long, narrow winding entrance channel, so the salinity varied from winter to summer from less than 3‰ to 45‰. After the Dawesville Channel was opened in 1994, the Peel–Harvey became a new environment. Maximum tides increased from about 15 per cent to 48 per cent of ocean tides in the Peel and from 17 per cent to 55 per cent in the Harvey. Currently a more marine state dominates, an environment more similar to that of 6,000 years ago.

TABLE 1.9 *Summary of the mollusc fauna living in the estuaries of Swanland: marine species (M), estuarine–marine species (EM), estuarine species (E), freshwater (F), and episodic presence (P).* Source: Hodgkin & Hesp, 1998. Data for Princess Royal Harbour and Oyster Harbour, Roberts & Wells, 1980. New Peel–Harvey, Whisson et al., 2004.

Estuary	Total molluscs	M	EM	E	F
Permanently open					
Swan marine channel	97	64	25		
Swan lagoon	13		6	7	
Swan riverine	8			7	1
Peel–Harvey	10	2	4	4	
New Peel–Harvey	12	4	4	4	
Hardy	14		9	5	
Nornalup	22	7	10	5	
Princess Royal Harbour	90	54	32	4	
Oyster Harbour	109	70	33	6	
Seasonally open					
Wilson	17 or 16		9	8	
Broke	11	1	3	7	
Irwin	4		4		
Normally closed	5–6		P	5–6	
Permanently closed	4		2	2	

Historically the Leschenault flowed through a narrow channel to open on the northern side of the basalt headland. This opening was closed (1951), and a new entrance opened through the coastal dunes opposite the river mouths allowing greater tidal exchange and more marine conditions year round.

On the south coast, the Hardy receives the largest river of south-western Australia. The entrance has migrated eastward twice over the last 130 years but appears to have only closed once in that time, although the entrance is often shallow. As a result salinity in the estuary reflects river flow, and salty marine water flows upstream as river flow diminishes.

Similarly, the Nornalup although open to the sea has a narrow shallow tidal channel. River flow in winter creates stratified conditions where fresh water overrides salty denser water, but in summer marine water dominates and moves up the riverine reaches.

In the north the Murchison receives more irregular flow (about 3–5-year intervals) from the large arid catchment. The shallow estuary never appears to have closed completely, and dredging is now used to ensure a navigable channel that is only inaccessible during high seas.

All the permanently open estuaries experience a wide range of salinity depending on the volume of river flow, size of the lagoons and tidal exchange. As a result the fauna in different areas of the estuary contains plants and animals reflecting the dominant salinity from freshwater areas near the river entrances to marine species near the coast. Lagoon areas are generally dominated by a few euryhaline or estuarine types (E species), species able to withstand variations in salinity. Generally more marine (M) species enter over summer, and those common in marine and estuarine environments are classed as 'estuarine–marine' (EM) species.

Seasonally open–closed estuaries

Seasonally open estuaries generally receive less river flow than permanently open estuaries, their bars break each winter, or are broken artificially, and close again after river flow ceases. The larger seasonally closed estuaries generally follow the salinity pattern of the PO estuaries, varying with winter river flow and the intrusion of marine water while the bars remain open. While the bars are closed, water level and salinity fluctuate with evaporation and river flow. In Wilson Inlet with one of the largest lagoons (48 square kilometres), water level varies from about mean sea level (MSL) to + 1 metre at which height the bar is mechanically breached. Salinity is seldom less than 10‰ or more than 35‰, and the water is generally well mixed although the 5 metre deep basin is sometimes stratified. In contrast Irwin Inlet, which is smaller in area and much shallower, can become hypersaline when closed and the even smaller Parry Inlet sometimes dries out completely.

As the bars close over quite quickly, tidal exchange is limited and the chances for recruitment of marine animals is more restricted. For example, Wilson Inlet which is open for longer contains a larger range of species than the smaller Irwin which becomes hypersaline, however even in Wilson Inlet there are fewer marine species than in the PO estuaries.

Normally closed estuaries

Conditions in normally closed estuaries are more extreme than the seasonally open–closed estuaries. Generally only a few types of estuarine animals live permanently in the three estuaries that retain some water, the Beaufort, Wellstead and Stokes, although some estuarine–marine species enter when the bars are open where they may flourish for a time before the waters become hypersaline or before they dry out completely. Recruitment from coastal waters is therefore limited to the unusual periods when bars are open, and persistence of the fauna depends on how long conditions remain favourable. Some species may persist in the deeper parts or riverine reaches from which they are available to recolonise when the rivers flow and the estuaries contain water again. The presence of the salt-lake snail *Coxiella striatula* in estuaries like Oldfield and Barker Inlets is indicative of the extreme conditions where the estuary is becoming lagoon or lake-like.

Permanently closed, former estuaries

There are even fewer species in permanently closed estuaries like Jerdacuttup and Culham. Inclusion of Culham in this category is debatable, as it has opened several times in the last decade. However its inclusion is instructive as it illustrates biological differences between open and closed estuaries. Until 1993 the mound-forming tube worms *Ficopomatus enigmaticus*, a species also found in normally closed estuaries, dominated along with some estuarine molluscs; however after the 1993 opening the number of species increased slightly due to recruitment of species that live in marine and estuarine waters.

HABITATS AND ORGANISMS

The size of an estuary is often ill-defined, the lateral extent of the water body varying with the water level but indicated by the characteristic fringing vegetation of samphires such as *Sarcocornia* and *Halosarcia*, sedges and rushes such as *Juncus kraussii*. The fringing trees include sheoak *Casuarina obesa* and paperbarks or tea-trees *Melaleuca*, the saltwater species *Melaleuca cuticularis* and freshwater or swamp species *Melaleuca raphiophylla* and, in the Leschenault Estuary, the mangrove *Avicennia marina*.

Although individual species vary in their tolerances, these plants are generally able to tolerate periods of desiccation or waterlogging and extremes of salinity. However longer periods of inundation, drying due to draining or increased tidal influence gradually bring changes in the communities. For example, the changes induced by the Dawesville Channel in the Peel–Harvey have brought higher water levels and more regular inundation with salt water with the result that some trees are now dying. Although the exact causes are unknown, the waterlogged soils are probably becoming oxygen deficient and roots are no longer accessing the layer of fresh water floating on top of saline ground water. In addition, bores have probably depleted the freshwater layer that has been replaced by salt water. Fringing vegetation also declines with disturbance and weed invasion.

Whatever the mechanism of change, the great loss when fringing vegetation declines is the loss or alteration of function that is likely to follow. Fringing vegetation is important for trapping unwanted nutrients and buffering erosive forces. Fringing vegetation also contributes to the estuarine production of organic matter and creates habitat for terrestrial and aquatic animals. (See **Chapters 2–4**.)

Life in estuaries—living in extreme conditions

A large range of physical, chemical and biotic features affects organisms living in the estuaries of Swanland. These features vary between the different estuaries, within the estuaries, as well as over daily, seasonal, yearly and decadal time scales. They also vary with the relative riverine and marine influences, and in Swanland with the progression of estuarine evolution from open to closed state. Estuaries also provide a range of different habitats, including the fringing marshes, water column and estuary bottom.

Organisms living in the water column are pelagic while those on the bottom are termed benthic. Pelagic organisms form the plankton (floating forms) or the nekton (swimming forms). The distinction between the two terms, however, is often blurred as many planktonic algae (phytoplankton) migrate through the water column or are bottom dwellers resuspended by waves that disturb the sediments in shallow areas. Phytoplankton may be near the surface where light is abundant during the day and move to the bottom at night where they can take up nutrients. The movement of phytoplankton may be passive due to buoyancy of gases and compounds made during photosynthesis, or active through the use of flagella (tiny whip-like threads).

Habitats for bottom-dwelling organisms, referred to as the benthos, include sediments of various grain size (fine mud, sand and gravel), rocks, as well as other organisms both dead (organic matter or detritus) or living. Plants living on other plants but not as parasites are called epiphytes. A common usage of the term epiphyte is algae living on other algae or seagrasses. Small colonial encrusting animals, such as bryozoans and hydroids, are also on occasion referred to as part of the epiphytic community (Box 1.8).

Box 1.8 THE NAME EXPLAINS THE HABIT

Many terms used in biology are derived from Greek words. **Epiphyte** is a term derived from the Greek *epi* meaning on or upon and *phyto* from *phuton* meaning a plant or vegetation. Hence the term means a plant that grows upon another plant, but not parasitically as it does not obtain food, water or minerals from it.

In the water column, the term **pelagic** is from *pelagos* meaning sea, **plankton** from *planktos* meaning wandering, **nekton** from *nekton*, a swimming thing. The planktonic organisms generally float or drift and are usually small.

Euryhaline is from the Greek *eurus* meaning wide and *hals* salt, hence it describes an organism able to tolerate a wide range of salinity. Similarly, **stenohaline** from the Greek *stenos* meaning narrow refers to organisms able to exist only within a narrow range of salinity.

Microalgae—phytoplankton small in size but often abundant

The small size of the algae in the phytoplankton, and hence the large surface area, allows rapid uptake of nutrients so that they grow and divide rapidly forming scums or blooms. Although blooms can be a nuisance, they are an indication of the high productivity of estuaries that forms the base of food webs leading to the fish resources we value and the birds we love to watch. The microscopic algae, larger algae and plants produce oxygen-creating conditions favourable for animals, however as they use oxygen at night in respiration large blooms can create other problems (Box 1.9).

The phytoplankton are composed of green algae (Chlorophyta), diatoms (Bacillariophyta), Chrysophyta, Cryptophyta, Euglenophyta, Raphidophyta, dinoflagellates (Dinophyta formerly known as Pyrrhophyta) as well as cyanobacteria (Cyanophyta) that form blue-green algal blooms (Fig. 1.10). The microalgae, like the larger macroalgae, are classified according to the light harvesting pigments (photosynthetic pigments), storage compounds and structures of the cells (Table 1.10). Many occur as single cells, while others form colonies or aggregations. Because different types of algae respond to the environment in different ways and their roles may be quite different, the abundance and diversity of different types can used to gauge how an estuary is functioning.

Variety and form of microalgae

GREEN ALGAE The single-celled or colonial green algae may have small tails or flagella that enable them to move. Some species such as *Chlamydomonas* are common in fresh water. They may form blooms in the upper reaches of estuaries such as the Swan–Canning after a freshwater flush. In contrast, *Dunaliella salina* grows in the very saline conditions found in the Hutt Lagoon where it is cultivated for beta-carotene production. (See **Box 5.4 Super science meets the natural world**.)

DIATOMS Diatoms are tiny yellow-brown plants enclosed in a glass-like skeleton made of silica. The tiny cells are made of two parts, are often highly sculptured, and covered with pits through which parts of the cell protrude, but they do not have flagella. Diatoms are found in all types of water bodies. Some forms are free floating, while others attach to firm surfaces, sand grains or macroalgae. After death, the shells sink to the bottom where they can accumulate in deposits known as diatomaceous earth or diatomite. These deposits are mined for use in polishes, filters, paint thickeners and kitty litter. As diagnostic details in the silica shells are preserved in fossil deposits, diatom deposits can also be used to indicate changes in environmental conditions.

Blooms of diatoms turn the water a murky yellow-brown. Although many such blooms are harmless, some forms are toxic, while others damage fish gills. In the Swan–Canning blooms of *Skeletonema costatum* are quite common, particularly as the winter run-off slows and salinity increases. *Skeletonema* is an important part of the diet of small animals in the water column as well as the filter feeding mussels that live attached to hard surfaces.

Box 1.9 PROCESSES FOR LIFE

Photosynthesis is the process by which organisms convert the inorganic compounds carbon dioxide (CO_2) and water (H_2O) using light energy absorbed by pigments such as chlorophyll *a* to carbohydrates (CH_2O) (organic molecules sometimes referred to as sugars) and oxygen (O_2).

Organisms that photosynthesise are termed **autotrophs.**

The rate at which organic molecules are formed is the **primary productivity.**

Some energy is used to maintain life and make new compounds in the processes of respiration and metabolism.

Respiration. 1. The process of living organisms of taking in oxygen and releasing carbon dioxide (breathing). 2. The chemical breakdown of complex organic compounds such as carbohydrates and fats that takes place in cells of plants and animals during which carbon dioxide is produced and energy is released. It can be either aerobic or anaerobic.

Aerobic respiration is the process by which carbohydrates (CH_2O) are oxidised in the presence of oxygen (O_2) to produce carbon dioxide (CO_2), water (H_2O) and energy.

Anaerobic respiration is the breakdown of molecules without oxygen. This can be glycogen to lactic acid in vertebrate muscle or glucose to ethyl alcohol and carbon dioxide by yeasts. Anaerobic respiration does not release as much energy as aerobic respiration.

Metabolism. The chemical processes within an organism that involve the breakdown of organic compounds (catabolism) and release of energy, or the build up of organic compounds (anabolism) and use of energy. Energy may come from the breakdown processes or from the use of external energy such as light energy in photosynthesis. The rate of metabolism is usually referred to as the amount of oxygen consumed or carbon dioxide produced.

Energy not used in metabolism and respiration may be stored as **biomass.** The build up of biomass through time is called the **net primary productivity.** The **gross primary productivity** includes the storage (net primary productivity) and that used in respiration.

Heterotrophic organisms cannot manufacture their own organic compounds and depend on plants for their energy. They may be detritivores and consume dead material (detritus), or grazers consuming living material. Biomass built up over time by heterotrophs is referred to as **secondary production.** The gross secondary production includes the parts lost in faeces and urine, and metabolism. Further up the food web other heterotrophs consume the detritivores or grazers.

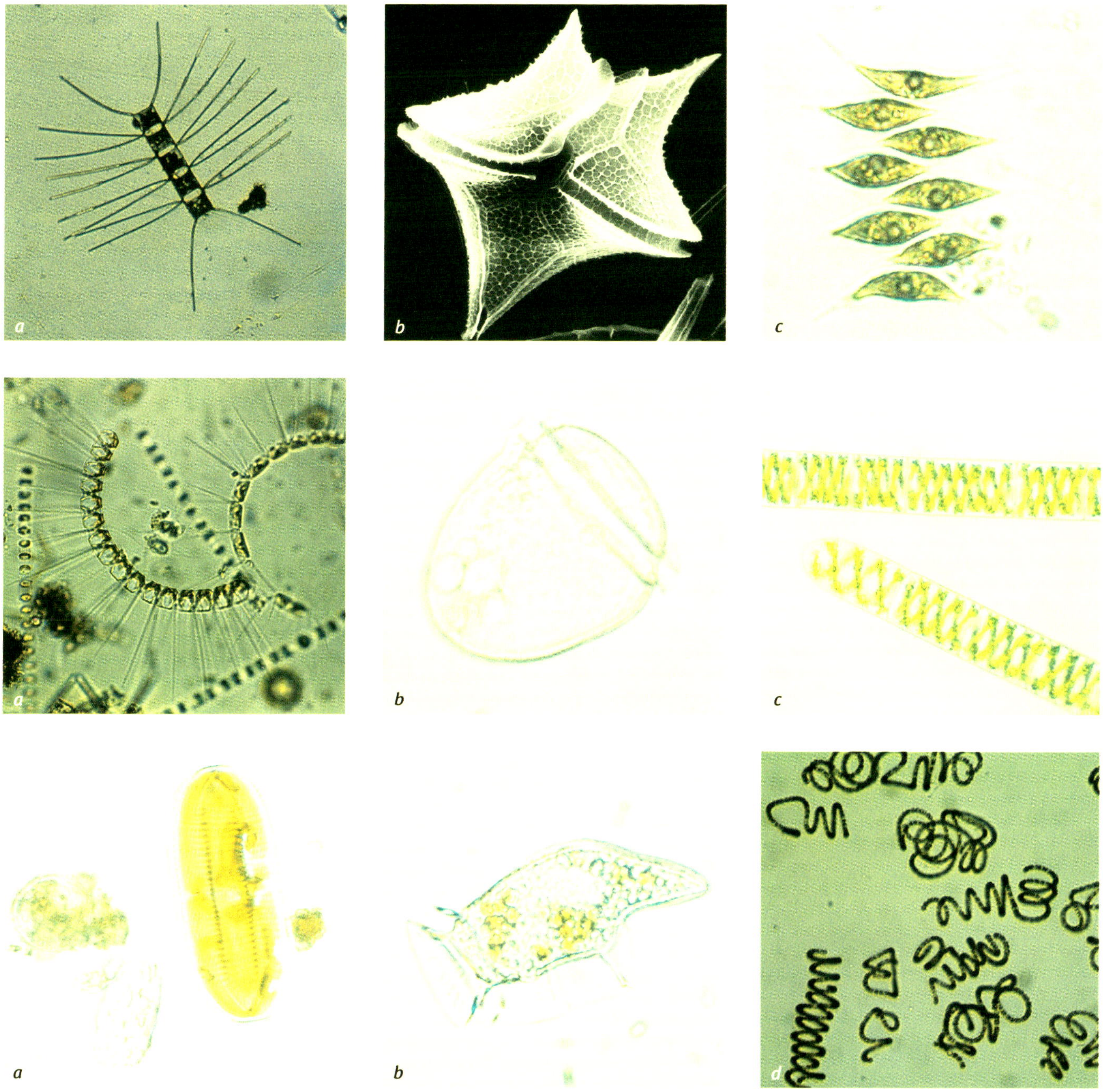

FIG. 1.10 *Different types of phytoplankton: (a) diatoms with shell made of silica—two types of spined* Chaetocerous, *and pennate* Hitzschia; *(b) dinoflagellates—*Protoperidinium, Phalachroma, and Dinophysis; *(c) Chlorophyta* Scenedesmus *and* Spirogyra; *and (d) blue-green algae* Anabaenopsis circinalis. Courtesy Aquatic Science Group, WRC.

DINOFLAGELLATES Dinoflagellates are single cells or colonies of cells with two thread-like strands called flagella, which are used for movement. The name is derived from the Greek word *dinos* meaning whirling and Latin *flagrum* a little whip. There are two groups: 'naked' with a thin shell or armoured with a thick shell. When they aggregate they colour the water red-brown, hence the name 'red tides' used to describe dinoflagellate blooms.

Dinoflagellates such as *Scrippsiella* sp. and *Gyrodinium* sp. are a very important part of the diet of mussels and other organisms such as copepods, and the base of many food webs in estuaries. When they are numerous some species can form scum annoying to swimmers. Others such as *Alexandrium minutum* produce nerve toxins (neurotoxins) that cause gastric or nerve illnesses such as Paralytic Shellfish Poisoning (PSP). In May 2003 a bloom of *Karlodinium micrum* caused the deaths of tens of thousands of fish in the Swan–Canning estuary.

TABLE 1.10 *Characteristics used to define some of the common types of phytoplankton and macroalgae in the estuaries of Swanland. The term haptophytes refers to the threadlike extension between the flagella that can be used to capture prey. Coccolithophorids, a major group of haptophytes, are covered with sculptured calcite plates. Fossils of these are a large component of chalk deposits such as Dover Cliffs in the UK.* Source: adapted from Bold & Wynne, 1985; Tomas, 1997; with assistance from A. Begum & W. Hosja, Aquatic Science Branch, DoE.

	Division/Phylum	Major pigments	Storage	Examples
Phytoplankton				
Green	Chlorophyta	Chlorophyll *a*, *b*, β-carotene	Starch (amylose + amylopectin)	*Chlamydomonas, Dunaliella, Tetraselmis*
Diatoms	Bacillariophyta	Xanthophyll, chlorophyll *a*, *c1*, *c2*, β-carotene, diadinoxanthin, fucoxanthin	Chrysolaminaran	*Skeletonema, Chaetoceros, Melosira, Cyclotella, Thalassiosira*
Dinoflagellates	Dinophyta	Xanthophyll, chlorophyll *a*, *c2*, β-carotene, peridinin	Starch	*Gymnodinium, Prorocentrum, Karlodinium, Scrippsiella, Gyrodinium, Alexandrium, Heterocapsa*
Cryptomonads	Cryptophyta	Xanthophyll, chlorophyll *a*, *c2*, β-carotene, alloxanthin	Starch	*Cryptomonas, Rhodomonas*
Haptophytes	Prymnesiophyta	Chlorophyll *a*, *c1*, *c2*, *c3*, β-carotene, 19'–HF xanthin	Chrysolaminaran, Paramylon	*Prymnesium*
Yellow-brown algae	Chrysophyta	Chlorophyll *a*, *c2*, *c3*, β-carotene, 19'–BF xanthin	Chrysolaminaran, fat and oil may be present	*Apedinella, Pseudopedinella, Synura*
Raphidophytes	Raphidophyta	Chlorophyll *a*, *c1*, *c2*, β-carotene, fucoxanthin	Starch	*Heterosigma/Fibrocapsa*
Euglenophytes	Euglenophyta	Chlorophyll *a*, *b*, β-carotene	Paramylon, occasionally Chrysolaminaran	*Euglena, Eutreptiella*
Cyanobacteria–blue-green algae	Cyanophyta	Phycocyanin, phycoerythrin, chlorophyll *a*, β-carotene	Myxophycean starch (glycogen-like)	*Anabaenopsis, Anabaena, Microcystis, Nodularia, Oscillatoria*
Macroalgae and other aquatic plants				
Green	Chlorophyta	Chlorophyll *a*, *b*, β-carotene	Starch (amylose + amylopectin)	*Ulva, Chaetomorpha, Cladophora, Enteromorpha, Rhizoclonium, Caulerpa, Polyphysa*
Brown	Phaeophyta	Xanthophyll, chlorophyll *a*, *c*, β-carotene	Laminaran	*Hormophysa, Sargassum, Hincksia, Cystoseira*
Red	Rhodophyta	Chlorophyll *a*, phycocyanin, phycoerythrin, β-carotene	Floridean starch (amylopectin)	*Gracilaria, Polysiphonia, Mazoyerella, Ceramium, Gelidium, Chondria*
Stoneworts–charophytes	Charophyta	Chlorophyll *a*, *b*, β-carotene	Starch (amylose + amylopectin)	*Lamprothamnium*

CYANOBACTERIA Cyanobacteria, also called blue-green algae, are an ancient group of photosynthetic bacteria without a true nucleus. Cyanobacteria dominated the earth for millions of years forming the structures known as stromatolites. They produced carbohydrates from solar energy and released oxygen as a by-product, creating conditions suitable for the evolution of other life forms. (See **Box 1.2 Stromatolites and blue-green algae—early forms of life**.)

The ability of some cyanobacteria to utilise atmospheric nitrogen allows them to grow more quickly than other algae when nitrogen is limiting. Potentially toxic blue-green blooms of *Anabaena*, *Microcystis* and *Nodularia* occur regularly in Western Australian wetlands. *Anabaena* forms long chains in rivers and lakes, spherical colonies of *Microcystis* are common in lakes and reservoirs, and the filaments of *Nodularia* occur in fresh and brackish lakes, rivers and estuaries but only rarely in more saline waters. In the early 1980s, massive blooms of *Nodularia* in the Peel–Harvey were linked to the presence of phosphate-rich freshwater run-off events in winter. Blooms occurred when the salinity increased during spring, when light increased and the waters warmed. Since the opening of the Dawesville Channel in 1994 and the creation of a more marine environment, *Nodularia* blooms have ceased in the Peel–Harvey. They are however common in the estuarine reaches of the Serpentine, where the layering of salt and fresh water creates anoxic conditions along the riverbed and favours release of phosphate to fuel the blooms. (See **Box 3.3 Blue-greens—ancient life a modern day nuisance**).

Macroalgae

Like the single-celled plants of the water column, the larger macroalgae are classified according to the type of light-capturing pigments that are present, the forms of carbohydrate they store and how they reproduce (Table 1.10). Along the Australian coast there are about 2,000 green algae (Chlorophyta), 350 brown algae (Phaeophyta) and 1,300 red algae (Rhodophyta), and the southern coast has the highest diversity of species in the world. Compared to the oceans however, comparatively fewer types of algae live in estuaries with diversity decreasing with distance from the ocean entrance. The green algae are the most common and are usually the types that cause a nuisance. The estuaries that are permanently open to the sea often have more types of brown and red algae than estuaries with more restricted flow (Fig. 1.11).

Macroalgae are found as filaments, clumps, balls and leaf-like forms, but unlike the higher plants, the seagrasses, they do not have leaves, rhizomes or roots, and they take up nutrients directly from the water. Macroalgae live attached to rock, wood, seagrasses, other algae or they may be free floating. Under different conditions however, they may grow quite differently. For example, in the 1970s the green alga *Cladophora* in the Peel–Harvey grew as a loose carpet of small dark balls rolling along the bottom and stranding on the shore, while in the Swan another *Cladophora* generally grows in small tufts on hard substrate. *Rhizoclonium*, another green alga, which is a filamentous form, is found as blooms in the still waters of the upper Canning River estuary. Along with *Enteromorpha*, it was very common in the Freshwater Bay–Peppermint Grove area of the lower Swan in the 1970s and 1980s. Although blooms of green algae are not toxic, the dead cells use up oxygen in the decay process and even emit nauseating odours.

Like phytoplankton, macroalgae grow quickly when nutrients are available. The particular species vary in the different estuaries and parts of estuaries depending on the different proportions of nutrients, salinity, the timing of freshwater flow, nutrient concentrations, as well as depth and water colour factors that affect light. For example, the green alga *Caulerpa* only grows in the estuaries open to the sea such as the Swan–Canning and the Peel–Harvey, and is most abundant near the entrance channels. *Caulerpa racemosa* and *Caulerpa distichophylla*, common species along the coast, have become more abundant in the Peel–Harvey since the Dawesville Channel was opened.

In the past, dense mats of green algae were common in estuaries such as the Swan and Peel–Harvey, and Albany Harbours where nutrient concentrations had increased due either to fertiliser from the surrounding land and rivers, or from industrial pollution. Here, however, lies the conundrum. Macroalgae are great food and habitat for many tiny animals that are in turn eaten by fish and birds, which are abundant in estuaries and even more common when there is higher algal growth. Algal growths become a problem, however, when they smother other plants or when they die and decay, an unpleasant smelly process. The decay process uses up oxygen and changes the water and sediment chemistry, which leads to release of nutrients from the sediments to fuel further algal growth and formation of compounds unfavourable for animals. (See **Chapter 3 The Peel–Harvey Estuarine System: Fish and weeds—a complicated picture**.)

The clear association of animals and algae, including high fish catches, brings up the question of why the animals cannot control the algal blooms. Perhaps the type of algae have changed and the species that grow quickly are not suitable food for animals, or perhaps small animals that eat algae are no longer present or are not reproducing as one would expect when food supplies are abundant. Have they been affected by poisons used to control pests in crops and gardens, the mosquitoes that live in the surrounding marshes, chemicals from septic systems, rubbish dumps along the foreshores or run-off from roads? The presence of banks of algae or scums of microalgae

FIG. 1.11 *Green algae (a)* Ulva, *(b)* Cladophora, *(c)* Enteromorpha, *(d)* Caulerpa distichophylla *and (e)* Caulerpa racemosa; *brown algae (f)* Hincksia *(formerly* Giffordia*), (g)* Sargassum, *(h)* Hormophysa *and (i)* Colpomenia; *red algae (j)* Gracilaria *and (k)* Ceramium. See also Huisman, 2000. Courtesy: J. Huisman, Murdoch.

e

h

i

f

j

g

k

may be due to increased nutrients, but they indicate that the ecosystem has changed and that it now functions in a different way. We can ask the question: is this estuary change death by a multitude of slow changes?

The lessons for the future obviously lie in trying to maintain the attributes we treasure: the water we can swim in and boat on, and the fish still present for the catching that can be eaten with safety. How little we know of how the ecosystem processes intertwine.

Seagrasses

Seagrasses are true plants; they have leaves, rhizomes, roots and flowers. Like land plants they take up nutrients through roots, but like algae they also take up nutrients from the water through leaves. There are many different types of seagrasses and southern Australia has the highest diversity of types, about seventeen species. Seagrasses live in habitats that range from the wave swept ocean down to 40 metres to calm shallow water, and include the turbid estuaries where salinity fluctuates. As with the algae, the marine influence determines the types of seagrasses present in an estuary (Fig. 1.12).

The large ribbon weeds *Posidonia australis* and *Posidonia sinuosa*, and the sea nymph also called wireweed *Amphibolis antarctica* and occasionally *Amphibolis griffithii*, species that are usually found in sea water, are found in Oyster and Princess Royal Harbours and Waychinicup where ocean influences are strong. The estuaries with either permanent or seasonal connections to the sea are colonised by seagrasses such as the paddle weed *Halophila ovalis*, *Zostera mucronata* and *Zostera tasmanica* (formerly *Heterozostera tasmanica*) in the Swan, Peel–Harvey and Leschenault. Swan grass *Ruppia megacarpa* grows in most estuaries throughout Swanland, including the normally closed eastern estuaries such as Beaufort and Wellstead Estuaries, and is particularly luxuriant in the Wilson Inlet. (See **Chapter 2 The Swan–Canning Estuary: Large macrophytes—algae and seagrasses; Box 8.3 This thing called *Ruppia*** and **Box 9.3 Shrinking seagrasses versus growing algae**.)

Other aquatic plants

In the more saline eastern estuaries such as the Beaufort and Wellstead, that remain closed for long periods, the charophyte *Lamprothamnium* also grows abundantly along with *Polyphysa*. *Lamprothamnium* was common in the Deadwater and Swan Lake, part of the Hardy Inlet during the 1970s, but with the migration of the entrance to the east this habitat has now changed.

A number of other aquatic plants, such as the native duckweed *Lemna* and introduced species *Salvinia* and *Hydrocotyle*, often grow abundantly in the upper reaches of estuaries such as the Canning and move downstream during freshwater flow. In some cases these may shade a water body inhibiting the growth of phytoplankton, however the rapid growth results in a large amount of organic material that as it decays depletes oxygen and releases nutrients.

Invertebrates and foraminifera

The high productivity of estuaries, due to vegetation around the margins, in the water and sediment, or attached to hard surfaces creates rich feeding grounds for invertebrate animals such as worms, molluscs (gastropods and bivalves) and crustaceans (crabs, prawns, amphipods, isopods and barnacles). (See **Chapter 2 The Swan–Canning Estuary: Invertebrates.**)

Another group of organisms, the foraminifera often called forams, are also common in estuaries. These are single-celled protists with shells (tests). The tests (chambers, spheres or tubes) are made of organic material, calcium

FIG. 1.12 *Seagrasses in the estuaries of Swanland.* Adapted from a painting courtesy E. Rippey.

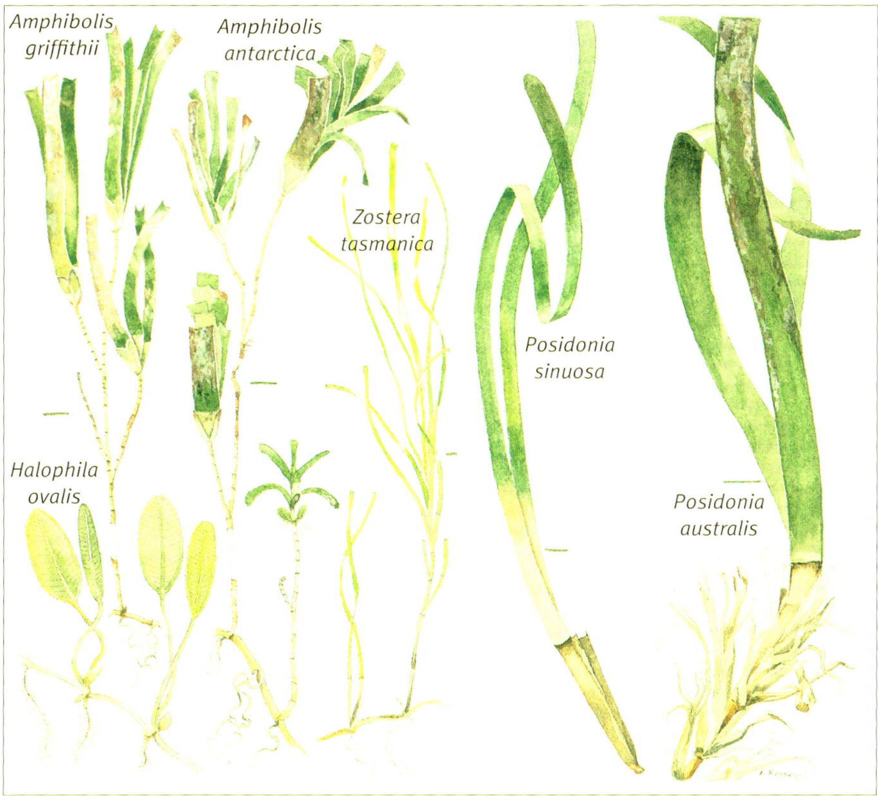

carbonate (calcite), silica or particles of sand or other material cemented together, and range in size from 100 micrometers to 200 millimetres. As the tests preserve well, fossil forams are particularly useful for inferring information about past environments. They feed on dissolved organic molecules, bacteria and diatoms, and some contain algae that live within the shell in a symbiotic relationship. They may be extremely abundant and are an important food item for invertebrates and fish.

Characteristically, although the populations are large, they usually consist of only a few species, less than those found in nearby freshwater and marine habitats. This low diversity of species is probably due to the stressful and changeable conditions in estuaries, indicated by the greater diversity of fauna in the less changeable estuaries where marine influences are most obvious.

The most important of the factors affecting animals is salinity, although other factors such as temperature and oxygen, which fluctuate widely in shallow estuarine waters, turbidity, different substrata and interactions between organisms, are also very important.

Dealing with salinity—types of organisms and ways of coping

Many animals are only found in particular parts of the estuary, depending on the gradients of salinity from river to mouth. Those tolerating a narrow range of salinity, either marine or fresh water, are termed stenohaline while those tolerating a wide range of salinities are termed euryhaline. In the estuaries of Swanland, particularly those east of Albany where the bars close for many months, estuarine animals also need to cope with hypersaline conditions where salinities are higher than sea water.

Some animals are migrant species that pass through an estuary on the way to the river or sea. Others enter for a short period to feed when conditions are favourable. The most tolerant species, those able to cope with the changing salinities, have physiological adaptations to control movement of water and salts into or out of the body tissues, or behaviours to limit exposure to unfavourable salinities. For example, some prosobranch snails in low salinity excrete dilute urine, while some fish take in salts through the gills. The shell of crustaceans such as the crab limits the surface of soft tissue exposed to the water. Shells also protect snail-like animals, such as the dog whelks (*Nassarius*), that can retreat into their shells and close the operculum or 'trapdoor' over the foot, while mussels and barnacles can close the valves of their shells. Other types of animals, such as bivalve shells and worms, survive by burrowing into the sediment where changes are less extreme or occur more slowly allowing time for the animal to adjust.

Saline layers—limiting oxygen

Estuarine animals often need to cope with low oxygen conditions due to the low solubility of oxygen in warm water, consumption of oxygen by algal blooms at night, the decay of organic matter, or isolation of the bottom waters. Low oxygen conditions due to stratification of the water column with light fresh water at the surface and dense salty water at the bottom can be quite common in estuaries. To overcome oxygen deficiencies, burrowing animals in particular often have enlargements of the body surface to increase gas exchange. Others possess microscopic hairs or cilia on the gills, or use a pumping action of their soft body to create water currents through the burrows. Some have special respiratory pigments to facilitate oxygen uptake when concentrations are low, and many have low oxygen requirements. More active animals, such as crustaceans and fish, respond to low oxygen by increasing ventilation currents over the gills or moving to more favourable areas. When conditions are extreme or widespread and the options of migration are limited, however, mortality of fish and other animals can be high.

Living in the oozy bottom

Characteristically estuaries contain layers of sediments of sand, clays, silt and organic material brought by rivers, and organic material produced in the water body. Despite the thick sediments, animals are usually found in the top 1–2 metres, their depth limited by the supply of oxygen in particular. The depth of the oxygenate layers varies with rate of exchange with overlying waters, water movement, the amount of organic matter and the size of sediment particles.

In sheltered waters, where fine particles of clay, silt and organic matter (typical of estuaries) settle out, oxygen exchange is limited and bacterial decomposition of the organic material creates a high biological demand for oxygen, so that the aerobic layer is thin and the anaerobic layer thick. This type of habitat, however, is particularly suitable for deposit feeders, which feed on organic matter. In contrast, in areas of higher water movement, where sediments are sandier and there is less organic matter, the oxygenated layer is thicker and the fauna is often dominated by suspension or filter feeders that collect organic material from the water column.

Complex interactions—animals driving the sediment and water chemistry

Bioturbation or the burrowing activity of organisms also affects sediment chemistry through increasing exchange with the water column, disrupting and redistributing sediments, and the production of faeces and mucus. Some larger organisms overcome problems of low oxygen by pumping water through their burrows, and some oxygen passes though into the sediment forming an oxygen rich zone, characteristically lighter in colour than the underlying anaerobic or deoxygenated layer. This in turn affects the distribution of microbes and small animals, which are often found deeper in sediments than they would normally be, clustered around burrows where they feed on organic material released by the burrowing organisms.

The oxygen content of sediment is often very obvious around the edges of estuaries, indicated by a change in colour from brown to grey or black. The change in colour reflects the presence of different types of bacteria. Generally surface bacteria use oxygen as an energy source, while those deeper down use other molecules such as nitrate, sulfate, carbonate and iron although these are less efficient in terms of the amount of energy produced. One of the most obvious of the deeper bacteria are those that use sulfate (SO_4^{2-}) and produce hydrogen sulfide (HS^-), the bad-egg odour; a problem that plagued people living around the Peel–Harvey during the algal bloom years. The hydrogen sulfide combining with iron causes the characteristic black-grey colour of the sediment. Sulfides are toxic to many animals so that the depth of the black deoxygenated (anoxic) layer determines how deeply many animals live.

Altered oxygen concentrations also influence other chemical processes such as the process of nitrification, the conversion of ammonium (NH_4^+) to nitrite (NO_2^-) and nitrate (NO_3^-), and denitrification, the conversion of nitrate (NO_3^-) to nitrogen gas (N_2), and the release of phosphate (PO_4^{3-}). (See above **Additional nutrients and the onset of eutrophication**.)

Adding extra organic matter

The sediments in productive environments such as estuaries contain large amounts of organic matter. Organic material arrives from the land in the distant catchments, from the nearby fringing marshes, from microalgae in and on sediments, together with bacteria, microscopic plants of the water column (phytoplankton) and larger plants (macrophytes) including algae and seagrasses. In estuaries there are also inputs from industry (Swan–Canning and Albany Harbours), sewerage (early years of the Swan and Albany Harbours), and feedlots and piggeries (Swan–Canning and Peel–Harvey).

When large amounts of organic matter are present, bacterial communities are active and oxygen demands are high. Large increases in organic loads therefore lead to more stressful conditions and a change in the communities of animals. For example, the larger animals disappear and large numbers of small deposit feeding worms such as *Capitella* and *Scololepis* dominate the fauna.

Utilising the rich feeding grounds—feeding strategies and types of food

In addition to the abundance of food in estuaries, there are differences in the types of food and how food is utilised. Some animals change feeding mode depending on material available. For example, some grazers of living plants also feed on detritus and some polychaete worms can switch between a suspension and deposit feeding mode (Fig. 1.13).

Suspension or **filter feeders** filter or trap organic matter, bacteria, microscopic plants and animals that are suspended in the overlying water. The filters can be tentacles on polychaete worms, hairs on mouthparts or legs of crustaceans such as crabs, and barnacles and smaller amphipods. Some animals, like sponges, fan worms, bivalve molluscs and ascidians, use cilia (microscopic) beating hairs to create currents and pump water through chambers in their bodies. Other animals secrete mucus that traps pieces of organic matter and small animals. Some like the mussels attach to hard surfaces while fan worms live in burrows.

Deposit feeders ingest the sediment and other small particles of organic material. Some animals select or sort inorganic and organic digestible material or the size of particles that they ingest, while others simply ingest the sediment. The dead organic material or detritus also includes bacteria, fungi, microscopic algae and animals living on the detritus.

Detritivores also feed off dead organic material and include some of the more obvious estuarine fish such as sea mullet and Perth herring.

Grazers or **herbivores** consume plant material. They vary in size with large types consuming whole plants or removing large pieces, while others feed on smaller algae such as epiphytes growing on other algae or seagrass. Some grazers are also browsers and vice versa, and move over the surface of the sediments or plants consuming fine organic material. Gastropod molluscs use a toothed radula to rasp material from the surface, while small amphipods pick and chew small pieces. The group characteristically produces small particles of faeces that accumulate in the oozy bottom of estuaries. Some of the large fish in estuaries such as garfish and striped (six-lined) trumpeter are herbivorous. While others such as black bream and yelloweye mullet are omnivores feeding on both plant and animal matter.

Predators or **carnivores** eat other animals. Some actively search for prey, while others live in burrows ambushing other animals as they pass by. Predatory animals include the small crustaceans such as amphipods and crabs, and vertebrates such as fish and even dolphins.

Scavengers feed on dead animal matter. The dog whelk *Nassarius* is often one of the most obvious scavengers in estuaries as they gather around dead animals such as fish stranded in the sandy shallows.

Parasites live on or within the body tissues of other animals from which they obtain their food.

Large differences in abundance

We have discussed the large changes in populations of estuarine animals throughout a year and from year to year, changes that usually relate to river flow and tidal flows, salinity and temperature and, in some estuaries of Swanland, with isolation from the sea. The numbers and types of invertebrates living in an estuary vary greatly throughout an estuary, at scales of less than a metre to kilometres, even on the same day and daily within the same area.

Such variability makes the understanding of biological communities and processes, and detecting changes, very complicated. Changes can be due to water and sediment movement, differences in larval settlement, and interactions between different types of animals including competition for resources and predation. They may also be due to historical or recent human (anthropogenic) influences, such as disturbance from dredging, dumping of spoil, influx of organic matter, poisons and pollutants, extra nutrients or different nutrients, water movement due to boat wash and impediments of jetties, groynes, walls or anchors. Changes can also be very subtle following changes to the less obvious primary producers, the bacteria, algae and plants that affect the larger and more complex plants and animals.

For example, during the years of high algal growth in the late 1970s through to the early 1990s, the benthic invertebrates in the Peel–Harvey were dominated by a few types of generally small, fast-growing species that produce many young. Typical species of these eutrophic (nutrient enriched) habitats included the worm *Capitella capitata*, the amphipods *Corophium*, *Tethygenia*, *Melita*, the bivalve *Arthritica semen*, and the gastropod *Hydrococcus brazieri*. This was a contrast to the less extreme conditions of the Swan–Canning Estuary, where the invertebrate community was more diverse consisting of larger, longer lived species with a larger biomass. In 2000 however, one of the most abundant and significant contributors to the production of molluscs in the 1970s, *H. brazieri*, was no longer present or drastically reduced in number.

FIG. 1.13 *Characteristic community in the estuaries of south-western Australia. Large floating and attached algae and seagrasses such as swan grass and paddle weed create a three-dimensional habitat as well as a food resource. Note burrowing worms disrupt the boundary between the oxygenated surface sediment and buried low oxygen layers. 1. Hardyheads (Atherinids) feeding and laying eggs in floating algae such as* Chaetomorpha *sp. 2. Worms (*Ficompomatus *sp.) buried in the sediment filtering water for food including microalgae, tiny animals such as copepods and organic material. 3. Spionid worms casting mucous 'nets' to catch tiny plants, animals and organic matter. 4. Mud mussels filtering water for food and also providing habitat for other animals. 5. Small crustaceans feeding on the tiny algae growing on the seagrass fronds (*Ruppia*). Large fish feeding on the smaller animals, living and dead plant material that grows so abundantly in shallow, calm nutrient-rich estuarine waters.* Courtesy T. Rose, WRC.

Exotic imports and introductions

Since the beginning of European settlement new algae, plants and animals have been brought to our shores. Stories of escapees overgrowing native plants or out competing and preying on native animals are well known, but the threats of tiny algae are less obvious and perhaps even more insidious. Freshwater fish introduced to control mosquitoes, as sports fish or aquarium escapees now live in many rivers where they compete with or eat the native organisms and wash downstream with freshwater flows. Ballast water, taken on board in one port and carried worldwide, can be dumped in or near other ports and may carry organisms to areas free of the natural competitors and predators that keep them controlled. We know that some larger organisms such as the Asian bag mussels *Musculista senhousia* in the Swan–Canning and European fan worms *Sabella spallanzanii* in harbours near the Swan–Canning, Leschenault and Princess Royal Harbour have been introduced but so far these have caused little trouble. The potential damage to aquaculture in Tasmania by the dinoflagellate *Gymnodinium ctenatum* is a warning of the type of problem that could be lurking undetected in Western Australia.

Fish and fisheries

At the interface of river and sea, estuarine fish populations vary with the degree of fresh and salt water influence, tolerance to fluctuating conditions, the connection between the estuary and sea, and the fish present in the adjacent offshore areas.

Fish in the estuaries can be grouped according to their dominant habitat or where they breed. The groups include the Freshwater group that are sometimes flushed downstream from the river reaches; Semi-anadromous group that includes species such as the Perth herring *Nematalosa vlaminghi*, which also lives in the ocean but spawns in the upper estuary reaches; Anadromous group, the only representative in the South-West being the pouched lamprey *Geotria australis* that uses estuaries as a route from the ocean feeding area to breeding areas in the distant inland river reaches; Estuarine group that breed and complete their life cycle within the estuaries e.g. black bream *Acanthopagrus butcheri*, yellow-tailed trumpeter *Amnitataba caudivittatus*, some species of hardyheads or atherinids, and gobies; Estuarine and Marine group, species that can complete their life cycles either in the estuary or in sheltered marine waters outside the estuary, e.g. cobbler *Cnidoglanis macrocephalus*; Marine–estuarine Opportunist group that breed in the ocean and use the sheltered and productive estuary as a nursery area, e.g. yellow-eye mullet *Aldrichetta forsteri*, sea mullet *Mugil cephalus*; and the Marine Straggler group that enters estuaries infrequently and usually in low numbers, e.g. giant herring *Elops machnata*, pilchard *Sardinops neopilchardus*, buffalo bream *Kyphosus sydneyanus*, sea trumpeter *Pelsartia humeralis*, various types of goat fish, flatheads and leatherjackets (Tables 1.11, 1.12, 1.13, 1.14 and Fig. 1.14). (See also Table 2.9 for growth rates and longevity, and Table 2.10 for food items of common estuarine fish.)

The connection to the sea and the marine influence are the key to the diversity of species found in the various estuaries. Estuaries such as the Swan–Canning, the New Peel–Harvey, Walpole–Nornalup, Princess Royal and Oyster Harbours, where the marine connection is permanent and salinity is similar to the ocean and more stable, have the greatest diversity due to marine opportunists and marine stragglers that enter to feed in the sheltered productive waters. In the estuaries where the entrances are shallower or close seasonally, fewer species have the opportunity to enter and recruitment depends on the length of time the bar is open, the timing of the bar opening, and the fish along the coast available for recruitment. In the estuaries that remain closed for long periods, the normally closed and 'permanently' closed estuaries, conditions are more extreme and the fish die as conditions deteriorate. While conditions are favourable however, the fish populations may remain high although the marine species do not reproduce. Fish numbers are reduced as many of the estuaries are fished by commercial and recreational fishers.

The fish populations also vary within the estuaries with salinity and depth. For example, on the south coast in the basin areas of Nornalup and Wilson Inlets, the shallows are dominated by estuarine spawning species while the deeper waters (where salinity is higher and more stable) have a greater number of marine species including snapper. In Nornalup these include five species of rays and sharks.

In terms of individuals, most of the estuaries of Swanland are dominated by the estuarine group of species that spawn and complete their life cycle in the estuaries and the marine–estuarine opportunists that enter as juveniles and use them as nurseries. As the name marine–estuarine opportunist suggests, fish in this group use the permanently open estuaries as an alternative to sheltered bays along the coast. Juveniles are abundant while feeding and growing in the sheltered waters along with adults that form a large part of the commercial and recreational fisheries. Species include sea mullet *Mugil cephalus*, yellow-eye mullet *Aldrichetta forsteri*, cobbler *Cnidoglanis macrocephalus*, tailor *Pomatomus satatrix*, and mulloway *Argyrosomus japonicus*.

Generally fish of the marine opportunists group enter estuaries in Swanland as juveniles rather than larvae. This pattern is also found in some estuaries along the east coast of Australia, but is in contrast to many of the estuaries in the northern hemisphere where the larvae of marine–opportunists are transported by tides into estuaries that are then used as nursery areas.

FIG. 1.14 *Representation of various ways that fish use the estuaries of Swanland.* Adapted from Potter & Hyndes, 1999; Courtesy I. Potter, Murdoch & G. Hyndes, ECU.

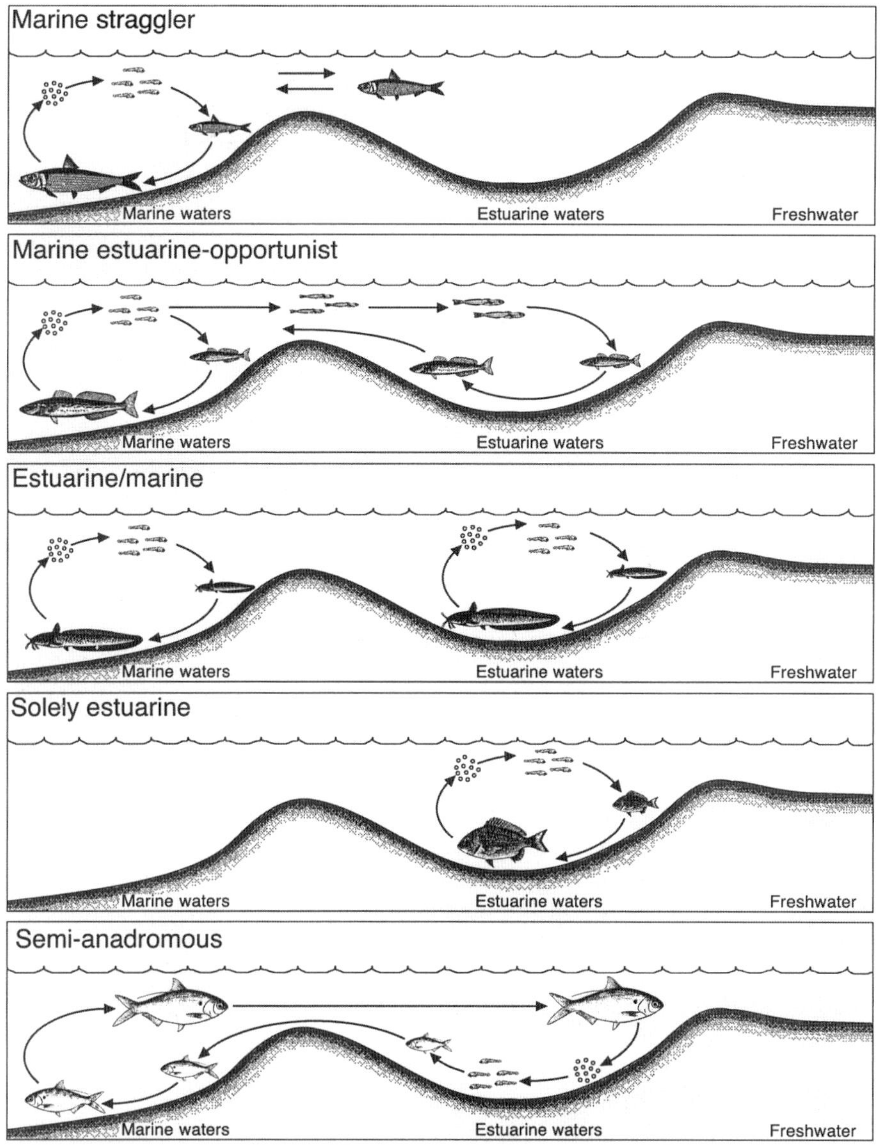

Fishing in the past—fishing for the future

Estuarine fish have sustained people living in the South-West for millennia. The Aboriginal people hunted with spears and kylies along the shallow banks or caught fish in traps of stone and brush erected on the rivers leading to estuaries such as the Murray and around margins of Broke Inlet, Wilson Inlet and Oyster Harbour. (See Fig. 3.7 and **Box 9.1 Stone circles—echoes of the past**.) After European settlement estuarine fish remained important, particularly in the early years before the colony was self-sufficient (Fig. 3.16).

Until 1899 no distinction was made between commercial and recreational fishermen. After that time commercial fishers required a licence. Initially the estuarine fisheries were small, supplying the local population, although around Mandurah some fish were canned or smoked for export. The development of a commercial western rock lobster fishery *Panulirus cygnus* in the 1950s and the need for bait brought changes to the estuarine fisheries providing a market for the most common species: yellow-eye mullet *Aldrichetta forsteri*, sea mullet *Mugil cephalus* and Perth herring *Nematalosa vlaminghi*. The estuarine fish combined with Western Australian salmon *Arripis truttaceus* and Australian herring *Arripis georgianus* provided the bulk of the bait for rock lobster until the 1980s when cheaper imported frozen bait undermined the local market. In response the estuarine fishers targeted more valuable table fish, including crabs, prawns and cobbler *Cnidoglanis macrocephalus*, escalating the conflict with recreational fishers. The conflict between recreational and commercial fishers is still increasing. In 1987 an estimated 284,100 recreational fishers participated at least once a year, but by 1999 the number had increased to 544,300 persons. With population growth and tourism, the level of recreational fishing using lines and nets is likely to increase still further. There are also indications that species such as black bream are already fully or over exploited in some estuaries. Although restocking is a possibility, there are however problems as each estuarine population of black bream is genetically distinct, and haphazard introductions may disrupt the gene pool. In addition, it is not known if restocking replenishes a breeding population or simply increases stock in the short term.

Since 1969 there has been a reduction in the number of commercial licences through various Voluntary Fisheries Adjustment Schemes following a study that suggested many of the operators were not viable. Three lower west coast estuaries are open to commercial fishing and are managed separately as restricted entry fisheries, although they are likely to come under one management plan in the future. The south coast estuaries are managed as a single commercial fishery, the South Coast Estuarine Managed Fishery. In June 2001, the west coast fisheries included the Swan–Canning (4 units), Peel–Harvey (Mandurah) (8 units) and Hardy Inlet (1 unit). The final six licensees in the Leschenault Inlet ceased fishing in 2000 (see also Table 2.11).

In 2001 the commercial catch in the west coast estuarine fisheries of 264 tonnes, down from 305 tonnes in 2000, consisted principally of sea mullet, blue manna or blue swimmer crabs, yellow-eye mullet, yellow sand whiting (also termed western sand whiting), cobblers, tailor, Perth herring, Australian herring, King George whiting, black bream and other species (Table 1.14 and Graphs 2.12 and 3.6). The target species of recreational fishers varies according to the estuaries but includes tailor, whiting other than King

TABLE 1.11 *Common fish in the estuaries grouped according to the main habitat and breeding areas.* Adapted from Loneragan et al., 1987; Potter & Hyndes, 1999.

Group	Habits	Popular name	Scientific name
Marine straggler	Enter infrequently and in low numbers only when conditions are favourable	fan-bellied leatherjacket giant herring sea trumpeter	*Monacanthus chinensis* *Elops machnata* *Pelsartia humeralis*
Marine–estuarine opportunist	Breed in the ocean and use the sheltered and productive estuary as a nursery area	yellow-eye mullet sea mullet banded toadfish	*Aldrichetta forsteri* *Mugil cephalus* *Torquigener pleurogramma*
Estuarine and marine	Complete their life cycles in the estuary or in sheltered marine waters outside the estuary	cobbler	*Cnidoglanis macrocephalus*
Estuarine	Complete their life cycle within the estuaries	black bream yellow-tailed trumpeter hardyheads, certain gobies	*Acanthopagrus butcheri* *Amniataba caudivittatus* atherinids
Anadromous	Lives and feeds in ocean, passes through estuaries to breed in inland	pouched lamprey	*Geotria australis*
Semi-anadromous	Lives mainly in ocean but spawns in upper estuary	Perth herring	*Nematalosa vlaminghi*
Freshwater	Live in river reaches, washed downstream when river flow is strong	native minnows freshwater cobbler red fin perch mosquito fish goldfish carp tilapia	*Galaxias* *Tandanus bostocki* *Perca fluviatilis** *Gambusia holbrooki** *Carassius auratus** *Cyprinus carpio** *Tilapia mossambicus*

**Introduced species.*

TABLE 1.12 *Number of fish species in estuaries along the west and south coast according to estuary type. Note that the patterns remain the same although the numbers of fish species differ slightly according to the season and methods used to collect samples.* Adapted from Lenanton & Hodgkin, 1985 and updated; West Coast, Murchison and Chapman R., Lenanton (Fisheries WA), unpubl. data; Moore, Young et al., 1997; Swan, Loneragan et al., 1989; Peel–Harvey, Potter et al., 1983; Leschenault, Potter et al., 1997; Hardy, Hodgkin, 1978; Valesini et al., 1997; South Coast, Walpole–Nornalup, Potter & Hyndes, 1994; Wilson, Potter et al., 1993; Wellstead, Young & Potter, 2002; Hamersley, Culham and Stokes, unpubl. data, courtesy S. Hoeksema & I. Potter, Murdoch.

Estuary type	West coast	South coast
Permanently Open	≈31 Murchison, 74 Swan, 55 Peel–Harvey, 42 Leschenault, 57 Hardy (1974–75), 49 Hardy (1994)	29 Walpole–Nornalup
Seasonally Open	6 Chapman, 14 Greenough, 27 Moore; most unknown	38 Wilson
Normally Closed	–	9 Hamersley, 13 Wellstead, 12 Stokes
Permanently Closed	–	11 Culham

George, King George whiting, Australian herring, black bream, flathead and skipjack trevally. Although the magnitude of the recreational fishing has probably been underestimated, recreational and commercial catches of the top five recreational targets in the Swan–Canning were similar (Tables 2.12 and 2.13). In 1998–1999 recreational fishing in the Peel–Harvey was about 17 per cent of the combined commercial and recreational finfish catch, but 80 per cent of the blue swimmer crab catch.

The South Coast Estuarine Fishery is located between Cape Beaufort and the Western and South Australian border (Table 1.15). The licences or units, thirty-one in April 2000, have access to each of the individual estuaries that are open to commercial fishers except for the Beaufort Inlet where only three licensees are granted access each year. A review of the fishery suggests that to stabilise the commercial catch at historical levels, the optimum number of fishing units should be only fifteen. In 2001 the commercial catch in the South Coast Estuarine Fishery was 274 tonnes, slightly higher than the previous year of 252 tonnes. The annual catch consists principally of cobbler, black bream, Australian herring, sea mullet, King George whiting, and blue swimmer crabs although over forty species of finfish, sharks, rays and invertebrates are represented. Although the importance of black bream and cobbler to the south coast estuarine fisheries is obvious (Graph 1.4), each of the estuaries harbours different breeding populations of both these species with different growth and maturity rates so that each of the estuaries has different management issues. The source of the black bream catch varies from year to year with the Stokes, Beaufort and Oldfield contributing most of the catch in the five years to 1999, while in 2000 the Gordon and Wilson Inlets were also important. The particularly high catch in early 1992 was due, however, to the 64 tonnes from Culham Inlet before the inlet opened to the sea in 1993 after being closed for many years. In many estuaries cobbler have declined due to overfishing and habitat changes, but in Wilson Inlet they appear to have increased in number as a result of the maintenance of appropriate closed areas and the growth of the swan grass *Ruppia*. (See **Box 8.4 Cobbler: spines to fear—great to eat.**)

Recreational net fishing is permitted in some estuaries but the Donnelly and Warren Rivers and Parry Inlet are closed to this activity. Some of the estuaries however, can only be fished during particular times of the day or year, and there are restrictions on the type and use of nets. The rules governing net fishing are subject to change, and current information should be sought from the Department of Fisheries. Generally angling is permitted in all the rivers and estuaries, however there are bag limits and regulations for some species and a Freshwater Angling Licence may be required in the upper freshwater areas.

TABLE 1.13 *Life cycle characteristics of fish caught in estuaries of south-western Australia, Moore River to Wilson Inlet (Moore, Swan–Canning, Peel–Harvey, Leschenault, Hardy, Nornalup–Walpole and Wilson). In addition, the anadromous pouched lamprey migrates very rapidly through the estuaries of the lower west and south coasts.* Adapted from Potter & Hyndes, 1999.

Life category	Number of species
Marine	
stragglers (MS)	52
estuarine opportunists (O)	24
Estuarine	
estuarine and marine (E&M)	11
solely estuarine (E)	9
Semi-anadromous (A)	1
Freshwater (native and introduced) (F)	6
TOTAL	103

The effects of fishing and the removal of herbivorous or carnivorous fish from the estuaries has never been examined in Western Australia and only rarely elsewhere. There is however potential that fishing pressure on either category could have modified the ecology of some estuaries in what may be termed a Top Down Effect or Trophic Cascade.

Given that the fish and fisheries are highly valued by many people as a recreational or commercial resource, the role of fish and their response to estuarine changes, which are principally eutrophication, nutrient additions and algal growth, are a major question for managers.

We know that fish are abundant in estuaries because estuaries are productive, there is abundant food, the waters are sheltered, and there are few large predatory fish-eaters (piscivorous fish). An abundance of algae, however, also creates problems as haul nets cannot be deployed if the fish cannot be seen, and hauling nets loaded with macrophytes is difficult. In addition, some blooms are poisonous and in the Peel–Harvey fish died and appeared to move away during the toxic blue-green *Nodularia* blooms. Despite these problems fishers generally thought that fish were more abundant in the Peel–Harvey during the macroalgal blooms of the 1970s.

TABLE 1.14 *Principal fish in the west and south coast estuarine fisheries 2000 and 2001.* Adapted from Ayvazian & Nowara, 2001a & b, 2002a & b.

Species	West coast (tonnes) 2000	West coast (tonnes) 2001	South coast (tonnes) 2000	South coast (tonnes) 2001
Sea mullet	88.8	69.4	21.8	24.7
Blue swimmer crabs	73.0	93.0	16.9	38.9
Yellow-eye mullet	48.3	51.4	9.5	5.9
Western sand whiting (yellowfin)	36.2	16.4	0.5	0.2
Tailor	9.6	6.2	0.3	0.5
Perth herring	9.3	*	0	0
Australian herring	8.5	4.9	28.5	22.8
King George whiting	7.0	*	19.4	9.1
Cobbler	5.0	*	62.0	77.1
Black bream	5.0	*	30.0	27.4
Leatherjacket	0	0	16.2	17.5
Silver bream (tarwhine)	*	*	10.0	9.5
Pink snapper	0	0	3.8	9.3
Flathead	*	*	8.6	9.1
Other species	14.3	22.4	24.2	22.0
TOTAL	305.0	263.7	251.7	274.0

**Catches not reported separately as there were fewer than five fishers, but were included in other species.*

TABLE 1.15 *Commercial access to south coast fisheries 2002.* Government Gazette, WA, 2002. *The plan commenced operation on 1 July 2002.*

Open to commercial fishing	Closed to commercial fishing
Broke Inlet	Donnelly River
Irwin Inlet	Warren River
Wilson Inlet	Gardner River
Princess Royal Harbour	Walpole–Nornalup
Oyster Harbour	Parry Inlet
Waychinicup Inlet	Torbay Inlet
Beaufort Inlet	Taylor Inlet
Gordon Inlet (Gairdner River)	Normans Inlet
Hamersley Inlet	Cheynes Inlet
Culham Inlet	Wellstead Estuary
Jerdacuttup Lakes	Saint Marys River
Oldfield Inlet	Fitzgerald Inlet
Stokes Inlet	Dempster Inlet
	Barker Inlet
	Torradup Inlet

A comparison of Catch Per Unit Effort (CPUE) for the Swan, Peel–Harvey and Leschenault (Table 1.16) indicates that fish catch rates were higher in the more eutrophic Peel–Harvey and Leschenault than the Swan. This raises questions as to whether this was due to faster growth rates or if there were more fish? Information on fish biology including growth rates, the types of fish, as well as abundance are obviously necessary to answer these questions. It would appear that the main species grew at a similar rate in the Swan and the Peel–Harvey. However details of the major components of the catch are also interesting (Graph 4.3). Catches of yellow-eye mullet were higher in the Peel–Harvey and Leschenault in the 1970s and 1980s when algae were abundant. This is exactly the type of fish that could be expected to do well in eutrophic algal-dominated systems. Yellow-eye mullet are omnivores and feed on a variety of food, they grow quickly, and have a high reproductive rate. The higher catches in the 1970s in the Leschenault may indicate a response to earlier growth of algae and seagrasses following construction of The Cut. Another interesting feature of algal growth in the Leschenault lies in the type of species: the Leschenault is dominated by brown algae and seagrasses, while growth in the Peel–Harvey was due principally to the green alga *Cladophora montagneana*.

In the Leschenault, the decline in cobbler catch did not reflect a response to algal growth, but intense fishing pressure and slow recovery due principally to their low reproductive rate. In the Peel–Harvey the increase in cobbler was probably due to increased food due to high numbers of invertebrates associated with the algae. Similarly the increase in catches of sea mullet, a species that feeds on dead organic matter in the Peel–Harvey but not in the Swan, is also related to algal growth. In contrast, the decrease in western king prawns in the Peel–Harvey may have been due to the smothering mats of *Cladophora* that covered the sandy bottom where they burrowed.

Obviously algal abundance negatively affects some fish communities, and we know that fish die in estuaries when oxygen concentrations are low during blooms or after their collapse during decay (Table 4.11 Vasse–Wonnerup wetland system). The problem seems to be linked to the extent of the blooms, the degree of oxygen depletion and opportunity for fish to escape to other areas of the estuary as they did in the Peel–Harvey or, in the case of the Vasse–Wonnerup, out of the estuary. For example, we know that fish often die in the seasonally and normally closed south coast estuaries as water quality declines and salinity rises. This indicates that the overall water quality needs to be considered and that the different estuaries have different problems. Blooms of phytoplankton are a particular problem as they develop rapidly, the bars are generally closed, and conditions are already stressful when they occur. Growth of macroalgae, while not particularly pleasant when they wash

GRAPH 1.4 *South Coast Estuarine Fishery 1980–2001 (a) The annual catch, effort and catch per unit effort (CPUE); (b) Contribution of black bream to the South Coast Estuarine Fishery.* Ayvazian & Nowara, 2002a. Courtesy DoF.

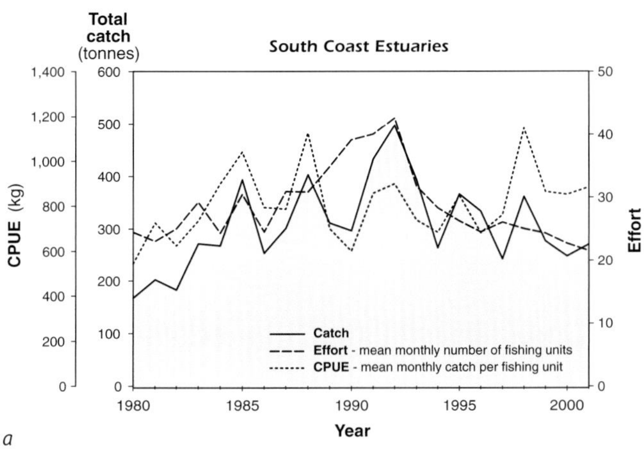

a

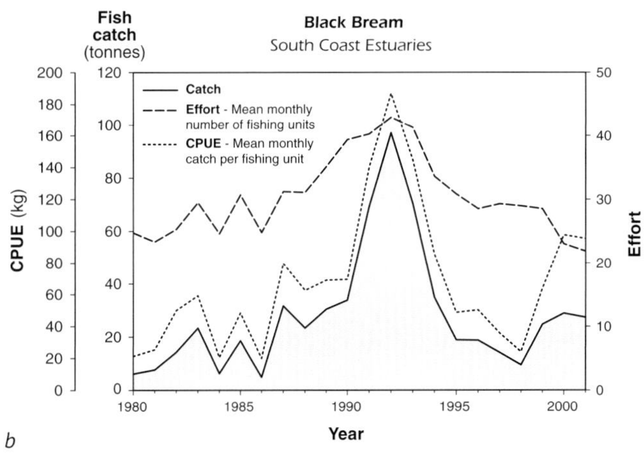

b

up on beaches and decay, generally take longer to develop and are usually patchy in distribution so there is some opportunity for escape; and they have the added attributes of providing habitat and protection.

When enough becomes too much

The importance of algae in estuaries is clear: they oxygenate the waters and form the base of food chains as well as providing protection and places for other organisms to live. Excessive algal growth due to nutrients has become more common as populations of people increase and land uses change. More frequent algal blooms upset the natural balance of ecosystems, other plants are smothered, waterways are clogged, and rotting material decays producing offensive odours that reduce oxygen so that animals die. Conservation, recreation and scenic values deteriorate to a stage that fisheries and all users of the estuary are affected. Around the Peel–Harvey rotting algae was such a problem that over 20,000 tonnes a year were harvested off the beaches, with over $60,000,000 spent to harvest weeds and build the Dawesville Channel. This is a large monetary price to correct a problem, with a solution that may still precipitate ongoing problems, and may require equally costly efforts to solve and maintain. In addition, changes brought by the Dawesville Channel have reduced some problems but this may be short term; described by some as a 'bandaid approach', in other respects perhaps it has 'swept things under the carpet' because the causes of the problem remain. The causes, the diffuse nutrient sources, still persist requiring far greater, long-term efforts to solve. This is a lesson for management of the Swan–Canning and all the estuaries in Swanland as problems cannot be solved quickly and, in the case of the Swan, there is no short-term cure as a modified and deepened marine access already exists. Similarly, *Cladophora* smothered about 90 per cent of the seagrass in Oyster Harbour and Princess Royal Harbour between the 1960s and mid 1980s despite the good marine exchange.

Changes to ecosystems also mean that conditions favour different types of organisms so that the dominant species change. In many urban areas, eutrophic conditions in estuaries and wetlands lead to algal blooms that create favourable conditions for midge larvae that swarm over the nearby houses.

Over the top—the birds

Estuaries are wonderful places for bird watchers; the range of habitats of trees, reeds, marshlands, mudflats and open water provides for a huge variety of birds that frequently gather in great numbers to feed on the bountiful plants, invertebrates and fish. Even estuaries like the Swan, surrounded by urban development with only small pockets of natural vegetation, have

	1960s	1970s	1980s
Yellow-eye mullet			
Swan	45 ± 14	78 ± 17	110 ± 45
Peel–Harvey	259 ± 56	555 ± 22	829 ± 23
Leschenault	379 ± 87	444 ± 12	823 ± 16
Sea mullet			
Swan	198 ± 46	189 ± 24	252 ± 39
Peel–Harvey	196 ± 36	426 ± 11	656 ± 70
Leschenault	145 ± 42	195 ± 64	254 ± 71
Cobbler			
Swan	113 ± 43	103 ± 30	67 ± 15
Peel–Harvey	169 ± 13	284 ± 12	212 ± 13
Leschenault	163 ± 15	91 ± 39	27 ± 90

TABLE 1.16 *Catch per unit effort (in kg) (mean per boat per month per year ± 95% confidence limits) of yellow-eye mullet, sea mullet and cobbler in the Swan–Canning, Peel–Harvey and Leschenault Inlets. Algal growth was high in the Peel–Harvey in the 1970s and 1980s and in the Leschenault from the later 1950s. Fishing techniques also changed during the 1970s with the adoption of monofilament nets and trailerised dinghies.* Adapted from Steckis et al., 1995.

wonderful populations of birds although some of the icon species like the Black Swan are now much fewer in number.

Many estuaries like the Swan, Peel–Harvey, and Vasse–Wonnerup have received international recognition as habitats for migratory species that fly from nesting grounds in the Northern Hemisphere to feed in summer on the shallow sandy and muddy shores of our estuaries. (See **Box 1.10 Global travellers: living a double summer life—transequatorial migrant birds; Chapter 4 Leschenault Estuary and the smaller estuaries of the Geographe Bay area: Wetland of significance** and **Box 2.18 Turn of the tide and a passing parade**.) On a more local scale, birds like Black Swans, pelicans, ducks and cormorants range widely across the South-West but use the estuaries for summer feeding grounds when other watercourses are dry. As a result bird populations on any estuary vary greatly from year to year. Estimating the size and health of populations is difficult, and depends on comprehensive wide ranging surveys, consistent methodology and careful interpretation of conditions throughout the birds' natural range.

In the Swan, gatherings of black cormorants still fascinate as they did years ago. (See **Box 2.19 Black cormorant feeding frenzy**.) The cormorants, although unwelcome visitors on anchored boats, are a sign that the waterways are full of fish, particularly schools of hardyheads or silver atherinids that feed on tiny planktonic plants and animals in the surface waters.

The shallow banks, however, support the greatest variety and number of birds. Species with short bills pick up surface dwelling animals, while those with longer beaks can penetrate deeper into the sediment. Along all the estuary banks, pelicans cruise by probing banks of algae for hidden fish, head flicking as they swallow but watching all the time to see if they have been noticed, ever ready to gather at boat ramps for a hand-out. The shore is also the territory of silver gulls that proliferated in the years of open rubbish dumps. The natural habit of gulls, scavenging and clearing the beaches, is now misdirected to stealing picnics and serves as a reminder that we can upset a natural balance. (See **Chapter 2 The Swan–Canning Estuary: Birds of the Swan—memorable icon of the environment** and **Box 3.5 Types of birds using the Peel–Harvey; Box 3.6 The Banded Stilt;** and **Box 3.7 The Pelican**.)

The story of the Black Swan, the icon of Swanland, is perhaps the most illustrative for showing how some estuarine habitats have been changed. Once in awe-inspiring numbers, Black Swans are now found infrequently on the Swan River. Here the once shallow water with gentle sloping banks is now lined with steep limestone walls rising from deeper water. Black Swans, however, still inspire visitors to shallow estuaries like the Leschenault and Vasse–Wonnerup on the west coast and Wilson Inlet, Beaufort and Gordon on the south coast. (See **Chapter 2 The Swan–Canning Estuary: Recreating Swanland—bringing back the Black Swan**.)

The wider communities of animals—including mammals

The estuaries of Swanland are also visited by dolphins that enter to feed on fish, however as they move upstream behind sandbanks they can be trapped. The sight of dolphins in estuaries like the Swan–Canning and Peel–Harvey is a healthy sign showing that fish populations are high and they are not too threatened by water traffic.

As the estuarine habitats not only include the waters but the fringing vegetation where the boundaries between land and water are blurred, estuaries also create habitat for a far wider range of animals including birds, lizards, frogs and mammals. In many estuaries this fringing vegetation is a small enclave in a much altered landscape.

Box 1.10 **GLOBAL TRAVELLERS: LIVING A DOUBLE SUMMER LIFE—TRANSEQUATORIAL MIGRANT BIRDS**

Every summer about two million wader birds of about seventy species visit the coastal areas of Australia along with another eight species of resident waders such as the Banded Stilt and Hooded Plover. The migrant species breed in the northern summer and feed in the southern summer when the north is frozen. The epic transequatorial migration of about 25,000 kilometres each way is based on the need for a year-round source of food. Living for about 20 years, many of these tiny birds weigh less than 30 grams but travel over 400,000 kilometres in a lifetime.

On the migrations, in what is termed the East Asian–Australasian Flyway from the Arctic Circle through eastern and south-east Asia to New Zealand and Australia, they make three to four feeding stops each of two to three weeks at wetlands, swamps, coastal lakes and estuaries. This makes conservation of habitat a very complex issue. Australia shares responsibility for these habitats with Russia and a number of countries in Asia, and is signatory to two important treaties: CAMBA, the China–Australian Migratory Bird Agreement and JAMBA the Japan–Australia Migratory Bird Agreement. In Australia there are major staging grounds at 80 Mile Beach near Broome, in the Gulf of Carpentaria, and in southern Australia at Corner Inlet and Wilsons Promontory, where tides uncover vast areas of mudflats inhabited by marine life—rich picking for the travellers.

The breeding grounds in the Arctic tundra are characterised by low vegetation of lichens, mosses and shrubs. The swampy land is frozen in winter and in the long days of summer covered with swarming insect life that provides an abundant food supply for raising young. Here adults spend about eight weeks raising the young, leaving again before the young are fully fledged on the first stage of the great migration to the Southern Hemisphere where they spend the summer feeding—their winter grounds. The young birds follow later.

In Australia the birds frequent inland waters, estuaries and protected parts of the coast. Here they move onto sand and mud flats as the tide retreats, moving to higher ground as the water rises again. Generally they feed for about eight hours then rest for four hours before feeding again; habit attuned to exposure of mudflats with the tides.

In Australia the feather colours are generally muted, but at the end of the southern summer they undergo a moult and change to more vibrant breeding plumage before the adults begin the flight north. While the adults spend about five to six months in the Southern Hemisphere, the young remain for at least one winter before making the first breeding flight back to the Arctic.

The northwards migrations occur in response to day length, completion of the flight feather moult, and the amount of fat stored—fuel for the long flight. The body weight increases by about a third, even up to double the usual weight, while some non-essential organs may shrink in size. Then when weather conditions are right the journey commences. It is a hazardous journey taking about two months, and if flying conditions deteriorate they may turn back to the wintering grounds as they must arrive in good condition to breed. Flights between resting–feeding stops may be 4,000 kilometres long, sometimes twice as long. The birds usually fly at heights of 3,000–8,000 metres according to wind conditions, avoiding gales and using tail winds to cover about 30–60 km/h. Red-necked stint, one of the smallest and most numerous of our visitors, flies further than the distance from earth to the moon in its lifetime.

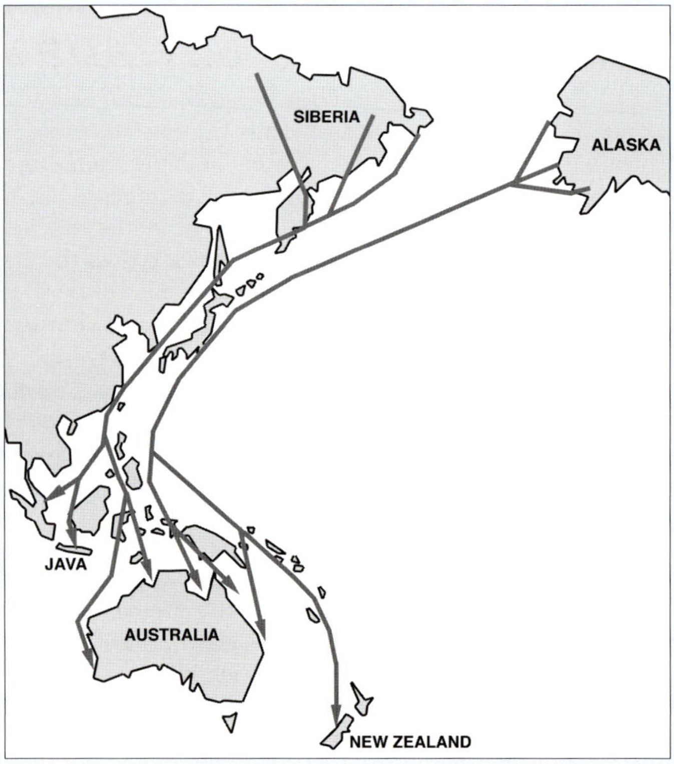

MAP 1.14 *Flight paths of the transequatorial migrant wader birds.*

Sources: Pizzey, 1985; Rowe, 2002; Saunders & Rebeira, 1985. Map adapted Rowe, 2002; *The West Australian*, 15 Dec 1998.

Chapter 2 The Swan–Canning Estuary

The Swan River is the heartland of Perth. It is not only a wonderful setting for a city, but has been put to practical use as a source of water and transport from the time of the city's founding. As the population grew, drains and waste began to degrade the waterways and, with concern for the health of the river and estuary, the Swan River Reference committee was initiated in 1943. Further pollution and degradation led to the proclamation of the Swan River Conservation Act in 1958 with provision of statutory power for the management of the river by the Swan River Conservation Board, later the Swan River Trust. Increasing algal blooms and community concerns about the declining health of the river brought the Swan–Canning Cleanup Program, and three years of research 1996–1999 to develop an Action Plan to cope with increasing degradation of the catchment and waterways. It was, however, the sudden appearance of a toxic bloom in February 2000 that brought the unthinkable to reality: the Swan River was closed to boating, swimming and fishing. Although the bloom resulted from a combination of unusual events, it was a very dramatic reminder of the vulnerability of our estuary. This chapter describes the Swan as the first explorers found it, and traces changes that dictate how the estuary functions or is likely to function in the future.

THE SWAN–CANNING ENVIRONMENT

The Swan River Estuary extends for about 60 kilometres from Fremantle to the junction with Ellen Brook near All Saints Church in the upper Swan, and for 6 kilometres up the Canning River to Kent Street Weir. The water of the estuary passes through the major landforms of the Swan Coastal Plain; the alluvial soils of the Guildford Formation; and sands of the Bassendean Dunes, Spearwood Dunes and Quindalup Dunes. The estuary enters the sea through the limestone cliffs of the Spearwood system, bordered to the north and south by the young Quindalup Dunes although these are poorly developed in this part of the coast.

The estuary can be subdivided into three main areas separated by shallow sills of sand and mud that affect water flow and the biological processes. Heirisson Island and the shallows around Maylands separate the Upper Estuary, where river flow and fresh water are major influences, from the wide main basin or Middle Estuary. The Middle Estuary is essentially the main basin of Melville Water. The Lower Estuary is downstream of Karrakatta Bank and Point Walter and influenced by marine processes.

In the Upper Estuary around Middle Swan the channel is about 2–3 metres deep except for some deeper areas to about 5 metres, before becoming shallower to about 1 metre in the area known as Perth Water. The main basin is 12 kilometres long and 2 kilometres wide, and about 21 metres deep in Mosman Bay before it passes through a narrow cliff-lined channel, Blackwall Reach, where it is only 5 metres deep. The channel then deepens in the Harbour that is dredged to about 11 metres (Fig. 2.1 and Box 2.3).

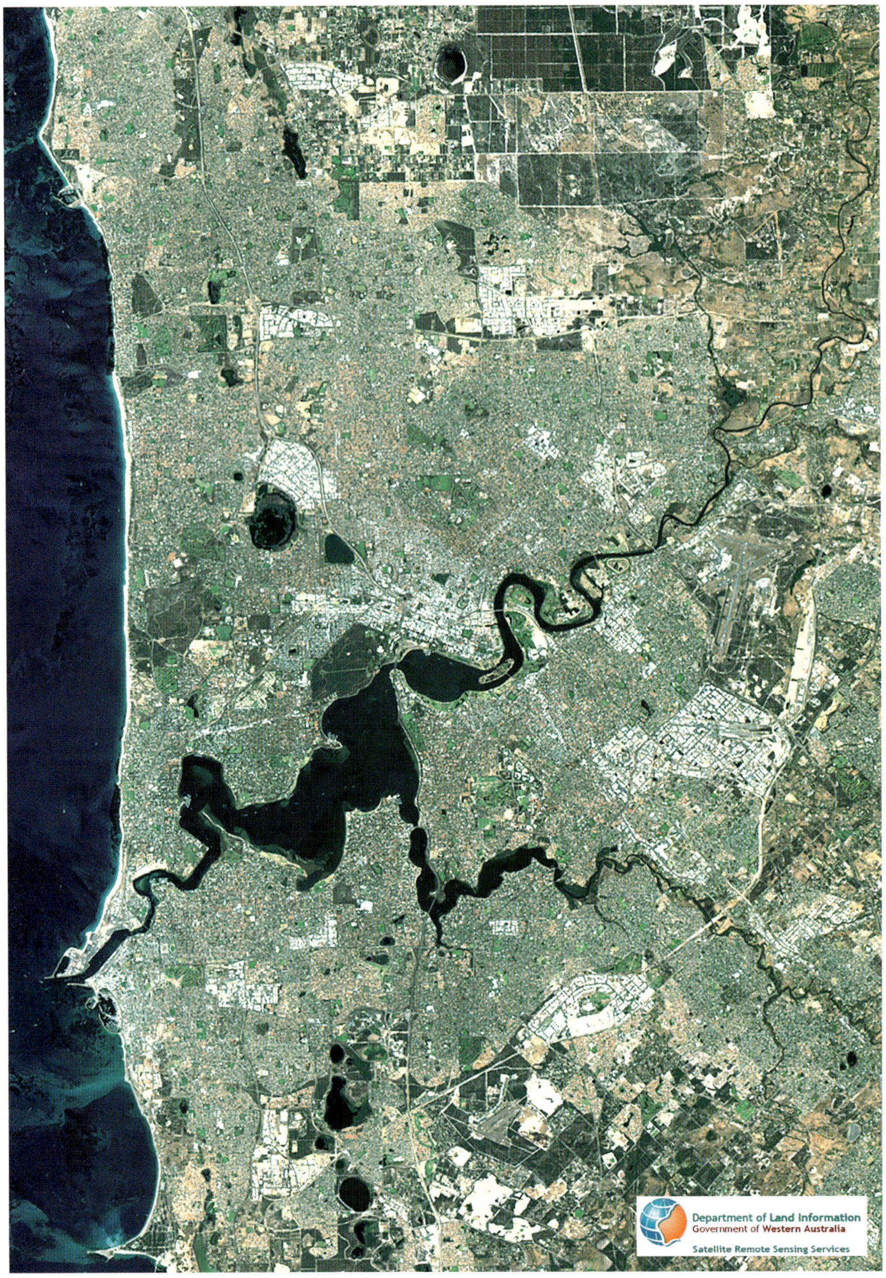

FIG. 2.1 *View from space of the Swan–Canning Estuary: the narrow reaches of the river and Upper Estuary meander across the flat coastal plain before widening into the shallow basin of the Middle Estuary, where it is joined by the Canning (on the left). The estuary narrows again in the lower reaches as it passes through the limestone gorge of Blackwall Reach.* (Land content–16 December 2001, Water–join of 16 December 2001 and December 1988. Enhanced SRSS–October 2003.) Landsat imagery provided by Australian Centre For Remote Sensing (ACRES), Geoscience Australia, Canberra and digitally enhanced by Satellite Remote Sensing Services, DLI, WA.

Winter to summer a changing state—fresh to salt

Salinity in the estuary changes dramatically with the seasons from fresh to salt. In winter, rainfall in the catchment brings river flow, which, depending on the strength and duration of flow, can completely flush the estuary with fresh water. As flow diminishes salty water moves upstream from the ocean, and in late summer the water in the Lower Estuary is as salty as the sea. Many marine animals enter to feed in the sheltered waters but, apart from the similarity of salinity, the estuarine waters are quite different to those on the open coast. Estuarine waters of the Swan–Canning contain ten times more silica than sea water, an important factor for microscopic algal (diatom) growth. Other nutrients are also more concentrated in the estuarine waters and blooms of algae are quite common. The sediments also contain larger amounts of plant debris and nutrients, and are more siliceous than the calcareous marine sediments.

Formation of the estuarine environment

The estuary of the Swan lies in a river valley drowned by the sea about 6,000 years ago when the sea level rose to slightly above its present level (0.5–1 metre). The catchment of a much older river system is marked by salt lakes on the Yilgarn Plateau, which probably converged on the ancient Swan–Avon and flowed to the sea through the 1,000 metre deep Perth Canyon millions of years ago (see Maps 1.5, 1.6 and 1.10). Sediments from this era lie beneath the present river and are known only from boreholes (Fig. 2.2). The Kings Park Shale marks the presence of an older (60 million years) estuarine system of Palaeocene or Eocene age, which covers even older marine sediments of Cretaceous age. These old sediments only outcrop to the north on the Dandaragan Plateau near Gingin and further south between Dunsborough and Augusta. (See **Chapter 1 Understanding Swanland: Flat coastal landscape of water and wind** and **Chapter 5 Estuaries north of the Swan: Gingin catchment—a different landscape**.)

Since the formation of the river channel, the sea level has risen and fallen many times due to changes in glaciation at the poles (eustatic changes). The changes are marked by dune lines on the coastal plain left as the coastline moved progressively westwards and seawards forming what is termed a prograding shoreline. Changes in sea level also modified the shoreline, shape and size of the estuary and location of the mouth. Shell beds from four of these transgressions are recognisable around the shores of the present estuary. The oldest deposit is at Point Resolution (Dalkeith), the next at Peppermint Grove, followed by Minim Cove (Mosman Park) and Caversham, and the youngest (Middle Holocene age) and best preserved at Point Waylen (Alfred Cove), a deposit that also extends below the surface upstream to East Guildford and along the older Canning to Ferndale–Higham. (See **Box 2.1 An older estuary hidden under Guildford** and **Box 2.2 Seashells at Peppermint Grove**.)

A remnant of the old river valley from the time of the Last Glacial Maxima is the deep channel of Blackwall Reach. In the last 6,000 years the old channel has also filled with sediment, up to 30 metres deep in places, varying from dark mud to quartz and carbonate sands. Silt and organic material brought by floods generally flocculate (settle-out) when the fresh

Box 2.1 AN OLDER ESTUARY HIDDEN UNDER GUILDFORD

Channel dredging, about 26 kilometres upstream at Guildford near the Helena River confluence in 1969–1970, unearthed fossil shells in fine grey mud that were radiocarbon dated at 6,660 ± 120 years BP or middle Holocene age. They are similar in age and composition to fossil beds at Heirisson Island, Perth and Melville Waters, and Cannington, Beckenham and Ferndale on the Canning arm of the estuary. Fossils from clay pits at Caversham are different and are thought to be slightly older, of Pleistocene Age, deposited when the sea level was several metres higher during the Last Interglacial Period 130,000–120,000 years ago.

During the last glacial maxima and extending into the middle Holocene, the climate of south-western Australia was more arid than at present; a condition that lasted until about 4,500 years ago. During that time, the Swan Estuary received less winter flooding than at present, and the slightly higher sea level allowed greater tidal exchange.

The fossils of filter-feeding bivalves, together with various herbivorous algal feeders, scavenging, carnivorous and parasitic gastropods, sea anemones and feather sabellid worms, are typical of those found in estuarine seagrasses such as *Halophila* and *Ruppia* today. Some like the mud oyster *Ostrea angasi* no longer live north of Cape Leeuwin. The presence of this community indicates the location of the seaward part of the palaeoestuary.

A few freshwater mussels *Westralunio carteri*, a species that inhabits the freshwater tributaries such as the Helena and Bennett Brooks, also indicate that the river flows were fresh. These freshwater mussels were once common throughout the South-West. In the Swan–Avon they have become extinct due to rising salinity.

Source: Kendrick, 1977, Yassini & Kendrick, 1988.

water meets the more saline water, and accumulate in deeper water along with faecal material produced by animals burrowing in the rich ooze or living in the water column.

Sandy and shell material has accumulated in sand bars and spits of the Lower Estuary such as Karrakatta Bank in Freshwater Bay, the Point Walter Spit and others in the entrance channel near Rocky Bay. Many were formed early in the estuary's development where sediment accumulated due to slowed water movement, or where sediment banked up against rocky outcrops like the ridge below Point Walter Spit. Others are younger, created by wave action and tidal currents that continue to move and reshape them.

THE RIVERS FLOWING INTO THE SWAN – CANNING ESTUARY

The Swan–Canning Estuary receives water from two main sources: the Swan–Avon and the Canning Rivers draining an area of about 121,000 square kilometres. Although the inland Avon Basin with sub-catchments the Avon, Yilgarn and Lockhart covers the largest area, most of the winter flow of the Swan originates in the 2,117 square kilometres higher rainfall area nearer the coast in the Darling Ranges (Graph 2.1).

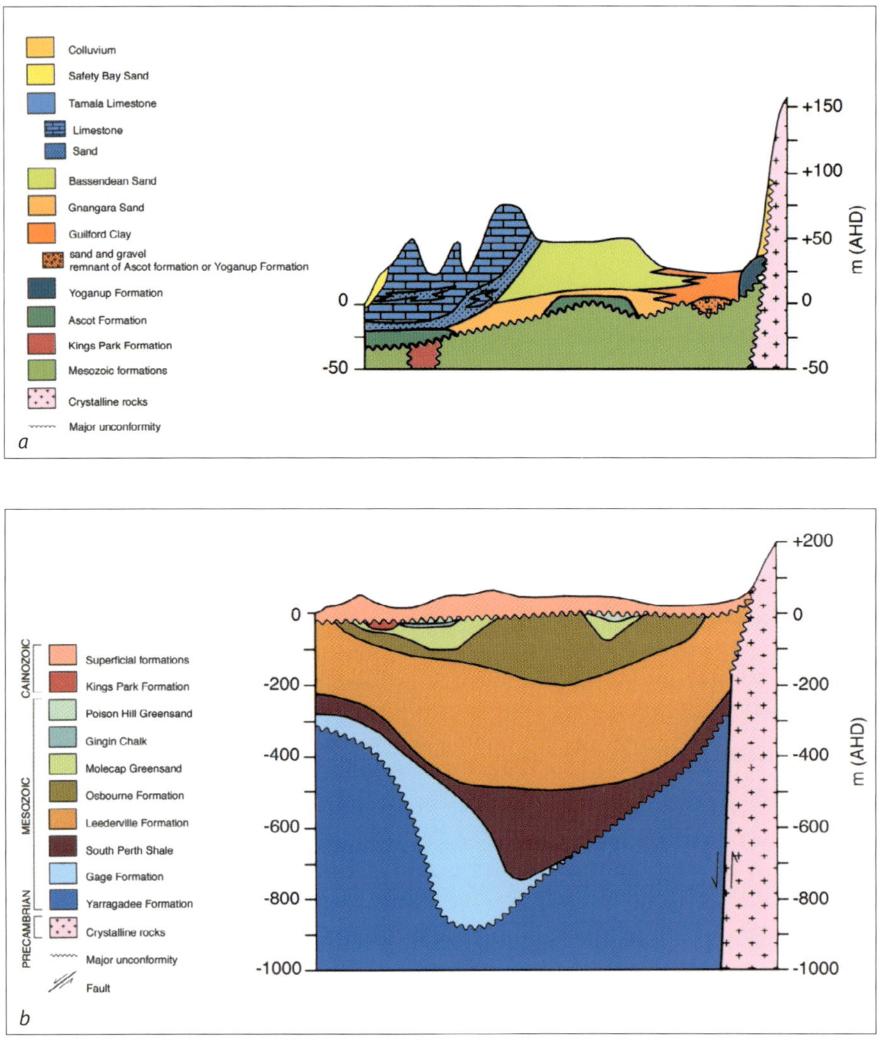

FIG. 2.2 *Major landforms and sediments underlying the Perth area of the Swan Coastal Plain. The configuration and depth of strata varies greatly from north to south and from east to west: (a) superficial formations and (b) older deposits of Cainozoic and Mesozoic eras. There are unconfined ground water aquifers in the superficial sediments, the upper surface of which reflects the watertable, fluctuating seasonally with winter rainfall and summer pumping. Older waters at the base may be about 2,000 years old. Large ground water aquifers, the north and south Gnangara Mounds and the Jandakot Mound, form a major part of the Perth urban water supply. Activities that affect water supply and quality such as variations in rainfall, overuse and contamination are of concern. There are also deeper ground waters in the Rockingham, King Park, Mirrabooka, Leederville and Yarragadee Aquifers that contain much older water. Water in the deepest aquifer, the Yarragadee, is generally about 36,000 years old although it is probably only 600 years old in the intake area south of Perth.* Modified from Davidson, 1995.

Box 2.2 **SEASHELLS AT PEPPERMINT GROVE**

Shells embedded in the limestone cliffs at Peppermint Grove, Minim Cove and East Fremantle provide a key to understanding the sea level changes in the last two million years. These water-laid deposits of the Tamala Limestone are of middle to late Pleistocene age.

Rocks at East Fremantle, about 3 metres above sea level, contain shells characteristic of exposed and sheltered rocky and sandy shores. The abalone, limpets, turban and whelks indicate an open reef environment, while the bivalve shells are more typical of open sandy beaches. Other bivalves and moon shells are typical of sheltered sandy and muddy substrates in bays or estuaries. Those of open reef and sand are similar to species living along the coast today, while some characteristic of sheltered bays are no longer present in the general area. The fauna represented by the East Fremantle deposit was moved around by waves and currents after death and cast up on a shoreline.

At Minim Cove, shells are found in two concentrations, seven metres and two metres above present sea level. The species are characteristic of a shallow marine bay with sand and mud. Some are species that live in the estuary today, and in sheltered marine areas like Cockburn Sound. Most, about seventy per cent, live today along the lower west coast off Fremantle while about twenty per cent live along the south coast east of Cape Leeuwin, and a few others are found today only north of Geraldton.

The fauna found in the rocks at Peppermint Grove seven metres above present sea level is also characteristic of a shallow marine gulf. Unlike the fauna at Minim Cove, only half live today along the lower west coast, about forty per cent live along the south coast east of Cape Leeuwin and a few are found north of Geraldton. Thus compared to Minim Cove, the Peppermint Grove assemblage has greater affinity with the southern region and weaker affinities with the area to the north.

This intermingling of species of southern (Flindersian–Peronian Province), northern (Damperian Province) and west coast distributions follows in a general way the patterns in the fauna living in the area today. At the time that the shells were living, part of the Swan was probably similar in form to Cockburn Sound where many of these species live today.

During the time over which the fossils were deposited, the sea retreated to beyond Rottnest and rose again, re-establishing marine conditions that then changed again with regular winter flooding of the Swan. Some of the more marine species were no longer able to live in the estuarine conditions, but today live in coastal waters further south.

The estuarine fauna today is not static: it changes seasonally, annually and, as the fossils indicate, over longer time periods. Some species move into the estuary and upstream in summer, particularly after dry winters, and move downstream and out of the estuary, or are killed, with the winter freshwater flows. The marine influence has probably been greater since the removal of the rocky bar and harbour construction in 1896, mimicking conditions during periods of high sea level in the past.

Source: Kendrick, 1960; Kendrick et al., 1991.

The Swan–Avon River

The Swan rises as the Avon about 370 metres above sea level and 200 kilometres inland near Lake Yealering and the Yenyenning Lakes south-east of Beverley (Map 2.1, Figs 2.3 and 2.4), part of a larger and ancient drainage system that extends eastwards into the arid western goldfields. Flow from the upper part of the catchment is generally low, but heavy rainfall particularly in summer after cyclonic activity in the north occasionally saturates the inland catchment and reactivates the old drainage system bringing vast quantities of water to the estuary, as happened in January to February 2000.

Following clearing of the deep-rooted natural vegetation of eucalypts and acacias, flooding became common around the towns of Brookton, Beverley, York and Northam with over twenty-seven floods between 1830 and 1955. After disastrous floods in 1955, the Avon was 'trained' (straightened and deepened) to increase flow away from the towns, a strategy now under review as the accumulating silt is removed and meanders are recreated. As a result of the river training, overgrazing, the rising watertable and salinisation, the catchment is in a particularly poor state and water from the Dale River, and Spencer, Wongamine and Toodyay Brooks is brackish to saline.

Below Toodyay, the Avon descends from the Darling Scarp in a series of rapids through a large forested valley lined with towering hills 150 metres

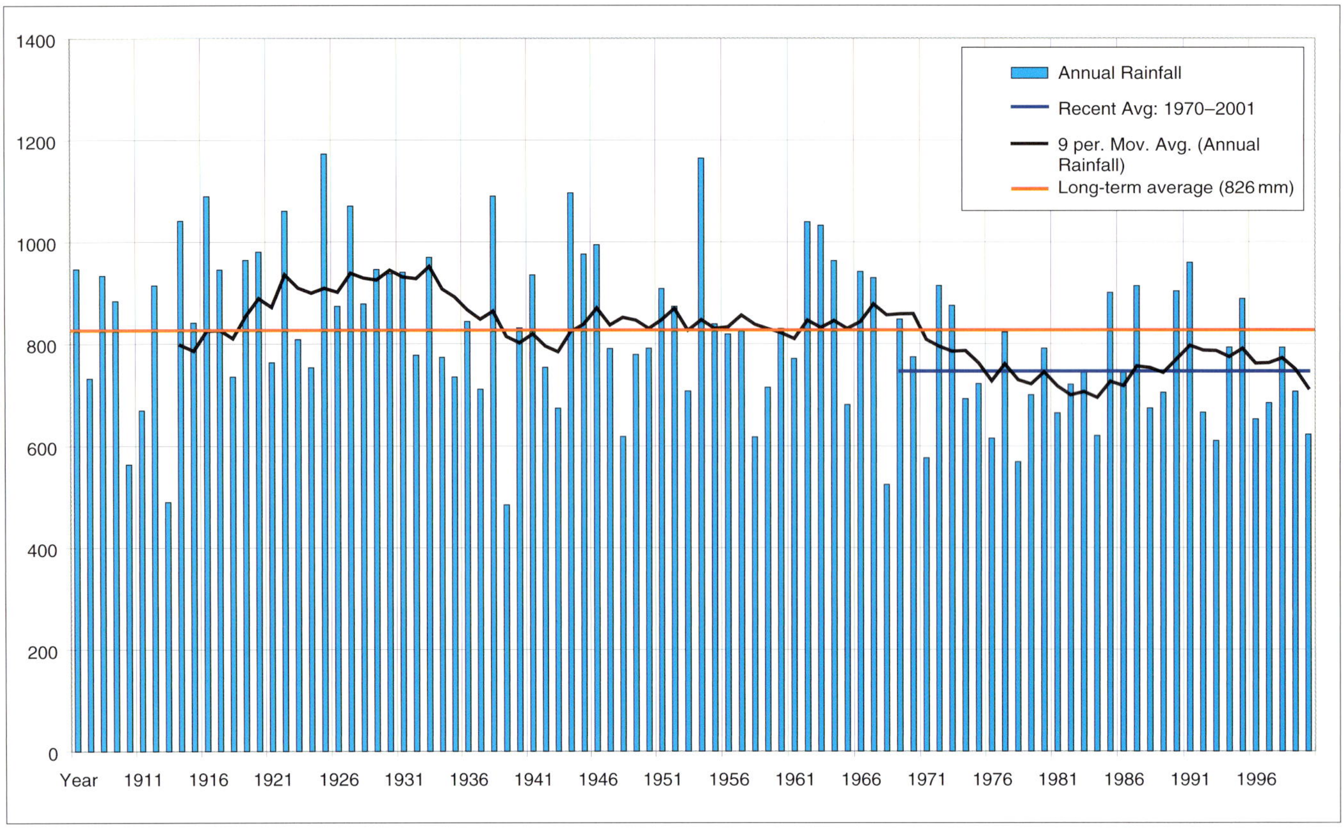

GRAPH 2.1 *Rainfall at Perth Airport 1907–2001. Note the long-term average of 826 mm and the decline in rainfall since the 1970s to 746 mm, a trend that appears to be occurring across south-western Australia.* Courtesy WRC.

high. Although rainfall is 600 to 900 millimetres, the tributaries of the Brockman and Gatta Rivers and southern Wooroloo Brook are brackish, so despite some water flowing from the forested part of the ranges the Avon flow is saline. The Avon emerges from the plateau in the Walyunga National Park, changing in name to the Swan.

Near the base of the Darling Scarp, Jane, Susannah and Ellen Brooks join the Swan. The alluvial soils (Guildford Formation) of the Ellen Brook catchment, deposited by another older and larger river system, have been extensively cultivated and are now a major source of nutrients flowing into the estuary (Map 2.2).

The Swan turns south then south-west crossing the coastal plain in wide meanders, passing through farmland and vineyards. Bennett Brook enters the main channel at Success Hill and the largest tributary, the Helena River, joins about a kilometre downstream near Guildford.

Mundaring Weir dams the Helena River, which drains the northern jarrah–marri forest. The weir captures about 50 per cent of the river flow, a feature common to most of the rivers between Perth and Bunbury. Under a scheme designed by C. Y. O'Connor in the late 1890s and completed in 1902 (enlarged in 1951 and 1959), Mundaring Weir supplies water to the wheatbelt towns and the Coolgardie–Kalgoorlie goldfields. Much of the catchment in the less than 900 millimetre rainfall zone is prone to salinisation that follows clearing and logging, so that the forested country reserved for maintenance of quality water also provides a dual function of conservation reserve.

Although tidal effects extend another 35 kilometres upstream to the junction with Ellen Brook in Henley Brook, Heirisson Island marks the junction of the riverine Swan–Avon and the main basin of the Swan. Here dredging has transformed the original river delta of mudflats, shoals, bars and salt marshes to islands surrounded by narrow channels.

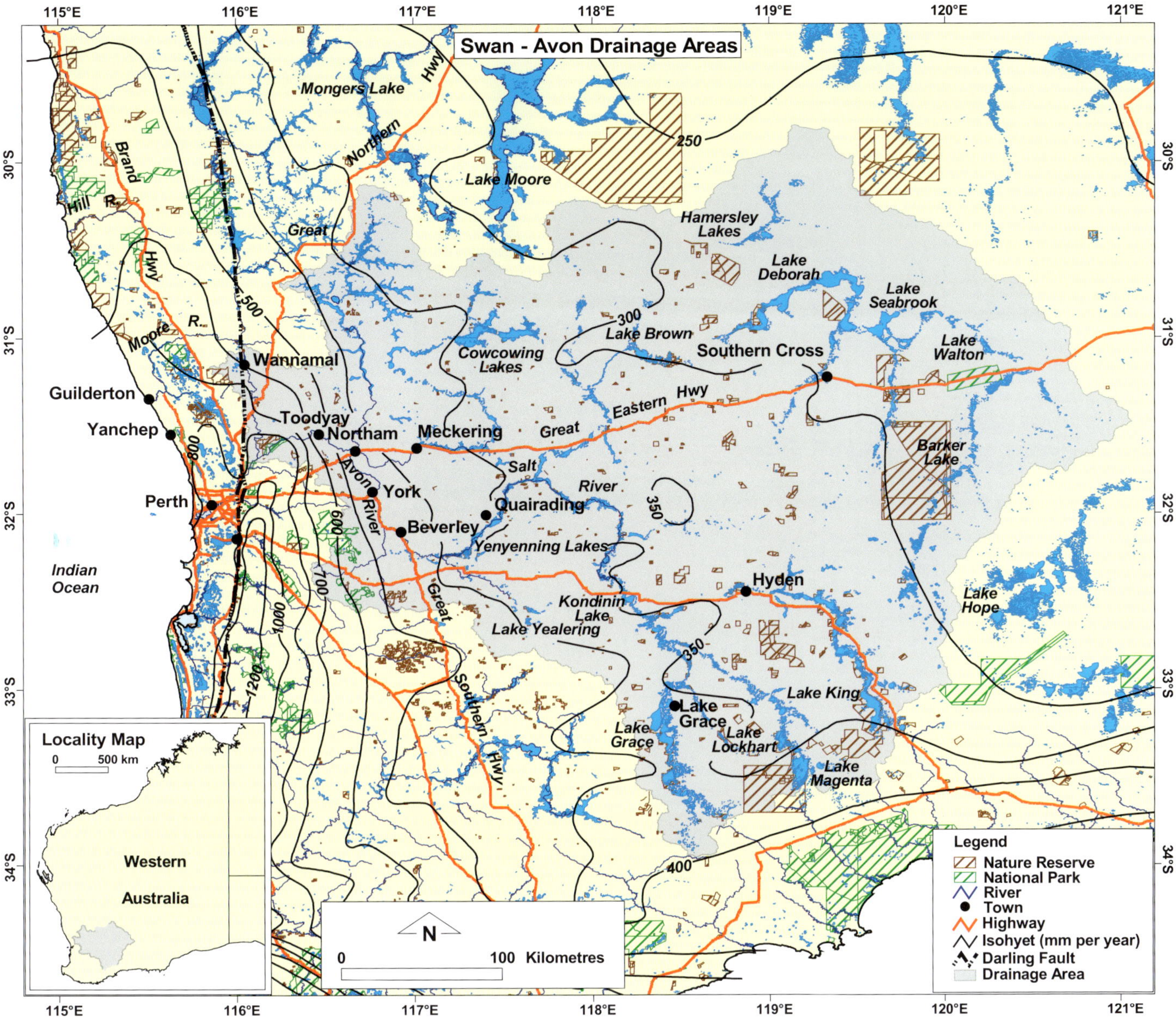

MAP 2.1 *Main tributaries and rain isohytes for the catchments of the Swan, Avon and Canning Rivers. The catchment area covers 121,000 km², however much of the water in the estuary originates in the area of the Darling Ranges and coastal plain.* Courtesy DoE.

FIG. 2.3 *Yenyenning Lakes at the source of the Avon River. Many of the wheatbelt lakes, such as Lake Yealering, were fresh in the early days and the water was used in the boilers of steam engines. The lakes became salty as the watertable rose after the land was cleared. The dead tree stumps bordering the creek lines are a stark reminder of the original freshwater vegetation.* L. Pen, 1999; Courtesy WRC.

FIG. 2.4 *The Avon River, Gwambigyne area, March 1997. The Avon has been 'trained', deepened and straightened, to increase storage and flow. Sediments eroded from the catchment now fill much of the river course, and deep pools once a feature of the river and a refuge for wildlife over summer have disappeared. Recently restoration programs to remove sediment, contour the banks and plant riparian vegetation have been initiated. Historic records and museum collections indicate that freshwater animals once lived in the now saline waters.* Courtesy E. Rippey.

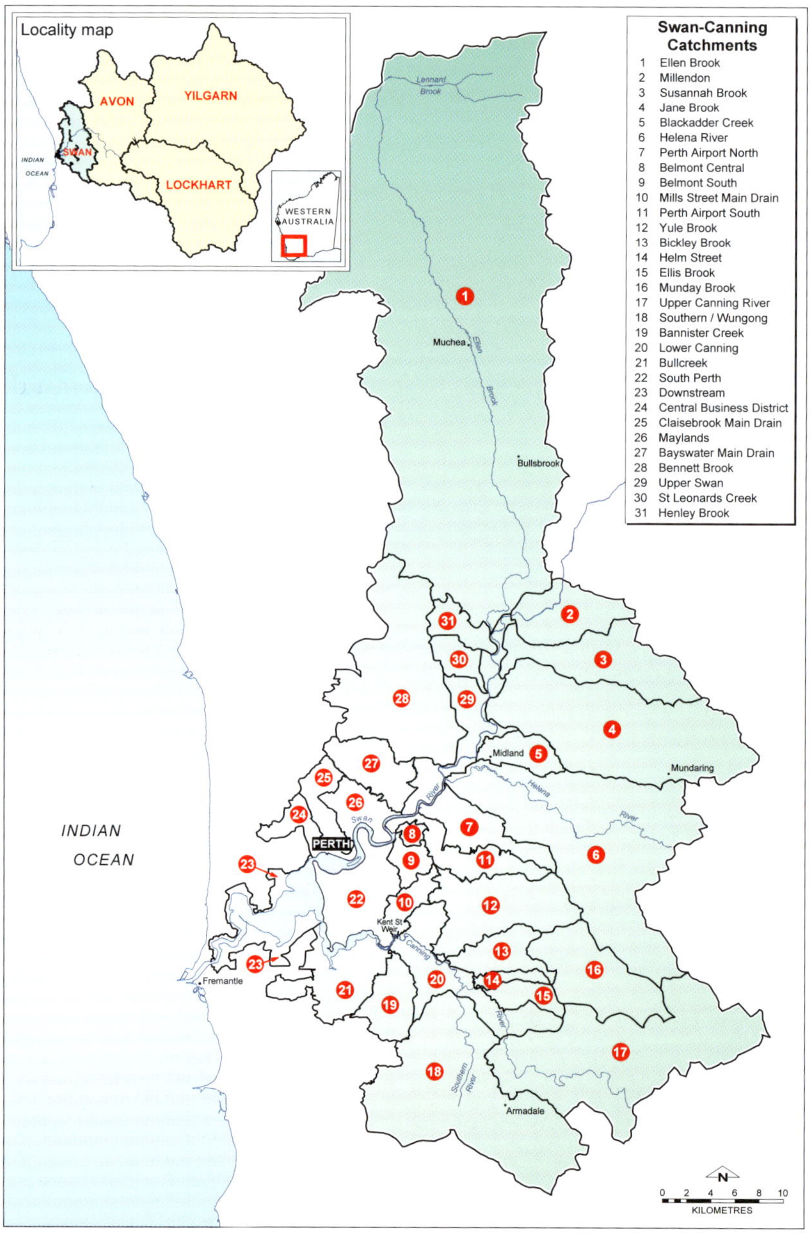

MAP 2.2 *Catchments of the Swan–Canning system on the Swan Coastal Plain: there are 31 sub-catchments within the coastal plain, while many drain urban areas, others drain rural areas. Flow reaches a peak in mid to late winter, with flow from the Swan River and its tributary the Avon about fifteen times greater than flow from the Canning. Inflow in summer is mostly due to ground water gradually moving through the sandy soils surrounding the estuary. In 1996, the total surface run-off entering from the Avon was about 715.9 million m^3, and ground water discharge between Guildford and the Causeway was about 80,000 m^3 per day or about 29.2 million m^3/year, about 4% of the total river flow.* Source: SCCP Action Plan, 1999, Courtesy SRT.

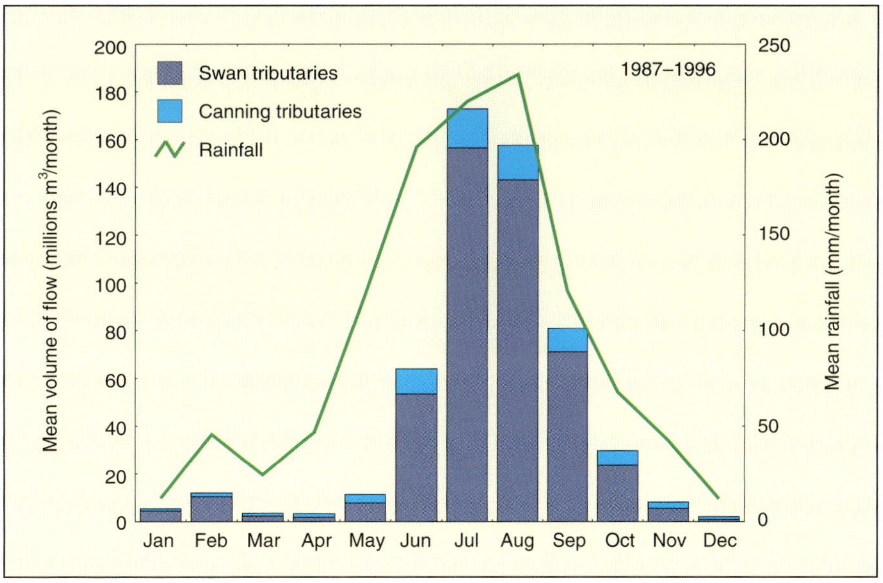

GRAPH 2.2 *Rainfall and river flow of the Swan–Canning Estuary 1987–1996: rainfall peaks in winter and early spring. Flow from the Swan River, which receives water from the undammed Avon River, is fifteen times greater than that from the Canning River. Flow from the Avon is saline due to its source on the Darling Plateau where salt has been mobilised by rising ground water following clearing of the deep-rooted natural vegetation.* Source: SCCP Action Plan 1999, Courtesy SRT.

Canning River

The Canning River catchment is much smaller than the Avon catchment and lies in forests receiving 700–1100 millimetres of rainfall on the Darling Plateau. After descending to the coastal plain at Kelmscott, the main channel is joined by other small tributaries flowing from the scarp. Water flow is now much reduced by dams on the Canning River (built in 1940) and the largest tributaries, the Southern–Wungong River, the Munday (Victoria Reservoir 1891) and Churchman Brook (Graph 2.2).

Flow is also reduced by a weir at Kent Street, built in 1927 to hold water for agriculture by trapping fresh water as the winter flows diminish and preventing the salt water penetrating upstream in summer. When the river flows, water joins the main body of the Swan at Riverton.

THE SWAN — AS IT WAS

Noongar Aboriginal people were drawn to areas around the Swan–Canning Estuary and surrounding wetlands, the Derbal Yarrigan, for camping, fishing and hunting particularly over summer (Map 2.3). (See **Box 9.1 Stone circles—echoes of the past** and Fig. 3.7.) Legends and stories telling of the rivers' formation and of tribal histories are part of the rich tradition passed down through the generations. Fish and waterfowl were easily accessible at Pelican Point. The Mooro tribe of Yellagonga used the Pelican Point area, while Midgegooroo, the father of Yagan, moved with the Beeliar Tribe through the Alfred Cove area in the summer. The area between the Swan and Canning, known as Beeloo, was the territory of Munday. The name Milyu, an Aboriginal term for samphire the low-lying succulent vegetation characteristic of estuarine wetlands, is now used for the foreshore nature reserve along the Kwinana Freeway in South Perth.

MAP 2.3 *Tribal districts of the Aboriginal people around the Swan–Canning at the time of European settlement. Although there are many different spellings of the name, 'Noongar' is preferred, Whitehurst, 1997.* Modified from Hasluck, 1961. Courtesy N. Hasluck.

Just visiting — first impressions

Although traders in the sixteenth century knew the north coast of Australia, mapping of the south-western part of The South Land really began with the discovery of a shorter, faster route to the Dutch East Indies. The route along the high latitude westerly winds before turning northwards was hazardous given the lack of reliable methods for calculating longitude. Not surprisingly miscalculations brought the first visitors and shipwrecks, the first by Dirk Hartog in the *Eendracht* in 1616. Dirk Hartog landed on the island that bears his name in Shark Bay, where he left a now famous plate nailed to a stake.

Crews of the Dutch vessels *Dordrecht* and *Amsterdam* probably sighted the area around the Swan River as early as 1619, before they sailed northwards. The next visitors were not so lucky: the *Batavia* was wrecked on the islands of the Houtman Abrolhos in 1629 and the *Vergulde Draeck* (Gilt Dragon) near the coast between the Moore River and Ledge Point in 1656. Only the few survivors of the wrecks and of the massacre on the Abrolhos ever left the Southland to tell the tale. One of the survivors from the *Vergulde Draeck*, Abraham Leeman van Santwitz (Leeman), returned on a rescue expedition in 1658 led by Samuel Volckertszoon (Volkersen) in the *Waeckende Boey* and Aucke Pieters Jonck in the *Emeloort*. Volkersen sailed past Rottnest Island and discovered wreckage of the *Vergulde Draeck*. Hampered by foul weather, the search for survivors resulted in yet another mishap in which Leeman, survivor of the wreck, was marooned, again making his way to Batavia in an open boat. In an expedition plagued by unsuitable and mismatched boats and personalities, the resulting maps were, however, quite detailed, showing the dangers of the coast but contributing little to the knowledge of the land.

Loss of yet another ship, the *Ridderschap van Holland* between the Cape of Good Hope and Batavia (Jakarta) in 1694, heralded the next expedition led by Dutchman William de Vlamingh, commander of the ships *Gellvinck* (Gold Finch), *Nijptangh* (Nipper or Pincers) and *Weseltje* (Little Weasel). Instructed to also search for wreckage of the *Vergulde Draeck*, Vlamingh made landfall around Rottnest Island, variously referred to as 'Fog Island', 'l'Isle des filles' (Isle of Girls) and ('t Eylandt 't Rottenest') (Rats Island), before moving to the coast in early January 1696 (Fig. 2.5 and Map 5.1). Initially, members of the expedition probably landed near Swanbourne Beach and walked inland to the large river 'Swane Rivier', where they were met by others of the crew who had entered from the sea. Some days later, three boats were dragged over the rocky bar and Vlamingh himself joined the explorations upstream, probably as far as East Perth. Unlike the situation today with the estuary salt in summer, Vlamingh's party recorded the water as 'brackish', 'somewhat

FIG. 2.5 *Vlamingh's expedition entering the Swan River in 1696. Although the elevation of the landscape has been exaggerated, Arthur Head is visible to the right of the river mouth. Points of land projecting across the river were excavated when the harbour was built; the inner point the site of the limestone sill across the river and location now of the Fremantle Traffic Bridge. It is likely that Vlamingh originally intended to name the river the Witsen after Nicolas Witsen, a director of the VOC (Dutch East India Company) and promoter of the expedition, but the name Swan River was used by Victorszoon the cartographer. (See Map 5.1 & Playford, 1998.)* Picture accompanying a map by Johannes van Keulen. Swartte Swanne drift op het Eyland Rottenest, from Verhandeling by Francois Valentyn. Dordech 1726. MAP NK 2110, Courtesy National Library of Australia.

brackish' or a 'little brackish', nevertheless Vlamingh noted that it was an estuary influenced by tides.

Impressions of the new land were mixed; some men became ill after eating *Zamia* palm nuts, the flies were troublesome, and despite efforts to meet the natives, only fires and footprints were seen. Vlamingh was, however, impressed by the scenery and weather, referring to one area (possibly the Crawley–Dalkeith foreshore) as 'Belvedere' (beautiful view) and Rottnest Island as paradise on earth. The river's Black Swans became the first export; three were captured and taken to Batavia where they died. In general, discoveries by the Dutch East India Company along the Western Australian coast were not conducive to settlement, revealing little in the way of useful products and a native population regarded with fear and suspicion.

> *As regards to the Country, it is sandy, and, in the place where we have been were a great number of Trees, among which were some almost three or four fathoms (Sic) thick but bearing no fruits; in short, full of prickles and thorns. Several of these Trees yielded a kind of resin nearly like wax, of a brownish red colour. All were very shy, the Men, the Birds, the Swans, the Brent-geese, the Crammed-gees, the Cockatoos, the Parakeets etc. The best of it is no vermin are found here: but during the day one is terribly tormented by the flies.*
>
> *Nijptangh* diarist from the translation by Robert, 1972.

The British era of exploration brought William Dampier to the Kimberley in 1688, and later in 1699 to the Shark Bay area. For much of the eighteenth century however, although a colony was founded in New South Wales, and Europeans such as Captain George Vancouver in 1791 visited the southern coast, Vlamingh's report generally discouraged interest in the west coast. Much of the Great South Land slumbered on until the French

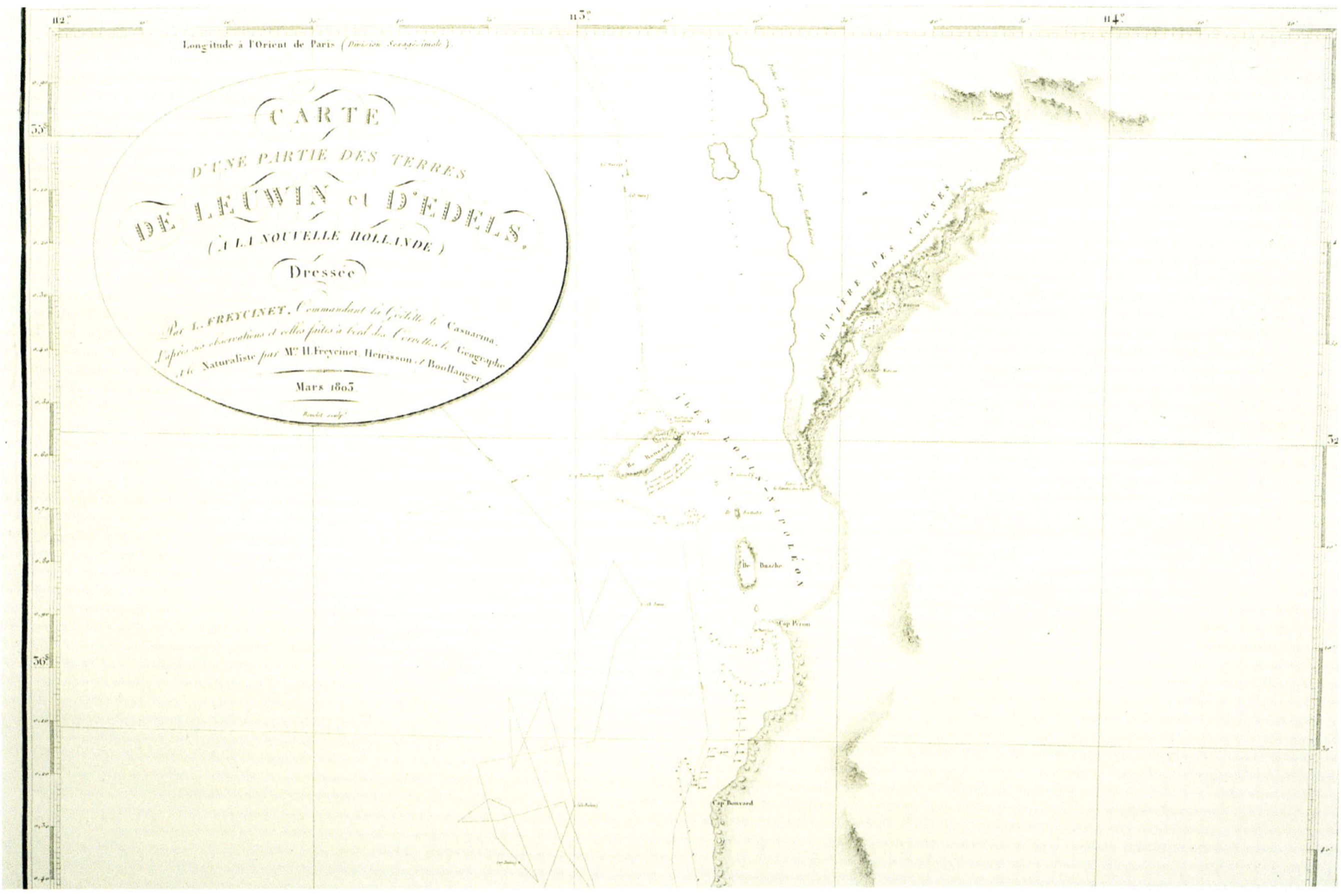

MAP 2.4 *Map drawn by Freycinet and Heirisson of the west coast from the Mandurah area to the north of the Swan River showing the route taken by the 1803 Baudin expedition. Carte d'une Partie des Terres, de Leuwin et d'Edels. (À la Nouvelle Hollande).* See Marchant, 1982. Courtesy T. Burrows, Scholars' Centre, UWA.

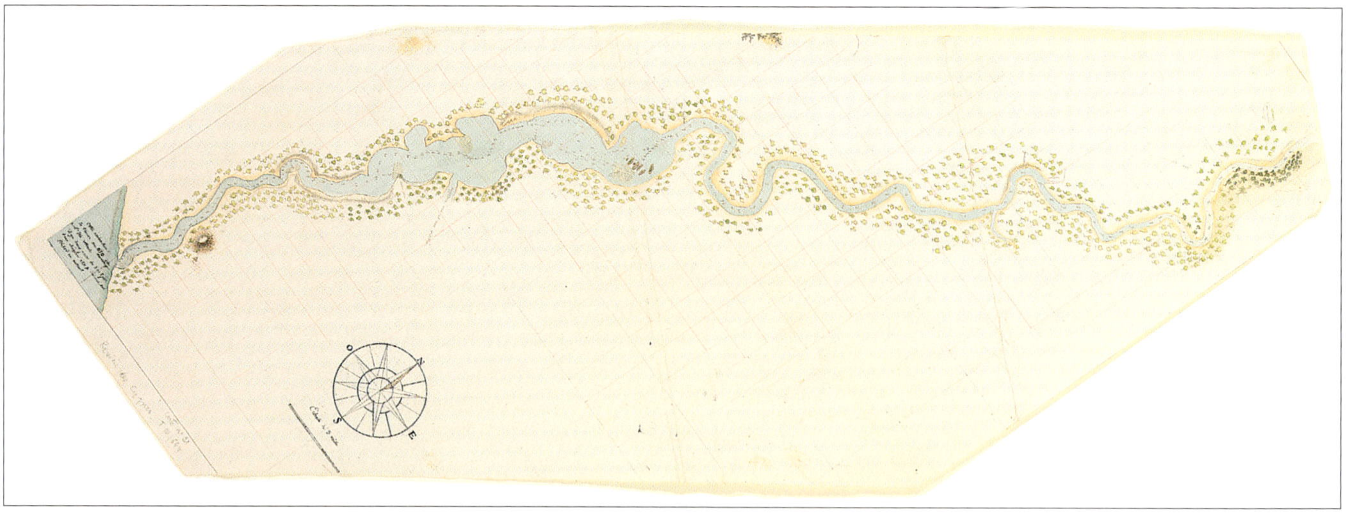

expedition of Baudin in 1801–1804, when England and France were still at war, restimulated interest in the west coast.

The two voyages of the Baudin Expedition (1801–1804) were well equipped, supported by the French government and scientific societies, and its crew carefully chosen. Under command of Nicolas Baudin on the *Géographe* were an astronomer, engineer–geographer, botanist, mineralogist, three zoologists including François Péron, and assistants, while the *Naturaliste* commanded by Felix Hamelin also had a team of scientists. Although reports of the expedition (archived in Paris) were kept by most of the scientists, the published accounts of Péron (with the artist Charles Alexander Lesueur), and Louis Desaules de Freycinet, sub-lieutenant aboard *Naturaliste* and later commander of the *Casuarina*, provide the first details of the west coast (Map 2.4).

After exploration around Cape Naturaliste and Geographe Bay, winter storms separated the boats. (See **Chapter 4 Leschenault Estuary and the smaller estuaries of the Geographe Bay area: Visitors—mapping and naming the landscape**.) The *Géographe* and Baudin were blown further north, and only those on board the *Naturaliste* spent time around the Swan River. Although lacking the observant eye of Péron, Sublieutenant Heirisson, Joseph Bailly, and François Collas recorded some details of the river and surrounding land. Bailly, a mineralogist, noted the calcareous cliffs along Blackwall Reach concluding that an earlier sea deposited them, a fact now confirmed by studies of the fossils. Passing by the mouth of the Canning River ('Entrée Moreau', named after Charles Moreau a midshipman with the exploring party), the boats proceeded upstream past high ground (Mount Eliza) before grounding on shallow areas upstream, naming the small islands after the commander Heirisson (Map 2.5). Upstream they noted the low-lying land, tree-lined riverbanks and the hills beyond, although they did not venture far from the river. Rejoining other members of the crew at the river mouth they sailed around Rottnest before departing for Timor where they met up with the *Géographe*. The *Géographe* again visited Geographe Bay and the area around the Swan River during the second voyage in 1803, but although the weather was more favourable reports of the land were still not encouraging. Baudin recorded Hamelin's comments made in 1801:

> *All that can be said of it, judging by the report of Captain Hamelin and his officers who stayed there quite a long time, is that the anchorage is very bad, insecure and open to the winds from the South-West to North-West, and that at such times as they are blowing, the sea is very heavy there. The Swan River, which was explored for 18 to 20 leagues, offers no resources at all. Here and there some fresh water was found in the streams that enter it, but they are very small.*

MAP 2.5 *The Swan River drawn by Heirisson on the Baudin Expedition in 1801. The chart shows details of the surrounding country including the bar, Point Walter spit, a hill near Bicton, the entrance to the Canning River* Entrée Moreau, *Heirisson Island and a freshwater stream, Claise Brook. Further upstream, the Helena River and Bennett Brook are shown, the chart ending near Henley Brook. The chart also shows the water depth and vegetation. Heirisson estimated he travelled 18 leagues (about 55 miles) along the river. Note the compass rose indicates the true orientation of the chart.* Purchased by the Western Australian State Library Custodians in 2002, Catalogue 5901A/1, Courtesy Battye Library.

MAP 2.6 *Evaluation of the Swan River by James Stirling and Charles Frazer, March 1827. The part coloured green marks land intended for settlers and public purposes; yellow land to be granted to Mr T. Peel on condition of his landing 400 persons before 1 November 1829; and purple land to be granted to Captain Stirling RN. In the foreground the text reads 'The white line of the Coast for as far as Geographe Bay is a Limestone Ridge varying from 20–200 feet in height and extending inland to the distribution of from One to Five Miles'. To the right are 'Fine Undulating Grafs Plains' and on the left 'Undulating Grazing Country Thinly Wooded'. Along the river notations include 'Ellens Bk' and 'Henley Park'; 'Swan River'; 'Canning River' and 'Melville Water'; River water fresh here; Abundance of fresh water; Many fresh springs; Mineral Springs. Notations along the hills include 'An undulating open Country extending to the Base of the Range; General Darling's Range; St Anne's Hill about 3,000 feet high; Mt William 3,060 feet high & very remarkable'. Mountains from 1,200 to 1,500 feet high.* Source: British Parliamentary Papers, 1829. Courtesy T. Burrows, Scholars' Centre, UWA.

Interest anew

British interest in the western coast of Australia was rekindled by Captain James Stirling, who was commissioned to take supplies to Sydney and the Melville Island in 1826. An experienced sailor but no longer in naval service, Stirling was looking for opportunities. His interest in the Indian Ocean was stimulated by family connections. Stirling like the Dutch before him saw the advantages of a port on the west coast, and approached Governor Darling with a plan to investigate the area around the Swan. Stirling's plan obviously fell on fertile ground, as a British presence on the south coast was already of interest to Darling, who in fear of French intentions had sent a small garrison to King George Sound in December 1826.

Accompanied by Charles Frazer, Colonial Botanist of New South Wales, Stirling left for the Swan River in the *Success*, arriving in March 1827. After exploring around Carnac and Garden Islands and discovering a protected anchorage, 'a Magnificent Sound', which was his first objective, Stirling prepared to voyage up the Swan. Unlike the Dutch, he appears to have had no trouble crossing the bar, recording the headlands on either side and naming the southern one Arthur Head after Sir George Arthur, Governor of Van Diemen's Land, and the northern Rous Head after a friend Captain H. J. Rous. Proceeding upstream he was charmed by the landscape describing it as extremely beautiful and picturesque. The mudflats near Heirisson Island made travel difficult, as they had done to the earlier visitors, but once passed the surrounding land looked even more promising (Map 2.6). Like Vlamingh before him, Stirling noted that water levels in the river were influenced by tides and waters were estuarine:

> *I must protest here against the term "River" as applied to the Estuary in which we are now entering. It is a misnomer which leads to confusion of Ideas; and I shall therefore designate the various ramifications of the Sea within the two heads just mentioned by the general name Melville Water, limiting the use of the name Swan River to that Stream, which joining the Sea at the Islands below Frazer's Point concludes its career as a River.*
>
> Narrative of Operations enclosed with a letter from Captain James Stirling to Governor Darling, 1827, pp. 554–5.

FIG. 2.6 *View taken at the commencement of the fresh water, Swan River 1827. A gentle river abounding with wildlife and surrounded by large trees was taken as evidence of fertile soils. Frederick Garling, pen and ink and watercolour and pencil.* Courtesy Art Gallery of Western Australia, 957/W5.

Frazer, a person already familiar with the Australian bush, its uniqueness and variety (Fig. 2.6) echoed the enthusiasm of Stirling. Acknowledging the obvious proximity to the Cape of Good Hope, India, the Spice Islands to the north and as a port of call for ships travelling to East India and China, Frazer listed the attributes:

> *In giving my opinion of the Land seen on the Banks of the Swan River, I hesitate not in pronouncing it superior to any I ever saw in New South Wales east of the Blue Mountains, not only in its local character but in the many existing advantages, which it holds out to Settlers; these advantages I consider to be.*
>
> First.—*The evident superiority of Soil.*
>
> Secondly.—*The facility with which a Settler can bring his Farm into a state of immediate culture, resting upon the open state of the Country, a state which allows not a greater average than 10 trees to an Acre.*
>
> Thirdly.—*The general abundance of Springs, producing water of the best quality, and the consequent permanent humidity of the Soil, two advantages not existing on the Eastern Coast, and*
>
> Fourthly.—*The advantage of Water carriage to his Door, and the non-existence of impediments to land carriage.*
>
> Frazer's report inserted by Stirling into the Admiralty document and appended to the report to Governor Darling, 1827, pp. 581–2.

Stirling and Frazer were to be proved wrong on many accounts. They underestimated the extent of the good land, the infertility of the remaining land, the difficulties of transport on the sandy soils, the overabundance of water in winter leading to waterlogging and flooding, and the change to drought over summer when freshwater supplies diminished. Finally, they failed to appreciate problems of the inhospitable coast as a refuge for ships.

Stirling ended his letter to Darling with the suggestion: 'The Name "Hesperia" indicating a Country looking towards the Setting Sun, would be descriptive of the Situation of the Country in question; it would not interfere with any Name previously given, nor would it be subject to the imputation of Nationality.'

Observations to advertising—Swan River mania

Satisfied that the two requirements for settlement, a safe anchorage and fertile land, were assured, Stirling set sail for Sydney to plead his case with Governor Darling. Perhaps not surprisingly, Darling supported the idea as the garrison already established at King George Sound was more isolated and reports of the land less favourable. Initially the British Government was slow to agree to a new colony, but with renewed interest by the French and with Stirling's arrival to champion his cause, the Colonial Office reversed its decision with two conditions. First that it should not be a penal settlement, and second that the Government would not be required to spend money in establishing the colony.

The enthusiastic but somewhat scanty and naive reports of Stirling and Frazer were now reinterpreted by the press and embellished by speculators. The policy of providing free land grants to all settlers arriving before 1830 at a rate of 40 acres for every three pounds of capital aroused public interest to a level dubbed 'Swan River Mania'.

Living in a strange land

The proclamation of British Rule on 2 May 1829 encompassed the remaining third of the South Land, which was all the land not included within the boundaries of New South Wales. It was, however, a long way from home. For the most part the settlers were small investors, mostly educated, with the manners and customs of an English Society, but generally ill prepared for the conditions that awaited them. (See **Box 6.1 Georgiana Molloy—lady of flowers: a change in perception**.)

The safe anchorage that so impressed Stirling on his first visit proved deceptive. Reefs and sandbanks surrounding the entrance claimed the *Parmelia*, whose passengers were offloaded onto Garden Island. With little shelter and no storage for their goods, the conditions for the new arrivals were appalling, and as surveying and allocation of the land proceeded a more bitter truth emerged: that the extent of good land was limited.

Exploration to the south and east quickly followed, and people requiring larger grants were sent further afield to the south around Augusta on the Blackwood River, Peel–Harvey, Leschenault and Geographe Bay, and finally further north to Champion Bay, Greenough and Geraldton areas. Around the Swan, settlers had the choice of land along the Swan or Canning Rivers or further inland over the ranges in the Avon District. Along the Swan, allocations were limited to half a mile on one bank so the lots were long and narrow stretching outwards into poorer soil (Map 2.7).

Changing the landscape

The present landscape around the Swan–Canning reflects not only the natural landscape but the past and present values of the community, both urban and agricultural. Early development around the Swan centred around three nodes, the Fremantle Port, the Perth Capital and Guildford further upstream near the end of navigable waters adjacent to the fertile soil of the upper Swan Valley. Alluvial land along the Canning was also settled early, with Kelmscott established in 1830. Initially the river provided the main access because of sandy soil, but tracks gradually became more reliable being covered with limestone or wooden planks. The first, from Fremantle to Perth, were along the southern shore to Canning River crossing to Perth via the Causeway, and a northern track along the route of the Stirling Highway with a ferry crossing the river at Preston Point, until a bridge was built at Fremantle in 1866.

Change in the estuary was under way when the moles were built in 1897 to create a harbour at the river mouth. The arrival of convicts in 1850 heralded major changes to the Swan River colony in an era of public works and buildings. Buildings included the Fremantle Goal and Courthouse, the Customs House, the Lunatic Asylum (now a branch of the Western Australian Museum), the

Perth Courthouse (1856) and Old Perth Boys School (1852–1854). The most easily sourced material for building was limestone from quarries along the river banks at East Fremantle, Rocky Bay and Mosman Bay. Quarrying at Rocky Bay, once 70 metres high, levelled the landscape creating areas suitable for another generation of industry including the State Engineering Works, automotive manufacture, sugar refining, fertiliser manufacture and rubbish disposal. These in turn left a legacy of contamination for the new era of urban development in the 1990s (Fig. 2.7). Quarrying at nearby Minim Cove went on much longer into the 1920s, with rocks used for the Fishing Boat Harbour and walls along the South Perth foreshore and Riverside Drive.

Growth of a city—a landscape of wetlands, lakes, rivers all gone

The siting of the capital Perth (12 August 1829) along the river bank accessible by boat downstream to the port and upstream to the best alluvial land, determined its character as a city on a river. Originally the town was laid out east–west along a sandy rise from St Georges–Adelaide Terrace to Hay Street at the top (Map 2.8). Cross streets William and Barrack ran down to jetties built over the shallow margins of the river to service boats and cargo

MAP 2.7 *Land grants around the Swan and Canning Rivers: once the importance of river access and the limits of the fertile land were recognised, land grants along the Swan and Canning River were surveyed in long strips with only a narrow river frontage. Detail from Roe's map published by J. Arrowsmith, October 1839, showing the growth of settlement.* Courtesy Battye Library, Map Rare 24/1/4.

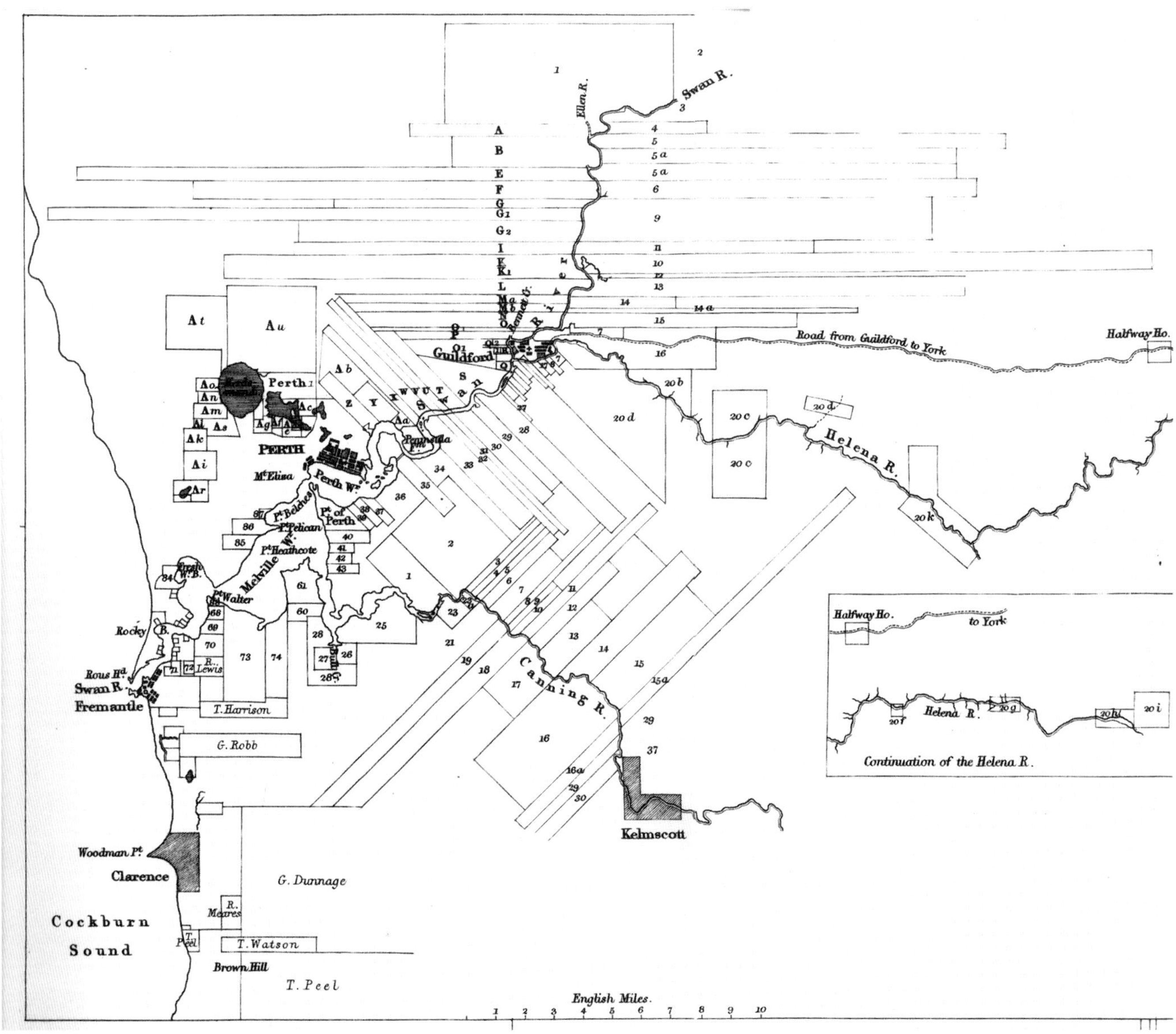

transport along the river. To the north a chain of lakes limited development, and flooding became a problem as the watertable rose after land clearing and the city grew across the low-lying land.

Drains dug to alleviate flooding of Murray Street and Forrest Place in 1847 heralded the beginning of a massive drainage program around Perth that would change forever the character of the land. By 1854 Lake Kingsford was drained, cleared of rushes and the land sold, while others such as Queens Park remain today much altered by clay mining for brick making. Drains around the Railway Station were open until 1899 when pipes replaced them. Drainage across the plain continued as the population increased. Low-lying areas around Osborne Park were drained in 1912. Herdsman Lake was drained in 1925 but reverted to bulrushes and dairying when found unsuitable for market gardens. Although generally viewed as progress, there was some regret. W. E. Bold, Town Clerk of Perth 1900–1944, expressed the wish that more of the swampy areas had been kept as parkland.

Other problems emerged more slowly. Nutients from garbage dumped as fill in low-lying areas gradually started to leach into water supplies, along with others from septic sewage systems that worked so well in the sandy soils of the Bassendean Dunes. The enriched ground water flowing to the Swan Estuary led to stinking rotten masses of algae as early as the 1900s.

FIG. 2.7 *Peppermint Grove, Freshwater Bay with Tuarts and Rottnest Island Pines, April 2003. Only a few areas of natural vegetation remain along the Swan River; invasion by weeds, such as pepper trees in the foreground, is a problem.* Photograph C. Brearley.

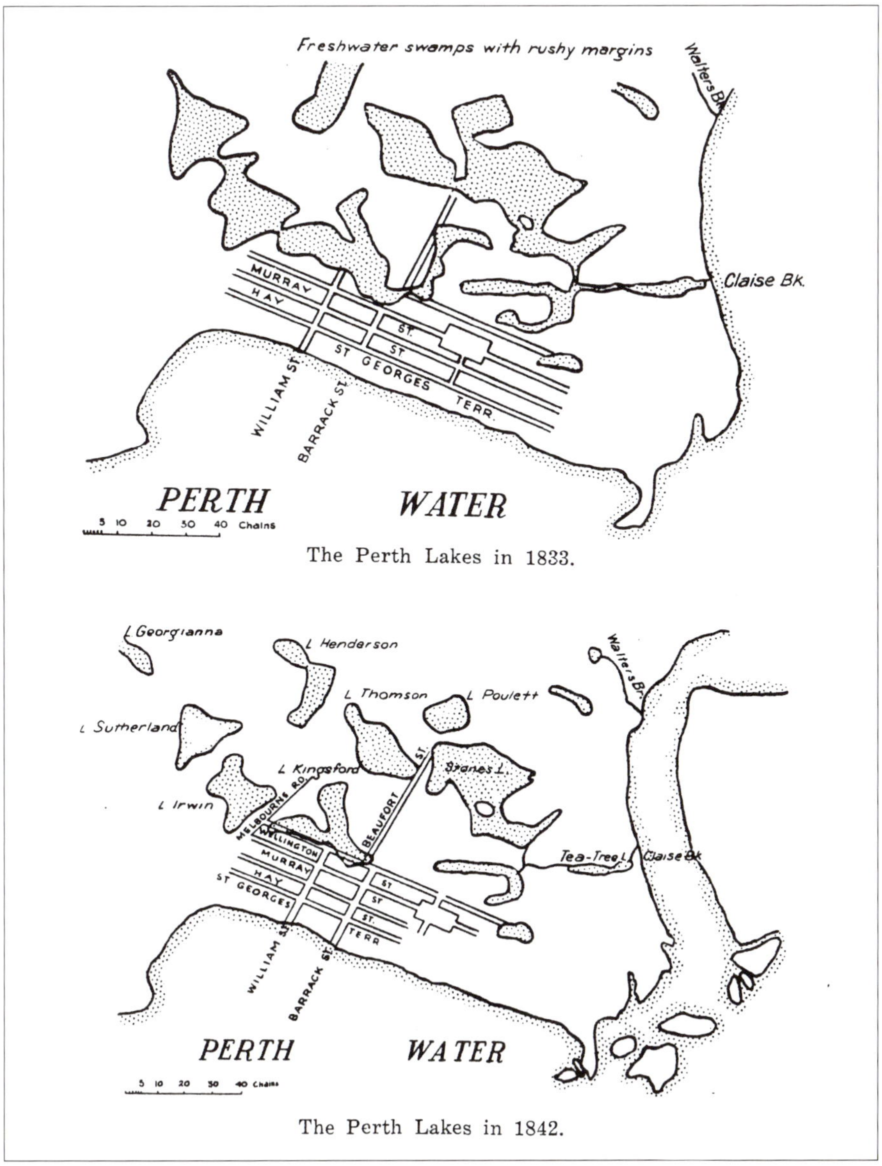

MAP 2.8 *The Perth Lakes 1833 and 1842 before the area was drained and filled. The north-eastern lakes drained through to Claise Brook, one of the areas described so picturesquely by Frazer and now the site of the East Perth development. Lakes Kingsford, Irwin, Sutherland and Henderson were used to work a mill marked today by Mill Street. The eastern end of St Georges Terrace becomes Adelaide Terrace and links with the Causeway across the islands (now Heirisson Island). The lake system also extended westwards to Lake Monger and Lake Herdsman (originally the Great Lake). A drain completed in 1925 currently takes water from Lake Herdsman, now a compensating basin for an area stretching 10 km to the north, to the Indian Ocean at Floreat.* Modified from Serventy 1948, based on the maps of Arrowsmith 1833 and 1842. Courtesy of Emu, journal of Birds Australia.

FIG. 2.8 *View of South Perth from Kings Park, 1900–1910. Note the low-lying land, the long jetties across the shallow sand banks and the 'mill pond' at the end of Labouchere Road. In response to flooding the area was extensively filled and walled, and is now covered with a freeway and high rise apartments.*
Courtesy Battye Library, Catalogue # 9366P.

Floods—damage and response

Chronicled since the early days of settlement, flooding following heavy rain increased as development spread. Drains were unable to cope with sudden heavy showers, or did not function on the flat landscape when water levels downstream were high due to earlier run-off, tides and onshore westerly winds (Fig. 2.8).

The floods of 1926 were particularly spectacular and memorable. Floodwaters spread over five kilometres wide at Guildford, and covered large areas of the Perth Esplanade and South Perth, where the residents boated down Labouchere Road or were transported across to the ferry depot in horse drawn drays. The rush of water demolished the North Fremantle Railway Bridge and 12,729 million litres cascaded over Mundaring Weir.

Although periodic floods still damage buildings and roads, they are less extensive as dams on the Helena and Canning Rivers now capture much of the run-off. Deeper channels dredged at the Causeway and in the Fremantle area, and river training in the Avon catchment, allow water to drain to the sea more quickly although this is still limited by low gradients on the coastal plain. Total run-off has, however, increased with the higher watertable after clearing.

Although devastating, floods also rejuvenate. The fertile soils that so attracted the early settlers around the upper Swan and Canning are alluvial, deposited by floods. Accumulations of silt and organic material are swept away and redeposited where they fuel new growth, and the freshwater conditions often extend growing and breeding seasons for animals into the summer months.

By 1966, more than 50 per cent of all wetlands between Yanchep and Rockingham, 31 per cent of those between Rockingham and Harvey, and 96 per cent between Harvey and Dunsborough had been drained. In 1991 it was suggested that the total might be as high as 70–80 per cent for the whole coastal plain. Where lake areas were not drained, the surrounding areas were cleared leaving little of the marginal vegetation for wildlife. In summer as the waters dry areas of wetland are smaller still, and become the foci of waterbirds from a much wider area that extends to the land over the ranges. The future for the remaining wetlands is poor as many landholders are opposed to plans for conservation rather than development.

River transport and river changes

As the population grew and agriculture spread across the land, the estuary too was modified. The extensive shallows hampered river traffic and the rocky bar at Fremantle limited access between the port and city. The port, hardly a port, was exposed to the full force of the Indian Ocean.

Access to the sea

Provision of safe anchorage became a greater problem as the number of trading ships increased. With the mouth of the river blocked by a rocky bar 1.8 to 2.5 metres deep and 15 metres across, vessels anchored in Gage Roads, the safer refuges of Owen Anchorage and Cockburn Sound only accessed over the treacherous shallow areas of Success and Parmelia Banks (Fig. 2.9 and Map 2.9). While early attempts to remove the bar had failed, the first in 1849 due to a lack of blasting powder, the innovative scheme of C. Y. O'Connor, Engineer-in-Chief, solved the problems of access and harbour anchorage. The bar was removed and the harbour created by building stone breakwaters (north and south moles). Opened in 1896, the original 130 metres long and 9 metres deep channel was increased to 11 metres between 1914 and 1922, and Fremantle remained the only port until 1956 when large ships started using Cockburn Sound. The estuary was now permanently open to the sea, floodwaters could flow outwards more quickly and salty water could move upstream relatively unimpeded, although the Fremantle Traffic Bridge built in 1866 still acted as a weir, this final obstacle was removed in 1968 (Figs 2.9 and 2.10).

Too shallow—deep channels

The sandy shallows that had hampered the progress of the first visitors were an even greater hindrance to travel for the colonists, and by 1831 a channel was cut through the banks of Heirisson Island and across the neck of Burswood Peninsula. In 1844 another channel was cut through the Point Walter Spit to shorten the distance between city and port, and a toll was charged for its use. Dug close to shore to a depth of about 1.3 metres, which was the draught of a ferry and a depth limited by the underlying limestone, the channel required regular dredging and was finally abandoned. Little evidence of the channel remains, although it is shown on maps drawn in 1896.

The shallow water was also a problem on the Canning River. In the early days only flat-bottomed barges could move upstream of Mount Henry due

FIG. 2.9 *View of Fremantle, Western Australia, from Church Hill east of the Town Hall, Richard Morrell, 1832. This early view of the river mouth shows small waves breaking on the bar across the entrance. Access to the river was originally limited by the bar and sandbanks, which were removed during harbour construction. The headland and 'Round House' gives a reference point to the location of low-lying swampy ground covered now by the main streets of Fremantle.* Courtesy AGWA 1940/00W2, Art Gallery of Western Australia.

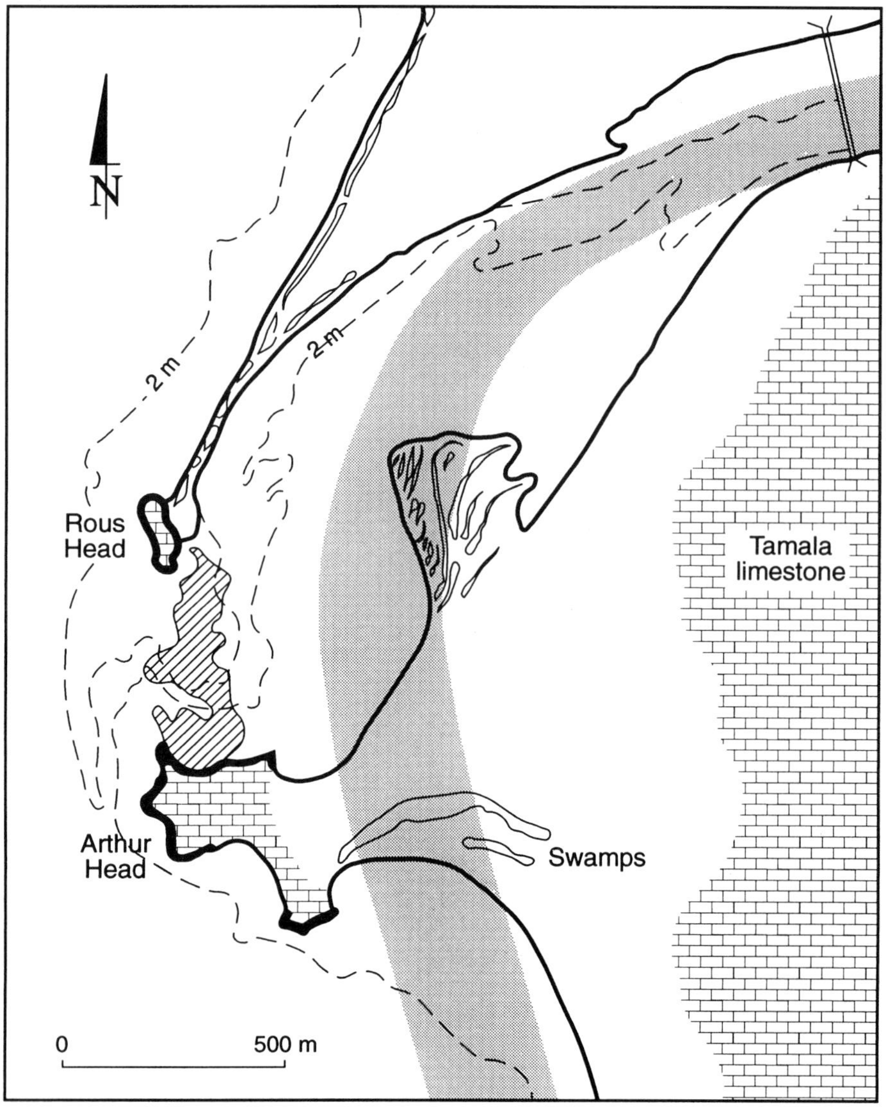

MAP 2.9 *The entrance to the Swan Estuary drawn from charts of c. 1833 and 1873, superimposed with the sea level coral limestone reef (hatched) and the approximate line (shadowed) of the river channel during the last glacial low sea level about 10,000 years ago. Later coastal sands accumulated in a flood-tide delta and a tombolo closed the channel and the entrance moved northwards. At the time of European exploration and settlement, the entrance between Rous Head and Arthur Head was barred by the reef and travellers had to negotiate a passage across the rocky bar and the shallow delta. The swamps and low-lying areas in the vicinity of the old channel behind Arthur Head created problems for the developing Port of Fremantle (Fig. 2.9). Prone to flooding, they were soon contaminated by urban waste.* Data from R. Gordon, Fig. 10; Adapted from Hodgkin & Hesp, 1998. Courtesy Marine and Freshwater Research, CSIRO Publishing, Victoria, Australia.

to snags, sandbanks and rocks. Logging in the jarrah forest on the Darling Range near Cannington during the 1860s increased the need for river transport. Landing docks were built and channels deepened. Initially the channels were dug by hand using convict labour, but after transportation ceased in 1869, the channels quickly silted up and the dredge *Black Swan* was used to keep the channels open. After the timber industry closed at the end of the century the channels silted up again.

Dredging of river channels and the reclamation of the foreshore using dredge spoil became a feature of the Estuary (Table 2.1 and Fig. 2.11). Grab dredges were used in Canning River and Perth Water around Barrack Street, Mends Street, Coode Street and upriver through the Causeway. Downstream the bucket dredge *Parmelia* and suction dredges *Premier* and *Governor* were used between 1892 and 1896 to create Fremantle Harbour. The spoil was used to create Victoria and North Quays, and the rock dumped on Hall Bank opposite Leighton Beach.

In 1925 the Swan River Improvement Act provided money to improve river flow, eliminate the mosquito problem and canalise the river. Reclamation of Burswood Island began in the late 1930s using slag, dredging material, fly-ash from the nearby power station and demolition refuse. The new dredge *Stirling* was purchased in 1928. Between 1929 and 1950 the *Stirling* and later the *Throsby* worked almost continuously, except for times during the Great Depression and World War II. In 1939 Millers Pool at Mill Point, South

FIG. 2.10 *Fremantle Harbour, aerial view, March–April 2002. See also Map 2.9. This a wild sandy coast exposed to oceanic swells offering little shelter to ships that anchored originally to the south of Arthur Head around the three small harbours used by the fishing industry and private boats. River access was improved by blasting the rocky bar, dredging sands from the river delta, and constructing breakwaters–moles to provide sheltered waters. More recently the northern shoreline has been extended seawards for additional port structures. This image is part of a series of high quality photographs of benthic habitats along the coastline from Quinns Rock to Singleton, and has been georeferenced to GDA94.* This section courtesy DALSE.

FIG. 2.11 *Contrasting views of Perth from Kings Park. (a) An extensive view of Perth, water colour and pencil with body colour over pencil by George Nash, 1851.* Courtesy Art Gallery of Western Australia 1992/0248. *Jetties, wharfs and warehouses originally built across the shallow margins of the estuary to service the river traffic gradually fell into disuse as the roads and tracks improved. Later shallow areas were filled with dredge spoil or rubbish to create easy access to deep water. (b) In the 1970s there was more infilling for roadways leading to the Narrows Bridge and freeway systems shown in the 2003 view.* Courtesy T. Alford, UWA.

Perth (Fig. 2.8) and the Nedlands Foreshore were filled. Between 1954 and 1959 the area for the Narrows Interchange and Kwinana Freeway were filled (Mounts Bay Road 28 hectares, Mill Point to Canning Bridge 17 hectares) with another 4 hectares around the Applecross foreshore. In 1960–1967 the Mitchell Freeway brought more infilling. Between 1960 and 1964 the Canning River area was dredged and the spoil bought by the Shire for use as landfill for urban subdivision.

In the early to mid 1960s the dredges also worked on programs to improve yacht clubs and other recreational facilities at Preston Point, Point Walter, the foreshore from Point Walter to Point Waylen, at North Fremantle and the British Empire and Commonwealth Games rowing course near the Narrows. Some of these areas were also used for rubbish disposal, which has left a legacy of contaminants (Table 2.2).

In the late 1960s the emphasis moved upstream past the Causeway 'to clean out' the river and alleviate flooding. In 1968 the area between Bunbury Bridge and East Street, Maylands was dredged to create deeper water for yachting. Between 1968 and 1973 dredging upstream of Garratt Road Bridge was used to create recreational ground on the left bank, a navigational channel and Garvey Park Island. Sand removed from the mouth of the Helena River was used to create playing fields. Work was also done at Belmont Park area (1968) and at Maylands Peninsula (1971–1973) where 14 hectares of the left bank was removed.

In the 1980s and 1990s dredging occurred around developments: in the Upper Estuary for Ascot Waters, East Perth Claisebrook Cove, and Barrack Square, and in the Lower Estuary for Pier 21. There was also maintenance dredging for yacht clubs, Royal Perth, Royal Freshwater Bay, Perth Flying Squadron, Nedlands, Swan, South of Perth, Claremont and East Fremantle, and this continues today.

TABLE 2.1 *Summary of dredging on the Swan–Canning Estuary 1892–1976, based on files Public Works Department, Harbours and Rivers Branch.* Adapted from Riggert, 1978.

Category	Area of dredging (m^3)	Period of activity
Navigation	11,651,583	1892–1976 Barrack St and South Perth area
Construction	8,488,337	1922–1976 Narrows interchange
Flood control and foreshore protection	6,773,544	1946–1973 Upstream of the Causeway
Mosquito control	3,636,929	1935–1964 Canning River and Rivervale
Recreation	2,424,541	1950–1976 Rowing course on Canning
TOTAL	32,974,934	

TABLE 2.2 *Areas of foreshore and wetlands used for landfill and rubbish disposal in the 1950s–1970s. About 80% of the low-lying wetland areas in the metropolitan area were also used for the same purpose.* Adapted from Riggert, 1978.

Authority	Location	Duration of use	Area (ha)
City of Stirling	Deeley St, Maylands	1960s	1.2
City of Stirling	Queen St, Maylands	1961–1963	2.4
Shire of Bayswater	Slade St, Bayswater	1960s–1973	18.2
City of Perth	Rivervale	1952–1972	101.0
City of Perth	Trinity College, East Perth	1972	1.5
City of South Perth	Swanview Tce, South Perth	1959–1969	32.4
City of Melville	Canning Highway, Attadale	1952–1969	60.7
Shire of Belmont	Stoneham St, Belmont	1952–1973?	8.0
Town of Canning	Wendouree St, Wilson	1962–1973?	14.2
Town of Canning	Kent St, Cannington	1956–1959	0.8
Shire of Swan Guildford	Third Ave, West Midland	June 1958–1973?	8.0
Town of Mosman Park	Queen St, Mosman Park	1951–1973?	0.8
TOTAL			249.2

The end of gently sloping sandy banks

Dredging removed the shallow sandbanks creating deeper channels, while the remaining shallows overhung with trees and bounded by rushes were gradually overrun with weeds such as bamboo. Walling of the river before World War II made harsh division between land and water, replacing the sandy beaches with little consideration for wind and waves. Along the Nedlands and South Perth foreshores, the large areas that were walled during the 1930s were gradually extended with the establishment of roads, freeways and playing fields. Today there is some reversal of policy with efforts to stabilise remaining areas without walls, introduce a subtle outline and restore communities of fringing vegetation.

THE SWAN—AS IT IS

Fringing vegetation

Only a few pockets of the vegetation around the Swan–Canning that appealed so favourably to Stirling and Frazer in 1827—a landscape of salt marsh, rushes and fringing forest—remains. (See **Chapter 3 The Peel–Harvey Estuarine System: Plants on the edge; A different sort of ecosystem, Around the edges**.) Now fragmented and disturbed, the remaining vegetation is increasingly vulnerable to invasion by exotic weeds (Map 2.10, Figs 2.12 and 2.13).

Fringing vegetation provides habitat not only for waterbirds but also for land birds such as rails, crakes, warblers and grassbirds which nest in the rushes, shrubs and trees and feed on seeds or small animals. Other birds such as ibis, egrets and herons feed on the adjoining mudflats, and use the fringing vegetation for nesting and roosting.

The fringing vegetation also protects the estuary margins: the dense root mat binds sediment, while the leaves and stems reduce water flow and the erosive power of river flow and boat wash. Acting as a baffle to inflowing water, fringing vegetation reduces erosion and allows sediments to accumulate, slowing the entry of nutrients and other pollutants such as heavy metals, hydrocarbons and pesticides. Although most effective with gradual flows rather than the sudden flushes associated with storm run-off, sedge-like vegetation has been shown to remove as much as ninety per cent of nitrogen and phosphorus entering a waterway.

Costs to the estuary due to lost fringing vegetation are difficult to estimate. Habitats at the interface of land and water are gone. The filter between the land and water for sediment and nutrients means food chains are disrupted, with a proliferation of algal blooms due to the added nutrients. Toxic pollutants are also a potential health hazard to the food chains. In addition, the plant community itself has an intrinsic value.

The slow moving waters that are such a wonderful habitat for birds and fish are, however, often smelly and home to mosquitoes. This is undesirable for nearby urban areas and leads to conflict of interest, expressed as:

> *Between Riverton Bridge and Nicholson Road Bridge there are some 750 acres [300 hectares] of low lying, swampy, mosquito-breeding areas that will require a lot of effort and finance to improve before scenic roads and river beautification can take place. Until then that section of the river is hardly likely to be used for any purpose.*
>
> Cardon, 1968

The value of low-lying areas and fringing vegetation to the estuary is now more widely known, and state and local government agencies along with community groups are undertaking restoration and landscaping projects to reduce bank erosion, minimise nutrients entering the riverine and estuarine waters, and restore natural aesthetic values.

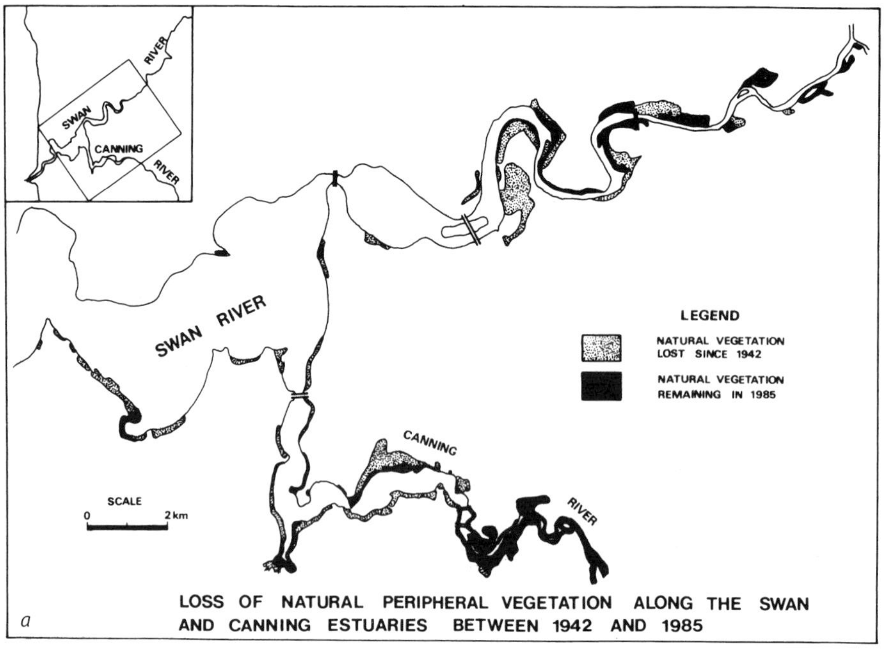

MAP 2.10 *(a) Loss of the natural fringing vegetation along the Swan–Canning Estuary 1942 and 1985. Although some of the fringing vegetation was cleared before the 1940s, clearing continued after World War II. Details of change in vegetation between 1942–1945 and 1980–1981 at (b) Maylands and (c) Alfred Cove. At Maylands salt marsh replaced the paperbark and rushes (*Melaleuca–Juncus *complex) growing in an area flushed by fresh water after drains were dug channelling flow directly to the river allowing the incursion of salt water. At Alfred Cove stormwater reduced salinity favouring colonisation of the salt marsh by rushes and areas of saltpan by samphires.* Modified from Pen, 1987.

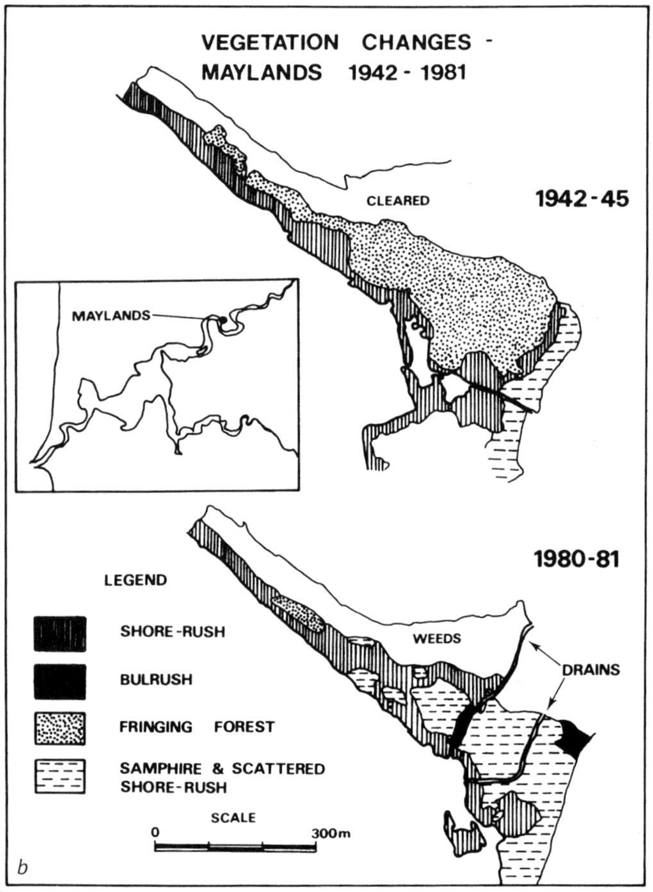

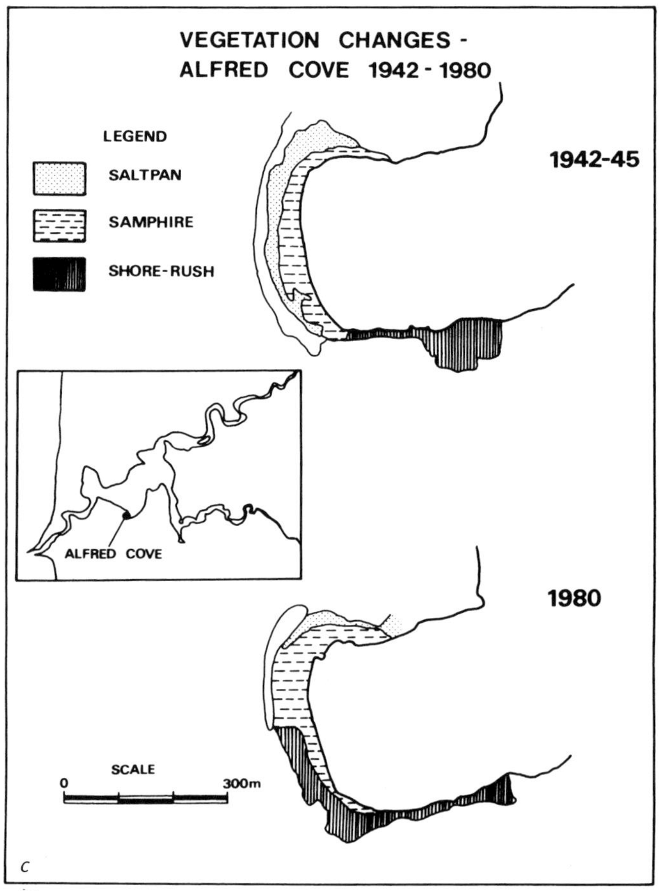

FIG. 2.12 *Alfred Cove, aerial views showing the development of a city, (a) 1945 and (b) 1988. Much of the fringing vegetation in the Swan–Canning Estuary has been removed and filled with dredge spoil. This is only remaining area of largely unmodified sand and shell beds of Middle Holocene age (6,000–4,000 BP), a catalogue of the estuary's evolution. Note in 1945 the area was sparsely populated and bordered by extensive sand flats with Burke Drive as a reference mark. By 1988 the suburbs of Melville and Attadale have grown and the area along the river had been filled with dredge spoil.* Courtesy DLI.

FIG. 2.13 *Narrow band of fringing vegetation: aesthetically pleasing, baffle for erosion, filter for water run-off and habitat for water and bird life. In some areas digging for worms for use as bait, a practice that is now discouraged, is a threat to remaining fringing vegetation. Holes adjacent to the rushes or in the root mat increase the risk of washaways and undermining of the whole fringing community.* Courtesy SRT.

The waters—flowing river or salty bay

The extreme seasonality of river flow causes great differences in the water regime of the estuary between summer and winter (Fig. 2.14). Rainfall in winter brings fresh water that flows outwards to the sea. At the beginning of winter or when run-off is low, the fresh water flows seawards over the more dense salty water and the water column is stratified or layered. Oxygen concentrations are low in the vicinity of the Narrows Bridge. In the main water body, winds gradually mix the layers of fresh and salt and a layer of salty water persists only in the deepest parts, about 5 metres, below the level of the rocky sill at Fremantle. In years of high run-off, the whole estuary can be flushed with fresh water and the plume of coloured floodwaters can be seen stretching way out to sea as far as Rottnest Island (Fig. 1.8). Episodes of complete flushing have incredible effects on the water life, and species that can usually tolerate lower salinities are 'killed off', which resets the biological clock when more marine animals recolonise. This run-off, particularly from the Avon, also contains high concentrations of soluble nutrients particularly nitrate, but much of this is flushed to sea. Soluble phosphorus also enters from Ellens Brook but is diluted by the larger turbid flow coming down the Avon, and much of it binds with the sediment and settles to the bottom.

As the run-off slows in spring, tidal forces become more obvious gradually pushing or pumping dense salt water, the salt wedge, upstream along the bottom, while the outward lighter flow continues. As the flow diminishes wind mixes the two water bodies; the flow progresses further upstream gradually replacing all the fresh water until the summer saline state is established. In summer, the salinity of the Lower Estuary to about Perth Water is similar to sea water (35‰), decreasing gradually towards the head of the estuary in the upper Swan (Graph 2.3). Unusual summer rains like those of January 2000 disrupt this salinity pattern, creating a freshwater summer state. During the usual summer state, the gradual progression of the salt wedge upstream also brings a change in the location of water low in oxygen, and in summer this is usually confined to the upper reaches. (See **Box 2.3 Dividing up the Swan Estuary**.)

The temperature of the waters also follows the seasons. In the Lower Estuary surface temperatures are 12°–14°C in winter and 22°–24°C in summer. Temperatures also fluctuate more in the shallow areas and, like salinity, the variations have large implications for water life. The rise in temperature in spring and sunlight create favourable conditions for phytoplankton growth (indicated by a rise in chlorophyll *a* concentrations), fuelled by the nutrients brought by winter flow. The slower river flow also reduces flushing to the

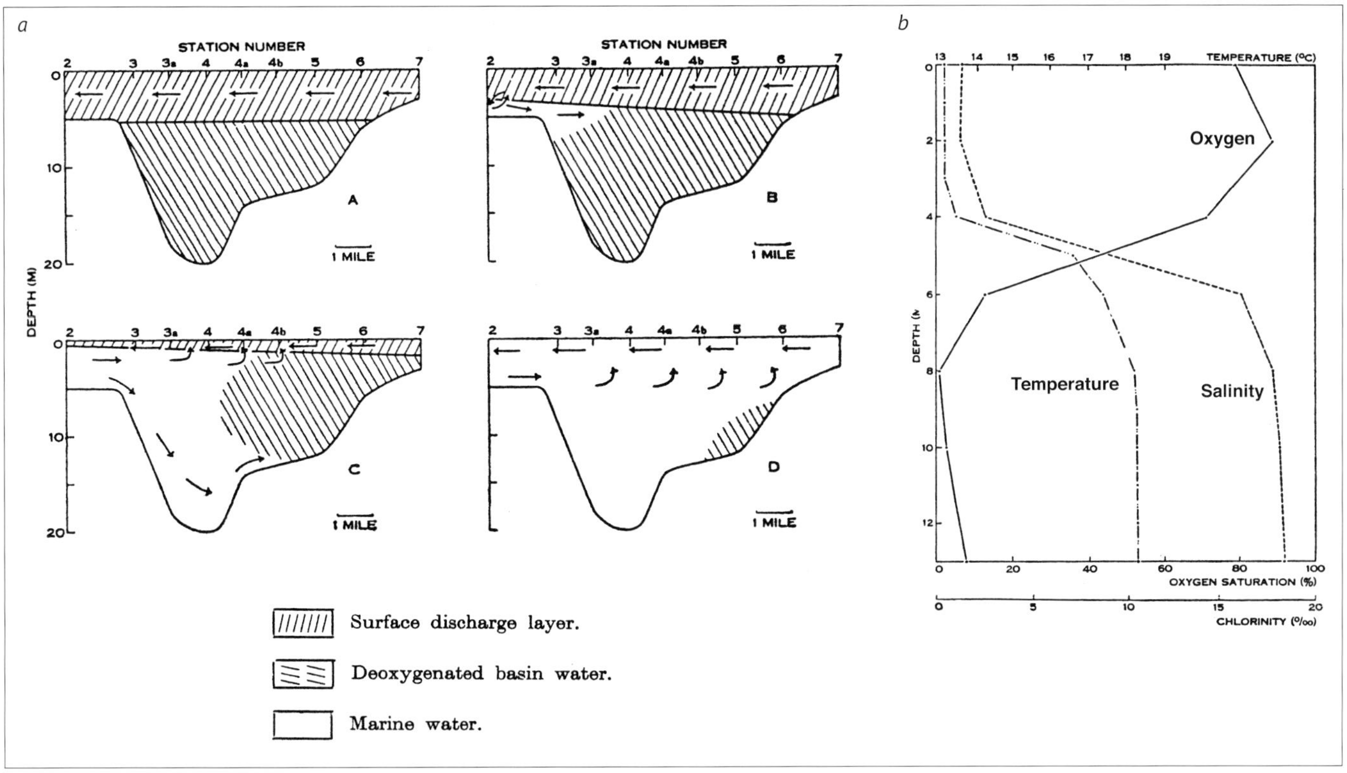

FIG. 2.14 *(a) The dynamic seasonal cycle of salinity in the Swan estuary first documented by Spencer in 1956. Station numbers from the left, downstream to the upper reaches: (A) winter freshwater flow over the saline bottom water; (B) as the freshwater flow diminishes, saline water starts to intrude; (C) ocean water penetrates further upstream; (D) saline marine conditions in summer. (b) Temperature of the deeper water does not fluctuate as much as the surface water, which warms quickly as the temperatures rise in spring.* Adapted from Hodgkin, 1987.

ocean so that blooms have time to develop. Spring is the time of a succession of blooms; as the cells in one bloom die, they settle to the bottom, decay and the nutrients released fuel another bloom.

The summer state is also ideal for phytoplankton blooms particularly in the Upper Estuary. With progression of the salt wedge into the upper estuary, water low in oxygen has also moved upstream. Although the stratification is not as strong and the waters become brackish, there is sufficient isolation to maintain low oxygen concentrations favouring release of ammonium and phosphate that fuel blooms of algae, particularly species that can migrate between the nutrient-rich bottom and sunlight at the surface. Phytoplankton growth is also rapid in the salty Lower Estuary, dependant on the cycling of nutrients but generally limited unless there are sudden additions of nutrients with summer floods.

The change from summer to autumn throughout Swanland varies greatly; rainfall may be early or late and heavy or light. By now nitrate concentrations are low, but ammonium and phosphate may be high in areas of low oxygen. Sudden freshwater flows bring the nutrients that have accumulated over summer, and if the weather is still warm and sunny, conditions are perfect for algal blooms. Conversely, with a later onset of rain, conditions may be less favourable and nutrients may be swept to sea before phytoplankton respond.

Seasonal changes in salinity in the lower Canning River are similar to those in the Swan River, but the Kent Street Weir limits the extent of marine influence. In winter, the upper boards of the weir are removed to reduce flooding and allow river flow downstream. In summer, the boards are replaced to retain fresh water for longer and prevent salt water moving upstream. As a result the upper reaches are more lake-like in behaviour, and any stratification is due to temperature rather than salinity.

Layers of water

In winter to spring (July to October) the estuary waters are termed 'stratified'. The surface waters are fresh, and the deepest water is salt and below the influence of usual winter flushing and blowing winds. The often abrupt difference between the two layers is termed the 'halocline'. This stratified condition means that the bottom waters are isolated from the surface. Little light penetrates the dark floodwaters, plants cannot grow, no oxygen is produced, and the bottom layer becomes deoxygenated, which is an unfavourable condition for many animals. Occasionally the deep deoxygenated water moves to the surface, resulting in fish deaths.

The best documented 'overturn event' of this kind occurred in 1955, when unseasonal summer rain in February created a fresh layer over the salty water, sealing it off from the surface. In April 2001 upwelling of deoxygenated water caused a major fish kill like those that occur in the southern estuaries. (See **Chapter 10 Estuaries to the east of Albany: Beaufort Inlet: Portrait of a fish kill** and **Chapter 4 Leschenault Estuary and the smaller estuaries of the Geographe Bay area: Vasse–Wonnerup wetland system: Dead fish—symptoms of a distressed environment**.)

Similar overturn events also occur in shallow areas of the Upper Estuary where rain in autumn and early winter creates a cold fresh layer over the warm brackish water. Termed 'meromixis', it is an unusual state as cold water is usually heavier and therefore at the bottom, but this is a common condition in water bodies throughout Swanland.

Water movements—tides, currents and winds

In the typical west coast pattern, changes in water level in the Swan–Canning Estuary are quite striking despite the relatively small astronomical tides. These include changes due to atmospheric pressure on a scale of days, ocean

Box 2.3 DIVIDING UP THE SWAN ESTUARY

Different parts of the Swan Estuary have different physical characteristics (depth, salinity), nutrient sources, plants and animals, and biological processes. Hence problems vary in the different areas, so the areas need to be managed accordingly.

Early workers, such as Spencer interested in hydrology in 1956 and Wilson in bivalve mussels in the 1960s, used the terms Upper and Lower Estuary with Heirisson Island as the defining barrier. In the 1970s a study of estuarine fauna defined the estuary into three zones: Lower, Middle, Upper.

The Upper Estuary corresponded to that of earlier workers: above Heirisson Island, a tidal river with meandering channel and fluvial aspect including the Canning River to the Riverton Weir. It is chiefly alluvial sediment but with some quartz sands from old dunes, and increasingly muddy downstream. Water depth varies from 6 metres at Maylands to 2 metres at Middle Swan with some deeper holes. There are fresh water inflows in winter with a gradual increase in salinity as river flow ceases and salt water penetrates upstream.

The Middle Estuary basin area ranges from Heirisson Island to Blackwall Reach with lower sections of the Canning River to Salter Point. It is a drowned river valley with extensive shallows in Perth Water and a deep basin and sand flats in Melville Water. There are banks of quartz sands deposited during the Pleistocene and some deposits of carbonate coastal limestone. Shallow sandy areas are covered with the seagrass *Halophila* while deeper areas have accumulations of fine grey mud. River flows are less significant with marine waters persisting for longer.

The Lower Estuary ranges from Blackwall Reach to Fremantle Harbour with a narrow winding channel incised through the coastal limestone. The channel is 16 metres deep near Blackwall Reach but less than 5 metres deep near the bridges. In the channel are sediments of fine grey mud. Shallow sand flats downstream are covered with seagrasses *Halophila* and *Zostera*. The Lower Estuary is only affected by fresh water when river flow is strong or prolonged. Marine water enters when the river flow is low so salinity is higher and stable for longer periods. Flora and fauna are typically marine although not as diverse.

Studies of macroalgae and fish have also revealed distinct communities in different parts of the estuary. Similarly, the zones are also useful when considering nutrient additions, chemical processes and algal blooms; however the area around The Narrows and Perth Water is considered a major barrier and water quality is often defined in terms of the U. (Upper) and L. (Lower) Estuary. The nuisance algae generally bloom in response to nutrient-rich freshwater flows in the Upper Estuary, moving downstream if the blooms persist.

It is interesting that James Stirling in 1827 defined the Heirisson Island area, at the time much shallower, as the upstream extent of the estuary.

currents on an annual scale, and continental shelf waves on an inter-annual scale. Winds also generate movement, moving water along the estuary and stirring the water column. At any given time water level results from a combination of all these factors.

Tidal changes, now much greater since the rocky bar and sandbanks were removed, bring marine water into the estuary. After the winter freshwater flow diminishes, dense salt water moves along the bottom as a wedge, the salt wedge under the fresher lighter layer. In winter, the salt wedge is usually situated 5–20 kilometres upstream from the ocean depending on the strength of the river flow. As the outward flow decreases still further, the saline layer moves further upstream, about 50 kilometres by the end of summer. Although the water column is well mixed in some areas, the Upper Estuary is stratified with little vertical mixing and the bottom waters become deoxygenated; the low oxygen conditions leading to release of nutrients.

Non-tidal influences due to atmospheric pressure have a large effect on the Swan. Low pressure allows water to rise to greater levels, and high pressure depresses water levels still further. These longer term effects, usually in cycles of five to ten days, also affect salinity when low pressure associated with a winter rain-bearing front causes a 'backup' of freshwater run-off. The subsequent rise in pressure as the front passes allows the water level to fall rapidly, flushing out to sea, often with a greater effect than the astronomical tide. In this situation the change from summer salt to the winter fresh condition with a drop of 20‰ or more may only take a few days (Fig. 2.15).

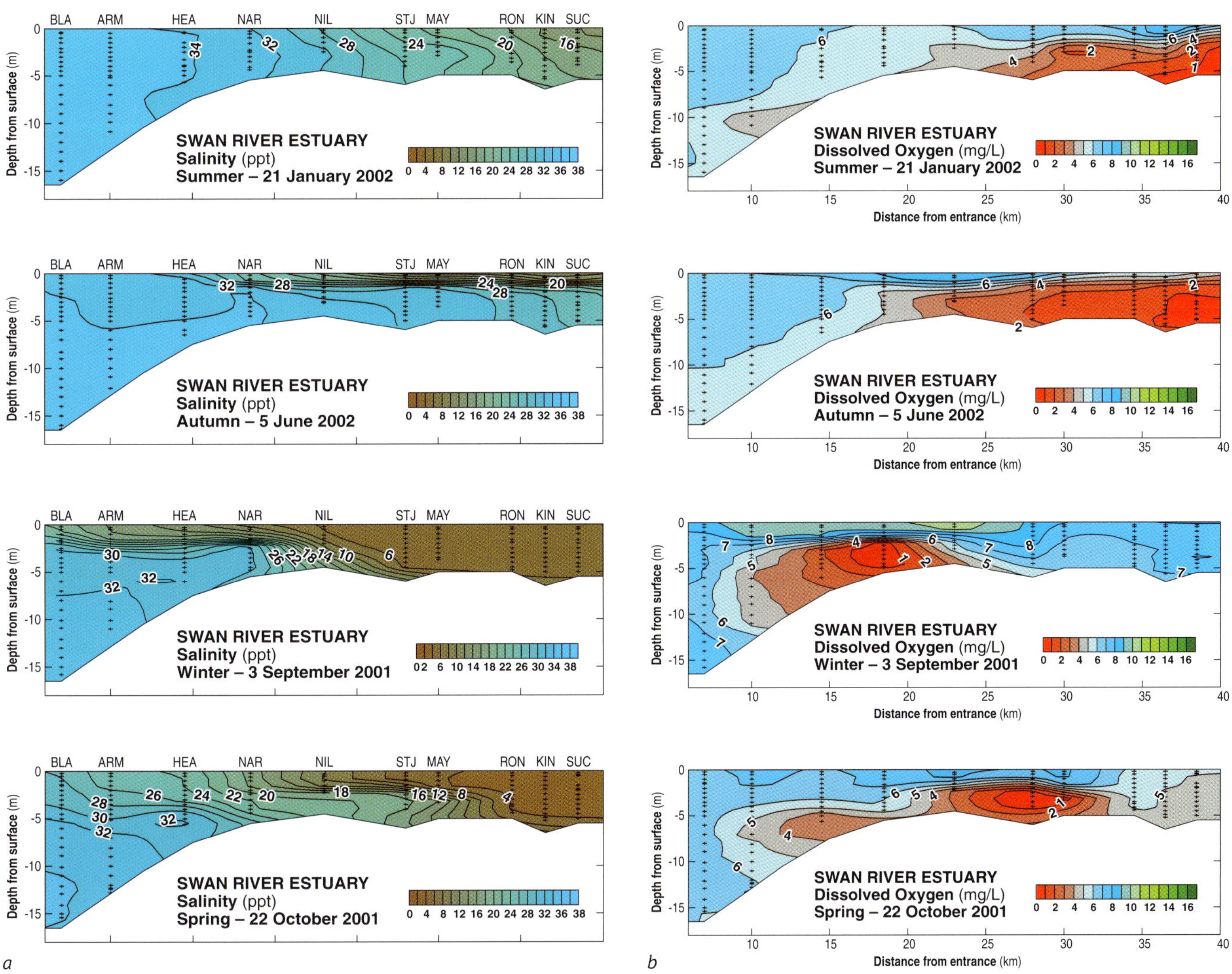

GRAPH 2.3 *Profile of (a) salinity and (b) oxygen concentrations in the Swan River Estuary throughout a yearly cycle: downstream areas on the left and upstream areas on the right of each graph. Water in the upper reaches is not completely fresh; water from the Avon is salt about 5‰ although some of the other tributaries are less salt at about 0.5‰. Note the progression of the salt wedge and low oxygen water upstream as freshwater flow declines in spring. In summer the lower reaches near Blackwall Reach contain sea water, the salinity of the water decreasing further upstream. Suburbs adjacent to sampling stations indicated on graphs: BLA Bicton, ARM Dalkeith, HEA Heathcote, NAR Narrows Bridge, NIL East Perth, STJ Rivervale, MAY Maylands, RON Ashfield, KIN Bassendean, SUC Guildford.* Source: *River Science 8*, WRC 2002a, Courtesy WRC.

FIG. 2.15 *(a) High water levels in the Swan, July 2001.* Courtesy West Australian Newspapers Ltd. *(b) Matilda Bay, May 2003.* A. Brearley. *A combination of river flow, high tides, low pressure, onshore winds and swells brings flooding to the flat landscape.*

Water levels in the estuary are also affected by global weather cycles. In winter, the Leeuwin Current present along the Continental Shelf brings less salty and warmer water from the tropics, and raises water levels by ten to twenty centimetres (Map 1.9 and Fig. 1.4). This can be either in synchrony or antagonistic with astronomical tides and air pressure. In summer, surges or continental shelf waves generated by tropical summer cyclones off the north-west coast, moving down the coast at a rate of 500 to 600 kilometres a day, can also influence water levels in the estuary over intervals of about two weeks. Typically sea level falls as a cyclone crosses the coast, then rises before declining again. The effect of *Rosita*, a Category 4 Cyclone, took about ten days to reach Perth in April 2000.

On a longer time scale, water levels fluctuate with the formation and melting of the polar ice sheets. Higher water levels due to global warming, regardless of whether this is induced by humans or part of a more natural cycle, are of great importance to estuaries particularly where they are surrounded by urban developments on low-lying land like those around the Swan–Canning. Much of this land is at or below the current Australian Height Datum (AHD), which is mean sea level (zero) based on thirty tide gauges around the county in 1966 and 1968. (See **Chapter 1 Understanding Swanland: Water movements—the tides and currents of Swanland.**)

With the water flow—nutrients

Run-off from surrounding natural bushland generally contains only small quantities of nutrients. Organic material from native plants is generally nutrient poor and decomposes slowly. Changes in land use since the first days of settlement have altered this pattern so that today, nutrients such as phosphorus, nitrogen and carbon enter the Swan–Canning from urban, industrial and agricultural sources. The highest concentrations of dissolved phosphorus and nitrogen occur in the Ellen Brook, the Southern–Wungong River and the South Belmont and Mill Street main drains (Map 2.11). Soil-bound phosphorus appears to originate mostly from the Avon River catchment and the hill areas along the Darling Scarp. Estimates in the 1990s suggested that since the 1940s, 72.5 tonnes of phosphorus and 750 tonnes of nitrogen have entered the estuary each year from catchments of the coastal plain.

As in other estuaries on the Swan Coastal Plain, such as the Peel–Harvey, most of the phosphorus originates from superphosphate applied to farmland and enters the Swan dissolved in the run-off or attached to soil. While the water is oxygenated, the phosphorus accumulates in the sediment bound to metal ions of iron, manganese, aluminium and calcium. When oxygen concentrations are low, both nitrogen and phosphorus are released back into the water along the bottom of the river and estuary. The metal-bound

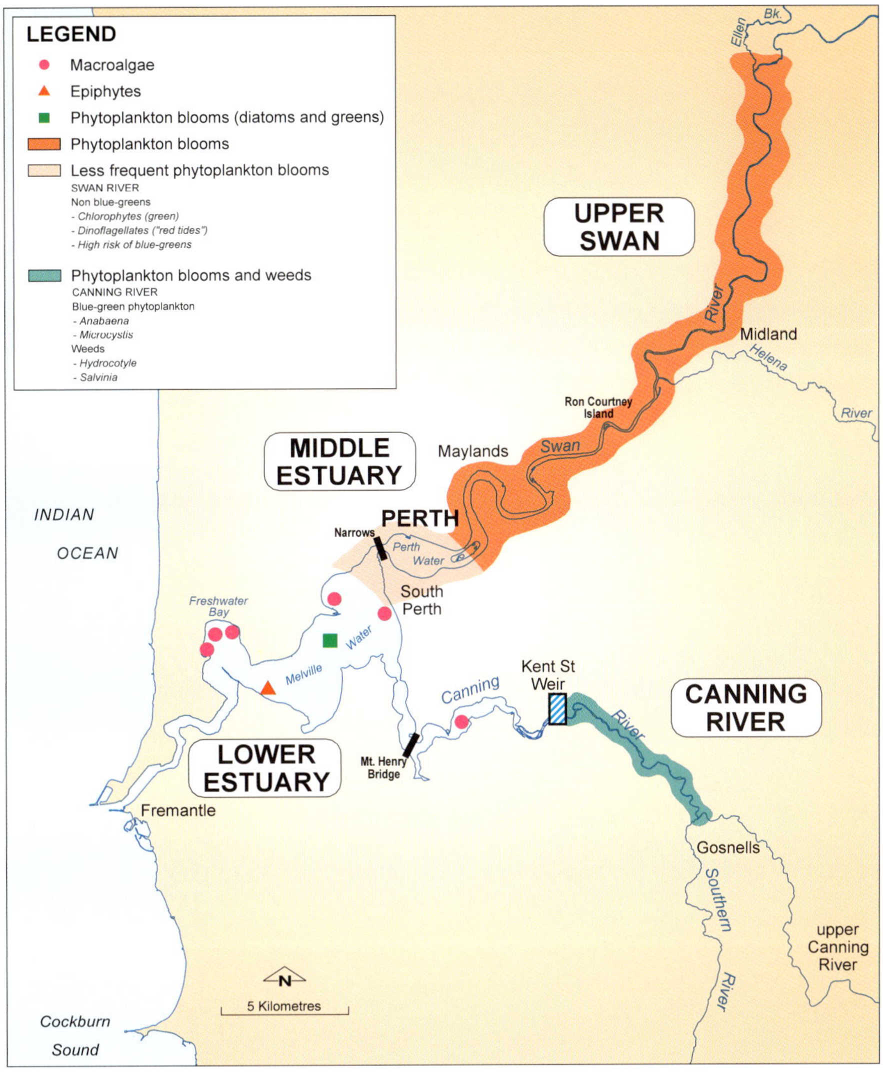

MAP 2.11 *Location of problem areas around the Swan–Canning River and estuarine system. The map shows the general water quality problems associated with each area. The annual flow from Ellen Brook is only about 10% of that from the Avon but it brings most of the dissolved phosphorus and high concentrations of nitrogen. Algal blooms are common in the Upper Swan but are considered nitrogen 'limited', however both nitrogen and phosphorus concentrations need to be reduced. Toxic blue-green blooms and outbreaks of aquatic weeds are common upstream of the Kent Street Weir in the freshwater section of the Canning River. The estuary areas of the upper Swan and lower Canning experience a range of algal blooms, but extensive blooms of toxic blue-green cyanobacteria were not recorded until January 2000.* Source: SCCP Action Plan, 1999, Courtesy SRT.

phosphorus becomes more soluble, while bacteria facilitate release of nitrogen through nitrate to nitrogen gas. Despite the large amounts of nitrogen entering the rivers and estuaries, nitrogen concentrations in the estuary water column are generally low, indicating that the denitrification process is occurring or that algae have been taking it up. In the Swan, algal growth in spring and autumn is considered nitrogen limited. (See also **Box 1.6 Terms used for water quality**.)

Disturbance of sediments by wind, waves and currents also releases nitrogen and phosphorus back into the water column. The now more salty river also stratifies the water column, isolating the bottom layers and encouraging development of low oxygen nutrient releasing conditions.

In the Upper Swan ground water flows also play an important role in the release of nutrients from the sediments. Ground water flows move nutrients from the spaces or pores between the sediments accounting for about ten per cent of nutrients, especially nitrogen, entering the water column. Pore waters are also disturbed as the salt wedge moves upstream.

The fuels for algal growth—nitrogen and phosphorus

Blooms of microalgae (phytoplankton) in the Swan–Canning Estuary develop in response to the amount of nitrogen available in spring, summer and early autumn. A large amount of dissolved nitrogen in the run-off fuels the spring algal bloom. The bloom converts inorganic nitrogen to organic nitrogen that accumulates in the sediment as the algal cells die. In the sediments, decomposition converts the organic nitrogen to ammonia and then to nitrate, which are then available for other algal blooms in summer. Dinoflagellates, which are motile algae that dominate the summer blooms, are able to move down into these nutrient-rich deeper waters at night to take up nutrients and up into the sunlight during the day to photosynthesise. This is a cycle that repeats through summer while conditions are suitable (Graph 2.4).

It is interesting to note that concentrations of nutrients in the Swan in the 1980s were similar to those in the eutrophic Peel–Harvey, but there were fewer blooms of nuisance algae in the Swan. The absence of similar algal blooms in the Swan Estuary was probably due to the variability of salinity over the year and along the length of the estuary, which provided conditions suitable for a variety of algae rather than proliferation of a single species. In addition, the saline state in summer and autumn is generally less favourable to nuisance species. This situation changed in 2000, when widespread floods in February brought nutrient-laden fresh water to the estuary, where the sheltered shallow, warm waters created a perfect environment for massive blue-green algal blooms in the hot conditions. Blue-green blooms have been annual events in the upper Canning River since 1994.

In the past larger macroalgae had caused problems, however in the early 1990s new problems emerged as the organisms causing the blooms changed. Blooms were mostly due to microalgae. Although the more obvious sources

GRAPH 2.4 *Surface and bottom concentrations of total nitrogen, total phosphorus and chlorophyll* a *between winter 1995 and spring 1997 were consistently high in the upper estuarine areas of the Swan River. Nitrogen concentrations are highest during periods of river flow, and when algae are abundant in late summer indicated by chlorophyll* a *concentrations. Total phosphorus in contrast is highest in late summer when the river is stratified and oxygen concentrations decline, which favours release of phosphorus in the sediment. In the Swan–Canning, deepening of river channels and removal of the shallow tidal flats have probably altered the capacity of the sediments to release nitrogen gas. The deeper areas are also more likely to become anoxic favouring phosphorus release.* Source: SCCP Action Plan 1999, Courtesy SRT.

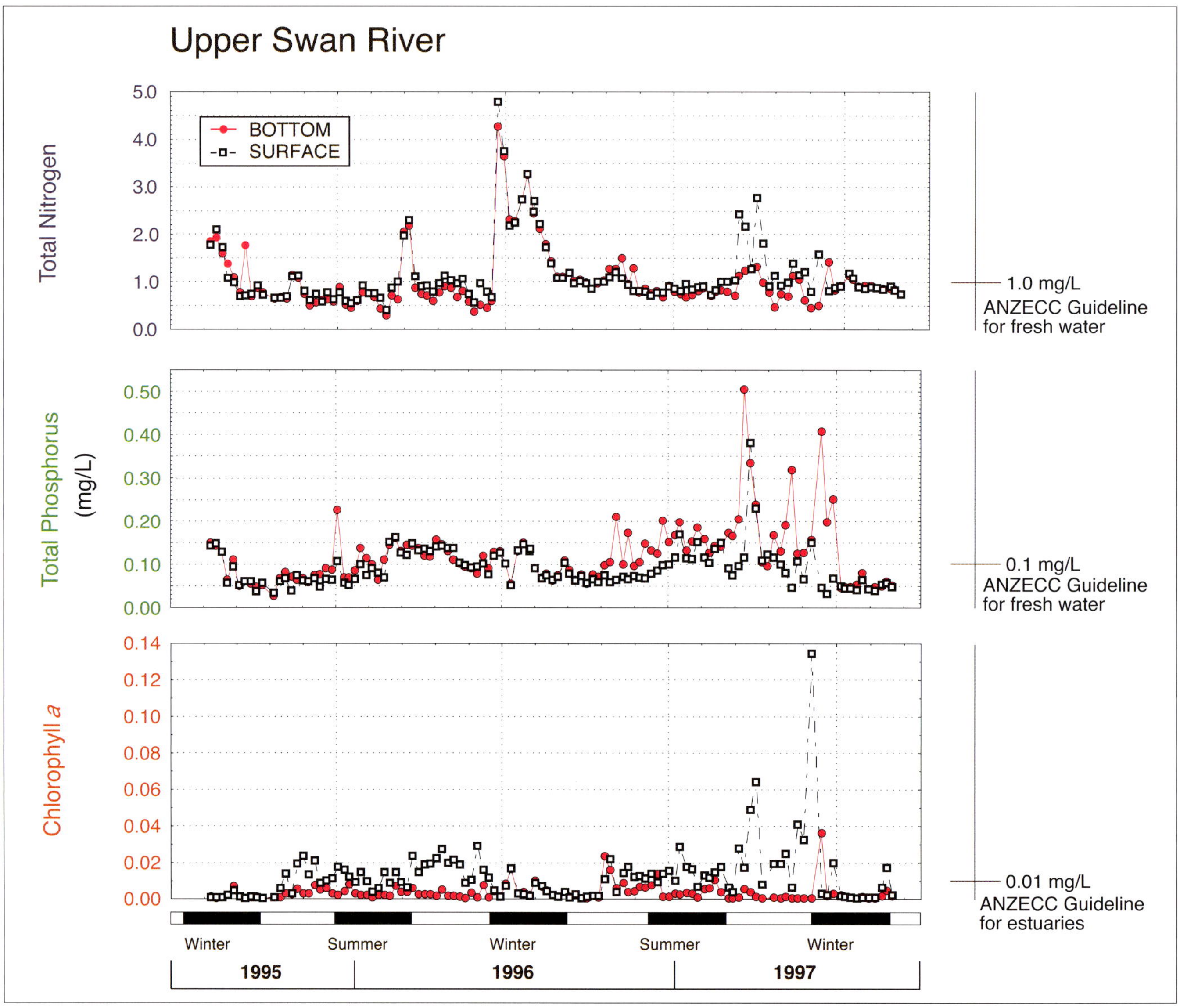

of pollution, untreated sewerage and industrial discharges, had been stopped nutrient concentrations were still high, and the eutrophic state had increased.

Australian waters tend to have low ratios of nitrogen to phosphorus so blue-green algae (cyanobacteria), including toxic species that can utilise or fix atmospheric nitrogen, often dominate. Though blue-green algal blooms have occurred in the Swan–Canning Estuary they are more common in the upper Canning River, many urban wetlands and in pools of the Avon River and Ellen Brook (Tables 2.3 and 2.4). The background presence of these blooms, in particular those in the Avon, were the inoculants that set the scene for the massive outbreak in the main estuarine water body in February 2000 (Figs 2.16–2.18, Graphs 2.5 and 2.6, Map 2.12). (See **Box 2.4 Tiny cells on the rampage take a city's playground off limits**.)

TABLE 2.3 *Blooms in the Swan–Canning system.* Source: Aquatic Science Branch, DoE.

Date	Type	Location
Mar 1978	Blue-green scum	Middle Swan Bridge
Nov 1981	*Microcystis*	Riverton Bridge
Jan–Mar 1994	*Anabaena–Anabaenopsis*	upstream Kent St Weir
Feb–Apr 1994	*Microcystis*	upstream Kent St Weir
Apr 1996	*Anabaena*	upstream Kent St Weir
Jan 1997	*Anabaena*	Kent St Weir, Bacon to Wharf St, Kent St, Greenfield St
Feb 1997	*Anabaena*	Bacon to Wharf St
Mar 1997	*Anabaena*	Castledare
Jan 1998	*Anabaena*	Kent St Weir to Greenfield St. Blooms all year at Castledare.
Feb 1998	*Anabaena*	Kent St Weir to Mason Landing 2 weeks, Bacon St
Mar 1998	*Anabaena*	Wharf St, Ellison Drive
May 1998	*Microcystis*	Wharf St
Jan 1999	*Anabaena*	Wharf St, Kent St Weir
March 1999	*Anabaena*	Kent St Weir
Feb 2000	*Microcystis*	Swan River estuary
Jan 2001	*Anabaena*	Canning River, Kent St Weir to Hester Park
Jan 2002	*Anabaena*	Canning River

TABLE 2.4 *Common nuisance phytoplankton, cyanobacteria 'blue-green algae' and aquatic plants in the Swan–Canning system.* Adapted from SRT, 1999.

	Criteria for health alert (cells/mL of water)	Comments
Heterosigma phytoplankton	> 15–20,000	Slimy layer—nuisance warning and potential to kill fish
Scrippsiella, Gyrodinium dinoflagellate phytoplankton	15–20,000	Red tide—nuisance warnings, some adhere to body
Rhizoclonium macroalgae	No health threat	Unaesthetic appearance and temporary loss of recreational amenity
Anabaena blue-green phytoplankton	> 20,000	Blue-green—unsafe for recreation; prevalent in the upper Canning River
Microcystis blue-green phytoplankton	> 20,000	Blue-green—unsafe for recreation; prevalent in the upper Canning River
Anabaenopsis blue-green phytoplankton	> 20,000	Blue-green—unsafe for recreation; prevalent in the upper Canning River
Hydrocotyle plant	No health threat, any growth eradicated	Introduced—gazetted, noxious plant; occasional outbreaks in the Canning River
Salvinia plant	No health threat, any growth eradicated	Introduced—gazetted, noxious plant; most common in the Canning River

Box 2.4 **TINY CELLS ON THE RAMPAGE TAKE A CITY'S PLAYGROUND OFF LIMITS**

FIG. 2.16 *(a) Bright green slicks of toxic algal cells along the beaches of Matilda Bay, February 2000.* Courtesy West Australian Newspapers Ltd. *(b) Sign showing a health warning.* Source: *River Science 2*, WRC, 2000c; Courtesy WRC.

In February 2000 the almost unthinkable happened: the Swan River estuary was closed to all activity, the danger of algal poisoning considered a risk to all. Unseasonal heavy rain in early January soaked the normally dry catchment. A week later another 12,000 million tonnes (12,000 gigalitres) of rain fell on the already saturated Lockhart subcatchment in the Yenyenning Lakes area of the Avon River catchment sending 270 gigalitres downstream. This was enough water to fill the rivers and estuary five-and-a-half times. The water flushed an area that had not flowed in summer for over forty years.

The flow brought an accumulation of plant material, soil, nutrients and fertilisers containing nitrogen and phosphorus. Total nitrogen concentrations were 7 milligrams per litre and total phosphorus concentrations 0.2–0.3 milligrams per litre. Concentrations of nitrogen were seven times higher than those considered acceptable in freshwater flow to estuaries, and phosphorus concentrations two to three times higher than acceptable.

In total, about 800 tonnes of nitrogen and 35 tonnes of phosphorus in fresh water entered the wide, shallow, protected estuary at a time when the waters would normally be salty. The warm sunny conditions of summer were a perfect place for algal growth, the unusual freshwater state (5–10‰) allowing growth of species found in stagnant wetlands and rivers.

The species responsible for the bloom *Microcystis flos-aquae*, originally identified as *Microcystis aeruginosa*, had several features accounting for its success. These were buoyant cells and rapid growth. The gas vacuoles of *Microcystis* help it float near the surface in the sunlight where it produces sugars through photosynthesis. The pressure of the

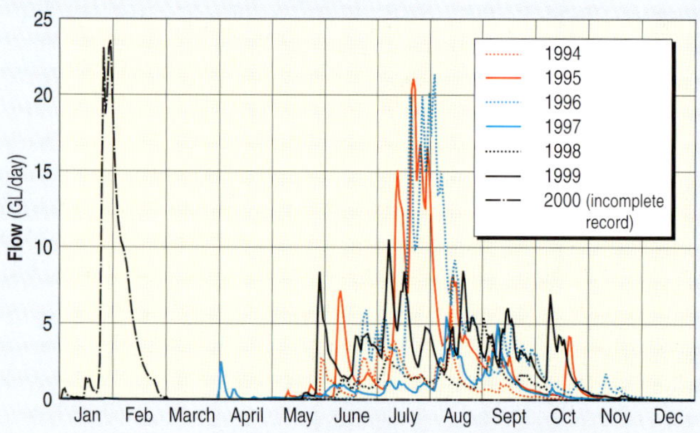

GRAPH 2.5 *Flow from the Swan–Avon River measured at Walyunga Canyon showing the normal summer and winter flows 1994 to 1999 and the unusual summer flow 2000.* Source: *River Science 2*, WRC, 2000c; Courtesy WRC.

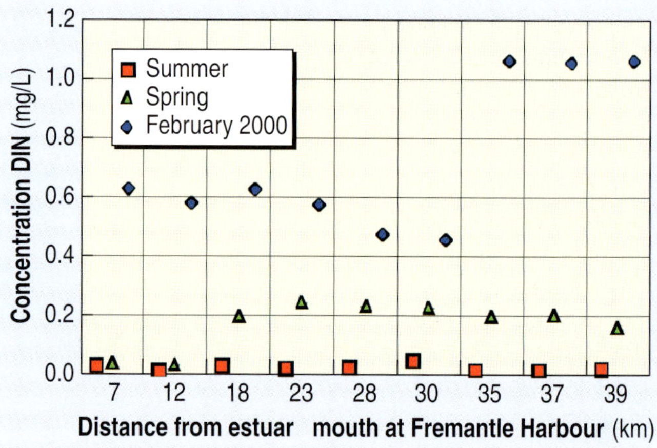

GRAPH 2.6 *Median concentration of dissolved inorganic nitrogen (DIN) during the* Microcystis *bloom in 2000 compared to those usually observed in spring and summer.* Source: *River Science 2*, WRC, 2000c; Courtesy WRC.

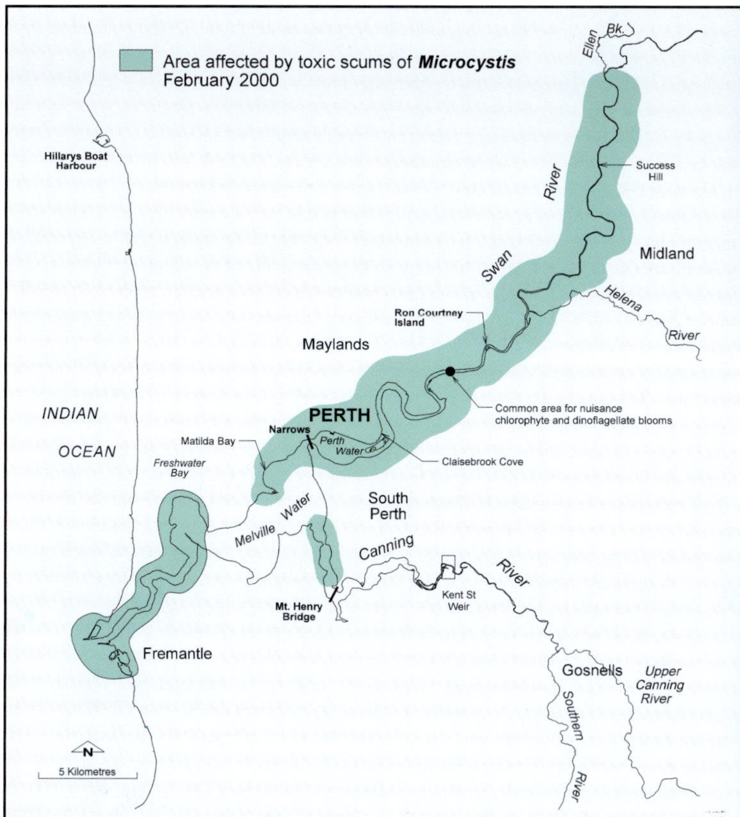

MAP 2.12 *Extent of the Swan Estuary affected by the toxic* Microcystis *bloom in February 2000.* Source: *River Science 2*, WRC, 2000c; Courtesy WRC.

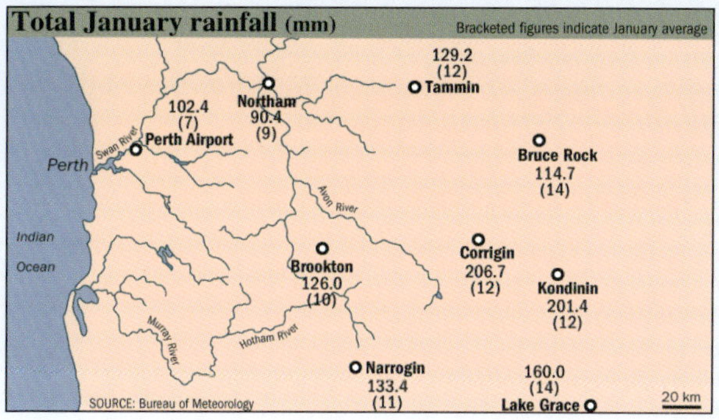

FIG. 2.17 *Rainfall in the Swan–Avon catchment in January 2000 with the average January rainfall in brackets.* Source: *River Science 2*, WRC, 2000c; Courtesy SRT and WRC.

FIG. 2.18 *Skimming was a highly effective method of removing concentrations of algal cells during the bloom in February 2000.* Source: *River Science 2*, WRC, 2000c; Courtesy WRC.

sugars cause the cells to sink below the surface where they consume the sugar overnight, rising again the following day. When conditions are favourable the cells divide every day, so bloom growth is rapid and exponential. About 20,000 cells per millilitre are considered a health hazard; in some areas surface concentrations reached 130 million cells per millilitre in bright green scums. In the more open areas where water movement was higher, counts were generally well below those considered a hazard.

The mass *Microcystis flos-aquae* formed thick green scums along shorelines at Maylands, Claisebrook Cove, Barrack Square, Matilda Bay, the Nedlands–Dalkeith foreshore and Freshwater Bay. The bloom was observed flowing out to sea at Fremantle. The bloom concentrated on the northern shoreline because of wind and tide, which were not sufficient to push cells down into the water where they would die, starved of light. For twelve long days from Success Hill down to the port, the river was closed to all summer activities: swimming, wind surfing, rowing, water skiing, fishing and prawning.

Microcystis flos-aquae produces the toxin microcystin. At the height of the bloom toxin concentrations were 7,000 micrograms per litre, that is, 7,000 times higher than the World Health Organisation (WHO) standard for drinking water and 700 times higher than concentrations considered safe for recreational activities. While the toxicity was high, the risk to humans was quite low because the risk of contact or ingestion was considered low due to the bans placed on activities where the blooms were concentrated.

Efforts to control the bloom included spraying sea water, application of salt and clay slurry, as well as skimming off high concentrations using oil spill equipment. Although many areas could not be treated successfully, the trials with fine bentonite clays showed promise and could be useful if blooms occur again, particularly if it could be applied early.

The bloom collapsed as the freshwater flow subsided at the end of February and the natural summer salinity regime re-established with intrusion of salty marine water. Once salinity reached about 10‰ the bloom started to die, first in the Lower Estuary and then progressively upstream in the Middle Estuary and Upper Estuary as the wedge progressed. By the end of February, the last of the isolated pockets of *Microcystis* in the upper reaches had collapsed. Nature, that unpredictable initiator of the bloom, also provided the cure.

Memories of the summer flood and bloom linger: stories of brown jellyfish flushed way out to sea, washed kilometres up the coast; the silence of deserted beaches; the sight of a river without sails; the ominous fluorescent green scum approaching the beach; and the dead fish lining the sand. A warning of the fine line we walk when we change our landscape.

Source: WRC, 2000a.

LIVING IN THE ESTUARY

Changes in the estuary water are closely followed by changes in the plants and animals. The seasonal salinity changes in particular, restrict the flora and fauna to those species able to cope with the changes or able to re-establish between seasons. The seasonal changes create three environments suitable for freshwater species, estuarine species and marine species.

During the freshwater phase the algae and other organisms move downstream with the river flow, and establish while the fresh water persists. In the Swan–Canning Estuary, the short freshwater state in winter slows algal growth and limits the activity of some animals.

As the river flow diminishes, the advancing salt wedge brings more saline conditions upstream, which is the estuarine phase. Although only a few species are able to tolerate the salinity changes, many are present in low numbers year round, proliferating as the temperatures rise in spring and summer, the algal growth fuelling a food chain of invertebrates, fish and birds.

As summer progresses and the salinities reach oceanic concentrations, the Lower Estuary enters a marine phase, when plants and animals that are limited to stable marine waters visit the sheltered productive water to feed, and leave again or die when freshwater flows commence in winter. These are termed the opportunist and marine stragglers.

The abundant fish and birds that so intrigued the first visitors to the Swan provided a welcome variety to the diet of the first European settlers, as it had done for Aboriginal people. This abundance is typical of estuaries where the primary producers of fringing vegetation, microscopic algae and larger macrophytes of algae and seagrasses all thrive in the shallow, well lit, nutrient-rich and calm waters, contributing directly or indirectly through detritus to the food web (Fig. 2.19).

In times past nutrients for plant growth changed with the seasonal cycles, now, however, the types and abundance of particular nutrients also depend on less predictable water flow, and nutrients applied in the catchment. The change of nutrients promotes the growth of some plants and a gradual change in plant communities with the potential to influence the animal communities.

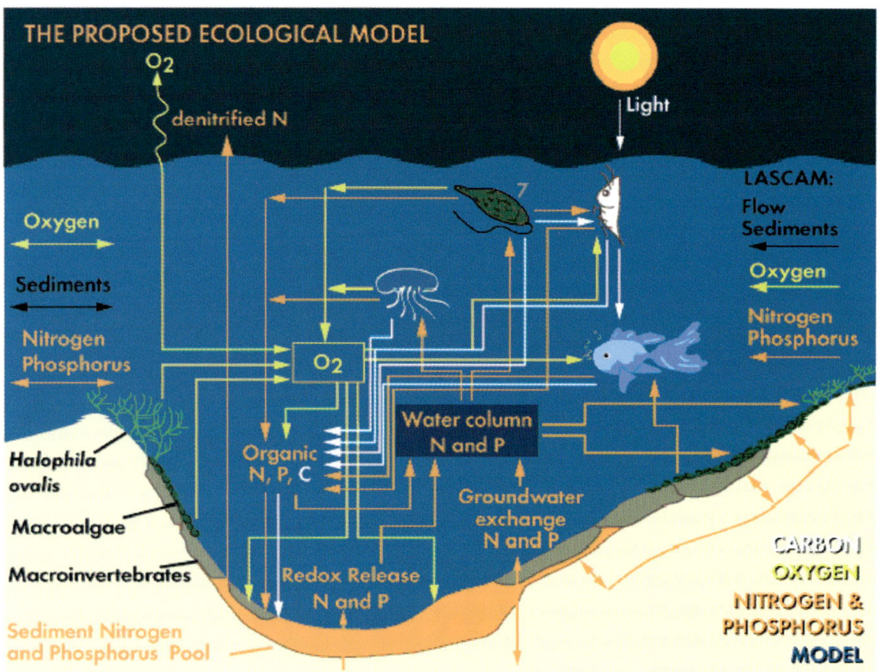

FIG. 2.19 *Ecological model for the Swan–Canning: the model illustrates how the physical and biological components of the estuary interact and influence each other. Plant growth in the estuary is dependent on light, nutrients and salinity. Primarily water depth influences light so that the deeper waters are populated by phytoplankton, while benthic plants, both micro and macroscopic, occur where light levels are high in shallow water. The nutrient cycles in turn are mediated by plant growth. Animal life similarly depends on the physical conditions but ultimately on food derived from plants. The activity of the animals also influences nutrient cycles and therefore plant growth.*

Source: SCCP Action Plan 1999, Courtesy SRT.

The tiny phytoplankton— the good guys and bad guys

In general the seasonal changes in salinity determine the types of phytoplankton, but nutrients determine the size of the populations or blooms (see above). Rainfall run-off brings fresh water loaded with nutrients that fuel algal blooms. The nutrient concentrations then decline as they are taken up by the phytoplankton. Other factors also affect the amount of phytoplankton: light and temperature affect the rate at which algae grow and divide, and grazing by small animals such as copepods, mussels and barnacles reduce the number of cells in the water column (Graphs 2.7 and 2.8).

As eutrophication progresses, the microscopic algae in blooms change from smaller to larger forms, which are more difficult for zooplankton to eat. As a result, the blooms of large ungrazed algae do not enter the food chain but collapse when the nutrients are exhausted. The dead cells fall to the bottom adding to the sediment nutrient store, and remove oxygen as they decompose initiating nutrient release to fuel further blooms.

Microscopic variety

The community of microscopic water plants or phytoplankton consists of chlorophytes (microscopic green algae), diatoms and other organisms such as dinoflagellates responsible for 'red tides', as well as cyanobacteria that cause blue-green algal blooms (Table 1.10). The microscopic green algae or chlorophytes are most common in fresh water. Blooms of *Chlamydomonas* are stimulated by nutrients in the freshwater flows and occur in the upper reaches of the Swan in spring or down as far as Perth Water after wetter winters. Although blooms of green algae are not toxic, the dead cells use up oxygen in the decay process.

In early spring the diatoms dominate the blooms in the lower reaches. The blooms turn the water a murky yellow-brown but generally they are not toxic. More than 400 diatom species have been recorded in the Swan–Canning Estuary. Blooms generally start in Melville Water in spring as the winter run-off slows and salinity increases, and then move upstream past Maylands with the advancing front of salt water. The blooms are often dominated by *Skeletonema* a species important in the diet of small animals in the water column as well as the filter-feeding mussels.

In late spring as the water turns brackish in the middle reaches, blooms of dinoflagellates (*Scrippsiella* sp.) develop along with the blooms of diatoms (*Skeletonema* and *Chaetoceros* spp.) and the water is a brownish red. The dinoflagellates bloom is a response to the nitrogen released into the brackish water from the sediments. Dinoflagellates are a natural part of the food web and very important in the diet of mussels and other organisms, however they can form annoying scums and some species produce toxins.

In midsummer when the water is salty, blooms consist of dinoflagellates (*Gymnodinium–Prorocentrum* sp.), raphidophytes (*Heterosigma–Fibrocapsa)* and cryptophytes (*Cryptmonas* spp.). The size of the blooms largely depends on the amount of nutrients released from sediments. The blooms colour the water red or brown and some species are regarded as a nuisance.

In autumn the blooms are composed of diatoms, dinoflagellates and marine algae. The species depend on the salinity, either the summer saline state or fresh water if there are early rains. Early rains also bring a flush of nutrients to the warm estuary, and nutrients may also be released from the sediments. Some species are important to the food chain, while others are nuisance species and a potential risk to human health.

The summer and autumn are also the time of potentially toxic blue-green algal blooms. Blooms of *Anabaena*, *Microcystis* and *Nodularia* have occurred regularly in Western Australian wetlands and in low numbers in the Upper Swan. Short-lived blooms of *Anabaena* in 1978 and *Nodularia* in 1981 highlighted concerns that algal blooms could become more common—a fear well demonstrated by the February 2000 bloom of *Microcystis*.

In the lower reaches of the Canning River, phytoplankton also respond to the yearly cycles of salinity as they do in the Swan River Estuary. In contrast, upstream of the Kent Street Weir saline water has a minor influence, with light and temperature playing a more important role in determining the

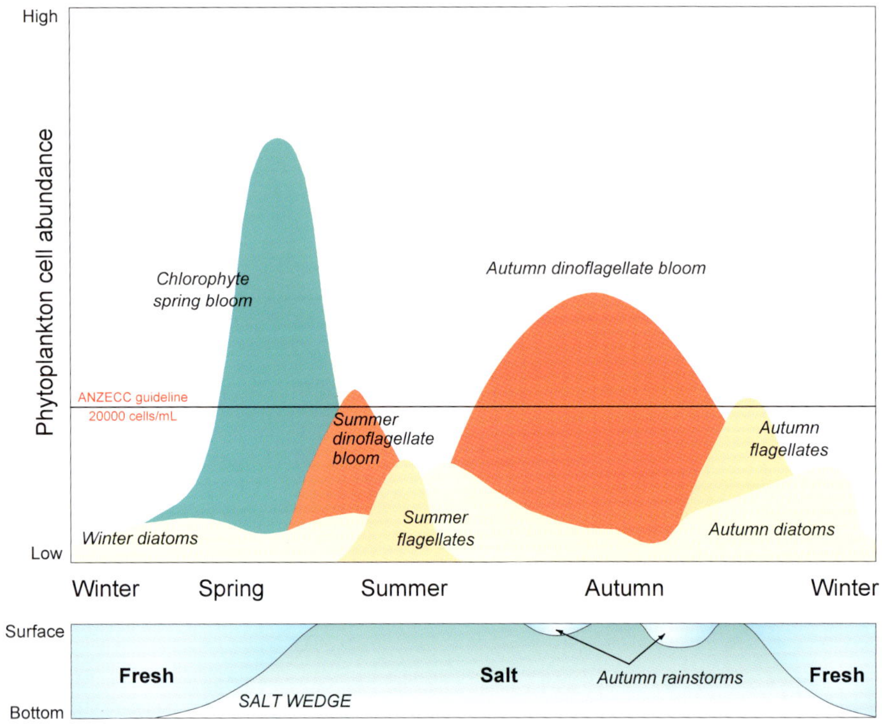

GRAPH 2.7 *Phytoplankton succession and relative abundance in response to salinity and timing of the salt wedge. In winter freshwater flow allows the establishment of freshwater species. The change to more saline conditions with the upstream advance of the saltwater wedge allows other species to establish. The salinity stratification isolates the bottom water creating stagnant low oxygen conditions favouring release of sediment bound nutrients, which promotes algal blooms.* Source: SCCP Action Plan 1999, Courtesy SRT.

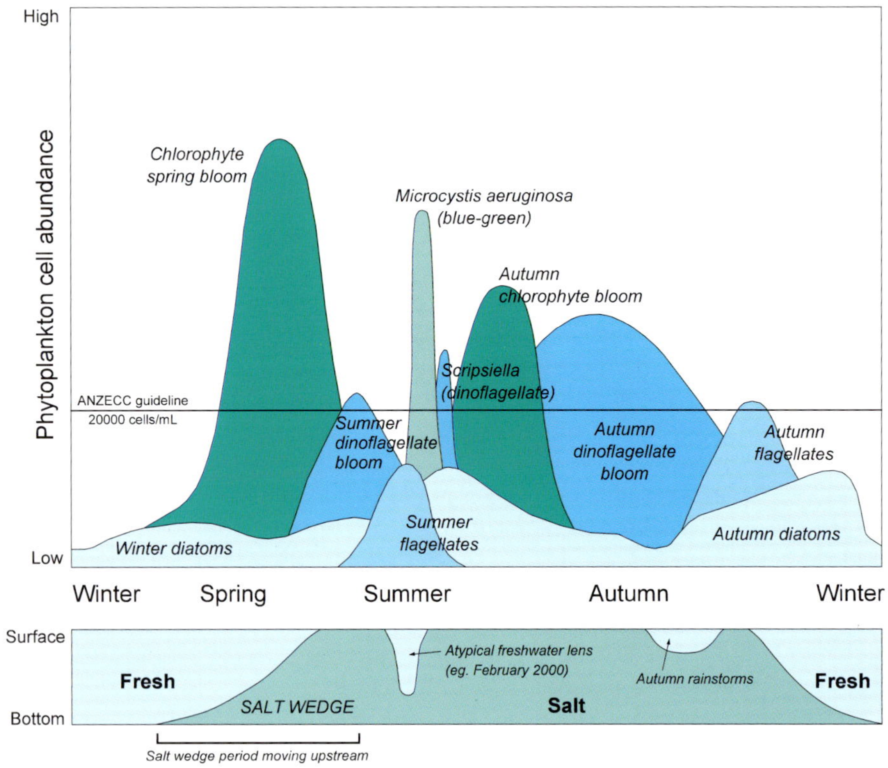

GRAPH 2.8 *Algal blooms following a flush of fresh water in late summer and autumn. The bloom of the blue-green alga* Microcystis aeruginosa *in February 2000 was a response to fresh water and nutrients in warm calm conditions.* Source: *River Science 2*, WRC, 2000c; Courtesy WRC.

succession of phytoplankton. Prolific growths of large aquatic plants also compete with phytoplankton for nutrients in the Upper Canning River. The subsequent decay of organic material appears to provide the low oxygen environment favouring sediment nutrient release. In the Canning, green algae bloom in spring as they do in the Swan, but blooms in late spring and summer when the water is still fresh and warm often contain blue-green alga such as *Anabaena circinalis*.

The Canning also has a larger variety of freshwater algae, and larger aquatic plants or macrophytes. The macrophytes include the native species *Hydrilla verticillata*, *Potamogeton crispus*, duck weed *Lemna* and floating water fern *Azolla* as well as on occasion the introduced species, aquarium escapees *Salvinia* and *Hydrocotyle*. These shade the water, compete for nutrients and probably helped to control blue-green algal blooms, however they also add substantial amounts of organic material to the waterways that eventually decompose releasing nutrients, the fuel of algal blooms.

Tiny plants on the surface of the sand

As well as microscopic plant cells in the water column, tiny algal cells also live on the sediment surface where they have the appearance of a green or brownish discolouration. The film disappears and reappears quite quickly as the cells move downwards or upwards to the light. The film consists of a community of diatoms, green algae, blue-green algae and dinoflagellates that are either free moving forms (epipelic) or attached to sand grains (episammic). Benthic microalgae (BMA), sometimes referred to as microphytobenthos (MPB), grow rapidly and contribute to the overall production of a water body (Graph 2.9).

The small size of benthic microalgae means that they contain little carbon, but production due to the fast growth is high, much greater than seagrass. However when respiration is taken into account, a process that goes on through the day and night, when photosynthesis stops, the net production is only small, about 44 grams of carbon per square metre per year (Table 2.5). When the respiration of the bacteria, tiny animals and the decay of organic matter are included, the amount of carbon produced is still large but not as great as the seagrasses. The benthic microalgae also serve an important function in oxygenating the sediments, are of great importance to the fauna, but also control the release of nutrients.

GRAPH 2.9 *Chlorophyll* a *in the top 5 mm of sediment and the overlying water column from different ecosystems in Western Australia. In all environments the chlorophyll* a *in sediments was greater than in the overlying water column. In the Swan Estuary, benthic microalgae were more abundant in the coarse-grained sediments of the Lower Estuary where they contributed more to the total ecosystem productivity than in the fine sediments of the Upper Estuary where the water column phytoplankton were more important. Chlorophyll* a *is a relatively easily measured compound used in the process of photosynthesis where carbon dioxide is made into carbohydrates and oxygen released, and can be used to derive estimates of carbon biomass.* Modified from Masini 1987, Courtesy R. Masini, DoE.

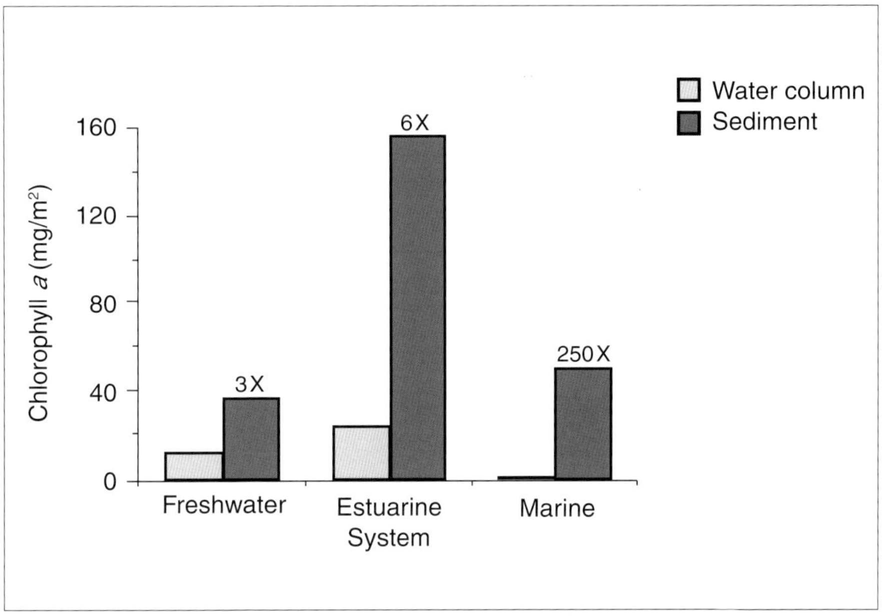

	Seagrass	Intact core
Plant biomass (g C m^{-2})	15	2.5
Production rate (g C m^{-2} yr^{-1})		
Gross	350	570
Net	300	44
% cover (800 ha)	60	60
Annual production (tonnes carbon)		
Gross	1,750	2,850
Net	1,500	220

TABLE 2.5 *Comparison of production on an areal basis of bare sediment and the dominant seagrass* Halophila ovalis *in the Swan River Estuary. The community is limited to areas less than 2 m deep.*
Source: Masini, 1987.

Large macrophytes—algae and seagrasses

Compared to other estuaries, the Swan–Canning has a much higher diversity of macroalgae, about sixty-six species, probably attributable to the range of habitats and the open connection to the sea. The range of species is more limited, however, than in the nearby open ocean where more than 350 species occur. The diversity also decreases upstream. In the less marine waters the green algae predominate. Encrusting blue-green alga, *Calothrix parietina*, grow in a band high on rocks as far upstream as Perth Water. Lower in the water, green sea lettuce *Ulva* and *Enteromorpha* form another band with the red alga *Polysiphonia*, *Gelidium* and *Gracilaria* occurring slightly deeper. Although the red algae often germinate and start growing attached to solid stones and shells, they also continue to grow as they float around in the water. Downstream of Preston Point, brown alga such as the bubble weed *Colpomenia peregrina* and the fucoid brown algae *Hormophysa triquetra* and *Cystoseira trinodis* become more common and create a habitat for seahorses and other marine life (Fig. 1.11).

Like phytoplankton, macroalgae also grow quickly when nutrients are available. Between 1912 and 1936 macroalgae were a problem around Burswood. Dense macroalgal accumulations of *Rhizoclonium* were present in Freshwater Bay in the 1970s and 1980s, and from the 1990s in the lower Canning River near Riverton Bridge. It was particularly abundant in 1996, a wet year, when it extended downstream to Mount Henry Bridge. (See **Box 2.5 Unacceptable weeds on beaches—pressure for controls** and Table 2.15.)

Seagrasses

Like the large algae, seagrasses in the Swan–Canning are only represented by a few species, the small paddle weed *Halophila ovalis* and fine eelgrass *Zostera mucronata*, while more than eight species are found along the coast of Western Australia (Fig. 1.12).

The paddle weed seagrass *Halophila ovalis* (Fig. 2.20) is the most common species of seagrass in the Swan–Canning Estuary, covering about 500 hectares or 20 per cent of the main estuarine basin. *Zostera* is most abundant in the more marine waters of the Lower Estuary.

Salinity and light are important in determining the distribution of *Halophila* in the estuary, while temperature in combination with light are the most important factors for growth. Laboratory test and field observations in the 1980s showed that *Halophila* grew best at a salinity of about half to over marine concentrations (*Halophila ovalis* also grows in the ocean), and remained quite healthy in water of about 25 per cent marine salinity (≈ 9‰) for about four weeks. It died, however, when exposed to fresh water for longer periods (Graph 2.10). In some years when winter run-off is low, growth only slows for a short time, but in years of high flow and lower salinity *Halophila* dies. This salinity response probably accounts for the

Box 2.5 **UNACCEPTABLE WEEDS ON BEACHES—PRESSURE FOR CONTROLS**

The cause and solution to algal blooms in the Swan–Canning have been debated since the early days. Complaints of stinking mud and algae around Mill Street and the Causeway first appeared in newspapers in the 1870s, and by the 1900s were common mostly around the reclaimed Esplanade in Perth Water and in Freshwater Bay at Claremont. Opening of the sewerage works at Burswood Island in 1912, however, moved the focus upstream where *Cladophora*, *Chaetomorpha* and *Rhizoclonium* accumulated and rotted.

Dr F. Stoward of the Department of Agriculture in 1913 conducted the first investigation. Stoward suggested if the extra nutrients from the sewerage plant were aggravating the problem, the shallow slow-moving river conditions favoured algal growth so that changes to flow rather than treatment with copper sulfate were likely to have a greater effect. Complaints continued, and other scientists examined the issues in 1919 and 1922. Professor G. E. Nichols commented in 1922 that although nutrients were being added near the outfall, the black silt was likely to be contributing nutrients. Nichols noted that the problem was widespread with weeds also accumulating on the banks of the Lower Estuary and in the Canning. This view was shared by Dr E. Atkinson of the Public Health Department and Mr F. W. Lawson, an engineer with the Metropolitan Water Supply and Sewerage Department, who suggested an ocean outfall was a desirable option.

Advice from Colonel F. F. Longley, Advisory Expert in Sanitary Engineering to the Commonwealth Department of Health 1922–1924, confirmed concentrations of nitrogen and phosphorus in the river were sufficient to support algal growth without additions. Longley suggested a costly ocean outfall would not completely remedy the situation and reiterated that improved flow would be equally effective with the added advantage that dredging could be used to reclaim low-lying areas around the river. Up to this point most reclamation had been for flood control.

In 1928, after 400 tons of algae were removed from near the Bunbury Traffic Bridge and chemical controls failed, calls were made for a Swan River Conservancy Commission 'to control the points at which discharges to the river are made and ensure adequate purification before discharge'–although no action of this sort was implemented.

During the debate, a letter to the *West Australian* from G. P. Stevens noted the algal problem was present before the Burswood Plant commenced and weeds had been removed from Freshwater Bay earlier in the century. Commenting that more complaints did not mean more problems, rather an increasing population of people to complain about the problem; 'for every person holding his nose and anathematising the authorities 25 years ago, there are at least a thousand now and the volume of the complaint is correspondingly greater'.

The complaints continued, of course, as the population produced more waste and the number of voices grew. Eventually with the filter beds at Burswood needing considerable additions, money was allocated for an ocean outfall–creating employment during the depression. The filter beds closed in December 1936, but the problem continued and new calls were made for scientific research despite the disregard for scientific opinion over the previous 24 years.

During World War II, complaints diminished although the problems continued and a Swan River Reference Committee was formed in 1943 and reconstituted in 1948 with a sub-committee formed in 1952 to define pollution and suggest methods of control. The 1955 report including one by R. D. Royce who concludes 'by far the greatest pollutant is the alluvial mud brought down from agricultural areas which is rich in nutrient fertilizers such as nitrogen and phosphorus'.

An independent Swan River Conservation Committee with wide representation was formed in 1946. This committee had representation on the Reference Committee and input to the drafting of the Act in 1958 for the establishment of the Swan River Conservation Board, a statutory body charged with maintaining the 'purity and cleanliness of the waters'. The Board functioned until 1977 when, with the establishment of the Waterways Commission, it was replaced by the Swan River Management Authority and in 1989 by the Swan River Trust.

In the 1950s the Public Works Department Harbours and Rivers Branch cleared weed from the beaches, at first by hand raking but later a bulldozer was used where ground was sufficiently firm. About 400 tonnes of wet weed was removed in 1967 and 200 tonnes the following year. Although algae and seagrass are still removed from beaches, the problem of large weeds is now minor and there are other more pressing concerns.

The origin of the problem is still debatable. The sewerage works certainly added nutrients but probably only about 16 tonnes per year, compared to 200 tonnes per year entering from all sources in the 1980s. The eutrophic conditions in that region of the river probably existed before European settlement, part of the natural cycle of estuaries where nutrients accumulate; the sewerage works probably adding a little more. The social response is interesting, with algae deemed a health hazard when eyes and noses are offended. It would seem we have conveniently short memories about conditions when we choose to live alongside nature at the edge of the estuary.

Source: Hodgkin & Vicker, 1987.

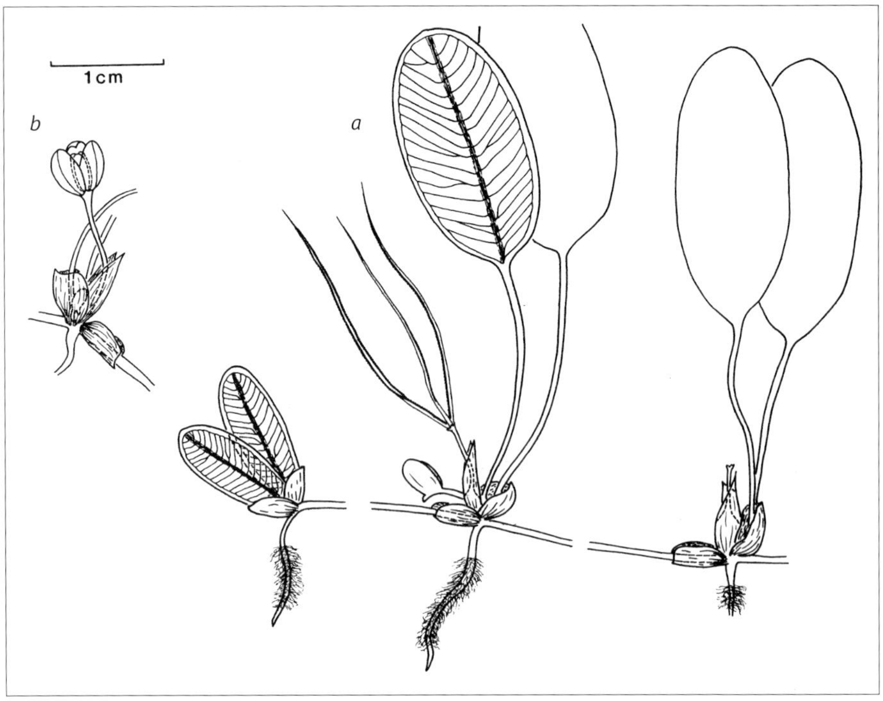

FIG. 2.20 *The paddle weed seagrass* Halophila ovalis*: the seagrass is a true plant with leaves, roots, rhizomes, flowers and seeds. The flowers are quite small and inconspicuous lying between the leaf bases. The plants grow from seed and small pieces of rhizome. (a) Flowering plant with female flower; and (b) male flower. The amount of plant material in the leaves above the sediment and below the sediment in the roots and rhizomes is very similar.* Halophila *in the Swan is both a sink and a source of nutrients. It takes up nutrients for growth, some of which are released when the plant dies back and tissues decay. The roots also release oxygen creating a favourable environment for burrowing animals and nutrient cycling.* Adapted from Womersley, 1984; from original drawings by Enid Robertson.

ephemeral nature of *Halophila* in the Canning River area and its absence upstream of the Narrows Bridge.

In the Swan–Canning Estuary, *Halophila* growth is higher when water temperatures are higher, and is only limited by unfavourable low temperatures for a short time during the winter months when temperatures may decline to 15°C. Like salinity, low temperatures alone are not extreme enough to account for the low growth rates in winter (Graph 2.10).

Light conditions are the most significant factor influencing *Halophila* in the Swan River Estuary, the same factor known to be important to seagrasses growing in other areas and accounting for their decline. (See **Box 9.3 Shrinking seagrasses versus growing algae**.) Most seagrasses need about 5 per cent of full sunlight and live to depths where this level of light occurs, which is called the photic zone. In the Swan–Canning Estuary this is about 2.5 metres except in winter, and about 1 metre in the Upper Estuary. Not surprisingly this corresponds to the distribution of *Halophila*. In experiments, no growth occurred at two per cent full sunlight but increased up to 15 per cent sunlight, indicating that *Halophila* at 15 per cent was receiving all the light it could use. A combination of low light and low salinity probably excludes *Halophila* from the Canning and Upper Estuary.

Although estimates based on the distribution of *Halophila* and average nutrients vary greatly, in summer about 350 tonnes dry weight biomass, 4,100 kilograms nitrogen and 630 kilograms of phosphorus are bound up in the seagrass tissue; thus seagrass can be termed a 'nutrient sink'. In contrast, in winter about 280 tonnes of biomass, 2,800 kilograms of nitrogen and 460 kilograms of phosphorus are released when the seagrasses die, so they may be termed a 'nutrient source'.

At the end of the growing season in late summer to early autumn, seagrass biomass can be about 120 grams dry weight per square metre. At this time the tissue contains about ten times the amount of carbon, and twice as much nitrogen and phosphorus as the phytoplankton in the water body. Despite the high biomass of seagrass and its fast production, restriction of the seagrass to the shallow areas means that, overall, the amount of carbon and productivity by phytoplankton is still much greater (Table 2.6).

Production of *Halophila* in the Swan–Canning amounts to about 1,500 grams dry weight per square metre annually (500 grams of carbon per square

TABLE 2.6 *Total biomass and nutrient concentrations of nitrogen and phosphorus in tissues of* Halophila ovalis *in the Swan–Canning Estuary during summer.* Source: Hillman et al., 1995.

	Area occupied ($\times 10^6$ m^2)	Biomass (t DW)	Nitrogen (kg)	Phosphorus (kg)
Summer 1976	5.63	346	4,066	623
Summer 1982	5.93	351	4,214	632
Winter 1981	5.93	76	1,316	172

Summer: 18 mg, 8 mg and 5 mg N g^{-1} DW and 2.3 mg, 1.3 mg and 0.8 mg P g^{-1} DW of leaves, rhizome and roots respectively; assuming proportions of total biomass 2:1:1. Biomass summer 1982, 100 g DW m^{-2}. Winter: 24 mg, 18 mg and 5 mg N g^{-1} DW and 3.5 mg, 2.5 mg and 0.8 mg P g^{-1} DW of leaves, rhizomes and roots respectively; assuming proportions of total biomass 1:1:1. Biomass maximum winter 1981, 20 g DW m^{-2}.

metre per year), about 100 grams of carbon per square metre per year when estimated over the entire estuarine basin. In contrast, phytoplankton is about 300–500 grams of carbon per square metre per year. It is in the shallow areas, however, that seagrass is most important. Here combined seagrass and phytoplankton production can be up to 1,700 grams of carbon per square metre per year, with 80 per cent due to seagrass. Almost half this productivity is available to food chains as dissolved organic matter while the remainder, which is particles of detritus colonised by micro-organisms, accounts for the abundance of invertebrates and fish found in the shallow areas.

Like other seagrass habitats, meadows of *Halophila* also create places for other plants and animals to live as they stabilise sediments and release oxygen through photosynthesis. (See **Chapter 9 Albany Harbours and the smaller estuaries of the Albany region: Why seagrasses? What's wrong with algae?**) The high primary production and low biomass result in turn over of leaf material every eleven to twenty-four days. This is a turn over rate similar to other small seagrasses with the ability to establish and grow quickly, which is a useful attribute for a coloniser species. The high productivity may, however, account for the small load of algal epiphytes (unlike the case of *Posidonia* in Oyster Harbour). As algae also have fast turn over rates however, the role of epiphytes in the ecology of the estuary may still be significant.

A productive ecosystem

In the Swan as in other estuaries, the high abundance of crustaceans (prawns and crabs) and fish is based on detritus (dead plant material). The detritus originates from the fringing vegetation, the macroalgae, or benthic microalgae on the sediment or growing on other plants (epiphytes) including seagrass, and to a lesser extent the phytoplankton in the water column. Determining the plant items that contribute most to food webs is very difficult: while fish

GRAPH 2.10 *Laboratory results on the effect of: (a) salinity and (b) temperature on the growth of* Halophila ovalis. *Marine salinity equals 35‰. Although* Halophila *grew over a wide salinity range from low (5‰) to high (45‰) for a few weeks, growth eventually slowed and the plants died under extreme conditions. This explains field observations.* Halophila *did not grow at temperatures below 10°C and grew more rapidly at higher temperatures. Production was calculated as the average increase in dry weight per day throughout the experiment.* Adapted from Hillman et al., 1995.

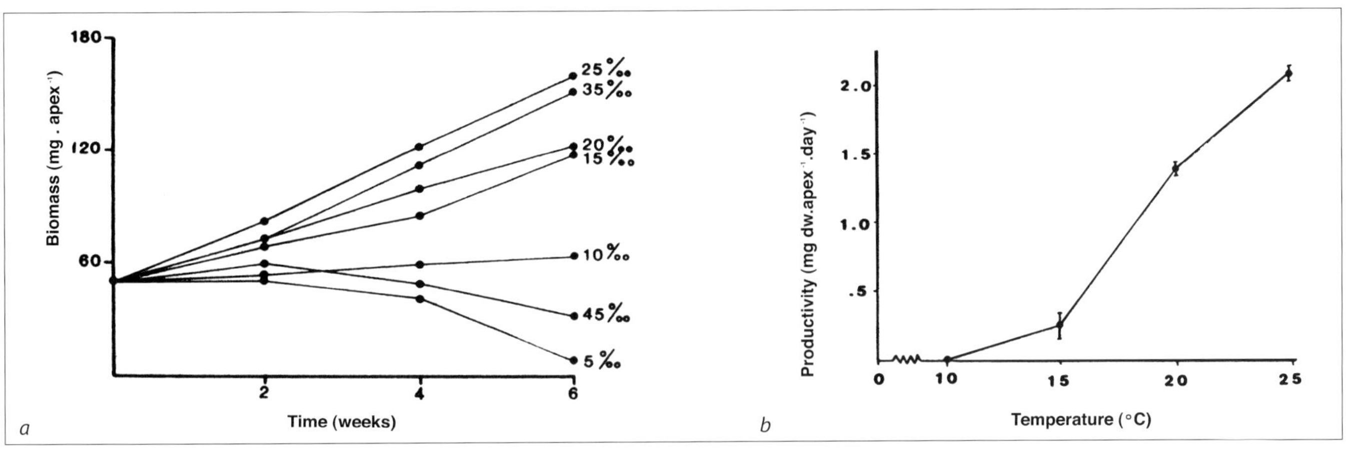

behaviour and stomach contents can be examined, for the smaller animals that eat smaller food items it is almost impossible. The use of naturally occurring stable isotopes, however, has greatly enhanced this type of study (Graph 2.11). (See **Box 2.6 Using stable carbon isotope ratios to disentangle food webs.**)

Wildlife of the Swan

Despite substantial changes to the Swan–Canning landscape over the last 170 years, the estuary still retains some of the natural features that attracted the early explorers and settlers. Letters and journals tabulated by Riggert in 1978 record an abundance and diversity of wildlife that both fascinated and provided sustenance: a resource that had in the past sustained the Aboriginal inhabitants during their summer migrations to the coast.

All water birds are still present, although some like the Black Swan and waders are reduced in number. Similarly, some of the land birds have disappeared along with the surrounding natural vegetation. Fish are still abundant supporting recreational and commercial fishing.

Perhaps one of the most noticeable animals in the water are the swarms of white and brown jellyfish (Fig. 2.21). The bane of swimmers, jellyfish were recorded by one of the earliest visitors, James Backhouse in the summer of 1837–1838. 'Numerous shoals of fish were sporting in the sunshine, and multitudes of Jelly-fish of great beauty, were floating just below the surface of the water.' (Backhouse, 1843) (Fig. 2.22). Backhouse noted two types of jellyfish, the common white saucer jellyfish *Aurelia aurita* and the brown jellyfish *Phyllorhiza punctata*. For many years this early record was forgotten, and following the occurrence of particularly large numbers in the summer of 1947–1948 it was suggested they had probably been introduced on the U.S. Navy Catalina flying boats based at Crawley during World War II. The jellyfish may, however, have been introduced much earlier by the first explorers and colonists, the scyphistome stage from which the young medusae (ephyra)

FIG. 2.21 *Swarms of jellyfish up to a square kilometre in area are a feature of the river; this swarm of* Aurelia aurita *in November 2001. Often accumulating downwind, they can also swim by contraction of muscle around the bell.* Courtesy V. Hosja, WRC.

Box 2.6 USING STABLE CARBON ISOTOPE RATIOS TO DISENTANGLE FOOD WEBS

Carbon (C) is found in two forms, stable isotopes, ^{12}C and ^{13}C. About 98.9% of carbon exists as ^{12}C while 1.1% is ^{13}C. Different plants have different ratios of ^{13}C to ^{12}C depending on how the plant obtains carbon or how it is incorporated into tissues. Carbon can be derived from carbon dioxide or bicarbonate. Animals eating plant tissue take on the ratios of their food source with only a small increase in ^{13}C, so that food sources can be inferred by comparing the ratios on an array of plants and animals. Values are expressed in parts per thousand (ppt) as deviations δ (delta) from a world standard, and are usually negative since this standard is very high in ^{13}C. In the Swan–Canning Estuary the ratios of carbon in *Halophila* (–12 ppt) and phytoplankton (–20 to –23 ppt), which are the main primary producers and sources of carbon, are quite different. Most of the fauna in contrast showed intermediate values, indicating that they obtained carbon from both seagrass and phytoplankton with most using 5–50 per cent of seagrass in their diet. Based on fish abundance, 10–60 per cent of the estuary food web is based on carbon derived from seagrass. Seagrass health is therefore a major concern for maintaining fish stocks.

Source: Hillman, 1987.

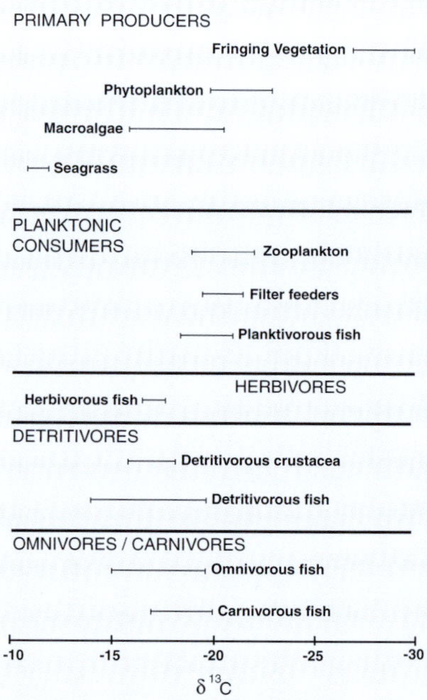

GRAPH 2.11 *^{13}C:^{12}C ratios of selected flora and fauna in the Swan–Canning Estuary.* Source: Hillman, 1987, Courtesy J. John, Curtin.

FIG. 2.22 *(a) Drawings of jellyfish made by James Backhouse in the summer of 1837–1838 during a visit to the Swan River Colony.* Backhouse, 1843. *(b)* Phyllorhiza punctata *in the Swan.* Courtesy T. Wernberg, UWA.

bud off having been carried on the bottom of the ships. The jellyfish, like many other organisms including fish and molluscs, could have also arrived at the Swan River via the southward flowing Leeuwin Current.

Aurelia and *Phyllorhiza* are found throughout south-eastern and south-western Australia. They also appear to be widespread in the Indo-Pacific, in the tropical Atlantic and in the eastern Pacific, particularly in shipping harbours. Their occurrence in the Swan is unusual as they are not found in other estuaries along the West Coast. Absence from the Peel–Harvey is thought to be due to the almost complete winter flushing with fresh water occuring in that estuary. Although the Swan is also flushed with fresh water, small pockets of warmer salty water generally remain in the deepest areas: laboratory studies have shown these are suitable for survival of the scyphistome stage. The more marine conditions in the Peel–Harvey since the channel was constructed may prove to be suitable for jellyfish.

Aurelia is a predator–grazer species that increases in abundance as the waters become more saline in spring (Fig. 2.23). *Phyllorhiza* is most abundant through the summer when the salinity exceeds 25‰. The brown colour of *Phyllorhiza punctata* is due to a dinoflagellate unicellular alga *Symbiodinium* sp. living within the jellyfish tissues in a symbiotic relationship. In corals, similar symbiotic algae (zooxanthellae) release organic matter to the host tissue and the environment. Although the concentration of chlorophyll *a* in large patches of jellyfish is unknown, it is probably similar to low concentrations of phytoplankton. The life cycle of *Phyllorhiza* is similar to *Aurelia*. Although the scyphistomae stage of *Phyllorhiza* has never been found in the estuary, they have been grown in the laboratory. Most planulae settle on the underside of horizontal surfaces and the scyphistomae produce buds as they grow. The scyphistoma divides and a single ephyrae is released as the water temperature rises and day length increases during November and December. Medusae that develop from the ephyrae are large enough, about thirteen centimetres, to reproduce sexually within six weeks. The population expands rapidly due to the production of ciliary buds and ephyrae.

Studying the fauna

The first professional scientific report on the fauna of the Swan was made by W. Michaelsen and R. Hartmeyer during the Hamburg South-Western Australian Research Expedition of 1905. Published as *Die Fauna*

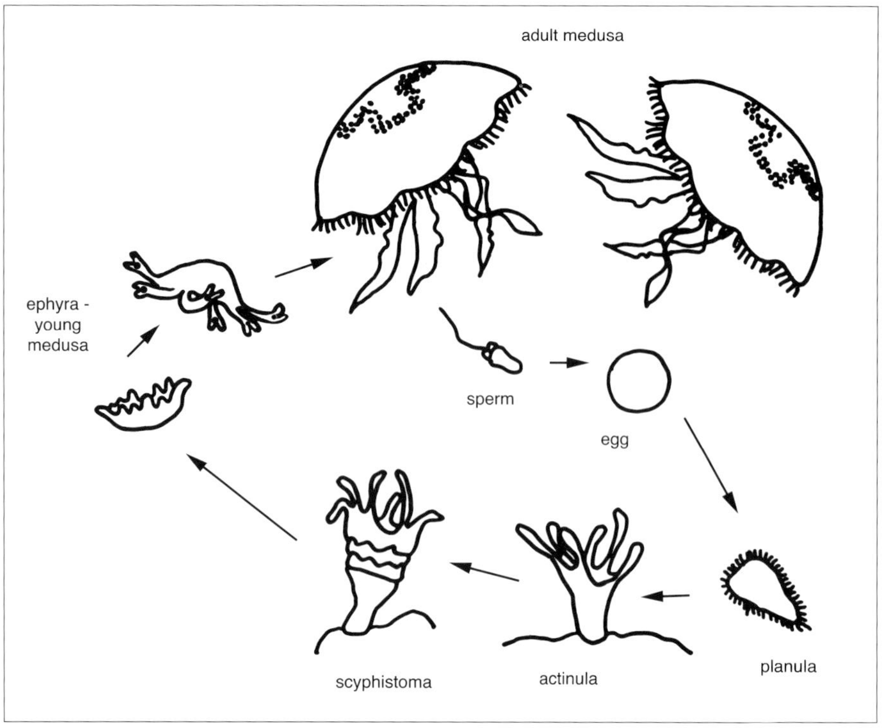

FIG. 2.23 *Life cycle of the saucer jellyfish* Aurelia aurita. *The adult medusa produce eggs or sperm in horseshoe shaped gonads (coloured a light brown or pink) located between the oral arms. After fertilisation the eggs develop into larvae, planulae that attach to algae or rocks. These grow into polyps that divide horizontally producing a layered structure in a process called strobilation. Later the layers are released forming small medusae or ephyrae about 1 mm in diameter. The adults feed on small planktonic algae and animals that are passed from the frilled oral arms to the central mouth and onto the gastric pouches for digestion.* Adapted from Knox et al., 2001. See also Rippingale & Kelly, 1995.

Südwest-Australiens in five volumes from 1907 onwards, the text and species descriptions, particularly of the invertebrate groups Polychaeta, Crustacea and Mollusca, are still relevant today.

Local studies followed the appointment of a trained zoologist, W. B. Alexander, to the Western Australian Museum in 1912 and Professor W. J. Dakin to the first chair of biology in the newly founded (The) University of Western Australia in 1913. Alexander expanded the Museum collections, an example followed by the next curator Ludwig Glauert. Dakin collected the juvenile stage or (nauplii) of the school prawn, commenced a study on the life cycle of the common jellyfish finding that the ephyrae stage appeared in June. Dakin also described the single celled ciliate responsible for phosphorescence in the estuary and Cockburn Sound. Although much of the early research ended when Dakin and Alexander left the state, Dakin's influence on naturalists and marine biologists Australia wide has been substantial. His popular book, *Australian Seashores* published in 1952, has been reprinted and revised nine times, and fully rewritten and illustrated in 1987 by Isobel Bennett. Perhaps his experience around the west coast was pivotal to his appreciation of the estuarine and marine habitats.

Later bird life and hydrology was extensively studied by Dom Serventy, a scientist instrumental in nature conservation throughout Western Australia and in the establishment of many national parks. The more watery world was largely the realm of Ernest Hodgkin and his students at the Zoology Department, The University of Western Australia (Table 2.7).

Studying the processes

Over the years efforts to understand how the estuary functions have included collections of fauna, and studies of the biology of some of the larger animals including some of the fish. The studies have enabled changes in the fauna to be documented on seasonal and longer time scales.

The extremes of salinity mean the mobile species that are unable to cope with low salinities leave the estuary when conditions are no longer suitable, while sessile species attached to jetty piles and rocks or buried in the sand die. At times when flushing is strong or sudden, fish, crabs and prawns usually found in deeper water also die. Seasonal changes restrict

TABLE 2.7 *Studies, books and articles initiated by Dr E. P. Hodgkin during his appointment at the Zoology Department, The University of Western Australia (1945–1973). This list focuses on studies by students and primarily those conducted in the Swan River Estuary. It does not include Ernest Hodgkin's own numerous papers dealing with other estuaries or advice given to many researchers.*

Year	Subject	Author
PhD		
1964	Breeding biology of some Western Australian marine and estuarine Mytilidae (Molluscs: Bivalvia)	Wilson, B. R.
1966	The biology of certain planktonic organisms of the Swan River Estuary with special reference to *Sulcanus conflictus* Nicholls (Crustacea: Copepoda)	Bhuiyan, A. L.
1968	The biology of the whelk *Dicathais* in Western Australia	Phillips, B. F.
1971	The larval stages of some Australian species of *Halicarcinus* white (Crustacea, Brachyura)	Lucas, J. S.
1971	The ecology of the crab *Portunus pelagicus* (Crustacea: Portunidae) in south-western Australia	Meagher, T. D.
1972	Biology of the copepod *Gladioferens imparipes* Thomson	Rippingale, R. J.
1975	Strategy and zoogeographic implications of the persistence of the estuarine catfish *Cnidoglanis macrocephalus* (Val.) (Plotosidae) in estuaries of south-western Australia	Kowarsky, J.
1978	The biology of the estuarine shrimp, *Palaemonetes australis* (Decapoda, Palaemonidae) in South-Western Australia	Bray, D. M.
MSc		
1955	Ecology of the Western Australian limestone reefs	Marsh, L. M.
1970	Biology of the commercially fished whiting (*Sillago* spp.) in Shark Bay, Western Australia	Lenanton, R. C. J.
1975	The estuarine ecology of two species of nassarid gastropods in south-western Australia	Smith, P. R.
Honours		
1964	Benthic fauna of the Cockburn Sound sand flats	Jones, R. (ed.)
	Sedimentary environment and distribution of foraminifera on the southern flats of Cockburn Sound	Nunn, R. M.
	Prosobranchiata of the southern flats Cockburn Sound	Jones, A.
	Faunal distribution	Meagher, T. D.
1969	Macrofauna in the Swan River Estuary chapters	Blackwell, J. (ed.)
	Arthritica helmsi	Ashman, R.
	Balanus amphitrite	Bedo, D.
	Plotiopsis australis	Blackwell, J.
	Fluviolanatus subtorta	Carrie, D.
	Parcanassa burchardi	Kowarsky, J.
	Polychaete Worms	Joll, L.
	Palaemonid Shrimps	Monzu, N.
	Gammarid Amphipods	Prince, J.
1971	Gametogenesis and the annual reproductive cycle of *Fluviolanatus subtortus* in the Swan–Avon River system	Chiffings, A. W.
1973	The response of two estuarine molluscs (*Velacumantus australis* and *Notospisula trigonella*) to seasonal hydrological change	Hughes, J.

longer-lived fauna to those species able to cope with extremes of salinity, temperature and oxygen, or short-lived species either with resistant eggs or the ability to recolonise from the sea. We know that some marine animals occasionally enter the lower reaches when conditions are favourable and that some of these include tropical species brought south by the Leeuwin Current. The studies also highlight recent pest introductions, bivalve shells that probably arrived in ballast water. Studies of fossils show that 4,500 BP the estuary was at that time a stable arm of the sea with less winter flooding and a more marine fauna than now, allowing interpretation of what changes we could expect if the estuary waters became more marine.

Invertebrates

Many invertebrates, except perhaps prawns and crabs, are not obvious or striking in appearance although their importance is unquestionable: forming the base of food webs for larger, more spectacular fish and birds.

The zooplankton of the Swan

Tiny animals or zooplankton in the water column of the estuary include the larvae of many bottom-dwelling worms, and crustaceans (crabs, prawns and barnacles) as well as the true plankton such as calanoid and cyclopoid copepods. Like the tiny plant cells or algae of the phytoplankton community, the species of zooplankton change with the winter river flow and the summer inward movement of salt water, but the phytoplankton and zooplankton communities also interact. Phytoplankton, as a primary food source, as we have seen determines zooplankton abundance and in turn the zooplankton reduce phytoplankton abundance. In winter there are generally few phytoplankton in the dark fresh waters. As the river flow decreases at the end of winter, salinity increases, temperatures rise and the algal cells multiply, as do the zooplankton.

The copepod, *Gladioferens imparipes*, is one of the most abundant species of the zooplankton in the Swan–Canning as well as in other estuaries throughout Swanland (Fig. 2.24). (See **Box 2.7 Tiny algal eaters**.) Studies during the 1970s demonstrated that in the Swan–Canning Estuary, *Gladioferens* persisted in the Lower Estuary over winter. Adults had a wide salinity tolerance, congregating in the more saline deeper waters (30‰) during the day, but migrating over 20 metres at night to the surface where salinity was only 5‰. Larvae of *Gladioferens* appeared less able to cope with the salinity changes and remained at the surface during the day. As summer saline conditions established, the population of *Gladioferens* moved upstream into the middle and upper reaches at night ahead of the saline water, gradually becoming restricted to the extreme head of the estuary where low salinity water persisted. In the Lower Estuary, as *Gladioferens* populations decreased, populations of another copepod, *Sulcanus conflictus*, gradually increased in the more saline Middle Estuary, with a third species, *Acartia clausii*, dominating

FIG. 2.24 *Calanoid copepod from the Swan estuary:* Gladioferens imparipes. Gladioferans *generally feeds by filtering small plant cells (phytoplankton) and organic debris while swimming, but spines on the dorsal surface allow it to continue feeding while attached to a hard surface.* Courtesy R. Rippingale, Curtin.

Box 2.7 TINY ALGAL EATERS

The copepod *Gladioferens imparipes* has a wide salinity tolerance, living in almost fresh to hypersaline waters. As a filter feeder removing tiny phytoplankton and debris from the water, *G. imparipes* is able to feed over winter when there is little phytoplankton growth. Females produce thirty-five to fifty-five eggs in batches about every second day depending on temperature. The eggs mature in about twenty-eight days at 15°C and twelve days at 25°C, and these growth rates allow the species to exploit the detritus in the winter run-off. The tiny hairs on the back allow the animals to attach to the bottom and still feed. It is likely that they move upstream with the tide when they are planktonic at night, and then attach during daylight and falling tides so that they retain position to be moved further upstream on the next cycle.

Populations of *Sulcanus conflictus* follow the cycle of *G. imparipes* in space and time, but generally require more saline water. They appear to be entirely planktonic and omnivorous, possessing unusual feeding appendages developed for detection and capture of prey as well as filter feeding. They prey on juvenile *G. imparipes*, possibly accounting for the decrease in numbers of this species. Studies in 1995–1996 suggest there has been a substantial decline in the population of *Sulcanus conflictus* in the last thirty years, however it is unclear if this is indicative of long-term change in zooplankton in populations of the Swan–Canning.

Acartia clausii is widespread in coastal marine and estuarine waters around the world but only lives in water of salinities more than 20‰. In the Swan–Canning it overwinters in deeper saline water and dominates the Lower Estuary and Middle Estuary in summer when water is 25‰. An omnivore, it feeds on material filtered from the water and on copepod larvae.

Oithona nana is never abundant and little is known about its biology. Like *A. clausii* and *O. nana* it enters from the ocean.

Source: Rippingale, 1987.

Box 2.8 **BARNACLES: ALGAL EATERS— TROUBLESOME FOULING ORGANISMS**

Barnacles feed on tiny algae and animals in the water, and in turn are eaten by fish and crabs. They are part of the community that keep the estuaries healthy. They are however a curse: along with mussels, sponges and other organisms they form the fouling community on boats and river structures. Efforts to reduce settlement have seen development of antifouling paints that contain toxic compounds some of which, such as tributyltin (TBT), have long-term effects on other organisms. They are now banned for use on vessels less than twenty-five metres. In many coastal areas worldwide, a legacy of TBT remains in sediments around harbours and marinas where mussels, a favourite food for people, can be contaminated.

There are three species of barnacles in the Swan–Canning. *Tetraclitella purpurascens* is only found in the lower Swan near the Fremantle Traffic Bridge. An Australian endemic it occurs in temperate waters in Western Australia from Dongara southwards, and is recognised by its depressed cone shape, its membranous base and multipored shell plates that give a honeycomb appearance when viewed from below. *Balanus amphitrite* is very tolerant of varying salinity and turbidity, and is common in estuaries and harbours throughout Australia and worldwide. It is easily recognised by the white shell with purplish-brown stripes, rounded shape and calcareous base. *Balanus variegatus* is an Australasian species ranging from Queensland around the south coast to near the Abrolhos. Found in wave-sheltered shores, harbours, bays and estuaries from low tide to about four metres on rocks, wharf piles and shells, it is tolerant of silt and other suspended material and varying salinity. It is recognised by the crosshatched purplish-brown plates, the toothed opening and the calcareous base. In the Swan–Canning the two *Balanus* species often occur together, although *B. variegatus* extends further upstream to Garrett Road Bridge on the Swan and *B. amphitrite* is often found at higher water levels. Both are stressed when salinities fall to 15‰ and prolonged salinities of less than 10‰ cause death. *Balanus trigonus*, easily recognised by its white and pink striped appearance, is a cosmopolitan species found throughout temperate and tropical regions. They occur near low tide level and are common in fouling communities on boats and structures in sheltered water such as harbours and estuaries.

Source: Jones, 1987; Jones & Morgan, 1994.

a

c

b

d

FIG. 2.25 *Barnacles of the Swan–Canning Estuary: (a)* Tetraclitella purpurascens, *base ≈ 25 mm; (b)* Balanus amphitrite, *base ≈ 13 mm; (c)* Balanus variegatus, *base ≈ 18 mm; and (d)* Balanus trigonus, *base ≈ 18 mm.* Source: Jones, 1987; Courtesy D. Jones, WAM.

the most saline areas above 25‰. Later in summer a cyclopoid type (*Oithona nana*) appeared that overlapped the range of *Acartia clausii* and *Sulcanus*.

Although there has been some information on the feeding habits of various copepods with respect to the type and size of the preferred food, the effects of grazing has, however, not been known. The role or potential role of copepods in controlling algal blooms has become more important with deteriorating water quality and the increase of blooms. A computer modelling study in the late 1990s using data collected in the field and laboratory demonstrated that copepod grazing reduced the number of dinoflagellates in phytoplankton and the size of blooms in the Upper Estuary. This new knowledge, coupled with the role of copepods as food items for other animals higher in the food web, illustrates that the health of zooplankton although appearing somewhat insignificant is an important issue for maintaining ecological function.

Box 2.9

SUMMER SPORT—PRAWNING PARTIES

Prawning parties on the banks of the Swan–Canning and the Peel Inlet were a feature of balmy summer evenings in Western Australia. Unfortunately this favourite pastime has become less successful as the prawn population has declined in recent years. The prawns are caught as they emerge from the sand to feed, particularly when the nights are darker at new moon. Most common are the river prawn, also called the greasy backed or western school prawn *Metapenaeus dalli*, and the larger western king prawn *Penaeus latisulcatus*. In the South-West, as the name suggests, river prawns spend their entire life within the rivers and estuaries. Further north they are found in coastal waters especially around Shark Bay and Exmouth Gulf through to the Kimberleys, Northern Territory and Indonesia. Similar prawns are found in estuaries along the eastern Australian coast. King prawns are found in coastal areas from Madagascar to Japan and throughout Australia.

Prawns are an 'annual crop' as they breed and grow to a size worth catching in one year. The moon dominates their life cycle. When young, they moult each month with the full moon and high tides. At maturity growth slows and they moult less often. King prawns mate on the full moon, when their skins are soft after moulting. River prawns mate at another time in the moon cycle just before the females spawn and, unlike other prawns, mating occurs when the shell is hard. Female prawns are generally larger than males and distinguished by the genital opening on the third pair of walking legs, and the males by the genital opening on the fifth pair of walking legs.

River prawns live all year in the estuary reaching about five centimetres at nine to ten months in spring. They spawn when about one year old, during summer (Fig. 2.27). Although some live for two years in deeper waters of the Middle Estuary, most are caught by prawners or eaten by other animals in the first year. Two or more dry winters in a row provide the best conditions for prawns because more young prawns survive through the winter. River prawns are usually of a size (50 millimetres) between October and February, although the season may be limited to December and January in the upper reaches. River prawns can only be caught by hand-pulled nets of four metres width and a mesh size of 16 millimetres, or 25 millimetres in Mandurah Channel. In the Murray

FIG. 2.26 *The prawns we catch: river prawns are almost colourless and translucent with distinctive blue tips and a slippery almost greasy feel, while king prawns (figured) are cream in colour with brown body marks, blue legs and tail fins. King prawns generally grow to a larger size than river prawns, but are similar in size when caught early in the prawning season.* Courtesy C. Bryce, WAM.

and its tributaries, and the Yunderup Canals, only hand scoop nets no larger than 61 centimetres and mesh of 16 millimetres are allowed.

In 1987 an estimated 50,000 people went prawning in a season, predominantly in the Swan–Canning and Peel Inlets. In recent years catches of river prawns in the Swan have been poor and far fewer people are prawning. To ensure that breeding stocks are maintained, a closed season has been imposed in the Swan–Canning from 31 July to 1 November and in the Serpentine, Murray, Dandalup and Harvey Rivers from 1 May to a date in November announced each year. A catch limit of nine litres per person per day ensures that the prawns are available to many people.

King prawns spend only part of their life in the estuary (Fig. 2.28). In the South–West, they spawn in ocean in spring and summer when temperatures are high enough. Further north where waters are warmer they may spawn year round. After hatching, the young go through several moults in the ocean before settling in the shallow coastal bays or salty lower reaches of the estuaries where they feed on plankton and dead material. They reach a size to catch during January or February, but by late summer can be up to eight centimetres in length. They move back into the ocean between March and July particularly when the first rains begin to flush the estuaries. In dry years, when the freshwater flush is small, some king prawns stay in the lower reaches of the rivers and estuaries and grow larger. The largest prawn recorded in Shark Bay reached over 24 centimetres and weighed 99 grams.

King prawns in the estuaries reach the best size in late summer and autumn when they swim at the surface and are carried seaward on the outgoing evening tides. Attracted by bright lights, they can be caught from boats and jetties using a fine-mesh scoop net. More recently hunting for king prawns has become a popular sport of scuba divers operating in the Lower Estuary.

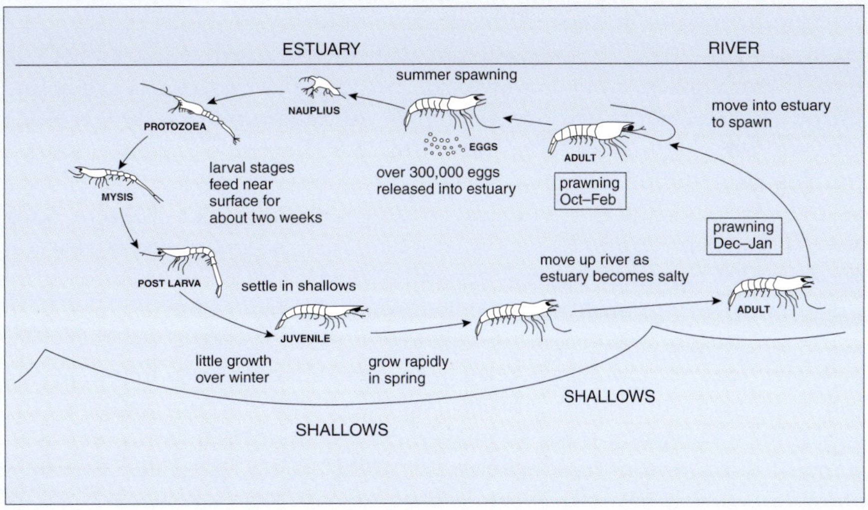

FIG. 2.27 *Life cycle of the river prawn: they breed and live within the rivers and estuaries. River prawns can produce 300,000 eggs and king prawn 1.2 million eggs at each spawning.* Courtesy Department of Fisheries Western Australia.

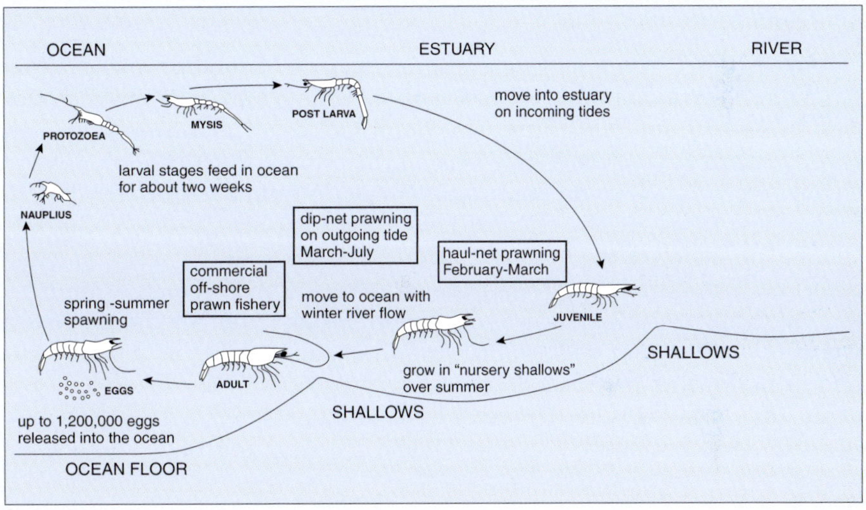

FIG. 2.28 *Life cycle of the western king prawn: they spend only part of their adult life in the estuaries. King prawns breed and the juvenile stages feed in the ocean before some enter the coastal bays or lower reaches of the estuaries.* Courtesy Department of Fisheries Western Australia.

King prawns are caught by commercial fishers in the Mandurah Channel using beam tide nets, and in the sea off Mandurah and Fremantle using larger boats and otter trawls during winter and spring. At Mandurah, however, the higher tidal flows through the Dawesville Channel have made the prawns more difficult to catch. Combined with other king prawn grounds in Shark Bay, Exmouth Gulf and Nickol Bay, the total catch is valued at about $40 million a year. To maintain this valuable resource strict regulations based on scientific research into the distribution, habits and breeding cycles cover the number of trawlers and when they can fish.

The prawns themselves are an important part of the food chain in estuary areas, providing food for black bream, mulloway, tailor, cobbler and the juveniles of many other fish species that inhabit the sheltered estuary waters.

Source: Fisheries WA, 1990a; Jones & Morgan, 1994; Potter et al., 1986.

Larger crustaceans

Amphipods, commonly referred to as sea-fleas or beach-hoppers, are more abundant in the Middle Estuary basin than in the riverine Upper Estuary reaches. Amphipods, isopods (pill bugs), shrimp *Palaemonetes australis*, and sea-spider crab *Halicarcinus* live among the rocks and mussel clumps along with the barnacles (Fig. 2.25). (See **Box 2.8 Barnacles: algal eaters—troublesome fouling organisms**). Of greater interest to many people are the river or school prawn *Metapenaeus dalli* and the larger Western King Prawn *Penaeus latisulcatus* as well as the blue manna crab *Portunus pelagicus*. (See **Box 2.9 Summer sport—prawning parties**.) Leader prawns (giant tiger prawns) that probably arrived as larvae in ballast water of ships from south-east Asia are also caught occasionally in prawning nets.

Seven isopods, pill bugs, related to garden slaters are found in the shallow subtidal areas of the Swan, showing a distribution reflecting the salinity regime. Several species found in other parts of the world appear to be recent introductions. One species, the wood boring *Sphaeroma quoyanum*, is found in eastern Australia, New Zealand, western USA and estuaries along the south coast of Australia. It is likely that *S. quoyanum* has been resident in the Swan since the 1920s and is now found in the middle estuarine reaches. In the saline Lower Estuary near the harbour, however, it appears to have been displaced by another recently introduced species.

Molluscs

A study in the 1970s found ninety-seven species of mollusc living either continuously or temporarily in the estuary, falling into four main groups (Figs 2.29, 2.30 and 9.9). First, sixty-four species of marine affinity that mostly entered as larvae and only lived in the estuary when conditions were suitable. Second, twenty-five species, also of marine affinity, which lived more or less continuously in the estuary. Third, seven species that are exclusively estuarine not living in the freshwater areas of the river or the ocean, and one species limited to fresh water with limited estuarine and no marine affinity.

Due to the marine influence, the Lower Estuary contained the greatest diversity of molluscs as well as other groups of animals such as echinoderms, and worms. Further upstream the fauna was progressively less diverse. Of the twenty-five species that lived more or less continuously in the Lower Estuary, only six lived continuously in the Middle Estuary, and only four moved

FIG. 2.29 *Common gastropod molluscs of the Swan–Canning. (a) The creeper snail* Batillaria *(previously* Velacumantus*)* australis *≈ 40 mm, one of the larger and more obvious species, is quite tolerant of salinity changes and lives in the estuary year-round. Common on the sandy shallows, the creeper snail is host to one stage of the life cycle of the schistosome bird parasite. When the parasite leaves the creeper snail to seek the next host, they have been known to burrow into human flesh causing a wheel-like rash. The dog whelks, (b)* Nassarius pauperatus *≈ 20 mm and (c)* Nassarius burchardi *≈ 8 mm, are often seen crawling across the shallow sandy areas: as scavengers they often congregate around dead fish.* Courtesy F. Wells & C. Bryce, WAM. See also Wells & Bryce, 1984.

Box 2.10 MUSSELS—LARGE AND SMALL, NATIVE AND INTRODUCED FEEDING ON THE TINY ALGAE

The mussels attach to hard surfaces with fine threads of silky material, the 'byssus'. The large edible black mussel *Mytilus planulatus* (60 millimetres) is commonly regarded as a subspecies of the Northern Hemisphere *Mytilus edulis*, hence the often-used name *Mytilus edulis planulatus*. It may, however, be more closely related to another species *Mytilus galloprovincialis* also from the Northern Hemisphere. If this proves to be the case, the name may be changed in due course. Regardless of any future name changes, the group has a remarkable worldwide distribution and may indicate a very early introduction of a marine organism to southern Australia. In the Swan it is most common in the Middle Estuary and Lower Estuary. Larvae of black mussels enter from the ocean in spring with the advancing salt water, and adults often die when salinity decreases.

The estuarine mussel *Xenostrobus securis* (15 millimetres) occurs throughout the estuary from Blackwall Reach to Guildford, its upstream limit set by the length of time the low salinity water (4‰) persists, when it must remain closed without feeding. A closely related species *Xenostrobus pulex* (≈ 20 millimetres), occasionally found in the Fremantle Harbour, lives along the coast where salinity is constant and occasionally enters the lower reaches when salinity is favourable.

The Asian bag mussel *Musculista senhousia* (25 millimetres) is a recent introduction, first found in the Swan in 1982. A native of the north-west Pacific, it was introduced into California in the 1940s and is now common near harbours throughout Australia and New Zealand. The shell is greenish in colour with radiating reddish lines, and small internal teeth on the dorsal margin distinguish it from native mussels. Like the other mussels it grows quickly and secretes byssal threads, but these usually form a web or bag around the mussel in which the faecal pellets accumulate. Unlike the other mussels, they often live in large masses on sand and the ecological impact of the invasion is still debated. The bag mussel is eaten by some fish, however the byssal bag provides a camouflage or makes it unattractive to other fish that ignore it, perhaps accounting for its success as an invader. In some countries where it has become established it was thought to inhibit seagrass growth, but other studies suggest that seagrass growth is enhanced around mussel clumps due to faecal accumulations.

The sunset shell *Soletellina* (previously *Sanguinolaria*) *biradiata* (≈ 50 millimetres) buries itself in sandy sediments of many of the estuaries and sheltered bays in south-western Australia. Often large numbers of the shells wash up on beaches. The trough shell *Spisula trigonella* (≈ 20 millimetres), although a native species common in many other estuaries in Swanland, was first found in the Swan in 1964. Documentation of this change in distribution was important, contributing to our knowledge of how natural changes occur and our understanding of the fauna found in fossil deposits.

Source: Chalmer et al., 1976; Slack-Smith & Brearley, 1987; Wilson 1967, 1968 & 1969; Wilson & Kendrick, 1968.

FIG. 2.30 *Common bivalve molluscs of the Swan–Canning. (a) The large edible black mussel* Mytilus edulis *> 60 mm; (b) estuarine mussel* Xenostrobus securis *≈ 15 mm; (c) little black mussel* Xenostrobus pulex*; (d) Asian bag mussel* Musculista senhousia *≈ 25 mm; (e) sunset shell* Soletellina *(previously* Sanguinolaria*)* biradiata *≈ 50 mm; and (f) trough shell* Spisula trigonella *≈ 20 mm.* Courtesy F. Wells & C. Bryce, WAM. See also Wells & Bryce, 1984.

into the Upper Estuary after long periods of low river flow. The six species that lived more or less continuously in the Lower and Middle Estuary, the sunset shell *Soletellina* (formerly *Sanguinolaria*) *biradiata*, trough shell *Spisula (Notospisula) trigonella*, *Tellina (Macomona) deltiodalis*, *Theora (Endopleura) lubrica*, the creeper *Batillaria (Velacumantus) australis*, dog whelks *Nassarius pauperatus* and *Nassarius burchardi*, are commonly found in other estuaries but occasionally occur in the marine enviroment. (See **Box 2.10 Mussels—large and small, native and introduced feeding on the tiny algae**.)

Typical estuarine species are the estuarine black mussel *Xenostrobus securis*, the tiny white bivalve *Arthritica semen*, the false mussel *Fluviolanatus subtorta* (*Anticorbula amara*), and the tiny gastropod snails *Tatea preissii*, *Potamopyrgus* sp., *Assiminea* sp. and *Hydrococcus brazieri* (previously *graniformis*). The first four may move into the Lower Estuary during floods. *Potamopyrgus* sp. and *Assiminea* sp. are only known from the Upper Estuary and *H. brazieri* has not been found in the Swan since 1840. *H. brazieri* fluctuates greatly in abundance and was the dominant mollusc in the Peel–Harvey during the years of algal blooms. (See also Fig. 3.21.)

Only four species of molluscs appear to move into the Upper Estuary probably reflecting the extremes of salinity and perhaps scouring river flow. Since the 1930s, the range of the false mussel *Fluviolanatus subtorta* has extended up the Avon River beyond York, reflecting the salinisation of the river water. The truly freshwater species *Plotiopsis australis* is found in the upper reaches of the Swan.

New arrivals

The fauna of the Swan varies from year to year depending on conditions in the estuary due to river flow and the persistence of saline conditions, or the supply of larvae in the offshore waters. Some of the most common species, such as the bivalves *Spisula trigonella* (1964) and *Theora lubrica* (1965) and the snail *Nassarius burchardi* (1973), while known from other south-west estuaries have only been present in the Swan–Canning for a relatively short time. In contrast, the Asian Bag Mussel *Musculista senhousia* (Fig. 2.30d), present since 1982, probably arrived from further afield as larvae in ballast water of ships or as fouling. Recent introductions also include a number of isopod crustaceans (similar to garden slaters) that are now common in the main estuary basin.

Worms

Burrowing worms are also abundant, particularly in the sandy and muddy shallows where they are an important food for fish and wading birds. Worms also play a role in nutrient cycling, the burrowing habit over-turning and aerating sediment in the process termed bioturbation. Other species, such as fan worms, live attached to rocks and other structures and filter small organisms from the water column.

Fish and fisheries—reaping the Swan's rich bounty

The fish community in the Swan–Canning Estuary is quite diverse with over 130 species recorded. Detailed studies report that the greatest diversity is due to those species classed as marine stragglers, species that enter infrequently and in low numbers when conditions are favourable (Fig. 2.31 and Table 2.8). Marine stragglers are more likely to enter in summer when the waters are more saline, and are most common in the Lower Estuary. Another group, the marine–estuarine opportunists, breed in the ocean and use the sheltered and productive estuary as an alternative nursery area to the coastal bays. The estuarine and marine category contains species that complete their life cycle in the estuary or in sheltered marine waters outside the estuary. The solely

FIG. 2.31 *Seahorse* Hippocampus *is a common visitor to the lower reaches of the Swan Estuary, particularly in summer when the waters are saline.* Courtesy C. Bryce, WAM.

estuarine group complete their life cycle in the estuary. In the Swan–Canning the semi-anadromous group is only represented by one abundant species, the Perth herring *Nematalosa vlaminghi*, that lives in the ocean but spawns in the Upper Estuary. Freshwater species are sometimes caught in the estuary when flushed downstream from the river reaches. An anadromous species, the pouched lamprey *Geotria australis*, uses the estuary as a route from the ocean feeding area to breeding areas in the distant inland river reaches. (See **Box 7.1 Swimming through**.)

The diversity of marine species entering the estuary is not surprising when the direct connection to the sea, the limited seasonal river flow, and intrusion of salty water are considered. Sharks and rays, giant herring *Elops machnata*, pilchard *Sardinops neopilchardus*, buffalo bream *Kyphosus sydneyanus*, sea trumpeter *Pelsartia humeralis*, various types of goat fish, flatheads, leatherjackets, box and puffer fish, all common along the coast, move through the harbour and into the estuary basin particularly over summer. Despite this diversity, some species such as King George whiting *Sillaginodes punctata*, yellowfin (western sand whiting) *Sillago schomburgkii* and sea garfish *Hyporhamphus melanochir* are not as abundant as expected when compared with their abundance in the Peel–Harvey. This has been attributed to the passage they must to take to enter the estuary through the narrow harbour, which is noisy and turbulent due to shipping. Other explanations for the different fish abundances in the two systems may be related to habitat, the Peel–Harvey having more extensive shallow areas than the Swan–Canning.

As the name marine–estuarine opportunist suggests, fish in this group use the Swan–Canning as an alternative to sheltered bays along the coast. Juveniles are abundant and feed and grow in the sheltered waters along with the adults that form a large part of the commercial and recreational fisheries. Species include sea mullet *Mugil cephalus*, yellow-eye mullet *Aldrichetta forsteri*, cobbler *Cnidoglanis macrocephalus*, tailor *Pomatomus satatrix*, and mulloway *Argyrosomus japonicus*. The major commercial species have been the sea mullet, cobbler, yellow-eye mullet and Perth herring and, to a lesser extent, the blue manna crab. The most sought after recreational species are the cobbler, black bream, tailor, mulloway, western school and king prawns, and the blue manna crab. Cobbler breed in the estuary and in the nearby marine waters, but the populations do not interbreed. (See **Box 8.4 Cobbler: spines to fear—great to eat.**) Maintenance of the tailor population, one of the most popular fish with recreational fishers, became a major concern in the 1990s. It was recognised that larger breeding fish weighing one to two kilograms or larger, and such a thrill to catch, had become rare and the traditional 'runs' of fish in summer were almost nonexistent. A campaign to ensure size and bag limits are honoured appears to be successful. (See **Box 2.11 Tailor limits: reducing the plunder—maintaining a resource**.)

Some of the most abundant estuarine species include the black bream *Acanthopagrus butcheri*, yellow-tailed trumpeter *Amnitataba caudivittatus*, Australian anchovy *Engraulis australis*, the hardyheads or atherinids, and the bottom-dwelling gobies. Most of the estuarine species spawn in late spring

TABLE 2.8 *Life cycle characteristics of fish in the shallows of the Swan–Canning Estuary. Note in each category the numbers of fish caught in deeper offshore waters of the estuary are slightly different but the pattern is very similar.* Adapted from Potter & Hyndes, 1999.

Life category	No. of species	% of species	% of individuals
Marine	**47**	**66.2**	**18.5**
Stragglers	28	39.4	0.3
Estuarine opportunists	19	26.8	18.2
Estuarine	**19**	**26.8**	**34.2**
Estuarine & marine	9	12.7	20.4
Solely estuarine	10	14.1	13.8
Semi-anadromous	**1**	**1.4**	**47.2**
Freshwater	**4**	**5.6**	**0.1**
TOTAL	**71**		

Box 2.11 **TAILOR LIMITS: REDUCING THE PLUNDER—MAINTAINING A RESOURCE**

In 1993 an estimated 86,000 anglers pursued tailor. Not surprisingly 70 per cent of the fishing occurred near the main urban areas between Bunbury and Jurien Bay. As predatory fish, tailor are fast moving with voracious appetites and easy to catch, characteristics that makes them vulnerable to overfishing.

The west coast estuaries of the Swan–Canning, the Peel–Harvey and the Leschenault along with the sheltered coastal bays are key nursery areas for young tailor. Young tailor about 35 to 45 millimetres in length enter the estuaries in late spring to early summer to feed on the abundant small fish such as anchovies. Over winter, they probably move to more salty deeper areas in the lower reaches or leave if the freshwater flush is strong. At one year they are about 15 to 18 centimetres in length. Those that remain in the estuaries leave towards the end of the second summer and school along beaches before moving offshore as the water temperature drops. Spawning takes place in spring although the location of the spawning grounds is unknown. Tailors take two to three years to reach maturity at about 30 centimetres length. Some tailor may survive over fourteen years and reach over 8 kilograms in weight.

The program to maintain tailor stocks and ensure *Fish for the Future* entailed research studies on the biology of the fish, surveys of recreational and commercial catch, an assessment of the survival of released fish, and a publicity campaign to encourage the speedy return to the water of all undersized fish and those in excess of the bag limit. The bag limit was also reduced from twenty fish to eight per fisher.

Further surveys and research will allow assessment of the success of these conservation measures, a greater understanding of how tailor populations vary over natural cycles, and levels of exploitation that will ensure the ever increasing recreational fishery remains viable.

Source: Fisheries WA, 1993b.

and summer when water movement due to river flow is low, and salinity is fairly high and stable. Many are found in particular parts of the Estuary. For example, Swan River Presbyter's hardyhead *Leptatherina presbyteroides* and the long-finned goby *Favonigobius lateralis* are found mainly in the entrance channel. Ogilby's hardyhead *Atherinosoma ogilbyi* is found in the lower reaches, the elongate hardyhead *Atherinosoma elongata* and gobbleguts *Apogon rueppellii* in the basin, while Wallace's hardyhead *Leptatherina wallacei* and a number of other gobies are found in the less saline reaches. Seahorses are usually found near pylons or the brown alga *Cystoseira* (Fig. 2.31). (See **Box 2.12 Quite a mouthful—a safe place to keep one's eggs**; **Box 2.13 A popular fish—the black bream** and **Box 2.14 An unpopular fish—the common blowfish or banded toadfish**.)

Life cycles of fish in the Swan–Canning Estuary

Four of the five species of hardyhead have a one-year life cycle, and reach a total length at maturity of forty-five to eighty-five millimetres. Ogilby's hardyhead and gobbleguts live for two years, and in the Swan–Canning reach 160 and 104 millimetres respectively. Yellow-eye mullet, sea mullet, Perth herring, cobbler and blowfish live for several years (Table 2.9). Species that live in the estuary generally produce relatively few eggs. Hardyheads attach eggs to seagrass leaves, and gobies build nests under stones attaching eggs to the inner side.

Growth for most fish species is seasonal. There is an increase in length between spring and autumn, which is generally faster for fish in the estuary compared to those that remain in the near shore marine areas. For example, one study found that cobbler in the estuary grew 200 millimetres compared to 160 millimetres in the ocean, and blowfish grew 85–100 millimetres compared to 65 millimetres, illustrating the value of estuaries as nurseries. The high growth is thought to be due to abundant food, calmer waters and osmotically favourable salinity. The absence of predatory fish is also thought to account for the abundance of juvenile fish.

Box 2.12 QUITE A MOUTHFUL—A SAFE PLACE TO KEEP ONE'S EGGS

Gobbleguts are found from Albany on the south coast along the west and north coasts across to Arnhem Land in the Northern Territory. They are one of the most abundant fishes around seagrasses along the coast and in estuaries. In the Swan they are often found around the seagrass *Halophila* and lurking around rocks and jetty piles (Fig. 2.32).

Gobbleguts generally live for one year and grow to about 60 millimetres, but some may survive a further year reaching a length of about 104 millimetres. Adults move into deeper water during winter and then move back into shallow water to spawn as the temperature rises in spring. After the female lays the eggs, the male collects them and holds them in his mouth until they hatch in summer. While brooding the mouth is greatly distended, and the sight of a fish with a mouth full of eggs probably accounts for the name 'gobbleguts'. Generally gobbleguts produce about 200 eggs, which is less than many other cardinal fishes. However the diameter of the eggs tends to be larger, over 2 millimetres, which is greater than those produced by other cardinal fish that are less than a millimetre. At hatching they are also much larger, about 7.5 millimetres, over twice the size of most other species in the group. This larger hatching size is probably an advantage for a fish living in unpredictable and stressful estuarine environments. Conditions in the Swan in comparison with other estuaries are however relatively stable, except for brief periods of heavy freshwater discharge in winter, and even then deeper areas of saline water provide a refuge.

Gobbleguts are carnivores, feeding on the abundant small animals that live in estuaries. They feed on tiny copepods when young and on larger crustaceans, bristle worms and small fish when larger. Adults also nip the tops of the siphons of buried bivalves. The prey of adults also changes over a daily cycle. More copepods are eaten during the day and more amphipods at night, probably reflecting the activity patterns of the prey. Copepods tend to move to surface at night, whereas amphipods move into the middle of the water column where gobbleguts tend to swim.

Source: Chrystal et al., 1985.

FIG. 2.32 *The gobbleguts* Apogon rueppellii. Courtesy B. Hutchins, WAM.

TABLE 2.9 *Life cycle duration and growth of some common fish in the Swan–Canning Estuary.* Adapted from Loneragan et al., 1987.

Common name	Species name	Length range after one year (mm)	Maximum length (mm) and age (years) in the Swan
One-year life cycle			
Presbyters hardyhead	*Leptatherina presbyteroides*	50–75	85 (1)
Elongate hardyhead	*Atherinosoma elongata*	50–80	85 (1)
Wallace's hardyhead	*Atherinosoma wallacei*	40–65	70 (1)
Mullet hardyhead	*Allanetta mugiloides*	50–70	70 (1)
Two-year life cycle			
Ogilby's hardyhead	*Atherinomorus ogilbyi*	40–80	160 (2)
Gobbleguts	*Apogon rueppellii*	45–80	104 (2)
Three or more years life cycle			
Perth herring	*Nematalosa vlaminghi*	80–150	361 (8)
Cobbler	*Cnidoglanis macrocephalus*	180–250	683 (6)
Yellow-tailed trumpeter	*Amniataba caudavittatus*	70–160	289 (> 4)
Black bream	*Acanthopagrus butcheri*	100–160	> 400 (4)
Yellow-eye mullet	*Aldrichetta forsteri*	100–200	380 (> 4)
Sea mullet	*Mugil cephalus*	140–260	580 (> 4)
Common blowfish	*Torquigener pleurogramma*	75–130	229 (6 +)

Box 2.13 **A POPULAR FISH—THE BLACK BREAM**

The black bream *Acanthopagrus butcheri* is one of the most popular fish with recreational fishers in the Swan–Canning. Although found from mid-west Western Australia to Victoria, each population is quite distinct as the estuarine life cycle limits interbreeding between populations. (See **Box 10.2 Black bream—a real estuary fish**.)

In the Swan–Canning Estuary, black bream move downstream during winter and only leave the estuary at times of heavy flooding. They move upstream in August and November to breed in the upper reaches, moving downstream and into deeper water as they grow. In the Swan–Canning they grow rapidly, particularly over summer, reaching maturity at 2.2 years when they are about 215 millimetres long.

Black bream use their peg-like teeth to prise mussels, barnacles and tubeworms from rock, piles and pylons and also eat crabs, prawns, river worms and small fish as well as algae, which is probably swallowed along with the small animals. The younger smaller black bream generally feed on smaller soft-bodied prey. In estuaries east of Albany (such as Wellstead Estuary) black bream eat a lot of algae, the food quality of which is lower, probably accounting for the slower growth rates observed in these estuaries.

A study in the 1990s found very few black bream older than five years in the Swan–Canning. This is probably due to the heavy fishing pressure from recreational fishers throughout the year in all parts of the system and by commercial fishers during winter and spring in the Lower Estuary.

Hatchery studies to raise stock for aquaculture and for restocking of depleted estuarine populations through Swanland are currently being conducted at Fremantle TAFE. In 1995 about 767 juveniles tagged with yellow 'spaghetti' tags were released in the Swan–Canning with about 13 per cent recaptured over the next three years, indicating that hatchery raised fish could survive in the wild. In 1998 250,000 juveniles about 30 millimetres long and five months old were also released into the Swan after the loss of large breeding stock due to a chemical spill. Success of the later release is unknown but the abundance of black bream in 2001–2002 certainly indicates that environmental conditions were suitable for survival.

Hatchery raised juveniles were also released into the Blackwood–Hardy in mid 2002. The first 60,000 released, about 20–25 millimetres in length, had been 'tagged' by chemical staining of the otoliths (a grain of calcium carbonate in the inner ear) with further releases of older larger fish marked with external tags planned over the next few years. Debate continues, however, as to the effect this will have on fish stocks in the long term.

Source: Sarre & Potter, 1999.

Diet of fishes

Many of the fish such as black bream are carnivorous, feeding on the abundant invertebrates. The anchovy *Engraulis australis*, the sandy sprat *Hyperlophus vittatus* and the Western Australian pilchard *Sardinops neopilchardus* are zooplanktivores, feeding on the tiny animals that feed on the even tinier algae floating in the water. Herbivorous fishes include the garfish and striped (six-lined) trumpeter, although they also consume some animal material particularly at night (Table 2.10). Perth herring and mullet feed on algae but also consume detritus, some of which is derived from the seagrass *Halophila* growing in the shallows. (See **Box 2.15 Perth's herring *Nematalosa vlamingbi*** and **Box 2.16 Fishing prize—the mulloway *Argyrosomus japonicus*.**)

Not always a good place to live—fish and fish kills

The fish populations of the Swan–Canning are one of the tangible ways for most people to assess the health of the estuary. The fish populations are remarkably diverse and support other fish and bird life. Despite the enveloping city with problems of nutrients and contaminants, we can still catch and eat fish from the river.

Fish kills are quite common in many estuaries; some are associated with low oxygen concentrations during algal blooms. Other fish kills are associated with the overturn of anoxic water and the presence of hydrogen sulfide and ammonia. (See **Chapter 10 Estuaries to the east of Albany: Beaufort**

Box 2.14 **AN UNPOPULAR FISH — THE COMMON BLOWFISH OR BANDED TOADFISH**

The common blowfish *Torquigener pleurogramma* is found in estuaries and bays along the southern half of Australia from Queensland to about Coral Bay on the west coast. On the west coast, blowfish often form large schools and have been known to attack spearfishers. Blowfish are a nuisance to line fishers, attacking baited lines and prawners when large numbers are caught in nets. Blowfish can reach 22 centimetres in length and about 220 grams in weight. The body is torpedo-shaped with a soft skin but with small spines and fused beak-like teeth. Banded toadfish have several vertical lines on the lower head and a horizontal line along the sides (Fig. 2.33). The common name of blowfish refers to the ability to inflate the abdomen by swallowing water or air.

Blowfish and pufferfish belong to the Family Tetraodontidae. Members of this family, along with others in closely related families that include boxfish and porcupine fish, are poisonous to eat as the skin contains the poison tetrodotoxin. Concentrations of the poison vary throughout the year, and when concentrations are high cleaning the fish can contaminate the flesh. Known as fugu, blowfish are eaten in Asian cultures, however despite the tradition of use and preparation the practice is still risky and many people die each year. In southern Australia people have died from eating the flesh, as have dogs foraging along beaches. Some seabirds such as seagulls and other fish, however, appear to be immune to the poison.

In the Swan–Canning, young fish less than 130 millimetres long feed mainly on polychaete worms and amphipods. Older fish feed on bivalve molluscs such as the sunset shell *Soletellina biradiata*, the introduced Asian bag mussel *Musculista senhousia* along with the black estuarine mussel *Xenostrobus securis* and the tiny gastropod *Hydrobia buccinoides*.

FIG. 2.33 *The banded toadfish* Torquigener pleurogramma. Courtesy S. Morrison, WAM.

Blowfish in the Swan–Canning are particularly abundant in the shallow water of the Middle Estuary and in the more marine area near the estuary mouth, a trend also observed near the entrances to the Peel–Harvey. As an indication of their abundance, in the summer of 1985–1986 an estimated 30,000 kilograms of blowfish were discarded by people prawning in the Middle Estuary. Blowfish are less common in the Upper Estuary but move upstream with the wedge of saline water, particularly in years when the winter freshwater flow has been light and more marine conditions establish earlier. Some blowfish stay in the estuary all year.

Blowfish leave the estuary to spawn in the sea at the end of the second year when they are about 125 millimetres long. Peak spawning occurs in early to midsummer. Young blowfish start moving into the estuary or coastal bays in July to August, a time when freshwater discharge is usually high. The newly spawned fish are then about six months old, 50–70 millimetres long and weigh 2–6 grams. At the same time older blowfish re-enter the estuary so that populations often contain several generations: in the Swan–Canning this can include up to seven age groups. In this respect blowfish are similar to yellow-eye mullet and cobbler. Populations in the estuary and along the coast are distinct and fish in the estuary grow faster, no doubt due to the large amount of suitable food. Populations of blowfish fluctuate greatly, for example in late 1980 and 1985 populations were particularly large due to the large recruitment of young in those years. Numbers were also very high in 1977, 1978 and 1984 but in contrast, due to weak spawning in 1983–1984, there was only a low recruitment in 1984–1985.

Source: Edgar, 1997; Hutchins, 1986; Hutchins & Swainston, 1996; Potter et al., 1988.

Inlet: Portrait of a fish kill and **Chapter 4 Leschenault Estuary and the smaller estuaries of the Geographe Bay area: Vasse–Wonnerup—Fish kills**). Fish kills in the Swan, however, appear to have been due to a number of causes. Strong river flow can stratify the water column so that fish die in the isolated deoxygenated bottom waters. In 1992 a large fish kill between Rivervale and Guildford was attributed to the low oxygen concentrations of the water column. In contrast, mullet deaths in the Swan between East Perth and Guildford in March 2000 were caused by a parasite *Myxobolus*, although there may have been additional stress from low oxygen conditions in the deeper areas isolated by the freshwater layer. There was also a large kill upstream of Perth Water associated with the dumping of poisonous herbicides and pesticides in June 1998. The most spectacular was in April 2003 when heavy rain brought fresh water loaded with nitrogen to the warm calm waters,

TABLE 2.10 *Food items in the diet of the common fish species in the Swan–Canning Estuary.* Adapted from Loneragan et al., 1987; Potter & Hyndes, 1999.

Common name	Species	Prey type	Food item
Detritivore			
Perth herring	*Nematalosa vlaminghi*	I	Organic debris
Sea mullet	*Mugil cephalus*	E/I	Detritus. Juv. also planktonic copepods and cladocerans
Herbivores			
Striped trumpeter	*Pelates sexlineatus*		Red and green algae
Sea garfish	*Hyporhamphus melanochir*		Seagrass *Zostera*, algae, diatoms
Omnivores			
Black bream	*Acanthopagrus butcheri*	E/I	Juv. detritus, amphipods, nereid polychaetes and algae. Adults bivalves (*Xenostrobus*) and crabs
Yellow-eye mullet	*Aldrichetta forsteri*	W/E	Juv. terrestrial insects, diatoms and planktonic copepods. Adults algae, amphipods, gastropods and bivalves, shrimp *Palaemonetes* and capitellid polychaetes
Carnivores			
Hardyheads	*Leptatherina presbyteroides*	E/W	Planktonic copepods, larval bivalves, forams and terrestrial insects
	Atherinosoma elongata	E/I	*Palaemonetes*, nereid polychaetes, small molluscs
	Leptatherina wallacei	E	Planktonic copepods, amphipods and nereid polychaetes
	Atherinomorus ogilbyi		Benthic crustaceans, polychaetes
	Allanetta mugiloides		Planktonic crustaceans
Cobbler	*Cnidoglanis macrocephalus*	I/E	Amphipods, bivalves, nereid polychaetes
Bar-tailed flathead	*Platycephalus endrachtensis*		Fish, shrimp and prawns
Yellow-tailed trumpeter	*Amniataba caudavittatus*	I/E	Juv. nereid polychaetes, amphipods, algae and bivalves. Adults nereid and spionid, polychaetes, bivalves, algae, crabs, prawns, and midge larvae
Gobbleguts	*Apogon rueppellii*	E	Amphipods, planktonic copepods, polychaetes
Yellowfin whiting	*Sillago schomburgkii*		Polychaetes, amphipods
Tailor	*Pomatomus saltator*		Fish, prawns and shrimp
Mulloway	*Argyrosomus hololepidotus*		Fish, prawns and shrimp
Common blowfish	*Torquigener pleurogramma*	E adults E/I	Juv. amphipods and nereid polychaetes. Adults amphipods and large bivalves

Prey type: I—infauna; E—epibenthic fauna and flora; W—throughout water column.

where high light favoured growth of yet another little known dinoflagellate alga *Karlodinium micrum*. Initially the low oxygen concentrations due to the layer of fresh water probably created stressful conditions for fish that then succumbed to toxins produced by the algae. Between April and June 2003 the *Karlodinium* bloom in the Swan and Canning Rivers killed thousands of fish of all sizes and many types including Black bream, Perth herring, mulloway, flathead, flounder, mullet and Swan River gobies. Another bloom of *Karlodinium* in late June 2004 killed about 30,000 fish, predominantly gobbleguts and juvenile blowfish in the Como foreshore and Matilda Bay.

Although most fish kills have been associated with low oxygen conditions or algal blooms following an influx of fresh water loaded with nutrients, particularly oxidised nitrogen, there is growing concern about a decline in fish health due to other pollutants. Fish appear to be affected by changes in pH (more acidic water), hydrocarbons or heavy metals that enter the river and estuary in stormwater from drains. Like the influx of nutrients these pollutants are particularly high in the first flush of water after a period of low flow, however in contrast to episodic fish kills associated with algal blooms they may herald a chronic decline in fish health.

Fishing in the estuary—an important resource

Fishing techniques in the Swan–Canning have been through many changes. At first the Aboriginal people armed with kylies, a type of small boomerang, and spears fished along the shallows in summer (Map 2.3 and Fig. 3.7). In the early days of European settlement anyone could catch, barter or sell fish; and fish were an important food or necessary resource while the colony

Box 2.15 PERTH'S HERRING *NEMATALOSA VLAMINGHI*

The Perth herring (boney herring) *Nematalosa vlaminghi* has a distinctive elongate filament at the rear of the dorsal fin and a black spot behind the gill (Fig. 2.34). Common in marine and freshwater environments along the west and north-west coast, Perth herring are very abundant in the Swan–Canning often contributing a greater biomass than any other species to the commercial fish catch. Generally regarded as poor eating fish, they are caught mainly as rock lobster bait. Perth herring belong to the same family (Clupeidae) as the scaly mackerel, pilchard and sprats, and largely feed on detritus.

The Perth herring usually enter the estuary in spring and migrate upstream to spawn in the Upper Estuary in late spring and summer, behaviour that is termed semi-anadromous. They grow fairly slowly to about 105 millimetres in length and 8 grams in weight at the end of the first year, not reaching commercial catch size until the third year when they are sexually mature. In contrast, sea mullet grow rapidly in the Swan reaching a minimum legal size (240 millimetres) at the end of first year. Although there is no minimum legal size, the legal mesh of 57 millimetres usually excludes immature fish shorter than 160–170 millimetres. Perth herring probably live about 8 years, although a population aged in the late 1970s was mostly less than 5 years old.

FIG. 2.34 *The Perth herring* Nematalosa vlaminghi. Courtesy B. Hutchins, WAM.

Source: Chubb & Potter, 1986.

Box 2.16 **FISHING PRIZE—THE MULLOWAY *ARGYROSOMUS JAPONICUS***

Near the city the riverbanks are often lined with anglers patiently waiting, sometimes all night, for a sudden rush of line that heralds the presence of a mulloway or kingy. What a sight, to catch such a giant; the fish twisting and turning through the city lights as the catch is brought ashore. The first and last quarters of the moon are the most popular times for mulloway fishing. Rising tides at dawn or dusk appear to be optimum times.

In Australia, mulloway range from Rockhampton in Queensland, around the south coast and up the west coast to near Exmouth. A closely related fish is found further north. In eastern Australia mulloway are known as 'jewfish' abbreviated from 'jewel fish', a name derived from the large ear bones or otoliths that are used in jewellery. In the west the name 'kingy' has been popular, derived from the river kingfish, but this is no kingfish like the yellowtail, amberjack or samson fish. Around South Africa and Madagascar they are known as kob. Mulloway are also known as croakers from the characteristic croaking sound produced by vibrations of the swim bladder.

Although mulloway can grow to over 60 kilograms and 1.8 metres in length along the west coast, a fish of 30 kilograms is considered large, and most of those caught along the banks of estuaries and beaches of south-western Australia are only 5–15 kilograms. Newly caught fish exude a smell reminiscent of freshly cut cucumber. Juvenile fish of about 1.5 kilograms are termed 'soapies', referring to their taste, and larger fish of 2–5 kilograms as 'schoolies'. Primarily caught by recreational anglers, a further 20–30 tonnes per year are caught as 'by catch', by beach seine and wetline commercial fishers. Often large fish congregate under nets of pilchard's 'mulies', but fishers are limited to a daily catch of two per person.

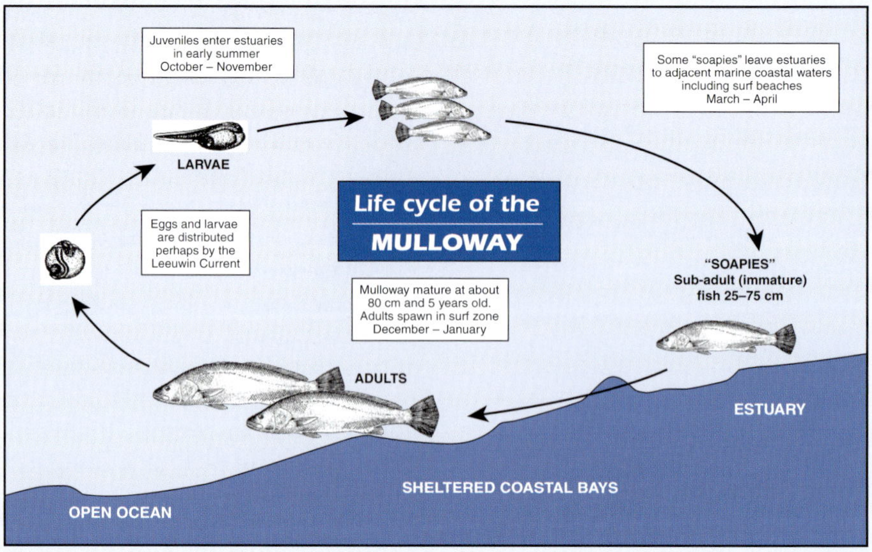

FIG. 2.35 *Mulloway use the Swan–Canning Estuary as both a nursery and a feeding ground for adults. Larger fish prefer more salty water and are found in the middle and lower reaches during spring and summer, generally leaving the estuary when salinity falls in winter. They may overwinter in Cockburn Sound and move back into the estuary in September–October to feed and build up reserves for spawning. They leave the estuary to spawn (probably behind the surf zone) in December–January and return again towards the end of summer. Juveniles found throughout the estuary appear to be more tolerant of fresh water and may remain in the estuary over winter when the freshwater flush is not too severe. Adult fish move in and out of the estuary.* Courtesy DoF.

Recreational fishers have a daily bag limit and there are restrictions on the minimum length of the fish.

Mulloway are covered with thick scales, and have a line of silver 'portholes' along the lateral line and fiery amber eyes. With a body more like a slow moving fish, the convex spade-like tail powers the fish through the water after prey. Mulloway are predators possessing a large mouth with small strong canine teeth. They are opportunistic, feeding on whatever they find, such as small crustaceans, worms, squid, octopus to large fish. In the Swan, like other estuaries, they favour the main resident fish such as bony herring, mullet, trumpeter and tailor. They hunt by smell and vibrations are sensed through the lateral line rather than sight, so they are well equipped to feed at night in turbid waters.

Mulloway appear to reach sexual maturity at about five years old when they are 70–80 centimetres long and 4 kilograms in weight (Fig. 2.35). Older and larger females produce higher numbers of more healthy eggs. The larvae probably grow at sea for some months before entering estuaries at 15–16 centimetres, then move outside to the open ocean in the following winter at about 35 centimetres. In Western Australia juveniles less than one year old live in nearshore marine areas, while fish one to six years old live in both marine and estuarine water. Adult fish generally live in the sea, although they are still common in and near estuaries and appear to stay in one area or return to particular areas. They often form large groups or shoals of similarly sized fish, although large mulloway also hunt alone.

Source: Fisheries Department of WA, 1993b.

relied on imported and sometimes irregular supplies before the adjacent land was brought into production. After 1899 and until 1905 licences were only needed by people selling fish or using a seine net, a rule that was extended to include simply possessing a net. The importance of the fish supply again assumed great importance during World War I and the Depression with up to 130 men fishing mostly for sea mullet, crabs and prawns. Until 1940 anyone could catch fish with nets other than seine nets as long as the fish were not sold. Licences were required for all net fishing from 1940 to 1949. After 1949 a formal licensing distinction was drawn between professional and amateur fishing.

During World War II canning commenced using Perth herring, mullet and cobbler, continuing at varying levels until the 1970s. Perth herring were also caught for rock lobster bait during the 1960s. The cobbler catch has fluctuated with demand for table fish.

Commercial fishery

Commercial net fishing is not permitted in the Swan upstream of Heirisson Island; near the Narrows Bridge; on the Canning upstream of Second and Sulman Avenues at Rossmoyne; within 100 metres of the jetties at Mends, Coode Street, Como, Applecross, Point Walter and Nedlands or the Canning Bridge; or between Point Walter and Point Resolution and the western ends of South and North Moles. In essence the fishery is restricted to Melville Water, a section of Perth Water and the lower Canning. There are also restrictions on the time and type of nets that may be used. There is little bycatch or discard of catch as all catch is marketed in the greater metropolitan area.

Total fish catch was 30 tonnes in 1952 increasing to 320 tonnes in 1973. Catch then declined to 120 to 140 tonnes in the early 1980s. The initial rise reflected a demand for yellow-eye mullet and Perth herring for rock lobster bait, and the decrease in the 1970s that continued through to the late 1990s reflects a decline in the market (Graph 2.12). The maintenance of Catch per Unit Effort (CPUE) indicates that in most instances commercially exploited fish have not declined in abundance, but the decrease in catch has been due to decreased fishing effort from thirty-two units in 1979 to only four in 2001 (Table 2.11). The reduction in fishing effort follows the voluntary surrender of licences through a Fisheries Adjustment Scheme. This reduction has been part of a management program to maintain fish stocks and reflects a shift to recreational fishing. (See **Chapter 1 Understanding Swanland: Fishing in the past—fishing for the future**.)

Perth herring, sea mullet, yellow-eye mullet and cobbler account for 80 per cent of the total finfish catch. The catch of Perth herring was 180 tonnes in 1968, decreasing to 60 tonnes in 1970, to 160 tonnes in 1973 and 50–90 tonnes in the early 1980s. Sea mullet catch varied annually between the 1950s and early 1970s but declined after 1973 as a result of a reduction in fishing licences. The CPUE for both Perth herring and sea mullet has remained at a similar level to that of the previous two decades indicating that fish are still abundant in the estuary. The yellow-eye mullet catch has always been lower than that of sea mullet, ranging from 2 tonnes in 1952 to 30 tonnes in 1970s, declining further after 1976.

TABLE 2.11 *Fisheries units operating in the estuarine fisheries; along the west coast fisheries units are limited to a single estuary.* Source: Lenanton, 1984; Fisheries Western Australia 1999; Ayvazian & Nowara, 2000, 2001, 2002, 2003.

	1979	June 1998	October 1998	April 1999	April 2000	May 2001	May 2002
Swan–Canning	32	9	6	6	4	4	4
Peel–Harvey	45	19	14	14	9	7	8
Leschenault Inlet	18	6	6	6	6	0	0
Hardy Inlet	4	4	2	2	1	1	1
South Coast	72	36	33	31	31	29	15

In contrast, the catch and the CPUE for cobbler and black bream followed a similar downward trend. Catches of cobbler were much lower in the 1970s and 1980s than in the earlier years. The decline continued from about 4–12 tonnes in the 1980s to less than a tonne in 1998. The cobbler fishery is regarded as over exploited in the Swan–Canning, as it is in the Peel–Harvey and the Leschenault. (See **Box 8.4 Cobbler: spines to fear—great to eat**.) The reported commercial catch of black bream in the Swan–Canning was only 2.6 tonnes in 1998 down from the record high catch in 1996 of 7.5 tonnes.

The commercial finfish catch has continued to decline from 115 tonnes in 1991–1992 to 43 tonnes in 1998–1999 (Table 2.12). The catch was dominated by sea mullet (17.4 tonnes) and Perth herring (16.1 tonnes) with the remainder consisting of yellow-eye mullet (3.0 tonnes) and other fish. The commercial catch of blue swimmer crabs was 24 tonnes in 1998–1999, a reduction from 31 tonnes in 1997–1998 but higher than 8.6 tonnes in 1991–1992. In 2001 with only four commercial fishing units in the Swan–Canning, information on catch became confidential. (See also **Box 4.2 Crabbing—a summer pastime**.)

Swan recreational fishing in the Swan–Canning Estuary Basin

The number of people competing for a share of the recreational fish catch has been growing steadily year by year while the commercial catch has declined. Despite wide acknowledgement of this trend however, the size of the recreational catch has been debatable. To address the lack of data and

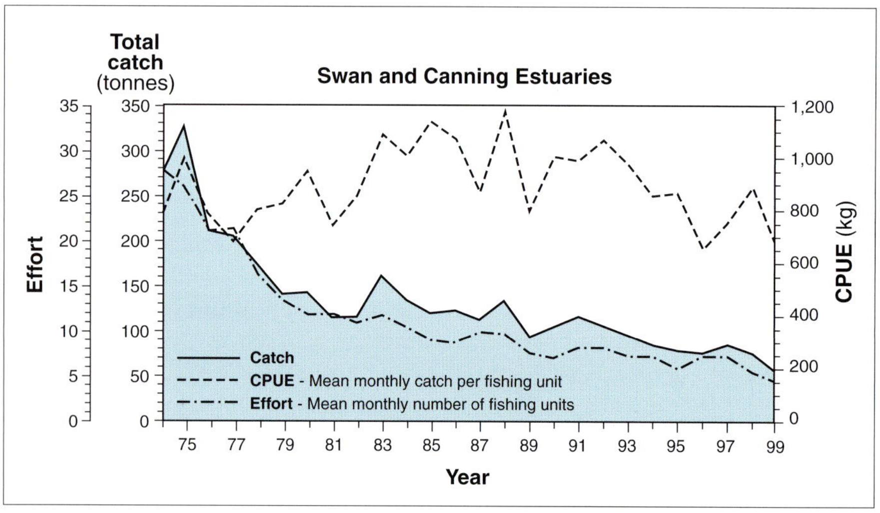

GRAPH 2.12 *Annual catch, effort and CPUE for the total fishery (finfish and crustaceans) of the Swan–Canning Estuary over the period 1974–1999. Post-1999 figures not available as there were only four licensed fishing units and the records are confidential.*
Source: Fisheries 1999, Courtesy DoF.

	1998			1999	
	Catch (t)	Estimated value to fishers ($)	Projection (t)	Catch (t)	Estimated value to fishers ($)
Swan–Canning	75.6	209,000	63–85	55.6	158,000
Peel–Harvey	325.8	711,000	240–420	256.7	763,000
Leschenault Inlet	88.1	194,000	67–108	59.8	140,000
Hardy Inlet	NA	NA	NA	NA	NA
South Coast	365	1,151,000	220–480	280	1,021,000

TABLE 2.12 *Commercial catches of finfish and crustaceans in the Swan–Canning Estuary 1998–1999 and comparison with other areas. Note that from 2000 with four units the weight of catch in the Swan was confidential.*
Source: Fisheries Western Australia, 1999; Ayvazian & Nowara, 2000.

FIG. 2.36 *Fishing.* Courtesy SRT.

to aid management, a twelve-month survey was conducted from August 1998 to July 1999. The survey team interviewed 1,300 people at boat ramps and 378 shore-based fishing groups at jetties and bridges, however it only covered fishing activity in the estuary basin downstream of Point Walter and Claremont (Fig. 2.36).

Of the people interviewed at boat ramps only 40 per cent were intending to fish and about half of these, although launching boats in the estuary, were planning to fish in the ocean. Nevertheless these boat fishers accounted for a total of 22,265 fisher days and shore-based fishers about 8,000 fisher days.

About 40 per cent of the boats and 9 per cent of the shore-based groups were targeting blue swimmer crabs. As expected, most of the crabs were caught in summer in upper areas of the basin when water temperature and salinity were high. The catch was less in winter with the decline in temperature and salinity, although some crabs were caught in deep areas of the lower reaches where salinity remained high. Most of the people fishing for crabs caught less than the daily limit of crabs, that is twenty-four crabs for a single person or forty-eight crabs for two or more people fishing from a boat. In total the recreational catch was estimated to be 7.3 tonnes or 20,176 crabs (Table 2.13). This was about 30 per cent of the commercial catch for that year. Examination of catches showed that most fishers had complied with regulations and released undersized crabs.

The recreational finfish catch is about 23 per cent of the combined recreational and commercial catch. Common fish kept are the common blow fish, tailor, whiting other than King George, Australian herring, black bream, flathead, trumpeter and flounder. In contrast, sea mullet, Perth herring and yellow-eye mullet have dominated the commercial catch. Both recreational and commercial fishers target black bream. The recreational catch of black bream is probably an underestimate as these are targeted in upper estuary areas, which were not surveyed. The survey found that recreational fishers also caught small numbers of wrasse (Labridae family), red gurnard *Chelidonichthys kuma*, yellow-eye mullet *Aldrichetta forsteri* and Ogilby's hardyhead *Pranesus ogilbyi*.

The fish catch depended on the season: black bream were most commonly caught in autumn, although more than half of these were released and were the dominant fish in winter. Flatheads were most common in spring.

TABLE 2.13 *Estimated recreational fish and crab catch in the Swan in the Swan–Canning Estuary Basin 1998–1999. The catch is based on interviews of boat-based fishers at boat ramps between Coode Street jetty in Perth Water to Mount Henry Bridge on the Canning River, and shore-based fishers on areas, jetties and bridges downstream of Point Walter and Claremont.* Source: Malseed & Sumner, 2001b.

Common name	Scientific name	Boat-based			Shore-based		
		No. kept	Weight (t)	No. released	No. kept	Weight (t)	No. released
Tailor	*Pomatomus saltatrix*	2,306	0.73	271	406	0.09	336
Other whiting	*Sillago* sp.	2,153	0.20	900	182		75
Black bream	*Acanthopagrus butcheri*	900	0.50	1,449	699	0.31	927
Flathead	*Platycephalus* sp.	849	0.25	1,223	302	0.12	370
Australian herring	*Arripis georgianus*	843	0.12	105	831	0.12	41
Common blowfish	*Torquigener pleurogramma*	793		14,086	3,544		24,801
Flounder	*Pseudorhombus* spp.	775		765			
WA butterfish	*Pentapodus vittata*	681		327			
Silver bream/Tarwhine	*Rhabdosargus sarba*	181		903			
Skipjack trevally	*Pseudocaranx* spp.	160		73			
Trumpeters	Tetraponidae family	107		400	1,049		1,887
Mulloway	*Argyrosomus hololepidotus*	107		0			
Blue swimmer crab	*Portunus pelagicus*	20,176	7.1	9,952	699	0.2	651

Whiting other than King George were caught in summer when salinity in the estuary was high and they moved in from the sea. The estimated catch rate was 0.25 fish kept/boat/hour for whiting and 0.17 fish kept/boat/hour for black bream during winter and tailor during summer.

Fishing pressure and retaining a valued resource

The black bream is arguably the most important fish in the Swan–Canning: it is prized by recreational and commercial fishers and, as a truly estuarine fish that lives and breeds in the Swan–Canning, it can be regarded as an indicator of estuarine health. Numbers have, however, declined over the years and like other estuaries in Swanland this has been due to overfishing, lack of breeding success and changes to the habitat.

Many of the species caught in the Swan–Canning use the estuary for only part of their life cycle, so catches in the estuary are not the only factor affecting the catch in following years. But as fishing in all estuaries and off the coast has increased, we all need to be mindful of the potential for excessive harvesting of these fish stocks. The ongoing success of the recreational and commercial fisheries depends on the spawning and recruitment success, larval survival and growth. As the average level of production from year to year cannot increase however, ongoing success is only assured if environmental conditions in the estuary remain favourable, that is if the sources of water in the surrounding wetlands, water catchments and shallow feeding areas remain healthy, and the oceanic sources of juveniles are maintained.

BIRDS OF THE SWAN— MEMORABLE ICONS

In the late 1980s an estimated 10,000 waterbirds used the Swan–Canning each year. The abundance of waterbirds is matched by a rich diversity of over eighty-seven species, 77 per cent of the waterbirds recorded in south-western Australia. In comparison, seventy-eight species have been recorded on the Peel–Harvey, seventy-four on Lake Argyle in the Kimberley, seventy-three on Vasse–Wonnerup and sixty-six on Forrestdale Lake. This diversity and abundance makes the City of Perth a very special environment, a privilege for those people living nearby and a responsibility for the future. This feature was recognised in the declaration of the Swan Estuary Marine Park and Adjacent Nature Reserves in 1999 (Map 2.15). (See **Box 2.17 Waterbirds**.)

The wetland system of the Swan Coastal Plain, although now highly modified, is part of the reason the bird population of the Swan is so diverse. Some species range widely across the coastal plain and throughout the South-West using many different water bodies; dependent on a variety of factors such as weather conditions and availability of food. Others, such as the Little Black Cormorants *Phalacrocorax sulcirostris* and smaller numbers of darters *Anhinga melanogaster*, Great Cormorants *Phalacrocorax carbo* and White Ibises *Threskiornis molucca*, feed in the estuary but nest each year in trees around wetlands such as Booragoon Lake.

Around the Swan–Canning are thirty-one of the bird species listed in the Japan–Australia Migratory Bird Agreement (JAMBA Treaty), under which the signatory nations are obliged to protect species. Twenty-eight of the listed species are trans-equatorial migrants that breed in the Northern Hemisphere above the Arctic Circle and Eurasia and move to the Southern Hemisphere to feed in mid-spring to late autumn, swelling the bird population of the Swan–Canning to about 15,000. In the past before dredging and walling changed the Swan–Canning the numbers of birds were probably much higher, as they are in other estuaries such as

Box 2.17

WATERBIRDS

Waterbirds are those birds dependent on wetlands for feeding, breeding or loafing (resting and roosting). They include grebes, pelicans, cormorants, herons, ibises, spoonbills, swans, ducks, rails, waterhens, waders (shorebirds), gulls and terns. Birds use all the areas of the Swan–Canning environment: the open water, tidal mudflats, beaches, tidal marshes, tidal pools and channels, riverine woodland as well as the artificial structures.

Box 2.18 TURN OF THE TIDE AND A PASSING PARADE

Bird counts are so important to understanding the health of bird populations, but counting birds is difficult, even more difficult if they use a number of different areas. A survey involving teams of 'watchers' placed around the Swan–Canning in January 1999 revealed details of habitat use and illustrates the problem. At the time there were an estimated 2,250 wading birds on the Swan–Canning, the majority at Alfred Cove. As the tide rose in the late afternoon, the first to move from Alfred Cove were 1,000 Red-necked Stints that flew to Milyu joining another group of over 400. As the water rose further the larger waders (Grey Plovers, Black-winged Stilts, Great Knots and Greenshanks) moved to flooded samphires at Alfred Cove, or across the river to Pelican Point. At sunset the medium sized waders (Curlew and Sharp-tailed Sandpipers) and a few remaining Red-necked Stints left in a large group, some stopping briefly at Point Walter before disappearing over the coast. At dark, the sandbars at Alfred Cove were almost flooded with only the Pied Oystercatchers and pelicans remaining.

Source: Bamford, 1999.

Vasse–Wonnerup where there are still extensive shallow feeding areas. It is, however, difficult to determine the extent to which bird numbers have declined as there have been changes in censoring methodology, the influence of conditions in other areas is often unknown, and preceding breeding seasons need to be taken into account.

Only three major habitats for wading birds remain in the Swan–Canning (Map 2.15): the Milyu Nature Reserve along Como foreshore (40 hectares), the tidal flats and salt marshes of Point Waylen and Alfred Cove (130 hectares), and Pelican Point (20 hectares). Although waders use all three areas on the Swan–Canning estuary, depending on the extent of the mudflats that vary with the tides, barometric pressure, wind and disturbance, Alfred Cove is the most important. It is not only the largest area but the birds are most abundant and there is a variety of vegetation complexes protected from disturbance. Although disturbance is a major factor affecting birds, we know little about other important factors such as food availability. The effects of nutrients or other pollutants on the plants and animals on which birds feed have never been investigated. (See **Box 2.18 Turn of the tide and a passing parade**.)

Wading birds

The Red-necked Stint *Calidris ruficollis* and the Curlew Sandpiper *Calidris ferruginea* dominate populations of waders (Fig. 2.37). Other common species include the Red-necked Avocet, godwits, plovers, sandpipers, stilts, dotterels and knots. Numbers of Red-necked Stints appear to be similar to counts made in the 1980s, however some of the other migratory waders such as the Curlew Sandpiper, Red and Great Knot, Bartailed Godwit, and Sharp-tailed Sandpiper may have declined. In the case of the Curlew Sandpiper, this may be part of a wider decline throughout the East Asian–Australasian Flyway, but for the larger species disturbance within the estuary is probably a factor. In contrast, one of the non-migrant waders, the Pied Oyster Catcher with a resident population of about 100, was relatively unknown on the estuary until the 1980s, a change that could indicate a greater marine influence. (See **Box 1.10 Global travellers: living a double summer life—transequatorial migrant birds**.)

Waders feed on a range of animals including worms, molluscs, crustaceans and insects, the type probably depending on the abundance and size of the organism. While some waders feed on swimming invertebrates stranded along the tide line, the length of bill may limit the depth to which they can probe the sand or the leg length may dictate the water depth in which they feed. The Curlew Sandpiper with longer legs is able to feed in deeper water than other waders, and selects a diet of insects (larvae and adults), worms, crustaceans, molluscs and bugs as well as plant material. The tiny Red-necked Stint feeds on mosquitoes, weevils, bees and wasps, beetles and possibly seeds.

FIG. 2.37 *(a) Red-necked Stint and (b) Curlew Sandpiper, wader birds of the Swan–Canning Estuary.* Babs and Bert Wells Collection, Courtesy CALM.

Swans and ducks

Among the most common are the Black Swan *Cygnus atratus*, Grey Teal *Anas gibberifrons*, Australian Shelduck or Mountain Duck *Tadorna tadornoides*, Black Duck *Anas superciliosa* and Blue-winged Shoveler *Anas rhynchotis*. Many spend the winter months in wetlands throughout the South-West and move to the permanent waters of the Swan as more ephemeral sources dry out. Numbers of ducks and swans around the estuary appear to have increased since the late 1990s possibly due to the years of lower rainfall and subsequent drying of smaller wetlands. They may also move during the course of a day, feeding on the estuary in the morning and in the freshwater lakes in the evening. Basically herbivores, swans and ducks feed by sifting algae and other aquatic plants along with the associated invertebrates from the bottom.

Birds of the reed beds

Birds of the reed beds include the Spotless Crake *Porzana tabuensis*, Buff-banded Rail *Rallus philippensis* and the Little Grass-bird *Megalurus gramineus*, species that shelter in the fringing vegetation and feed on insects among the vegetation or on mud. As a group they are probably the most threatened, their habitat being the fringing vegetation which is now severely reduced.

Fish eating birds

Fish eating birds include the White Egret *Egretta alba*, White-faced Heron *Ardea novaehollandiae*, Nankeen Night-heron *Nycticorax caledonicus*, Pied Cormorant *Phalacrocorax varius*, Little Pied Cormorant *Phalacrocorax melanoleucos*, Black Cormorant *Phalacrocorax carbo*, Little Black Cormorant *Phalacrocorax sulcirostris*, Darter *Anhinga melanogaster*, Australian Pelican *Pelecanus conspicillatus*, Hoary-headed Grebe *Poliocephalus poliocephalus*, Caspian Tern *Hydroprogne caspia*, Whiskered Tern *Chlidonias hybrida*, Fairy Tern *Sterna nereis*, Crested Tern *Sterna bergii*, and Silver Gull *Larus novaehollandiae*.

Herons and egrets feed on small fish in the shallow areas and around the edges of the reeds. Darters, grebes, pelicans, cormorants and terns catch fish in slightly deeper water, while gulls are scavengers. Cormorants generally feed by diving to catch small bottom-living organisms, but the Little Black Cormorant also uses community fishing to catch pelagic prey. (See **Box 2.19 Black Cormorant feeding frenzy**.) In the 1930s Dom Serventy commented that the cormorant population of the Swan–Canning was the most numerous he had seen anywhere, suggesting that this was due to the exceptionally large populations of easily-caught food organisms of a convenient size, including gobbleguts, gobies and hardyheads. Serventy's work, undertaken as a response to demands for a cull of the cormorants that were reputed to be depleting the number of edible fish, clearly showed that over 90 per cent of the species eaten were non-edible fish. There are indications that increased numbers of cormorants could be linked to the more frequent algal blooms and the increase in small fish that feed on phytoplankton.

Recreating Swanland — bringing back the Black Swan

> *The quantity of black swans, pelicans, ducks, and aquatic birds seen on the river was truly astonishing. Without any exaggeration, I have seen a number of black swans which could not be estimated at less than 500 rise at once, exhibiting a spectacle which, if the size and colour of the birds is taken into account, and the noise and rustling occasioned by the flapping of their wings previous to their rising, is quite unique in its kind. We frequently had 12 to 15 of them in the boats, and the crews thought nothing of devouring eight roasted swans in a day.*
>
> Charles Frazer 1827 from Hay, 1906.

Box 2.19 **BLACK CORMORANT FEEDING FRENZY**

Although not restricted to the west coast, large feeding flocks of Little Black Cormorants form one of the characteristic sights of the Swan River Estuary and other inlets in the South-West; a sight sufficiently noteworthy to be recorded by Captain C. H. Fremantle on the HMS *Challenger* in 1829, and Dr Dom Serventy in 1939 (Figs 2.38 and 2.39).

In contrast to the individualistic fishing methods of the other Cormorants, P. sulcirostris *gathers in large companies which may be seen flying low over the water, very close to the surface, in a long column. When a shoal of fish is encountered the leading birds wheel round and settle on the water, probably concentrating the shoal. The remainder continue coming up and the birds aggregate to a dense mass, harrying the fish. The flight may be represented by the figure:*

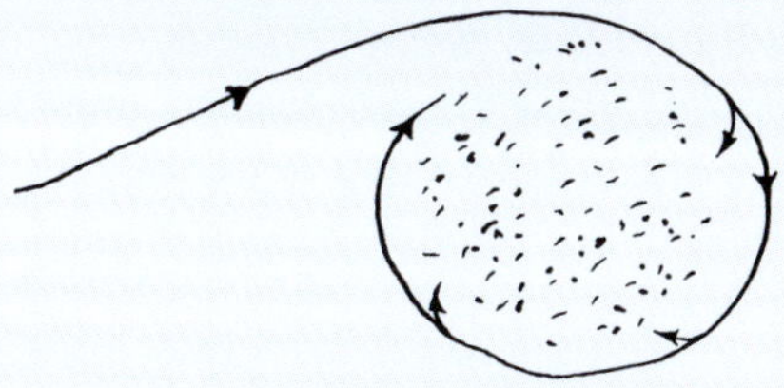

FIG. 2.39 *Feeding aggregations of Little Black Cormorants on the Swan–Canning Estuary described by D. Serventy, 1939.* Courtesy *Emu*, journal of Birds Australia.

FIG. 2.38 *Large flock of Little Black Cormorants resting on a sand bar after feeding.* Babs and Bert Wells Collection, Courtesy CALM.

After a feeding orgy the birds perch en masse on a sand spit or jetty.

... On the Swan River the flocks are accompanied by an occasional Little Pied Cormorant and a few Silver Gulls which share in the results of the systematic fishing of the Little Black Cormorants.

Source: Serventy, 1938a, 1939; Sassoon, 2003.

The Black Swans of the Swan River, once a presence so memorable and inspiring that it is an emblem of the State, The University of Western Australia and many local products, featured on colonial buildings, lamp posts and fountains, have now largely disappeared from the metropolitan area (Fig. 2.40). Some were no doubt eaten (Fig. 2.41). The spectacle recorded by Frazer in 1827 of a flight of 500 birds is long gone, and so too has the habitat of the shallow reed-backed foreshore that has now been removed, walled or infilled as the city developed. Although bringing Black Swans back to the river will require re-creation of suitable habitat, there is the added bonus of conserving other more secretive and less spectacular plants and animals.

Black Swans are nomadic, ranging across the South-West, often overwintering and nesting inland on natural and artificial lakes and waterways, moving to the more permanent water around coastal lakes and estuaries in summer. In 1992 a survey from Lancelin to Bunbury recorded 3,406 Black Swans on the Swan Coastal Plain (Table 2.14). Important features of habitat include: safe roosting areas; access to fresh water; access to shallow, fairly sheltered water where birds can graze on submerged aquatic plants, about 0.5 to 1.5 metres deep, at a depth where plants are abundant and within reach.

Many of the family groups observed on Lake Monger probably originated from Herdsman Lake where there is an abundance of nesting habitat. Black Swans were less abundant on the Swan–Canning Estuary. At Matilda Bay, Milyu Nature Reserve, and Perth Water sightings were generally of solitary birds. Upstream of Heirisson Island, Black Swans were more common on the nearby wetlands than on the river. About fifty were present around the Burswood foreshore, and breeding pairs were seen on wetlands adjacent to the river at Belmont, Maylands, Ashfield, Bayswater and Ascot as well as Guildford and Viveash. Up to eight Black Swans were present at Lucky Bay in Alfred Cove in 1998–1999 and eleven in 1999–2000, and over eighty birds were seen in the Waterford area of the Canning River.

Availability of food is a major factor determining the number of swans on a wetland (Graph 6.5). Swans feed on plants such as the paddle weed seagrass *Halophila*, pondweeds, *Potamogeton*, introduced water-couch grass *Paspalum distichum*, as well as young shoots of bulrush *Typha* and other rushes and sedges. Not surprisingly the eutrophic and weed rich lakes such as Lake Monger are prime swan habitat.

For nesting, water levels, materials for nest building, proximity of feeding areas, and fresh water (for young) are important (Fig. 4.14). Nest mounds are constructed in areas surrounded by water (0.3–1.0 metres deep) for protection but without the risk of flooding. Black Swans nest in colonies where the distance between nests may be only 2.4 metres, or as solitary pairs where the territory around a nest is about a hectare in area. Colonies are more likely where there is an abundance of nesting material or where space is limited on islands. Conversely, where food is adequate and the depth of water is suitable but nesting material is limited, single nests are more likely. They appear to nest in the same area each year, possibly using the same nest mound, and will use artificial nest sites—an important factor for re-introductions. Pairs usually remain together for many years, possibly for life. In captivity swans can live up to twenty years, although longevity is unlikely to be as long in the wild. Swans are prone to poisoning through pollution, botulism, fungal respiratory infections, parasitic flukes and worms; predation by foxes, hawks and eagles; and accidents including collision with power lines.

Black Swans generally breed in winter and spring. Eggs are usually laid in batches of five to six (although two to nine have been recorded) in July to August. The eggs are brooded by both parents and hatch after about forty days. The long incubation period means that late breeders produce young when supplies of fresh water are diminished, resulting in high mortalities.

Cygnets take to the water soon after hatching although they are sometimes carried on the parents' backs under the wing feathers. Initially the parents pull up aquatic plants for the young to feed. Cygnets grow very slowly taking about five months to fledge at about four to five kilograms. As the cygnets develop, grey-brown feathers replace the light brown-grey down, the bill turns reddish-purple and the eye colour changes to brown with a black iris. The wing feathers develop last, the primaries and secondaries at first white tipped, gradually changing to pure white at maturity in about two years. Although family units are usually maintained for about six months and the adults, particularly the males, are aggressive towards other birds and animals, cygnet mortality is quite high; at least one per brood does not survive. Black Swans are most vulnerable when the young are unfledged, and adults are flightless while undergoing the annual moult.

Examination of areas where Black Swans are abundant provides insights into causes leading to their disappearance and their requirements in the future. We know they are quite amenable to living in proximity to humans as they are most abundant at Lake Monger in an urban area and they are now found in other parts of the world. About 100 Black Swans were introduced to New Zealand in the 1860s and by 1972 an estimated 200,000 birds were present throughout the country. In the 1960s about 5,000 pairs of Black Swan were recorded nesting at Lake Ellesmere in New Zealand, which is a much larger nesting population than anywhere in Australia.

The most favourable areas on the Swan–Canning for recreating Black Swan habitat are:

FIG. 2.40 *Crests using the Black Swan image.* Reproduced with permission of the Government of Western Australia, The University of Western Australia, City of Swan, City of Melville, City of Perth, City of Subiaco, City of Canning and Swan Brewery.

- Attadale Reserve (Lucky Bay–Alfred Cove)
- Clontarf Bay, Waterford
- Como Foreshore (the Narrows to Como Jetty)
- Sir James Mitchell Park, South Perth
- Ascot Waters

All are areas where Black Swans are known to congregate, where the habitat already contains features required, and where these can be extended with the least effort and cost or conflict with existing land use. In many areas projects to 'Bring Back the Black Swans' can be linked to other rehabilitation and beautification plans. Perhaps the greatest problem is acknowledgment that the requirements of swans are not compatible with some other land uses: human access for playing fields, dog exercising, jogging, cycling, golf practice and on the river, boating, water skiing, fishing and jet skiing.

The photograph (Fig. 2.42a) was printed in the *West Australian* on 6 August 1932 with the title 'Mount's Bay Road 30 years ago' and the caption 'Mount's Bay Road, near the Swan Brewery, early in this century. The Old Men's Home was then situated there, and the inmates are seen fishing from their favourite rocks, while others are along the shore, the black swans were always to be found in the locality.' The accompanying articles and letters detailed earthworks being undertaken—infilling the bay (where smelly algae accumulated) to improve the road between Fremantle and Perth.

FIG. 2.41 *Icon and sustenance: Mrs Beeton's recipe for Black Swan, 1936.* Courtesy Orion Books, UK.

MRS. BEETON'S

EVERYDAY COOKERY

WITH ABOUT

2,500 PRACTICAL RECIPES

AND SECTIONS ON

LABOUR-SAVING	CARVING AND TRUSSING
HOUSEHOLD WORK	THE ART OF "USING-UP"
SERVANTS' DUTIES	TABLE DECORATION
LAUNDRY WORK	TABLE NAPKINS
MARKETING	MEALS AND MENUS
RENOVATIONS	BEVERAGES
Etc.	Etc.

WITH 16 PLATES IN COLOUR
AND NEARLY 300 ILLUSTRATIONS

WARD, LOCK & CO., LIMITED
LONDON AND MELBOURNE

BLACK SWAN, ROASTED OR BAKED.

This bird should be quite young, otherwise it is better cooked in hashes and stews. Very frequently it is cooked just in the same way as goose, but it is very nice stuffed with farce made from minced beef-steak in the same manner as sausage farce, substituting beef for the pork. Truss as for goose, squeeze lemon-juice over and wrap in bacon rashers. Bake for about 2 hours in a moderate oven, remove the rashers ½ an hour before serving, and baste well and dredge flour over till crisp and brown.

TIME.—About 2 hours. SUFFICIENT for 8 persons.

TABLE 2.14 *Wetlands in the Perth metropolitan area with the largest populations of Black Swan (mean numbers counted during the survey period 1990–92). Note the survey area covered wetlands over a much larger area of the Swan Coastal Plain from Lancelin to Bunbury.* Source: Storey et al., 1993. See also Bamford, 1999; WRC, 2000b.

Lake	Largest populations	Nest and families
Lake Monger	404.4	51
Thomsons Lake	241.9	7
Bibra Lakes	189.5	17
Forrestdale Lake	182.4	19
Lake Herdsman	122.1	8
Lake Cooloongup	104.3	
Lake Yangebup	49.5	1
Big Carine Swamp	32.9	14
Lake Joondalup	32.7	
Perry Lakes	23.0	6
Lake Kogolup	21.1	14
Lake Claremont	16.4	5
Lake Gwelup	9.6	2
Maylands Claypits	9.2	

FIG. 2.42 *(a) Home of the Black Swan as it was in the early 1900s, near the brewery at Mounts Bay.* Courtesy Battye Library, Catalogue 225,657 P. *(b) View of the brewery apartments and restaurant, 2003.* A. Brearley 2003.

CONSERVING AND PROTECTING THE SWAN–CANNING ESTUARY

Additions to the water—pollutants from a city

Perhaps it is not surprising that the estuary in the heartland of metropolitan Perth, with approximately 1.5 million people, has shown symptoms of stress. Soon after settlement, pollution from the city started to have undesirable effects on the river: the small streams were contaminated by human waste and with products from the first industries. As the population grew development altered the catchment, and the estuary became a drain and a conduit to the ocean of all the products of the rural hinterland and urban surroundings.

Complaints started early and led to the formation of the Swan River Reference Committee in 1943 and the Swan River Conservation Act in 1958 with the provision of statutory power for the management of the river and the formation of the Swan River Conservation Board. (See **Box 2.5 Unacceptable weeds on beaches—pressures for controls**.) The influence of the Board spread to other waterways conservation agencies throughout Australia and led to the proclamation of the Waterways Conservation Act of

1976, followed later by other initiatives in 1988 and 1995 (Table 2.15). (See also **Chapter 3 The Peel–Harvey Estuarine System: No longer a local issue**.)

Despite the best intentions and management programs, the problems of suitable waste management continued for many years. One of the first industries, a brewery established in 1837 at Mounts Bay Road still had no sewerage outlet in 1926; its output including diatomaceous earth from filters and solution from a bottle wash went into the river. This large organic load continued even after the opening of modern premises at Spring Street in 1966 until the brewery was connected to Perth Main Sewer.

Even when Perth's first sewerage treatment plant was established in the early 1900s near where the Burswood Casino now sits, the problem of nutrients from sewage still plagued the river. This was because the waste water from the system was filtered over large open trickle beds, and inevitably some of the nutrient-rich water entered the river. The result was large algal blooms stretching from Maylands to South Perth, although the Public Works Department of the day denied the trickle beds were the cause.

In the 1970s a zinc plating industry contributed cyanide to the river. Stockyards and abattoirs on the Swan and Helena Rivers used the rivers for effluent disposal, and stock access to the riverbanks caused erosion. Household and industrial refuse was used in landfill as part of a program to control mosquito breeding areas, but leached contaminated nutrients and toxic material into nearby water bodies. At North Fremantle effluent from wool scourers was still entering the river in the 1970s although it was usually discharged on a falling tide. In addition to these major point sources of pollution, there have been additions of stormwater from roads, urban and industrial areas, and the slow leaching of septic tanks.

Although the established agencies dealt with major point source pollution from industry and foreshore rubbish dumps and gradually reduced discharges to the river, nutrients from diffuse sources in the surrounding urban areas and rural catchment of the hinterland were ever increasing. Not only were the incoming loads increasing, but nutrients were also building up in the sediments. These hidden diffuse sources were and still are much harder to manage and control.

TABLE 2.15 *Management agencies of the Swan–Canning Estuary.* Sources: Riggert, 1978; SRT, 1999 with advice from B. Hamilton.

	Regulatory bodies and programs	Composition, roles and aims
1943	Swan River Reference Committee (4 members increased to 15)	Voluntary and advisory to co-ordinate works and deal with purity of water and cleanliness of foreshore.
1952	Sub-committee: Deputy Commissioner of Public Health, the Chief Engineer of Metropolitan Water Supply, Sewerage and Drainage Department and Deputy Government Analyst	Define pollution and suggest controls.
1955	Public concern; industrial waste and other pollution	
1958	Initiation of systematic sampling by Public Health Department and Government Chemical Laboratories for the Swan River Conservation Board	To assess chemical and bacterial concentrations.
1961	Swan River Conservation Act (1958) establishes Swan River Conservation Board	First of its kind in Australia, wide powers subject to Minister for Works and Water Supply, and to appeal to Courts.
1964	Boundaries extended Two additional members: (1) Biologist nominated by Minister for Fisheries and Fauna; (2) Representative of Local Government in the Canning River area	(1) Swan River to Fremantle Traffic Bridge upstream to Middle Swan Bridge (2) Canning River to Nicholson Road Bridge (3) Helena River to Scott Road Bridge Parliamentary approval required for reclamation in excess of 0.8 ha (in response to outcry over Narrows Bridge interchange).

	Regulatory bodies and programs	Composition, roles and aims
1971	Boundaries extended to include agricultural areas upstream	(1) To the confluence of the Swan, Avon and Gatta Rivers (2) Lower Helena River Diversion Weir (3) Brookton Highway Bridge on Canning River
1977	Waterways Conservation Act (1976) proclaimed and Swan River Conservation Act repealed. Swan River Management Authority replaces the Swan River Conservation Board	Provided for the Waterways Commission and three statutory authorities for Swan–Canning, Peel–Harvey and Leschenault Inlet. Also for the establishment of other Waterways Management Authorities in Western Australia.
1986	Environmental Protection Act 1986 (WA)	
1989	Swan River Trust Act (1989) proclaimed to establish the Swan River Trust (SRT)	Swan River Trust to work with state and local government and the community to ensure the Swan and Canning river system is conserved and managed to enhance its environmental quality and amenity. Additional powers to control and manage development in and around the Swan and Canning Rivers.
1994	Western Australian Estuarine Research Foundation (WAERF)	Integrated predictive ecological model of catchment hydrology and water quality based on 12 projects to investigate a particular component of the physics, chemistry or biology.
1995	Water and Rivers Commission Act 1995	Waters and River Commission (WRC) from 1 January 1996 to ensure effective management of Western Australia's water resources. Waterways Commission merged into WRC but SRT remains a separate statutory body.
1995	Swan–Canning Cleanup Program (SCCP) Task Force established to prepare a comprehensive 'cleanup' program for the Swan and Canning Rivers	SRT is responsible for involving state agencies, local government, universities and research institutions to develop an action plan based on research and the work of the WAERF.
1996	SCCP–WAERF Communication Forum	Groups involved in care of waterways including the Swan–Avon Integrated Catchment Management Program (SAICM).
1998	Environmental Protection (Swan and Canning Rivers) Policy gazetted (SCEPP)	To establish standards and develop a Comprehensive Management Plan to protect the Swan and Canning Rivers.
1999	Swan–Canning Cleanup Program Action Plan	Strategies to improve the health of the Swan–Canning to maintain public health and safety, ecological function, and reduce nutrients and other contaminants from urban and rural sources while maintaining productivity and profitability based on the SCCP.
1999	Swan Estuary Marine Park and Adjacent Nature Reserves	Declaration of Marine Park: estuarine lands and waters within the Swan River at Alfred Cove, Milyu (Como) and Pelican Point.
2002	Department of Environmental Protection (DEP) merged with the WRC to form the Department of Environment (DoE)	This merger brings environmental management pollution control and water resources management together.
2003	Launch of Riverplan: Environmental Protection (Swan and Canning Rivers) Policy 1998 for Public Consultation	Environmental management framework for the Swan and Canning Rivers; first step of the 'Comprehensive Management Program' under the SCEPP.

The Swan – Canning Cleanup Program

In 1994, in response to community concern for the health of the estuary, the Swan River Trust was commissioned by the Western Australian Government to prepare an action plan to reduce algal blooms. Thus the Swan–Canning Cleanup Program and Western Australian Estuarine Research Foundation (WAERF) investigated water quality and ecological processes to develop ways of controlling blooms. Management strategies ranged from engineering programs targeting blooms, to methods of reducing nutrient inputs in the rural and urban catchments and along waterways, together with development of strategies promoting individual responsibility and community action. (See **Box 2.20 Goals of the 1999 Action Plan**.)

Box 2.20 **GOALS OF THE 1999 ACTION PLAN**

Public health and amenity: Algal blooms are kept at a level where there is no threat to public health and amenity, and they are not a nuisance to the community. Toxic algal blooms are kept to a minimum.

Ecological function: Water quality in the Swan–Canning system is suitable for maintaining a healthy ecosystem. People can swim or catch fish at any time.

Catchments and targets: Contaminants in stream run-off leaving the catchments are within set targets. Rural catchments are managed to be productive and profitable, and are attractive and affordable places to live.

New urban and industrial areas: Areas are designed so that discharges have reduced contaminant levels and meet set targets before entering rivers, while they still remain attractive, affordable places to live and productive and profitable places to work.

Older urban and industrial areas: These areas are modified over time to reduce contaminant levels so that drainage discharges meet set targets before entering the rivers.

The studies

The program funded a number of projects including river and catchment monitoring; and mapping and research on nutrients, algae, invertebrates and fish. Other studies included modelling of estuarine dynamics, stormwater design, water quality management and Catchment Management Plans for key areas. The initial studies and development of the Action Plan(s) drew on expertise from government agencies, scientific organisations, local government and community groups.

The studies reached some very important conclusions:

- The main pathways by which nutrients enter the system are from run-off in winter and spring, and from urban drains, ground water and sediments in summer.
- It is nitrogen rather than phosphorus that normally limits algal growth; it is known that there is usually enough phosphorus in the water.
- Controlling both nitrogen and phosphorus is essential for effective management.

(Dr Des Lord, Taskforce Chairman, The Swan–Canning Cleanup Program).

To reduce the nutrients responsible for the blooms, it was recognised that improved water quality would require a comprehensive program for reducing nutrient loss from the catchment: a program that would probably take up to twenty years or more before it was effective. Reiterating the old adage: 'there's no quick fix'.

Box 2.21 **ACTION PLAN OBJECTIVES 1999**

1 Support Integrated Catchment Management to reduce nutrient inputs.
2 Improve planning and land-use management to reduce nutrient inputs.
3 Modify river conditions to reduce algal blooms.
4 Monitor river health, fill critical gaps in knowledge and report progress to the community.

Issues, options and actions

The 1999 Action Plan highlighted the need for a whole-ecosystem approach to the problems in the Swan–Canning, and the need for increasing community awareness to achieve ongoing change. This required a combination of reforms in land-use planning, statutory controls, improved land management and restoration practices, river intervention and catchment management.

The Action Plan recognised that some controls, such as catchment management and foreshore restoration, might take up to twenty years before changes were apparent. In the shorter term (three to five years), implementation of some specific actions could reduce algal blooms by treating sediments or maintaining high oxygen concentrations to remove or lock up nutrients. Short-term solutions, although relatively costly, essentially required only immediate monetary support. In contrast, the longer term solutions would need ongoing commitment by community, local, State and Commonwealth government for ongoing resources. (See **Box 2.21 Action Plan Objectives 1999**.)

Catchments

Nutrients and other contaminants enter the catchment from surface and sub-surface drainage. These can include 'point source' inputs from industrial and commercial activity, or intensive agriculture such as piggeries, poultry farms, feed lots, market gardens, horticulture areas, turf farms, vineyards, tree cropping and golf courses as well as grazing: all of which produce nutrient-rich waste or apply large amounts of fertiliser to the soil. Although many point sources arise from activities approved by local government or licensed by the Department of Environmental Protection, (it should be noted that in many cases) their discharges are not always effectively controlled. In addition there are diffuse sources from rural and urban areas.

Rural issues

Diffuse sources from rural areas around the Swan–Canning are similar to those from catchments around other estuaries, and are usually due to applications of fertiliser and pesticides, soil losses, and organic waste or carbon in faeces. Ellen Brook sub-catchment in particular (Map 2.13) contributes the highest nutrient concentrations to the Swan–Canning. Options for reducing nutrient

MAP 2.13 *Classification of the sub-catchments of the Swan–Canning on the coastal plain based on total nitrogen and total phosphorus concentrations (mg/L) between 1995 and 1997. Nitrogen concentrations have increased in recent years (since classified in 1987–1988) and are only low in the Avon River, Helena River, Jane Brook and Upper Canning River, the exception being during floods (e.g. summer 2000) when the water contains large amounts of nitrogen. Eleven of the sub-catchments have low concentrations of phosphorus, only Ellen Brook, Mill Street Main Drain, Southern–Wungong River and South Belmont Main Drain exceed guideline concentrations. In some areas, management programs appear to be effective and phosphorus discharge concentrations in some catchments such as Bayswater Main Drain decreased between 1987 and 1997.* Source: SCCP Action Plan, 1999, Courtesy SRT.

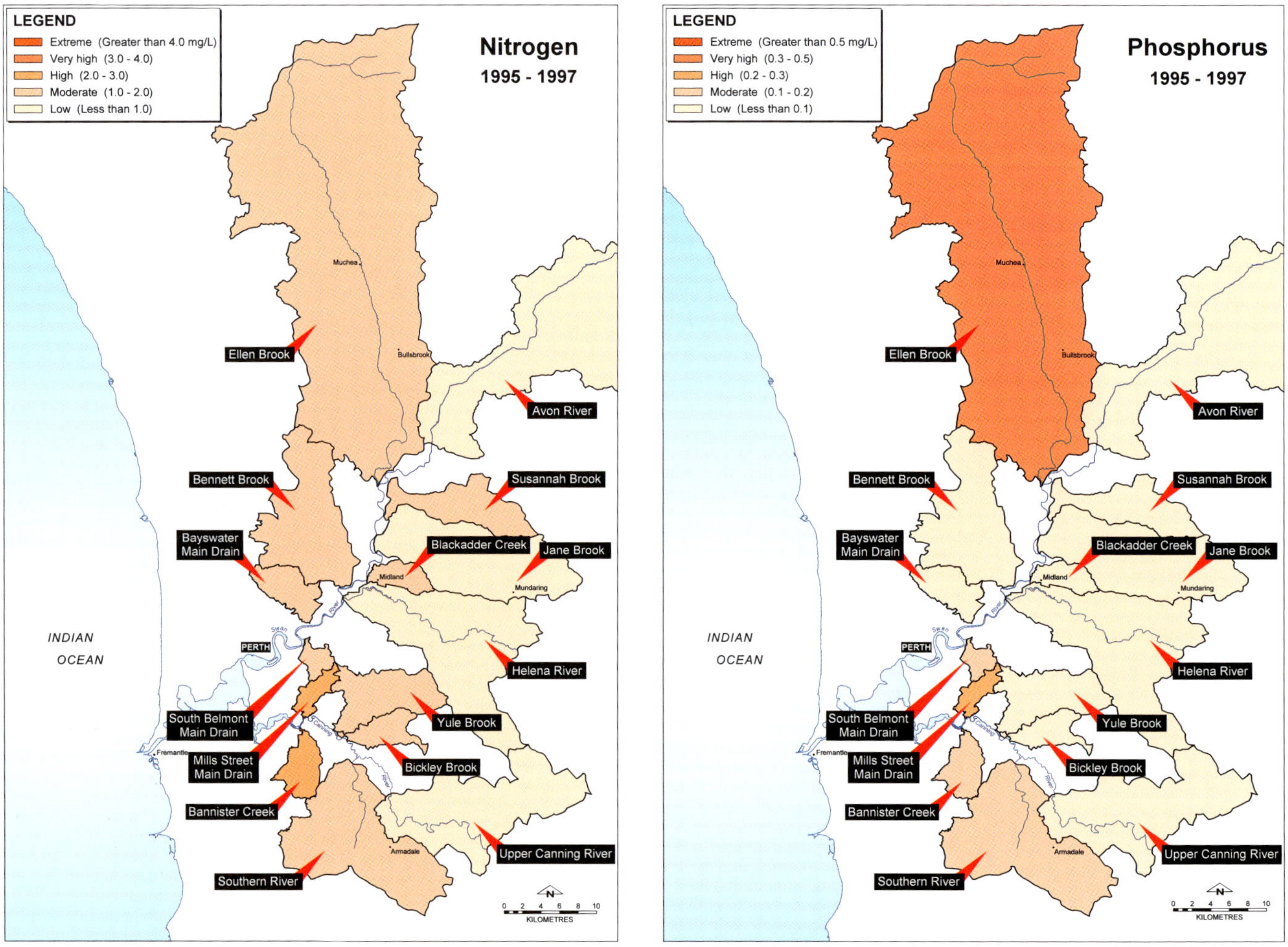

loads include legislation, controlling nutrient generating and land degrading practices, increasing use of slow release fertilisers, restoration of water courses, construction of artificial wetlands to strip nutrients, clearing controls in areas not yet developed, and appropriate planning to reduce erosion and nutrient losses.

Urban issues

The larger Perth area has a long history of contamination from waste disposal and industry: many of these areas are no longer obvious but the legacy continues to be a problem (Table 2.2 and Map 2.14).

In residential areas there is nutrient input from vehicles, domestic animals, septic systems, atmospheric fall out, garden fertiliser run-off and spills; most entering the rivers through ground water or more directly from road drains. Obvious sources include dog faeces that account for 10 per cent of nitrogen and phosphorus in urban drains, and lawn and garden fertiliser. About 40 kilograms of total phosphorus is added to each hectare of urban gardens each year, possibly as much as 5,000 tonnes in Perth metropolitan Area, or double the application rate in agricultural areas.

Obvious management options in urban areas include smaller gardens where application of fertilisers is easier to control, and street sweeping that removes nutrients and contaminants before they enter drainage systems. More effective use of water is required to reduce flows for irrigation of parks and recreational areas. In the later 1990s 75 per cent of urban water was used for non-domestic purposes (gardens) in summer, and 70 per cent of this may move through the root zone straight through to ground water together with nutrients. Currently two-thirds of Perth's public water supply comes from water trapped by reservoirs and dams on the rivers flowing to the Swan–Canning Estuary. Water sourced from these rivers can be regarded as fully utilised, as further diversion would affect estuarine processes (Table 2.16). Use of the other major water sources in shallow and deeper ground water is likely to rise, however the effect of pumping to the watertable, on overlying vegetation and seepage to the estuary are not fully known and still debated.

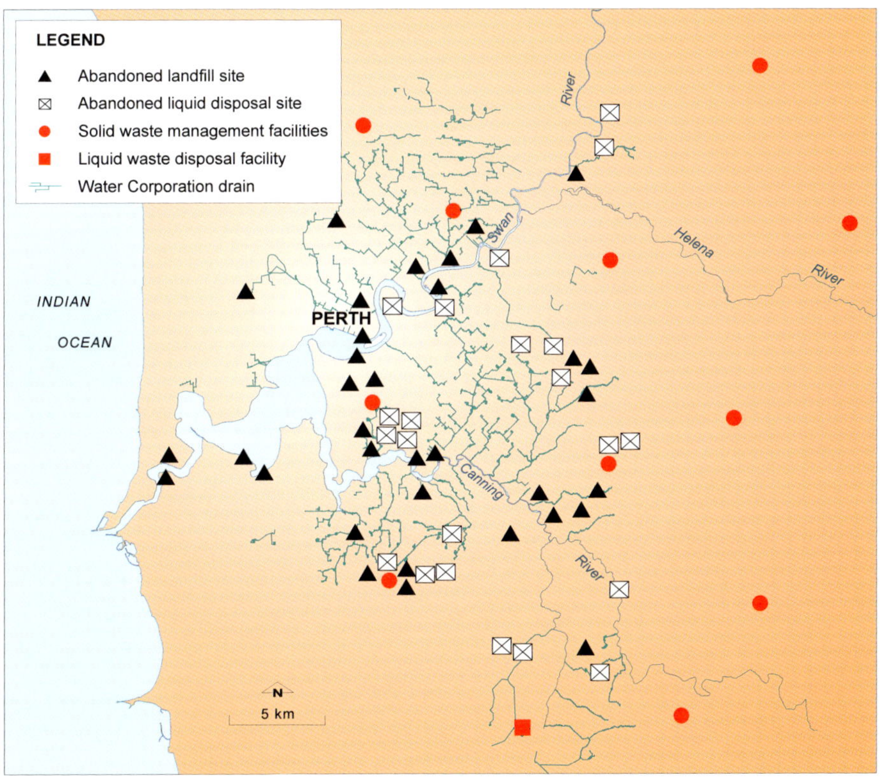

MAP 2.14 *Landfill and waste disposal sites and sources of nutrients and other pollutants entering the ground water around the Swan–Canning Estuary. There are also a number of industrial areas along with abandoned waste disposal sites nearer to the coast with potential to contaminate ground water. Surrounding these are large areas where septic systems are still in use or were in use until quite recently, and numerous parks and suburban gardens that also contribute nutrients. Although some areas can be linked directly to contaminated flows and algal blooms, relatively little is known about the pathways or rates of ground water flows reaching the estuary. However the numerous drains, shallow watertables, and porous soils make contamination of ground water a serious threat to estuarine health.* Source: SCCP Action Plan, 1999, Courtesy SRT.

Surface water sources	Mean yield (gL/annum)	Ground water sources	Mean yield (gL/annum)
Mundaring Weir and Lower Helena Pumpback	36.0	Deep artesian bores ground water schemes	18.5
Churchmans Brook Dam	3.7	Mirrabooka	20.0
Wungong Dam	20.6	Wanneroo	45.0
New Victoria Dam	6.0	Jandakot Stages 1 and 2	9.8
Canning Dam	37.7		
Other importing into the Swan–Canning system	106.1	Other importing into the Swan–Canning system	26.0
Total surface yield	210.1	Total ground water yield	119.3
		TOTAL SYSTEM YIELD	329.4

gL = gigalitres.

TABLE 2.16 *Water for a thirsty land: sources of water adjacent to the Perth metropolitan area.* Adapted from SRT, 1999. *Additional sources of water are under investigation and include damming Samson Brook and Wellesley Creek in the south-west, and greater extraction of ground water from aquifers such as the Yarragadee. However the effect of these measures on river flows is still unclear and could be influenced by a decline in rainfall. There are also plans for the desalination of sea water and reuse of waste water. A pipeline to bring water from the Kimberley has also been debated for many years.*

Conversion of remaining unsewered households to the sewerage system has been another major strategy. In the past septic systems were major sources of nitrogen and sulfate with the added complication of illegal inappropriate disposal of material removed from septic tanks. In addition to expansion of the sewerage system has been the development of more efficient methods for stripping nutrients from the sewage, and the development of contingency plans for preventing overflows of untreated sewage to waterways due to equipment failures.

Controlling the blooms

Although the most effective control of algal blooms depends on control of nutrients entering the system, there are a number of techniques with the potential to control conditions that favour nutrient release from the sediments in the estuarine waters. A reduction in nutrients would reduce uptake by algae, or growth of blooms. As we have seen, conditions favouring release are layering of salt and fresh water (stratification) of the water column that leads to low oxygen conditions favouring release of nutrients. Destratification using a 'bubble curtain' has been found to be ineffective, but two other techniques showing promise are:

- oxygenation of bottom waters to enhance removal of nitrogen, phosphorus and organic carbon, and
- sediment remediation with modified clays for phosphorus removal.

A bubble curtain destratification

The proposal to create a curtain of bubbles to mix the water was based on techniques used in lakes and reservoirs to overturn the water column and remove temperature stratification, increase oxygen concentrations, and move blue-green algal blooms below the surface away from the light. In the Upper Swan, the technique aimed to introduce oxygen and turbulence into the bottom of the water column to break down the salinity and temperature stratification, which would allow wind mixing of atmospheric oxygen, creating conditions that reduce sediment nutrient release and increase nitrogen loss to the atmosphere. Unfortunately the stratification redeveloped upstream, and it would seem that the salinity stratification due to the advancing salt wedge was more difficult to disrupt than modelling showed (Fig. 2.43).

Oxygenation of the bottom waters

Aeration, the mixing of air and water to increase the oxygen content, is employed in many circumstances and on many scales from fountains and small ponds to larger lakes and water bodies as well as in sewage treatment.

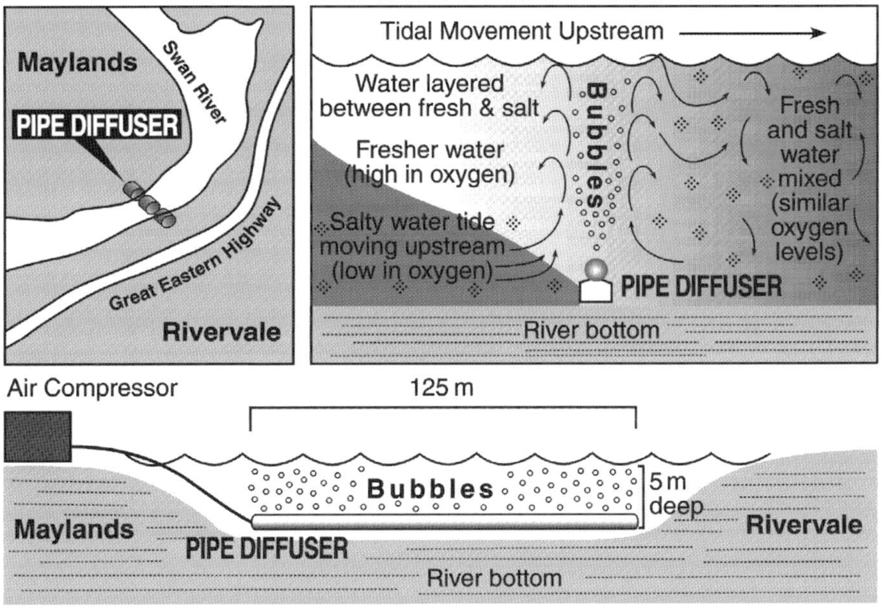

FIG. 2.43 *Trials at Maylands in spring 1997 using a 'curtain of bubbles' to mix water from the salt wedge with the overlying fresh river water were partially successful. The water column was destratified for 40 m upstream and there were some secondary effects 400 m away, although the stratification re-formed upstream of the trial site.* Source: SCCP Action Plan, 1999, Courtesy SRT.

Larger water bodies generally require more direct oxygenation techniques of injecting pure oxygen or compressed air from which the nitrogen has been removed. In the Swan–Canning direct injection of oxygen has been trialled and may be useful to complement other treatments to reduce nutrient release from sediments before the salt wedge advances in summer, and during periods of strong stratification after the salt wedge has moved up river (Figs 2.44 and 2.45).

Sediment remediation

Techniques to immobilise nutrients so that they are not released to the overlaying water column have also been researched, but most have been unsuitable. Development of modified clays Phoslock™ by the Swan River Trust, the Water and Rivers Commission and CSIRO, however, appear to be more promising. These were tested in the Upper Swan in December 1996, in Lake Monger summer 1997–1998, and in the Canning River in early 1998 and January 2000 (Fig. 2.45). In the Canning River, phosphorus was rapidly adsorbed onto the clay particles as they fell through the water and settled on the bottom. On the bottom, the clays also appeared to bind phosphorus dissolving out of dead algal cells and other organic matter, while areas outside the trial were releasing phosphorus in forms available for blue-green algal growth. The modified clays may also remove 95 per cent of phosphorous from the water column when broadcast on the surface after rainfall, and could also be used in traps in drains and wetlands to remove phosphorous in flowing waters.

At this time it appears that it will be possible to bind phosphorus, but not nitrogen, in brackish or saline water and nitrogen in fresh water. The technique appears suitable for use in constructed wetlands, compensation and nutrient stripping basins, and drains and rivers where it can be applied to sediment or the water column.

Physical capping

Capping or isolation of contaminated sediments with a layer of sand or other material has been used in coastal areas of Japan and in US harbours. This technique, however, is not feasible for the Swan due to large nutrient-rich ground water flow; continued inputs of nutrient-rich sediments and organic material. There are also problems with accuracy of application, short life of capping, the cost relative to the area to be covered, and the effect on benthic habitat and organisms.

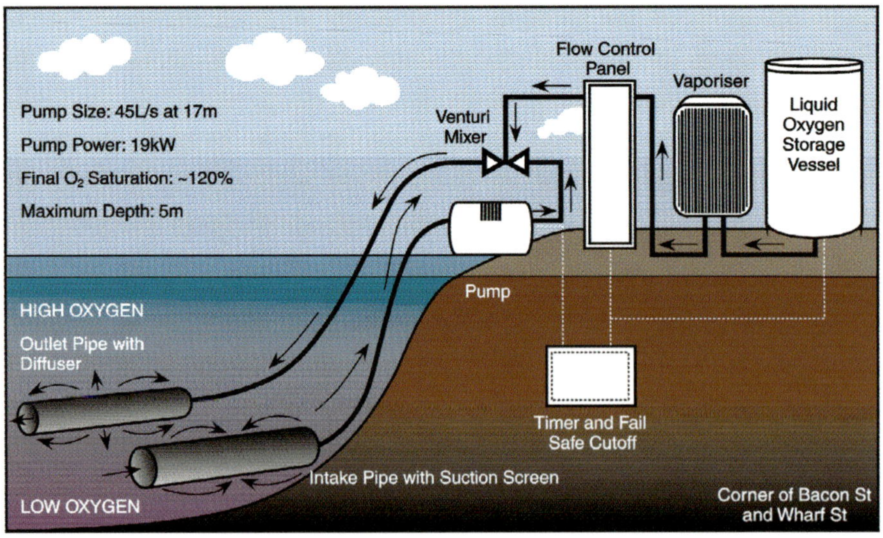

FIG. 2.44 *Oxygenation trial in the Canning River, 1997–1998. Low oxygen water was pumped from the bottom of the river to a mixing plant where oxygen was added. The oxygenated water was returned to the bottom of the river without disrupting the natural salinity and temperature layering. In the Thames Estuary (UK), oxygen is injected directly to the bottom of the river to improve water quality after high organic inputs due to storms and pollution.* Source: SCCP Action Plan, 1999, Courtesy SRT.

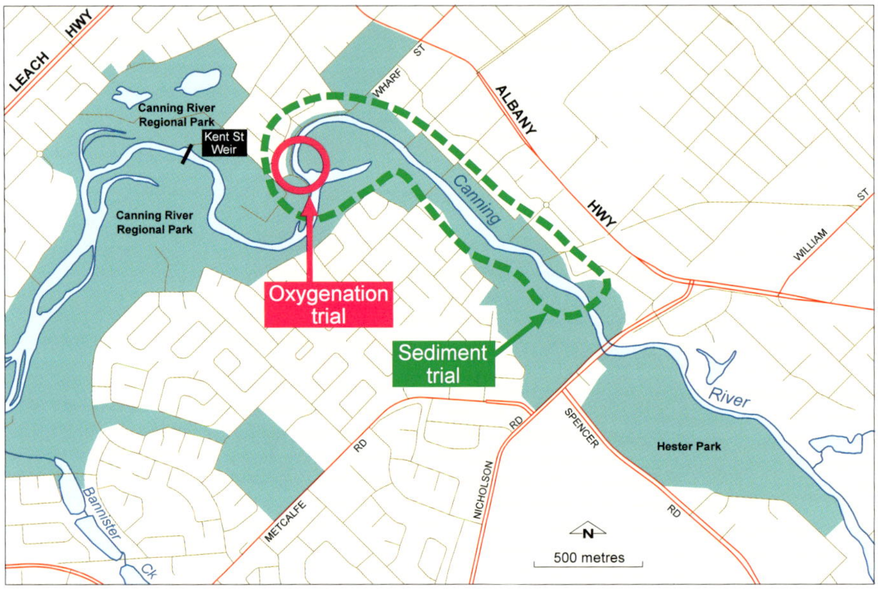

FIG. 2.45 *Location of oxygenation and sediment remediation trials in the Canning River 1997–1998. A granular form of the modified clay Phoslock™ is now manufactured in China under a joint venture with a public company listed on the Australian Stock Exchange in 2003. The product could be used in natural watercourses, reservoirs for drinking water and irrigation, sewage treatment plants and areas of turf.* Source: SCCP Action Plan, 1999, Courtesy SRT.

Sediment removal

Although dredging has been used to remove contaminated sediments from rivers, estuaries and bays in Canada, USA and England, this is not considered a suitable option for the Swan–Canning. There is a problem in targeting the generally thin mobile layers, and equipment to deal with the fine easily suspended sediments is not readily available or likely to be cost-effective. In addition, there are likely high environmental impacts due to problems of appropriate disposal of large quantities of nutrient- and sulfate-rich material, and with the continuing inflow and redistribution of sediments. Habitat disturbance or even destruction could also be a problem.

Channel modification

Although some degree of success in the Peel–Harvey has been achieved with the Dawesville Channel and in the Leschenault Estuary with the 'Cut', this is not feasible for the Swan. In the Lower Estuary, there is already marine flushing greater than that produced by the Dawesville Channel since the removal of the limestone sill in the late 1800s, and the wooden piled railway bridge in the 1960s. Further, the configuration of the Swan–Canning is different as there is not the same large estuarine basin parallel to the coast. In addition, algal blooms in the Swan–Canning occur in the upper regions where increased oceanic exchange is already limited. The tides are naturally reduced (damped) along the long estuary channel, with the range at Barrack Street only 80 per cent of the ocean tidal range; this is better than Peel–Harvey with the Dawesville channel where tides are 48 per cent and 55 per cent of the ocean range.

Generally it is considered that the cost of river bed and channel modification is very high relative to its effectiveness, and it would be necessary to remove large sandbars at Point Walter and in Perth Water. The scale of work would cause severe environmental impacts and problems of disposal of material. Such works are still being modelled.

Bathymetric smoothing

Until recently it was postulated that nutrients were derived from organic rich material that had accumulated in deep holes created by dredging. Recent surveys, however, have shown that these are generally limited in extent covering only 10 per cent of the estuary bottom, and that most were very wide depressions less than one metre deep with gently sloping sides. River flows also appear to remove sediment accumulations. Nutrient-rich sediments are probably spread in a thin layer across the riverbed and renewed annually with the winter rains, rather than as deep deposits in 'problem' areas.

Canning River options

The problems in the Canning River differ to those in the Swan River. The Kent Street Weir creates a freshwater environment with still water that is high in nutrients, particularly phosphorus, when light levels and temperatures are high. Suggestions to remove the Kent Street Weir and return the river to its natural state are problematic. The weir has been in place for a long time, which means that attitudes and current use reflect the freshwater conditions with considerable sentiment attached to the environment.

The return to an estuarine state over summer is not preferred because it would create a stratified water column conducive to blue-green algal blooms such as *Nodularia*. It might create less favourable conditions for the freshwater weed *Hydrocotyle*, however it would not be favourable to the riparian vegetation, which has adapted to the freshwater environment. Salinisation of ground water would be another problem for users, and lower water levels would expose large areas of mud flats.

Another option is to increase river flow but the potential for this is limited as there are dams on the Upper Canning and Wungong, and water allocations for irrigation and commercial uses. However there may be potential for agreements to release water at critical times that, together with lowering of the weir water levels in spring, would allow some saltwater intrusion to discourage exotic weeds.

Biological intervention to combat the weed problems is also considered unlikely to be useful. Strong seasonal changes and a lack of substantial area of hard surfaces for attachment make it a difficult environment to manipulate. Native zooplankton grazers are generally unable to cope with the large algal cells typically responsible for blooms. In addition, evidence from overseas suggests that low levels of contaminants such as oil, grease and herbicides similar to those in the Canning may affect small invertebrates. It is therefore likely that some native invertebrates have already been affected by contaminants resulting in populations that are now much reduced.

Water and Rivers Commission

The Water and Rivers Commission operated under the Water and Rivers Commission Act 1995 and was responsible to the Minister for the Environment and Heritage, and Water Resources (Table 2.15). The Commission worked with all levels of government, industry and communities to protect and manage Western Australia's water resources, including surface water (wetlands, rivers, estuaries and inlets), and underground water (ground water). The Commission set the framework for how water should be used throughout the state, and the balance between the need for water by the community and environment.

This required assessment of how much surface and ground water there is in a particular area and the quality of the water (including nutrients, salinity, pollutants and algae). This information was then used to gauge the requirements for maintenance of environmental features such as wetlands, and the level of human use for public water supplies, industry, mining, agriculture and horticulture.

The Water and Rivers Commission provided advice, planning and education to all sectors of the community and assistance to communities in managing their water resources through community-based bodies such as the Geographe Bay Catchment (GeoCatch). In 2004 and during the writing of *Swanland*, the Government has restructured the Water and Rivers Commission and the Department of Environmental Protection to form a new Department of Environment (DoE) (Table 2.15). The vision of the restructure was to provide better management of the State's estuaries and waterways by integrating all aspects of catchment management.

Swan River Trust

The Swan River Trust is responsible for coordination of work necessary to balance use and protection of the waterways and shorelines, and to restore degraded environments within a gazetted management area. It is responsible to the Minister for the Environment and Heritage, and Water Resources. The management areas include the Swan–Canning River system and adjoining parks and recreation reserves from the Fremantle Traffic Bridge to Moondyne Brook on the Avon River, to the lower diversion dam on the Helena River, along Southern River to the Allen Road crossing, and the Canning to its confluence with Stinton Creek.

The Trust assists with development of management plans and provides advice to the Minister, the Western Australian Planning Commission and local government authorities about developments that affect the river and estuary. It also works with other agencies, landowners, community groups, boat owners and anglers to control shoreline erosion and pollution. In addition, the Trust also maintains foreshores and clears pollutants and rubbish (Table 2.17). The decrease in rubbish cleared from the estuary from 1995 to 2000 appears to indicate that community education had a positive effect.

The Trust conducts water quality monitoring to detect pollution from nutrients and other chemicals. This allows identification of sources of nutrients that fuel algal blooms and development of strategies to reduce their occurrences. Excessive algal blooms are a response to excessive nutrients and an indication of a biological response, but we need to remember that some algal growth is a necessary part of estuarine function. We need to be able to distinguish between bad smells that may be unpleasant but not a danger, and

TABLE 2.17 *Material collected by the Swan River Trust.* Source: SRT, 2002.

		95–96	96–97	97–98	98–99	99–00	00–01	01–02
Domestic rubbish collected from beaches	tonne	123	80	88	85	46	56	77
Logs and timber from fallen trees	tonne	253	152	109	150	109	101	121
Rotting weed removed from foreshores	tonne	348	460	612	350	197	390	529
Dead fish left by prawning parties	tonne	12	11	11	10	7	5	4
Dead birds	number	26	31	176	120	123	158	83
Derelict/abandoned boats salvaged	number	4	5	6	1	3	3	7
Syringes on beaches and public places	number	109	146	242	118	232	169	144
Dead animals (cattle and sheep)	number	5	9	8	6	4	11	7

amounts of algae necessary for sustaining estuarine processes. This includes the food webs that lead to prolific fish and bird life.

Excessive amounts of algae indicate that the balance has been tipped, and the natural processes are no longer operating in harmony so the environment is degrading. The question then remains if we are happy with the changes, or if the balance can be corrected and what monetary and convenience cost that might incur?

The role of assessing the state of the Swan–Canning, managing past influences and planning for the future lies with three main bodies: the Department of Environment, the Swan River Trust, and the Department of Conservation and Land Management that liase with all groups in the wider community.

The Swan Estuary Marine Park and Adjacent Nature Reserves

The Swan Estuary Marine Park and Adjacent Nature Reserves cover estuarine lands and waters within the Swan River at Alfred Cove, Milyu and Pelican Point (Map 2.15). The Marine Park and Pelican Point are vested in the Marine Parks and Reserves Authority (MPRA) and the two other land areas adjacent to the Marine Park are vested in the Conservation Commission of Western Australia. All lands and waters are managed by the Department of Conservation and Land Management (CALM). A Management Plan 1999–2009 (No. 41) has been compiled by the Department of Conservation and Land Management.

The three foreshore areas are important feeding grounds for trans-equatorial migratory wading birds protected under agreements Australia has with Japan and China. The sand, mudflats and beaches provide habitat for diverse assemblages of aquatic and terrestrial flora and fauna, and are the only significant areas remaining where wading birds can feed and rest within

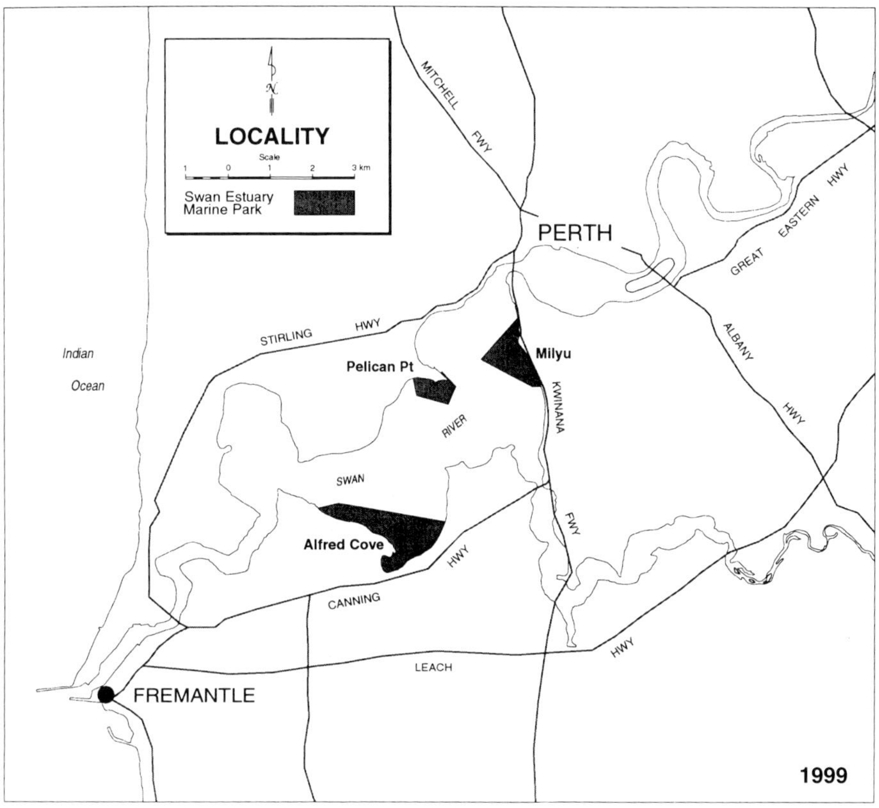

MAP 2.15 *The Swan Estuary Marine Park and Adjacent Nature Reserves.* Source: Management Plan 1999–2009, Courtesy CALM.

the Swan Estuary. Within the Park, birds move between the three areas on a daily basis. The Park is not only managed for conservation but for recreation, education and scientific study as well as commercial and recreational fishing. Pressures of a modern city on the areas are increasing so that continued conservation (status) of the areas is most important.

Keeping the jigsaw complete

The Ecological Model for the Swan–Canning Estuary (Fig. 2.19) has provided a framework for how the estuary works and an indication of the key processes required for maintaining estuarine health with a growing city in the centre of 2,000 square kilometres of urban development housing approximately 1.5 million people at the beginning of the year 2000.

The Swan–Canning has been subjected to an accelerating rate of change for 175 years and, as we have seen, some problems have been obvious for many years. The opening to the sea, and the salty marine influence, has been increased by dynamiting the bar. The channels have been dredged and deepened to facilitate access to the estuary and progress along its river course, creating areas where the water column can stratify promoting low oxygen conditions. The spoil from dredging has been dumped on the foreshores reducing the extent of the shallow sandbanks where many of the chemical and biological processes necessary for healthy estuarine function occur. We have used the foreshores and low-lying parts of the landscape for the disposal of large amounts of rubbish and pollutants that contaminate ground water that flows eventually to the river and estuary. Drains from factories, rubbish dumps, septic tanks and houses have brought nutrients and poisons to the estuary waters. Dams have been built across the rivers leading to the estuary, reducing the flushing capacity with the result that nutrients and organic material are not flushed to sea but now accumulate in the sediments.

It is clear that excess nutrients are a key in fuelling unwanted noxious and toxic algal blooms, even though the estuary has always been productive. An obvious management issue is therefore control of the delivery of additional nutrients from the catchment, which is dependent on the various levels of government regulation and enactment. Of even greater importance are community attitudes for complying with and supporting the regulatory authorities. An obvious part of supporting regulations is in the understanding and knowledge about the issues at stake and mechanisms needed to deal with the issues.

The problems of how to deal with already large nutrient stores lies in 'locking up' the nutrients, and maintaining or creating conditions less favourable for blooms. Clearly large shallow areas are needed for the natural processes of degassing of nitrogen. These are also the areas where seagrass and algae grow, and where the tiny animals live that provide the rich feeding grounds for fish and birds.

As we have seen, the margins of the estuary are now very different to those encountered by the first developers. We now have beautiful playing fields and roads as well as dredged boating channels with only small remnants of the earlier landscape. These small undisturbed areas fringed by rushes and reeds create a backdrop to the urban landscape, are key areas for wildlife and the chemical processing to keep the estuary healthy, and also a buffer to nutrients.

The reserves set aside are obviously key areas not just 'museum exhibit pieces' of what was, but an intrinsic part of the currently relatively healthy estuarine system. Declaration of the reserves has been timely as there are few areas around the Swan that could be reserved in the future. We can, however, protect them from undesirable change by creating or maintaining buffers; revegetating nearby areas; and reducing pollutants of nutrients and chemicals poisonous to the plants and animals, which we now know are also likely to affect human health. Again this depends on the regulatory authorities but also on community support.

Chapter 3

The Peel–Harvey Estuarine System

Situated only 80 kilometres south of Perth and Fremantle, the oldest and major centres of population in Western Australia, it is perhaps not surprising that changes to the hinterland around the Peel–Harvey commenced soon after European settlement. In the early 1800s low-lying coastal land was cleared and drained for agriculture, with major effects to water flow and water quality.

Originally aquatic plants (seagrasses) such as swan grass *Ruppia megacarpa* and paddle weed *Halophila ovalis* grew in the shallow waters of the Peel–Harvey. An increase in nutrients in run-off from surrounding agricultural land following the development of superphosphate to supplement poor quality soils, however, led to eutrophication. In the Peel Inlet this was massive green macroalgae growths, and in the Harvey Estuary blooms of microscopic algae (diatoms) and blue-green algae. The algae changed over time, with the first blooms in the 1970s of the green *Cladophora montagneana*, replaced in the 1980s by *Chaetomorpha linum*, *Ulva rigida* and *Enteromorpha intestinalis*. The blue-green cyanobacteria *Nodularia* problem developed later in 1978, mostly in late spring and early summer.

To alleviate the algal problems, management adopted three major strategies: reducing nutrient inflows; removing algae from the water body by harvesting; and increasing tidal flow to flush nutrients to the sea through a 2.5-kilometre long channel cut across the dunes near Dawesville. This huge undertaking has resulted in more marine conditions—clearer, more salty water that fluctuates with the tides—creating a New Peel–Harvey Environment.

This chapter describes the Peel–Harvey as it was and follows the changes that occurred, which caused concern about the problems and to scientific research, political lobbying, decision making and formation of policy that led eventually to the massive engineering works and birth of a new and different ecosystem.

LANDSCAPE OF THE PEEL–HARVEY —THE LARGEST BODIES OF WATER

The Peel–Harvey Estuarine System lies on the western edge of the Swan Coastal Plain, separated from the sea by a line of dunes formed less than a million years ago in the late Pleistocene. Peel Inlet lies between the Spearwood Dunes to the west and the slightly older Bassendean Dunes to the east, while the Harvey Estuary lies in a depression between the Spearwood Dunes. The natural but narrow 5-kilometre long channel, now deepened and partially walled, enters the sea at Mandurah (Fig. 3.1 and Map 3.1). Tidal flow through the channel was slow, and tides were only a minor influence in the estuary until the Dawesville Channel was opened in 1994.

FIG. 3.1 *Landsat false colour composite of the Peel–Harvey Estuarine System, 4 December 1980. The red colour in the Harvey Estuary and the NW Peel Inlet is due to a massive bloom of the blue-green algae* Nodularia spumigena. This view was used on covers of the Peel–Harvey reports with acknowledgements to H. Houghton, Department of Land and Surveys and Dr F. Honey, CSIRO for computer analysis and scene generation.

The Peel Inlet (75 square kilometres), and the Harvey Estuary (61 square kilometres) are the largest inland water bodies of south-western Australia. Separated by a shallow sandy sill at Point Grey, a deep channel allows water some flow between the two water bodies. Peel Inlet is roughly circular, about 10 kilometres across and up to 2 metres deep, bordered by wide gently sloping shallow areas less than half a metre deep that cover about half the basin. In contrast, the Harvey Estuary is elongate, slightly deeper with a narrow fringing border, only 2–3 kilometres wide and 20 kilometres long, oriented NNW–SSE in line with the prevailing winds.

In winter, the estuaries are even larger as the main basins, the ill-defined river courses and the drains connect with what is still a vast wetland covering the low-lying areas on the Bassendean Sands behind the hard core of the Spearwood Dunes. In summer, river flow slows to a trickle or ceases and the lower reaches become more saline as the marine waters gradually move upstream.

MAP 3.1 *The Peel–Harvey Estuarine System showing the main drainage areas; the coloured areas indicate the major sources of nutrients.* Hodgkin et al., 1980. Courtesy DoE.

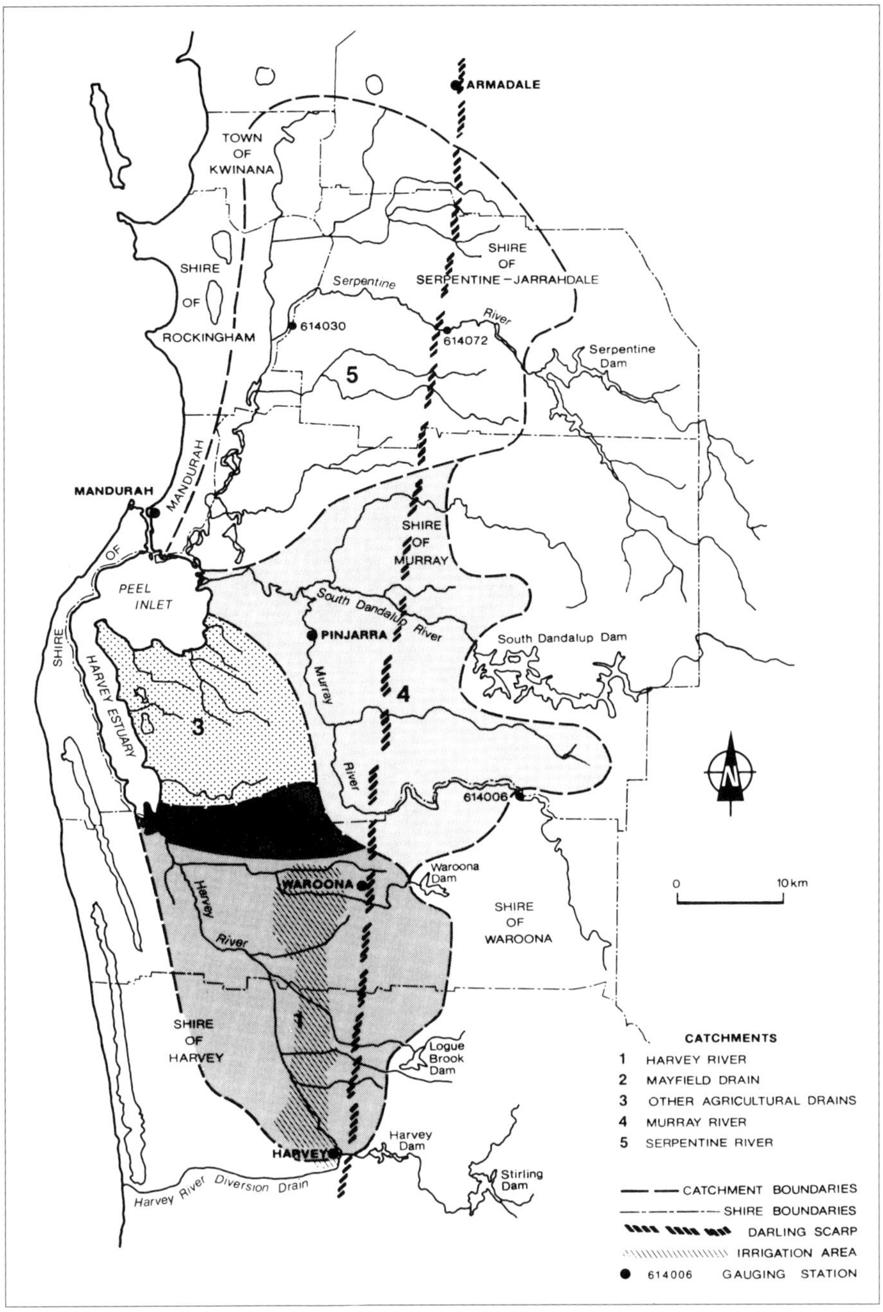

Sources of the water—river flow

The Peel–Harvey system receives water from three rivers, the Serpentine, the Murray and the Harvey, and fifteen or so major drains from agricultural areas. The largest river, the Murray, with its tributaries the Hotham and Williams Rivers, rises 150 kilometres inland near Narrogin on the Yilgarn Block (Map 3.2). The upper reaches flow through the broad valleys of the wheatbelt where rainfall is only 500 millimetres a year. Nearer to the western margin of the Yilgarn Block, where rainfall is about 1,000 millimetres a year, the Murray flows through the forested valleys of the Darling Range where the Serpentine, Harvey, North and South Dandalup Rivers as well as a number of other small streams rise. Here laterite or gravel, a brown capping of iron and aluminium oxides supports jarrah and marri forest, vegetation with deep penetrating roots well adapted to the infertile soils and dry summers (Fig. 3.2).

FIG. 3.2 *Murray River, Baden-Powell Water Spout in the Darling Range, May 1985.* E. P. Hodgkin.

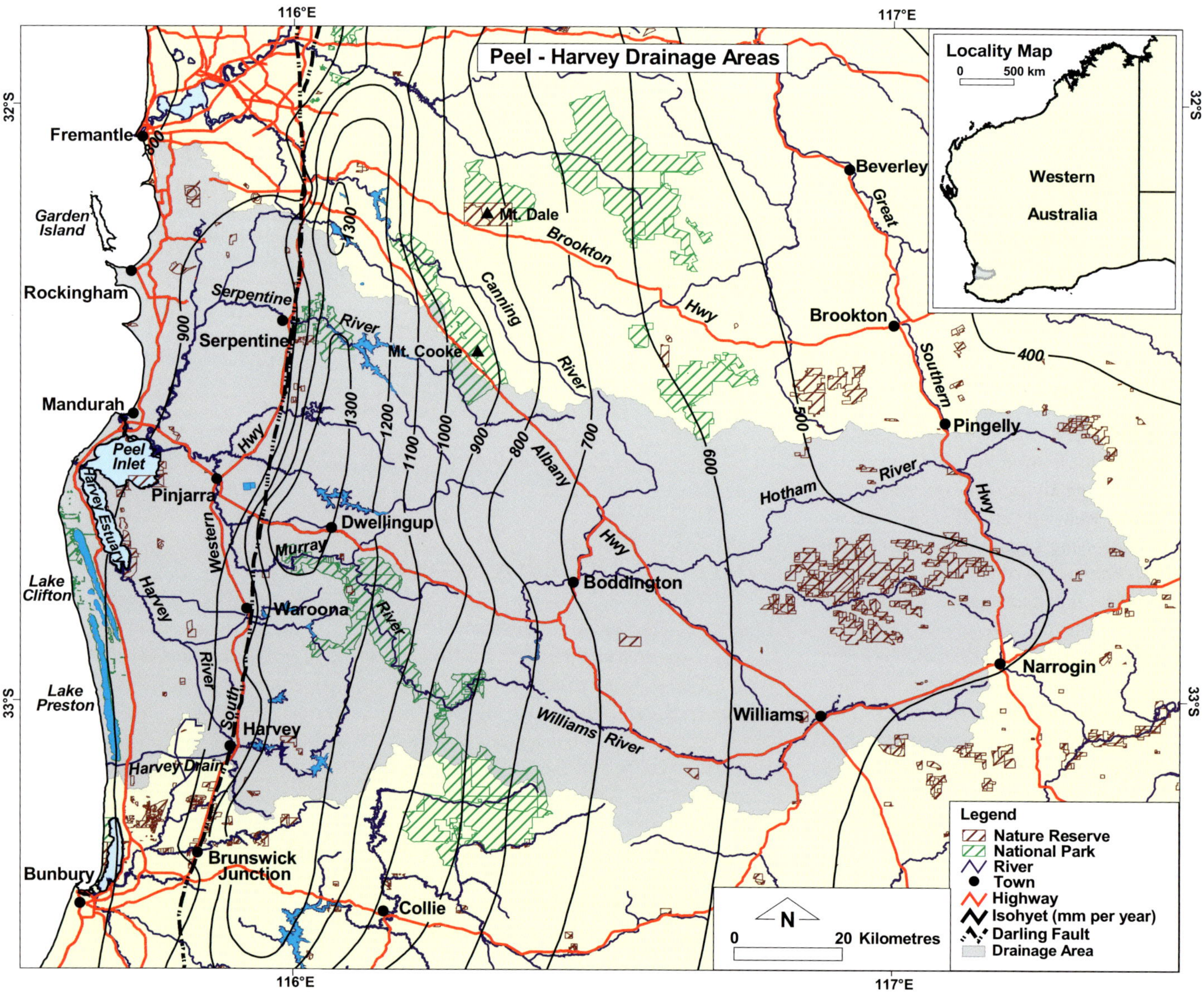

MAP 3.2 *Catchment areas and rivers of the Peel–Harvey Estuary. The longest river, the Murray, rises in the agricultural region on the ancient plateau where rainfall is less than 500 mm a year. In the lower reaches the Murray, Serpentine and Harvey Rivers flow through eucalypt forest before descending to the sandy coastal plain where the river courses were once a maze of swamps.* Courtesy DoE.

The rivers descend in steep-sided valleys with rapids and waterfalls before spreading out across the Swan Coastal Plain in what was once a network of swamps and small rivers, a huge continuous wetland in winter. The Serpentine flows southward between the Bassendean and Spearwood Dunes through Lake Geogrup to the northern side of the Yunderup Delta. The Murray, tidal to the weir at Pinjarra, flows through a more defined course entering the Peel by the twisting channels of the southern Yunderup Delta through which a channel has been dredged (Fig. 3.3). The Harvey River flows northwards across the plain to the southern end of the Harvey Estuary entering through a large shallow 'crowsfoot' delta area (Fig. 3.4). Numerous drains flow into the system from the east and small deltas now cross some of the mouths.

The Harvey and Murray catchments have been extensively cleared for agriculture. In the south-east of the Harvey, irrigated pastures support a dairy industry and some intensive horticulture such as citrus orchards. In

FIG. 3.3 *The Murray River Delta at Yunderup showing the Yunderup canals and the Murray Lakes development, aerial view, 30 October 1996. The soils in low-lying swampy areas around the Peel–Harvey and other estuaries are rich in sulfate, iron and organic material, and this makes them prone to acidification. When exposed to air during clearing and dredging the iron sulfides (iron pyrites) oxidise producing sulfuric acid that may release toxic concentrations of aluminium, iron and arsenic from the soils. The acid and metals can cause fish kills and contaminate ground water as well as damaging equipment and buildings. Some canal and reclamation areas around the Peel–Harvey contain dangerous concentrations of acidic water.*
Photograph courtesy DALSE.

FIG. 3.4 *The Harvey River delta, aerial view, November 1980. The vegetated reaches of the river, fringing marshes and open sands are a haven for waterbirds. The seclusion, year-round fresh water, protection from strong winds, safe roost sites, and the close proximity of shallows for feeding and loafing all combine to make this part of the Peel–Harvey a near-ideal dry season refuge for waterfowl. Sand eroded from the catchment has formed a 'crowsfoot' delta around the river mouth.* E. P. Hodgkin.

the central and western areas, nitrogen-fixing clover pastures support beef cattle, sheep and hay production, while the Murray catchment is mostly used for sheep and wheat. Although the Serpentine catchment has been less cleared, water flow and nutrient inputs have changed. The flow from pristine forested areas is trapped by a dam to supply drinking water, and cleared areas are used for agriculture, horticulture, grazing and hobby farming (Map 3.3).

Arriving in the water

Water flow to the estuary is seasonal, with 90 per cent of water reaching the estuary in the winter and spring months between May and October. Sources of water also vary from year to year depending on how much rain falls and on which part of the catchment it falls. In most years, the Murray with its large catchment brings about 50 per cent of the water to the system and the Harvey about 30 per cent, but in other years most of the flow comes from the shorter rivers in the higher rainfall area near the coast. Up to 1989 stream flow to the estuary ranged from 370×10^6 to $1,200 \times 10^6$ cubic metres (370 to 1,200 gigalitres) each year, with an estimated total phosphorus input of 95–237 tonnes. The Harvey River contributed only 30–40 per cent of annual stream inflow and brought 50–75 per cent of the annual phosphorus load. In contrast the Murray, bringing most of the water, brought relatively smaller amounts of phosphorus and other nutrients, while the Serpentine with only a small flow had much higher nutrient concentrations and contributed a larger amount of phosphorus. More recently during the 1990s nutrients coming to the estuary from the Harvey River declined but nutrients coming in waters from the Serpentine appear to have increased.

Where does the estuary begin, where does it end?

The amount of water in the estuary varies with river flow, tide and atmospheric pressure. At times the water body extends over the low-lying margins; at other times a broad band of sand surrounds the shallow water body. In this area between the extreme high tides and mean sea level water, salt-tolerant plants form the fringing or salt marsh community (Fig. 3.5).

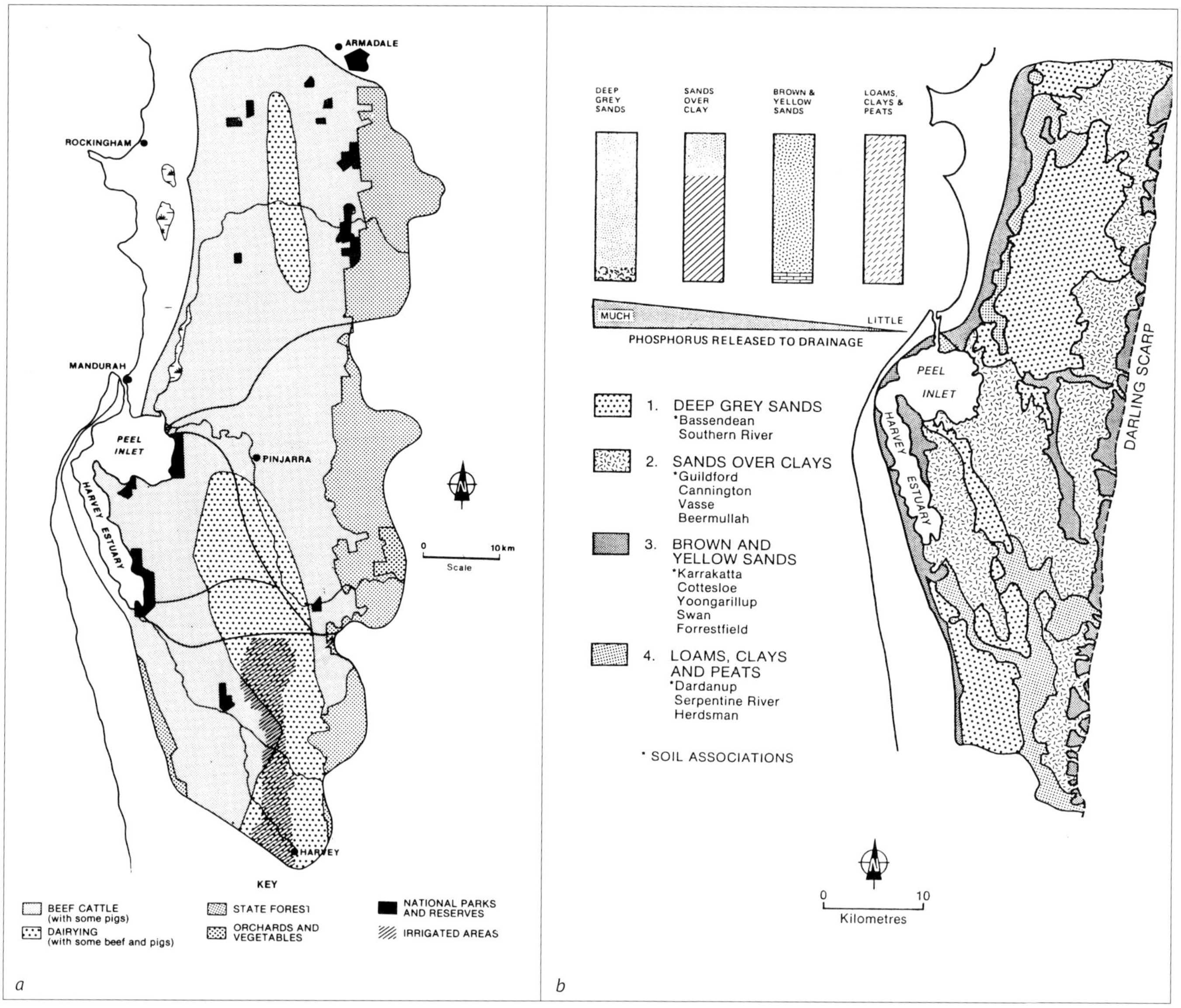

MAP 3.3 *(a) Land use on the Swan Coastal Plain: the upper river reaches pass through forest before descending to the coastal plain where most of the land has been cleared for beef and dairy cattle as well as more intensive agriculture and horticulture.* Hodgkin et al., 1980. *(b) Soils on the Swan Coastal Plain: the soils have little capacity to retain fertilisers applied to overcome the naturally poor soils.* DCE 1983a. Courtesy DoE.

Plants on the edge

Subjected to periods of flooding and drying, these plants on the edge of the water body are also on the physiological edge. They are exposed to waterlogging and accompanying anoxic conditions, and extremes of salinity from fresh waters brought by rain and river run-off to salty marine water brought by tides. They are also subjected to erosive wave action and to sediment burial. A difference in tolerance to these conditions and the duration of these conditions result in zonation of plants and a mosaic of plant communities. Even apparently bare areas are colonised by highly productive microscopic algae, fungi and bacteria.

The plant communities are characterised by three main zones: low-lying flats or pans regularly immersed by tides; a higher transition zone with both low-lying and higher elevation plant types and infrequent immersion; and a zone of samphires, rushes and salt-tolerant trees where immersion is very irregular. This most elevated zone is where storms, tides, higher water levels during floods and waves have deposited sediment in ridges leaving behind evidence of higher water levels.

The positions of the different plant communities and the topographies of the marsh and ridges provide a record of past water levels and processes of erosion and accretion. The sediments below the plants also record changes

FIG. 3.5 *(a) Fringing vegetation with samphire, the sea rush* Juncus kraussii *and sheoak* Casuarina obesa *trees at Coodanup, 1985. (b) Heron Point, Harvey Estuary with* Melaleuca cuticularis, *November 1977. (c) Salt marsh pan in low areas of the salt marsh, Falcon, May 1985. These still, protected areas, although used occasionally for dumping rubbish and as breeding grounds for mosquitoes, are also important feeding grounds for fish when flooded. As buffer areas they protect the margins of the estuary during flooding.* E. P. Hodgkin.

TABLE 3.1 *Estimates of fringing marshes in the Peel–Harvey dominated by samphires in 1994.* Source: McComb et al., 1995.

Location	Area (ha)
Austin Bay area	38
Creery Wetlands	140
Harvey Delta	145
Goegrup Lake area	45
Roberts Bay	81
Other	181
TOTAL	630

Areas with less than 30% samphire cover and areas dominated by terrestrial shrub and trees were excluded in the 1994 calculations. The area of marsh cover in 1985 was estimated at 13.9 km², recent re-evaluation revised the 1985 figure to 9 km² giving a discrepancy of 5 km². In the future, monitoring of Permanent Transects will allow assessment of changes to the fringing vegetation.

in water movement with coarser sands indicating more vigorous conditions and reworking of sediment, while finer fractions such as silt, mud and detritus derived from riverine inputs and in situ processes indicate calmer conditions.

The most extensive marshes in the Peel–Harvey are on either side of the Mandurah Channel including the area known as the Creery Wetlands, along the eastern side of the Peel Inlet and around the Harvey Delta (Table 3.1). In other areas around the estuary, and the lower part of the Serpentine River and Lake Goegrup, this fringing plant community is often present as a narrow band. Salt marsh communities are naturally dynamic, changing with the natural variation between wet and dry years, cycles of accretion and erosion, and maturity of the biological community. Algal accumulations that occurred in the 1970s brought other changes: algae smothered some areas of marsh while in other parts marsh plants grew over the algal banks, no doubt thriving on the rich organic mulch.

Mapping that has used historic aerial photographs and computer assisted techniques for eliminating geometric and scaling problems, indicates that salt marsh areas decreased in the northern part of the estuary between 1965 and 1986. The first decline occurred immediately after construction of a permanent opening to the Mandurah Channel in 1977 that increased the effects of flooding tides. Other losses particularly in the Creery Wetlands were due to disturbance from human and vehicle access. This area became the site of canal developments in the early 21st century. Marsh land also declined around the lakes area between 1986 and 1994. In contrast, the marsh around Harvey Delta and Austin Bay increased between 1977 and 1994. (See below **The New Peel–Harvey: Around the edges**.)

PART OF THE SEA — THE PAST, AND NOW?

The estuary has gone through many changes since its original formation: from a river system flowing to a distant ocean, a part of the sea, isolation to an estuarine environment and, with recent engineering, part of the sea again. Changes are recorded in the shape of the valleys and the fossil sediments (Map 3.4 and Fig. 3.6). (See also **Chapter 1 Understanding Swanland** and **Chapter 2 The Swan–Canning Estuary: Formation of the estuarine environment**.)

About 18,000 years ago when the sea level was 150 metres lower, the Murray River after joining with the Harvey flowed past Halls Head and across the coastal plain to the sea. The estuary first formed when it was flooded by rising sea levels about 8,000 years ago. The extent of the estuary gradually increased as the sea rose still further until about 4,000 years ago when it was between 0.5 and 3 metres higher than present. At this time the low lands of the coastal plain were part of the estuary; the boundary marked today by the Vasse soils under the marshes and lakes. After that time the sea level fell to about its present level and the estuaries took on their present size and shape.

The Peel Inlet is covered with about 3 metres of sediment overlying the Pleistocene valley. Sediments brought by the rivers, sea and winds were topped by a layer of black mud with a high carbon content from algal growth and animal faecal pellets. Compared to other estuaries such as the Hardy Inlet (5 metres of sediment), the sediment layer is quite thin indicating that the accumulation rate has been quite slow at about 0.3 millimetres per year.

Fossils of marine shells in the lowest sediments show that the system was more marine between 5,000 and 4,000 years ago. At this time there were about forty different types of molluscs living in the estuary and the fauna was similar to that found in Oyster Harbour at Albany today. A change

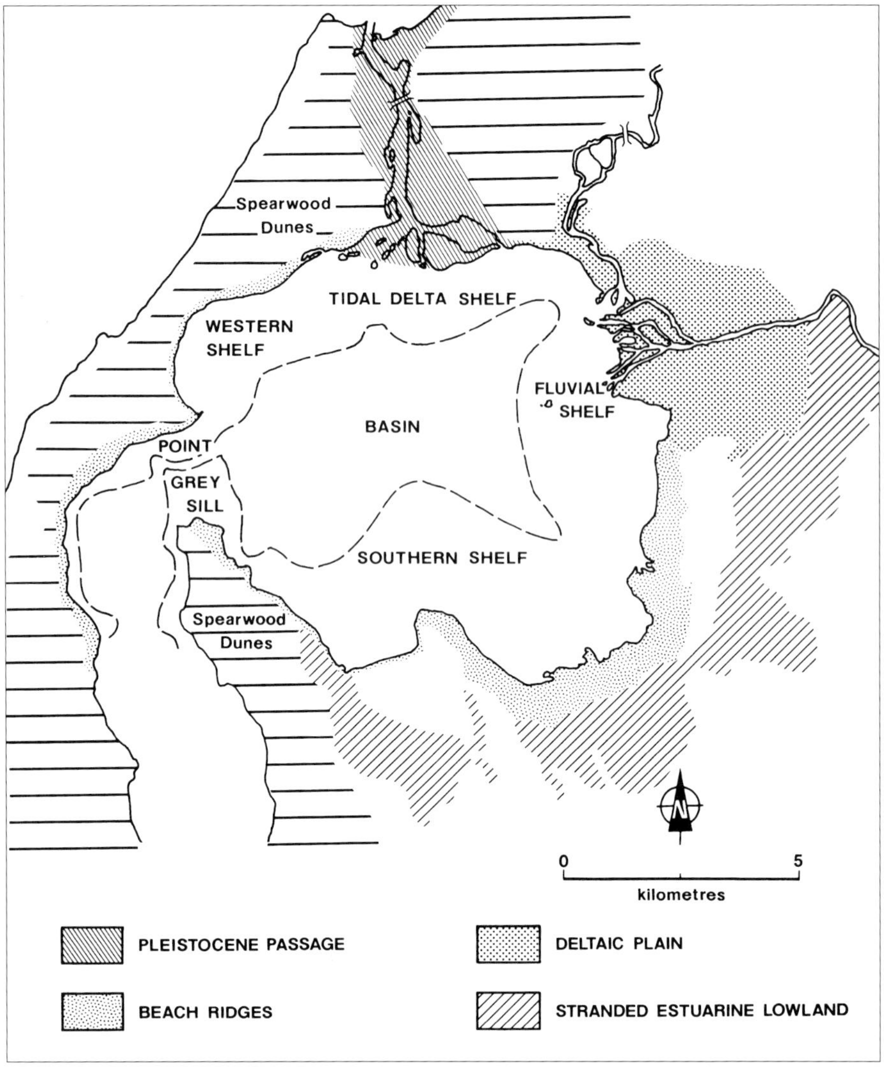

MAP 3.4 *Peel Inlet showing the extent of sediments deposited when the estuary first started to form about 8,000 years ago. The estuary continued to grow extending over the shaded area of 'stranded estuarine lowlands' until about 2,000 years ago, when the sea level started to fall again leaving the wide flat ground that surrounds the estuary today.*
Hodgkin et al., 1980. Courtesy DoE.

in the fossils indicates that the water body was gradually isolated from the sea. In the 1970s only twelve species were present; those able to tolerate the fluctuating salinity and temperature typical of an estuarine environment. Opening of the Dawesville Channel initiated another marine phase in the estuary's history.

Isolation from the sea and development of more extreme conditions was probably caused by a number of interconnected events. The low areas of the basin and inlet channel gradually shallowed with sediments brought in by winds and rivers. The opening to the sea closed when water outflow was low. Climate change affected river flow and brought lower sea levels that moved sand onshore to further block the entrance.

In the past, the mouth of the Peel–Harvey opened and closed depending on the build-up or erosion of sand at the bar. The movement of sand varied with the outward flow of water (influenced by the amount and timing of rainfall) and with the northwards longshore drift of sand in the ocean. When the bar closed in winter, floodwaters banked up until a fresh path scoured through the bar. In summer with no river flow and with evaporation the water level fell over 40 centimetres exposing sandflats, and salinity increased with evaporation to 50‰.

In the case of the Peel–Harvey we can say 'past', but which past? Modifications to the processes controlling water flow have been made since the first days of European settlement when efforts were made to accelerate the exit of floodwaters from the low-lying hinterland. Old records indicate

FIG. 3.6 *Sediments in the Peel Inlet show the different stages in the estuary's development. The bottom sediments contain marine shells showing the type of environment 4,000–5,000 years ago, while the surface sediments have been deposited as the sea level fell again.*
Hodgkin et al., 1980. Courtesy DoE.

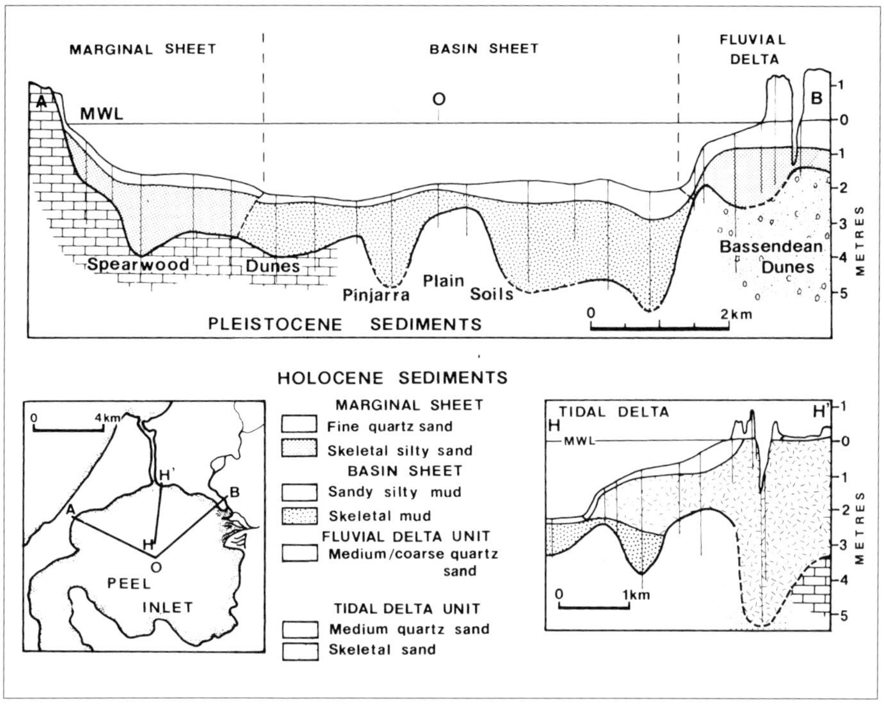

that the bar closed in the summer of 1914–1915 and again in 1935 and 1958, but interference to keep the bar open was by then more frequent and substantial.

Ever increasing speed of change—explorers, settlement and start of change

The quiet waters of the Peel–Harvey were explored in the quest for fertile land soon after British settlement around the Swan in 1829. In July 1829 Captain Currie and the botanist James Drummond, the first Europeans to visit, entered the estuary and named the Serpentine River. Later in the year Lieutenant Preston and Mr Collie passed through on their way to and from Leschenault Inlet. Once reports of the fertile soils reached Perth, settlers soon followed. The reports were so promising that even Governor Stirling cancelled his earlier land grants around the Swan River and applied for land from the Harvey River to Leschenault Inlet. Development around Stirling's grant was slow however, due to the swampy nature of the land and property speculation following Stirling's departure from the colony.

Thomas Peel, one of the first to arrive, was allocated land extending from Woodman Point to the north bank of the Murray River and inland to the Darling Scarp. Peel first settled some of his followers and indentured labourers at Clarence just north of Rockingham early in 1830, but later established a small settlement at Peeltown (Mandurah). Another small outpost further up the Murray River at Pinjarra, the limit to which a boat could travel, was soon abandoned due to hostile natives and floodwaters.

By the 1840s farming had started in a small area of river flats where the Serpentine River spreads out over the plain, known today as Lowlands. It is interesting to note that the area was described 'as fine grass in summer, water-covered in winter' and 'very rich wheat land', hinting at the problems soon to follow.

Very soon there was conflict between the Aboriginal people and the new arrivals, particularly in the fertile areas around Pinjarra. Conflict escalated to the 'Battle of Pinjarra' in October 1834 and the killing of members of a Noongar camp: the details shrouded by time can still stir great sorrow today. Five months later a truce between settlers and the Murray and Swan Noongars opened the way for settlement on the river flats and coastal plain around the Peel–Harvey (Fig. 3.7). (See **Box 9.1 Stone circles—echoes of the past**.)

Wet or dry–farming in an upside down land

The environment of the Swan Coastal Plain presented two strikingly different seasons to the new settlers. In summer there was little fresh surface water, but dense undergrowth covered much of the landscape. In winter heavy rains over a few short months filled the low areas of the landscape, and the rivers spread out across the sand plain in a maze of swamps marked by paperbarks, flooded gums and sedges. Crops planted to grow in spring and summer, following European traditions, failed through lack of water over the long dry summer, while other crops rotted in the waterlogged ground in winter.

The new environment presented other dilemmas. The cattle soon ate out the native grasses, and stock losses were high as starving cattle strayed or died of nutritional deficiencies. Following observations that recaptured strays were healthy, the practice of moving herds between coastal and heavier loam areas near the hills gradually developed. Although this required the use of large tracts of country, the cattle were healthier. This practice was later verified when deficiencies in cobalt, zinc and other elements in the fodder were recognised.

In the earliest days livestock were also moved to higher ground during the wet winter, and buildings were set on higher land. Later as the human populations increased and the land was subdivided for more intensive farming, low-lying areas were utilised and winter flooding became more difficult to deal with. At the same time water run-off was increasing as the native deep-rooted vegetation was removed. (See also **Chapter 1 Understanding Swanland: Vegetation and clearing leading to changes**.)

Initially development centred on small pockets of fertile land and the spread of grazing–homesteading was slow. The Murray River was used for transporting produce to the coast on the way to Fremantle. Later roads were cut through the bush, wheat growing increased, and small mills were established to supply the growing markets in Perth and Fremantle until cheaper wheat was produced further inland. The rate of change accelerated

FIG. 3.7 *Barragup Weir on the Serpentine River, photographed ca 1900. This wooden weir, used by the Aboriginal people to trap fish during the historical period, was destroyed by Marine and Harbours in the 1930s.* Courtesy Anthropology Department, WAM. Copyright WAM.

with the start of convict transportation in 1850. Convicts supplied labour for clearing and new roads such as the Old Coast Road to Bunbury, and in turn increased the market for produce. The construction of the railway linking Perth and Bunbury in 1893 and more efficient transport brought more settlers, land holdings were sub-divided, more trees were cleared and the land became wetter. The railway line also added to problems of water flow.

Around the Harvey problems of waterlogging increased as the old pastoral runs were divided for fruit growing and dairying, emulating similar projects at Renmark in South Australia and Mildura in Victoria. By 1898 properties around Coolup were abandoned and calls for draining the land were growing.

By spade, horse and cart—the drains begin

'Desnagging' or the clearing of trees and logs from the network of tea-tree swamps converting them to faster flowing drains, initiated in the early 1900s, started another sequence of events. Where was the water to go? Swampy areas without a defined river course presented other problems; so channels and finally larger drains were dug, again with the problem of where the water was to go (Fig. 3.8).

FIG. 3.8 *(a) Swamp in Meredith Road Reserve with dense tea-trees typical of the ill-defined watercourses that once spread across the Swan Coastal Plain, February 1982. (b) Meredith Drain created to move water across the low flat landscape, January 1982.* E. P. Hodgkin.

Major works to remove the water started with construction of the Coolup Main Drain to the Harvey and the Greensland Road Drain into a large swamp at Austin Bay, but winter waterlogging remained a problem. When the downstream areas were soaked and the estuary was full, the water did not flow away across the flat plain. There were also new problems of erosion and silting in the channels that highlighted questions of maintenance responsibilities and labour costs.

Following World War I, a new push for development saw renewed interest in the land to the north of the Peel Inlet in the area around the original Peel Estate. This, the beginning of the Group Settlement Scheme, brought urban migrants from England to establish farms throughout the South-West. The scheme was doomed to failure through the inexperience of farmers and bad management, which resulted in two Royal Commissions and heartbreak for many.

Drainage around the Peel Estate brought the same problems—where was the water to go? Drainage around Mandogalup sent water southwards to the Spectacles Wetlands where water levels rose 4 metres; so drainage works were continued further south along what is now know as the Peel Main Drain.

With ongoing dramas of flooding and levees bursting, the water moved southwards; more drains were dug creating the Serpentine River Diversion. Then the bends were straightened, and the river pools cleared to create uninterrupted flow from Baldivis to Geogrup Lake, the original main fish nursery of the Peel–Harvey into which the new drain was designed to discharge at a rate of 45 cubic metres per second. Even so, the water was not draining fast enough; the land was too flat. Although good crops produced in dry years gave a sense of achievement, others years of flooding and crop losses followed. By the end of the 1920s, many farms were abandoned or holdings amalgamated. The cycle of drain building escalated as silt from the eroded banks caused the waters to bank up again.

Despite enthusiasm for draining and opening up more land, land-holders were reluctant to meet the costs and government was increasingly involved with providing money and machinery for the work. A way of providing labour came with the Great Depression of the 1930s, and the desire to keep unemployed men occupied and away from populated areas.

The Harvey Diversion Drain

During the Depression, large groups of men were used to clear the riverbeds and dig a major diversion drain from the Harvey to the sea. In the beginning, most of the work was done by hand using axes, shovels and wheelbarrows, with horse and cart used to transport heavy loads (Fig. 3.9). First the Wokalup was diverted to the drain, and by 1934 water from the Harvey no longer spread across the land but ran straight to the sea. At one time there was even a proposal to link the Leschenault and Harvey with Lake Preston to provide a clear waterway from Bunbury to Mandurah.

Clearing of streamlines was also carried out along the North and South Dandalup and the Serpentine Rivers, where some trees had started to regrow. This, the final destruction of native vegetation that had survived the previous clearing, led to the disappearance of the freshwater fauna such as freshwater catfish or cobbler *Tandanus bostocki*, mussels *Westralunio carteri* and native water rat *Hydromys chrysogaster*. Recognition that stream-side vegetation reduces erosion and filters nutrients has brought a reversal of policy with revegetation 'streamlining' and 'resnagging' of streams now promoted to reduce flow and help in the re-establishment of stream communities.

Post World War II expansion

With the return of peace, commodity prices rose and new research to increase production on the infertile Bassendean and Spearwood sands using trace elements and new fertilisers was paving the way for an agricultural boom.

FIG. 3.9 *Diverting the Harvey River to the sea to reduce flooding on the Harvey flats and remove silt that was filling the main drain. Between October 1931 and December 1932, 2,500 men based at a campsites at Myalup shifted 1,377,000 cubic metres of earth and dug a trench 25 kilometres long, 10 metres wide at the top and 6–8 metres deep. Mechanical diggers were only used further inland near the Wokalup River during final stages.* Courtesy Battye Library, 5297B/13.

The rate of land clearing increased, and with it the need for yet more drains. The need to drain also became 'entrenched'; the need for more drains in many cases unquestioned, with old drains widened and deepened. The demand for land and clearing continued into the 1980s with government departments at odds over land use. The soils of the Meredith area on the middle reaches of the Harvey River were described in 1971 by the Public Works Department as 'rich swamp lands' and by the Department of Agriculture as 'poor grey and coastal sands'. Where was the overview, the insight? Within six years the Meredith Drain was identified as a significant source of nutrient pollution affecting the estuary.

Collection basin – what comes in stays

Estuaries occupy low points in the landscape where water collects, bringing sediments and nutrients. At these points they meet the ocean and, in exchanging water, allow both fresh and marine organisms to mix. As a result estuaries are usually extremely productive, although the communities contain only a few types of plants and animals; those able to survive fluctuating and extreme conditions.

The Peel–Harvey is no exception, and although the plants and animals have changed with seasons or larger climatic cycles, the system has supported large populations of fish and birds since its formation thousands of years ago. By the 1970s, however, the changes were different and people were starting to notice and complain about the amount of algae building up on the beaches and fouling fishing nets.

Hindsight is wonderful; the cause of problems can be so obvious. One trouble is that as we change our environment, the consequences and effects can be small and remain relatively unnoticed for some time. Conditions can also change quickly when a series of events trigger a larger response. Looking back, both these scenarios occurred in the Peel–Harvey: initial changes were infrequent and small, the indications were there but an understanding of events was lacking.

Algae were a concern as early as 1946 and 1947 when a pink slime weed *Mazoyerella australis* previously called *Corynospora australis* was fouling fishing nets. In fact the greatest concern for the health of the Peel–Harvey system was related to the effects on fishing. Weed infestations in the Serpentine Lakes in the 1950s were linked to changes in drainage, but as with flooding, the solution was to increase the flushing from lakes and inlet. Despite these weed problems, and in fact probably due to the increased algal growth, the 1960s and 1970s were a boom time for fishing. Sea mullet and yellow-eye mullet were abundant, prices were high, and there was increased commercial and recreational fishing.

Apparently the aquatic plants, the seagrasses, swan grass *Ruppia* and paddle weed *Halophila* were abundant during the 1960s. Although dead leaves accumulated on the beaches, these were often gathered for fertiliser and were not regarded as a problem. Perhaps the accumulations were small, perhaps they decomposed quickly, or the smaller population at that time meant that fewer people were affected.

Algal growth continued to increase and the types of algae changed. In the late 1950s and 1960s, the green alga *Cladophora* or goat weed that grew as unattached balls was floating to the surface and accumulating in shallow areas around Coodanup and Dawesville. Probably about this time swan grass (the seagrass *Ruppia*) along the western shore of the Harvey Estuary started to disappear, with little remaining by the mid 1970s. The changing weed communities led to changes in bird life and a decline in Black Swan numbers. Weed growing on the bottom and surface also affected fishing, with net hauling no longer possible across large areas in the southern end of the Harvey.

The smell of rotting algae was unpleasant and hydrogen sulfide gas was affecting people's health. Although there had been a few blooms of the toxic blue-green alga *Nodularia* in the Serpentine River in 1970 and small blooms in the Harvey Estuary, *Nodularia* blooms in the Harvey were becoming larger and more frequent. The changes were now more difficult to ignore, and concern about the estuaries' health was spreading to the wider community.

NO LONGER A LOCAL ISSUE

Until the 1970s, the inlet and its problems were dealt with by state government agencies such as the Fisheries and Wildlife Department and Public Works Department, together with the local council authorities. The growing weed problem highlighted the need for a more coordinated approach, which eventually culminated in 1971 with establishment of the Peel Inlet Conservation Advisory Committee, a forum for local government to work with relevant government agencies to ensure that the area retained its general attractions. This became the Peel Inlet Management Authority (PIMA) in 1977. (See Chapter 2, Table 2.15.)

Other environmental problems were also causing concern at this time, and the Environmental Protection Authority (EPA) was established in 1971 along with the Department of Environmental Protection to oversee state wide response to environmental problems and new developments. In 1973 the EPA established the Estuarine and Marine Advisory Committee (EMAC) of five members experienced in dealing with pollution along the coast.

Problems emerging in other rivers and estuaries throughout the South-West and the need for environmental assessment of mining and dredging claims in the Blackwood River at Augusta in 1975, highlighted the lack of information about estuarine processes. In many respects, although the Blackwood River Estuary Environmental Study showed that it was a different sort of estuary to many others, it illustrated the importance of our estuaries and the need for understanding processes in the catchment as well as the estuarine basins. (See **Chapter 6 Hardy Inlet and estuaries of the Naturaliste–Leeuwin Ridge**.)

As far as the Peel–Harvey was concerned, some of the problems were now clearly recognised. A preliminary study published in 1975 (Final Report to the Metropolitan Water Supply, Sewerage and Drainage Board on the Peel Inlet Study) conducted by researchers at the Western Australian Institute of Technology (WAIT), established that the estuarine system was nutrient rich or eutrophic. Drainage and dams had altered water flow and there was political pressure to dam the Murray. The Mandurah Channel did not flush the estuary sufficiently to cope with changes in the catchment, and chemical and animal fertilisers were polluting the inflowing water. Where there had been flooding, there was erosion, the trained rivers and channels were silting up; and then there were the algae. Efforts to control the algae were largely cosmetic, and the cause or causes needed to be identified to assist in formulating long-term solutions. Public outcry, both local and in the larger Perth area, had increased: but who was to deal with it?

In 1976, the EPA asked EMAC to undertake a major study of the Peel–Harvey Estuary to be led by Dr Ernest Hodgkin. Since his retirement in 1973 as Associate Professor of Zoology at The University of Western Australia, 'Hodge', as he was known to his students, had led the Blackwood study. At the Blackwood, he had coordinated teams of scientists from different disciplines producing an integrated study of an ecosystem the like of which had not been done before in Western Australia. The Peel–Harvey Study that was based on similar principles ran from 1976 to 1980 with the objectives:

> *Specific: To determine the causes of the excessive growth and accumulation of green algae in Peel Inlet and if possible to propose methods for its control.*
>
> *General: To gain an understanding of the working of this estuarine ecosystem so that environmental problems can be foreseen and decisions made about its management on the basis of sound knowledge.*
>
> Hodgkin et al., 1980.

The study used expertise from government and educational institutions. Studies included hydrodynamics with an emphasis on freshwater input, evaporation and other meteorological factors, as well as the sources and quantities of superphosphate. Particular attention was paid to understanding the biology of life in the estuary with studies on aquatic plants, including the microscopic and larger forms that were causing the problems, on molluscs, fish and waterbirds. Understanding how the plants used nutrients coming into the estuary was crucial to understanding the whole ecosystem. Professor Arthur McComb of The University of Western Australia and later Murdoch University was responsible for leading the research on the Peel–Harvey and other wetland environments that led to an understanding of ecological processes throughout the South-West and along the coast.

Something's wrong—untangling the causes

Along with the natural fluctuations, the waters of the Peel–Harvey had been changing for 170 years but the speed of change accelerated with increased mechanisation. While clearing of the land led to higher watertables, swamp drainage, river diversion and dam construction changed water flows. Sandbars were dredged from the rivers and at the sea mouth, and then there were the extra nutrients. While legume pastures and fertilisers added to the nitrogen input following the first rains, the large-scale clearing of the deep grey Bassendean Sands on the coastal plain catchment in the 1960s and 1970s and application of phosphate fertilisers brought the final blow. Phosphorus readily leached from the sandy soils into the rivers, particularly the Harvey, and accumulated in the estuarine waters. Although phosphorus coming down the rivers in the winter flow had been noted in 1957 and associated with superphosphate, it was the large-scale application through the 1960s when fertiliser prices were low that really tipped the balance. The Serpentine also brought nutrients from piggery effluent and feedlots.

Too much of a good thing—nowhere to go

The other problem was poor outflow to the ocean that allowed nutrients to accumulate. This made the system particularly vulnerable, as the shallow waters with abundant sunshine provided ideal conditions for plant growth.

Over the years phosphorus accumulated in the sediment, and the microscopic algae and larger macroalgae formed organic ooze on the bottom. This ooze reduced the oxygen content of the underlying sediment, a chemical state that favours release of phosphorus fuelling further plant growth. The living plants also reduced the oxygen content of the water at night, again favouring phosphorus release and fuelling plant growth in an increasing and self-perpetuating cycle. Gradually the sediment store of phosphorus built up to concentrations where nutrients were available year round and every year, not just with the first flush of winter rain or when rainfall and river flow was large. (See **Box 3.1 Ingredients of a green soup** and **Chapter 1 Understanding Swanland: Understanding water health—components we measure**.)

Two different problems—the Peel and the Harvey

When the studies of the Peel–Harvey began, it was obvious that although the two water bodies were connected they were quite different environments. The Peel had massive populations of green macroalgae *Cladophora montagneana*, which formed floating rafts and smothering mats, whereas the Harvey was dominated by microscopic plants or phytoplankton.

Large water plants—the macroalgae

The high nutrient state of the estuary in the 1960s to 1980s favoured growth of macroalgae. As in other estuaries, the algae were restricted to fast growing species able to cope with the changing salinities, usually red algae (Rhodophytes) and green algae (Chlorophytes). The brown algae (Phaeophytes) are usually less tolerant of low salinities, favouring marine water with low nutrients. As algae generally grow attached to a hard surface, the sandy and muddy bottom of the Peel–Harvey further limited the types to those able to grow in the water column. During the 1960s, the main algae were the green algae, *Cladophora*, *Chaetomorpha*, *Ulva* and *Enteromorpha* and occasionally the red alga *Chondria*, species that do not need to be attached to firm surfaces.

STINKING MASSES IN THE PEEL In the Peel, the most obvious species in the 1960s and 1970s was *Cladophora*, which formed small balls of branched filaments (Figs 3.10–3.14). The algal balls formed mats in deeper water over the sediments where nutrients were high but later formed floating masses buoyed up by gases produced during photosynthesis. The extent of the problem algae was not due to particularly rapid growth (only 3.4 grams dry weight per square metre per day), but because the floating mass could move to colonise new areas. (See **Box 3.2 Understanding the system—test tubes, cages and computers**.)

Growth was fastest in spring and summer when nutrient concentrations were low but temperature and light levels were high. Where the banks of *Cladophora* were extensive they became self-fuelling; the lower shaded layers dying, releasing nutrients for growth in the upper layers. The light requirement may indicate why *Cladophora* was not as abundant in the Harvey Estuary where conditions were more turbid due to phytoplankton and sediment suspended by winds. In contrast, the red alga *Chondria* was more abundant in Harvey Inlet, possibly better able to cope with low light in the deeper and turbid waters. *Chondria* occasionally grew in the Peel when conditions were turbid, as they were in the winter of 1978.

Box 3.1 **INGREDIENTS OF A GREEN SOUP**

Increased nutrients, particularly nitrogen and phosphorus, stimulate the growth of some plants so that the dominant species may change. Eventually the system has fewer species and the community is simpler. Once ecosystems are modified, changing them back again can be difficult, if not impossible; problems tend to recur so that plants and some of the animals they harbour become a nuisance.

The amount of nitrogen in the estuary depends on the amount of nitrogen in the catchment and the flow rates of surface and ground water. Total nitrogen (TN) includes all forms of nitrogen including dissolved forms and compounds in the plants and sediments. Organic nitrogen in plant and animal tissue is less mobile than soluble forms, such as nitrates (NO_3^-) and nitrites (NO_2^-) (oxidised forms) common in cultivated soils and flowing water, while ammonium and ammonia (NH_4^+ and NH_3) (reduced forms) are common in surface run-off and stagnant water. All of these can be converted or denitrified to nitrogen gas (N_2) so that as little as 20 per cent of nitrogen added to the sandy catchment may reach the estuary. The tiny phytoplankton, larger plants and bacteria usually take up ammonium and nitrate, although ammonium is often taken up more easily and quickly.

The amount of phosphorus reaching the estuary also depends on water flow but, as it adsorbs to iron and aluminium oxides, soil properties are important. In the upper Murray catchment on the Darling Plateau, phosphorus moves only about a metre in centuries because it binds to the iron and aluminium oxides in laterite. Most of the export occurs in particles following heavy rain when soils erode. In the sandy soils of the lower Serpentine on the Swan Coastal Plain, where there is no binding iron or aluminium, dissolved phosphorus leaches quickly to the surface and ground water.

Phosphorus occurs in organic form in tissues of organisms, or in inorganic forms in sediments as soluble non-apatite and insoluble crystalline apatite-phosphorus bound to calcium or fluoride. In crystalline form it is stable and not available for plant growth, a feature important for assessing the impact of the phosphorus nutrients washed to sea where they readily bind with calcium. The non-apatite phosphorus is usually bound to iron (Fe). The concentration of oxygen (related to the redox potential) plays a critical role in the binding and release of phosphorus. When oxygen is high (aerobic) the Fe^{3+} form of iron binds to phosphorus, whereas when oxygen is low (anaerobic) the Fe^{3+} is reduced to Fe^{2+} and phosphorus is released: some of this may enter the water column.

Concentrations of total phosphorus in the sediments of Peel Inlet have increased about fivefold since 1950 and dissolved inorganic P about twofold. In the Harvey Estuary these have both increased about sevenfold in the same period. These increases amount to about 25 kilograms per hectare in the sediment with slightly more flushed to sea. At the height of the Peel–Harvey 'problem', algal booms trapped about 50–70 per cent of the phosphorus arriving in the river water.

The more oxygenated state of the water after the Dawesville Channel was expected to decrease the release of phosphorus from the sediment and increase the amount moved out to sea where it would become inert by binding to calcium.

Source: Hodgkin et al., 1980; DCE, 1983; Gerritse et al., 1998.

The Peel supported 5,000–20,000 tonnes dry weight of *Cladophora* each year during the 1970s reaching 60,000 tonnes in 1979 at the height of the problem. Calculations showed that this amount of growth would have taken about 12 years to develop, and fits with the timing of the first large blooms.

Cladophora dominated the Peel Inlet until 1979 when it declined dramatically. After this the amount of macroalgae fluctuated with different species dominating at different times, probably due to changes in light and nutrients. *Chaetomorpha* was more competitive in lower nutrient conditions and *Ulva* and *Enteromorpha* during times when nutrients were abundant. There were also occasional blooms of the *Enteromorpha*, a green ribbon-like algae, in late winter and early spring. Like the *Cladophora* blooms, these drifted onto beaches especially around Coodanup and Austin Bay. *Chaetomorpha* was often found mixed in with *Cladophora* in the eastern area of Peel Inlet where the algal accumulations were most offensive.

Tiny water plants in great numbers—phytoplankton

In the Peel–Harvey and other south-western Australian estuaries, most of phosphorus and nitrogen entering the estuarine system arrives with winter run-off. Most of these nutrients are taken up by microscopic plant cells,

Box 3.2 **UNDERSTANDING THE SYSTEM— TEST TUBES, CAGES AND COMPUTERS**

Studies were made of how light, temperature and salinity controlled algal growth in the laboratory so that these could be related to field observations and assessment of what could be done about the algal growths.

The studies showed that the *Cladophora* was tolerant of a wide range of salinities (euryhaline), with salinity having little effect on photosynthesis and growth. High light and temperatures, however, increased growth and the uptake of phosphorus and ammonium nitrogen. As the actively growing algae never contained the amount of phosphorus at which growth was fastest, these laboratory studies also indicated that phosphorus was more important than nitrogen in controlling growth.

With the detailed knowledge of algal growth obtained in the laboratory, experiments were made in the inlets to verify the findings. All the results were then to be used in computer models in which conditions were changed to predict what was likely to happen during different seasons, different weather events, and with the larger engineering changes. Modelling was also important because measuring growing algae in the estuary was almost impossible; the algae moved around with the winds and growing conditions within the mats varied.

Source: Gordon & McComb, 1989; Gordon et al., 1980.

FIG. 3.10 *(a) Professor Arthur McComb, Murdoch University, viewing an accumulation of* Cladophora *on the north-east shore of Peel Inlet, August 1978. Professor McComb supervised many of the studies investigating the algae and nutrients. (b) Dr Gary Kendrick and Dr Dave Gordon (students at the time) collecting samples from the bank of algae. Gas produced during photosynthesis made the balls buoyant and they floated to the surface where they were caught by winds and waves, and accumulated upwind in large floating banks or on the sandy beaches. In Peel Inlet* Cladophora *was perennial, the balls and fragments breaking up and regrowing rather than being recruited from spores. This bed was 140 m long with an estimated dry weight of 75 t containing 230 kg of phosphorus and 2,000 kg of nitrogen.* E. P. Hodgkin.

FIG. 3.11 *Harvesting algae at Falcon, Peel Inlet, August 1979. Front-end loaders only operated along the beaches and in water less than 10–20 cm deep; offshore harvesters, able to operate in water deeper than 45 cm, came into use in 1983.* E. P. Hodgkin.

FIG. 3.12 *(a)* Cladophora *bed: decaying material created conditions favourable for nutrient release from the sediments that fuelled further algal growth. (b) Main areas where* Cladophora *accumulated.* Hodgkin et al., 1980. Courtesy DoE.

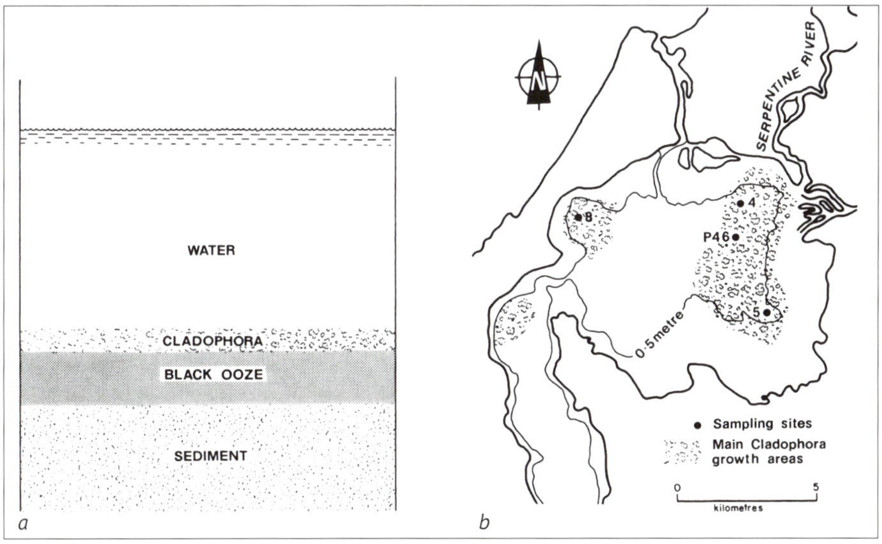

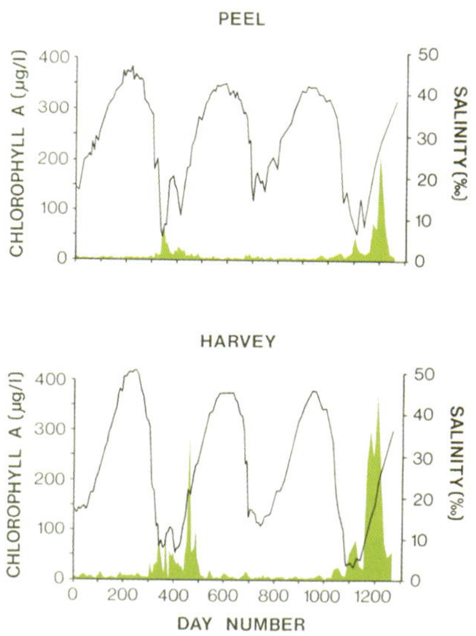

GRAPH 3.1 *Chlorophyll* a *pigments and salinity. Note the peaks in chlorophyll* a *correspond with low salinity due to inflow of nutrient-rich fresh water from the catchment.* Courtesy A. McComb, Murdoch.

the phytoplankton. If conditions are favourable cells in the phytoplankton multiply and accumulate into blooms (Graphs 3.1 and 3.2).

Phytoplankton communities change as conditions of temperature, light, salinity and nutrients fluctuate throughout the year, or as the cells are eaten by tiny zooplankton. Diatoms usually dominate with dinoflagellates becoming more abundant as the temperatures rise and the weather calms (Graph 3.3). As the diatoms die or the faeces of zooplankton that feed on them settle to the sediment, the decay process releases nutrients to fuel growth of larger macroalgae and the blue-green alga *Nodularia* in spring (Graph 3.4 and Fig. 3.15). *Nodularia* fix their own nitrogen from the atmosphere and grow quickly when phosphorus concentrations are high and the water is brackish, a common condition in the Peel–Harvey in spring and early summer. (See **Box 3.3 Blue-greens—ancient life a modern day nuisance.**)

At the height of the algal problem, blooms of *Nodularia* were confined to the Harvey and to areas around the Serpentine River where the low salinity and high phosphorus conditions in late spring and early summer were particularly favourable to germination and growth of spores. The bloom size was linked to the amount of phosphorous entering the previous winter. In the Harvey, blooms were also fuelled by phosphorus resuspended from the

FIG. 3.13 *(a) Balls of* Cladophora *algae blown ashore, November 1980. (b) & (c) Structure of* Cladophora *under low and high power magnification (length of strand ≈ 40 μm).* A. L. Huber, E. P. Hodgkin Collection.

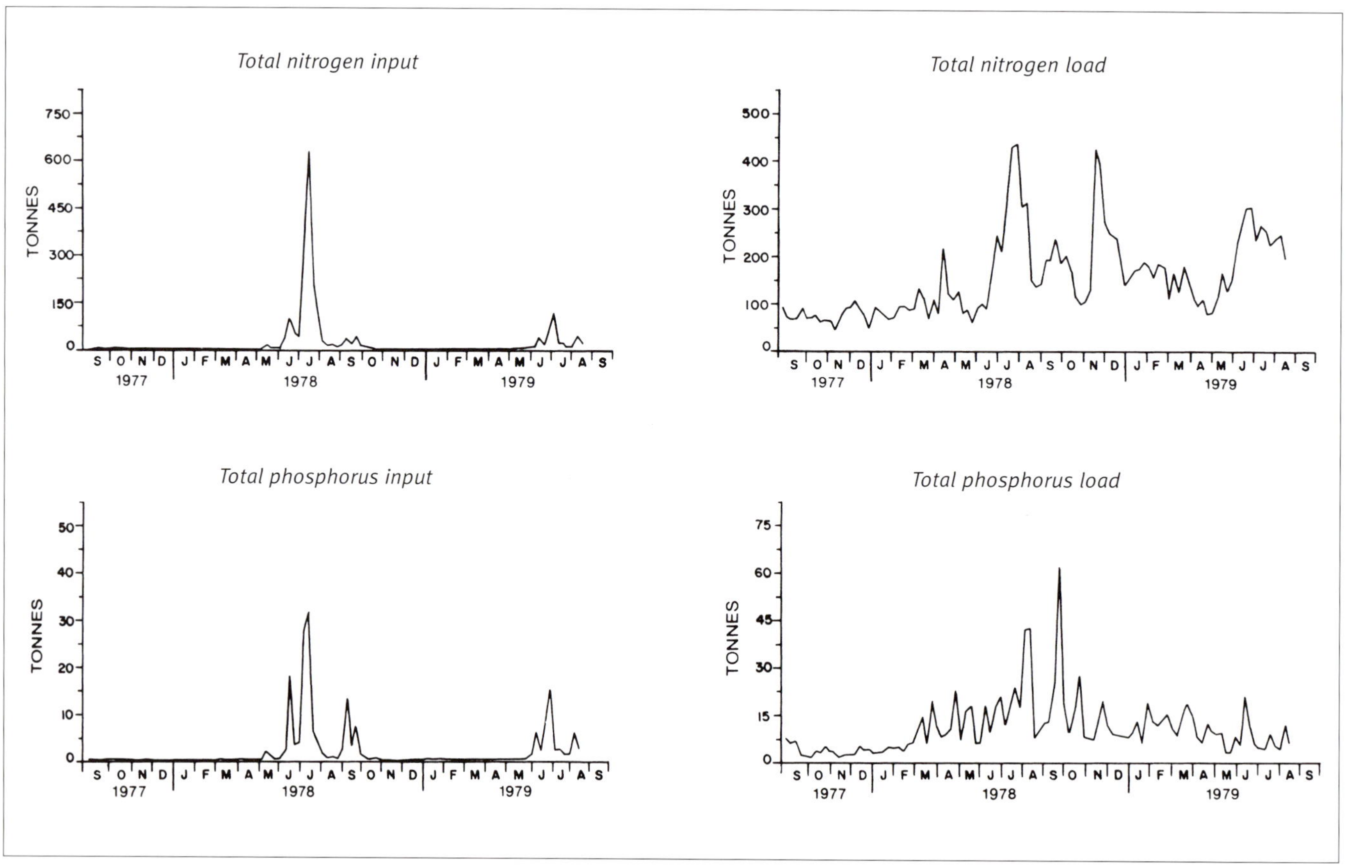

GRAPH 3.2 *Total nitrogen and total phosphorus: inputs and loads to the Peel–Harvey. Note the peaks correspond to river flow.* Hodgkin et al., 1980. Courtesy DoE.

Winter–Spring (June–September)

Each winter a load of nitrogen and phosphorus is delivered to the estuary in the river water. Some goes straight out to sea, and some of the nitrogen is lost to the atmosphere. Much of the phosphorus, however, stays in the estuary and is used by microscopic plants such as diatoms, or binds to suspended sediment and settles to the bottom.

Spring

The diatoms grow rapidly, then die and settle to the bottom where they decompose. In the process of decomposition, bacteria reduce the amount of oxygen creating conditions favourable for phosphorus release from the sediments.

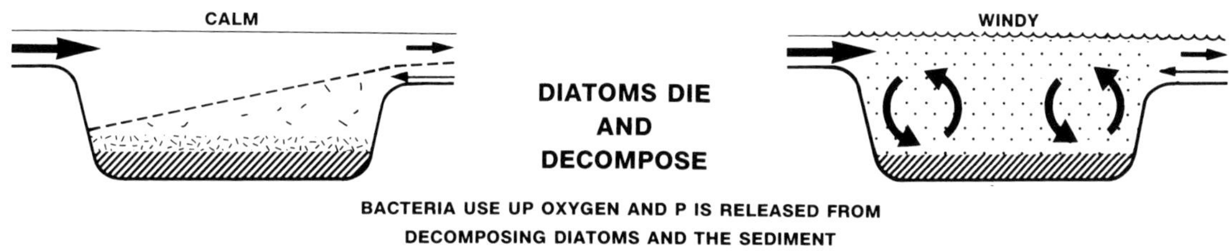

Spring–Summer (September–December)

As the water warms, the phosphorus stimulates growth of the dormant Nodularia *spores in the sediment.* Nodularia *multiplies rapidly and by October blooms are so dense that other plants are shaded. The* Nodularia *can outgrow other algae because it is only limited by phosphorus as it can fix nitrogen from the atmosphere in a similar way to legumes such as clover.*

Summer–Autumn (January–May)

In about December Nodularia *starts to die, probably as the water is too salty and the phosphorus supply is running out. In late summer the* Nodularia *blooms decompose releasing phosphorus to fuel another crop of algae.*

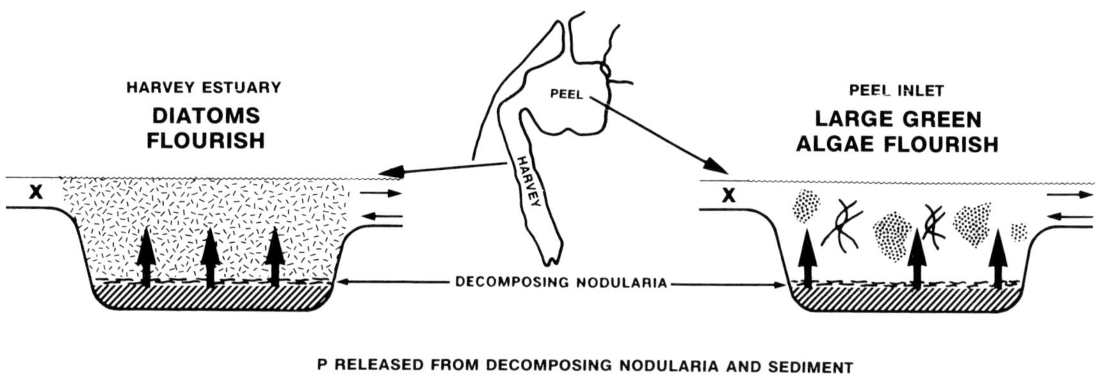

FIG. 3.14 *Nutrient cycling in the Peel–Harvey at the height of the algal problem.* Modified from DCE, 1983b.

sediments by winds blowing along the main south to south-west and north to north-west axis. As the blooms increased in size they spread into the lower part of the Peel, where *Nodularia* was generally not as abundant due to the lower nutrients. The blooms collapsed when salinity reached about 30‰ as summer progressed, although other factors such as nutrient depletion and photo oxidation may have also been important.

TINY PLANTS ON THE SURFACE OF THE SEDIMENT In addition to the obvious blooms of microalgae in the water column, studies during the early 1980s found that there were also significant amounts of microalgae in the top centimetre of sediment of the Peel–Harvey. In fact, except for the winter diatom blooms and the *Nodularia* blooms, the biomass due to benthic microalgae was forty times greater. Production by these tiny plants was also thought to be higher than the floating blooms of microalgae, particularly in low rainfall years when nutrient inflows were low, as benthic microalgae could use the sediment-stored nutrients. The other interesting finding was that except during the winter–spring diatom bloom and the

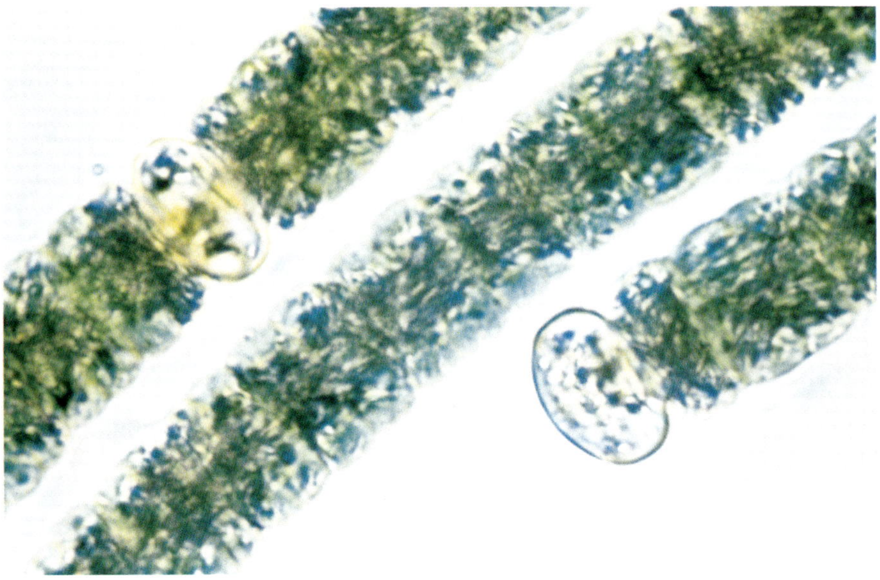

FIG. 3.15 *Blue-green algae* Nodularia spumigena. Photograph Ann L. Huber, E. P. Hodgkin Collection.

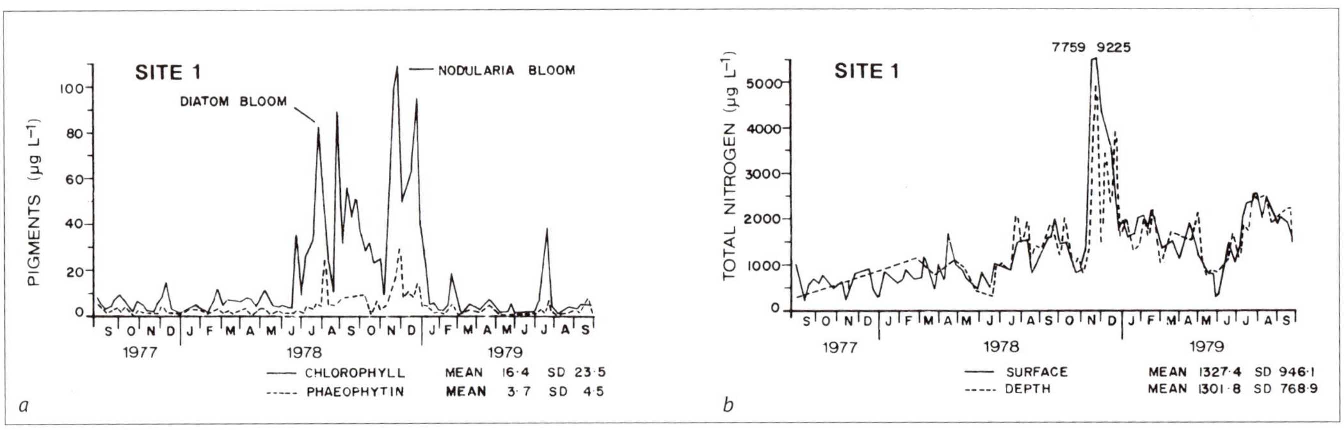

GRAPH 3.3 *(a) Chlorophyll* a *pigments in the Harvey Estuary near Mealup Point (Site 1). The abundance of cells in a volume of water is usually estimated from the concentration of chlorophyll* a*, the green light harvesting pigment used in photosynthesis. High points in the graph show the initial diatom bloom when nutrients rise with winter flow, and collapse of the blooms followed by* Nodularia *blooms in spring. (b) Total nitrogen in the Harvey Estuary near Mealup Point (Site 1). The high levels in late 1978 were caused by a bloom of* Nodularia. Hodgkin et al., 1980. Courtesy DoE.

following *Nodularia* bloom, most of the phytoplankton were benthic diatom species that had been suspended by wind driven waves, again emphasising the importance of the less well known benthic microscopic plants.

SEAGRASSES Seagrasses are marine plants and can die quickly if salinities fall below about three-quarters sea water. They are also affected by reduced light. High nutrient conditions favouring the growth of algae have negatively affected seagrass growth in many marine environments. (See **Box 9.3 Shrinking seagrasses versus growing algae**.) A few species, such as the paddle weed *Halophila ovalis*, eel grasses *Zostera mucronata* and *Z. tasmanica*, and swan grasses *Ruppia megacarpa* and *Lepilaena cylindricarpa*, are more tolerant of low salinity and light and grow in many south-western Australian estuaries. In the Peel–Harvey, *Ruppia* and *Halophila* were the most common types, with *Ruppia* found on the shallowest parts of the fringing shelf particularly around Austin and Robert Bays in the Peel, and *Halophila* in the slightly deeper area of the fringing shelf. The *Zostera* species grew mainly in the more marine conditions of the inlet channel and *Lepilaena* was only found in the eastern part of the Peel Inlet.

Seagrasses generally grow in spring and summer. *Halophila* generally grows from the rhizomes of previous years' growth that are buried in the sediment (see Fig. 2.20) but *Ruppia* grows mostly from seed in very shallow water and from rhizomes in deeper water. The leaves of both species detach and float but never form extensive accumulations like the algae.

In the Peel–Harvey, the large algal drifts of the 1960s, 1970s and 1980s are reported to have smothered some of the areas previously colonised by the small seagrasses, so that they disappeared in the Harvey and declined in the Peel.

Nuisance weeds— the base of a productive system

The phytoplankton, macroalgae, seagrasses and marsh vegetation around the estuary edges provide the basic building blocks for complex food webs supporting large numbers of animals. Some animals, the zooplankton of small crustaceans and fish larvae, spend most of the time in the water column, while others, bottom dwelling or benthic species, such as worms and molluscs live in the sediments. In turn, these support larger animals such as the crabs, prawns, fish and waterbirds that make the area so popular with people.

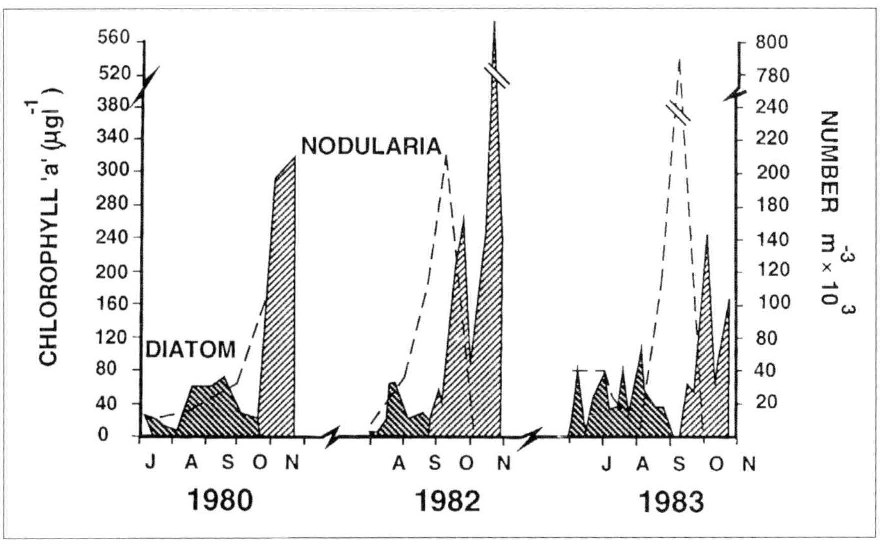

GRAPH 3.4 *Chlorophyll* a: *concentrations and numbers of cells in the water column of the Peel–Harvey. The initial blooms consist primarily of diatoms; the numbers decline as nutrients, primarily nitrogen, in the water are depleted or as the diatoms are eaten by copepods in the phytoplankton (numbers of copepods shown as a dotted line).* Nodularia *blooms capable of fixing atmospheric nitrogen follow; the growth is fuelled also by the release of nutrients after the diatom blooms collapse.* Courtesy A. McComb, Murdoch.

Box 3.3 BLUE-GREENS—ANCIENT LIFE A MODERN DAY NUISANCE

Nitrogen is an important nutrient, one of the building blocks of proteins. Although nitrogen is the most abundant gas in the atmosphere, about 78 per cent, most is not in a form that can be used by organisms. Most plants and bacteria require ammonium or nitrate ions, formed when nitrogen is dissolved in water or changed by bacteria. In contrast, many blue-green algae, also called cyanobacteria, can fix atmospheric nitrogen allowing them to grow rapidly and giving them a competitive edge over algae.

Cyanobacteria occur as single cells or in filaments of cells joined end to end. Like plants, they contain the pigment chlorophyll *a* and some contain the blue phycolobins that give them the blue colour and their name. Cyanobacteria are the organisms that help create **stromatolites**. Cyanobacteria are more highly differentiated than other bacteria in having specialised cells called akinetes and heterocysts.

An **akinete** is an enlarged food-filled cell or spore. Akinetes of the blue-green algae *Nodularia* lie dormant in the sediments until conditions are favourable when they germinate, such as after several days of exposure to low salinity warm water. They grow well in relatively low light of 16–30°C and in salinities of 3–30‰, but not in marine waters. In the Harvey, conditions in late spring and early summer of high phosphorus and low salinities were particularly favourable to *Nodularia*.

Heterocysts, along with other cells, are sites for nitrogen fixation and may be involved in asexual reproduction. The thick wall of the heterocyst creates favourable chemical conditions for nitrogen fixation by limiting the amount of oxygen that can enter.

The amount of nitrogen fixation can be quite staggering. Calculations based on nitrogen measurements in the Harvey Estuary during 1978 and 1979 suggest that *Nodularia* fixed 270 tonnes of nitrogen, almost as much as the Harvey River input.

Blooms form when phosphate concentrations are high, allowing fast growth, and when water movement is slow and cells can stay at the surface in the sun. Blue-green algal blooms are found where run-off from over fertilised pastures and crops or sewage enters slow moving water as found in lakes, dams and estuaries. In some places, such as rice paddy fields, the nitrogen-fixing ability of cyanobacteria is utilised as a fertiliser.

Some species such as *Anabaena* and *Microcystis* can produce toxins—**toxic algal blooms**. Their toxins can cause illness, skin irritation and even death. The poisons, small chain amino acids, have their greatest effect on the nervous system and liver function. *Nodularia* can be toxic to humans if ingested in contaminated shellfish such as mussels, cockles and oysters.

In estuaries, like plants, animals are limited to those that can tolerate the extreme conditions. The abundance of estuarine animals can change rapidly with seasonal changes of fresh river or tidal marine water flow, and throughout the estuary due to gradients of dissolved oxygen, sediment and water depth. Animals are most abundant in beds of vegetation fringing the estuary, macroalgae and seagrasses, where some dead plant matter accumulates. Too much dead material, however, can lead to low oxygen conditions and production of poisonous hydrogen sulfide gas.

Fish and fisheries in the Old Peel–Harvey

About sixty species of fish have been recorded from the Peel–Harvey. Some such as the black bream, yellow tail trumpeter and school prawns are true estuarine or euryhaline species that spend their entire life cycle in estuaries. Most are marine species, such as blue manna crabs, king prawns, yellow-eye mullet, cobbler, whiting, mulloway and tailor, which enter estuaries as juveniles and mature there during the saline summer phase. Before the Dawesville Channel was cut through the dunes, a few species such as school whiting entered very occasionally, usually in low numbers when conditions were essentially more marine.

The first industry and holiday fun

The Peel–Harvey provided a rich harvest of fish for many years. During the early years after settlement, the waters supported the most important commercial estuarine fishery in Western Australia as well as recreational fishers. In the 1870s, fish such as sea mullet and yellow-eye mullet (pilchard)

FIG. 3.16 *Canned fish from Mandurah.* Courtesy Western Australian Maritime Museum.

were dried and exported to Singapore or canned and smoked for the Perth and goldfields markets. 'The Chimneys' area today, near the old entrance channel, marks the site of a busy packing depot used when fish were abundant but distance to markets was a problem (Fig. 3.16).

By the 1890s the industry had declined and fish were scarce and smaller in size. Overfishing, disturbance, taking of small fish and disease had taken their toll. The first fishing regulations were introduced in 1898 when fishers and boats were licensed; limitations on net and size of fish were set; and rivers, deltas and the entrance channel were closed to all netting. Closure of the riverine nursery areas was unpopular and policing was a problem. One of the first fisheries inspectors was based at Mandurah in 1898 and the Fisheries Department was established the following year. From 1898 to the 1950s there were as many as fifty, rarely less than forty licensed boats on the Peel–Harvey. Catches ranged from 100 to 200 tonnes per year although composition of the catch varied from year to year.

Interest in prawns and crabs developed later with the commercial harvesting of crabs starting in the 1930s. River prawns were important until the 1980s when they practically disappeared, although they reappeared again in catches in the 1990s. These prawns, living entirely in rivers and estuaries, were caught by hand dredges, attracted by light with poultry pellets used as bait. The decline in prawn numbers was blamed on the Yunderup canal development, but probably reflects a more general decline in estuarine health. River prawns also migrated out of the system when salinities were very low. Western king prawns were also abundant in the early days and caught as they passed through the narrow entrance channel to breed in the ocean. (See **Box 2.9 Summer sport–prawning parties**.)

There has always been competition for prawns between recreational and commercial fishers, but for many years the blue manna crab was regarded as a nuisance by commercial net fishers as the crabs ate dead and dying fish and damaged nets. In the past, 'plague crab' years for commercial fishers were bumper years for recreational fishers, but more recently the commercial catch has become a source of antagonism between the two groups. Commercial fish catches have varied with market demands, changing from table fish to bait fish for the rock lobster industry, though imported baits have now reduced demand. The industry also changed with the creation of the dredged entrance channel in 1967 that allowed boat fishers offshore to moor within the estuarine system.

In the 1960s through to the 1980s, the bulk of the commercial catch was made up of five species: yellow-eye mullet *Aldrichetta forsteri*, sea mullet *Mugil cephalus*, western king prawn *Penaeus latisulcatus*, blue manna crab *Portunus pelagicus* and estuarine catfish or cobbler *Cnidoglanis macrocephalus*. Recreational fishers also targeted blue manna crabs, western king prawns and cobblers, as well as King George Whiting *Sillaginodes punctata* and tailor *Pomatomus saltator*.

Fish and weeds—a complicated picture

The eutrophic condition of the estuary affected fish populations in different ways. The abundance of macroalgae favoured sea mullet, cobbler and yellow-eye mullet, providing shelter from predators and food due to high numbers of invertebrates living in the drifting mats. Conversely, the numbers of western king prawns decreased, probably due to the accumulations of *Cladophora* and *Ulva* covering the sand where juvenile prawns burrowed. Toxic *Nodularia* affected some types of fish and crabs. Many species avoided blooms and fewer species were caught when blooms were intense. At times, anoxic conditions also led to fish deaths.

The blooms of algae also affected fishing practices. The weight of algae made net hauling difficult and schools of fish were hard to find under banks of algae, so most fishing activity was concentrated in Peel Inlet rather than Harvey Estuary. Whether the blooms affected fish numbers is unclear. *Nodularia* blooms certainly coincided with lower catches of sea mullet and cobbler although catches of yellow-eye mullet were not affected. As the blooms were localised in the Harvey, overall fish abundance throughout the system was probably unaffected. Numbers of crabs, prawns and fish that entered from the sea probably varied more with conditions in the breeding areas offshore. The slightly lower catches during bloom years were influenced by higher river flows and the less favourable lower salinity.

KNOWING WHAT'S WRONG—THE STUDY REPORT

Investigations undertaken during the Peel–Harvey Estuarine System Study (1976–1980) added greatly to understanding Western Australian estuarine systems. Many reports published in scientific journals addressed fundamental biological processes of international interest. The final report, together with recommendations for future management, published in December 1980 was, however, largely ignored and not released until August 1981. A major problem remained, because there was little provision for funds to implement the management recommendations.

Managing the problems

All the studies indicated that the main cause of the algal blooms was the excess of phosphorus coming into the estuary, the bulk of which was retained within the system with very little flushed to sea: '...the only satisfactory long term solutions to the problem will be to substantially reduce the availability of nutrients especially phosphorus, to algae.' (The Peel–Harvey Estuarine System Study (1976–1980), Department of Conservation and Environment 1980, Report No 9.) It was clear that solutions would not be simple and would take a long time. Phosphorus entering from the catchment had to be reduced without disruption to agriculture on the poor soils; phosphorus in the estuary had to be reduced by stopping recycling from sediments, or the phosphorus store removed from the estuary. There was also the problem of how to improve the water quality conditions in the short term.

Harvesting

Algal harvesting, primarily to reduce odours and clean the beaches, was also used to decrease the recycling of nutrients from rotting algae that fuelled further algal growth. Harvesting started on a small scale using rakes and pitchforks, but the equipment became mechanised and more sophisticated as the problems grew (Fig. 3.17).

Harvesting was mainly conducted along the Coodanup and Novara–Falcon foreshores, areas adjacent to houses and used most heavily for recreation. In the late 1980s harvesting removed 15,000 to 20,000 cubic metres each year. By the 1990s the amounts harvested had declined to 5,000 to 10,000 cubic metres between 1990–1991 and 1993–1994. Complaints about rotting algal masses appear to have decreased since the Dawesville Channel opened, and harvesting has been much lower at less than 1,500 cubic metres in 1994–1995 and 1,300 cubic metres in 1995–1996. However, as these were years of below average annual rainfall, the effects of wet years on algal growth are yet to be seen.

Despite the success of harvesting in improving aesthetic and amenity values, its value as a management tool has been questioned. Harvesting each year only removed 0.1–5.0 per cent of the total algal biomass in the estuary. This was orders of magnitude less than the amount that could end up on the beaches. On a system the size of the Peel–Harvey, the cost was high and

FIG. 3.17 *Larger algal harvesters were used during the 1980s.* A. Brearley, 2001.

there were concerns about effects to the environment mainly through beach erosion and animal by-catch.

A comparison in the 1990s of sediments and animals on beaches regularly harvested since the 1970s versus beaches left unharvested with masses of algae, and beaches where algae did not accumulate, has allowed an evaluation of the effects of harvesting. Organic matter in the sediment was higher along unharvested beaches. The faunal communities were also different, probably due to animals being more abundant in the oxygenated surface layers. Where algae was abundant, the fauna were restricted to those able to cope with low oxygen conditions, or able to recolonise after extinctions in the rotting masses. The study also found that populations of small invertebrates recovered quickly after harvesting, usually over days or weeks. Although the study demonstrated the negative effects of rotting algae on fish and crab numbers, it also demonstrated the positive role of algae in providing food and shelter and in supporting a productive ecosystem.

Reducing fertiliser application in the catchment

With phosphorus recognised as the key to algal blooms, research into its use and a quest for better management continued through the late 1970s and early 1980s. Examination of application rates combined with soil testing showed up to 30 per cent of the phosphorus was lost from farmland, and more than 50 per cent was unaccounted for in surface soil, pastures or run-off. Phosphorus concentrations in 90 per cent of the clay loams and 40 per cent of the deep sands were found to be high enough for adequate paddock plant growth, needing no further applications. Soil testing also indicated that many soils were deficient in sulfur, potassium and some trace elements with superphosphate (containing about 11 per cent sulfur) often applied to overcome these other deficiencies. Like phosphate, these nutrients were leaching from the soil or were removed when the products left the farm with the sale of milk or livestock. Early applications of superphosphate in the growing season resulted in improved growth on soils that had sufficient stores of phosphorus. This effect was due to the sulfur in the superphosphate being in granules large enough to dissolve slowly and to supply just enough sulfur for the pasture in spring.

Clearly better fertiliser applications to decrease downstream effects on the estuary would reduce costs on the farms. The key in many cases was soil testing, and the development of new fertiliser products and strategies that continued through the 1990s and to this day. One of the requirements was for a range of 'slow release' fertilisers for specific locations. Although one of the new products, the highly promoted 'Coastal Superphosphate', failed to meet expectations, tailored mixes of superphosphate and muriate of potash (potassium chloride) are now available to meet phosphate and potassium needs, and coarse rock-gypsum (calcium sulfate) or a mix of iron, manganese and calcium sulfate to meet sulfur requirements. Soil testing has determined the optimal timing of applications at 3–4 weeks after germination when the root systems have developed, with further additions of sulfur and potassium later in spring to reduce leaching and maximise plant uptake. A slow release form of superphosphate is still unavailable despite production being relatively simple. At present the only slow release fertiliser widely available is for sulfur and it works on the simple principle of providing larger particles of gypsum that dissolve at a slow rate, making the fertiliser available well into spring when it is needed.

Although application of superphosphate declined from 27,000 tonnes in 1974 to 17,000 tonnes in 1979 and 9,500 in 1987, continued reductions will depend on the development, cost and use of suitable slow release fertilisers. Unfortunately reductions in fertiliser applications in the 1980s were offset by increased output from point sources such as piggeries, feedlots and sewage treatment plants. (See also **Box 9.2 The P problem—can it be controlled?**)

The effectiveness of fertilisers has been hampered by the acidification of the soil caused by the growth of legumes and the use of some nitrogenous fertilisers. Acidic conditions exacerbate phosphorus loss by reducing plant growth and the take-up of phosphorus that is (eventually) exported in farm product.

Reducing loss of phosphorus from the farmland

Methods of retaining phosphorus and other nutrients in the soils and reducing the amount entering the waterways were also investigated. These included the addition of binding agents such as red clay from bauxite mining, use of retention basins, replacement of shallow-rooted annual pasture crops with deeper rooted perennial species, growth of trees for shelter belts, fodder and timber, and improving or replacing streamline vegetation. Many of the strategies were developed and adopted by Landcare and Catchment groups (Fig. 3.18).

SOIL CONDITIONER—'RED MUD' FROM BAUXITE MINING The use of bauxite residue has been trialled for many years. After refining to extract the aluminium, two-thirds of the bauxite, more than 14 million tonnes per year, is left behind. Used as a soil, pigment and clay, the iron-rich residue also binds phosphorus, making it a suitable additive for phosphorus leaching soils. The bauxite residue is alkaline due to the presence of sodium carbonate (washing soda), which has the added advantage of rapidly liming the acidic soils. The sodium carbonate originates from the caustic soda used in alumina refining, but it rapidly reverts to the carbonate in contact with air negating the need for neutralisation when applied at less than 100 tonnes per hectare.

Phosphorus retention rates after applications of bauxite residue have been impressive with logistical and handling issues being solved using conventional equipment. Pasture production on light sandy soils can be almost doubled with significant reductions in phosphorus use. Field trials have proven that less than 20 tonnes per hectare can reduce phosphorus loss by 50 per cent with most farmers opting for 5–10 tonnes per hectare, nearer to lime application rates. Applying 5 tonnes per hectare is approximately 0.5 of a millimetre thick, and applications of as much as 20 tonnes per hectare are very difficult to see on the ground within a year.

Extensive research over more than 10 years into the safety of bauxite residue in soil amelioration has covered issues of leachates, dust, radioactivity, and human and animal health. The issue remains controversial, however, and while its use has been approved by a Public Environmental Review, the manufacturer removed the product from the market in the 2002–2003 season.

FIG. 3.18 *Farmers' Field Day, Meredith Drain, August 1982.* E. P. Hodgkin Collection.

Projections indicate that use of bauxite residue in soil amendment would only equate to about one day of yearly production, which would have little impact on stockpiles of residue while access costs would be significant.

The use of bauxite residue is now commonplace around leach drains in catchments sensitive to nutrients. It has also been used in compensating basins to bind heavy metals from road run-off, and trials have been conducted to assess use in composting to reduce the availability of heavy metals.

TREES—PLANTING MORE AND KEEPING WHAT'S LEFT Efforts to establish forestry to reduce run-off and nutrients in the streams were also beset by problems of communication with landholders and farmers. In the late 1980s, the Department of Conservation and Land Management (CALM) was developing a major reafforestation proposal for the whole South-West linked to proposals for a pulp mill: combining these plans with the Peel–Harvey catchment rehabilitation appeared strategic. A proposal to buy land, particularly those with soils contributing to the phosphorus problem, about 13,500 hectares at a cost of about $13.5 million, was however viewed as a threat to farming. Failure of the Coastal Superphosphate was remembered, and tree planting was viewed as government interference. Many considered that the issues had not been properly addressed and were sceptical of the profitability of the scheme.

Problems soon emerged when large scale planting commenced in 1987. Seedling losses were high due to waterlogging on the Bassendean Sand areas and the Pinjarra Plain, and the effort to break up the underlying 'hardpan' or 'coffee rock' was too expensive to use on a large scale. By 1991 it was obvious that extensive tree plantations were not a management option. Gradually perceptions have changed however. The success of smaller tree belts and concern about the decline of the remaining native vegetation from waterlogging, storms, compaction of earth and ringbarking by animals, has been the foundation for farmers and Landcare groups interested in preserving and extending native vegetation. Planting to improve roadsides, reduce erosion on creeks and maintain animal habitat have now extended to multi-purpose agroforestry plots for windbreaks and timber production.

The use of streamline vegetation to reduce erosion and filter nutrients has been part of the Landcare focus, and the effectiveness of remnant vegetation, particularly sedges, for retaining phosphorus has been demonstrated but the effects may not be obvious for 5–10 years after implementation. Modelling also showed that the extent of the Peel–Harvey problem was so great that a wetland rehabilitation area extending from Pinjarra to Harvey would be needed to effectively reduce nutrients entering from the Harvey River and Mayfield Drain. Clearly other options were required.

Increasing flushing—plans to dig

During the early 1980s, various plans were considered to hold nutrient-rich waters in the Harvey that would then be diverted to the ocean through Lake Clifton and Lake Preston in a drain or tunnel. The option of flushing the nutrients to the sea was taken more seriously after a conference in late 1982 when a Public Works engineer drew up plans to dig a ditch through the hills at Dawesville.

Political interest in weeds

Although some funds were allocated to continue the studies in December 1981, there was a growing gulf between the local people, including the Peel Inlet Management Authority, the 'researchers' and the policy makers located in Perth. The poor state of the estuary over summer fuelled the debate at a state political level, and made good headlines in the press. With an election looming, the Peel–Harvey problems in the marginal seat of Mandurah were of great interest to the sitting Liberal member and the Labor opposition. The Government purchased new floating harvesters to reduce the amount of weeds reaching the beaches. Political debate became even more vigorous and pledges more dramatic; the Government promising to continue the programs in place, the Opposition pledging funds for a program to buy properties polluting the estuary and to make the use of slow-release fertilisers compulsory. Into the ring came an independent local candidate with a platform to dredge the Mandurah Channel, a proposal generally discounted by scientists and engineers involved in the studies, although creating an exchange between the estuary and ocean at Dawesville was under consideration. The February 1983 election went to Labor with wider policies for dealing with problems in the catchment, but frustration in the Mandurah community about the estimated time taken to improve the estuary was still high.

Seeking further solutions

In 1983 the engineering plan to dig a ditch was one of many (more than 100) options being considered. Although in many ways these were a continuation of programs already initiated, by 1984 these were narrowed down to eight for detailed consideration.

Three main strategies were:

- expansion of weed harvesting
- reducing phosphorus input to the estuary through modifying agricultural practices
- increasing the loss of phosphorus to the sea.

A further five supplementary measures were also proposed: application of algicides; amendment of leaching soils with bauxite residue; treatment of rural sources of concentrated nutrients; changes in land use; and effects of controls on clearing and on drainage construction (DCE, 1984).

While many facets of the management program had commenced and future management plans were being drawn up, the option of creating a new channel was gaining popularity. Premier Brian Burke announced that modelling studies had shown the problem to be so serious that major engineering works might be needed. This was a popular and politically astute statement in a marginal seat and one the local community had been advocating for years. A new entrance would not only increase flushing, but also create a new and more reliable boating channel. A major drawback was the cost.

Options to increase flushing, to reduce the nutrients and create more marine conditions were also supported by evidence from the biological studies. Salinity was now known to be a major factor controlling *Nodularia* blooms. The spores germinated in early spring when waters were relatively fresh (< 15‰), developed as the water became saltier in late spring, but when the blooms collapsed in summer the water was too salty for further germination of spores. A second bloom in midsummer of 1982 confirmed this link with fresh water, when rain after cyclones Errol and Bruno sparked a second germination of *Nodularia* spores.

It was also recognised that phosphate had built up in the estuarine sediments and these stores were fuelling the blooms. Early projections of how much the flow of phosphorus needed to be reduced were now known to be too low for any short-term reduction in the frequency of blooms. The damage was too large for any expectation of a rapid recovery.

Actions had not only to improve conditions in the short term, but also to provide a cure that would continue to meet the needs of a continually growing population. Following the Project Team's report, in August 1984 the government presented two options to the Mandurah Shire Council for consideration: a $26 million Dawesville Channel (depending on a $900,000 feasibility study) or an $8 million dredging of the Mandurah Channel (only partially solving the problem in the Peel, leaving the Harvey unchanged). The Government wanted a decision in two weeks.

After much debate and dissatisfaction with the lack of information, the Council supported the feasibility study of the cut option. The communication problems loomed again. The Murray Council was excluded from the discussions despite the implications for the catchment area, and the local community was angered over the lack of consultation about the planned path of the channel. Private individuals held some land but most was undeveloped, held since the 1970s by investors waiting for future subdivision potential when the area regained its popularity.

Feasibility studies presented a dilemma. The agency most suitable for the task of reporting to the EPA, the newly created Department of Conservation and Environment, was the agency proposing the channel. To overcome the problem a new coordinating structure, the Peel–Harvey Steering Committee, was formed with the newly formed Department of Conservation and Land Management as chair. This Committee was charged not only with reviewing the channel proposal but all the other management strategies, so that the engineering proposal to increase flushing could be considered in the context of management to reduce phosphorus through fertiliser modifications.

The first stage of the Environmental Review and Management Program (ERMP) determined that engineering works would only succeed if carried out in conjunction with the fertiliser efficiency program, and that weed clearing would be needed for many years. By August 1985 an action plan to continue with the fertiliser program and weed clearing, dredge the Mandurah Channel and conduct a detailed study of the new channel option by the following year was announced. Plans to start construction of the channel in January 1987 were under way when another election loomed, this time with both major political parties supporting the concept of a channel at Dawesville. The Government retained the seat helped along by the promise of creating a permanent opening through the Mandurah Channel. The Peel–Harvey was going to be linked to the sea again, even if the Dawesville Channel was some years away.

More reviews—the clean up continues

The second stage of the review process proceeded slowly. Disquiet was brewing again, the Mandurah Shire Council felt left out of the process and farmers thought they might be pressured to change their land use. The ERMP was finally released for public comment in May 1988, with the EPA finding the proposals environmentally acceptable in November. Within weeks, the Government approved funding for management, the purchase of land surrounding the channel, and construction of the channel at a predicted cost of $54 million. The timing was right and the decision was popular as the estuary experienced yet another *Nodularia* bloom, the worst ever with toxins at concentrations dangerous for swimmers, fish and shellfish. The bloom collapsed quickly due to increasing salinity, a pattern repeated over the next few years—blooms were intense and toxic but short-lived (8–9 weeks).

Under the approvals, the Ministers for Transport, Agriculture and Waterways (also Minister for the Environment) were jointly responsible for the implementation of conditions under which building could proceed. These included requirements for phosphorus management, a moratorium on clearing and drainage, appropriate waste disposal systems in new developments, suitable waste spoil disposal, and monitoring of impacts (such as mosquito populations) of the channel.

Approvals for the channel in 1989 were given in a climate of uncertainty with the Government embroiled in controversy surrounding business dealings—later referred to as 'WA Inc.' and the subject of a Royal Commission—when the date of another election in February was announced. The state of estuary was not a major part of the election debate except for 'green candidates' who regarded the channel construction as a 'bandaid technical fix' that did not address the biological and social causes of pollution.

Political issues around the Peel–Harvey were diverse; the Labor Government won the election and the new seat of Murray, but lost the seat of Mandurah to the Liberals. Although the new government reaffirmed commitment to the channel, the call for tenders was delayed as the Minister for Environment wanted a number of new options considered. These included dosing the estuary with powdered limestone, a successful technique used on smaller water bodies but found to be unsuitable for the huge Peel–Harvey system.

Creating a new environment— the Dawesville Channel

The decision to go ahead with the channel was announced in July 1990, with construction scheduled to start in 1991 and expected to take about five years. Although no funds were allocated, various interests were anxious to hurry the process along. The Department of Marine and Harbours had been investing time and personnel on the proposal for some years, and Wannunup Developments that owned adjoining land had been sitting on their asset since 1977. Over the years various proposals had been suggested to link the developments with the channel, but a new partnership with a Malaysian electronics firm provided the finance to undertake major development. With the cost escalating, and the Government in financial limbo, overtures by the developers were more attractive.

Proposals from the developers and Thiess Construction were agreed in principle in July 1991. With the amendments to the plans proposed by Marine and Harbours, including increasing the height of ocean retaining walls to protect the channel from ocean storms and sand drifts moving along the coast, cost estimates were now $76.8 million, three times the 1984 estimate. Wannunup and Thiess Contractors proposed to construct the channel, plus adjacent canal and residential developments, in one major integrated work program for $52.9 million. With some additional small costs, mainly for road works, the overall cost to the Government for the channel would be $56.7 million, considerably less than for the government proceeding alone.

After consultation with State Treasury and an independent financial consultant, the proposal was viewed as a sound benefit to the State, and Cabinet agreed to the proposal in November 1991. Lengthy negotiations of design and approvals were followed with broad concept plans finalised in early 1992. In a climate of economic recession and unemployment, particularly in the areas around Mandurah, the project offered employment, additional investment and tourism potential.

The final design was a channel 2.5 kilometres long, 200 metres wide, 6.5 metres deep at the ocean end and 4.5 metres in the north Harvey Estuary to the south of Point Grey (Map 3.5). The channel was to be spanned by a four-lane traffic bridge that would also carry pipes and cables (Fig. 3.19).

At the ocean end, it was expected the longshore processes that move sediment and create the bar at Mandurah would also operate at the new channel. To protect the channel from sediment build-up, a spur groyne was constructed to act as a sediment trap. A small groyne to the north was also built to reduce infilling during storms from the north-west. To stop a bar forming and to ensure that the beaches north of the channel would not be eroded as sand moved north, the infrastructure for a sand pumping system capable of dealing with about 60,000 cubic metres of sand a year was installed.

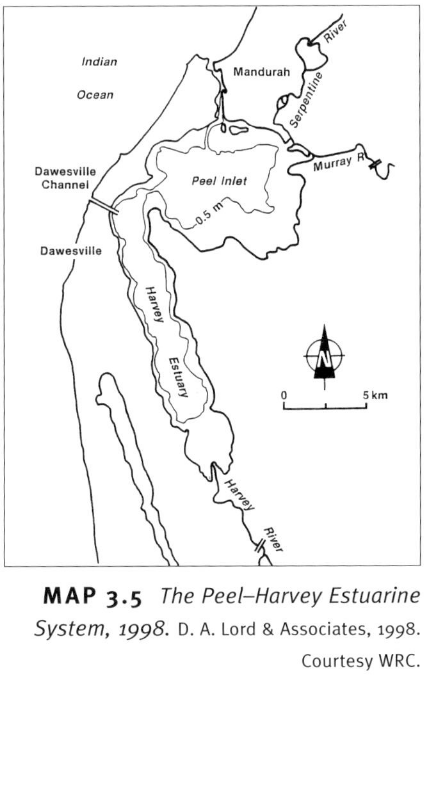

MAP 3.5 *The Peel–Harvey Estuarine System, 1998.* D. A. Lord & Associates, 1998. Courtesy WRC.

At the estuary end, the channel was extended through tidal flats into deeper water (2 metres) to maximise tidal flushing, with spoil from the channel used to fill the shallows behind the groynes. Although the original plan was for reclamation of 10 hectares (6 hectares to the north and 4 hectares to the south), the area was increased to 25 hectares to use extra spoil from the sea groyne extension, creating further areas for canal development. In total 500,000 tonnes of sand, mud and silt were removed and placed behind the retaining wall in the largest single dredging operation undertaken in the region.

Opening of the channel

The channel was completed within budget and more than nine months ahead of schedule. Timing of the opening was important. Removal of the bund was delayed because of concerns that the rush of sea water would stir up sediments and lead to algal blooms when light was strong in late summer. Removal was also timed for a period when the water level inside and outside was similar to minimise flow and reduce erosion while the gap was widened. Finally the channel was breached on 5 April 1994 with the official opening of the Dawesville Channel and festivals on 23 April, during which the historic boats *Endeavour* and STS *Leeuwin* passed through into the estuary. A Liberal Premier Richard Court performed the opening, the sixth premier to be involved since the project was first considered. A quotation from the Premier's opening address in which he acknowledged the efforts that had gone on for so many years serves as a reminder of the fragility of places like the Peel–Harvey and the importance of a healthy environment.

> *We should not expect miracles overnight. The problems of the Peel Harvey Estuary built up over several decades and likewise the solutions implemented may take at least five years before the full effects are realised but we have to start somewhere. I think we should look upon the Dawesville Channel as a symbol of care. The Channel will serve as an excellent daily reminder to us that we must take more care of our environment.*
>
> Premier Richard Court, 23 April 1994.

The cost of the Dawesville Channel should be a reminder of the price to be paid when our actions set in train a series of system-wide changes, each setting the stage for other unforseen events. Additionally we have now changed the Peel–Harvey again, back to a younger more marine environment.

The winter plume of estuarine water and lightly elevated levels of nutrients have been detected 2 kilometres offshore and 3 kilometres north of the channel. It was estimated that 100 tonnes of phosphorus and 900 tonnes of total nitrogen per year moved through the channel in 1994–1995, similar to amounts predicted during the planning. In the ocean phosphorus is expected to be converted to the insoluble apatite form and, although no build-up of nutrients or sediments has been detected, the long-term effects on marine life are unknown.

Despite recognition that the channel could recreate the marine environment of 6,000 years ago, and the view of the Auditor General that the government had a legal obligation under the Environmental Protection Act to monitor and manage changes, few funds were allocated for monitoring and management.

The water flows in

The connection established through to the sea brought some immediate changes. A plume of muddy water from the final stages of the excavation flowed into the Harvey. Water levels rose and salty water flowed up some drains on the eastern side. Although tidal levels had long been expected to play a major role in the estuarine processes, particularly in the long term, the initial high water levels were due to a low pressure system under the influence of a tropical cyclone in the north. The high water levels, however, indicated that the estuary waters were now clearly influenced by ocean processes.

FIG. 3.19 *Dawesville Channel, aerial view, 30 October 1996. The bridge over the channel built of precast sections was jacked over the span, supported by columns high enough to allow passage of boats with a mast 18 m tall and strong enough to withstand the impact of a 300 t vessel travelling at 8 knots.* Courtesy DALSE.

THE NEW PEEL–HARVEY

With the new millennium we have a new ecosystem. The catchment has undergone over 170 years of modification by people through draining, damming and nutrient enrichment. The soils hold a history of nutrient additions and now we have opened a new path to the ocean, but are we still treating the Peel–Harvey as a sewer? After so many studies, so much effort in the catchment, so much money, it is time now to assess how successful we have been in solving the algal blooms and gauging the effects of change.

Nutrients from the catchment—still too much?

Nutrients entering the Peel–Harvey were and continue to be from both diffuse and point sources. Reducing point sources can produce almost immediate improvements to water quality, but diffuse sources are more difficult to manage and contribute most of the problems in the Peel–Harvey. In the decades since the nutrient problems were first recognised, the distinction between point and diffuse sources has become more difficult. Activities in the catchment have intensified, and many animals previously housed in yards and responsible for point source pollution are now managed in large paddocks and contribute to diffuse pollution. Improvements in nutrient loading to the estuary system will not be obvious until nutrients in the ground water are reduced.

Nutrient monitoring is complex, time consuming and costly. In the past many samples have been collected to establish a picture of what is going on. Today we need to assess trends in the present nutrient status to determine if management programs are improving conditions or if the situation is deteriorating (Map 3.6). (See **Chapter 1 Understanding Swanland: Understanding water health—components we measure**.)

Subtle trends that are real and meaningful changes in nutrient concentration are difficult to detect however, as natural variations of water flow lead to natural wide variability in nutrients. Interpretation of nutrient trends and their environmental significance requires background knowledge of current and historic water flows, vegetation, land use and phytoplankton blooms. Interpreting large changes may also be difficult as they may be due to unusual events such as storm flow, or a change in sampling location or

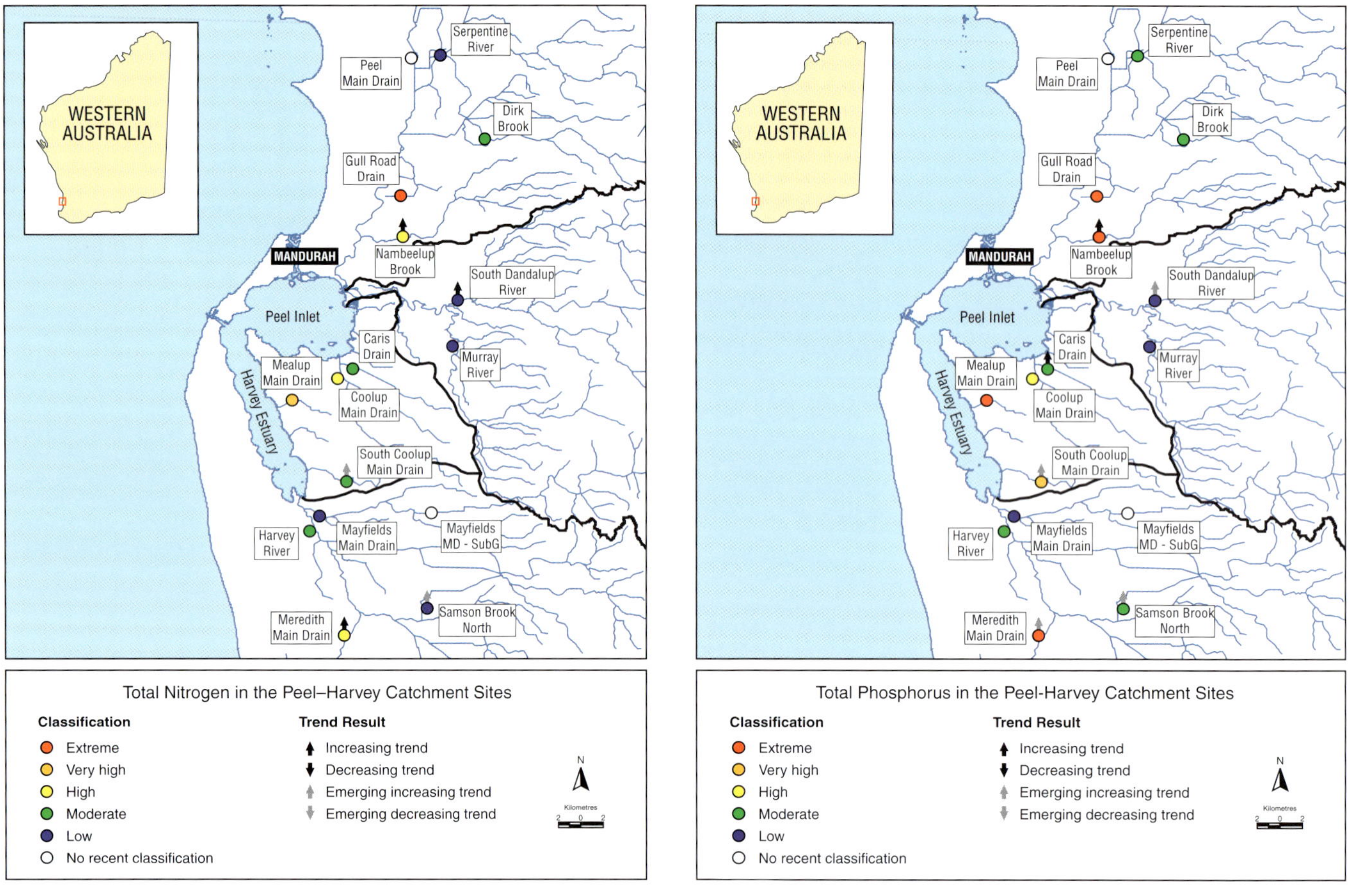

MAP 3.6 *The status and trends of total nitrogen and total phosphorus to 1998 in the Peel–Harvey catchment.* Source: Jakowyna, 2000, Courtesy WRC.

timing. They may also reflect a sudden influx due to a high concentration application of fertiliser, for example, recent superphosphate applications before a rain event. To determine a real trend or change rather than natural variation between years due to unusual events, nutrient concentrations are usually examined statistically using the median concentration over three consecutive years.

SERPENTINE The Serpentine River accounts for about 12 per cent of water entering the Peel–Harvey. In the past it was one of the major sources of nutrients entering Peel Inlet. Nutrients from the Upper Serpentine appear to be stable, although phosphorus concentrations are slightly elevated. Nutrient concentrations are higher further down the river from point sources, from fertilisers and re-cycling of phytoplankton blooms in the lower reaches of the river, and appear to be increasing (Table 3.2).

The tributary Nambeelup Brook brings about 6 per cent of the river flow to the Serpentine. It is nutrient enriched with total nitrogen ranging from 1.85 (1990) to 3.0 (1997) milligrams per litre, increasing at a rate of 0.133 milligram per litre per year. Total phosphorus ranged from 0.37 (1990) to 0.68 (1995 & 1997) milligram per litre increasing by 0.042 milligram per litre per year. Around the Nambeelup, stock grazing and pasture development with poultry farms, dairies, plantations and horticulture all contribute nutrients, and the sandy and duplex (sand over clay) soils favour leaching of nutrients.

In the middle reaches of the Serpentine, the Peel Main Drain accounts for about 8 per cent of the flow. While nutrient inflows from the Peel Main Drain may be decreasing, there are an increasing number of potential nutrient sources such as piggeries, poultry farms, horticulture, industry and stock holding yards. In 1999 it was noted that flow from one sheep lot alone accounted for about a third of the nitrogen in the drain. Dirk Brook is considered mildly enriched from diffuse and point sources. Nutrients in the

TABLE 3.2 *Trends in quality of water entering the Peel–Harvey.* Source: Jakowyna, 2000.

Waterway	N status	N trend 1985–98	P status	P trend 1983–98	Factor
Serpentine region					
Serpentine River	Low	None	Moderate	None	
Peel Main Drain	NA	None	NA	None	
Dirk Brook–Punrack Drain	Moderate	None	Moderate	None 1990–98, no data 1994–95	Some evidence of a step change; source not detected
Gull Road Drain	Extreme	None	Extreme	Steady	Point source
Nambeelup Brook	High	Increasing 1990–98	Extreme	Increasing	Point source
Murray region					
Murray River	Low	Decreasing?	Low	Decreasing but technique change	
South Dandalup River	Low	Increasing variation; increased run-off with removal of remnant vegetation	Low	Increasing variation?	Needs confirmation
Drains					
Caris Drain	Moderate	None	Moderate	Increasing	Gradual rise diffuse sources
Coolup Main Drain	High	None	High	None	
Mealup Main Drain	Very high	NA till 2001	Extreme	NA till 2001	
South Coolup Main Drain	Moderate	Increasing?	Very high	Increasing?	
Harvey region					
Mayfields Main Drain	Low	None	Low	None	
Mayfields Main Drain–Sub G	NA	None	NA	None	
Harvey River	Moderate	None	Moderate	None	
Samson Brook North	Low	Increasing?	Moderate	Increasing	
Meredith Main Drain	High	Increasing	Extreme	Increasing	

Gull Road Drain have, however, been high for over ten years due to a piggery, and the drain is considered one of the most nutrient enriched waterways in Western Australia although the overall flow to the lower Serpentine is small.

Studies of phytoplankton are particularly useful as they indicate how the environment has responded to nutrients. In the lower reaches of the Serpentine there were summer blooms of dinoflagellates and blue-green algae (mainly *Nodularia* and *Anabaenopsis*, *Merismopedia*, *Oscillatoria*, *Anabaena* and *Microsystis*) as well as dinoflagellates in 1994–1995, 1995–1996 and 1996–1997. Since then (autumn of 1997 and 1999) there were blooms of the haptophyte *Prymnesium*, a type known to cause fish kills, which may signal a change in phytoplankton diversity.

MURRAY REGION The Murray River accounts for 60 per cent of flow to Peel–Harvey, the single largest freshwater influence. Generally nutrients in the Murray River are low and the quality of water entering the estuary is good. Recent changes, however, in the amount and types of phytoplankton with a shift from summer–autumn diatom blooms to more dinoflagellates such as *Alexandrium* and chlorophytes may indicate that there are some negative changes in nutrient availability.

Nutrients also appear to be increasing in the South Dandalup River, the Murray's main tributary, probably due to clearing of the remnant vegetation and accompanying rise in the watertable and ground water flow.

DRAINS All the drains flowing to the Peel–Harvey contain high nutrient loads originating from the sandy soils. Although the flow volumes are low, these diffuse sources represent a large input on a regional scale.

Changes in Caris Drain are probably due to fertiliser application or a plume of phosphorus moving through sandy duplex soils, while the slight changes in South Coolup Main Drain may be reflecting long-term changes in ground water.

HARVEY REGION The Harvey River contributes the second largest flow, about 20 per cent of the annual flow to the Peel–Harvey. In the past the Harvey contributed more than half of the annual phosphorus load. Nutrient concentrations are elevated in the river and phytoplankton concentrations are probably similar to those in the Serpentine where *Nodularia* blooms are now common. In the Harvey Region, Meredith Drain accounts for about 3 per cent of the Harvey River volume but is nutrient enriched and shows evidence of further deterioration, probably due to the long history of stock grazing, lack of fencing and degraded riparian vegetation as well as fertiliser application. Recent rehabilitation programs for 'streamlining' and application of bauxite 'red mud' residue to reduce leaching of phosphorus appear to be having some effects on surface leaching but it will take some years to affect ground water nutrients.

Nutrients in Samson Brook also appear to be increasing and may reflect mining in the upper catchment. In contrast, water quality in the Mayfields Main Drain is better than adjacent waterways although the land use is similar. The differences may be due to the higher clay content of the soils, improved land management and farm practice.

The quality of water flowing from the catchment in 2004 indicates that nutrient concentrations remain a concern. Some of the rivers still contain high concentrations of nitrogen and or phosphorus, and in some the concentrations appear to have increased while in others they have decreased. Further analysis is obviously needed to clarify if the trends can be related to changes in land use or management.

Water in the estuary—is it different?

Exchange between the estuary and sea is three times greater than it was before the Dawesville Channel was constructed. Flood tides at 2.17 metres per second and ebb tides at 2.09 metres per second flush about 10 per cent of the estuary volume to sea through each day. The channel also affects flow throughout the estuary and in the old Mandurah Channel so that the previous residence times of water of about thirty days in Peel Inlet and fifty days in Harvey Inlet are now about ten days and seventeen days respectively. Calculations suggest that, as predicted, the higher flows remove about 280 kilograms of phosphorus on a tidal cycle (100 tonnes per year) and about 2,400 kilograms of nitrogen (900 tonnes year).

Despite the greater ocean connection, salinity in the estuary still fluctuates. Salinity throughout the water column is still lower in winter than the open ocean, particularly near the rivers where it may be < 1‰ and near the channel about < 10‰. In summer, salinity is usually higher than the sea at about 40‰, although the concentrations are not as great as before the channel when salinity could reach 50‰. During spring, however, once a critical time for *Nodularia* blooms when salinity was usually < 10‰, salinity is now higher, about 20‰ in most areas, too high for *Nodularia* to germinate.

Before the channel was opened the freshwater run-off in winter overlay the dense saline marine waters in the main basin, effectively isolating the bottom water that became deoxygenated. Since the channel opening these stratification events have decreased and oxygen concentrations have been more stable, and the super-saturated surface conditions sometimes associated with algal blooms in the Harvey have also disappeared. In the lower estuarine reaches of the rivers however, a 'salt wedge' now moves further upstream and the stratification isolates the bottom. As a result anoxic conditions in the deeper river channels are not uncommon leading to the release of nutrients from the sediment.

Although particulate organic nitrogen and phosphorus associated with algal blooms in spring and summer have decreased in the main basin since the channel opened, concentrations in winter are still high particularly near the river channels. Inorganic phosphorus concentrations are also high near the Serpentine, and inorganic nitrogen (nitrate–nitrite) near the Murray exceeds guideline concentrations for control of phytoplankton growth in estuaries in winter and at some sites in spring. The high concentrations probably indicate that nutrient flow from the catchment may have increased, so that in areas beyond the flushing influence of the channel conditions in the Peel–Harvey are still favourable for blooms of phytoplankton.

The oozy bottom

Organic matter in the sediments does not appear to have decreased since the channel opened despite the decrease in macroalgae and absence of *Nodularia* blooms. The amount of algae growing and decaying is probably still high enough to replenish any matter flushed to sea. Alternatively, there has been little flushing of organic matter to sea or inputs from the catchment are still high. Certainly large amounts of organic matter have been deposited at the mouth of the Harvey River since 1989.

Sediment nutrients are similar to concentrations before the cut except for areas adjacent to the channel, where water currents scouring the sediments are greater. In 1998 phosphorus concentrations were higher near the river mouths than before the channel opened, although the amounts of biologically active non-apatite phosphorus had decreased. In general, information on sediment nutrients before and after the channel was opened is insufficient to allow firm conclusions to be drawn about changes in the sediment store.

A different sort of ecosystem

Microscopic algae after the channel—gone but lurking in the background

Concentrations of chlorophyll *a* have decreased in the estuary basins since the channel was opened and there have been no *Nodularia* blooms or large winter diatom blooms, indicating that the numbers of microscopic algae have decreased. The saltier water in spring is not as favourable for growth, and nutrients that once fuelled the blooms have washed out to sea. The occurrence of the marine non-toxic dinoflagellate *Heterocapsa triquetra* phytoplankton bloom in July 1997 could indicate a greater marine influence.

The decrease in algal blooms and amount of chlorophyll *a* has resulted in generally higher light levels in the water except in turbid areas near the rivers or where wind waves re-suspend sediments. The decrease in algal blooms has also stabilised the pH. In blooms pH had reached about 10 when photosynthesis used up the carbon dioxide in the water or declined to about 2 when blooms collapsed and bacteria in the organic material released carbon dioxide. The more stable pH and higher oxygen concentrations have reduced the potential for release of nutrients from the sediment.

The situation in the rivers, particularly in the lower estuarine reaches, is not as good. The bottom sediments isolated by the salty water from the estuary become anoxic, and nutrients stored for many years are released to fuel blooms of *Nodularia* and occasionally *Anabaena*. The presence of these species in the river channels is a great concern. The presence of blue-green algal blooms supports the information on water quality that indicates catchment conditions remain poor and sediment nutrient loads are still large. As river flows have not been high since the channel was opened, a wet winter could bring a flush of nutrients downstream into the estuary starting another period of algal blooms in the basins.

Larger algae—less abundant but more species

Before the channel was opened, estuarine conditions were only suitable for a few types of macroalgae. The saltier, clear water from the sea was expected to bring a greater number algal species through the channel. Green *Chaetomorpha* was still abundant after the channel opened until spring of 1996. Other nuisance green algae, such as *Ulva* and *Enteromorpha*, have decreased in the Peel, but red algae and the brown alga *Hincksia* (previously named *Giffordia*) may be more abundant. The marine green algae *Caulerpa racemosa* and *Caulerpa distichophylla* have established in the Harvey where they now form a large part of the algal biomass.

The timing of algal growth has also changed with the more marine conditions. The greatest biomass now occurs in spring with the increasing temperatures and light, rather than in summer and autumn following the collapse of the diatom and *Nodularia* blooms.

The amount of macroalgae in the Peel–Harvey has fluctuated greatly over the years and the lowest biomass was recorded in 1993–1994 before the channel opened. This reduction was attributed to lower nutrient concentrations, due in part to the increased flushing following the 1987 dredging of the Sticks Channel that increased flow through the Mandurah Channel, and to reduced phosphorus inflow from the catchment.

The amount of macroalgae continued to decrease after the channel was opened (Graph 3.5 and Map 3.7) particularly in the northern Peel near the channel where the marine influence is greatest. In the late 1990s, the largest macroalgae accumulations occurred along the south-eastern shores of the Peel and lower reaches of the Serpentine and Murray Rivers, areas furthermost from the influence of the channel where eutrophic conditions persisted.

Time will tell if the reduction in nuisance algae is a real change due to the influence of the channel, or if the low rainfall and slightly lower nutrient inflows of the past few years have had a greater affect. Ongoing reductions in macroalgal biomass and a significant change in macroalgal diversity are unlikely until nutrient concentrations across the 'whole' water body including the drains, rivers and the estuarine basins are reduced. The clear calm shallow waters still provide suitable conditions for algal growth. The future interplay between algae and seagrasses will be interesting to watch! Areas of fast growing seagrass, such as *Ruppia*, could potentially increase under the marine influence as they have in Wilson Inlet. As yet they have not changed markedly, though there have been some increases particularly in the Harvey.

Around the edges

Despite their resilience, plants around the edge are affected by changes within the water body of the estuary, particularly inundation with water and salinity, the very features that the channel was designed to produce. The increased tidal range after the channel was expected to submerge a larger area of land but for a shorter time, and expose some of the lower intertidal areas more often but for shorter times. Salinity was expected to be less variable, generally more marine, and no longer hypersaline in late summer and autumn. The amount of algae accumulating over the marsh was also expected to change, initially increasing then decreasing as the system stabilised and became less eutrophic. This could depend on the amount of drift alga growing in the post-channel estuary.

As expected, since it takes a long time for plant communities to change, there were no immediate consistent patterns of changes between 1994 and 1995 in fringing vegetation despite the higher than expected water levels and increased salinity. Monitoring of transects in areas of particular concern showed little change from 1957 to 1999 at Point Grey, Serpentine Cove and Coodanup. Although sediment appeared to be accreting at Waters Edge on the western side of the Harvey, there was some erosion and vegetation loss at Creery Wetlands (Fig. 3.20), and on the eastern side of the Harvey at North and South Kooljerrenup. Comparison with earlier records indicated that about 10–20 centimetres of terrestrial vegetation was lost, replaced either by algal or bare substrate. Substrate changes were also evident, with a reduction in fine sediment where there was increased exposure to tidal flow and waves. The yellowing and loss of *Melaleuca* along the western shore of Harvey Estuary could be due both to insects and saltwater intrusion.

GRAPH 3.5 *Comparisons of seasonal mean (± standard deviation) total macroalgal biomass in the Peel Inlet and Harvey Estuary before (1985–1991) and after (1994–1999) the Dawesville Channel construction. Note that the scale for the Harvey Estuary is an order of magnitude lower than that for the Peel Inlet. The * represents a statistically significant difference in pre- and post-channel years (p < 0.01). Note the variability in biomass.* Source: Wilson et al., 1999. Courtesy S. Wilson, Murdoch.

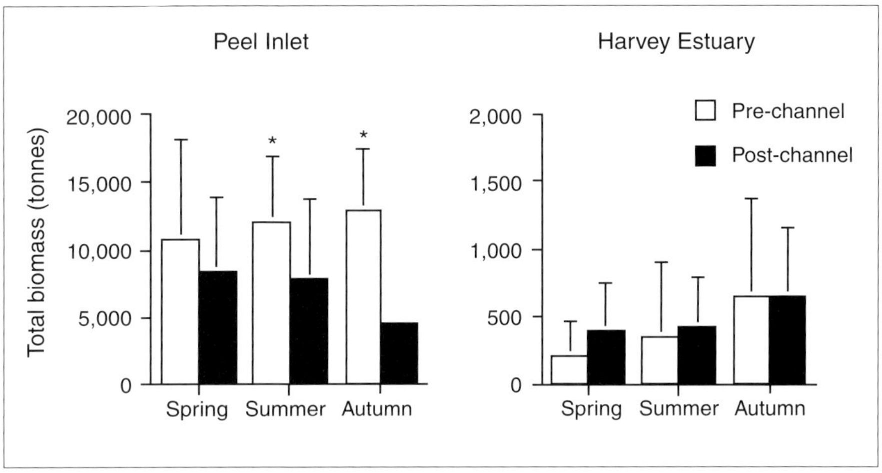

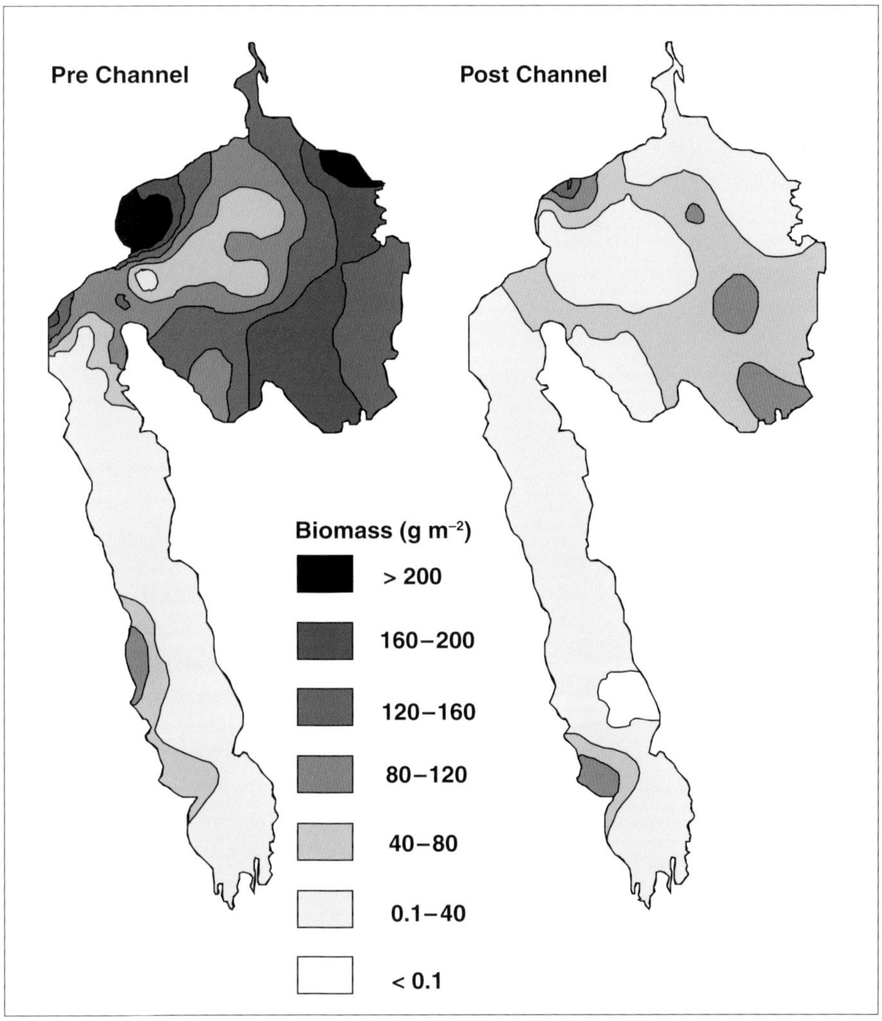

MAP 3.7 *Mean distribution of total macroalgal biomass during autumn, before and after the opening of the Dawesville Channel. Autumn differences are most dramatic with a mean biomass across the whole system of 120 g dry weight m^{-2}. The highest biomass is found on the western shore in the south of the Harvey Estuary and on the north-western shore of the Peel Inlet. Years of comparable rainfall were chosen for comparison: pre-channel 1985–1986, 1986–1987, 1987–1988, 1988–1989, 1990–1991; post-channel 1994–1995, 1996–1997, 1997–1998, 1998–1999.* Source: Wilson et al. 1999, Courtesy C. Wilson, Murdoch.

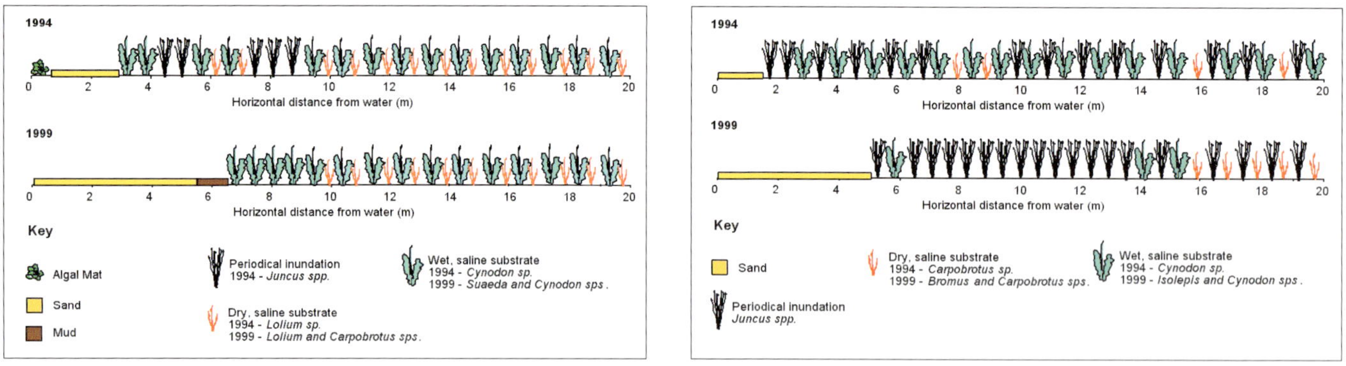

FIG. 3.20 *Comparison of pre- and post-Dawesville Channel vegetation transects A and B, Creery Wetlands. The area of bare substrate has increased and there has been a loss of fringing vegetation, exposure of underlying shell beds and steepening of the beach profile. The belt of* Casuarina *increased after 1957 but the wide* Juncus *fringe in 1977 had disappeared by 1999. Note symbols represent a vegetation type and are not indicative of actual species. Species names indicate dominant species.* Kobryn et al., 2000. Courtesy H. Kobryn, Murdoch.

Different sorts of animals

The channel was built to change conditions in the Peel–Harvey. As well as flushing nutrients out to sea to decrease algal blooms, the increased connection with the sea was expected to increase the marine state and provide a greater opportunity for animals to move into and out of the estuary. As a result, it was expected that a more diverse estuarine community of larger, longer-lived species such as are found in adjacent, less eutrophic estuaries like Leschenault Inlet and the lower reaches of the Swan–Canning would develop.

In the water—zooplankton

In the early 1970s and 1980s, typical estuarine calanoid copepods accounted for 80 per cent of zooplankton in the Peel–Harvey. By the late 1980s, however, these types accounted for less than 7 per cent of zooplankton, and other types of animals such as amphipods and nematode worms normally found on the bottom were more abundant. The capacity of amphipods and nematodes to grow and reproduce rapidly, and their high mobility meant they were able to move away from unfavourable conditions of low oxygen or toxins in *Nodularia* blooms and recolonise areas when conditions improved. These are traits well suited to eutrophic waters. Another unusual feature of the animal community was its seasonal abundance. In most other south-western Australian estuaries, peak abundance of zooplankton occurs in spring and summer. In the Peel–Harvey, peak abundance was in autumn, possibly as a response to the macroalgal growth at that time, although the low numbers in spring and summer may have been due to anoxic conditions or *Nodularia* toxins. The types of zooplankton living in the post-channel Peel–Harvey have not been investigated.

Living in the oozy bottom

Little is known about many of the bottom dwelling invertebrates before or during the massive algal blooms of the 1970s and 1980s. The molluscs, worms and amphipods appear to have been species common to other south-west estuaries. Molluscs were also abundant, particularly the tiny snail *Hydrococcus*, over 12,000 per square metre, and the bivalve *Arthritica*, 3,400 per square metre (Fig. 3.21). The larger trough shell *Spisula trigonella* was abundant in the Harvey and the bivalve *Fluviolanatus* attached to wood in the lower reaches of the rivers. The biomass however was quite low as most of the shells were small, less than a centimetre.

A few marine molluscs were found in the old Mandurah channel area with a total of eighteen species in the Peel Inlet and seven species in the Harvey Estuary. Compared to other estuaries such as the Swan (97 species) and Oyster Harbour (109 species), the number of species was quite low; the high diversity in these other estuaries due to a larger number of marine species. (See also Table 1.9.)

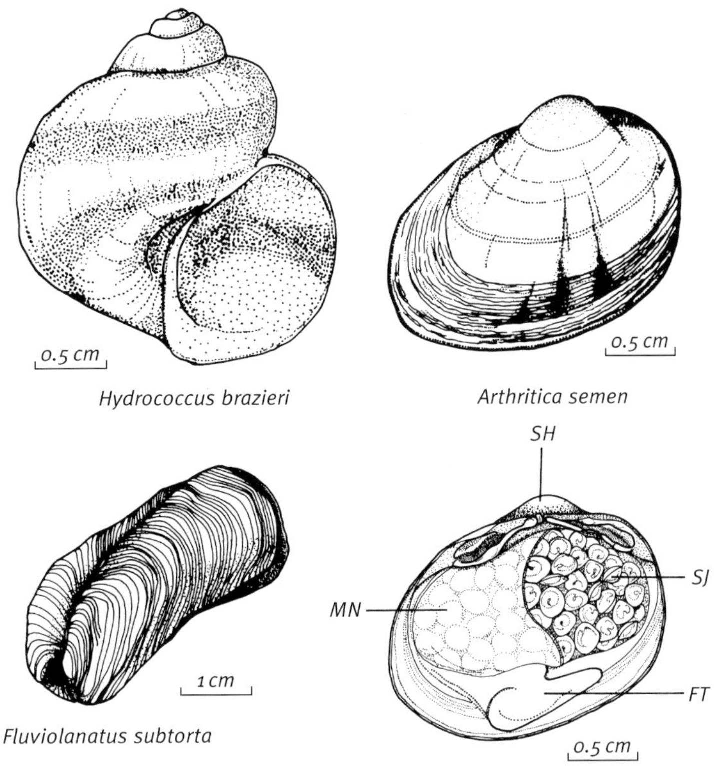

FIG. 3.21 *The bivalve* Arthritica semen *lives in the sand.* A. semen *are protandric hermaphrodites, initially maturing as males and after a brief male stage continuing to develop as females. The young are brooded within the female parent shell, a strategy that protects the vulnerable young from the changeable estuarine conditions. Shell (SH), mantle (MN) partially removed to show the shelled juveniles (SJ) (embryos), foot (FT).* Hydrococcus brazieri *are most obvious crawling over macroalgae, but numbers decline when the algae accumulates and decomposes, and many are eaten by birds. The mussel-like bivalve* Fluviolanatus subtorta, *previously known as* Anticorbula amara, *attaches to wood and coarse material.* Modified from Wells & Threlfall, 1982 and Wells et al., 1980. Courtesy F. Wells, WAM.

During the years of high algal growth in the late 1980s and early 1990s, the dominant benthic invertebrates in the Peel–Harvey were a few types of generally small, fast growing species that produce many young. These included the worm *Capitella capitata*, the amphipods *Corophium*, *Tethygenia*, *Melita* and the bivalve *Arthritica semen*, species typical of eutrophic (nutrient enriched) habitats. Interestingly one of the most abundant and significant contributors to the production of molluscs of the Peel–Harvey in the 1970s, *Hydrococcus brazieri*, was no longer present or drastically reduced in number. Although abundant, their small size meant that they contributed little biomass to the system although their production was high, about 30–75 tonnes per year.

Like the small zooplankton in the water column, most of the larger invertebrates were also more abundant in the Peel–Harvey in autumn and winter. Invertebrates are usually more abundant in spring and summer in other south-western Australian estuaries, but in the years of algal blooms conditions in the Peel–Harvey at this time of year were particularly stressful to small animals.

The most obvious and important invertebrates were the school prawns that spend their entire life in the estuary, the king prawns that visit during the summer and autumn months, and crabs that enter as young juveniles to feed in the productive waters. (See **Box 2.9 Summer sport—prawning parties** and **Box 4.2 Crabbing—a summer pastime**.)

Since the channel was opened, the dominant types of larger worms and molluscs have changed and a number of new species have appeared in the estuary. Although species lists can vary according to the methods used to collect the data, the increase in species from twenty-five species in the early 1980s, thirty-six species in late 1980s and fifty-six species in 2000, certainly indicates that there is now a wider range of invertebrates living within the less variable and less stressful conditions of the estuary.

The once abundant bristle worms *Capitella* and *Ceratonereis* have decreased in abundance. *Capitella* is however still abundant near river deltas where there are high concentrations of organic matter. In contrast, some larger, longer-lived species that prefer oxygenated, more marine conditions, such as *Australonereis eherlsi* and *Leitoscoloplos normalis*, have increased in abundance.

The dominance of small crustaceans has also changed since the channel opened, with a decline in deposit feeders such as *Corophium minor*, *Tanais dulongi* and *Melita zeylanica*; species typical of organic rich sediments and abundant macroalgae. The higher salinities also appear to have favoured long thin isopods such as *Cruranthura simplicia*.

The small bivalve *Arthritica semen* that decreased in numbers during the years of large algal blooms appears to have increased again, except for areas near the rivers where high concentrations of organic matter still persist. Similarly, numbers of the tiny gastropod *Hydrococcus* also appear to have increased particularly where macroalgae is growing and healthy. Other molluscs such as *Tellina deltoidalis*, *Soletellina* (previously *Sanguinolaria*) *biradiata*, *Nassarius burchardi* and *Spisula trigonella* also appear to have been favoured by the more saline marine waters.

The lower numbers of invertebrates near the river deltas, and presence of species that are more tolerant of nutrient rich habitats with large amounts of organic material, support the finding of recent water quality surveys. These indicate that nutrient conditions in the catchments and river are still high and a threat to estuarine health.

Animals of the marshes

The fauna living in the fringing marshes at the interface of water and land, a habitat so important for foraging birds, also changes with seasons and water levels. When water levels are high, the marsh fauna is similar to that of the permanently submerged estuary floor and fish move among the plants. When water levels are lower or the surface dry, aquatic animal numbers decline and the area is dominated by terrestrial species such as spiders and insects. Then there are the mosquitoes, particularly abundant in the marshes where they breed in shallow, sheltered ponds. Mosquitoes are not just a nuisance, they also carry the Ross River and Barmah Forest viruses. (See **Box 3.4 Mosquitoes—a nuisance, a danger and part of the natural web of life**.)

Mosquito breeding grounds have increased since the opening of the Dawesville Channel as the fringing vegetation is flooded more often, particularly when the sea level is high. To combat the mosquitoes, application of sprays to kill the larvae has increased. As the chemicals can affect other invertebrates however, there could be a long-term cost to the environment and the more popular creatures such as crabs, fish and birds. Other management controls are being investigated: to remove water from the marshes by 'runnelling', and allow predator species to enter from the main water body. They appear to be successful, but longer-term effects on the salt marsh are unknown.

Fish—marine visitors

The increased marine exchange through the new channel and more favourable salinity for oceanic fishes was expected to increase the types of fish found in the estuary, and decrease the numbers killed by toxins or low oxygen concentrations associated with the eutrophic conditions. Estuarine species, with the exception of juvenile mullet, were not expected to be greatly affected because of their tolerance of a wide range of salinities. Although the macroalgal contribution to the food chain and to the range of habitats was expected to decrease, the gradual increase in seagrass was expected to compensate for any losses.

Comings and goings

The short distance to the sea through the channel has resulted in the earlier entry and higher numbers of western king prawns and young crabs in Harvey Estuary. On the downside, the prawns and crabs now leave the Peel Inlet at

Box 3.4 MOSQUITOES—A NUISANCE, A DANGER AND PART OF THE NATURAL WEB OF LIFE

Salt marshes around the Peel Region are breeding grounds for mosquitoes that carry Ross River Virus (RRV) and Barmah-Forest Virus (BFV), diseases that are potentially debilitating. RRV causes polyarthritis in humans with symptoms of joint pain, swelling and lethargy that can last for many months or years. Symptoms of BFV are similar but generally milder.

The mosquito *Aedes camptorhynchus* is the major vector in the Peel Region. It breeds in brackish water from winter to spring. Another species *Aedes vigilax*, the summer salt marsh mosquito, breeds in warmer months. In the past, *A. vigilax* generally occurred in low numbers but it was abundant in February 1989 and March 1990, both periods associated with heavy rain following cyclonic activity in the north. Although it can carry the viruses, *A. vigilax* has been more of a problem as a nuisance species because it bites viciously, travels long distances (10 to even 100 kilometres), and is active even during the day.

Mosquito control is carried out by the Health Department of Western Australia and the Peel Region Contiguous Local Authority Group. Mosquito breeding areas are sprayed with Abate™ 50SG and more recently with Altosid™, larvicide-coated sand granules dropped from a helicopter one to four times between spring and autumn (September and April), when mosquitoes and virus are most abundant and when human exposure to biting mosquitoes is highest. As the larvae take about two weeks to develop in cooler months, and three to four days in warmer months, numbers of mosquito larvae in the marshes are monitored every two to three days to determine when the larvicides should be applied. Adult mosquito populations are also monitored and virus infection rates examined. The agencies also conduct public awareness campaigns promoting a 'cover the skin' and 'avoid being bitten' approach to disease management.

Opening of the Dawesville Channel was expected to create more favourable conditions for mosquitoes because of the frequent and widespread tidal inundation of salt marsh that would extend the breeding season for *A. camptorhynchus*. It appears the fear of more mosquitoes has been realised. Before the channel, larvicide was applied four times in year, now it is applied fifteen to eighteen times per season. With the increase of application and success of killing larvae (70–80 per cent), adult mosquitoes have generally decreased although there were large numbers of *A. camptorhynchus* in summer 1995–1996, and an increase in adult *A. vigilax* in 1994–1995, 1995–1996 and 1996–1997. Despite the success of the spraying, the summer mosquito species is now a major disease carrier.

One major problem with mosquito spraying is the effect on other non-target organisms, the small animals forming the basis of the food web that leads to fish populations. There is also the possibility of the mosquitoes developing resistance to the chemical. Other methods of mosquito control are being investigated. One is the use of 'runnels' (small, shallow, spoon-shaped channels designed to increase flow into and out of the salt marsh, and increase access of fish and other predatory animals to mosquito breeding areas) (Fig. 3.22). Runnelling appears to be successful in decreasing mosquito breeding areas, particularly in spring and summer when the channels need little maintenance. There appears to be no immediate effect on marsh vegetation although longer term effects, particularly those associated with acid sulfate soils, are not known.

FIG. 3.22 *A runnel or spoon-shaped drain dug through the salt marsh to increase flow to and from the estuary, 1995. Increased drainage decreases breeding areas for mosquitoes and effects on the health of the marsh appear to be minor, however there could be problems with acid sulfate soils in some situations.* Courtesy H. Kobryn, Murdoch.

Source: D. A. Lord & Associates, 1998; Jasinska et al., 1997, Lindsay et al., 1992.

a smaller size, probably due to the ease of exit through the new channel and the stimulus of tidal water movement. Crabs are still abundant in the estuary, as are king prawns, but the prawns are less concentrated in number and more difficult to catch in the stronger tidal flows through the channel.

Although fish populations in the system have always changed from year to year, the effects of the channel are now more apparent (Table 3.3). In this survey comparing fish populations before and after the channel opening, the numbers of semi-anadromous estuarine (1), estuarine (8), marine-estuarine opportunists (17) species remained the same, but the estuarine and marine increased from seven to nine and the marine straggler group increased from six to twelve species. The changes in abundance of particular species and overall decline in the number of fish are probably related to changes in macroalgal growth. As with the crustaceans, fish more commonly found along the coast can now enter the estuary more easily. The Peel–Harvey of today provides stable marine conditions, protected waters and readily available food, however many fish species occur only in low numbers and their abundance varies throughout the system; they are often most abundant in and near the channel. The common blowfish (banded toadfish) *Torquigener pleurogramma* was far more abundant after the channel opening in 1996–1997 and was present further south in Harvey Estuary. However as blowfish numbers fluctuate greatly from year to year, their increased abundance after the channel was opened may be a coincidence. (See **Box 2.14 An unpopular fish—the common blowfish or banded toadfish**.)

In contrast, the striped (six-lined) trumpeter *Pelates sexlineatus* and gobbleguts *Apogon rueppellii*, species favouring dense macroalgae, are less abundant since the channel was opened. The trumpeter and gobbleguts may, however, have declined earlier in the Peel Inlet with the disappearance by the early 1980s of the dense *Cladophora* stands rather than with the opening of the Dawesville Channel.

Catches of King George whiting in 1998 were the highest they had been since the 1970s. Numbers however declined in 1999, and it appears there was above average recruitment in the late 1990s accounting for the previous year's high catch.

The Peel–Harvey still supports the largest estuarine fishing industry in Western Australia, although the number of units actively fishing has declined from sixteen in June 1998 to seven (eight units available) in 2001. In 1999 a catch of 256.7 tonnes of fish had an estimated value of $763,000. In 2000 there were only nine units holders fishing in the Peel–Harvey and the catch, 192 tonnes, was much lower than the previous year reflecting much lower fishing effort. In 2001 only seven fishing units caught 196.5 tonnes, similar to the 2000 catch. The catch per unit effort (CPUE) declined sharply in 1989 but has remained relatively stable despite the decline in catch and effort since 1997, suggesting that the species contributing to the catch has remained constant although lower than in 1975–1985. In 2001 most of the finfish (54 per cent) were sea mullet and yellow-eye mullet and 37 per cent of the catch blue swimmer crabs (Graph 3.6).

Cobbler, a fish favoured by recreational and commercial fishers, is over-exploited in Peel–Harvey, Leschenault and Swan–Canning estuaries. In the Peel–Harvey the 2000 catch was about half of the 1999 level, and the 2001 catch even lower. The decline was probably due to many factors such as a decrease in bottom nesting sites with the decline in estuarine water quality, as well as overfishing, with many fish caught before they reach sexual maturity. As cobbler populations are discrete with virtually no breeding connections between adjacent areas, each stock needs to be managed separately. Restrictions on catch during the breeding season are being implemented, and closed seasons during spawning and an increase in minimum size may allow stocks to recover. In the Peel–Harvey, new types of breeding areas such as the rock walls of the channel may be replacing

TABLE 3.3 *Life cycle category, rankings by abundance (R), number (N) and percentage contribution (%) to the total catch of each fish species obtained seasonally from five regions of the Peel–Harvey Estuary in 1980–1981 and 1996–1997, i.e. prior to and after the construction of the Dawesville Channel, respectively. In this table the catches for 1980–1981 were adjusted so that, as in 1996–1997, they corresponded to 24 samples per region.* Source: Young & Potter 2003.

Common name	Species	Category	Pre-Dawesville Channel			Post-Dawesville Channel		
			R	N	%	R	N	%
Striped (six-lined) trumpeter	*Pelates sexlineatus*	O	1	17,765	21.8	10	250	1.0
Gobbleguts	*Apogon rueppellii*	E&M	2	15,670	19.2	7	863	3.4
Roach	*Gerres subfasciatus*	O	3	8,015	9.8	6	1,062	4.1
Sandy sprat	*Hyperlophus vittatus*	O	4	7,269	8.9	2	4,371	17.0
Long-finned goby	*Favonigobius lateralis*	E&M	5	5,311	6.5	8	641	2.5
Yellow-eye mullet	*Aldrichetta forsteri*	O	6	5,293	6.5	5	1,082	4.2
Common blowfish	*Torquigener pleurogramma*	O	7	4,666	5.7	1	10,643	41.4
Ogilby's hardyhead	*Atherinomorus ogilbyi*	O	8	3,043	3.7	14	144	0.6
Elongate hardyhead	*Atherinosoma elongata*	E	9	2,459	3.0	4	2,415	9.4
Presbyter's hardyhead	*Leptatherina presbyteroides*	E&M	10	2,269	2.8	3	2,442	9.5
Southern school whiting	*Sillago bassensis*	S	11	2,020	2.5	14	144	0.6
Soldier/devil fish	*Gymnapistes marmoratus*	O	12	1,530	1.9	12	223	0.9
Yellowfin whiting	*Sillago schomburgkii*	O	13	1,101	1.4	18	88	0.3
Sea mullet	*Mugil cephalus*	O	14	782	1.0	23	27	0.1
Wallace's hardyhead	*Leptatherina wallacei*	E	15	625	0.8	28	15	0.1
Blue spotted goby	*Pseudogobius olorum*	E	16	595	0.7	20	52	0.2
Yellow-tailed trumpeter	*Amniataba caudavittata*	E	17	430	0.5	32	5	< 0.1
Small-toothed flounder	*Pseudorhombus jenynsii*	O	18	309	0.4	17	111	0.4
Bridled goby	*Amoya bifrenatus*	E&M	19	287	0.4	26	22	0.1
Western river garfish	*Hyporhamphus regularis*	E	20	285	0.4	27	19	0.1
Mullet hardyhead	*Allanetta mugiloides*	E	21	243	0.3	30	13	0.1
King George whiting	*Sillaginodes punctata*	O	22	239	0.3	16	115	0.4
Prickly pufferfish	*Contusus brevicaudus*	O	23	199	0.2	23	27	0.1
Tarwhine	*Rhabdosargus sarba*	O	24	187	0.2	9	303	1.2
Tailor	*Pomatomus saltatrix*	O	25	126	0.2	19	62	0.2
Trumpeter whiting	*Sillago burrus*	O	26	123	0.2	13	168	0.7
Large mouthed goby	*Afurcagobius suppositus*	E	27	122	0.1	42	1	< 0.1
Southern sea garfish	*Hyporhamphus melanochir*	E&M	28	119	0.1	21	35	0.1
Blue weed whiting	*Haletta semifasciata*	S	28	119	0.1	31	7	< 0.1
Cobbler/estuarine catfish	*Cnidoglanis macrocephalus*	E&M	30	74	0.1	35	3	< 0.1
Pilchard	*Sardinops neopilchardus*	S	31	54	0.1			
Flathead goby	*Callogobius depressus*	O	32	27	< 0.1	32	5	< 0.1
Southern anchovy	*Engraulis australis*	E&M	33	25	< 0.1	11	237	0.9
Perth herring	*Nematalosa vlaminghi*	A	34	19	< 0.1	35	3	< 0.1
Elongate flounder	*Ammotretis elongata*	S	35	15	< 0.1	22	33	0.1
Rough leatherjacket	*Scobinichthys granulatus*	S	36	6	< 0.1	32	5	< 0.1
Hairy pipefish	*Urocampus carinirostris*	E	37	2	< 0.1	42	1	< 0.1
Western Australian Salmon	*Arripis truttaceus*	O	37	2	< 0.1			
Pygmy leatherjacket	*Brachaluteres jacksonianus*	S	39	1	< 0.1	42	1	< 0.1
Australian herring	*Arripis georgiana*	O				23	27	0.1
Western school whiting	*Sillago vittata*	S				29	14	0.1
Blue spotted flathead	*Platycephalus speculator*	E&M				35	3	< 0.1
Blue sprat	*Spratelloides robustus*	S				35	3	< 0.1
Six-spined leatherjacket	*Meuschenia freycineti*	S				39	2	< 0.1
Wood's siphonfish	*Siphamia cephalotes*	E&M				39	2	< 0.1
Skipjack trevally	*Pseudocaranx dentex*	S				39	2	< 0.1
Old wife	*Enoplosus armatus*	S				42	1	< 0.1
Port Jackson shark	*Heterodontus portjacksoni*	S				42	1	< 0.1
Bar-tailed goatfish	*Upenichthyes tragula*	S				42	1	< 0.1
	TOTAL NUMBER OF SPECIES			39			47	
	TOTAL NUMBER OF FISH			81,426			25,692	

A–semi-anadromous; E–estuarine; E&M–estuarine & marine; O–marine–estuarine opportunist; S–marine straggler.

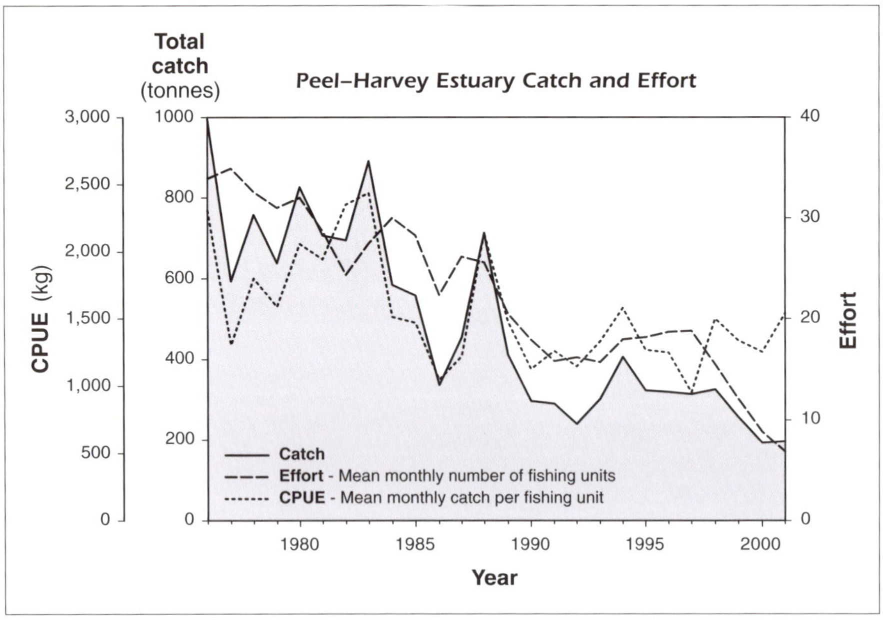

GRAPH 3.6 *The annual catch, effort and catch per unit effort (CPUE) for the total fishery in the Peel–Harvey Estuary, 1976–2001.* Ayvazian & Nowara, 2002b. Courtesy DoF.

those lost in degraded areas of the estuary. (See **Box 8.4 Cobbler: spines to fear—great to eat**.)

The clear waters coming through the Dawesville Channel have made the Peel–Harvey a more attractive area to visit, and, with more people living in the area, the number of recreational fishers has risen increasing pressure on the fish stocks. As well there are the more 'subtle' effects from boat noise, fuel spills, increased swash and boat landings. What impact these will have is anyone's guess, but we need to ensure we don't love the area to death.

Recreational fishing

A survey of recreational fishing in the Peel–Harvey conducted over the twelve-month period August 1998 to July 1999 found that boat fishing effort was 126,000 fisher days with 88 per cent of the effort focused on blue swimmer crabs. Shore-based fishing occurred over 191,00 fisher days with 83 per cent of the effort focused on blue swimmer crabs. The recreational catch of blue swimmer crabs was about 1,360,000 crabs or about 289 tonnes (Table 3.4), which was five times the mean annual commercial catch since 1991–1992. The commercial crab catch in the same period was 65.5 tonnes.

As expected from studies of crab biology, the recreational crab catch peaked in summer with the high abundance of crabs. Boat-based crabbers using drop nets mostly caught male crabs, with a ratio of females to males of 1:5.1. In contrast, shore-based crabbers (mostly scoopers) caught proportionally more female crabs with a sex ratio of 1:1.5. This bias was probably due to the abundance of female crabs in shallow areas (Fig. 3.23).

FIG. 3.23 *Crabbing: a most popular sport in the Peel–Harvey and other estuaries of Swanland.* C. Bryce, WAM.

The recreational finfish catch was relatively small (≈8 tonnes) in comparison to the blue swimmer catch; the most commonly caught and kept being the Australian herring, whiting other than King George (*Sillago* spp.), tailor, skipjack trevally, trumpeters, King George whiting, silver bream (tarwhine) and black bream (Table 3.5). Estimates of recreational finfish catches may have been underestimated as surveys were conducted only from November to April and did not include fishing from the banks of the Dawesville Channel or the rivers. In addition, recreational net fishing that was permitted on Wednesday nights from 1 November to 31 July was not included. Recreational net fishers would generally have caught mullet species that make up the bulk of the commercial catch.

The survey found that boat fishers generally kept to regulations with respect to undersized fish; only 5.8 per cent kept undersized fish and

TABLE 3.4 *Comparison of the commercial and recreational fish catches in the Peel–Harvey, 1998–1999. The blue swimmer catch by recreational shore-based scoopers was 90 t with 17 t from bridges and jetties.* Source: Malseed & Sumner, 2001a.

Common name	Species	Commercial weight (t)	Recreational boat weight (t)	Recreational shore weight (t)
Blue swimmer crabs	*Portunus pelagicus*	65.5	182	107*
Sea mullet	*Mugil cephalus*	109	NA	NA
Yellow-eye mullet	*Aldrichetta forsteri*	61	NA	NA
King George whiting	*Sillaginodes punctata*	16	0.3	NA
Other whiting	*Sillago* sp.	11	0.2	0.7
Australian herring	*Arripis georgianus*	5.8	2.3	2.0
Tailor	*Pomatomus saltatrix*	5.3	0.9	0.6
Black bream	*Acanthopagrus butcheri*	0	0.5	NA
Skipjack trevally	*Pseudocaranx* spp.	0.1	0.6	0.6
Silver bream	*Rhabdosargus sarba*	NA	NA	0.2

TABLE 3.5 *Estimated recreational fishing catch in the Peel–Harvey Estuary, November 1998–April 1999. Shore-based estimates for fish are from surveys of bridges and jetties in the Mandurah Entrance Channel. Catches from the area of Dawesville Channel were not included. Blue swimmer shore-based estimates also included catch by people wading and using scoop nets, and boat-based included both private trailed boats and hire boats.* Based on Malseed & Sumner, 2001a.

Common name	Scientific name	Boat-based		Shore-based	
		No. kept	No. released	No. kept	No. released
Australian herring	*Arripis georgianus*	21,551	1,060	25,989	549
Other whiting	*Sillago* sp.	3,210	1,076	12,194	1,767
Tailor	*Pomatomus saltatrix*	3,051	2,960	2,937	6,676
King George whiting	*Sillaginodes punctata*	2,659	602	NR	NR
Mussels	*Mytilus edulis*	2,630*	0	NR	NR
Black bream	*Acanthopagrus butcheri*	1,556	2,440	NR	NR
Trumpeters	*Tetraponidae family*	855	738	2,517	3,321
Skipjack trevally	*Pseudocaranx* spp.	495	166	4,513	898
Common blowfish	*Torquigener pleurogramma*	421	3,064	2,025	18,410
Australian salmon	*Arripis truttaceus*	237	309	NR	NR
Silver bream (tarwhine)	*Rhabdosargus sarba*	219	2,387	2,264	10,142
Flounder	*Pseudorhombus* spp.	86	0	NR	NR
Mulloway	*Argyrosomus hololepidotus*	42	335	NR	NR
Pink snapper	*Pagrus auratus*	0	293	314	1,590
Blue swimmer crabs	*Portunus pelagicus*	831,964	966,898	528,385	399,475

**Mussel catch in litres. NR – not recorded in survey area.*

6 per cent undersized crabs. Shore-based crabbers were less compliant; 13 per cent of crabbers kept undersized crabs. About 30 per cent of the boats with one person on board caught the daily limit of twenty-four crabs, while only 7 per cent of those with two or more people on board achieved the daily limit of forty-eight crabs. More than a third of boat-based fishers were local Mandurah residents living within a 20-kilometre radius of the estuary.

The combined boat-based and shore-based fish catch illustrates the importance and use of the estuary by recreational fishers. The recreational fishers generally sought the marine-opportunist species that breed in the ocean and move into the estuary to feed. The survey results showing the large number of undersize fish 'thrown back' into the estuary provide a good indication of the abundance of immature fish, illustrating the importance of the protected and productive waters of the Peel–Harvey as a nursery area. However as fish stocks in estuaries and all along the coast are not infinite, the escalation in recreational fishing is a problem.

A special place for birds

The Peel–Harvey, together with the Yalgorup, McLarty and Mealup Lakes, is recognised as a significant waterbird habitat and a Wetland of International Importance under the Ramsar Convention. More than eighty species have been recorded in the area including twenty-seven species listed in the Japan Australia Migratory Birds Agreement (JAMBA) and the China Australia Migratory Bird Agreement (CAMBA). In 1977 more than 100,000 birds were recorded on the estuary. Each year in summer around 20,000 migratory birds visit the estuary to feed on the rich pickings of the sandy flats before flying off to breed in the following Northern Hemisphere summer. The area is also important for species resident in Australia such as pelicans, swans, ducks, grebes, stilts and avocets. (See **Box 1.10 Global travellers: living a double summer life—transequatorial migrant birds** and **Box 3.5 Types of birds using the Peel–Harvey**.)

The Peel–Harvey environment with its wide sandy shallows, low-lying fringing marshes, and abundant food supply has been a waterbird haven. Although extensive areas remain undisturbed, the increasing urban development along the northern and western shores at Mandurah and Dawesville and the recent subdivisions to the east are increasing the fragmentation of suitable habitat and disturbance to bird populations.

The greatest numbers of birds occur in the north-central and south-eastern areas (Austin and Robert Bays) of the Peel Inlet and the southern part of the Harvey Estuary. The southern part of the Harvey is a particularly important refuge for waterfowl that congregate on the lower reaches of the river and the extensive delta. While the eastern shores and outer delta are reserved for nature conservation with a narrow reserve for 'drainage and conservation' on both banks of the lower river, most of Harvey Estuary is not officially protected as a waterfowl refuge area. Land to the west is privately owned, and the population is increasing as farmland is subdivided on both sides of the estuary.

The continuing value of the area for birds is unclear. The fringing vegetation, at the limits of physical tolerance, is particularly vulnerable to higher tides and more saline water. Since the Dawesville Channel was opened, some trees and shrubs appear to be showing signs of stress, the paperbarks are paler and more yellowish in late summer–autumn with some dead branches and a few trees have died. River salinities also appear to be increasing.

Box 3.5 TYPES OF BIRDS USING THE PEEL–HARVEY

The main types of birds using the Peel–Harvey and other estuaries in south-western Australia are:

- migratory waders (e.g. sandpipers, plovers and stints), which feed on benthic invertebrates in the intertidal areas at low tide
- resident waders (e.g. Banded Stilts, Black-winged Stilts, Red-necked Avocets and Red-capped Plovers). Banded Stilts are usually found in shallow inland non-tidal wetlands and rarely in estuaries with marked tidal influence
- long-legged waders (herons, egrets, ibises and spoonbills), which feed on fish and benthic invertebrates in shallow waters
- fish-eating birds (pelicans, cormorants, darters and terns), which occur all year round and fish in deeper water than waders, along with grebes that also feed on invertebrates
- waterfowl (ducks, swans, coots and swamphens). Black Swans feed mainly on large aquatic plants (especially the seagrass *Ruppia* and some species of macroalgae) and are found all year round providing there is access to low-salinity water for drinking
- gulls, which are omnivorous scavengers.

Most of these groups roost on the sandy spits and cays, or in the fringing vegetation (salt marsh and trees) of estuaries. Only a few species breed in estuaries: pelicans, swans, Pacific Black Ducks and Grey Teal, which usually build their nests on low-lying islands in undisturbed fringing vegetation, and Australian Shelducks (Mountain Ducks), which generally breed in trees.

Adapted from D. A. Lord & Associates, 1998.

Ultimate vagabonds

In general, birds are highly mobile, typically using the estuary as a feeding ground and breeding elsewhere, for example in the nearby swamps or the distant Arctic tundra. Numbers may vary greatly from year to year depending on breeding success and conditions in other parts of their range, as well as changes in the estuarine environment of the Peel–Harvey or similar water bodies throughout the South-West. At least 44,939 birds were present on the estuary during surveys in 1998–1999, the most common being Red-necked Stints with 16,436 birds recorded (Table 3.6).

Australian resident species such as pelicans and cormorants are present in varying numbers all year round, while migratory waders arrive in spring and stay through summer. The resident waders such as the Banded Stilt are more characteristically found in non-tidal lakes (Fig. 3.24). It has been suggested that their presence in Peel–Harvey during 1976–1977 was linked to a recent, inland breeding event and the small tidal influence in the pre-channel era (Table 3.7). (See **Box 3.6 The Banded Stilt**.)

TABLE 3.6 *Forty most common species of birds recorded in surveys in the Peel–Harvey 1996–1997 and 1998–1999.*
Source: Lane et al., 2002a & 2002b.

Species	1996–1997	1998–1999
Red-necked Stint*	14,672	16,436
Grey Teal	12,612	4,581
Banded Stilt	6,954	186
Australian Shelduck	4,527	3,890
Silver Gull	4,405	3,383
Red-necked Avocet	2,443	1,399
Black-winged Stilt	2,293	1,400
Little Pied Cormorant	1,665	1,693
Little Black Cormorant	1,368	2,636
Sharp-tailed Sandpiper*	1,285	1,322
Pied Cormorant	1,096	727
Black Swan	1,052	1,130
Red-capped Plover*	876	1,754
Pacific Black Duck	749	693
Whiskered Tern	736	337
Australian Pelican	641	774
Australasian Shoveler	501	100
Common Greenshank*	314	319
Curlew Sandpiper*	313	566
Bar-tailed Godwit*	307	137
Fairy Tern	262	167
Hoary-headed Grebe	220	NR
Caspian Tern	180	298
White-faced Heron	134	126
Grey Plover*	116	81
Great Egret	114	88
Crested Tern	112	121
Darter	95	79
Little Egret	89	78
Australian White Ibis	48	70
Pacific Golden Plover*	54	29
Great Cormorant	39	30
Red Knot*	39	34
White-fronted Chat	33	28
Black-tailed Godwit*	28	NR
Pied Oystercatcher	27	20
Great Knot*	26	37
Eastern Curlew*	19	24
Little Grassbird	15	22
Musk Duck	4	48

** denotes transequatorial migrant species; NR – not recorded in survey.*

As bird numbers naturally fluctuate from year to year, the effects of the new environment created by the channel are difficult to assess. Larger tides will alter some roosting, feeding and breeding areas but will potentially create similar conditions elsewhere. Disturbance is potentially a greater threat due to increased urban development, access and the popularity of boating.

Little Egrets *Egretta garzetta* have become more numerous and widespread around the estuary than in 1976–1977 when a major survey was conducted, reflecting a general increase in numbers of this species in south-western Australia in the past twenty-five years. Small numbers (ten to twenty) of Eastern Curlews *Numenius madagascariensis*, rarely observed elsewhere in the southern half of Australia, are still present particularly around the northern end of the Peel Inlet. Curlew Sandpiper *Calidris ferriginea* are now less abundant due to

Box 3.6 THE BANDED STILT

For a time in the mid-1970s, Banded Stilts *Cladorhynchus leucocephalus* were the most abundant waterbirds on the Peel–Harvey (Tables 3.6 and 3.7). Their presence was probably due to the low tidal amplitude and extensive area of shallow water available for feeding and roosting. Generally they are more common on non-tidal wetlands, particularly saline lakes of the coast or interior. Numbers can fluctuate greatly: 8,800 were recorded in December 1976, increasing to 63,000 in the following February and decreasing to only 27 in June 1977. The largest numbers have been found in the eastern and southern sides of Peel Inlet and in the Creery marshes, with few in Harvey Estuary. Although stilts were not as abundant in the same survey areas in 1996–1997, there were still quite large numbers on other parts of the estuary.

The number of stilts in the area also varies over shorter time scales, for example, more than 2,000 birds were found in Creery Wetland in summer 1995 but most birds had departed by March. Similar departures have been observed on Rottnest Island. A count in the middle of March 1983 recorded 3,000 birds, with another count 24 hours later recording only 200. This sudden drop in numbers is now known to be a response to the filling of the saline lakes of the arid interior following exceptionally heavy rain. When the lakes fill the bird's main food source, the brine shrimp *Parartemia*, hatch from drought resistant eggs and the birds congregate to breed. Although the birds were first scientifically named in 1816, their nesting habits eluded investigators until 1930 when nesting was first observed at Lake Grace.

TABLE 3.7 *Numbers of Banded Stilt on the Peel–Harvey.* Source: Lane et al., 2002a & 2002b; Lane & Pearson, 2002.

Year	Aug	Oct	Dec	Feb	Apr	June
1976–77	761	1,503	8,806	63,000	5,948	27
1996–97	–	124	760	6,954	–	–
1998–99	–	58	15	186	–	–

– indicates no survey at this time.

Three to four eggs, chalky fawn to pure white with black to dark brown twisted lines or blotches, are laid on the ground. The young are covered with pure white down, an unusual colour for young wading birds. Young birds in the first year of life lack the chestnut band across the chest.

Banded Stilts are also known as Rottnest Snipe. In the early colonial days, when Rottnest Island was used as the Governor's summer residence, shooting parties were a favourite entertainment. Bush brakes were erected around the lakes and prisoners used as beaters to move the birds towards the shooters. Today the summer flocks are a great attraction as they swim and feed on the brine shrimps or crowd into 'rafts' for resting. With increasing development along the coast, the shores of the Rottnest lakes are likely to become even more important for large groups of Banded Stilts.

Sources: Lane et al., 1997; Saunders & de Rebeira, 1985; Serventy & Whittell, 1962; Coate & Johnstone, 2003.

FIG. 3.24 *Banded Stilts.* A. G. Wells Collection, Courtesy CALM.

year to year variation in numbers migrating to Australia from the Arctic, a longer term decline in abundance in Australia or change in the Peel–Harvey.

Although not popular, the cormorants are an important part of the estuary system. As fish feeders, their numbers in part reflect the abundance of fish such as the small hardyheads and the health of the system. In the mid 1970s Little Black Cormorants roosted in single flocks of several thousands birds on the sandbars of Herron Point Ford and thousands of Pied Cormorants roosted on high points of a low sand cay north of Point Birch and on Nirimba Cay.

Pelicans

Increased water levels and disturbance are quite likely to affect the pelicans that historically have nested on low-lying islands such as Nirimba Cay, Creery, Channel, Boodalan and Boundary Islands and an unnamed island in Boggy Flat. Nesting has often failed due to human interference, vandalism, fox predation or flooding of nesting sites. (See **Box 3.7 The Pelican.**)

Box 3.7 **THE PELICAN**

The Australian Pelican *Pelecanus conspicullatus* regularly nests at eight or nine locations in Western Australia. Most nesting sites are located in the northern part of the state, between Shark Bay and the east Kimberley. A few also breed in the south in the Peel–Harvey at Nirimba Cay since 1962, Boodalan and Boundary Island in the Peel–Harvey, on Green Island in Oyster Harbour, Albany since 1985, and more recently in Shoalwater Bay on Penguin Island since 1998, and Seal Island since 2001.

Pelicans nest in colonies, generally on small low islands near the coast and around inland lakes. The nests are placed pecking distance apart and are typically simple scrapes in the ground, often without nesting material, although they are sometimes lined with plant material. The two to three eggs are white, elliptical in shape, and the young hatchlings are featherless. In recent times, nesting in the Peel–Harvey has been most successful on Boundary Island, although the ongoing success is doubtful due to disturbance from boats using the channel, its jetty access and status as a recreation reserve.

Boundary Island, created in 1987 from spoil dredged out of the natural outlet channel, is now an important bird habitat as many of the traditional shallow grounds have been altered or are no longer quiet due to increased people pressure. The first nesting of pelicans was recorded in 1989, but the numbers of successful hatching are unknown. Since the channel was opened there have been four periods of breeding varying from four to eleven months between July 1994 and July 1997 producing 500–700 young. These numbers are probably sufficient to maintain pelican populations, but there is no clear relationship between the number of pelicans counted on Peel–Harvey and the number of young fledglings from Boundary Island. The fluctuating numbers are typical of this highly mobile species. Many newly fledged young probably dispersed to other water bodies, like the older pelicans that use a variety of water bodies and breeding grounds.

Source: Abbott, 2001; Lane et al., 1997; Lane et al., 2002a & b.

FIG. 3.25 *Pelicans at Falcon on the Peel–Harvey, July 1983.* E. P. Hodgkin Collection.

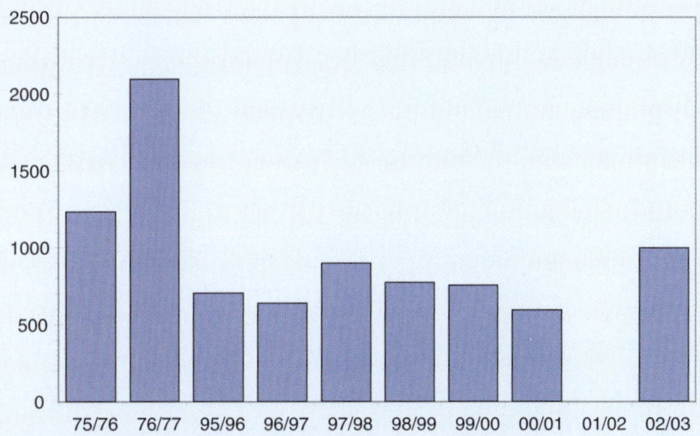

GRAPH 3.7 *Numbers of pelicans on the Peel–Harvey, February 1975 – May 1977 and October 1995 – February 2003. No counts were made in January 2001.* Courtesy J. Lane, CALM.

Australian pelicans are most abundant in the south-western estuaries in summer–autumn, and least abundant in winter–spring. Leg banding and monitoring of breeding colonies indicate numbers vary as adults leave to breed in the northern part of the state in winter, and with the return of both adults and young birds in summer. Birds staying in the South-West all year are either non-breeders, or in the case of the Peel–Harvey non-breeders and resident breeders (Fig. 3.25).

Counts of pelicans in the Peel–Harvey between 1995 and 2003 ranged from 268 in October 1995 to 998 in February 2003. These numbers are similar to those recorded from February to November 1975, but much lower than numbers recorded from October 1976 to March 1977 (Graph 3.7). The very high numbers of pelicans in 1976–1977 have been be linked to the filling of Lake Eyre in 1974. The Lake Eyre event supported a large breeding population that may have persisted into 1975–1976. As the waters receded and salinity increased, fish died and thousands of pelicans flew in search of food to other parts of the continent and New Guinea, possibly reaching the west coast including the Peel–Harvey. As 1976 was a drier than average year in the South-West, this would have also been an influence contributing to the high bird numbers in the permanent coastal waters.

Swans

The Peel–Harvey is also one of the most significant habitats for Black Swan in Western Australia. Populations fluctuate between years and throughout the year as swans range over a wide area of the South-West, but numbers appear to have declined since the late 1970s (Graph 3.8). This decline may be a short-term trend rather than a long-term change as numbers observed in the surveys of 1975 were similar to those in 1996–1997. Reasons for the decline are not known but could include differences in breeding success between years, of water habitat available across the South-West and possibly across the whole continent, food and differences in disturbance. A census in December 2002 at the height of an Australia-wide drought revealed a dramatic increase in swan numbers on Peel–Harvey, almost matching that of October 1976. (See **Chapter 2 The Swan–Canning Estuary: Recreating Swanland—bringing back the Black Swans**.)

Swans are abundant in the shallow areas of the Peel–Harvey, particularly over seagrass *Ruppia* sp. on the eastern and southern sides. The highest number recorded was 8,100 in October 1976. In some years they congregate in the southern part of Harvey Estuary in late summer and autumn where they have access to fresh drinking water that flows all year from the Harvey River. Small numbers also breed on the nearby freshwater wetlands and lead

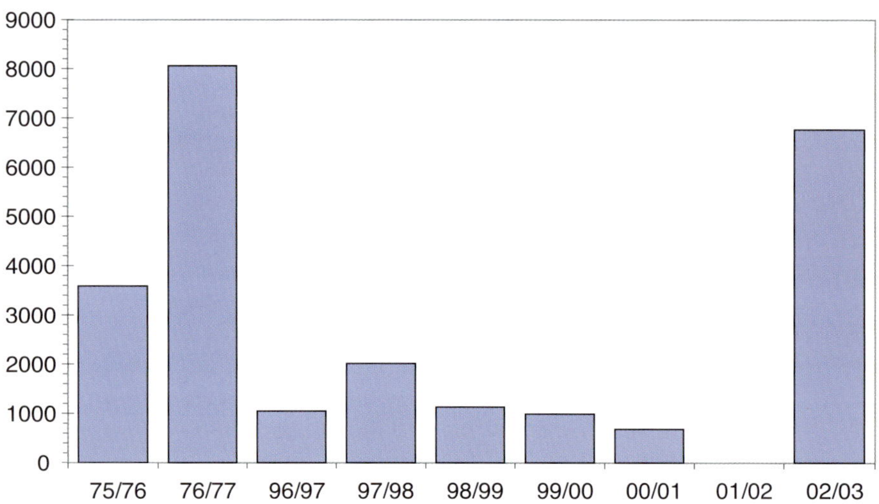

GRAPH 3.8 *Numbers of Black Swans on the Peel–Harvey, February 1975 – June 1977 and October 1996 – February 2003. No counts were made in January 2001.*
Courtesy J. Lane, CALM.

their young to the permanent waters of the estuary where they are reared. Other birds leave the area in winter to breed elsewhere on the Swan Coastal Plain or further inland, returning in summer as inland waters recede.

Like pelicans, swans are sensitive to disturbance from boats, and people or dogs wading in the shallows. They are particularly vulnerable for a period of four to six weeks when they undergo their annual moult of flight feathers. During this time they cannot fly and need large areas of undisturbed permanent open water with assured food supplies and fresh water for drinking.

As well as having great aesthetic appeal, swans are also an important component of the estuary food web playing a major role in cycling of nutrients. They graze extensively on submerged aquatic plants, particularly the seagrass *Ruppia*, hence the alternative name of swan grass. Their feeding also affects the distribution and composition of the plant community and the nutrient status of the sediments.

What of the future?

Problems in the Peel–Harvey emerged slowly as each of the small changes came together to create a huge effect, sending an ecosystem into distress in less than 150 years. The questions remain. Do the cures work? Do we like the result of our cures? Are they cures? And the most important question: have we learnt from the mistakes? The story of the Peel–Harvey is echoed throughout Swanland, throughout the state, the continent and the world: it was just larger, more visible in the Peel–Harvey being close to urban centres, and the cures more drastic and expensive.

If we had the chance to start again, we would probably go down much the same pathway. The land would have to be cleared, the watertable would rise, drains would be dug and we would add fertilisers to promote the pastures to feed the large animals we have added to the landscape. But there are some things we could do differently.

The first is probably accepting that small mistakes can add to a large whole and, conversely, adopting the idea that small remedies can make a difference. The birth of the Landcare movement and development of a stewardship for the land by communities as well as farmers has facilitated many small remedies through planting trees and revegetation of streamlines that buffer erosive water flows and filter nutrients, as well as creating a more pleasant landscape. The scale, however, is difficult to visualise. In 1995 an estimated 150 kilometres of the waterways had been revegetated, but there are about 2,000 kilometres of similar drains on the coastal plain.

Alternative types of fertilisers and reduced application rates will gradually reduce phosphorus stores and losses from the land. Certainly some of the land should never have been cleared, and clearing went on into the 1980s. Fertiliser use on farms has received most of the attention but the rivers flowing off the scarp are not those of 200 years ago, their flows having been impounded to provide water demanded by more distant communities. When the phosphorus application problems were being addressed, point source pollution from piggeries and sewage works commenced, and the growing population around the estuary used septic systems. Although some of the point source polluting industries have been 'cleaned up', some have become diffuse polluters with more intensive land use and stocking densities.

The waters flowing to the estuary still contain high concentrations of nutrients and these, together with those stored in the sediment, are the fuel for algal blooms that now plague the river reaches instead of the main estuarine basins. Then there are the acid sulfate soils that are likely to exacerbate problems. Although the masses of macroalgae have decreased, seagrasses along with algal epiphytes now grow profusely in the shallow well-lit waters indicating that nutrients are still abundant (Fig. 3.26). It is interesting that problems in the New Peel–Harvey continue to fuel political argument as they did in the 1970s and 1980s.

Today we have more knowledge at our disposal, and we should use it. Land clearing is no longer necessary: we have enough land, although it may not be in a state to make full use of its potential. The landscape needs to be viewed as a whole composed of many pieces. Large-scale engineering works have large downstream impacts, and we need to consider what these are or might be. We can use new products, or modify our practices. Drain flows can be regulated to accommodate irrigation and prevent flooding, but perhaps we do not need so many and perhaps they do not need to be so large or deep. It is in our own interests to maintain a healthy system, be it for the view (who wants to look only at buildings?), for the fishing (we all want to fish), for fun (there is no doubt that it is best to walk and swim in clean healthy water), or for the birds as they revive our spirits and show us that an environment is healthy and functioning.

FIG. 3.26 *Could this happen again? Ernest Hodgkin examining algae removed from the beach at Coodanup, June 1981.*
E. P. Hodgkin Collection.

Chapter 4

Leschenault Inlet and Estuary and smaller estuaries of the Geographe Bay area

South of Cape Bouvard near Mandurah and the Peel–Harvey through to Bunbury and Point McLeod, the coast lies in a north to south orientation backed by lines of dunes, ridges of limestone, wetlands and the estuarine lagoon of Leschenault Inlet. The Collie and Preston Rivers flowing to the Leschenault bring water from the 4,600 square kilometre catchment on the Darling Plateau.

South of Bunbury the coastline arcs westwards around Geographe Bay to Cape Naturaliste and the Leeuwin Naturaliste Ridge, where a change in the nature of the coast is marked by the start of granite outcrops. Here the rivers and creeks are short, flowing from the edge of the Darling Range, the Whicher Range and across the coastal plain to Geographe Bay. The Vasse–Wonnerup System and Toby Inlet, as well as the Broadwater, lie parallel to the coast just behind the mostly recently formed dune lines. In many ways the low-lying estuarine areas resemble the sandy gently sloping underwater landscape of the bay. Sediments and fossils underlying the coast in fact reveal the very recent separation from the sea. The landscape is now largely cleared of native vegetation, and many of the rivers have been modified for agriculture by a system of drains and tidal barrages that drain the low-lying areas.

FIG. 4.1 *Landsat view of the Leschenault Inlet and Estuary: the parabolic dunes are still migrating eastwards, spilling into the main estuary basin. 'The Cut' through the dunes can be seen opposite the Collie River delta in the south-east. The natural exit channel into Koombana Bay, now an isolated lagoon, lies on the left of the photograph.* Landsat imagery provided by Australian Centre For Remote Sensing (ACRES), Geoscience Australia, Canberra, and digitally enhanced by Satellite Remote Sensing Services, DLI.

Visitors — mapping and naming the landscape

Like much of the west coast, the area around the Leschenault and Geographe Bay was explored by the Dutch in the 17th century and the French in the 19th century, and settled by the British in the mid-1800s. The first visitors were searching for the lost treasure ship the *Vaergulde Dreack* or Gilt Dragon in 1658. Captains Samuel Volkersen and Aucke Pieters Jonck in the galjoort *Emeloort* and fluit *Waeckende Boey* mapped part of the west coast as far south as Bunbury. They found wreckage but mostly further north, and another 150 years passed before the next visitors.

The French scientific voyage in 1801–1803 led by Nicolas Baudin was the first to examine and map this area between Cape Leeuwin and the Swan River. After making landfall near Cape Leeuwin in 1801, the expedition proceeded northwards exploring the land around Cape Naturaliste and Geographe Bay and naming the areas after the expedition vessels *Naturaliste* and *Géographe*. Members of the expedition made numerous trips ashore around Geographe Bay and explored Wonnerup Estuary, originally named the Vasse River to commemorate the loss of a sailor of that name who drowned in the surf.

Impressions of the explorers varied from excitement at discovering new flora and fauna and aboriginal artefacts to disappointment at the lack of large trees, edible plants, apparent soil sterility, and scarcity of freshwater streams. Two years later in 1803 the area was visited again by members of the expedition including Louis de Freycinet in command of the *Casuarina*, a replacement for the *Naturaliste*, and Baudin on the *Géographe*. Freycinet entered Geographe Bay before Baudin, discovering Casuarina Point near the mouth of Koombana Bay before moving on to Rottnest Island, the agreed meeting place if the vessels were separated. Baudin followed a similar path two days later and anchored north-west of Casuarina Point, dispatching Midshipman Bonnefoi (Montbazin) to survey and chart the area. Bonnefoi charted Koombana Bay and discovered and partly mapped Leschenault Inlet to just beyond Middle Island, which he named. Baudin named the inlet Port Leschenault after the ship's botanist Jean Baptiste Claude Theodore Leschenault (Map 4.1).

LESCHENAULT ESTUARY

A detour off the busy South West Highway takes the traveller with time to spare along the shores of the Leschenault Estuary. In the north the narrow lanes pass through a paperbark cathedral bordered by rich green cattle pastures. A westwards glimpse through the trees reveals the estuary against a background of wooded dunes that border the Indian Ocean. The waters of the estuary are dotted with Black Swans feeding where in summer eager crabbers geared up with scoop nets and floating tubs harvest the estuary's rich bounty. Further south towering red mounds of wood chips, mineral sands and alumina wait to be shipped from the busy port of Bunbury, a major regional centre.

The Leschenault Inlet has been changed over the years. Damming of the rivers in the catchment, flood control programs, dredging of access

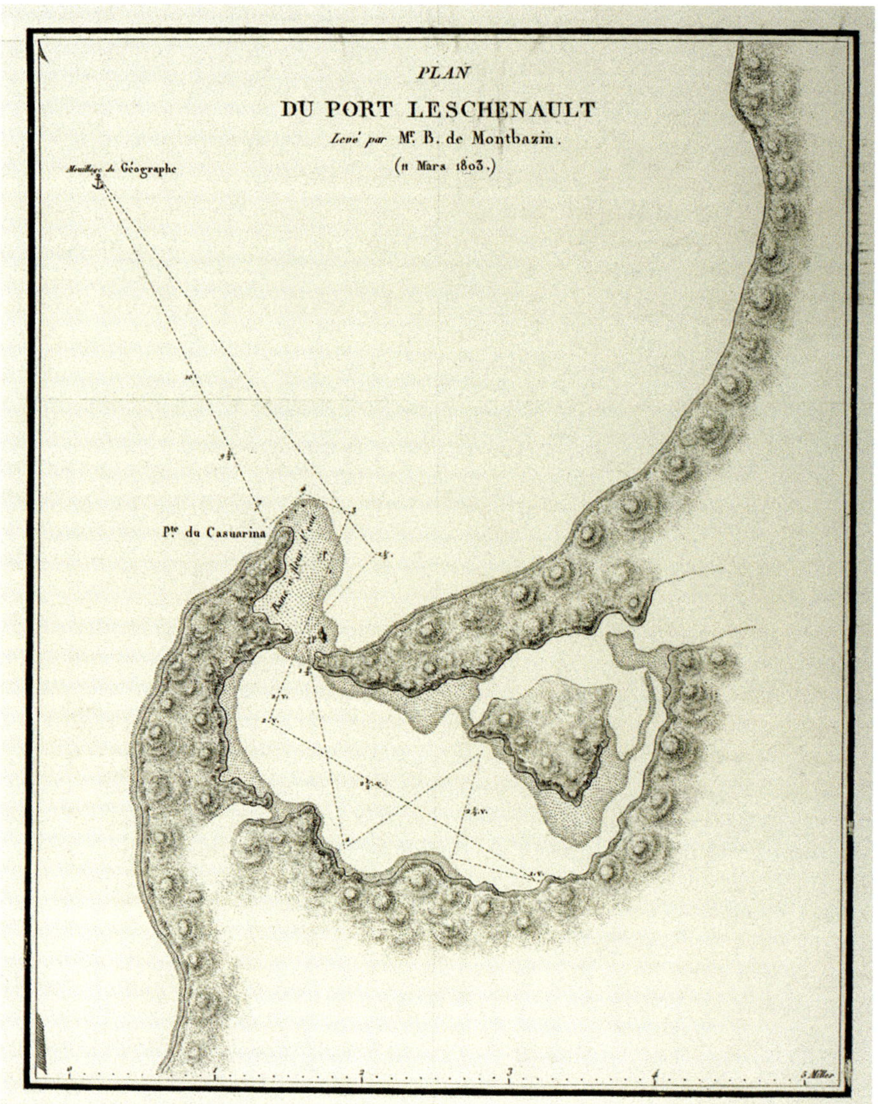

MAP 4.1 *The entrance to Leschenault Inlet in Koombana Bay drawn by Midshipman Bonnefoi during the second part of the Baudin expedition in 1803. The bay was considered suitable for small boats and called Port des Barques (Small Boat or Barque Harbour) and the visitors were impressed by the wildlife. The larger part of the inlet was not explored and the area around Middle Island is now isolated from the main estuary basin by port developments (compare with Figs 4.1 and 4.3).* See also Baudin, 1974; Marchant, 1982). Carte d'une partie des Terres de Leeuwin et D'Edels, F919.4042 1812 Voy (647); Courtesy T. Burrows, Scholars' Centre, UWA.

channels across the shallow shores and river mouths, barrier causeways, boating, fishing, and leakage of industrial wastes into the dunes, aquifers and inlet have taken their toll. Much of the low-lying swampy ground in the south is now covered with housing, roads and port facilities, and the original outlet to the sea is isolated from the main body of water. The main lagoon or water body of the Leschenault is now the estuarine system, and it has been suggested recently that the name Leschenault Inlet Estuary should be used when referring to this large body of shallow water, thus retaining the historic name but also including the current status of the water body. 'The Cut' made to the sea in the 1950s also changed the essence of the estuary, increasing the marine influence, although it probably allowed excessive nutrients from the catchment to escape so avoiding the problems encountered in the Peel–Harvey. The environment after The Cut, however, was different.

A different setting

The Leschenault is the estuary of the Collie–Brunswick and the Preston–Ferguson Rivers (Fig. 4.1). About 14 kilometres long but only 1.5–2.5 kilometres wide, the estuary lies parallel to the coast and covers an area of approximately 27 square kilometres. The elongate lagoon is separated from the sea by dunes, the Quindalup Dunes, laid down in the Holocene only 7,000 years ago, and bordered on the east by the sand and limestone of the Mandurah–Eaton Ridge of Pleistocene age (Fig. 4.2). Thus the Leschenault Estuary, like the Harvey Inlet, is a 'barrier estuary' lying between old dune lines in an inter-barrier depression, but the age of the dune barriers bordering the two systems is different. The Harvey Inlet is bordered on the west by the northern section of the Mandurah–Eaton Ridge, whereas this lies along the eastern margin of the Leschenault.

The estuary is very shallow (1.2–2 metres deep) and bordered by sand flats. The rivers enter the lower part of the estuary lagoon. As a result there are marked gradients in salinity from south to north. In the southern area the waters are influenced by freshwater river flow in winter and by marine waters brought by tides as flow diminishes in spring. The northern section, at a distance from the river mouths and marine influences through The Cut, becomes hypersaline over summer. Fresh water also enters the estuary in ground water seepage along the shoreline.

The geological history of the estuary and surrounding areas is reflected in the different landforms or geomorphology. The landforms, together with water movement and salinity, in turn create a range of habitats for plants and animals (Map 4.2).

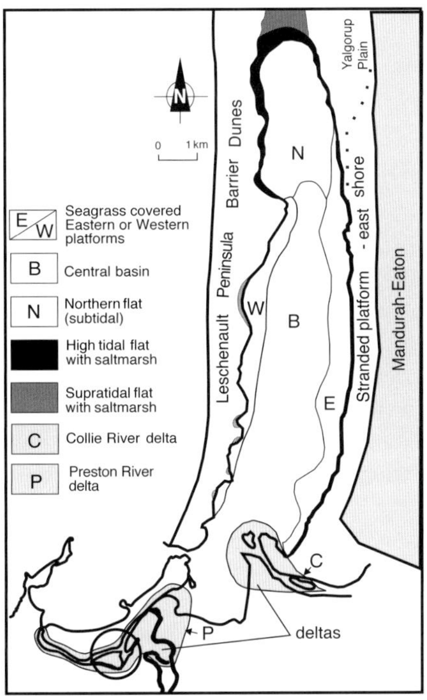

MAP 4.2 *Geomorphic units of the Leschenault Estuary and surrounding areas.* Semeniuk, 2000; Courtesy V. Semeniuk.

Rural outpost–growth of a port city and national park

Within months of European settlement at the Swan River in 1829, Dr Collie and Lieutenant Preston, ship's officers on HMS *Sulphur*, explored the area around the Leschenault discovering the rivers flowing to the estuary. The following year with a view to opening up the land for settlement, Stirling, accompanied by Collie and Preston, led another expedition primarily to map the area around the two rivers, which were now given their names. Although land was available from 1830 and Stirling took up a selection (Map 2.6), the land remained unsettled for a number of years, the area around the Augusta on the Blackwood Hardy being preferred.

The explorers and later the settlers including Lieutenant Henry William Bunbury, after whom Bunbury is named, refer to large numbers of Aboriginal people on the banks of the Leschenault (Bunbury & Morrell, 1930).

The numerous and well beaten paths near the banks of the estuary indicated the constant pressure of considerable numbers. Indeed, nowhere had I hitherto seen, even on the Murray [River], *where natives are numerous, such distinct paths as here or so many deserted huts, some of them made with some care of the paper bark, i.e. the bark of the tea tree, others of the leaves of the Zanthoriza, which afford excellent protection from rain but are not as warm.*

Bunbury went on to describe their fishing activities at night:

I soon discovered that the numerous lights I saw along the water's edge were for the purpose of attracting fish, which are speared by this means in great numbers both in the shoal water of the flats, where the native wade out carrying firesticks to spear Mullet and Cobblers, and also on the banks of the river, where fires on a large scale are made where

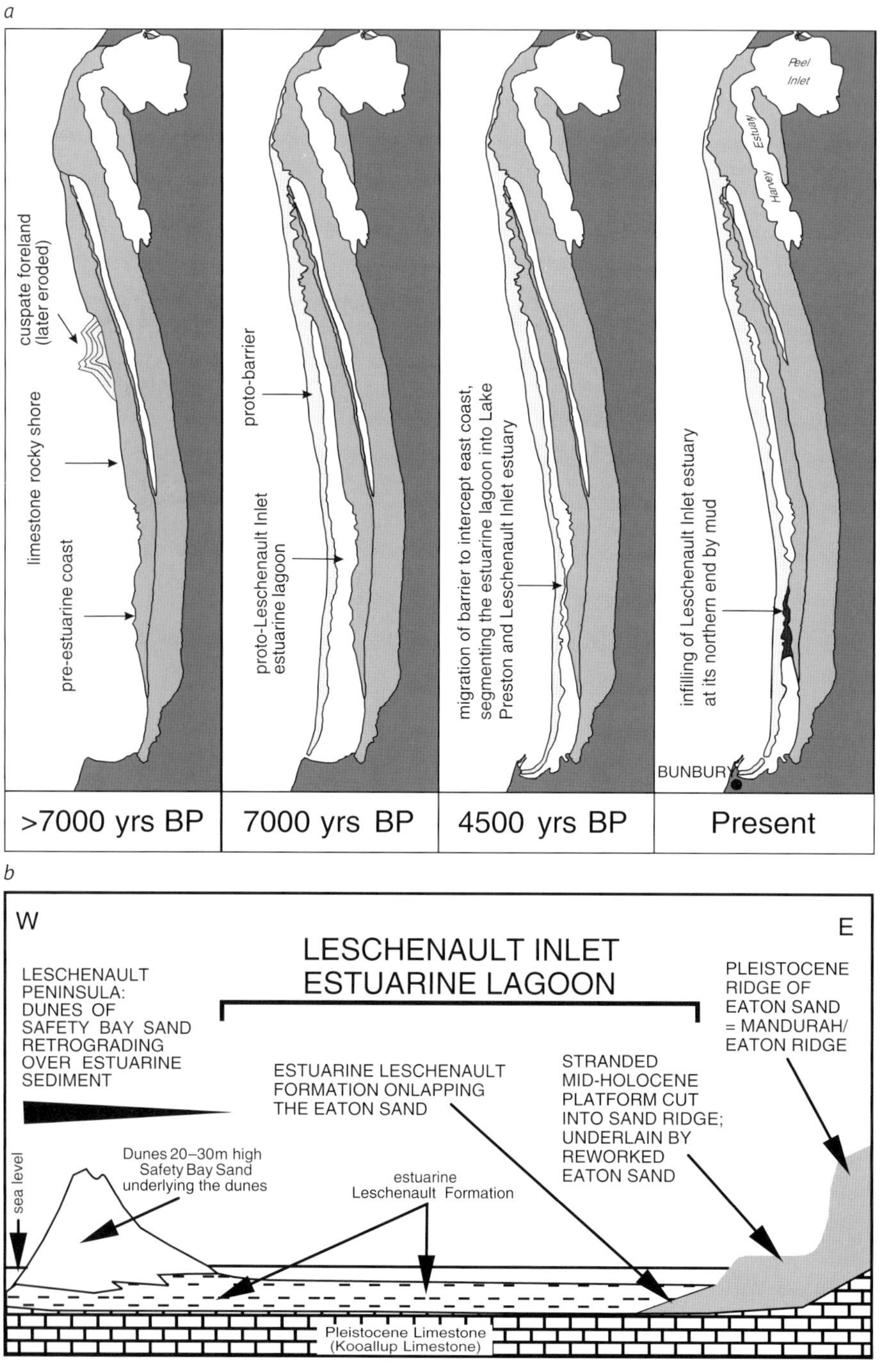

FIG. 4.2 *(a) Evolution of Leschenault Inlet through the Holocene. More than 7,000 years ago the coast between Mandurah and Bunbury was an open limestone shore with a distinct cuspate foreland to the north. Development of a dune barrier about 7,000 years ago formed a long barred estuarine lagoon (proto-Leschenault Inlet). This early estuary extended further north, included Lake Preston, and was much wider. About 4,500 years ago the eastward migration of the dune barrier narrowed the lagoon and separated it into two sections: Lake Preston and the estuary. After this time and continuing today the northern area gradually shallowed as mud accumulated. (b) Cross-section of the Leschenault Estuary.* Semeniuk et al., 2000, Courtesy V. Semeniuk.

water is deep close to and several natives watch the approach of the fish with their spears. Cobblers especially are thus taken in great quantities as well as larger fish such as Jew fish, Taylors and Black Snappers. [Note: Jew fish = mulloway, Taylors = tailor and Black Snappers = Black bream. R. Lenanton (DoF).]

How many Aboriginal people were in the area prior to 1830 is not clear, although the Leschenault area appears to have been a popular meeting place with abundant fish.

In late 1837 a family called Scott settled on land adjoining Lieutenant-Governor Stirling's grant on the Preston River, and the settlement of Leschenault now known as Bunbury was established on the shores of Koombana Bay the following year (Fig. 4.3). On the eastern shore of the estuary the town site of Australind marks the heart of the Western Australian Company's 1841 venture to establish a society based on the theories of Edward Gibbon Wakefield, champion of the colony in South Australia. The settlement at Australind under the leadership of Marshall Walter Clifton soon failed. Many of the land owners never arrived, the home company ran into serious financial difficulties, and the fine beach sand of the Koombana Bay coast proved unproductive, but settlers from the *Parkfield* and later *Trusty* spread out to pioneer the South-West.

In 1838 land on the Leschenault Peninsula was purchased by Thomas Little for Charles Robert Prinsep. Named 'Belvidere' after the Prinsep mansion in Calcutta, the estate was used to raise horses for the British army in India, and cattle. Two cattle herds were established with one herd of Bengali cattle or water-buffalo used for ploughing and as beasts of burden centred around Buffalo Homestead, known today colloquially as 'Buffalo Hut', at the northern end of the property on Bengal Station.

Success of the horse export trade based at 'Belvidere' is unclear. Thomas Hayward's diary noted that in the 1870s horses were sent to India 'all of which were sold to the Government for a good price' and 'several lots were taken to Singapore and Mauritius…but this trade was abandoned eventually as it was not very remunerative'. Henry Charles Prinsep (son of Charles Robert Prinsep) was later to record 'of course my main occupation was breeding horses for the Indian market, though I had to farm a good deal of land to support the employees, and also did a good deal of dairying'. The horse trade continued until 1869, together with export of jarrah railway sleepers to India (CALM, 1995).

In 1869 the Belvidere Estate was sold to William Henry Venn of Dardanup who rotated stock between the coastal and inland properties, and used the homestead area for stock-work and as a summer retreat (a strategy used throughout the South-West and until quite recently on the south coast).

FIG. 4.3 *A view of Bunbury and the entrance to the Leschenault Inlet published in the* Illustrated London News *1856–1857. Early records indicate that the shores of Leschenault were bordered by well-used paths by which the Aboriginal people accessed the rich estuarine waters.* Courtesy Battye Library, 2738P.

Over the years a series of fires destroyed the original house and replacements as well as a jetty. Remains of the jetty are still present near the date palms and fig trees probably planted in this period. The estate was sold a number of times, the final private owners using the site to foster alternative lifestyles in the late 1960s and 1970s. Purchased by the Government together with another grazing property to the south, much of the peninsula was used for acid effluent disposal until 1990 when it became a national park managed by the Department of Conservation and Land Management (CALM) (Map 4.3).

The Leschenault Peninsula contains natural vegetation representative of mobile and stable coastal dunes including mature tuart and peppermint forests, a narrow band of estuarine fringing vegetation and mangroves. The vegetation forms important habitat for terrestrial birds, reptiles and mammals, and water birds. The peninsula also has historic significance for Aboriginal people and early European farming.

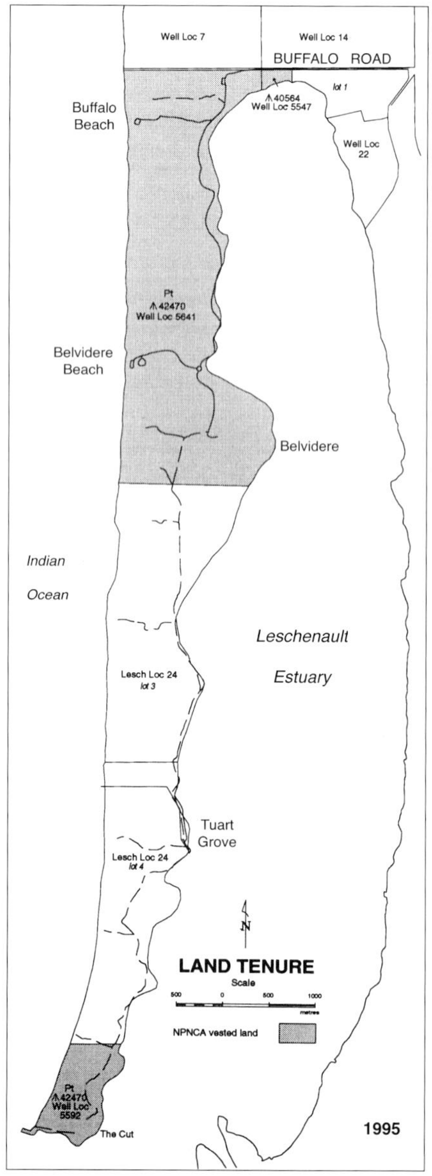

MAP 4.3 *The Leschenault Peninsula National Park and surrounding areas.*
Source: Draft Management Plan 1995 for the Leschenault Peninsula; Courtesy CALM.

Small port on a wild coast—changing world

At the time of discovery and settlement, the natural ocean outflow of Leschenault Inlet was through a 3-kilometre long channel to Point McLeod in Koombana Bay (Map 4.4). The inlet was probably always open to the sea, protected by the headland of basalt rock, but the channel was probably not deep enough to allow the passage of ocean-going vessels. There are reports, however, of small boats entering the inlet in the early days. Koombana Bay was a busy port of call for American whalers and coastal traders.

The area in the lee of Casuarina Point and surrounding the inlet was easy to settle as the land was flat with good access to the sea anchorage, but the low-lying areas were prone to flooding. Winter flooding of the Collie and Preston Rivers brought fine silts to the estuary that filled the exit channels, and run-off and sediment loads probably increased as the catchment was cleared. Not surprisingly change was on its way and, like the area around the Peel–Harvey, levee banks were built and drains dug (Fig. 4.4 and Table 4.1).

By the 1930s the harbour had shallowed, and a groyne ('North Shore Groyne') was constructed to trap littoral sand drift entering the harbour from the north. Access for construction of this groyne was by rail across a rail bridge spanning the estuary at Point McLeod. In 1949 the rail bridge was upgraded and the railway line extended northwards from the North Shore Groyne to the site of the northern groyne of The Cut. The Cut's northern groyne was constructed in 1950–1951, prior to excavation of The Cut. While the northern groyne of the channel was under construction, the North Shore Groyne was extended. The cutter suction dredge *Stirling* dredged the shallow areas of the estuary and then excavated a channel, breaking through to the sea in 1951. The new marine phase of the estuary was initiated with closure of 'The Plug', the original outlet.

TABLE 4.1 *Alterations to Leschenault Inlet since 1895.*

Alteration	Year completed
Original outlet to estuary filled (known as The Plug)	1951
'The Cut' opened	1951
Wellington Dam (Collie River) completed	1960
Laporte effluent pipeline to ocean*	1963–1992
Inner Harbour development	1967–1976
Reclamation of old inlet channel near Inner Harbour	1967–1976
Preston River channel redirected	1969–1970
New boat channel from lower estuary to ocean	1974
Parkfield Drain	1977

** The waste pipe from SCM Chemicals Ltd was supported by a 900 m rock causeway on the eastern margin and a trestle bridge. The bridge was removed July–Dec. 1992. A small section of the rock causeway remains; other sections were replaced by trestle walkways.*

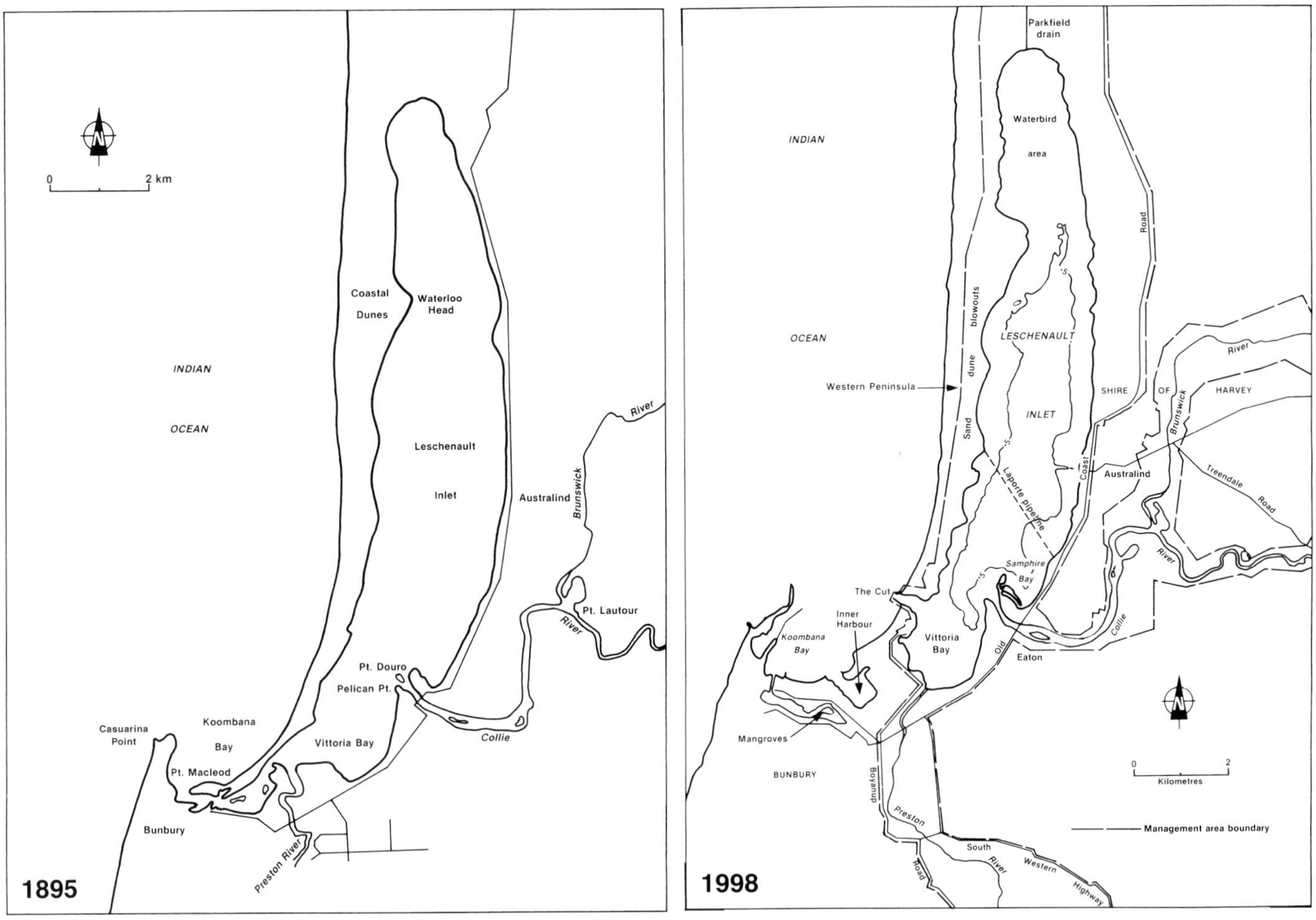

MAP 4.4 *Leschenault Inlet in 1895 and 1998. See also Fig. 4.4. Before the large changes of the 20th century, the long inlet channel opened to the sea in the lee of Point McLeod. Land reclamation, river diversion and construction of The Cut now take freshwater flow directly to the sea, allowing a greater marine influence when river flow diminishes. Note the dashed line marks the path of the Laporte effluent pipeline removed in the early 1990s.* Adapted from Leschenault Management Authority, undated.

Through the 1960s and 1970s, groynes and protective retaining walls were built along The Cut to reduce erosion. Low-lying areas to the north of the old entrance were dredged and reclaimed to create the inner harbour. The Preston River was redirected about 800 metres east to flow into Vittoria Bay, isolating the old entrance channel and transforming it to a coastal lagoon.

Port developments that changed the Leschenault landscape of the 1960s were part of many strategies to encourage industry to utilise the natural resources throughout the South-West. Among the industries was a mineral sands processing plant at Australind.

New industry— black sand and orange stains

The Swan Coastal Plain contains deposits of heavy-mineral sands that include titanium-containing minerals ilmenite (shiny black grains) and rutile. Mineral sands are derived from the igneous and metamorphic rocks of the Darling Plateau, and concentrated by the winnowing and sorting of wind and water over millions of years. The deposits are mined around the Bunbury region, at Eneabba to the north, and until recently near Augusta. The Augusta mine was closed following problems arising from oxidation of naturally occurring iron pyrite (FeS_2) that formed sulfuric acid in the mine pits (see Fig. 6.15).

Between 1963 and 1990 ilmenite was treated with sulfuric acid to produce titanium dioxide at Australind by Laporte Titanium (later SCM Chemicals Ltd and more recently Millennium Ltd). Effluent remaining after the refining process was very acidic (0.9–1.0) and contained large amounts of

FIG. 4.4 *Changes around the southern part of the Leschenault Inlet Estuary, aerial views: (a) 1941 before diversion of the Preston River and dredging of The Cut,* Map 1342, 150–6, Bunbury Run 4, 30.9.41*; (b) 1975 showing the diversion of the Preston, The Cut and development of the Inner Harbour,* WA 1565 (C) # 5010, 10.4.75. Courtesy DLI.

iron in solution, together with a suspension of fine mud containing other ions including heavy minerals (Table 4.2). Under the Laporte Industrial Factory Agreement Act 1961, the Western Australian Government was responsible for disposal of acidic effluent until the year 2011. This arrangement was part of a strategy to encourage industry to the isolated west coast where there was little infrastructure support and establishment costs were high.

Industrial waste disposal: out of sight out of mind—changing viewpoints

Originally effluent was carried across the estuary in a pipe supported by a causeway and trestle bridge to the Leschenault Peninsula, where it was discharged onto the ocean beach. This practice was terminated after a public campaign against the rust-coloured iron sulfate stains on the beaches (Fig. 4.5 and Map 4.4b). Between 1968 and 1991 when disposal of effluent ceased, effluent was deposited in the dunes where the calcareous sands neutralised the sulfuric acid forming gypsum (calcium sulfate), iron precipitate (ferrous sulfate) and carbon dioxide. Liquid reaching the underlying aquifer was therefore predominantly neutralised, although the gradual westward flow to the ocean periodically caused minor staining from (unneutralised) acid.

At first the effluent flowed into channels along the contour lines. Flow was later through a sprinkler network discharging onto the dune surface, and ultimately into a series of lagoons excavated into areas of exposed sand. At times the gypsum solids sealed off the lagoon floors and had to be removed to allow infiltration of effluent. In 1986 a central receiving reservoir was established where the solids were settled out, and the fluids pumped to infiltration lagoons.

TABLE 4.2 *Chemical characteristics of the effluent released from the titanium processing plant at Australind, 1982.*

Composition of effluent	Concentration (ppm)
Ferrous iron	6,000
Sulfate	35,500
Titanium	300
Calcium	37
Magnesium	39
Chloride	290
Manganese	230
Aluminium	80
Vanadium	13
Chromium	7
Solids	1,000–2,000
Barium, Copper, Nickel, Strontium, Tantalum, Thorium, Zinc	Traces

Source: Laporte Effluent Disposal Committee, 1982.

After lengthy research into alternative processing and disposal methods, the sulfate plant was replaced in 1989 by a chloride processing plant using hydrochloric acid produced from salt at Kemerton, 10 kilometres to the north. The sulfate plant at Australind ceased operating in October 1990, although the product continued to be finished and bagged there. Discharge to the Leschenault Peninsula ceased in January 1991. Brine produced during processing at Kemerton is discharged directly to the ocean near the end of Buffalo Road. Pending a permanent solution to its disposal by 2010, solid waste is trucked to a disposal area in sand dunes at Dalyellup, south of Bunbury.

During the years the factory operated, effluent periodically entered the estuary directly from breaks in the pipeline, or indirectly from seepage of partially neutralised or raw waste in the disposal lagoons. Fish mortality around spill areas was heavy but localised. There was initially a concern that fish could also be affected indirectly as small invertebrates at the base of the food chains were killed, but these were shown to recolonise polluted areas quite rapidly.

A legacy of the past

Radiation levels in disposal sites are slightly elevated, due mainly to thorium and its decay products. With the exception of two lagoons, radiation levels are similar to those in naturally occurring ilmenite strands towards the southern end of the Leschenault Peninsula. They are below the Radiological Council's 'maximum acceptable background' standard of 35 microrem per hour. Thus even if re-exposed in future, the deposits should not pose a hazard to human health. However, the deposits are aesthetically displeasing due to the grey colour and potential to cause milky discolouration on contact with the ocean. Burial under 500 millimetres of sand reduces radiation levels, and also provides a more suitable growth medium for initial seed germination and seedling establishment.

After closure of the sulfate plant and cessation of effluent disposal on the Leschenault Peninsula, approximately forty sites covering 100 hectares (about 10 per cent) of the peninsula required rehabilitation. Although the work started in the 1970s and 1980s, most of the rehabilitation was done between 1986 and 1992. Most of the programs aimed to reinstate communities of native vegetation by stabilising ground cover complemented by shrubs or trees. When responsibility for rehabilitation transferred to CALM in the mid-1990s, the lagoon disposal sites had been stabilised and required only follow up maintenance and monitoring. The legacy is by no means gone as natural erosion processes may re-expose unneutralised sediments that will require further burial.

FIG. 4.5 *(a) The Leschenault Peninsula showing effluent disposal and staining along the ocean beaches, aerial view, October 1970. Note the parabolic dunes migrating eastwards and spilling into the main estuary basin. The Cut through the dunes can be seen opposite the Collie River delta in the south-east.* E. P. Hodgkin Collection. *(b) Ponds used for effluent disposal, September 1982.* E. P. Hodgkin Collection.

The rivers and catchment

The Leschenault receives water from the Preston River with its tributary the Ferguson River, the Collie River with tributaries the Brunswick and Wellesley Rivers, and from numerous seeps around the margins particularly in the north. The rivers drain a catchment of 3,600 square kilometres but the Wellington Dam upstream traps much of the water flow leaving an effective catchment of 770 square kilometres. Water quality varies reflecting the amount and type of development in the catchment, rainfall and stream flow (Map 4.5, Graph 4.1 and Table 4.3). In the mid-1990s it was estimated that 51 tonnes of phosphorus and 610 tonnes of nitrogen entered the estuary in surface run-off from the rural catchment each year.

The Parkfield Drain (Map 4.4), flowing from low-lying horticultural areas in the north, also brings high concentrations of total nitrogen and moderate concentrations of total phosphorus, together with suspended solids and carbon to the estuary. At times the waters also contain high concentrations of phytoplankton including blue-green algae, with particularly high concentrations recorded in the autumn and late summer of 1997, 1998 and 1999.

Water quality varies along the length of the Brunswick and Collie Rivers. Nutrient concentrations are currently very low in the upper reaches, although salinity is a problem and there have been outbreaks of blue-green algae. Nutrients, organic carbon and suspended solids increase in the middle river reaches and are higher still in the lower reaches below the Wellesley River junction.

The Wellesley River passes through irrigated and drained areas used for dairy cattle. Although effluent management is being improved, water quality is poor with high phosphorus and moderate nitrogen concentrations, and periodic high loads of suspended solids, manures and organic matter. Phytoplankton blooms have not as yet been a problem, however the area is considered vulnerable. Diatom numbers are increasing and blue-green algae are sometimes present in moderate concentrations.

Below the Brunswick–Wellesley confluence the high concentrations of nitrogen and phosphorus together with high organic loads in the lower and estuarine reaches of the Brunswick River have not as yet caused major algal blooms. Phytoplankton numbers have increased since 1995 and blue-green algae are found in moderate densities in spring, summer and autumn, together with high densities of dinoflagellates and cryptophytes. Sea water intrudes a short distance up the Brunswick River and the river is not stratified except during times of freshwater run-off. Run-off in late summer or autumn can also bring large loads of organic matter that on decaying reduce oxygen concentrations in the river. This situation is thought to account for the death of fish and marron in a 10-kilometre stretch upstream of the Brunswick Caravan Park in March 2000, and a fish kill in 2002.

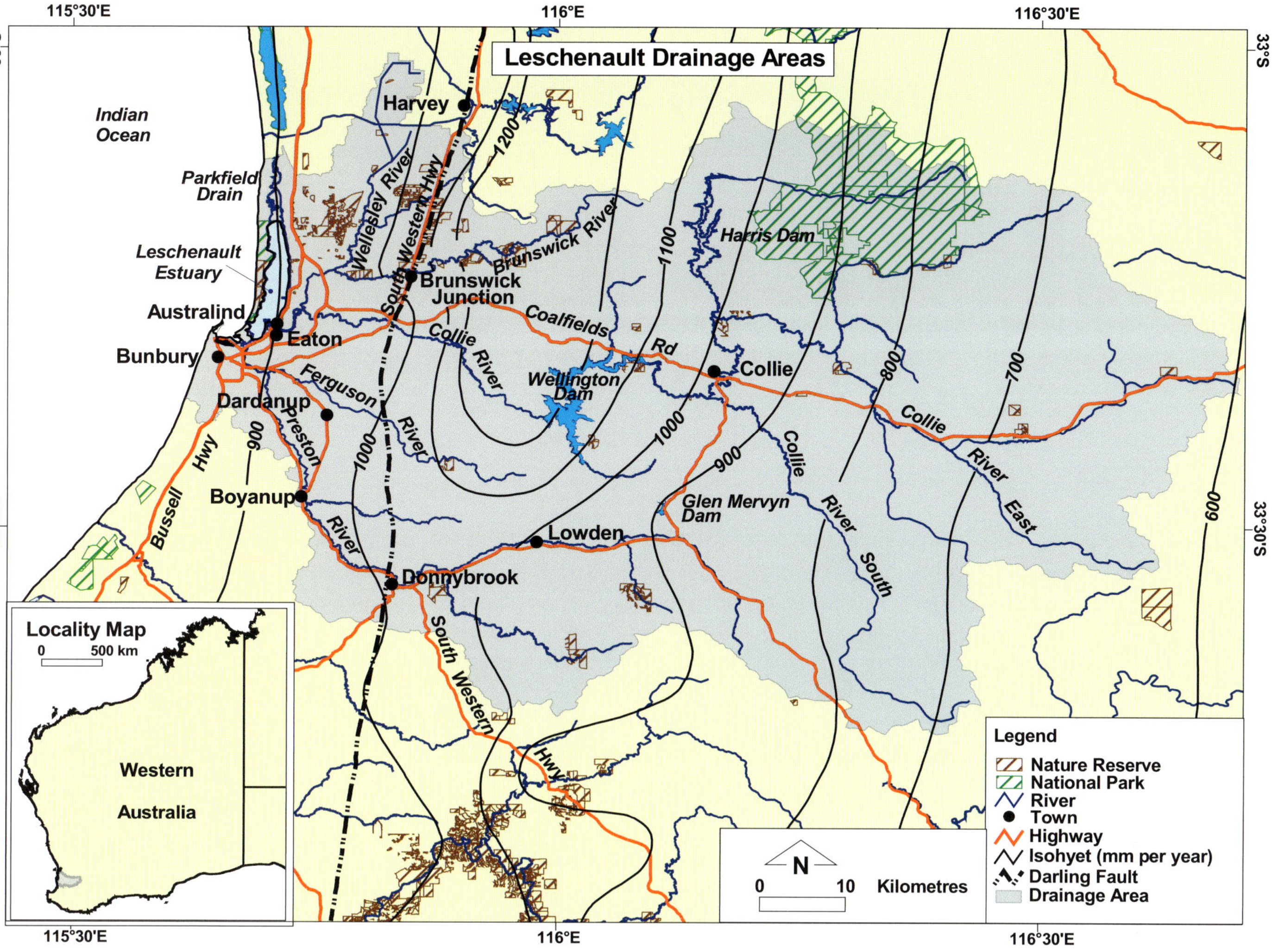

MAP 4.5 *The Leschenault catchment: the Preston River is dammed by Glen Mervyn Dam and the Collie River by Wellington Dam and by the Harris Dam on a tributary.* Drafted by DoE.

The lowest estuarine reaches of the Collie River are brackish during the winter with river run-off and more saline during summer (Graph 4.2). Salty water brought by tides intrudes for about 25 kilometres along the bottom producing a marked stratification. Dissolved oxygen concentrations are high in the surface waters, while the bottom waters often become anoxic when the salinity stratification is fully developed in summer and autumn. Nutrient concentrations in the estuarine reaches are higher than in the river reaches. Nitrogen concentrations are moderate and do not appear to have increased between 1995 and 1999. Phosphorus concentrations may have decreased slightly. Phytoplankton, dominated by diatoms, is more abundant in the estuarine part of the Collie River than in the main body of the estuary. A large bloom of diatom *Skeletonema* in January 1998 contained 368,246 cells per millilitre. Blooms of dinoflagellates often follow diatom blooms but blue-green algae are generally in low abundance.

Approximately half of the 603 square kilometres of the Preston River's catchment has been cleared, mostly along the valley, for grazing, orchards, potatoes and other vegetables, and dairying. Summer flow is controlled by Glen Mervyn Dam that releases water for irrigation. Nutrients in the upper reaches of the Preston River and its tributary the Ferguson River are generally low, although there may be periodic influxes of organic matter. Phytoplankton density is generally low in the upper reaches. Blue-green algae formed a bloom near Lowden in December 1997. In 1972 large numbers of fish, shrimps and marron died along a 5-kilometre stretch of the Preston River downstream from Boyanup, 40 kilometres to the north-east of Busselton. The deaths were possibly due to heat stress or pollution from a butter factory.

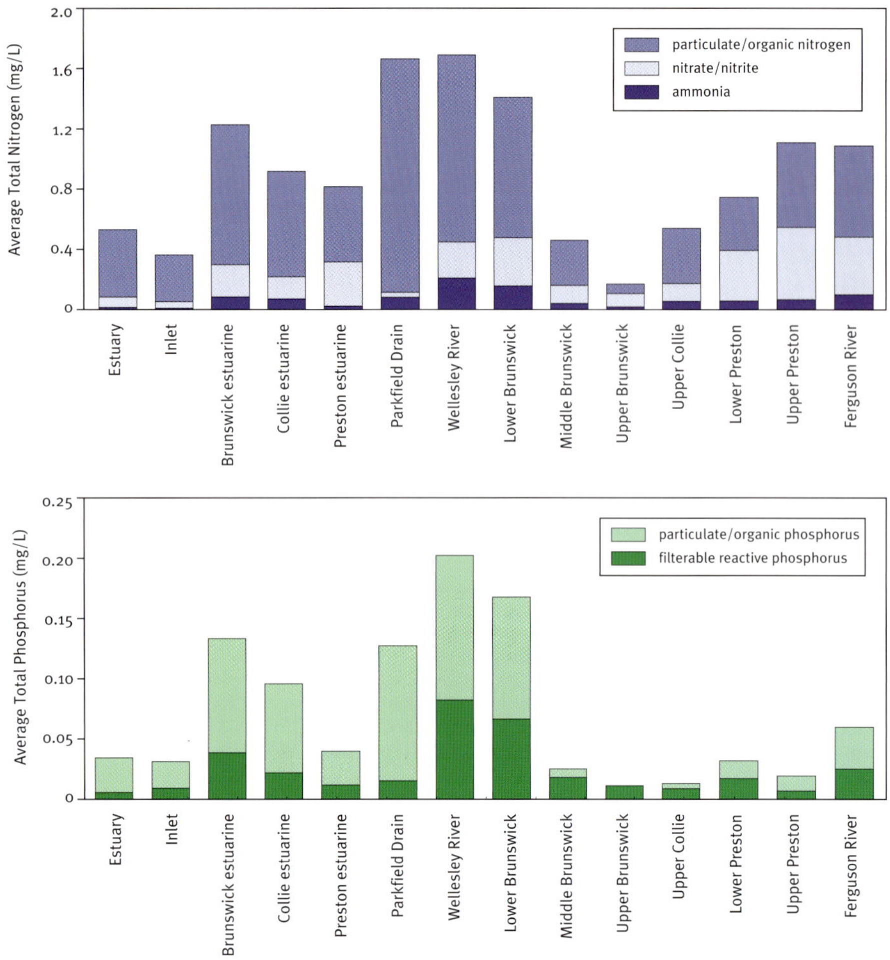

GRAPH 4.1 *Average nutrient concentrations in the Leschenault Estuary and catchment 1995–1999. ANZECC 2000 Guideline values for total nitrogen are 0.75 mg/L for estuarine and 1.2 mg/L for lowland rivers, and total phosphorus 0.03 mg/L for estuarine and 0.065 mg/L for lowland rivers. Parkfield Drain, Wellesley River, the estuarine reaches of the Collie River, and lower and estuarine reaches of the Brunswick River had the highest phosphorus and nitrogen concentrations. In the upper reaches high organic and sediment loads are due to manures and sediment washed into the waterways together with dissolved fertilisers. In the lower and estuarine areas, the high organic load is due to plant growth in summer taking up the dissolved nutrients from the upper catchment.* WRC 2001a.

Water in the lower Preston River (about 3 kilometres) is brackish in winter and more saline in summer. It does not appear to stratify, and oxygen concentrations are usually adequate for water life. Water quality does not appear to be declining but the amount of phytoplankton has increased since 1997. There have been occasional diatom blooms with a large bloom of 35,290 cells per millilitre of the marine diatom *Skeletonema* in autumn 1999.

Erosion is a problem in the Preston River catchment. Loads of suspended solids are high at times, particularly in winter (155 milligrams per litre), and the delta at the estuary mouth has grown because of erosion in the catchment.

Pesticides

Pesticide residues in the Preston River catchment have been a concern. Until the 1980s when their use was restricted, organochlorines such as aldrin, heptachlor, dieldrin, chlordane and DDT (note aldrin oxidises to dieldrin) were used to control African black beetle in potatoes. Chemically stable and not broken down by micro-organisms, enzymes, heat or ultraviolet light, organochlorines persist in soils, aquatic environments, plant and animal tissue. More recently, organophosphorus pesticides with equal or higher toxicity but with a shorter residual life have been used. However, while organophosphorus pesticides are less persistent in the environment they are more difficult to detect. Organochlorine concentrations were lower in 1985–1986 compared to 1980–1981, but reasons for the decrease are not clear. The highest concentrations occurred upstream of Donnybrook where they are still extensively used, and decreased in downstream areas where horticulturists had converted to less persistent organophosphorus pesticides. More current information about pesticide residues is not available.

TABLE 4.3 *Water quality in the Leschenault Estuary and rivers flowing into the estuary.*

Area	Total nitrogen			Total phosphorus		
	1995–97	1996–98	1997–99	1995–97	1996–98	1997–99
Estuary	low	low	low	low	low	low
Inlet	low	low		low	low	
Parkfield Drain	high	high	high	moderate	moderate	moderate
Wellesley River	moderate	moderate	moderate	high	high	high
Brunswick Estuarine	high	high	high	high	high	high
Lower Brunswick			high			high
Middle Brunswick	very low	very low	very low	very low	very low	very low
Upper Brunswick	very low	very low	very low	very low	very low	very low
Collie Estuarine	moderate	moderate	moderate	moderate	low	low
Upper Collie	very low	very low	very low	very low	very low	very low
Preston Estuarine	moderate	moderate	moderate	low	low	low
Lower Preston	very low	very low	very low	very low	very low	very low
Upper Preston		very low	very low		very low	very low
Ferguson River	low	low	low	low	low	low

Highlighted columns show areas where nutrients are higher than acceptable. Source: WRC, 2001a.

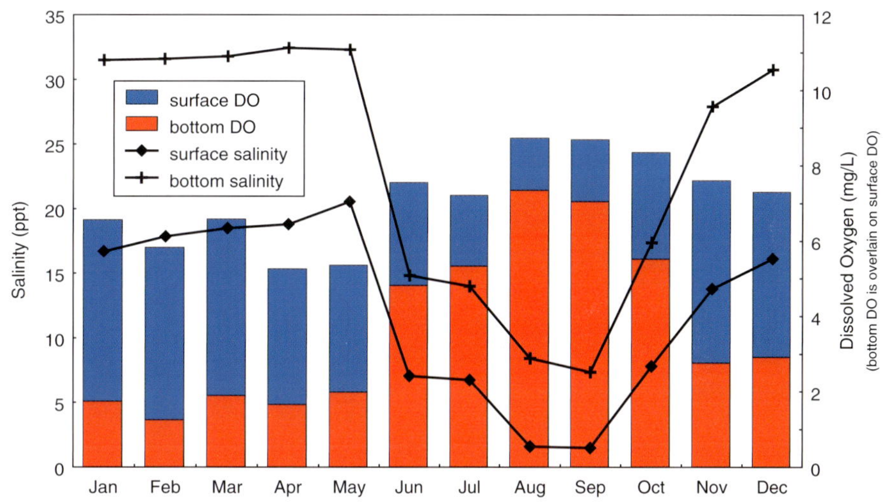

GRAPH 4.2 *Seasonal pattern of salinity and dissolved oxygen in the estuarine reaches of the Collie River.* WRC 2001a.

Waters of the estuary

Although the marine influence through The Cut is substantial, salinity still changes seasonally and with distance from The Cut. In winter, rainfall and freshwater run-off from the rivers and drains reduces salinity. In summer, the northern section becomes hypersaline due to the limited tidal influence and the concentration of salts with evaporation. Water temperatures also fluctuate daily and seasonally, generally following air temperatures. The temperature can rise to 25°C in summer and fall to about 14°C in winter.

Water quality in the Leschenault Estuary is generally good, and the estuary is well mixed and oxygenated. Although nutrient concentrations are slightly higher than in a natural state, they are considered acceptable for a developed catchment and do not appear to be rising. Many of the nutrients are flushed to sea through The Cut, and others are taken up by phytoplankton, macroalgae and seagrasses.

Not surprisingly, sediments in Leschenault Inlet are similar to those of the Peel–Harvey and the Swan–Canning. However surface sediments collected between 1988 and 1990 from the Leschenault contained more fine material than samples from the Peel–Harvey. They also contained more organic matter than the Peel Inlet, comparable with parts of the eutrophic Harvey Inlet. Sediments in the Leschenault have a slightly higher calcium carbonate content derived from shells and algae, probably due to the marine influence. Total phosphorus concentrations in the Leschenault sediments were slightly higher than in the Peel and slightly lower than in the Harvey. Organic phosphorus (19–31 per cent of the total phosphorus) was slightly higher than in the Swan–Canning and similar to proportions in the Peel–Harvey. Proportions of biologically inactive calcium phosphate, apatite phosphorus (37 per cent), were higher in the Leschenault compared to the other estuaries. Biologically active non-apatite phosphorus (36 per cent) was similar to the proportions in the Peel but lower than the Harvey and Swan–Canning. The potential for release of sediment-stored phosphate in the Leschenault appears to be relatively low, and problems from nutrient enrichment are perhaps a little less than in the Peel–Harvey. Phosphorus release from sediments, however, varies with environmental conditions that are characteristic of natural environments such as estuaries. The potential for higher concentrations of water column phosphorus either cycled from the sediments or from the river catchment remains a concern.

Although little is known about current phytoplankton and macrophyte populations, there have not been any large problem blooms since 1995. In summer and autumn, the marine diatoms *Cyclotella* and *Skeletonema* dominate the phytoplankton community. In much of the estuary blue-green algae have been present in very low densities (< 400 cells per millilitre) in autumn but these are not a health threat. The higher numbers in the northern part of the estuary near the Parkfield Drain, where circulation is restricted, are however of concern.

Around the edges

Leschenault Estuary lies between the Quindalup Dunes to the west and the older Spearwood and Bassendean Dunes to the east. To the west, the mobile, bare and vegetated dunes on the Leschenault Peninsula create a mosaic of vegetation types (Figs 4.5a and 4.6). In places, the migrating dunes have exposed earlier land surfaces and forest, or are engulfing vegetation in a gradual eastward migration. At the inlet, the fringing marshland gradually merges with the sandy shallow platforms that surround the deeper basin area.

Plant communities around the Leschenault Inlet are similar to those found around the estuaries of the Swan and Peel–Harvey. Five major categories of vegetation have been identified: salt marsh; fringing vegetation; fringing estuarine forest; freshwater; and sandy rise vegetation that can be further subdivided into twenty-eight plant communities. These vegetation patterns depend on variations in salinity, topography, tidal flushing and freshwater flushing.

SALT MARSH VEGETATION Salt marsh communities develop in saline areas that are inundated regularly with tidal water or where water is trapped behind a natural levee after periods of high water levels. Salt marsh plants are highly productive. They reduce the effect of water movement and create habitat for many animals. The salt marsh community forms a buffer zone between the main water body and the land. Aquatic invertebrates that remain in the salt marsh on the receding tide provide food for birds and fish.

The salt marsh community is dominated by herbs such as the samphire *Sarcocornia quinqueflora*, and sea rush *Juncus kraussii*, along with seablite *Suaeda australis*, sea heath *Frankenia pauciflora*, streaked arrowgrass *Triglochin*

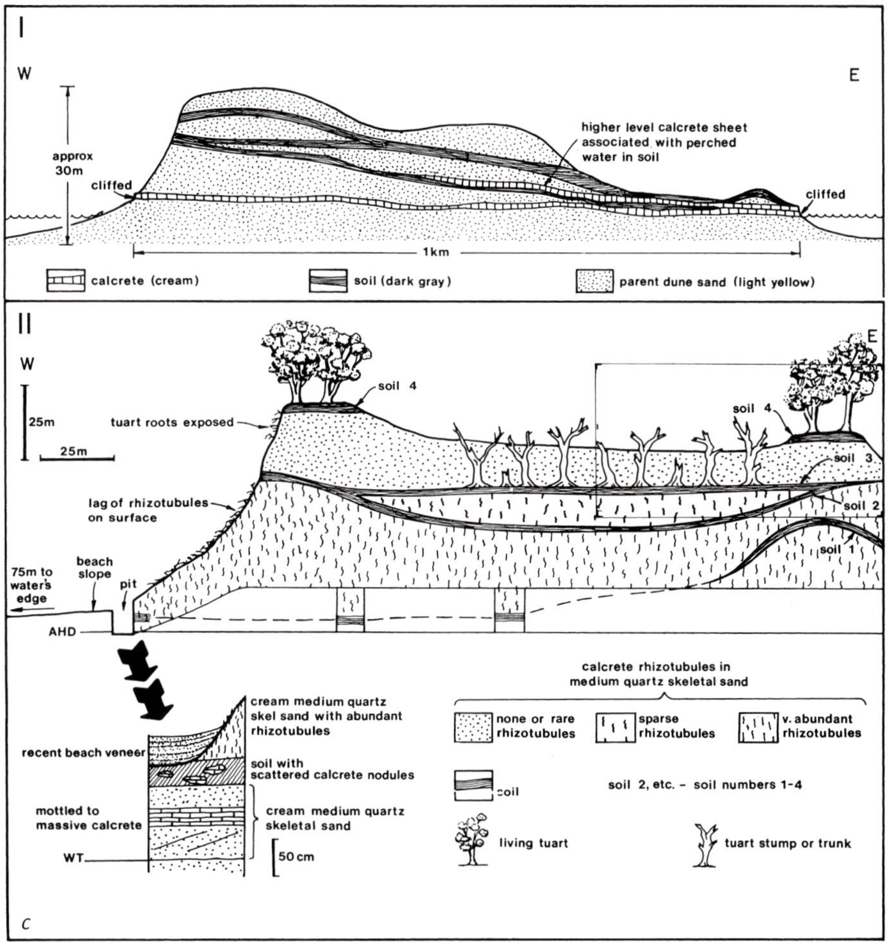

FIG. 4.6 *Landforms of the Leschenault Peninsula: (a) tuart trees exposed by wind erosion and (b) coastal erosion following cyclone Alby in April 1978. (c) I Diagrammatic relationship of dune sands, soils and calcrete. II Details of a transect in the south-west section of the Peninsula. Note the fossil tuart forest that was buried by migrating dunes. See also Figs 4.1 and 4.5a.* Source: Semeniuk & Meagher, 1981. Courtesy V. Semeniuk.

striata, creeping brookweed *Samolus repens*, shrubby samphire *Halosarcia halocnemoides* and the coastal sword-sedge *Gahnia trifida*. The species change in abundance and dominance with small changes in conditions and the history of disturbance. In highly disturbed areas, such as along the eastern shore of the Leschenault Inlet, introduced grasses such as couch *Cynodon dactylon*, kikuyu *Pennisetum clandestinum*, and pigface *Carpobrotus edulis* are invading the community.

FRINGING VEGETATION Fringing vegetation consists of species that live more or less permanently in shallow water. Typically it is dominated by the lake club-rush *Schoenoplectus validus* that forms a closed sedgeland up to 2 metres in height near the north-eastern end of the estuary.

FRINGING ESTUARINE FOREST Fringing estuarine forest is found where the ground water levels are high and salinity is moderate. The main species are small swamp sheoak *Casuarina obesa*, saltwater paperbark *Melaleuca cuticularis*, paperbark *Melaleuca viminea*, and swamp paperbark *Melaleuca rhaphiophylla*. They are found intermingled with samphires, herbs and sedges of the salt marsh and fringing communities. The white mangrove *Avicennia marina* also forms large stands of fringing forest in the area of Leschenault Inlet that was once part of the narrow channel connecting the estuary to the ocean, on the north-eastern estuary foreshore and on the Preston River delta (Fig. 4.7). The population of mangroves in the Leschenault is particularly interesting as it is the most southern occurrence on the west coast; the nearest population is at the Abrolhos. (See **Box 4.1 Mangroves**.)

FRESHWATER VEGETATION As the name suggests, freshwater vegetation is found close to the saline estuary but where freshwater inflow is higher, either from run-off or from ground water. It is typically found near the base of a ridge or dune. The twig rush *Baumea juncea* grows on the landward side of the sea rush *Juncus kraussii*. Only remnants of the community remain in the Leschenault Inlet as it was easily cleared for pastures.

Swamp paperbark *M. rhaphiophylla* and the peppermint *Agonis flexuosa* form a low open forest between areas prone to waterlogging and higher ground in regions near outflows of fresh water. Areas of fringing forest were once more extensive in the north-east, and today remnants of the former forest create the tree-lined arbor of Cathedral Avenue. Understorey species vary with topography. Estuarine rushes and sedges in low areas merge into drier land species typical of higher ground. Freshwater weeds such bridal creeper *Asparagus asparagoides*, soursob *Oxalis pes-caprae*, kikuyu *Pennisetum clandstinum*, irus, arum lily *Zantedeschia aethiopica* and buffalo grass *Stenotaphrum secundatum* are invading. Along the north-western edge of the estuary, woodland plants including wattle *Acacia cochlearis* and cutleaf hibbertia *Hibbertia cuneiformis* are more common in the understorey.

SANDY RISE VEGETATION Sandy rise vegetation is found where drainage is better on the ridges around the estuary and on the margins of coastal dunes. The stinkwood *Jacksonia furcellata* dominates in some areas with an understorey of sedges such as the common sword-sedge *Lepidosperma longitudinale* and knotted club rush *Isolepis nodosa* and woodland plants.

FIG. 4.7 *Mangroves on the Leschenault Peninsula near Belvidere. The air breathing roots, or pneumatophores, project above the mud and water on the estuary side of the tree. In the background the tree line along the Mandurah–Eaton Ridge slopes down to farmland along the margins of the estuary. This northern end of the estuary is a favourite haunt of Black Swans that feed on swan grass* Ruppia megacarpa *growing in the shallow water.* A. Brearley 2001.

Box 4.1 **MANGROVES**

The term mangrove is used to describe trees and shrubs that grow in the intertidal coastal areas. They are most common in tropical and subtropical areas. One species, the white mangrove *Avicennia marina*, is found throughout the Indo-Pacific and also in southern Australia in Western Port Bay near Melbourne, in St Vincent and Spencer Gulfs in South Australia and in the Leschenault Inlet. In Western Australia, the nearest populations today are found at the Abrolhos, along the mainland coast in Shark Bay at Carnarvon and further north. Mangroves in the Leschenault may be remnants of a much larger population from a time when the climate was warmer and estuarine habitats were more extensive, or recent arrivals. Evidence from fossil pollen studies indicates that mangroves were not present along the south-west coast in the early Holocene. The population in the Leschenault may have originated from seedlings brought southwards by a stronger Leeuwin Current that developed as climate became more arid.

Mangroves are adapted to cope with salt and waterlogging. The leaves of *A. marina* have epidermal cells that secrete salt. In fact, although mangroves tolerate high salt they do not require it for optimum growth. The tree has a large shallow root system, which supports upright air-filled roots or pneumatophores that reduce the effects of waterlogging. Other roots growing down penetrate the mud to collect nutrients (Fig. 4.8). The breeding strategy of many mangroves is also adapted to the watery habitat. After fertilisation the seedling develops on the parent tree until mature, when it falls to sediment and takes root, a feature termed vivipary. A seedling of *A. marina* does not develop on the parent plant to the same extent as some other mangroves, and resembles a large bean when shed.

Mangroves harbour rich invertebrate communities that encrust, bore and burrow into the trunks and roots of the trees (Fig. 4.9). Other animals live among the roots, where they feed on the decaying leaves and other material trapped by the root mat. Mangroves are nursery areas of juvenile fish, which are protected among the roots and feed on the rich plant and animal growth. The roots also slow water movement, bind the sediment and reduce erosion.

Sources: Hutchings & Saenger, 1987; Lear & Turner 1977; Morrissy, 1972; Semeniuk et al., 1978.

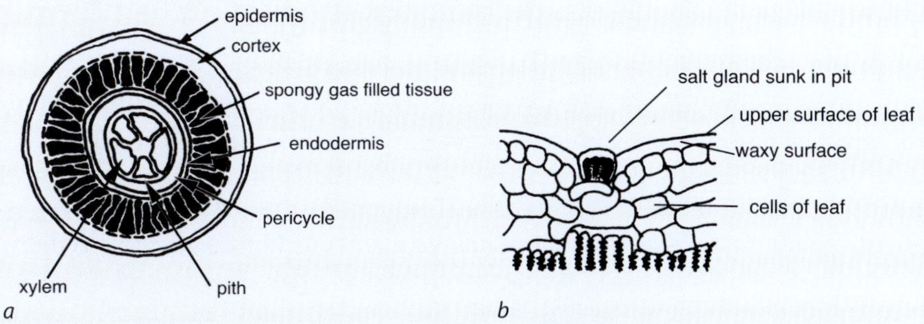

FIG. 4.8 *(a) Simplified cross-section of a pneumatophore of* Avicennia*: when covered by water, gas pressure in the spongy tissue falls as oxygen is used in the lateral roots. When the tide falls, exposing the pneumatophores, air is sucked into the root system. (b) Cross-section of a leaf showing the salt gland.*

Adapted from Lear & Turner, 1977.

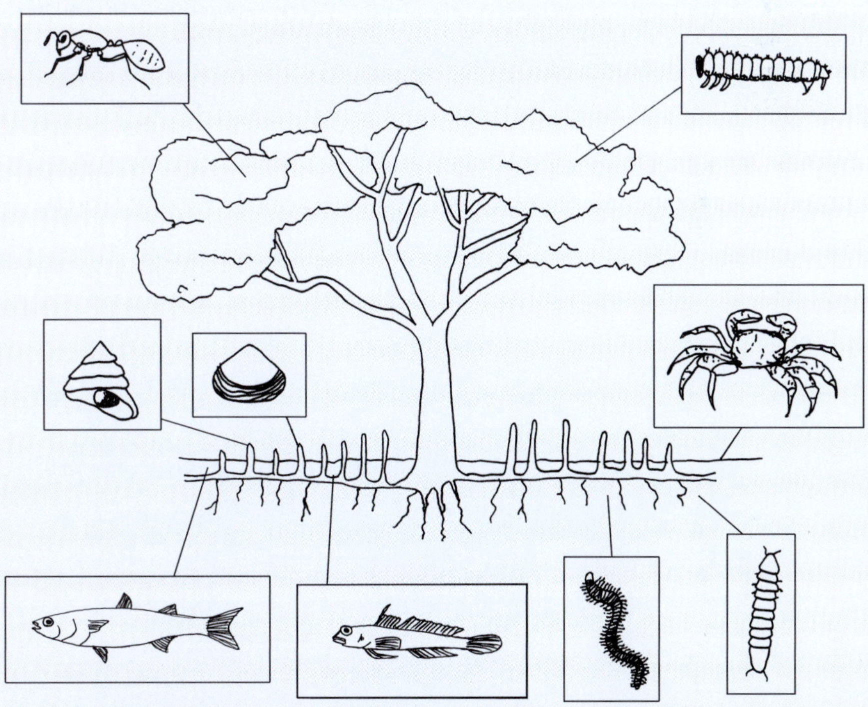

FIG. 4.9 *Mangrove community: mangrove trees create a rich and varied habitat for aquatic and terrestrial life. Mangroves can produce about 1 kg of leaf material per year for every square metre of mangroves. This plant debris is gradually broken down by a diverse community of microbes and animals forming an important part of the food web around the trees and in the wider estuarine environment.*

Courtesy G. Coupland, UWA.

On high ground near Pelican Point the flooded gum *Eucalyptus rudis* and swamp paperbark form a woodland. On the top of ridges, particularly where limestone outcrops, the tuart *Eucalyptus gomphocephala* woodland dominates. The tuarts around the Leschenault Inlet are representative of what was once a major vegetation unit along the coast, which has now become more restricted with urban and agricultural development. Weeds are common in disturbed areas, particularly the golden wreath wattle *Acacia saligna* in areas disturbed by fire.

Changes in fringing vegetation

Vegetation surrounding the inlet has changed since the area was first settled, when the land was cleared and pastures established. In the 1940s about 700 hectares remained, but clearing and reclamation over the next 50 years reduced the area to about 350 hectares in the early 1990s. The largest losses were to the north of Buffalo Road and in the south around the harbour. The fringing vegetation on the Leschenault Peninsula, while only narrow in extent, remains largely intact, however only a narrow strip of fringing vegetation remains along the eastern margin of the estuary. In the 1990s and 2001, the largest remaining area of fringing marsh was at Point Douro on the northern side of the Collie River delta facing into Samphire Bay. The fate of this area is under review; however development costs to meet environmental criteria may be very high due to high acid sulfate soil concentrations (Map 4.6).

Trees in the fringing forest, particularly saltwater paperbark *M. cuticularis* and swamp paperbark *M. rhaphiophylla*, have also died. Reasons for the decline are not obvious. Old aerial photographs indicate that tree deaths in the north-west area predate The Cut and subsequent increase in salinity and changes to the watertable. On the north-eastern shore, the change from slow year-round seepage to extreme and patchy salt or freshwater conditions through drainage probably accounts for losses of trees and shore rushes, and their replacement by samphire and bulrush *Typha*, species that are tolerant of salt and strong flushing.

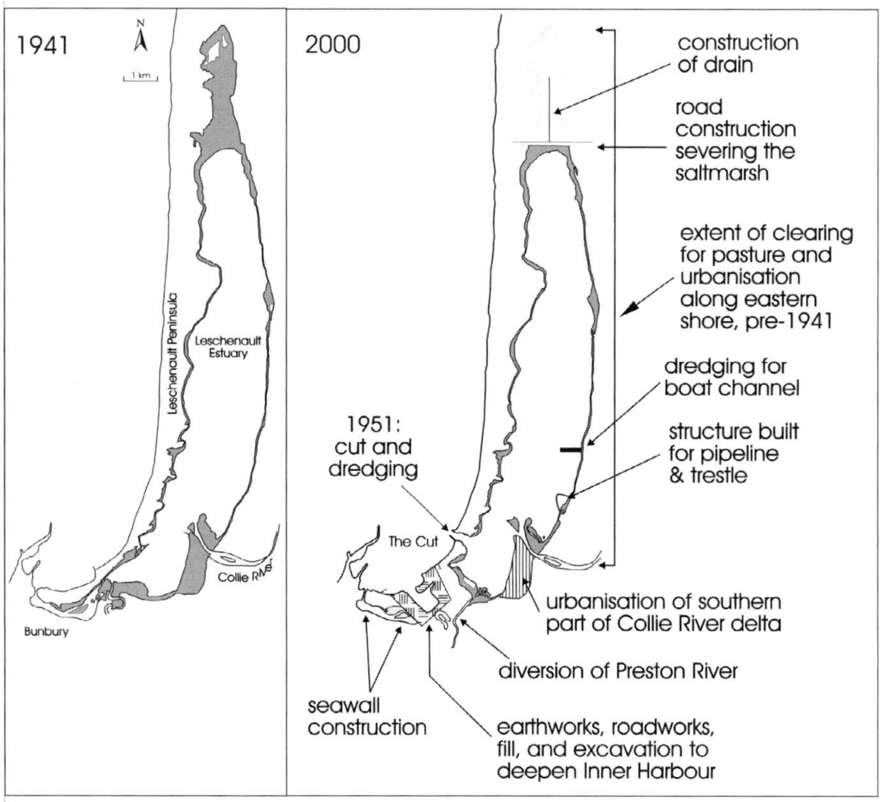

MAP 4.6 *Extent of fringing vegetation along the Leschenault Estuary and the later Leschenault Inlet in 1941 and 2000. The shaded area to the south of the Collie Delta has since been urbanised with housing and canal developments.*
Source: Pen, 1992a; Pen et al., 2000; Updated by V. Semeniuk.

Changes in the fringing community have been most dramatic in the Pelican Point area south of the Collie River delta where the paperbark community has been replaced by salt marsh. This was probably due to changes in drainage associated with urban development that have increased salinity. There has been some regeneration of paperbarks in this area and around Crimp Avenue.

Areas of remnant vegetation are also threatened by disturbances caused by clearing, fire or vehicles tracks. In disturbed areas, natural communities often fail to regenerate as the weed species often grow more quickly than native species.

The margins of estuaries are always dynamic with changes on land and in the water body affecting the fringing community. Growth of the fringing communities can also be seen in aerial photographs and along the shorelines in some areas. In many places around the estuary, the sea rush *Juncus kraussii* has encroached the samphire and areas of sand by 5–20 metres in the last 50 years. This growth is probably favoured by the decrease in winter water levels, as flood waters no longer rise as high as they once did but move through to the sea much faster than before The Cut. Samphires and shore rush have also colonised river deposits at Bar Island, off the mouth of the Collie River, the river delta around the Preston River, and mudflats at Point Mornington where a low open stand of *Casuarina* is developing on higher ground. Fringing vegetation communities are also changing along the southern part of the Leschenault Peninsula, where tidal lagoons formed by migrating dune sands are colonised by fringing vegetation.

Many water plants, but more marine

The aquatic flora in Leschenault Estuary is similar to the flora of other south-western Australian estuaries except that there is a larger diversity of red algae, an abundance of the brown alga *Hormophysa triquetra* and more seagrasses (Table 4.4).

Surveys of macrophytes conducted between 1984 and 1993 recorded total biomass of algae varied from 1,261 tonnes in May 1988 to 5,643 tonnes in October 1990. Maximum biomass occurred in spring, unlike the Peel–Harvey that typically reached maximum biomass in late summer–early autumn. On an areal basis, biomass in the Leschenault 68–258 (mean 144) grams dry weight per square metre was very similar to amounts in the much larger eutrophic Peel–Harvey 62–261 (mean 161) grams dry weight per square metre.

The major difference between the Peel–Harvey and Leschenault Estuaries was in the types of macrophyte contributing to the biomass. In the Peel–Harvey most of the biomass was due to green algae, whereas in the Leschenault brown algae and seagrasses accounted for most of the macrophyte biomass (Table 4.5). This high proportion of seagrass not only reflected the more 'marine' nature of the inlet, but suggested that the overall water quality was good compared to nutrient enriched systems like the Peel–Harvey.

Calculations based on concentrations of nutrient in algal tissue from the Leschenault Inlet and growth studies conducted in the Peel–Harvey indicate

TABLE 4.4 *Macrophytes in the Leschenault Inlet.* Adapted from Lukatelich, 1989. Adapted from Hillman et al., 2000.

Group	Species
Seagrasses	Paddle weed *Halophila ovalis*, swan grass *Ruppia megacarpa*, eelgrasses *Heterozostera* sp., *Zostera muelleri*
Charophyte	*Lamprothamnium papulosum*
Green algae	*Chaetomorpha linum*, *Enteromorpha* sp., *Cladophora* sp., *Caulerpa* sp.
Brown algae	*Hormophysa triquetra*, *Dictyota paniculata*
Red algae	*Hypnea episcopalis*, *Gracilaria* spp., *Chondria* sp., *Laurencia* spp., *Sypridium filamentosa*, *Ceramium* sp.

Macrophyte	Leschenault	Peel–Harvey
Green algae	11–43%	> 85%
Brown algae	14–49%	< 0.5%
Red algae	20–30% (spring), < 10% other times	seldom more than 10%
Seagrass	> 30%	< 15%

TABLE 4.5 *Macrophytes accounting for most of the biomass in Leschenault and Peel–Harvey Estuaries in 1984–1993.* Source: Hillman et al., 2000.

that algal growth in the Leschenault was not limited by either nitrogen or phosphorus. This is supported by the fact that the highest biomass did not correspond with the period of high nutrients. Growth of algae, particularly green algae in the Leschenault, may be limited by light; a similar situation to that found in turbid waters of the Harvey Inlet.

In the Leschenault Inlet the greatest biomass occurred in the very shallow (< 0.5 metre) and saline northern area where exchange with the ocean was limited. In this northern area plant biomass was dominated by the brown alga *Hormophysa triquetra*, the green alga *Chaetomorpha linum* and the charophyte *Lamprothamnium papulosum*. Water quality information shows that nutrient inflows from the Parkfield Drain are high. Therefore the presence of algae known to grow rapidly in response to nutrient enrichment is a concern.

The southern, essentially marine area of the Leschenault was dominated by seagrass (paddle weed) *Halophila ovalis*. Biomass, as in other estuarine systems, varied greatly through the year with the largest biomass in late summer–early autumn. *Halophila* was widely distributed throughout the estuary except in the deeper central basin, and most dense south of Waterloo Head on the sandy bank along the eastern margin. *Ruppia megacarpa* and *Heterozostera* sp. were generally confined to shallower sections and *Zostera muelleri* was found only around the entrance channel to the ocean. Many of the red algae were epiphytes growing on larger algae and seagrasses. Red algae, particularly *Gracilaria*, formed a significant part of the macrophyte biomass in spring. (See **Chapter 2 The Swan–Canning Estuary: Living in the estuary, Seagrasses** and **Box 8.3 This thing called *Ruppia*.**)

Invertebrates

Animals living in the Leschenault Inlet are typical estuarine species that tolerate changing conditions, the most obvious being the blue manna crabs *Portunus pelagicus*, western king prawn *Penaeus latisulcatus*, the school prawn *Penaeus dalli* and the shrimp *Palaemonetes australis*. The abundant but less obvious small bivalve and gastropod molluscs, crustaceans and polychaetes provide food for larger crabs, fish and birds that congregate in the productive and protected areas.

Records of invertebrates included thirty-one species of shelly molluscs and one nudibranch although only seven were common; twenty-one species of small crustaceans including two introduced species; fifteen species of polychaete bristle worms; and one species of brittle star or ophuroid (Table 4.6).

The small invertebrate animals living in the estuary fluctuate greatly in abundance throughout the estuary, and from year to year depending on conditions. Some species abundant for a short time may be present only rarely, while others are always present. Although tolerating the changeable estuarine conditions, the invertebrates do however show distinct habitat preferences related to salinity, water depth, type of sediment and aquatic vegetation of algae and seagrasses (Map 4.7). The tiny bivalve *Arthritica semen*, the polychaetes *Ceratonereis*, *Capitella*, *Nephtys* and *Scoloplos* and the crustaceans *Corophium*, *Melita* and *Tanais* were particularly abundant and

TABLE 4.6 *Common invertebrates in Leschenault Estuary.* Source: Semeniuk et al., 2000.

Group	Most abundant species
Molluscs	*Arthritica semen* *Tellina deltiodalis* *Nassarius burchardi* *Spisula trigonella* *Hydrococcus brazieri* *Acteocina* sp. *Bedeva paivae*
Crustaceans	*Tanais* *Corophium* sp.
Polychaetes	*Ceratonereis aequisetis* *Nephtys gravieri* *Capitella cf capitata*
Ophiuroid	*Amphipholis squamata*

MAP 4.7 *Distribution of the more common benthic invertebrates in the Leschenault Estuary.* Source: Semeniuk et al., 2000; Courtesy V. Semeniuk.

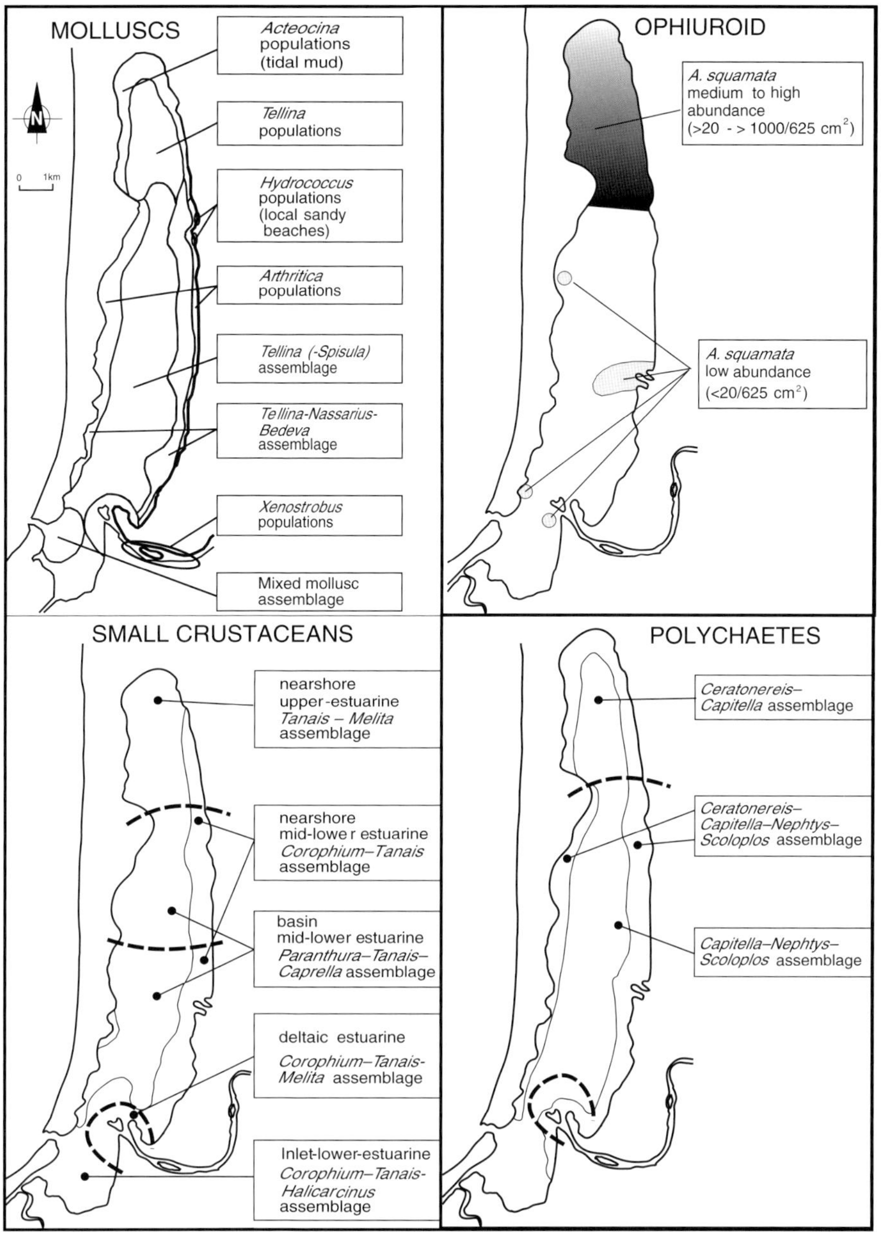

diverse on the sandy margins among seagrasses. Larger bivalves such as *Tellina* and *Spisula* were found more commonly in the deeper basin. The ophiuroid *Amphipholis squamata* is a species generally found in the ocean. In the estuary its abundance fluctuates, but it has generally been found in the middle to northern area often associated with seagrass or algae.

The rich bounty—fish and fisheries

Fish populations vary from year to year depending on conditions in the estuary and in the nearby marine areas. In the 1990s forty-two species were recorded (Table 4.7). Twenty were marine species using the estuary as a nursery (marine–estuarine opportunists), although collectively the group only accounted for 31 per cent of the fish in the shallow estuary area. The sandy sprat *Hyperlophus vittatus*, the mullets *Aldrichetta forsteri* and *Mugil cephalus*, and the King George whiting *Sillaginodes punctata* were among the more abundant fish. In nearby Koombana Bay the marine estuarine–opportunist group accounted for 45 per cent of the fish, demonstrating the importance of nearshore bay areas as alternative nursery areas to the estuary.

Life category	Estuary number	%	Bay number	%
Species				
Solely marine	0	0	6	17
Marine straggler (S)	9	21	5	15
Marine estuarine–opportunist (O)	20	48	19	56
Estuarine and marine (E&M)	7	17	4	12
Estuarine (E)	6	14	0	
TOTAL	42		34	
Individuals				
Solely marine	0	0	425	28
Marine straggler (S)	187	1	272	18
Marine estuarine–opportunist (O)	5,684	31	693	45
Estuarine and marine (E&M)	9,653	53	137	9
Estuarine (E)	2,745	15	0	0
TOTAL	18,269		1,527	

TABLE 4.7 *Life cycle strategy of fish species in the near shore shallow waters of Leschenault Estuary and Koombana Bay, February and December, 1994.* Source: Potter et al., 2000.

Thirteen species classed as estuarine or estuarine and marine accounted for 68 per cent of the total number of fish. These species live entirely in the estuarine waters, the most common being the long finned goby *Favonigobius lateralis*, the hardyheads and gobbleguts (Table 4.8). The abundance of estuarine species in the Leschenault Estuary is far greater than in the larger Swan River and Peel–Harvey but less than those in Wilson Inlet and Nornalup–Walpole on the south coast. Their dominance is probably due to the high and stable salinities over summer when spawning occurs. Only nine species classified as marine stragglers were recorded in the estuary basin despite the open connection to the sea. This was much less than in other west coast estuaries.

Only four species (cobbler *Cnidoglanis macrocephalus*, Australian anchovy *Engraulis australis*, black bream *Acanthopagrus butcheri* and yellow-tail trumpeter *Amniataba caudavittata*) complete their life cycle in the estuary, representing less than 3 per cent of the total fish caught during the survey.

The deeper basin of the Leschenault was dominated by larger fish, marine–estuarine opportunists such as yellow-eye mullet *Aldrichetta forsteri*, Perth herring *Nematalosa vlaminghi*, tailor *Pomatomus saltatrix*, yellowfin whiting *Sillago schomburgkii*, and Australian herring *Arripis georgianus*. The lower reaches of the Collie River were dominated by Perth herring (69 per cent), the second most common species in the deep estuary basin and by sea mullet *Mugil cephalus* (20 per cent). The dominance of these species is not unexpected as Perth herring, classed as semi-anadromous, breed in the low salinity upper reaches of the Swan River that is similar to the Collie. Similarly juvenile sea mullet are also known to migrate into lower salinity water. (See **Box 2.15 Perth's herring *Nematalosa vlaminghi*.**)

Depending on the productive estuary

The majority of fish in the Leschenault estuary depend on the abundant benthic fauna for food (Table 4.9). These in turn depend on the rich flora of microscopic and larger algae, seagrasses and dead plant material or detritus.

The cobbler, roach, yellow-eye mullet and banded toadfish eat large numbers of bivalve molluscs. Polychaetes worms are an important part of the diet of King George whiting, yellowfin or sand whiting, Australian herring and yellow-eye mullet. Small amphipods and isopod crustaceans are eaten by cobbler, yellow-tail trumpeter, yellow-eye mullet and banded toadfish; and larger crustacea, shrimp *Palaemonetes* and snapping shrimps *Alpheus* are eaten by trumpeter, yellowfin whiting, tailor and mulloway. Tailor and mulloway also feed on small fish such as southern anchovy, while yellow-eye mullet

TABLE 4.8 *Life cycle characteristics of fish in Koombana Bay and Leschenault Inlet compared with the Deep Basin of Leschenault and the lower Collie River. (S, marine straggler; O, marine–estuarine opportunist; E&M, estuarine and marine; E, estuarine; A, semi-anadromous). Note methods and time periods of catching fish in the estuary and bay (sampled with seine nets Feb.–Dec. 1994) were different to those in the deeper estuarine basin and lower reaches of the Collie River (gill nets sampled Sept.–July 1994) and may account for some apparent discrepancies of data, e.g. fish not sampled in estuary but present in deep estuary basin.* Source: Potter et al., 2000.

Common name	Species name	Life cycle	Percentage of catch per 100 m^2			
			Estuary	Bay	Deep Estuary Basin	Collie River
Long finned goby	*Favonigobius lateralis*	E&M	34	8		
Sandy sprat	*Hyperlophus vittatus*	O	23	1		
Swan River hardyhead	*Leptatherina presbyteroides*	E&M	17	< 1		
Hardyhead	*Atherinosoma elongata*	E	10			
Yellow-eye mullet	*Aldrichetta forsteri*	O	3	17	27	3
Gobbleguts	*Apogon rueppellii*	E&M	2			
King George whiting	*Sillaginodes punctata*	O	2		1	
Blue spotted goby	*Pseudogobius olorum*	E	2			
Perth herring	*Nematalosa vlaminghi*	A			18	69
Wallaces hardyhead	*Leptatherina wallacei*	E	2			
Large mouthed goby	*Afurcagobius suppositus*	E	1			
Sea mullet	*Mugil cephalus*	O	1	< 1	3	20
Soldier fish	*Gymnaspistes marmoratus*	O	1	1		
Common blowfish (toadfish)	*Torquigener pleurogramma*	O	< 1	< 1	1	
Prickly toadfish (puffer)	*Contusus brevicaudus*	O	< 1	15	< 1	< 1
School whiting	*Sillago bassensis*	S	< 1	15		
Sea trumpeter	*Pelsartia humeralis*	SM		2		
Flathead sandfish	*Lesueurina platycephala*	SM		25		
Tailor	*Pomatomus saltatrix*	O	< 1	< 1	12	4
Yellowfin whiting	*Sillago schomburgkii*	O	< 1	2	8	< 1
Australian herring	*Arripis georgianus*	O	< 1	3	6	
Silver trevally	*Pseudocaranx dentex*	O			5	
Striped (six-lined) trumpeter	*Pelates sexlineatus*	O	< 1	1	4	
WA Salmon	*Arripis truttaceus*	O		< 1	3	
Eagle ray	*Myliobatis australis*	O			3	
Cobbler/Estuarine catfish	*Cnidoglanis macrocephalus*	E&M	< 1	< 1	2	
Southern anchovy	*Engraulis australis*	E&M	< 1	< 1	1	< 1
Mulloway	*Argyrosomus japonicus*	O			1	2
Tarwhine	*Rhabdosargus sarba*	O	< 1	< 1	1	
Southern sea garfish	*Hyporhamphus melanochir*	O	< 1	< 1	1	
Roach	*Gerres subfasciatus*	O	< 1	< 1	< 1	1
Black bream	*Acanthopagrus butcheri*	E				1
Yellow-tail trumpeter	*Amniataba caudavittata*	E	< 1			< 1

also fed on midge larvae. In contrast, sea mullet and Perth herring mainly feed on organic detritus.

In the Leschenault, the blue manna feeds on a wide variety of small benthic fauna such as amphipod crustacea, polychaete worms, bivalve molluscs and small fish such as gobies. (See **Box 4.2 Crabbing—a summer pastime**).

Like other estuaries, the Leschenault environment is a nursery area where the high productivity of invertebrates provides rich feeding grounds for young fish, and the extensive beds of the seagrass *Halophila* provide protective habitat without the problems associated with large banks of algae. In addition, the estuarine environment is not suitable for many larger predatory fish.

The catch

The fishing industry of the Leschenault developed as the town of Bunbury grew. Originally much of the catch was used by the local residents, the problems of freight costs and access to market being similar to those of

Box 4.2 CRABBING—A SUMMER PASTIME

The blue manna crab *Portunus pelagicus*, or blue swimmer, lives throughout the Indian and Pacific Oceans. Although found in waters up to 40 metres deep, the blue manna is more common in shallow sheltered bays and estuaries near seagrass, where rich communities of small animals provide food. In south-western Australia, crabs are plentiful in areas such as Cockburn Sound, Geographe Bay and Peel–Harvey and Leschenault Inlets where they have been caught for generations (Fig. 4.10).

Blue mannas are continually on the move. Juvenile and adult crabs move into the estuaries when the waters are salty, peaking in abundance over summer. They move downstream and out to sea again when river flow begins and salinity drops. Females also move into the more marine downstream areas or into the sea to spawn.

In the estuaries, mating generally occurs in midsummer when the crabs congregate in deep water. After pairing, the male carries the female for four to ten days while he waits for her to moult. When she is still soft shelled after moulting, the male turns her over and deposits packets of sperm into the area of the abdomen flap. He then continues to carry the female for another three to four days until her new shell hardens. Males may mate with more than one female over a season. Most large females mate only once a year, although they may spawn up to three times during a season using stored sperm. The sperm are stored over winter until the eggs develop the following summer when they are fertilised. The young females reach maturity at about one year of age when some will mate.

Females are more active after moulting and feed to build up food reserves. The ovaries develop as the water temperature rises in spring. When ready to produce eggs, she settles into the sand with her abdomen extended, the eggs are extruded and attach to hairs on the abdomen, sperm are released and fertilisation takes place externally. Depending on her size, a 'berried' female may carry 180,000 to 2 million eggs. The newly spawned bright orange eggs gradually turn grey as they develop and use up the yolk. The rate of development depends on water temperature. It ranges from about eight days at 25°C to eighteen days at 20°C.

Females head to sea in 'spawning runs' only a few hours before the eggs are due to be released, usually at night in late summer. When spawning, the female raises her body, extends her abdomen, jerks her legs to break up the egg mass and combs the eggs out. Females may move back into the estuary or stay in the ocean between spawning events (Fig. 4.11).

In the ocean, the larvae moult about fifteen times growing larger and more crab-like with each stage. The first stage or 'zoeal' larvae appear to feed at the surface on tiny plankton in the morning. They may be carried offshore by the winds, sinking lower in the water before midday when the sea breeze starts blowing. The zoea have been found 80 kilometres out to sea after eight days, at which time they are in the second stage of development. At the third or fourth zoeal stage, they appear to start feeding near the surface in the afternoon and are transported back to the coast by the sea breeze. A round trip takes about sixteen days.

Next the larvae enter the 'megalopa' stage that lasts seven to eight days, when they live on the sea floor, and have been found 20 kilometres out to sea. The next eight stages are all crab-like, each lasting eight to fourteen days. During this time the animal increases from the size of a match head to about the size of a 10-cent coin. At this stage in mid to late summer some of the juvenile crabs enter estuaries.

The young crabs feed and grow rapidly, moulting as often as every two weeks depending on temperature.

FIG. 4.10 *Blue swimmer crab.* Courtesy C. Bryce, WAM.

Each moult brings an increase of 75–80 per cent in weight and about 20 per cent in width. They usually reach 76 millimetres just before winter, when growth slows. Some spend winter buried in the mud of the estuary.

The moult of males in October–November and females in December–January is probably linked to the mating cycle. At the end of summer, just over a year old, the crabs are about 100–115 millimetres across the carapace, although some may reach the minimum legal size of 127 millimetres. Some of these crabs may overwinter in the estuary but many appear to leave, accounting for the increases in commercial catches in Koombana Bay, Warnbro and Cockburn Sounds in winter.

At moulting the skin secretes a substance that dissolves the connection between skin and the old shell. The new shell forms below the old shell, which then begins to split. The crab then takes up water to fill out the new shell—accounting for the 'mushy' or poor eating state of newly moulted crabs. The old shell is eaten and incorporated into the gradually hardening new shell.

Crab numbers fluctuate from year to year due to the many different conditions they are exposed to during the long voyage of their larval life. The winds, currents, salinity and temperature directly affect the crabs and the food they feed on. Crab numbers are also affected by predators such as birds and fish, including stingrays, fiddler and gummy sharks, but the largest, most efficient predator is people.

There are usually equal numbers of males and females in the population, but catches are often biased towards one sex. Males are often more abundant early in the season when temperatures are cooler. Because males moult earlier than females, they are probably hungry and more easily caught in baited traps. Females, on the other hand, are generally less active before they moult and mate in midsummer. Later in the season, when more males have been caught, females dominate the catch.

Fishing pressure on crab populations in the estuaries is high, but recruitment from the sea is usually sufficient to maintain stocks. The ongoing success of our crabbing depends on a healthy offshore marine environment where the young crabs develop, and good conditions in the estuaries that provide the rich feeding grounds where they grow to catchable size. In the Peel–Harvey high crab numbers have been linked to an abundance of macroalgae, but crabs were reported to leave the system in spring during large blooms of the blue-green algae *Nodularia* and catches were generally poor the following summer.

Sources: Davis, 1988; Potter et al., 1983; Potter & de Lestang, 2000.

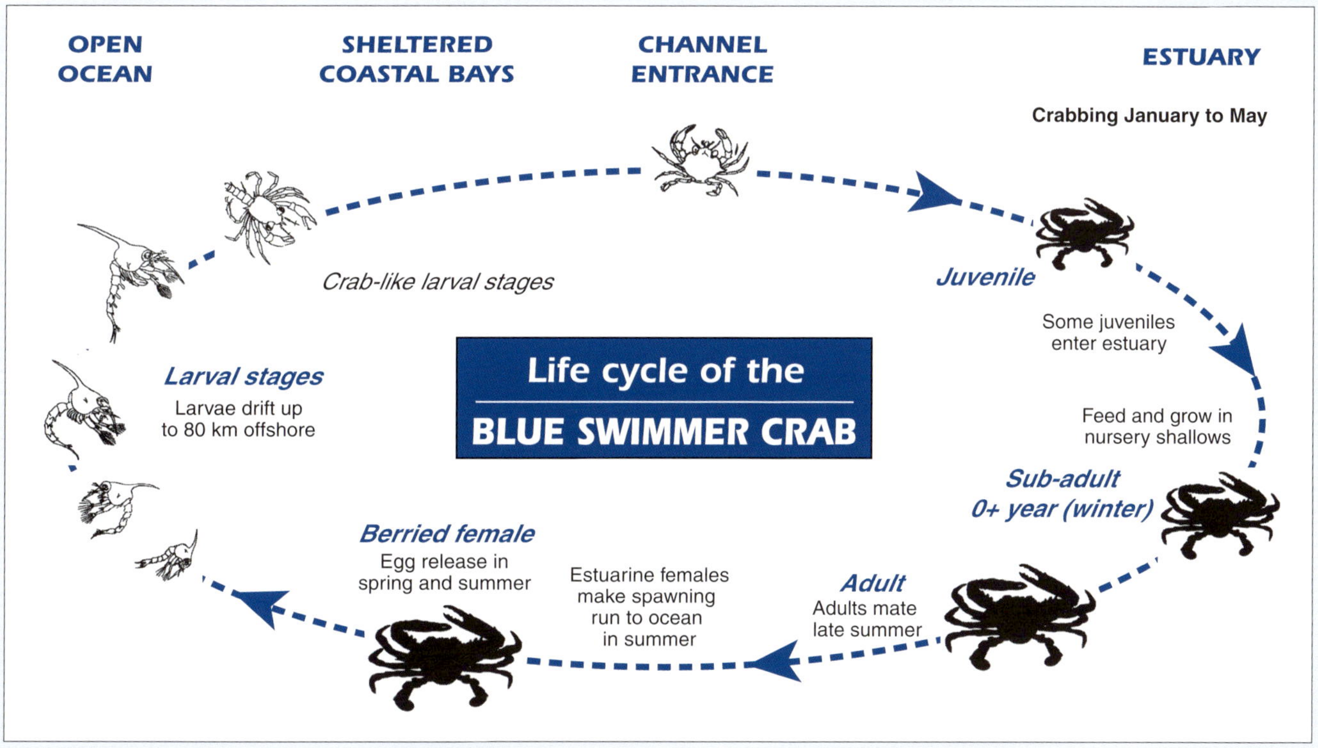

FIG. 4.11 *Life cycle of the blue manna (swimmer) crab. Female crabs spawn in the ocean in spring and summer. The first zoeal larval stages are spent at sea, the young crab-like stages entering the estuaries in summer. Young crabs are of catchable size at the end of the following summer.* Courtesy DoF.

other estuarine fisheries in the South-West. In 1979 there were eighteen commercial licensees operating in the productive waters, but by 1997 only six licences were still current as licences were not reallocated when a fisher ceased to operate. Commercial fishing ceased in the Leschenault in 2000. As the fishing effort declined the fish catch fell from about 150 tonnes in 1981–1985 to ca 100 tonnes in 1986–1992, ca 85 tonnes in 1993–1998 and 60 tonnes in 1999. In 1999 the catch was valued at $140,000. The catch for effort had, however, been relatively stable and fish populations healthy. Haul nets rather than gill nets were more commonly used in later years to reduce the bycatch and lessen interference with other estuary users. (See also Tables 1.16, 2.11 and 2.12.)

Commercial fishers caught yellow-eye mullet, sea mullet, yellowfin whiting (western sand whiting), King George whiting, cobbler, sea garfish and to a lesser extent tailor, Perth herring, black bream, skipjack trevally and Australian herring together with the blue manna crab, western king and river prawns (Graph 4.3). Yellow-eye mullet dominated the catch followed by sea mullet, whiting and Perth herring. Changes in the mullet catch reflected the decline in demand for bait fish (yellow-eye) and the increased demand for sea mullet as a table fish. The catch of King George whiting declined probably due to disposal of dredge spoil in nursery areas north of the estuary mouth. In contrast, the low numbers of cobbler probably reflect a trend in other estuaries where there has been strong fishing pressure with fish caught before the first breeding season. (See also **Box 8.4 Cobbler: spines to fear—great to eat**.)

Recreational fishing predominantly targets the three species of whiting, tailor, herring, juvenile salmon, skipjack trevally, and cobbler. A survey of recreational shore- and boat-based fishing from January to December 1998 estimated there were 33,392 boat-based fishing days with 86 per cent of the activity targeting blue swimmer crabs, and a further 5,610 shore-based fishers with 88 per cent targeting crabs. The total catch of 219,000 crabs was 45.7 tonnes (94 per cent of the total catch) or ten times the annual commercial catch over the previous five years. Over 70 per cent of the crabbers from the shore and in boats were from the nearby urban areas of Bunbury. During the survey period recreational fishers also caught 1,500 whiting other than King George whiting, 1,200 (0.24 tonnes) tailor mostly in summer, 800 (0.11 tonnes) Australian herring mostly in autumn, and 500 (0.06 tonnes) King George whiting mostly in spring.

The increasing pressure from the recreational fishing lobby brought the end to commercial fishing in the Leschenault in 2000. Ongoing success of the fishing resource in the Leschenault will depend on adequate management of recreational fishers both within the inlet and in the coastal waters nearby,

TABLE 4.9 *Mode of feeding and composition of the diet of the main fish species in the Leschenault Estuary.* Source: Potter et al., 2000.

Mode of feeding	Species	Composition of the diet
Herbivores	garfish, striped (six-lined) trumpeter	green and brown algae
Detritivores	sea mullet, Perth herring	bottom sediments, detritus
Omnivores	yellow-eye mullet, yellow-tail trumpeter	algae, bivalve molluscs, isopods, decapods, e.g. *Palaemonetes australis*; insect larvae, e.g. non-biting midge; bristle or polychaete worms
Lower-order carnivores	long finned goby, hardyheads, King George whiting, roach, common blowfish (banded toadfish), yellowfin whiting, cobbler	small crustaceans, e.g. amphipods; large crustaceans, e.g. *Palaemonetes australis*; snapping shrimp *Alpheus* spp., bivalve molluscs, polychaetes
Higher-order carnivores	mulloway, tailor	juveniles of other fish species, smaller fish species, e.g. southern anchovy; larger crustaceans, e.g. *Palaemonetes australis*

as target species are generally marine opportunists that visit the estuary when conditions are favourable. Maintenance of a healthy estuarine system will be crucial as conditions in the estuary also affect the visiting fish.

Over the top and in the shallow water—the birds

Over sixty species of wetland birds have been recorded on the open waters and in fringing tidal salt marsh of Leschenault Estuary and Inlet. Surrounded on the west by the Leschenault Peninsula National Park, this is the main wetland in the Bunbury region with up to 5,000 birds present at a time. Breeding colonies of egrets, cormorants, darters, spoonbills and herons commute to the estuary from nearby swamps; swans and ducks from the surrounding countryside seek refuge during the summer drought; and trans-equatorial wading birds escape the northern winter on the shores around the river mouths. Wader numbers are not as extensive as in the Peel–Harvey to the north, but include eighteen species listed in the Japan–Australia and China–Australia Migratory Bird Agreements. Freshwater seeps are a particularly valuable resource for water birds, but are increasingly under threat as the fringing marsh is cleared.

GRAPH 4.3 *(a) Commercial catch in the Leschenault Estuary basin 1981–1997; (b) main components of the finfish catch in the Leschenault Estuary 1981–1997.*
Source: Potter et al., 2000.

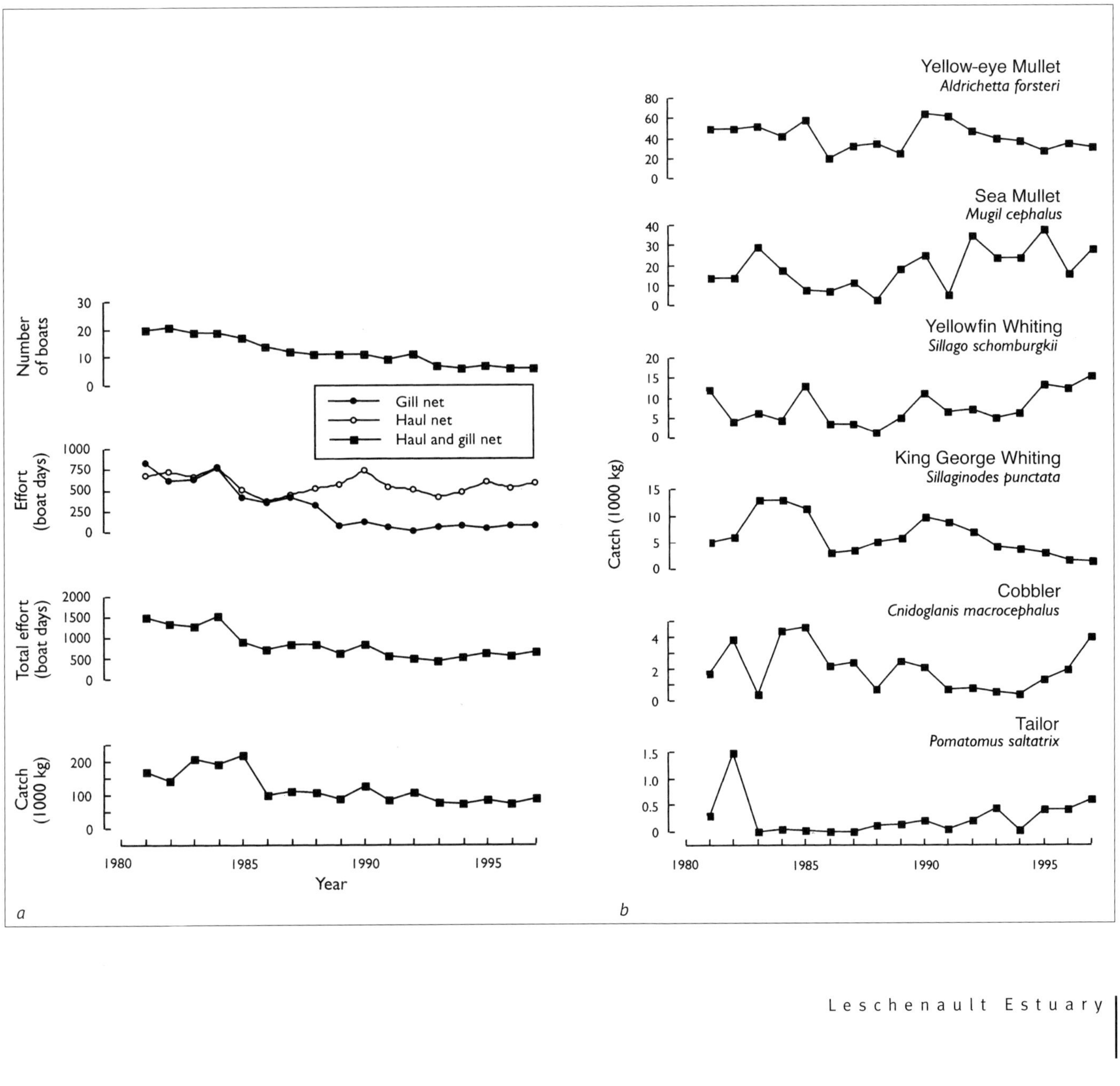

TINY ESTUARIES, INLETS AND LAGOONS OF THE GEOGRAPHE BAY AREA

The lagoons and wetlands of the Vasse–Wonnerup System north of Busselton and the Broadwater System, including Toby Inlet in the south, near Dunsborough (Map 4.8) are very new features of the landscape. They lie just behind the most recently formed dunes and in winter form extensive wetlands that are highly modified. A system of drains now channels water directly to the sea, bypassing some of the wetlands and inlets. Native vegetation has been cleared and replaced by pastures and urban housing. The soils are naturally poor with little capacity for retaining nutrients. As a result to maintain farm production fertiliser additions have been high, and nutrients, together with animal manure, leached to the waterways now fuel blooms of algae. Despite these impacts, these interdunal estuaries still retain estuarine features and vegetation that together create habitat for water birds.

MAP 4.8 *The original river systems were interconnected through a series of wetlands. Many still are but the channels have been highly modified and they now enter the sea through steep sided drains. Currently Toby Inlet receives water from a network of small rivers and wetlands. Originally other small rivers (Annie and Mary Brooks and Station Gully) as well as the Carbanup flowed to the sea.*
Drafted by DoE.

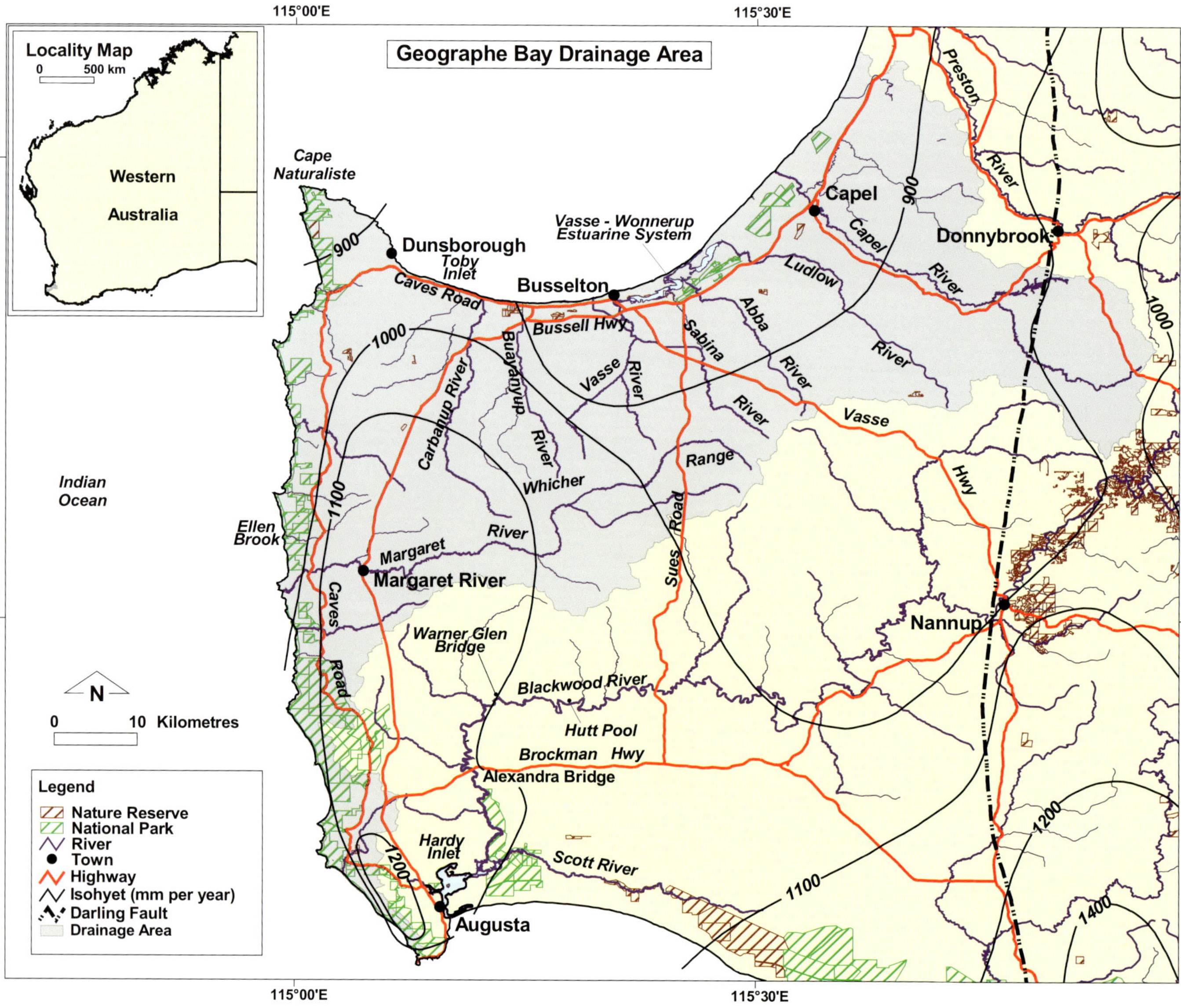

Vasse–Wonnerup wetland system

Settlement around the Vasse–Wonnerup occurred almost simultaneously with developments around the Peel–Harvey and Leschenault as settlers moved up and down the coast by land and boat. The whole area was attractive to Europeans seeking new farmland. There were sources of fresh water, fertile patches of soil, and the landscape was relatively flat unlike the area around Augusta where the country was hard to clear. The Bussells and Molloys from the failing settlement at Augusta were among the first to take up holdings around the Vasse, building homes *Cattle Chosen* and *Fair Lawn* on opposite sides of the river. The names make a statement about how the landscape was viewed. The Chapmans and Laymans settled near Wonnerup along the shores of the inlet. The description of the area around Wonnerup by Lieutenant Bunbury, however, illustrates other features of the landscape that were to prove a problem for the early settlers. The flat low-lying land was inundated with water during heavy rain (Bunbury & Morrell, 1930). The reasons for the drainage works that changed the landscape are the very factors that still plague the area today, when floodwaters invade the streets and houses that are only just above sea level.

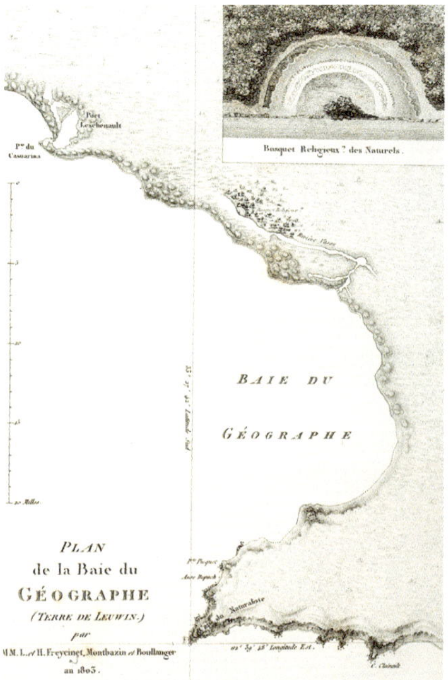

MAP 4.9 *Geographe Bay and Wonnerup Inlet drawn by Heirisson, a member of the Baudin expedition in 1801. The French spent several days exploring the area, describing the rivers, vegetation and wildlife and endeavouring to contact the local inhabitants. An Aboriginal settlement is marked on the map as* habitat de naturels. *At the time of discovery, the Aboriginal people had built a fish trap of stakes in the entrance. The top right illustration shows a bird's-eye view of a 'religious grove' found by Péron. A winter storm from the north-west hampered departure of the expedition and one of the party, Timothée Vasse, was drowned. Baudin later named the small bay at the entrance to Wonnerup Inlet Blunder Bay (Anse des Malaldroits).* Marchant, 1982. Map courtesy T. Burrows, Scholars' Centre, UWA.

> *I was sent to build barracks and form a township on a* presqu'isle *more fit for Dutchmen or frogs than British soldiers; where there were no settlers and no land to be taken up, and where in fact we were utterly useless as it was in the Governor's power to render us.*

Gradually settlement expanded and the town of Busselton grew from a whaling port visited by coastal traders to a major timber export centre, particularly after the completion of the kilometre-long jetty in the 1850s.

Place in the landscape

The Vasse and Wonnerup Estuaries are the largest part of the Vasse–Wonnerup wetland system that connects to the sea through the Wonnerup Inlet. The 9-kilometre long Vasse Estuary and the 5-kilometre long Wonnerup Estuary, together with the smaller Deadwater, cover about 900 hectares connected in winter through the Malbup Creek. In winter the Vasse–Wonnerup and associated wetlands spread out to cover about 1,500 hectares of low-lying land between the coastal dunes and the older Spearwood dunes. The opening to the sea at Wonnerup Inlet is shallow and often closed by a sandbar (Maps 4.9 and 4.10, Fig. 4.12).

Estuary to wetland

When the area was first settled by Europeans about 170 years ago the estuarine system, like others on the west coast, contained fresh water in winter. The bar opened when flow was strong and, as the outward river flow diminished, salty marine water entered with the tides establishing brackish conditions.

The Vasse Estuary received water directly from the Abba, Sabina and Vasse Rivers, and indirectly from Iron Stone Gully, Buayanyup River, Carbanup River and creeks to the west that flowed into the Broadwater–New River system to the Vasse. Wonnerup Estuary received inflow from the larger Capel and Ludlow Rivers.

Originally Wonnerup Inlet opened to the sea through two channels: the more western entrance, now covered by dune vegetation, closed sometime during the 1800s. Today only the eastern channel opens when water flow is high and conditions in the sea are favourable. More often the bar has been opened artificially to release floodwaters, a procedure repeated frequently throughout winter as seagrass, algae and sand clog the channel.

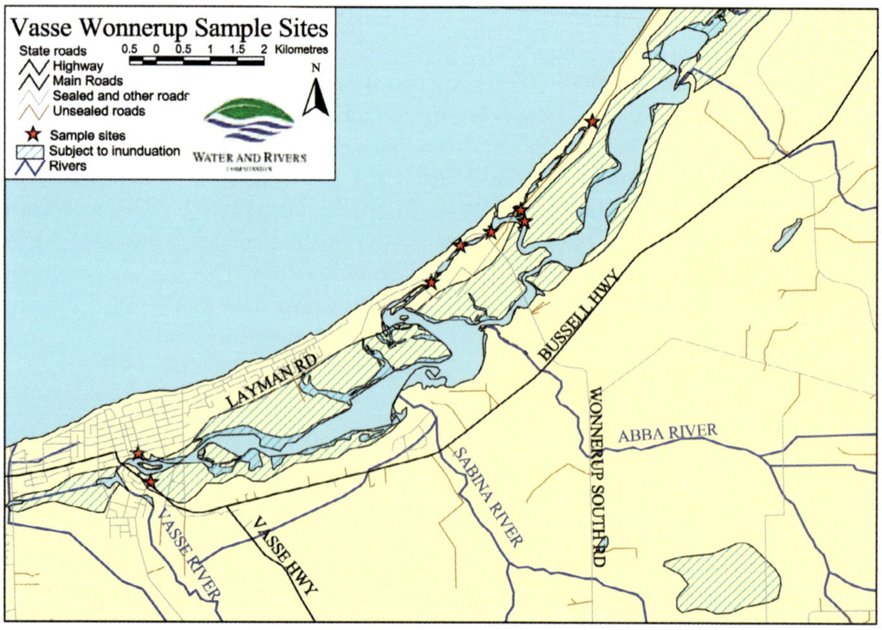

MAP 4.10 *Vasse and Wonnerup Inlets are fed by the Ludlow, Abba, Sabina and Vasse Rivers over about 1,500 ha of low-lying land to the east of Busselton between the coastal dunes and the older Spearwood dunes. Although the drainage system is now modified, water levels are controlled and the surrounding land has been extensively cleared for agriculture, the wetland system is still an important refuge for birdlife. A marina and breakwater called Port Geographe have been built in the middle section of the coast. The breakwaters interrupt longshore drift and trap seagrass, algae and sand. Sea water is now much closer to the wetlands and ground water, which are likely to become contaminated as the urban development expands.* Courtesy WRC.

FIG. 4.12 *The Vasse–Wonnerup System showing floodwaters across Ford Road, Busselton, aerial view, 30 June 1995.* Courtesy J. Lane, CALM.

In the early 1930s the channel was artificially closed at the end of winter to allow water to accumulate in the estuaries and prevent farmland from drying out too early in the summer. Closure of the bar stopped tidal exchange with the ocean and, coupled with reduced outward flow and intensive organic enrichment from farmland, water quality declined rapidly.

Drains

As settlement spread across the estuary flats and more of the low-lying land was occupied, winter floods that damaged crops and pastures became a problem: like the area around the Peel–Harvey pressure to drain the land grew. In the 1880s, the Capel River was diverted to the ocean via Higgins Cut, 5 kilometres to the north-east of Wonnerup Estuary. In 1908 floodgates were installed on the exit channels of both estuaries preventing the entry of sea water. In 1915 a cut was made to drain water from New River to the ocean, reducing its input to the Vasse.

Following World War I, drainage works started across the whole Busselton area. West of Busselton drains were dug to take water straight to the sea, and the 1920s brought extensive drainage networks in the Ludlow, Abba, Sabina, Vasse, Iron Stone Gully, Buayanyup and Carbanup River catchments. The drains increased the rate and volume of river flows and the need for further works to prevent flooding in the low-lying area farms and at Busselton. Following a particularly large flood in 1926 (1.14 metres Australian Height Datum (AHD)), a major drain was constructed to divert the headwaters of the Sabina and virtually all of the Vasse River's flow direct to the ocean. Soon afterwards Iron Stone Gully and the Buayanyup and Carbanup Rivers were also diverted to the sea.

Floodgates to control water levels and salinity

Floodgates were constructed on the Vasse and Wonnerup Estuaries in 1908. They were designed to open when the water level was higher on the upstream side releasing water downstream, and to close when the water level was higher downstream to prevent entry of sea water. The sill heights on both sets of gates were set at +0.01 metre AHD. This was the minimum level of discharge of the estuaries' water through the gates. Evaporation over summer however, frequently lowered water levels still further (Fig. 4.13).

The system changed as a result: the waters were now fresh or brackish at a lower level, and the estuary sediments were dry in summer. Although floods still occurred they were generally lower and of shorter duration; lesser threats to low-lying areas. The reduced water levels allowed an expansion of cropping and pastures onto areas that were previously too wet. The barring of sea water also allowed cropping on areas that would have been too saline. 'Wonderful results', but given the likelihood of changing conditions, the vagaries of seasons and malfunction of equipment could the bounty last?

GRAZING GROUNDS FOR CATTLE AND BIRDS By 1926 the floodgates installed in 1908 were in need of repair. They were no longer preventing the upstream movement of salt water that damaged the land, making it unsuitable for cattle. Landholders petitioned the Government not only to make repairs to reduce incursion of salt water, but also to lower the sills to enhance outflow when the sea was low. The floodgates were replaced in 1929 and the sills lowered 0.46 metre (–0.45 AHD). Initial response was favourable but it was found that the land was drying out too rapidly in summer. The next action was to close the sandbar at the end of spring to

FIG. 4.13 *Floodgates on the Vasse Lagoon, 1988, with a blue-green algal bloom and dead fish. The floodgates allow floodwaters to escape and prevent sea water flowing back into the low-lying agricultural land. The shallow, still, nutrient-rich water, however, promotes algal growth that reduces water quality, and the barrier limits the opportunity for fish to escape stressful conditions.* Courtesy K. McAlpine, DoE.

retain some water, a procedure that proved unsatisfactory. Another solution was needed.

Following years of debate, removable stop-gates were added to the floodgates in 1942. These could be added in four stages at the end of winter or in spring as the floodwaters subsided. In September 1944 all four boards were in place raising the level to 0.76 metre or +0.31 metre AHD as originally requested by most landholders. Although the timing of the placement and number of boards varied, boards were installed each year in September or October until the 1980s.

In the 1980s attempts were made to keep a higher and constant water level of 0.4 metre AHD for two months to create more favourable conditions for the Black Swan *Cygnus atratus*. When water levels were higher, nests were flooded. When they were lower, nests were exposed to predatory foxes. Maintaining a constant level required frequent removal and replacement of boards. This proved too time consuming and was discontinued after four years.

The original operation of the floodgates to prevent sea water intrusion was changed in 1988 following a fish kill. The gates were opened to allow dilution of the putrid waters and allow some fish to escape. Again the balance of interests was difficult to maintain as the clear salt water damaged the pastures and fringing freshwater vegetation. Since 1990 the gates have been opened when fish show signs of stress. The summer water level is maintained at –0.1 metre AHD. As a result water levels are usually higher in summer and autumn, and broad expanses of the estuary no longer dry out as often in summer. Although the water is sometimes saltier than the sea, most areas are flushed with fresh water in the following winter.

A changed world — grossly enriched

The Vasse–Wonnerup system is now highly modified. Most of the rivers and drains flow directly to the sea. Other parts hold floodwaters as part of a large 'compensation' basin to protect the town of Busselton. The trapping of fresh water has led to an increase of freshwater plants and animals. The waters are now eutrophic, nutrient rich due to livestock waste and fertilisers added to improve the poor soils.

The years of intensive agriculture have taken their toll. In 1987, a year of low rainfall, nutrient loads to Vasse by the Abba, Sabina and Vasse Rivers were estimated at 3.7 tonnes of phosphorus and 29.1 tonnes of nitrogen. In 1988, a high rainfall year, an estimated 19.6 tonnes of phosphorus and 148.4 tonnes of nitrogen entered the area, leading to assessment of the system as 'the most grossly enriched major wetland system known in Western Australia' (McAlpine et al., 1989).

Many of the nutrients flow to the sea with the winter floodwaters, although aquatic plants growing in and around the estuary take up large amounts. Some nutrients are retained by sediments and recycled to the water by chemical, physical and biological processes (Table 4.10).

In the 1990s the Vasse–Wonnerup entered a new phase with the development of a canal system for housing at Port Geographe. Sea water is now much closer to the freshwater wetland, as are gardens that require nutrient-rich soils. Nutrient enrichment of the canal areas together with algal blooms is likely given the porous soils and poor tidal flushing.

Green soup and dead fish

The rich green-coloured water in the lower Vasse River at Busselton is a 'fine example' of nutrient enrichment. Concentrations of nutrients are very high, different types of algae, including toxic blue-green species, are abundant and often accumulate in clumps (Graph 4.4). Although birds appear untroubled, permanent signs warn people against touching the waters.

TABLE 4.10 *Water quality in rivers feeding into the Vasse–Wonnerup estuarine system; much of the nitrogen is in the inorganic form, a state readily taken up by algal communities.* Source: WRC, 2001b.

System	Water quality	Total nitrogen (mg/L)	Total phosphorus (mg/L)	Salinity	Oxygen	Phytoplankton
Lower Vasse River	Poor — deteriorating	High to very high 1.7	High to very high 0.22	Fresh to brackish	High to very high but low when blooms collapse	Very high, persistent blooms 20 times ANZECC 1992 guidelines.
Vasse Estuary	Poor particularly where birds gather	Very high 1.5	Very high 0.15	Saline summer, brackish winter–spring	Variable: very low to high dependent on blooms	Blooms, summer and autumn of diatoms and chlorophytes but also blue-green algae. Permanent warning signs of danger.
Wonnerup Estuary	Poor — worse upstream	High 1.2	Moderate to high 0.1–0.2	Saline summer; brackish winter–spring	High but low where stratified near floodgates	Blooms usually diatoms, but blue-green *Nodularia* bloom in Jan. 2000 and filamentous green algae upstream.
Wonnerup Inlet	Acceptable	Low 0.49	Low to moderate 0.07	Saline — well flushed	Good	Dominated by a variety of diatoms. A marine species *Cyclotella* in Aug. 1999 reached 25,810 cells/mL. Highest blue-green counts 5–7,000 cells/mL in Jan. 2000.
Deadwater	Fair	Moderate to high 0.99	Low to moderate 0.06	Brackish winter, more saline in spring to hypersaline in summer	Adequate low to high with algal activity	Diatoms, some dinoflagellates and blue-green algae. Diatom bloom spread from inlet area in Aug. 1999, when there was a blue-green algal bloom of 50,000 cells/mL in northern area.

The quiet waters of the Vasse have collected nutrients from gardens, lawns, septic tanks, pastureland and factories for decades. The sediments alone contain enough nutrients to fuel blooms for years to come. Initiatives to improve water quality are expensive and time consuming. Generally they require removal or treatment of the sediment store, storm water management and revegetation to trap nutrients. Revegetation of streamlines also provides shade, which lowers light and temperature, creating less favourable conditions for algal blooms.

Further downstream in the Vasse Estuary, water quality is only a little better although the blooms are caused by a greater variety of algae. Toxic blue-green algae however pose a threat and, as fish are plentiful living off the abundant algae and small invertebrates, recreational fishers often ignore signs advising against fishing. Toxic algae, however, have generally not been the cause of the massive fish kills in the Vasse, which are usually due to low oxygen concentrations in warm weather. They are nevertheless a disturbing sight and are a warning of excessive nutrients and algal blooms. Although the dense algae produce oxygen by photosynthesis during the day, they consume oxygen through respiration at night so that many fish kills occur in the early morning when oxygen is low (Graph 4.5). To make matters worse, dying algal blooms also lower oxygen as they decay. In the areas around the floodgates, inflowing salt water can also form a layer along the bottom below the fresher water where it is isolated from the oxygenating effect of winds.

Water in the Wonnerup Estuary is generally poor, although it is slightly better near the floodgates due to tidal mixing. Planktonic algae are usually less abundant than in other areas and generally dominated by harmless diatoms. Blue-green algae have reached high densities at times, as in January 2000 when *Nodularia* was present throughout the estuary. Samples taken at

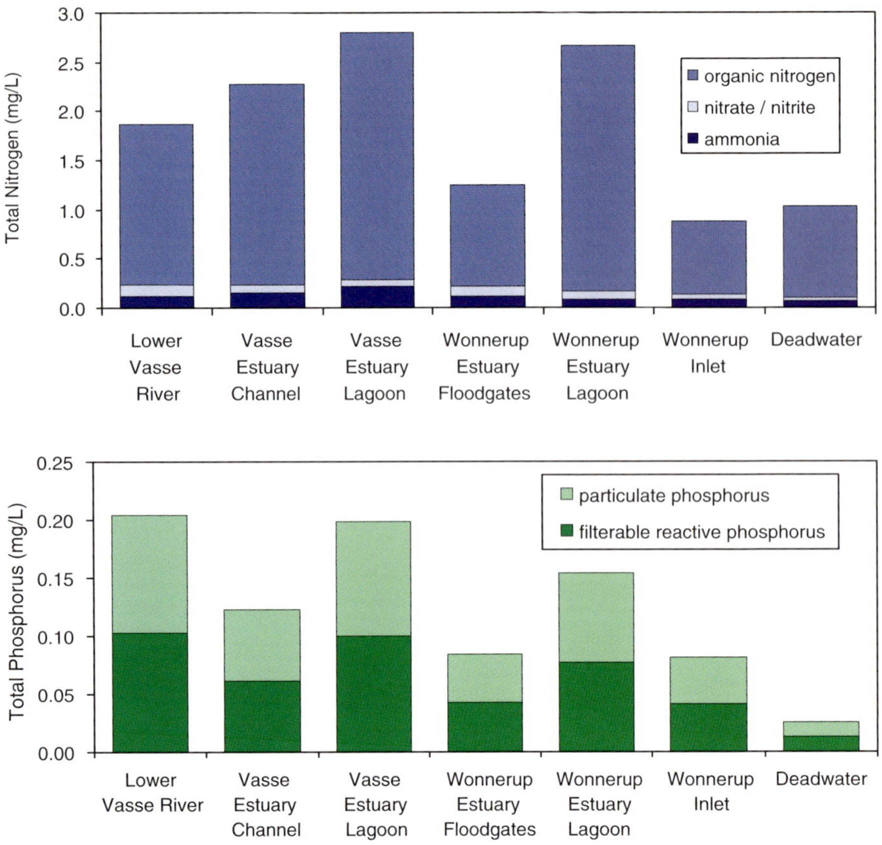

GRAPH 4.4 *Nutrients in the Vasse–Wonnerup Estuary: nutrient concentrations exceed ANZECC 2000 Guideline values for total nitrogen 0.75 mg/L for estuarine and 1.2 mg/L for lowland rivers, and total phosphorus 0.03 mg/L for estuarine and 0.065 mg/L for lowland rivers: the Vasse–Wonnerup system is considered eutrophic. There are similar amounts of particulate and dissolved phosphorus, and most of the nitrogen (organic nitrogen) is incorporated into plants.* WRC 2001b.

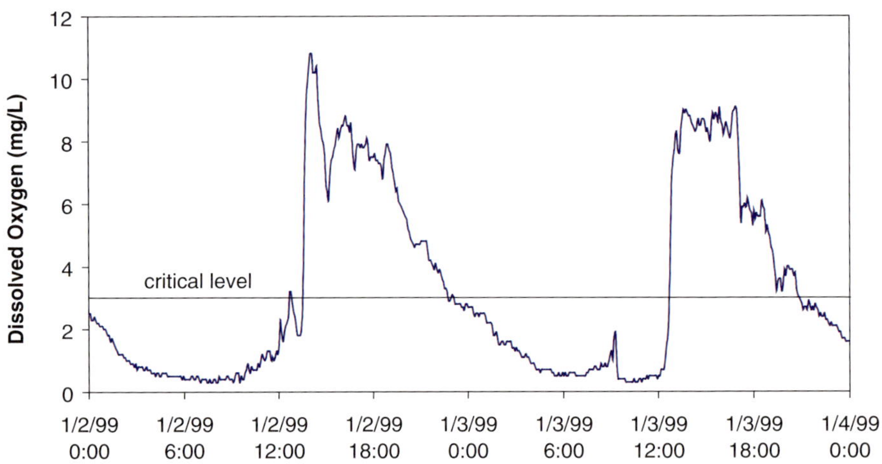

GRAPH 4.5 *Fluctuations in dissolved oxygen concentration at the Vasse Floodgates in January 1999. Methods of continuous oxygen monitoring are being developed to allow detection of low oxygen and potentially lethal conditions for fish to prevent fish kills.* WRC 2001b.

this time contained up to 99,005 cells per millilitre. Upstream areas can also contain large amounts of filamentous green algae.

Water quality is better and nutrient concentrations are moderate below the tidal gates in the Wonnerup Inlet where sea water enters. Algal growth is dominated by marine diatoms and is not excessive or a health risk. Concentrations of algae are usually below the guidelines of 20,000 cells per millilitre, but there was a bloom of the diatom *Cyclotella* in August 1999 with 25,810 cells per millilitre. Blue-green algae are present but usually in low numbers. The highest counts measured at two sites varied from 5–7,000 cells per millilitre in January 2000.

The brackish waters in the Deadwater area are in generally fair condition although they are isolated from the main flow and can become hypersaline in summer. Diatoms usually dominate the flora but dinoflagellates and blue-green algae are also present. In August 1999 there was a bloom of blue-green algae in the northern section (49,706 cells per millilitre). At the same time a diatom bloom in the Wonnerup Inlet extended into the Deadwater where concentrations were over 50,000 cells per millilitre.

A productive environment

Although floodgates restrict movement up and down the estuary, fish such as yellow-eye mullet, western sand whiting (yellowfin whiting), King George whiting and mulloway enter the Vasse–Wonnerup when conditions are favourable to feed on the abundant algae, the associated invertebrates and other small fish. Sea mullet that spawn in the ocean use the estuary as a nursery area.

Black bream live only in the brackish waters of the middle to upper estuary. As they do not live in the ocean, populations in each of the estuaries along the coast are distinct, adapted to local conditions. (See **Box 2.13 A popular fish—the black bream** and **Box 10.2 Black bream—a real estuary fish**.)

Dead fish—symptoms of a distressed environment

Mass fish kills are not a new problem, and similar events have occurred in other water bodies around the Busselton area at Toby Inlet and Quindalup, in the Brunswick River flowing into Leschenault Estuary and throughout the South-West. Although the first fish kills in the Vasse–Wonnerup were reported in 1905 and subsequently in the 1930s, reports of dying fish appear to have become more common in recent times (Table 4.11 and Fig. 4.13). Most fish deaths have occurred in warm weather and are generally attributed to lack of oxygen during times of high algal growth. One major fish kill however, followed a blue-green cyanobacterial bloom in the Wonnerup and may have been due to toxins. In other cases toxins may have been released by decaying seaweed.

Prevention or cure?

Various strategies have been trialled over the years to reduce the severity of fish mortality in the short term and to remedy some of the problems leading to algal blooms in the longer term. These have included opening the sandbar, netting or removal of fish (netting is now limited to commercial fishers for a short period) and opening of the Vasse Estuary floodgates. Monitoring of water levels, measurements of water quality (dissolved oxygen, temperature, salinity, nutrient and algal concentrations) together with observations on fish behaviour are conducted to determine more precisely the conditions leading to fish deaths, and assist in decisions to open the sandbar and floodgates.

Currently, the sandbar to the sea is opened through January and February to allow seawater into Wonnerup Inlet. The timing and frequency of opening depend on water quality and how often the bar is closing over. Better water quality should reduce fish deaths in the inlet, improve conditions upstream in the Vasse exit channel, and ensure the survival of fish released from the Vasse.

The entire system was open to commercial and recreational netting in 1960, but since 1992 only commercial netting has been allowed. Netting does not prevent fish kills, but it does reduce the number of fish dying when conditions deteriorate, and improve the aesthetics of the landscape.

Opening of the floodgates to allow fish to escape and relieve unfavourable conditions has to be balanced against allowing excessive seawater inflow that would threaten native vegetation, pastures and waterbird populations. Decisions to open the floodgates depend on observations of fish behaviour such as schooling near the floodgates or at the surface gulping air, the presence of algal blooms, and water quality sampling including monitoring of oxygen.

New guidelines developed by the landholders, government agencies and community have set summer levels at –0.1 metre AHD to reduce the salt threat. In reality however, the level has often fallen to –0.4 metre AHD due to evaporation.

The new regime, in operation since 1990, of opening the floodgates has had mixed success. At times limited openings have prevented fish kills but on other occasions, despite longer openings and higher water levels, fish deaths have still occurred. A fish kill occurred in February 1997 when the water level was +0.12 metre, but no fish died in the summer and autumn of 1994–1995 when the level was –0.1 metre. In February 2000 when 1,400 fish died, the new floodgate management allowed the release of more than 35,000 fish before they were affected by the low oxygen concentrations. Similarly, in

TABLE 4.11 *Summary of reported mass fish kills in the Vasse–Wonnerup system.* Source: Lane et al., 1997; WRC, 2001b, K. Hardcastle, pers. comm.

Date	Month	Location	Event	Likely cause
1905	January (late)	Vasse Estuary lower reaches (& west arm of Wonnerup Inlet?)	'Hundreds and thousands of fish of all kinds...rotting upon the banks'	'Dead and stagnant water' following closure of sandbar at mouth of Wonnerup Inlet
1933–34	'summer'	Vasse estuary	'A number of fish' died	Unspecified
Pre-1960	'summer'	'The estuaries'	'Heavy mortality during summer months'	Fish 'landlocked' by floodgates during summer
Pre-1960	February	Vasse Estuary channel	'Many dead fish lying around one February'	Unspecified
Pre-1960	'summer'	'Wonnerup estuary' (location doubtful)	'Not unusual for thousands of mullet to die' after waters subside	Unspecified
1966	April (late)	Wonnerup Inlet—between main road bridge and mouth	Estimated that nearly 4,000 fish of at least 5 species died	Lack of oxygen in water, fish gills clogged with sediment after sandbar artificially closed during floodgate repairs
1984	February (2nd week)	Vasse Estuary channel	Thousands of fish died including mullet and bream	Lack of oxygen when a heatwave coincided with a drop in the estuary's water level
1988	February (late)	Wonnerup Inlet	More than 1,000 dead fish strewn along the banks	High water temperature and lack of oxygen after bar closed for two months
1988	November	Vasse River, Rotary Park, Busselton	Carp *Carassius* sp. died	Bacterial infection 'fin rot', cause unknown, reports of stormwater pollution
1989	February (1st week)	Vasse Estuary channel	Estimated 1,000 pilchards and mullet died	Poor water quality
1997	February (2nd week)	Vasse Estuary channel	Many fish died including black bream	Algal breakdown, low oxygen, high temperatures
1997	June (mid)	Deadwater	Many fish died, at least 4 species	Sudden drop in salinity after heavy winter rains and run-off
1998	January (12)	Wonnerup Estuary	Estimated 770 mullet and 2 black bream died	Algal bloom decay, low oxygen
1998	January (19)	Vasse Estuary channel	232 fish died	Toxic phytoplankton bloom, 120,000 cells/mL of the haptophyte *Prymnesium* cf. *parvum*
1998	January (23)	Wonnerup Inlet (downstream of Vasse floodgates)	Juvenile sea mullet died	Low oxygen
2000	February (15)	Vasse Estuary channel	Over 1,500 fish died, mainly bream and yellow-eye mullet	Algal bloom decay, low oxygen
2003	December	Vasse Estuary channel	60 fish died and a further 6,000 released to Wonnerup Inlet	Algal bloom decay, low oxygen

December 2003 sixty fish died and a further 6,000 were released to the inlet. In recent years the floodgates have been opened for longer periods and the water level has risen +0.3 metre AHD.

New floodgates were installed on the Vasse and Wonnerup estuaries in 2004. The new structures incorporate the latest technology for continuous monitoring of water quality along with live video surveillance of fish behaviour. The gates will be automated and may be opened remotely when required to release stressed fish.

Controlling algal growth by reducing light and water temperature by planting of trees along the channels has been proposed. Meanwhile, some of the existing fringing vegetation of melaleucas and eucalypts is dying, salt concentrations in adjacent paddocks appear to be rising, hay production may be declining, and waterbirds may be affected. Many other procedures to maintain healthy conditions are costly and some have unwanted effects. Control of nutrients probably offers the best long-term solution.

In 2001 trials were conducted to determine if applications of the modified clay Phoslock™ developed by CSIRO in conjunction with the Water and Rivers Commission could be useful in reducing phosphorus available to algal blooms. The clay has also been tested in the Canning and Swan Rivers and in Lake Powell near Albany, and appears to be a useful tool if applied before blooms are initiated.

The effect of bringing salt water near to the wetland areas via the adjacent marina and canal developments (Port Geographe) remains unknown. The sandy soils are a ready conduit for water loaded with nutrients from the wetland entering Geographe Bay, or alternatively salt water from the bay moving into the freshwater areas of the wetland.

Wetland of significance

The Vasse–Wonnerup wetland system is an important refuge for wildlife, particularly birds that congregate to feed over the summer months when many other wetland areas are dry and barren. Surveys in the 1980s recorded over 30,000 birds of sixty species including migrants from other parts of Australia and the Northern Hemisphere. The species using the wetland vary from year to year, and eighty-five species have been recorded in total. The Vasse–Wonnerup was listed in 1990 as a Wetland of International Importance under the Ramsar Convention on Wetlands.

When the winter rains arrive and water spreads out across the land, the bird population around the Vasse–Wonnerup is at its lowest (Graph 4.6).

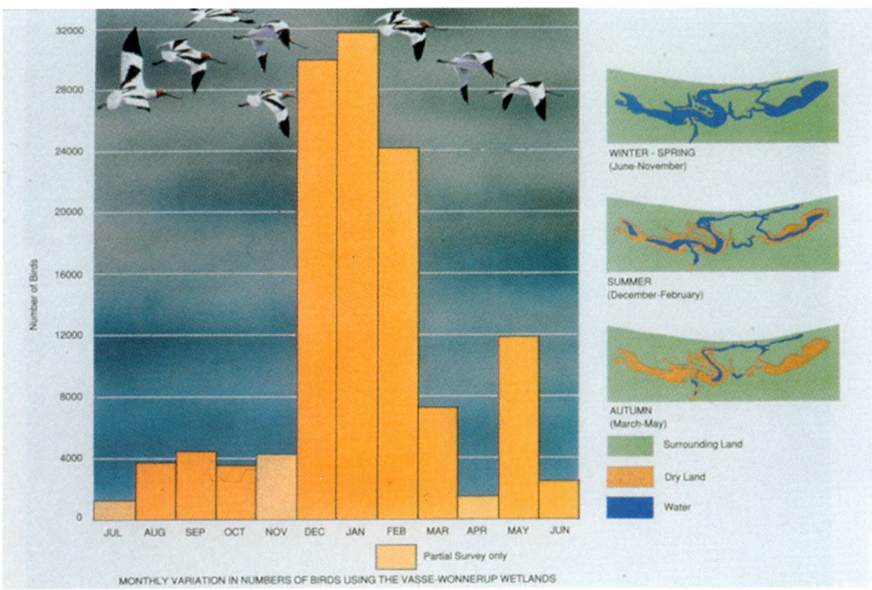

GRAPH 4.6 *Fluctuations in the number of birds in the Vasse–Wonnerup: these wetlands are an important habitat for birds particularly in summer when they congregate to feed in the productive shallow waters.* J. Lane, 1990, Courtesy Landscope, CALM.

FIG. 4.14 *(a) Flocks of Black Swan on the Vasse–Wonnerup.* Courtesy G. Chapman. *(b) Nest of the Black Swan, an island surrounded by water: eggs are laid early in the winter and incubated for about 40 days.* Courtesy Cliff Winfield.

Migratory birds replete with food and a fat store have left for breeding grounds in the Northern Hemisphere. Many waterbirds such as ducks and swans have spread out across the South-West to other swamps, lakes and dams replenished by the rains. Others including some ducks, Black Swans, coots (in great numbers), grebes, cormorants, heron and ibis use the sheltered waters over winter to feed or breed.

The breeding cycle starts with the pairing of Black Swans, in some years about 150 pairs. Old nests are repaired and new nests made, forming small islands as the waters rise in June (Fig. 4.14). Ducks also pair and build nests beneath the bushes in hollowed trees or tree branches.

With spring the waters begin to ebb, the days lengthen, temperatures rise, and sunlight promotes the growth of algae in the nutrient-rich waters. The algae provide food for tiny invertebrates and fish, and the birds return to feed. Other swans returning from their winter refuges join the newly emerged cygnets, and by November thousands of swans dot the inlet waters. Young ducklings also gather on the inlets, and the predatory ravens, Swamp Harriers, Osprey and Whistling Kites have rich pickings. (See **Chapter 2 The Swan–Canning Estuary: Recreating Swanland—bringing back the Black Swan**.)

With the drying summer conditions the waters retreat, and the shallow mud becomes a rich feeding ground for many species including about 12,000 teal, 5,000 stilt, 4,000 Black Duck, 4,000 Black Swan, coot and avocet. As the waters dry still further, many of the swans and other birds disperse to more permanent waters while some, such as pelicans, herons and egrets,

FIG. 4.15 *The Great Egrets and White Ibis feed on the fringing marshes and shallow waters of the Vasse–Wonnerup.* Courtesy Cliff Winfield.

arrive to feed on the stranded fish, and shelduck dabble in the rich oozy mud (Fig. 4.15).

The shorter days of mid autumn again bring the departure of the migrant birds that head for the beaches along the north coast, the first stage of their long flight back to the Northern Hemisphere. Winter brings rain and the promise of renewal, another water cycle, but the accompanying floods also show how little of the low-lying land is now left without developments where the need for flood control remains a major issue (Fig. 4.12).

TOBY INLET

Toby Inlet, about 5.6 kilometres long, lies parallel to the coast between a narrow line of dunes and a major highway, Caves Road. The inlet enters Geographe Bay at Station Gully Drain and the water body to the east of this point is known as the Deadwater (Maps 4.8 and 4.11). To the east-south-east lies a smaller 2.8 kilometre-long water body and to the south-west landscaped lakes around a golf course and housing estates form part of the same system. The lakes are evidence of flat land only slightly higher than sea level, which is waterlogged in winter (Fig. 4.16).

Originally the Carbanup River and Mary Brook flowed to the inlet, but as in other areas of the Swan Coastal Plain, drains were dug soon after farming began to reduce winter waterlogging and to speed the flow to seaward. In the early 1800s Jack Molloy diverted most of the Carbanup flow straight to the sea. The channel, known now as Molloy's Drain, changed the whole character of the western wetland structure and to some extent Toby Inlet.

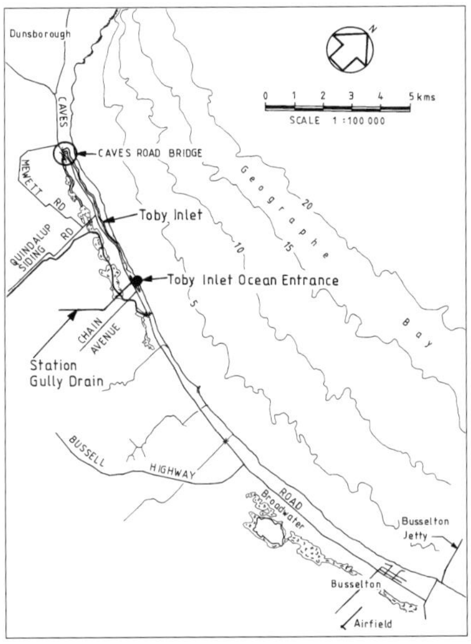

MAP 4.11 *In 1956 the Broadwater Wetland System was divided and flow to Toby Inlet was reduced when the culverts were removed during upgrading of Quindalup Siding Road. Previously the culverts had allowed water flow from east to west. Following their removal, water on the eastern side was diverted directly to Annie Brook Drain which joins with Mary Brook discharging to Geographe Bay through Station Gully Drain. Today most of the flow comes from about 5,100 ha to the south where unnamed tributaries form the Sussex sub-catchment area.*
M. P. Rogers & Associates Pty Ltd; Courtesy Sussex Land Conservation District Committee.

The ocean entrance has also changed considerably over time. Anecdotal evidence suggests that in the early 1900s the entrance was probably further east and connected to the Deadwater that would have been part of the main estuary body. The Deadwater formation and the channel in front of a large tourist complex mark the older channel (Fig. 4.17).

During the 1970s the Public Works Department maintained two separate exit channels. Cut perpendicular to the shore, these usually closed within a year after dredging. More recently, a single channel from Toby Inlet and Station Gully Drain has taken water to Geographe Bay. Although the combined flows are generally sufficient to keep the channel open, it has been manually opened three or four times. The channel was last excavated in 1996 and has since migrated 600 metres eastwards.

The state of the entrance after the bar is breached depends on the opposing forces to keep it open or closed. Generally if the scouring effect of inward and outward water flows exceeds the rate of sand deposition at the entrance, the inlet entrance remains open.

Out to sea—the coastal forces

Lying to the east of Cape Naturaliste, the shoreline near Toby Inlet is well protected from offshore swells from the south-west. Refraction around the cape, together with bottom friction and shoaling across Geographe Bay, change the direction and shape of waves so that most waves near Toby Inlet come from the north to north-west (Map 4.12). As a result, sand moves from west to east along the shore in a net annual longshore movement of approximately 25,000 cubic metres per year eastwards. Movement of sand, however, usually occurs in pulses and varies with the seasons and from year to year. The entrance channel is influenced by the outward water flows, longshore currents and sand movements, and varies by as much as 100–200 metres during a year.

Patterns of erosion and deposition in Geographe Bay are also affected by water levels. During episodes of high water, beach lines erode and the sand

FIG. 4.16 *Toby Inlet and wetland areas to the east, aerial view. The catchment, about 5,000 ha in area, is low-lying and was originally covered with paperbark woodlands and shrubs. Today the inlet is surrounded by grazing land, except for the most seaward section between the coastal dunes and Caves Road where peppermint trees hide residential developments. Toby Inlet was named after Captain Jacob Toby of the coastal schooner 'Ellen' that used to take on water at Meelup and trade with early residents along the Geographe Bay coast. The Inlet was once much deeper and small boats were able to sail upstream.* Courtesy B. Clay, Toby Inlet Catchment Group Inc.

moves with the longshore drift. Although tidal changes in the South-West are small, mean water levels in Geographe Bay are slightly higher in winter (0.1 metre). Water levels can increase by about 0.4 metre in a typical winter storm and about 1.0 metre in a severe storm. Storm surges created by cyclone Alby in 1978 were about 1.8 metres AHD at Busselton and peak flood level at the mouth of Toby Inlet about 2 metres AHD. Erosion due to high water levels and sediment movement has been a problem along Geographe Bay for many years. It is now complicated by the series of wooden barrages that trap sand and initiate erosion further along the shoreline. High water levels also reduce floodwater outflow, creating flooding conditions in Toby Inlet and Station Gully Drain for short periods of time.

Water flow in and out of the inlet is influenced by the stream flow from the catchment and the tides. During summer, stream flow is small and tidal flow dominates. During winter, increased water flow and tidal flows tend to act together. Stream flow will tend to reduce flooding tides and accelerate ebbing tides. As the channel migrates from west to east in response to longshore drift, it lengthens. The increasing length decreases the peak tidal currents limiting the ultimate length of the entrance channel. As Toby Inlet and Station Gully Drain migrate east, a barrier spit is formed by the accumulation of longshore drift sand. Under these circumstances, there is sometimes not enough sand to close the entrance.

Under normal wave conditions, sediment transport onshore is dominated by long period, low amplitude swell waves. These waves tend to slowly move sand onto the beach. Under modest swell conditions in summer, autumn and spring, sand supply is usually insufficient to close the entrance. During winter however, erosion of the dunes due to high water levels and storm waves increases the sand supply. The swell conditions transport more sand into the entrance where seagrass wrack can also accumulate and the bar may close.

Changing state — fresh to salt and nutrients too

Salinity changes over the year depending on the winter rainfall. When winter flow is strong the inlet waters are generally fresh, but as the freshwater flow decreases and tidal force dominates, salt water intrudes again beneath the

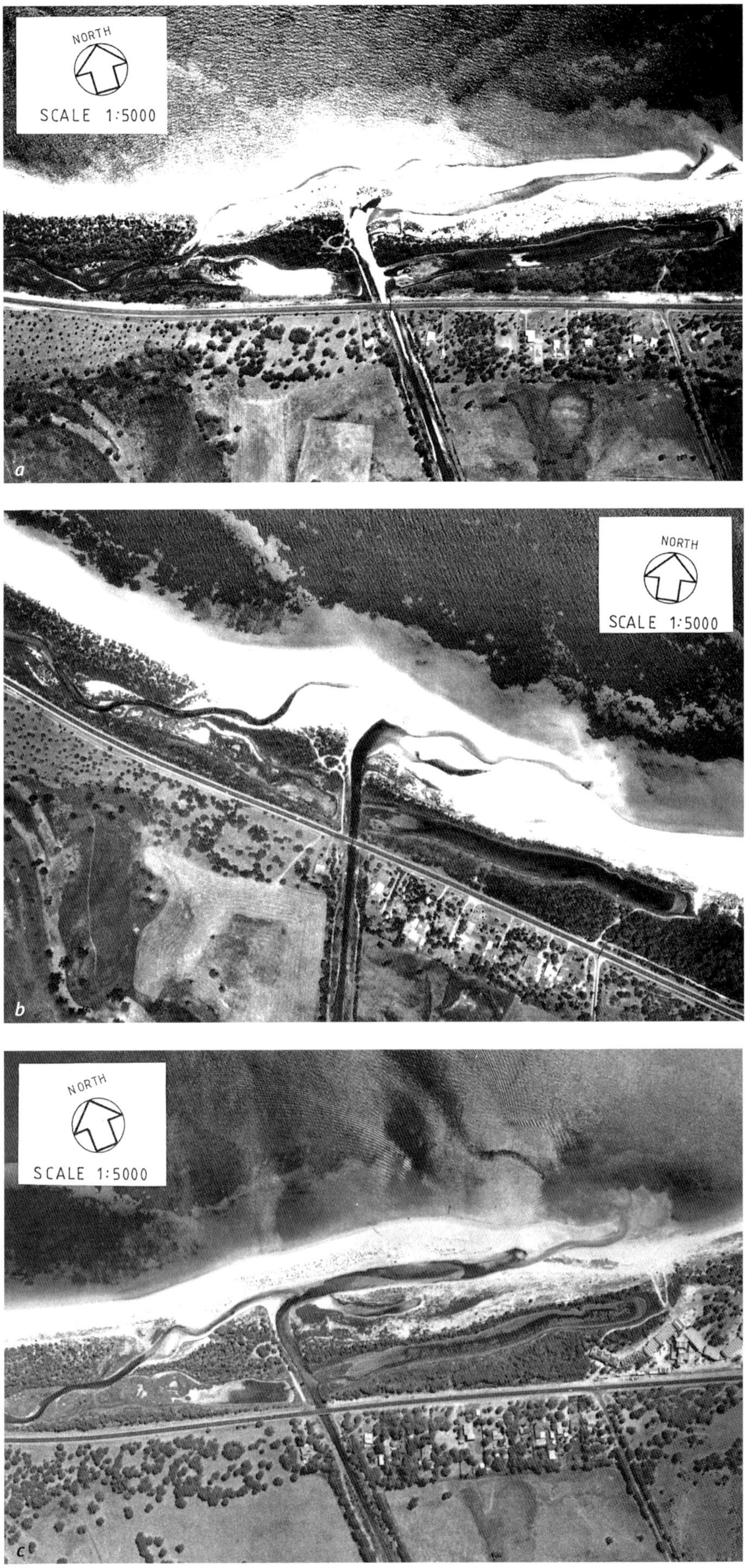

FIG. 4.17 *Migration of Toby Inlet entrance channels from west to east, the predominant direction of the longshore sand movement, aerial views. The entrance into Toby Inlet and Station Gully Drain can be two separate channels (1973, 29/1/1980, 20/11/1980) or a single entrance (1998). The remains of a boat buried beneath the sands are sometimes exposed when the sand moves. (a) 27/11/1973, (b) 20/11/1980 and (c) 3/2/1998. Note the new resort development that can be seen on the bottom right of (c).* M. P. Rodgers & Associates Pty Ltd, Courtesy Sussex Land Conservation District Committee.

fresh water in a wedge that moves back and forth with the tide. When rainfall ceases, the salt wedge penetrates further upstream to the vicinity of Quindalup Road, upstream of which the waters are fresher.

The winter rains also replenish the thin freshwater layer overlying the more saline ground water. Over use of the aquifer has caused some intrusion of salt water, a potential threat to the overlying deep-rooted vegetation that stabilises the inlet banks and creates the beautiful shady environment.

Nutrients from the surrounding catchment also arrive with the winter rains, and the inlet is eutrophic with high levels of algae in summer. Nutrients accumulating in the inlet appear to be derived from diffuse and point sources, and the sandy soils have little capacity to retain nutrients. In the middle river reaches grazing cattle are watered in the streamlines, and around the lower reaches the housing estates were until recently unsewered. Station Gully Drain also receives discharge from the Dunsborough Waste Water Treatment Plant, although this discharge should be reduced with the installation of a new treatment system. The new system is designed to remove over 80 per cent of the nitrogen and 90 per cent of the phosphorus, and the waste water used to irrigate a 47-hectare blue gum plantation during eight months of the year. Nutrient sources, however, are unlikely to decrease in the short term as the population grows and more land is occupied by houses, gardens and golf

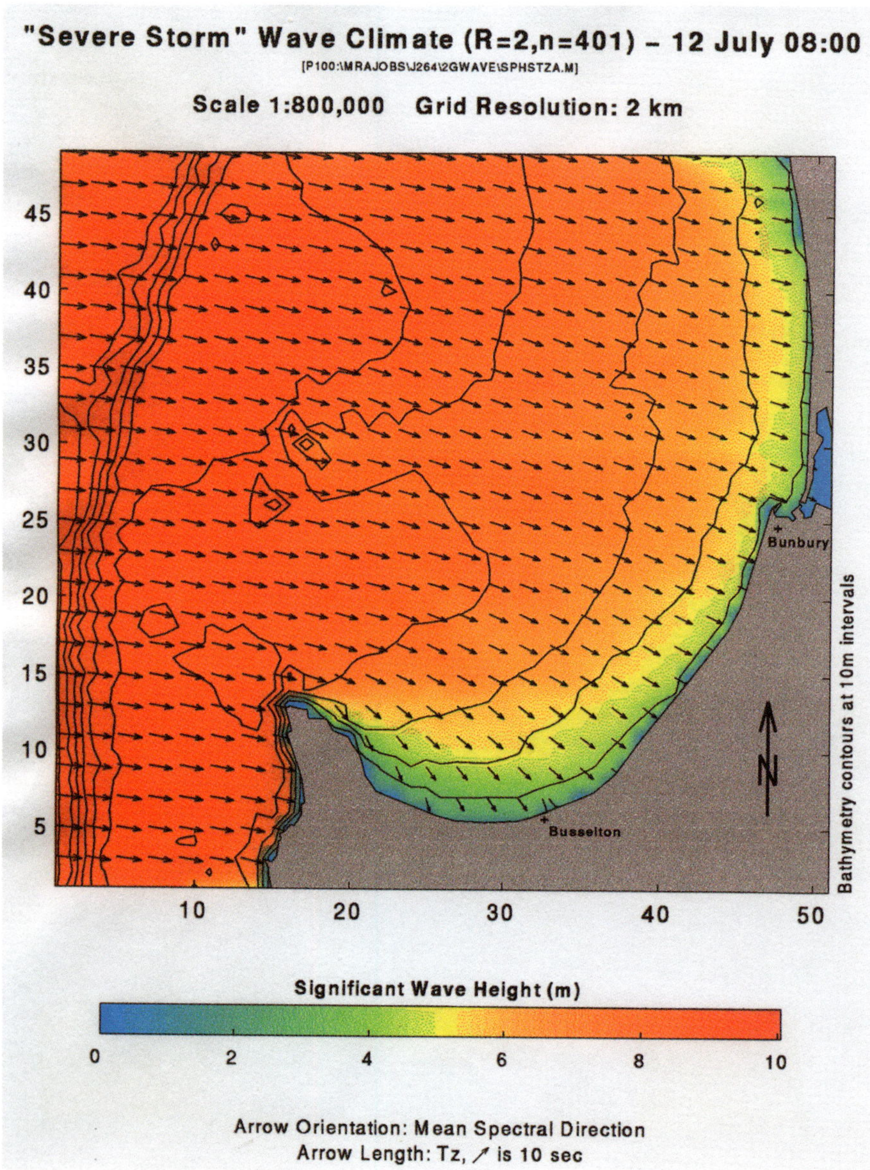

MAP 4.12 *Computer model of wave movement during a severe storm in Geographe Bay. Swell waves from the west are refracted around Cape Naturaliste and arrive along the beaches near Toby Inlet from the north-west.*
M. P. Rogers & Associates Pty Ltd; Courtesy Sussex Land Conservation District Committee.

courses. Despite the high nutrient loads, summer algal blooms have been relatively limited and short lived, probably due to the salty water and the maintenance of an open entrance channel.

Wildlife refuge and drain

Today the vegetation surrounding Toby Inlet and the associated wetland areas to the east is highly modified and only small areas of native vegetation remain. In the upper wetland areas, forest and woodland are now primarily grazing land. Nearer the coast, scrub and dune vegetation predominates. Typical estuarine fringing communities of sedges, samphires and paperbarks cover the wetter, more saline areas. About 140 species of plants have been found in the area, about fifty of which are introduced weeds. Although restricted in area and somewhat degraded, vegetation around the inlet has a conservation value as it represents plant communities that were once more widespread (Fig. 4.18).

The surrounding terrestrial vegetation provides habitat for eight frog species, thirteen reptile species, three native mammals and forty-two forest birds. Although most are common and widespread, the mammals include the endangered Western Ringtail Possum *Psuedocheirus occidentalis* and the Southern Brown Bandicoot *Isodon obesulus* (known by the Nyoongar Aboriginal name Quenda). Waterbirds also use the inlet during the summer months when other wetlands are dry, and Black Swans, Australian shelduck and wood duck nest in the area.

Water life in the inlet varies with salinity, reflecting to some extent distance from the mouth. In the upstream wetland areas, the relatively few invertebrate species appear to be affected by high nutrient enrichment and the absence of fringing or riparian plants. Generalist, cosmopolitan species, able to tolerate large amounts of organic matter and deoxygenated water dominate the communities. Downstream fish such as black bream, herring,

FIG. 4.18 *The scenic landscape of paperbark and sedges along Toby Inlet serves many purposes: a refuge for birds, a baffle reducing bank erosion, and a trap for nutrients before they enter Geographe Bay, May 2003.* A. Brearley.

whiting, mullet, salmon, trumpeter and crabs are still found in the inlet when salinity is favourable. Anecdotal evidence suggests, however, that numbers have declined in recent years. As in other estuaries, low water levels and algal blooms in late summer and autumn create stressful conditions for animals. Oxygen concentrations fall when algae are abundant and salt water lies along the bottom. A fish kill of trumpeter, pilchard and mullet in March 1983 was attributed to lack of oxygen during warm weather when the bar to the sea closed.

Threats to a little refuge

Lying in a landscape of grazing land and housing, the small areas of native bush around Toby Inlet are constantly under threat as people make paths to the beach; cut wood; trample, remove or burn the fringing vegetation; add fertilisers; use the ground water; or dump garden rubbish introducing competing weeds. Habitat decline is a major threat to native animals that are also under pressure from cats and other feral animals, fire and disturbance.

Toby Inlet is only a tiny water refuge and threats to its health are many. Initially cleared and managed for grazing and occupied by holiday homes, recent population growth is bringing further changes. Watertables are rising, competition for fresh water has increased and the risk of seawater intrusion is higher. Water is impounded upstream and nutrients, as well as sediment sources, are changing and increasing. Low-lying areas are used to control floodwaters that eventually find their way to sea, bringing nutrients from the catchment. In other areas such as Peel–Harvey, Cockburn Sound and the Albany Harbours, algae that grew where nutrient inflows were high and the waters poorly flushed smothered seagrasses. Although Geographe Bay is huge, there is concern that the health of Toby Inlet and the small streams and drains flowing to Geographe Bay could determine the future of the seagrass meadows that lie offshore.

Chapter 5 Estuaries north of the Swan

The northern estuaries of Swanland from the Murchison to the Irwin are all derived from rivers that flow with summer and winter rain, in contrast to the dominance of winter rainfall on those further south. Average rainfall is also much lower. River flow is often episodic following heavy rainfall during thunderstorms, but annual flows can be quite low either due to the small catchments or, in the case of those with larger catchments, because they flow though semiarid landscapes. As a result the rivers and estuarine valleys are limited in area. Many are filled with alluvium from the catchments, but in some this is only a shallow veneer over basement limestone rocks.

The estuarine parts of the rivers southwards of the Murchison, although smaller in area, have characteristics in common with other estuaries throughout the South-West, particularly those to the east of Albany. The bars are closed except when the rivers are flowing.

Compared to the southern estuaries, the northern estuaries have been less studied. This may reflect their smaller size, comparatively minor utilisation by fisheries and remoteness. The threats, however, are similar; much of the surrounding landscape has been cleared, and they increasingly receive more sediment, nutrients and contaminants than they would have in the past. They are, however, foci of attention; always pleasing to the eye and a fascination when the bar breaks and brown estuarine water moves into the blue ocean or when the tide turns and clear blue water enters. Remnant vegetation protects the banks and serves as a refuge for wildlife using the estuarine waters and nearby freshwater seeps.

LANDSCAPE OF THE CATCHMENTS

All the estuaries with the exception of the Murchison lie within the Perth Basin, but the Murchison spans both the Carnarvon and Perth Basins. The Carnarvon Basin extends from the Dampier Archipelago to the Murchison area. The oldest rocks, the Tumblagooda Sandstone of probable Ordovician age, outcrop along the spectacular gorges of the Murchison River. Further south outcrops also occur in headlands and cliffs, but in many places these are overlain by younger rocks and sands. The type section of the Tumblagooda Sandstone extends for about 70 kilometres along the Murchison River Gorge from near Hardabut Pool to Second Gully Point. The Tumblagooda Sandstone is over 1,000 metres thick near the Murchison, although it may be 3,000 metres thick in other areas. (See Maps 1.1 and 1.2.)

Inland the landscape adjacent to the rivers is dominated by outcrops of older rocks around Geraldton, Mingenew and Hill River on the Victoria, Darling and Dandaragan Plateaus. While estuaries like the Swan–Canning and Peel–Harvey further south lie on the coastal plain bordered by younger dunes representing old shorelines. Although the old shorelines generally decrease in age from east to west, as they do further south, in this northern section the coastal plain is narrower and the dune sequences are less developed.

In the Arrowsmith River region the Eneabba Plain replaces the Bassendean Dunes, a prominent feature of the coastal plain further south.

Steep dunes line the coast wherever the sediment supply is high, for example, around the Greenough. North of Port Gregory and the Hutt Lagoon, the coastal plain narrows still further. There are no offshore reefs and islands so the coast is exposed to high wave energy. Shore platforms and cliffs of Tamala Limestone line the coast, the most spectacular being the 200-metre high Zuytdorp Cliffs that extend for over 150 kilometres to the north of Kalbarri.

The estuaries

Most of the estuaries north of the Swan are riverine in nature and classed as 'seasonally open' or normally closed as the entrances are usually closed when river flow is low. The deepest part usually lies behind the flood tide delta or estuary mouth bar. They are also characterised by a 'flashy' pattern of discharge due to infrequent but intense flooding and more frequent but short-lived seasonal flow. Some such as the Hutt, Bowes, Oakabella, Oakajee and Buller may even dry out in summer (Appendix 1).

The smaller estuaries (Oakabella and Oakajee) lie in box-shaped river valleys filled with alluvium eroded from the Northampton Block (Complex). The larger Greenough Estuary, an inter barrier estuary, lies between dune lines where alluvium forms an extensive flood plain. The Greenough and the Moore Rivers empty into relatively wide basins before narrowing again at the entrance channel and bar. The largest, the Murchison, is permanently open although the entrance is often shallow due to sand accumulating behind the protective limestone reef at the mouth.

Vegetation

In the north, the vegetation typical of the more southern areas narrows to a thin band that gradually merges with the flora of the Desert (Eremaean) Vegetation Province reflecting the decline and change from winter to summer rainfall. The vegetation of the upper Murchison catchment consists of low woodland of mulga–wattle *Acacia aneura* with *Eucalyptus* spp, spinifex *Triodia* with ephemeral grasses and herbs (Fig. 5.1 and Map 1.12).

Further to the west where rainfall is higher and more reliable but more seasonal, falling mostly during winter, the vegetation is part of the South-Western Vegetation Province. Dominated by sandy soils and laterite, the vegetation is predominantly scrub heath kwongan near the coast with *Acacia–Casuarina* thickets further inland and other *Acacia* species, York gum *Eucalyptus loxophleba* and red river gum *Eucalyptus camaldulensis* on the more loamy alluvial soils.

Along the Swan Coastal Plain the vegetation reflects the various ages of dune formation as it does further south around the Swan River. The most seaward and mobile areas (Quindalup Dunes) are dominated by grasses *Spinifex longifolius* and *Spinifex hirsutus* and numerous introduced South African and Mediterranean species with native shrubs such as fan flower *Scaevola*, coastal daisy bush or rosemary *Olearia axillaris*, wattle *Acacia cyclops* and sword-sedge *Lepidosperma* more common where dunes are stabilising.

The vegetation of the older Spearwood Dune System varies from north to south reflecting the rainfall. In the northern part the Spearwood Dune System is a mosaic of scrub heath with *Banksia prionotes*, parrot bush *Dryandra sessilis* and one-sided bottlebrush *Calothamnus quadrifidus*. North of the Arrowsmith River the illyarrie *Eucalyptus erthrocorys* with its yellow stamens and bright red caps grows on rocky outcrops of limestone. Between Lancelin and Yanchep, banksia woodland dominates the vegetation with scrub heath on the limestone ridges. South of Yanchep and the Moore River, eucalypt woodlands replace

FIG. 5.1 *The upper catchment of the Murchison in the semiarid mulga woodland around Yalgoo. The rainfall, about 200 mm a year, is spread over summer and winter and supports a low woodland of mulga–wattle: (a)* Acacia aneura *with (b)* Eucalyptus leptopoda, *spinifex* Triodia, *ephemeral grasses and herbs. Good falls of rain bring annual grasses, while winter rains bring herbs such as 'everlastings' that create the famous wildflower carpets. This is the southern 'Station Country' where sheep range over large pastoral leases grazing on the palatable natural vegetation. Grazed since the 1870s, sheep numbers rose to 5.5 million in 1934 supported by underground water and good seasons, but as droughts, economics and grazing took a toll, sheep stocks declined. With the more palatable species reduced, the less palatable species such poverty bushes* Eremophila *dominate the vegetation. Where the soil is exposed to sun and weather, and trampled by hooves, rain no longer penetrates and erosion follows.* Courtesy (a) 3 miles south of Meekathara, D. Bell, 1980; (b) Steve Hopper, May 1986, BGPA.

the banksia community. The tuart *Eucalyptus gomphocephala* is dominant on the limestone ridges near the coast, with jarrah *Eucalyptus marginata* on the more leached soils further inland. Less commonly marri *Corymbia* (formerly *Eucalyptus) calophylla* occurs on deeper red soils in the swales, and red heart *Eucalyptus decipiens* on shallow patchy limestone.

Further inland on the weathered Bassendean Dune System, Banksia woodlands dominated by *Banksia attenuata*, *B. menziesii* and occasionally *B. ilicifolia* along with *Eucalyptus todtiana* woodlands occupy higher ground. Tea-tree *Melaleuca preissiana* and paperbark *Melaleuca rhaphiophylla* together with sedges and rushes occupy low swampy areas that used to create vast winter wetlands, particularly in the south around the Swan–Canning and the Peel–Harvey.

MAPPING THE WEST COAST OF THE GREAT SOUTH LAND

Exploration of the northern part of Swanland was both gradual and largely unplanned as the ships of the Dutch East India Company ('de Verenigde Oost Indische Compagnie' or VOC) on route to the Spice Islands miscalculated the longitude and so sailed too far east to the coast. The first or at least the first to tell the tale was Dirk Hartog in 1616 on the *Eendracht* (Accord) making landfall on Dirk Hartog Island, the seaward barrier of Shark Bay.

The first planned visit was made by Vlamingh in 1696; he was sent by the VOC to investigate the fate of the *Ridderschap van Hollandt* that had disappeared on the way to Batavia in 1694 carrying a prominent official of the VOC. Placed in charge of the *Gellvinck* (Yellow Finch), *Nyptangh* (Nipper or Pincer) and *Weseltje* (Little Weasel), Vlamingh was instructed to search for the wreckage and survivors of other ships such as the *Batavia* (1629) and the *Vergulde Draeck* (Gilt Dragon) (1656), and determine if the land was of economic value.

After making landfall around Rottnest Island in December 1696 and exploring and naming the Swan River, Vlamingh sailed northwards up the coast noting the low-lying hills (Map 5.1 and Fig. 2.5). Members of the expedition explored 'The Pinnacles' at Cervantes commenting that the barren land was without trees and the landscape 'an infertile desert' extending 'as far as the eye could see'. On another inland trek they probably entered the Greenough River before sailing northwards between the Abrolhos and the coast near Geraldton, where the flat topped hills were recorded to resemble those of the Cape of Good Hope.

On 23 January Vlamingh's party landed and walked inland to what is now known as Hutt Lagoon, describing the waters as 'blood red' and commenting that it was possible to collect 'as much salt as one would like'. Vlamingh deduced that sea water probably swept into the lagoon during very high tides and that 'through the power of the sun' the water would evaporate, leaving salt.

Proceeding northwards past a steep coast, Vlamingh reached a 'brave inlet' defined on the south by *Roode Houk* (Red Bluff) and anchored in the bay known today as Gantheaume Bay. Landing to explore on 25 January somewhere near Red Bluff, they apparently never saw the mouth of the Murchison, which remained undiscovered by Europeans for another 140 years until Grey was shipwrecked nearby. Thinking they would die of thirst in the thick scrub, Vlamingh's crew rediscovered the brackish Wittecarra Spring reputed to be the locality where Francisco Pelsaert marooned Wouter Loos and Jan Pelgrom in 1629 as punishment for their role in the massacre following the *Batavia* shipwreck. Vlamingh also noted the Zuytdorp Cliffs on his way northwards to Shark Bay, where he retrieved and replaced Dirk Hartog's plate. Commenting on the steep slopes and the fury of the waves crashing along the base, he found it an evil but fascinating place—almost a vision of the fate of the *Zuytdorp* in 1712.

Hamelin on the *Naturaliste* en route to his rendezvous with Baudin in Shark Bay in 1801 sighted Red Bluff (Point Rouge) and named Gantheaume Bay for the Vice-Admiral of the *Muiron* that rescued Napoleon from Egypt in 1799. However the coast north of the Swan through to Shark Bay remained largely uninvestigated, the land considered of little use. Perhaps too the stories of hazardous reefs and shipwrecks deterred more visits. The settlement of the Swan River colony by the British however, brought an impetus to explore the land further north.

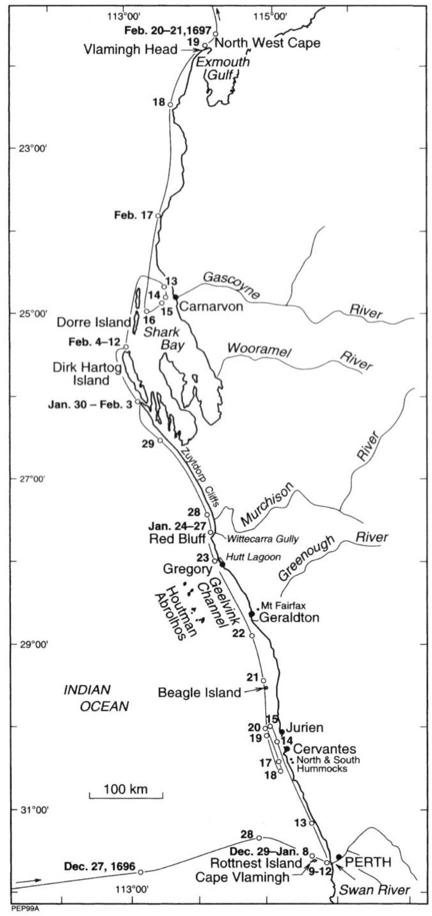

MAP 5.1 *Route of the three ships commanded by Willem de Vlamingh along the Western Australian coast, December 1696 – February 1697.* Adapted from Playford, 1998. Courtesy P. Playford.

Looking at the land

As the numbers of Europeans grew and settlement from the Swan spread to the south around Geographe Bay and eastwards towards Toodyay, Northam and York, the desire for grazing land brought increased exploration to the north. The first, George Grey in 1839, led a party of twelve in three whaleboats northwards as far as Dorre Island near Shark Bay and the Gascoyne area. After loosing one boat, short of provisions and suffering from exposure, an attempted landing at Gantheaume Bay resulted in the loss of the remaining two boats. An overland walk of 400 kilometres was the only way home to the Swan.

Walking home through fertile valleys, building castles in the air

In the days following the shipwreck, Grey explored the lower reaches of the Murchison and noted the dangerous seaward bar before setting a course to the south. Grey's journal, in part a long lament of the predicament of leading reluctant and inexperienced men through unchartered land, records details of the barren coast interspersed with patches of fertile alluvial soils. These soils supported the '*warran*' plants or native yams similar to sweet potato (*Dioscorea*) used by the Aboriginal people. This reliable food resource allowed a less nomadic lifestyle and hut-like mud plastered shelters were observed nearby. In places holes dug to harvest the *warran* were so dense that the explorers found travel difficult.

> *April 4 (1839) ...This was the first time we had yet seen this plant on our journey, and now for three and a half consecutive miles we traversed a fertile piece of land, literally perforated with holes the natives had made to dig this root; indeed we could with difficulty*

walk across it on that account, whilst this tract extended east and west as far as we could see.

It is interesting that Grey and his men were often short of food and yet they were passing through an area where the Aboriginal people were able to live in quite large numbers because plant and wildlife were abundant, as was fresh water. Grey's writing, although somewhat repetitive and showing a degree of hallucination probably due to worries and dehydration, reveals a little of the character of an explorer …

> *Sunday April 7… I turned to the north-eastward, and there burst upon my sight a most enchanting view. In the far east, that is, some twenty or five-and-twenty miles away, stretched a lofty chain of mountains, flat-topped. And so regular in their outline, that they appeared rather the work of art than of nature. Between this range and the nearest one, lay a large rich valley vying with the most fertile I have ever seen in an extra-tropical country. In front of us lay another valley that drained a portion of the large one, and in both rose gently swelling hills and picturesque peaks, wooded in the most romantic manner. Whilst I stood and looked at this scene, my woes were forgotten. Such moments as these repay an explorer for much toil and trouble. The distant range I at once named the Victoria, in honour of Her Majesty; and being now certain that the district we were in was one of the most fertile in Australia, I named it the 'Province of Victoria.'…*
>
> *Whilst I stretched my weary lengths along under the pleasant shade, I saw in fancy busy crowds throng the scenes I was then amongst. I pictured to myself the bleating sheep and lowing herds, wandering over those fertile hills; and I chose the very spot on which my house should stand, surrounded with as fine an amphitheatre of verdant land as the eye of man has ever gazed on. The view was backed by the Victoria Range, whilst seaward you looked out through a romantic glen upon the great Indian Ocean. I knew that within four or five years civilization would have followed my tracks, and that rude nature and the savage would no longer reign supreme over so fine a territory. Mr Smith entered eagerly into my thoughts and views: together we built these castles in the air, trusting we should see happy results spring from our present sufferings and labours,—but within a few weeks from this day, he died in the wilds he was exploring.*
>
> Journal of two expeditions of discovery in north-west and Western Australia during the years 1837, 1838 & 1839, George Grey Esq., (1841).

In the course of his incredible march, with the loss of only one man, Grey named the Gascoyne, Murchison, Hutt, Bowes, Buller, Chapman, Greenough, Irwin, and Arrowsmith Rivers, Mount Albert, Mount Victoria, Mount Fairfax, the Victoria Range and District, and the Gairdner Range. The journey revealed country suitable for agriculture and differences in rock formations to those further south, discoveries that served as an impetus to others like the Gregory brothers who followed, recording in greater detail the features of the land, rivers and mineral wealth. (See **Box 5.1 Murchison—coincidence of name?**)

The hidden wealth – coal and lead

One of the explorers to follow Grey, Assistant Surveyor Augustus Charles Gregory, found coal in the bed of the Irwin River in 1846. Two years later A. C. Gregory and his brother Francis Thomas Gregory discovered lead ore (galena) on the banks of Murchison River as they explored the inland regions.

The lead discoveries created great excitement in the south and prompted a visit by Governor Charles Fitzgerald whose approval was needed for mining. Landing at Geraldton (originally Port Grey) or Champion Bay, Fitzgerald trekked north over the Bowes and Hutt Rivers passing through the newly established pastoral leases. Following the Governor's approval, a lead mine at Geraldine became the State's first commercial mineral deposit in 1848. The first lead was shipped from Port Gregory, which was already in use as a whaling base. Although servicing the growing agricultural areas and used as a convict depot, the port was dangerous, surrounded by reefs, and was closed in 1861 when the railway linked the mines and new port to the south at Geraldton.

Reports on the mineral and agricultural potential by the Gregory expeditions brought settlers. Among the first was James Drummond, son of the colony's first botanist who settled on the Bowes River near Wanneranooka (now Northampton), where copper was found. Further finds of lead and copper around Northampton brought a shift in mining activity and development of a major mineral field. Cornish and Welsh miners together with ticket-of-leave convicts were brought to the colony, and by the 1860s the area was a major contributor to the colony's export earnings. The mining boom however was short lived, and low metal prices in the 1880s brought mine closures. Many of the miners became farmers, while others moved to the goldfields around Coolgardie and Kalgoorlie.

Lead mining revived for a short time when prices rose in 1909 and mines were brought back into production in response to the pre-World War I 'arms race', supported in part by the northwards extension of the railway in 1913 to Ajana near the Murchison River. In the 1950s another increase in activity coincided with the growing car industry and use of lead in batteries and fuels. The State Battery at Northampton (1950s–1983) around the upper reaches of the Bowes River, although mostly used for lead also processed tantalite, tin, zinc, tungsten and garnet. In total, an estimated production from the Northampton Mineral Field has amounted to lead 100,000 tonnes, copper 3,700 tonnes, zinc 40 tonnes and silver 15 kilograms (Fig. 5.2).

FIG. 5.2 *Copper mine at Galena on the banks of the Murchison River about 15 km north of Ajana, July 1929. The acacia trees were used for fuel in boilers and as props in the copper and lead mines.* Catalogue Number 225,625 P, Coutesy Battye Library.

A legacy of the mining days remains. Sediments with heavy metals, principally lead and copper and smaller amounts of cadmium, have moved from tailing dumps around the old mines into creeks and rivers and dissolved in the ground water of the Murchison, Bowes and Chapman Rivers. The degree of contamination is debatable as heavy metals also originate from undisturbed ore bodies throughout the area. In the late 1970s and early 1980s comparisons of sediments and water from upstream and downstream of mines and dumps showed that the type of heavy metal contaminants varied depending on the geology. They also decreased with distance from known sources due to dilution, adsorption onto sediments, precipitation and neutralisation by the natural alkalinity of the ground water. Lead, however, was present in sediments 20 kilometres from the tailing dumps along the Murchison. The headwaters of creeks around Protheroe draining to the Chapman contained zinc, copper, lead and cadmium, while contaminated sediment and water was found in the headwaters of the Bowes River in Udandarra Creek and in Nokanena Brook, downstream from the Northampton State Battery.

In the 1980s tailing dump pollution was considered to be relatively restricted in area, possibly due to the rather limited period of dispersion from ephemeral river flows. This type of flow, however, means that contaminants are dispersed during periods of peak flow, which were not investigated at the time. Threats to the human food chain were thought to be minimal, as water use by humans or stock was limited. Contamination of the aquatic food web was unknown. Sediments from rivers, however, accumulate in estuaries where human use increases and the estuarine bounty of fish and crabs is often harvested.

A new kind of settlement — minerals, wheat and sheep

Settlement of the Champion Bay district extending from the Murchison River in the north to the Arrowsmith River followed quite a different path to areas further south. Some of the earliest settlers brought experience from other parts of the colony, and the discoveries of economic deposits of minerals near fertile alluvial lands saw the development of wool, wheat and mining industries almost simultaneously.

Box 5.1 **MURCHISON—COINCIDENCE OF NAME?**

The Murchison River was named by George Grey in 1839 to honour Sir Roderick Impey Murchison, a great Victorian scientist and one of the first geologists to use fossils rather than rocks to define strata.

From his diary we know that Grey certainly recognised the ancient age of the rocks of the Murchison:

> *One remarkable feature in the Province of Victoria, is that the carboniferous series is here developed throughout a tract of Western Australia, extending in latitude from the bottom of Geographe Bay to near Cape Cuvier, and which I have carefully examined. The tract above alluded to, is the only one in which I have yet found the rocks belonging to this series: this circumstance, therefore, imparts a very high degree of interest to the district in question.*

But we do not really know if the name was proposed to honour a prominent scientist or more specifically for the ancient rocks.

Murchison was known as the 'King of Siluria', a name bestowed somewhat tongue-in-cheek when he was awarded the Royal Society's Copely Medal for his contribution to geology as the describer of the Silurian period. Apparently Murchison considered the Siluria a personal kingdom or empire that extended wherever these fossils might be found across the world. The Tumblagooda Sandstone was thought to be of Silurian age for many years, however it is now considered to be older, of Ordovician Age.

Murchison initially worked with another great scientist of the age, Adam Sedgwick. They were trying to untangle the sequence of strata in Europe more than 395 million years old, deposited when shallow seas covered most of the earth. Later however, the partnership developed into one of the great scientific feuds.

Working in Wales, Murchison began his study at the top of a sequence in the south-east, while Sedgwick began at the base in the north-west. Murchison's sequence contained many fossils that he used to document his sequence, naming it the Silurian after the Silures, a British tribe indigenous to the region noteworthy for resisting Roman invasion. Murchison identified himself with the Caractacus, the Silures' war leader. By comparison, Sedgwick's strata contained few fossils and he based his work on the rock types, naming it the Cambrian after another Welsh tribe. As it became clear that the two systems overlapped, the two collaborators fell out. The controversy was only resolved in 1879 by Charles Lapworth who recognised an intermediate division that he named the Ordovician System, again taking the name of an ancient Welsh tribe.

Before the great feud, Murchison and Sedgwick had already described the overlying system, the Devonian, after a study area in Devonshire where the fossils although few in number were quite distinct, intermediate between those of the Silurian below and the Carboniferous above. Murchison and Sedgwick also demonstrated that these fossils were also present (and richer in fossils) in Germany.

The fifth system of the Paleozoic strata, the Carboniferous System of coal-bearing beds in north England, was described by William Conybeare and William Phillips in 1822. Although recognised worldwide, in the United States the Carboniferous is often subdivided into the Mississippian and Pennsylvanian. The sixth Paleozoic System, the Permian, was also named by Murchison in 1841; a name derived from the Perm in Russia where thick limestone overlies the Carboniferous strata.

For many years Murchison was President of the Royal Geographical Society, which became a key national institution under his rule. Although a great forum some scientists, a number of Fellows (Charles Lyell, Charles Darwin, Joseph Hooker, Thomas Huxley and Alfred Wallace), shunned meetings considering that Murchison devalued science. However for most mid-Victorians Murchison was the Royal Geographical Society.

Recognition that the age of rocks could be estimated by distinguishing the fossils within them was a great step in geology, allowing other people to identify time periods and recognise similar deposits in different part of the world. The economic implications were enormous as strata likely to contain mineral deposits could be identified. Murchison championed the relationship between ancient strata and metal ores, and predicted gold discoveries in Australia. Although some of his theories were later disproved, his views served as a great impetus to exploration and he was a keen promoter of expeditions throughout the world and in Australia.

Source: Stafford, 1989.

In 1851 lots of 50 to 100 acres (20 to 40 hectares) were created around the Irwin and Greenough areas and the largest, The Cattle Company, was allocated 60,000 acres (24,000 hectares) of pastoral land that quickly increased to 145,000 acres (59,000 hectares). Generally cropping was not allowed on pastoral leases unless tillage leases fell within the pastoral boundaries, but many pastoralists grew grain for livestock feed. (See below **Irwin River and estuary: Changes to the landscape** and **Searching for explanations**.)

Changes to the landscape around the estuaries came as the native vegetation was removed: changes that accelerated after World War II when mechanisation increased the scale and speed of clearing. This is an old eroded landscape where flooding rains have sculptured the hills and filled the valleys. Rains now fall on bare earth in summer, erode the country, and fill the rivers with sediment accelerating the natural cycle of estuarine demise.

MURCHISON RIVER AND ESTUARY

The Murchison is the most northerly of the estuaries of Swanland, intermittently flowing through the desert gorges to enter the sea as an unbarred riverine estuary at Kalbarri (Fig. 5.3). Unlike the other estuaries of the area, the bar remains open to the sea and is more influenced by tides. The impressions of George Grey, the earliest European visitor after he and his twelve companions were shipwrecked near Red Bluff, still captures the essence of the estuary.

FIG. 5.3 *Mouth of the Murchison River, aerial view, 1990s. The town of Kalbarri can be seen on the southern bank. The estuary mouth is bordered by cliffs of Tumblagooda Sandstone to the south, and to the north by limestone of the Zuytdorp Cliffs that run for 200 km northwards to Shark Bay. When the port is inaccessible due to shoals in the entrance channel, the only sea access is across a reef near where Grey's Expedition was shipwrecked in 1839.* Courtesy DLA.

April 1 1839...On reaching the valley ...we found a small stream, and following this to the northward for about a mile, came out upon one of the most romantic and picturesque-looking estuaries I had yet seen: its shores abounded with springs, and were bordered by native paths, whilst the drooping foliage of several large sorts of Casuarina, the number of wild swans on its placid bosom, and the natives fishing in the distance, unconscious of our presence, imparted to the whole scene a quiet and a charm which was deeply felt by those who had now for so many days been either tossed about by the winds and waves, or had long been wandering over barren and inhospitable shores. We did not, indeed, find much good land about this estuary, but there were rich flats upon each side of it, whilst the nature of the rocks, and the lofty and peculiar character of the distant hills, gave promise of the most fertile region I had yet seen in extra-tropical Australia.

We followed the shores of the estuary to the northward and eastward until we saw a point, where it appeared to separate into two branches. The natives decamped as soon as they saw us coming, and Kaiber, who watched them with the most intense interest, indulged in various speculations as to the number they would bring back when they returned. We joined the party and traced the shores of the estuary to its mouth, which turned out to be the opening we saw in the morning: this mouth is completely sheltered by a line of breakers and reefs, which, although they present a most formidable appearance from the sea, can be doubled by keeping pretty close along the shore in approaching the mouth of the river. Owing to this reef there are no breakers in the bar, but its mouth is very narrow, and so shoal that I doubt if a boat could be got in at any other time than high water: some of the sailors with me, however, thought otherwise; but there is, at all events, convenient landing at this point under the shelter of the reef.

The rather romantic prose of Grey is perhaps not surprising considering the privations he and his followers had endured while exploring around Shark Bay and Carnarvon, however the long walk back to Perth was still before them. The Gregory brothers gave a more jaundiced version nine years later.

6 October (1848) ...Excepting the flats, and a narrow strip of land on each side, the country was very indifferent, the hills being composed of sandstone and sand, covered with coarse scrub and a gigantic species of grass, the leaves of which, instead of affording food for stock, were a source of great annoyance to our horses, being armed with sharp thorny points, and was somewhat appropriately called 'bayonet grass' by the party. The tide flows about five miles up the river, when it is obstructed by some slight rapids; although it seems shallow, and full of rocks and islands, I think it is navigable for small boats. Above the rapids the river is a succession of long reaches of water about 100 yards wide, and wide flats covered with reeds, the roots of which seem to form an important article of food with the natives. Many springs were seen on the left bank, but few on the right, the water of which was of excellent quality. After making observations of the bar, which appeared to be practicable for whale boats in moderate weather if the wind be south of west, we returned along the south shore of the estuary, which is about one and a-half mile long and half-a-mile wide; it does not appear to be of any great depth. My horse being quite knocked up, it was dark before we could reach a spot where we could obtain water and grass; having come to a convenient place, we bivouacked under a large overhanging rock, as it promised to be a wet night.

The Murchison River

The Murchison River drains a vast semi-desert hinterland of 82,300 square kilometres, the largest catchment after the Swan Avon (≈ 140,000 square kilometres). Rising near the Robinson Ranges north of Meekatharra about 400 kilometres to the east, the river flows along a circuitous 800-kilometre path to the coast (Map 5.2). Nearer to the coast the river flows through spectacular gorges cut deep into the Tumblagooda Sandstone (Fig. 5.4).

With over 80 per cent of the catchment in the semiarid region, discharge from the Murchison varies greatly from year to year. Flow occurs after winter storms from June to August or tropical summer cyclones from February

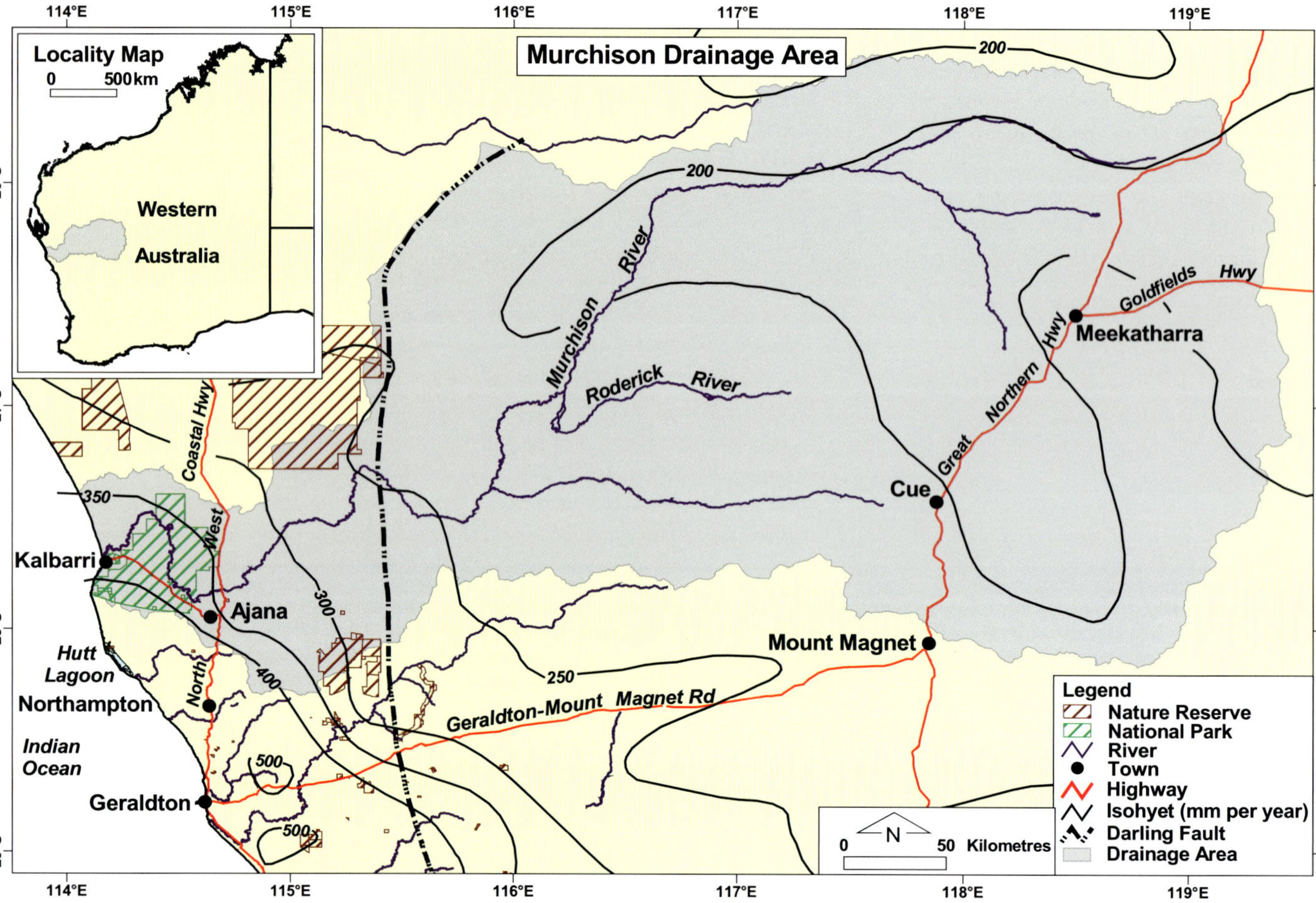

MAP 5.2 *Catchment of the Murchison River extends to the semiarid regions where rainfall is less than 200 mm per year.* Courtesy DoE.

to April (Graph 5.1). There were no significant flows in 1976, 1977, 1985 and 1993. In other years such as 1971, 1984 and 1992 multiple flows were recorded. The maximum flow occurred in March 1975 with other significant flows in July 1980 and 1974, June 1989 and April 1992. Peak flows of 220 cubic metres per second occur once in three years, 330 cubic metres per second once in five years, 460 cubic metres per second once in ten years and 680 cubic metres per second once in thirty years, although this could vary between twenty and fifty years. River flow needs to be about 500 cubic metres per second to affect the river mouth, a rate likely to occur every ten years.

Fresh water also enters the river and estuary from springs in the Tumblagooda Sandstone. These springs were used by the Aboriginal people, the early pastoralists and the first European holiday-makers at Kalbarri. Others such as Long Spring, 7 kilometres from the mouth, support small areas of swamp with dense rushes and other aquatic vegetation; an important refuge for wildlife when surface water dries up.

The estuary

The estuary is only four square kilometres in area and has been classed as an Inter-barrier Estuary, lying between the most recent dunes. As the basin development is limited and the lower river reaches form the estuary, it has also been termed a Riverine Estuary. The estuary is a sequence of shallow pools and elongate bars. The bars occur as thin veneer over cemented Pleistocene deposits. The estuary is lined with levee banks, overbank deposits, and point bars of sand and silt adjacent to some of the larger bars.

The Murchison is permanently open to the sea, and tide changes are present for about 18 kilometres to the Murchison House Ford. Salinity near

FIG. 5.4 *Gorges along the Murchison River carved through the Tumblagooda Sandstone, laid down in ancient seas over 430 million years ago, July 1980. As the land rose between 10 to 6 million years ago, the river cut down through the sandstone creating the spectacular gorges we see today at the Z Bend and The Loop. The flat top of the plateau has been eroded and overlying younger marine deposits of Cretaceous age have been removed.* Courtesy V. Ryland & K. McNamara, WAM.

GRAPH 5.1 *Discharge of the Murchison River (Emu Springs gauging station near Ajana) 1967–2000 (mean annual 162,800 and median annual 70,720 ML). Rain occurs in winter with the passage of cold fronts and in late summer associated with tropical cyclones.* Courtesy WRC.

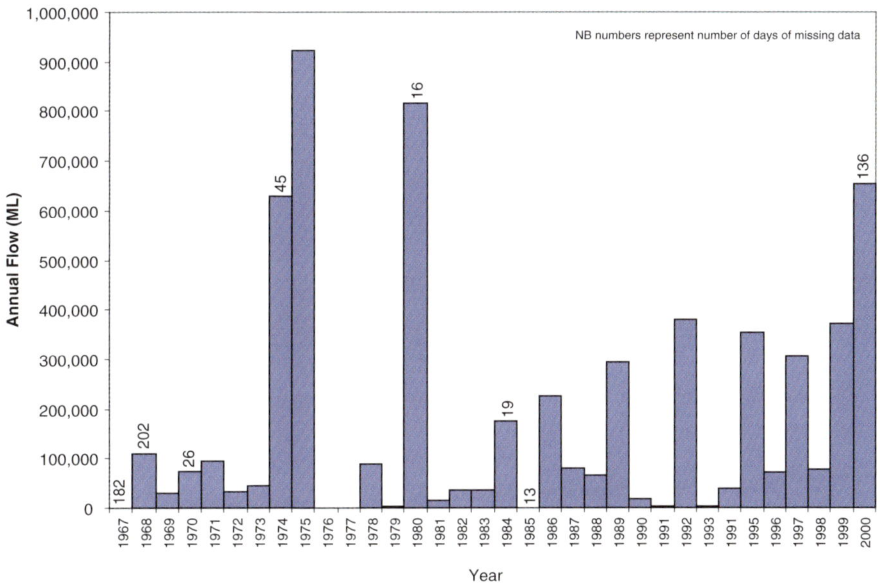

the mouth reflects the inflowing marine water, decreasing upstream to about 4‰ near the ford.

Riverine flows and marine processes of tides, waves, as well as wind influence the estuary. At low river flow, when littoral drift along the coast dominates, sediment accumulates, elongating and narrowing the channel. With heavy river flow, sediment is passed to the ocean and the channel broadens and deepens, but periods of high flow also bring sediments into the lower reaches and redistribute sediment. The waters are turbid due to the high sediment load of river flows and wind stirring.

The deepest part of the estuary lies immediately behind the flood tide delta at the mouth. The remainder, filled with sand and silt bars, is generally less than a metre deep and shallows even more further above Nannygoat Well about 15 kilometres upstream. The mouth of the estuary has probably never closed but the channel through the sea bar is very shallow and narrow, the sandbars forever changing the tidal and river flow. Although the channel is now dredged every year in October–November before the rock lobster season opens, passage through the estuary mouth is still difficult and dangerous when ocean swells are high. During these times boats are loaded and unloaded across the reef at Red Bluff (Fig. 5.5). The coast is exposed to high wave energy from the north-west in winter and south-west in summer. Based on estimates of longshore currents, littoral sand drift is estimated to be about 33,000 cubic metres northwards each year.

Living around and in the Murchison River Estuary

The vegetation of samphires and rushes lining the estuary, although characteristic of estuarine areas, consists of slightly different, more northern species. The river gorges and estuary surrounds are covered with jam *Acacia acuminata*, orange pea flowers *Jacksonia cupulifera* and rock sheoak trees *Allocasuarina huegeliana* with river redgum *Eucalyptus camaldulensis*, swamp sheoak *Casuarina obesa* and northern bottlebrush *Callistemon phoeniceus* (Fig. 5.6).

Little is known of the animals inhabiting the estuary, in part reflecting the general lack of studies along the northern section of the west coast. The estuary and fresh water springs from the Tumblagooda Sandstone provide refugia for birds and animals.

Collections made by Dr Rod Lenanton and Dr Ernest Hodgkin in 1977 (Table 5.1) show that the fish species are typical estuarine species tolerant of varying conditions. Popular fish include mulloway, tailor, King fish, sea and yellow-eye mullets and whiting that enter from the sea. The estuarine black bream breed in the upper reaches. Black bream are also caught near the mouth when the river flows but retreat up the estuary when sea water enters. Mud (mangrove) crabs are found upstream and blue swimmer (manna) crabs nearer to the mouth. Although there is no commercial fishing in the estuary and river, recreational fishing is a major tourist attraction and catches are quite substantial.

Settlement and changes

The advent of the pastoral industry brought changes to the area of the Murchison where Vlamingh and Grey had watched Aboriginal people fishing. One of the oldest and largest pastoral leases in Western Australia, Murchison House Station, was first settled in 1848. Ernest von Bibra the first lessee probably came overland from his other lease at Tamala Station. He used the

FIG. 5.5 *When the bar is impassable due to high swells, boats are moored down the coast at Red Bluff and Jacques Bay, where the catch is off-loaded and the vessels reprovisioned. This is almost as hazardous as entering the estuary through the bar in rough weather, and mishaps not unlike that of the colonial explorer Grey occur.* Source: The *West Australian*, 22 April 2000; Courtesy West Australian Newspapers Ltd.

FIG. 5.6 *The upper reaches of the Murchison River Estuary differ from the estuaries further south with turbid waters lined with outcrops of red flat-bedded Tumblagooda Sandstone overhung with flooded gums and Casuarina. Fish tolerant of the estuarine conditions are still abundant and a source of food for birds and the tourists that now flock to the area. Rod Lenanton and Ernest Hodgkin preparing to sample the estuarine fish, September 1977.* E. P. Hodgkin.

TABLE 5.1 *Fish and crab species caught by Rod Lenanton in the Murchison River Estuary, 21–22 September 1977. Salinity 1977: near the mouth 37‰ and in the upper reaches 25.7‰ and 27.6‰. M = mouth, U = upper tidal limit. Life cycle categories: S, marine straggler species; O, marine estuarine–opportunist species; E, species confined to estuaries; E&M, species represented by estuarine and marine populations; A, semi-anadromous.*

Fish name	Species name	Location in estuary	Life cycle
Black bream	*Acanthopagrus butcheri*	M, U	E
Yellow-eye mullet	*Aldrichetta forsteri*	M, U	O
Yellowtail grunter	*Amniataba caudavittatus*	M, U	E
Gobbleguts	*Apogon rueppelli*	M	E&M
Elongate hardyhead	*Atherinosoma elongata*	M	E
Eastern Fortesque	*Centropogon australis*	M	E&M
Honeycomb cod	*Epinephelus tauvina*	U	
Long-finned goby	*Favonigobius lateralis*	M	E&M
Roach	*Gerres baconensis*	M	
Striped (six-lined) trumpeter	*Pelates sexlineatus*	M	O
Southern sea garfish	*Hyporhamphus melanochir*	U	E&M
Flat-tail mullet	*Liza agentea*	M, U	E
Stripey	*Microcanthus strigatus*	M	
Fan-bellied leatherjacket	*Monacanthus chinensis*	M, U	S
Sea mullet	*Mugil cephalus*	M, U	O
Perth herring	*Nematalosa vlaminghi*	M, U	A
Wavy grubfish	*Parapercis haackei*	M	
Goatfish	*Parupeneus fraterculus*	M	
Short-headed sabre tooth blenny	*Petroscirtes breviceps*	M	S
Tailor	*Pomatomus saltator*	M	O
Ogilby's hardyhead	*Pranesus ogilbyi*	M	O
Swan River goby	*Pseudogobius olorum*	M, U	E
Small toothed flounder	*Pseudirhombus jenynsii*	M	O
Tarwhine	*Rhabdosargus sarba*	M, U	O
Western sand whiting	*Sillago schomburgkii*	M, U	O
Common blowfish (Banded toadfish)	*Torquigener pleurogramma*	M	O
Mulloway	*Argyrosomus japonicus*	U	O
Blue manna crab	*Portunus pelagicus*	M, U	
King prawn	*Penaeus latisulcatus*	M, U	
Smooth shore crab	*Cyclograpsus audouinii*	U	

well-watered river flats to raise horses for the Indian army and grow food for the convict lead miners upstream at Galena. The Indian connection was re-established in the 1980s when Prince Mukramm Jah, the eight Nizma of Hydrabad, held the lease.

Later the pastoral lease and a smaller freehold area were held by a number of people and used for grazing sheep and cattle, together with some wheat for use on the station. The lease originally extended across the current townsite of Kalbarri to the south of Red Bluff, but problems with stock control and the growing tourism traffic saw the lease reduced in area with a boundary about 2 kilometres north of the townsite. The most recent lessees have developed a tourism business around the old buildings with tours along the upper reaches of the estuary.

In the 1920s and 1930s families from the Galena mines and northern farming areas around Ajana and Binnu started to visit the lower estuarine reaches for holidays and fishing, a prelude to the current popularity of Kalbarri as a holiday destination and fishing port. Huts built during the war as a holiday camp for troops started the more formal period of settlement although no houses were built until 1949. One of the first in the early 1950s, a boarding house providing accommodation for men who came to fish, charged £10 ($20) per week. As part of the Northampton Shire, the town of Kalbarri in the surrounding Kalbarri National Park was gazetted in 1951 with a population of about 2,000 at the end of the 1990s. Tourism is still the principal source of employment and income, the population increasing during holiday periods and from September to October with visitors to the gorges and wildflowers in the Kalbarri National Park.

Box 5.2 **TUMBLAGOODA SANDSTONE AND GIANT SEA SCORPIONS**

The gorges of the Murchison are a spectacular feature of the national park, and these ancient rocks have a fascinating history (Figs 5.7 and 5.8). The age of Tumblagooda Sandstone has been widely debated. Until 1999 a Late Silurian age based on the presence of fossil tracks was widely accepted; a synthesis of evidence however now suggests a much older time of deposition. Currently the formation is thought to be no younger than Late Ordovician in age.

Outcrops of the Tumblagooda Sandstone occur patchily over an area of about 7,500 square kilometres between Greenough and the Murchison River, extending from the eastern margin of the Perth Basin to the coast. The sandstone also extends below the surface northwards into the Carnarvon Basin marking the path of a gigantic river system that drained the supercontinent of Gondwana more than 430 million years ago during the Ordovician Period. The Ordovician was a time of erosion when floods and winds swept sand and silt across the landscape before plants colonised the land. The age of the rock is revealed by the rare and poorly preserved microfossils, and the relationship between the younger rocks above and the older underlying rocks of the Northampton Complex. In some areas the Tumblagooda is overlain by chalky marine sediments of Cretaceous age, but in other areas these have been eroded leaving the flat-topped plateaus around the Murchison Gorges at Kalbarri.

The most common fossils, termed 'trace fossils', are burrows left by worms and trackways left by a variety of extinct arthropods. Many were very large, the largest being eurypterids or 'sea scorpions', animals that once lived in the shallow seas or estuary-like areas near river mouths.

Eurypterids, although resembling scorpions, are actually related to the living group of arthropods known as the king crabs. Eurypterids became extinct in the Permian about 260 million years ago. Scorpions have four pairs of legs but eurypterids have six pairs of appendages, three pairs at the front under the flat square head used for feeding and three pairs for walking or swimming, the hind pair generally larger and paddle shaped. The long body, three to four times the length of the head and divided into twelve segments, ended in a tail spine. The oldest forms were tiny but some grew up to 2 metres in length. Eurypterids were the main predator at that time and some fossils have been found with a stomach full of other small eurypterids.

In 1990 the rocks at Kalbarri revealed another ancient group of animals, the euthycarcinoids, freshwater arthropods, previously only known from France and the

FIG. 5.7 *(a) The layered rocks reveal different types of river flow: thick bands of coarse sand reflect strong flow during floods, thin layers of finer sediment often with ripples indicate periods of slower flow and very fine sediment transported by winds. (b) In some places the bedding has been disrupted or 'bioturbated' by organisms that lived buried in the sand. The same process occurs in the estuary today. The eurypterids and other arthropods left their footprints in coarse sediment that was later covered by fine wind-blown sand, the tracks exposed where the layers of rocks separated. The small depressions in the layers of rock mark small pools in what was once a sand flat. At this time in the earth's history, most animals were living in water and the great land invasion had only just begun. The occurrence of footprints around the pools indicate that the fossil tracks at Kalbarri are among the earliest record of animals walking out of water, possibly as they moved over the sand between waterholes as floodwaters subsided.* (a) Red Bluff, A. Brearley, July 1999 and (b) detail of burrows, M. D. Seale, Courtesy WAM.

United States in rocks 300 million years old. This group of cockroach-like creatures with eleven pairs of legs leave trackways with multiple imprints. Based on their shape euthycarcinoids were proposed as the ancestors of insects, but the age of the rocks in which they were found was too young and the theory discounted. The find at Kalbarri, which extends the fossil record of the group further back in time, means that this theory may be more likely. Scooped out areas where animals probably fed, together with the underlying U-shaped burrows, also allow speculation about what they were feeding on.

Source: Gorter et al., 1994; Haig, 2002; Hocking, 1991; McNamara & Trewin, 1993; Trewin & McNamara, 1994; Schmidt & Hamilton, 1990.

a

b

c

FIG. 5.8 *Fossils from the Tumblagooda Sandstone: (a) rock with an impression of a eurypterid* (K. Brimmell)*, (b) an artist's reconstruction and (c) tracks, 'trace fossils' made by a eurypterid walking over wet sand, now the Tumblagooda Sandstone in the Kalbarri National Park. The animal making the tracks had three pairs of legs and was about a metre long.* Courtesy K. McNamara, WAM.

Kalbarri National Park

The Kalbarri National Park, 186,000 hectares in area, was gazetted in 1963, with the boundaries changing as the popularity of the area and town have grown. In 1994 lands for future town growth, rubbish disposal and a new airport area were excised from the park and 30,000 hectares of unallocated Crown land to the north-east were allocated by State cabinet for inclusion in the park.

The national park covers an area of spectacular scenery of ancient rocks eroded by winds and waves; the multicoloured layers revealing secrets of Earth's history before plants and animals lived on the land. (See **Box 5.2 Tumblagooda Sandstone and Giant Sea Scorpions**.) The isolation and ruggedness of the river gorges offer great opportunities for adventure: rock climbing, canoeing when the river is full, bushwalking or simply quiet contemplation. The 20 kilometres of coastal cliffs offer a great contrast to the pounding ocean swells.

Kalbarri is renowned for its wildflowers with over 800 species representing the northernmost extent of the species rich South-West flora including the kwongan, that are replaced further to the north and east by the low diversity tropical and desert floras. The variety of soils from sand to heavy soils, rocky outcrops of sandstone as well as the coastal cliffs all host different plant communities. (See **Box 5.3 Kwongan**.) How changed is our perspective today to that of A. C. Gregory in 1848, the first explorer to traverse the country downriver to the mouth, who described the country as 'very indifferent'.

Management of the park is not without challenges. In the past sheep and cattle on pastoral leases grazed the vegetation. In the park area pigs and goats still graze and uproot vegetation competing for shelter with other animals, while feral cats and foxes prey on smaller native animals. Greater popularity has also brought problems of providing access while maintaining the wilderness experience and protecting natural habitats and native animals.

Fishing port on a wild coast

Kalbarri is also a fishing port principally for boats in the rock lobster industry, although wet-line fishers also use it. In 2000 Kalbarri was 'home port' for sixteen commercial rock lobster vessels, eight commercial wet-line boats, four charter vessels, one carrier boat and three hire-and-drive boats. A further sixty vessels used the port to off-load catch and for supplies during the Big Bank white crayfish run between February and March (Fig. 5.9).

Keeping the estuary open

With sediment accumulating in the entrance channel each year, maintaining access for boats has become a major issue with implications for changing the shallow nature of the basin and disposal of spoil. Between 1975 and 1983, dredging was undertaken by local fishers, and in 1984 by the Geraldton Port Authority, both with varying success. In 1985 (3,400 cubic metres) and 1987 (2,500 cubic metres) sediment was removed by the Department of Marine and Harbours creating a channel that persisted for about a month. Each year between 1988 and 1998 the entrance channel was dredged and about 20,000 cubic metres removed annually (maximum of 35,000 in 1997). A review in 1990 concluded that early efforts were unsuccessful due to the small volume of material and the 'sidecasting' of spoil, rather than removal from the system.

The volume of dredging was increased in 1989 and 1990, and used as infill for the river foreshore creating a channel that persisted for a little longer—about 3 months. Between 1991 and 1994 (minimum approximately 13,000 cubic metres in 1993 and maximum 21,500 cubic metres in 1994), spoil

Box 5.3 **KWONGAN**

The area around the Murchison River in the Kalbarri National Park is famous for its rich sandplain or kwongan vegetation. The name kwongan, derived from the Aboriginal language of the South-West, essentially meaning sandy country with open scrubby vegetation, first came into use in the mid-1970s although various forms of the term had been recorded as early as 1842. Other similar terms for areas with a mediterranean climate (medium temperatures with winter rain and summer drought) and now in common use include *maquis* in France, *macchia* in Italy, *chaparral* in California and *fynbos* in South Africa.

The kwongan characteristically contains sclerophyllous (hard-leaved) shrubs and heath including the families Myrtaceae (tea-trees and bottle brushes), Proteacea (banksias, grevilleas and dryandras), Leguminosae (pea-flowers) and the Epacridaceae (heaths and pearl flowers). Kwongan is famed for its diversity and includes many of the endemic species for which the South-West Botanical Province is renowned. The composition of the flora also varies on a small scale, with adjacent areas containing different species. This diversity on a small scale matches that of tropical rainforest. The diversity is probably due to many interacting factors, but the most important appears to be the development of strategies for acquiring nutrients from the poor soils.

Source: Beard, 1990; Pate & Beard, 1984.

FIG. 5.9 *A fishing boat in the sheltered water of the entrance channel to the Murchison Estuary, 1998. When conditions are favourable boats can pass through the narrow passage between Oyster Reef on the right and Chinamans Rock on the left into Ganthaeme Bay. The waters of the estuary are very shallow and turbid.*
Courtesy B. Hegge, DALSE.

was discharged to the ocean beach north of Oyster Reef. Sediment deposited after strong river flow in June 1995 (35,000 cubic metres) was removed from the south side of Oyster Reef and placed along the eroding Chinamans Beach. This practice is now controversial. Opponents of the practice suggest that sediments gradually move into the channel, while proponents argue that more sand is needed to reduce erosion.

In the past the channel was only dredged to 2.0 metres to cater for the shallow draft of most vessels operating in the Kalbarri Region, with larger vessels only having access during high tides. With larger boats now using the marina, the channel has been deepened to 3.25 metres (2.5 below datum) to cater for a nominal 20 metre-long trawler with a beam of 5.7 metres and a draft of 2.0 metres. This gives a 0.5 metre clearance with waves at low tide. The width of 34 metres allows two vessels to pass safely.

After considering a range of options, the Boating Access Improvement Committee in 1996 recommended that future dredging should at least remove the volume accumulating in the entrance each year, approximately 36,700 cubic metres in 1998 and 42,900 cubic metres in 1999. Between September and October 2000 about 30,000 cubic metres were removed, along with an additional amount of material from the channel upstream. The most suitable time for dredging was found to be October, a period with suitable weather conditions and least disruptive to boat access during the West Coast Rock Lobster Season from 14 November to the end of March.

Sediments accumulating in the channel are sand-sized carbonates and quartz with low concentrations of heavy metals such as arsenic, cadmium, chromium, copper, mercury, lead, nickel or zinc reflecting its marine and riverine sources. Underlying this at 2.5–3.0 metres is a layer of fine mud and clay. Earlier dredging operations disrupted the clay layer and created turbid conditions, while disposal of dredge spoil within the estuary allowed sandbanks to reform during periods of high flow. Changes to dredging protocols now limit the depth of dredging to less than 3.0 metres and require spoil to be disposed of on the seawards side of Oyster Reef about 900 metres north of the river mouth (Fig. 5.10). Generally littoral drift moves most of the material northward away from the mouth although material washes back into the estuary during high seas.

FIG. 5.10 *Discharge of sediment to the north of Oyster Reef, November 1999. Annual maintenance dredging in 1991–1996 removed 17,000 m³ in five years, but 1996–2000 the average volume dredged was approximately 35,000 m³ annually.* Courtesy Jesz Fleming and Associates; DALSE.

THE 'FOSSIL ESTUARIES': HUTT, BOWES, OAKABELLA, OAKAJEE, BULLER

South of the Murchison a series of small rivers, the Hutt, Bowes, Oakabella, Oakajee, Buller and the larger Chapman flow from the hills around Northampton and the Moresby Range across the narrow plain, a distance of about five kilometres to the coastal dunes. In most winters the streams rush down the sandy sediment-filled beds breaking to the sea through narrow gaps in the coastal limestone. Generally the bars close over quite quickly as flow slows, but when sea level is high sea water flows inwards. Pools of water along the rivers are probably maintained by springs from the Tumblagooda Sandstone. The estuaries are now little more than 'fossil estuaries', largely filled with sediments; those red in colour from the upper catchment and pale dune sands from the migrating dunes and ocean currents (Map 5.3).

Hutt River

The Hutt River, about 50 kilometres long, flows from the hinterland in the east. Near the coast the river generally flows to the north behind a large dune before entering the sea (Figs 5.11 and 5.12). After heavy rain however, heavy flows break straight through the dunes. Sea water enters when the bar is open forming a small riverine estuary. The river usually contains pools of water even over summer, possibly derived from springs, and the water is generally brackish 1.5–2.0‰.

In the past the river probably flowed through the Hutt Lagoon (about 24 kilometres) before infilling and formation of a bar to the north isolated the lagoon from the sea. Over time the salinity of the water increased as cycles of wetting and drying progressed, a feature noted by Vlamingh in 1697 and utilised in the early 1900s for a salt harvesting industry. The red colour of the water, also noted by Vlamingh, is now known to be a pigment made by a single-celled alga *Dunaliella salina* and used in the production of beta-carotene (Fig. 5.12). (See **Box 5.4 Super science meets the natural world**.)

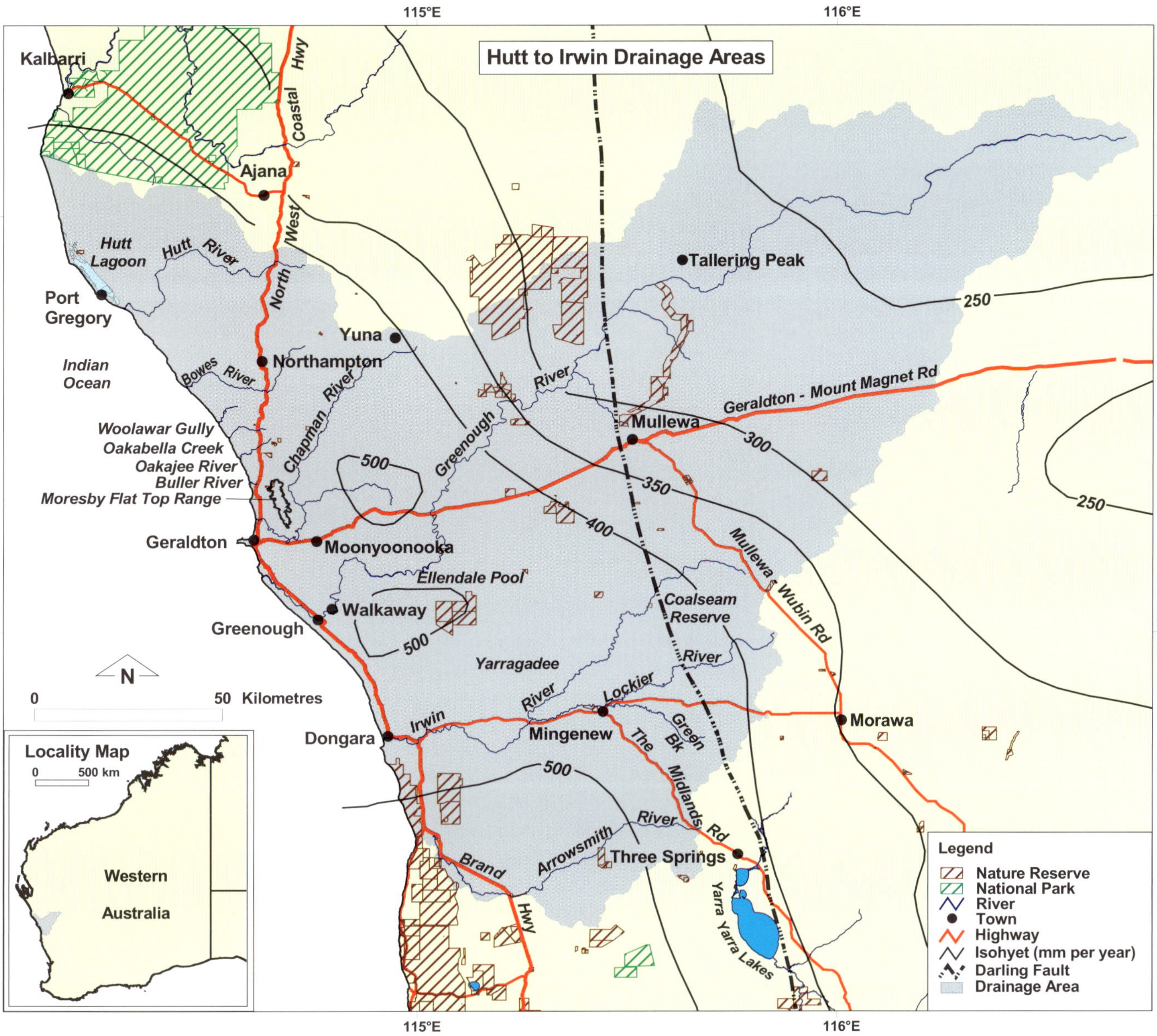

MAP 5.3 *The rivers and estuaries from Hutt River to Irwin River and Arrowsmith Drainage Area. The Greenough, the largest river in the area, flows from the Yilgarn Plateau about 180 km inland.* Courtesy DoE.

FIG. 5.11 *The mouth of the Hutt River, aerial view, April 1982. The entrance lies to the north of an extensive dune system. In 1999 massive flows broke the bar directly opposite the main channel in the right foreground. The white salt deposits around Hutt Lagoon lie to the left. The cleared cultivated land in the middle was settled in the 1850s and used as a convict-hiring depot, Lynton. Note the unvegetated migrating dunes near the coast and the winding course of the estuary.* E. P. Hodgkin Collection.

Box 5.4 **SUPER SCIENCE MEETS THE NATURAL WORLD**

Beta-carotene is a pigment, one that colours fruits and autumn leaves, and is found in chloroplasts or the light harvesting part of green plant cells. In plant cells, carotenes harvest particular wavelengths of light and pass these onto the chlorophyll molecule for the process of photosynthesis. They also protect the cell against photoxidation and are precursors of vitamin A: characteristics now utilised in the biotechnology industry. Beta-carotene is used as a food colour in products such as margarine. The antioxidant feature and ability to combine with free radicals produced by fats, stress and ultraviolet light also make it a valuable food additive for animal and human consumption, supplementing fresh fruit and vegetables.

The beta-carotene industry at Hutt Lagoon depends on the natural features of the area and its relative isolation uncontaminated by industry. The long hot dry summers maintain high evaporation rates, the near constant winds mix the water, and there is a source of clean brine from the ocean. Salinity is maintained near 25% NaCl, a compromise between the slightly lower salinity (18–21%) favourable for algal growth and the higher salinity (> 27%) favourable for beta-carotene production. Lower salinities also favour other non-carotenoid algae and grazing single-celled animals (protozoa) and brine shrimps (see also Table 1.10).

FIG. 5.12 *The ponds in Hutt Lagoon used for growing microalgae for β-carotene production, aerial view, ca 2000. Sea water is pumped into the large (100 ha) earth-walled ponds in the lagoon and the sun evaporates the water, increasing the salinity. Pond depth is maintained at about 20 cm for maximum light penetration favourable for algal growth. The strong south-westerly winds stir the ponds and mix nutrients. The brines are pumped to a processing plant where a combination of settling, flocculation and centrifuging are used to harvest the algae without breaking the cell membranes. The product is then dried or the β-carotene extracted with solvents and stabilised. The red colour was first noted by Vlamingh in 1697.* Lesley Borowitzka, Cognis Australia Pty Ltd.

Two types of brine shrimps that live in Hutt Lagoon prey on the microscopic green alga *Dunaliella salina*. The native *Parartemia*, preferring less saline water, are often abundant in pools around the fringing marshes after rain. The other type *Artemia*, an introduced species, is more salt tolerant. Now found in salt ponds or lakes used for salt production throughout Australia including those at Rottnest Island, *Artemia* was probably brought to Australia in salt harvesting equipment last century.

Source: Borowitzka, 1986; Borowitzka & Borwitzka, 1990; Geddes, 1981; Playford, 1998.

Bowes River, Wollawar Gully and Oakabella Creek

The upper Bowes River flows in a south-westerly direction through the old rocks of the Northampton Plateau then turns to the west-north-west before flowing west of south-west in the lower reaches. The other main tributary, Tomelton Brook, flows in a south-westerly direction parallel to the coast, while the lower Bowes flows in a valley cut through the old Jurassic rocks. Previously the Upper Bowes probably flowed southwards to the ocean though the channel of the current Buller River. The original Bowes, a shorter river, probably cut back to capture the Upper Bowes forming the present river system. The Oakabella, Oakajee and Buller Rivers today drain the old southwards flowing channel (Figs 5.13–5.15).

Near the coast, the Bowes River and Estuary are bordered by alluvial soils typical of the area that lies behind the dunes. The position of the river mouth may vary across the width of the bar.

FIG. 5.13 *Wollawar Gully looking eastwards, aerial view, April 1982. The flat-topped mesas in the background, remnants of the Victoria Plateau, are surrounded by cleared farmland: this is the landscape recorded in George Grey's journal of his travels in 1839.* E. P. Hodgkin Collection.

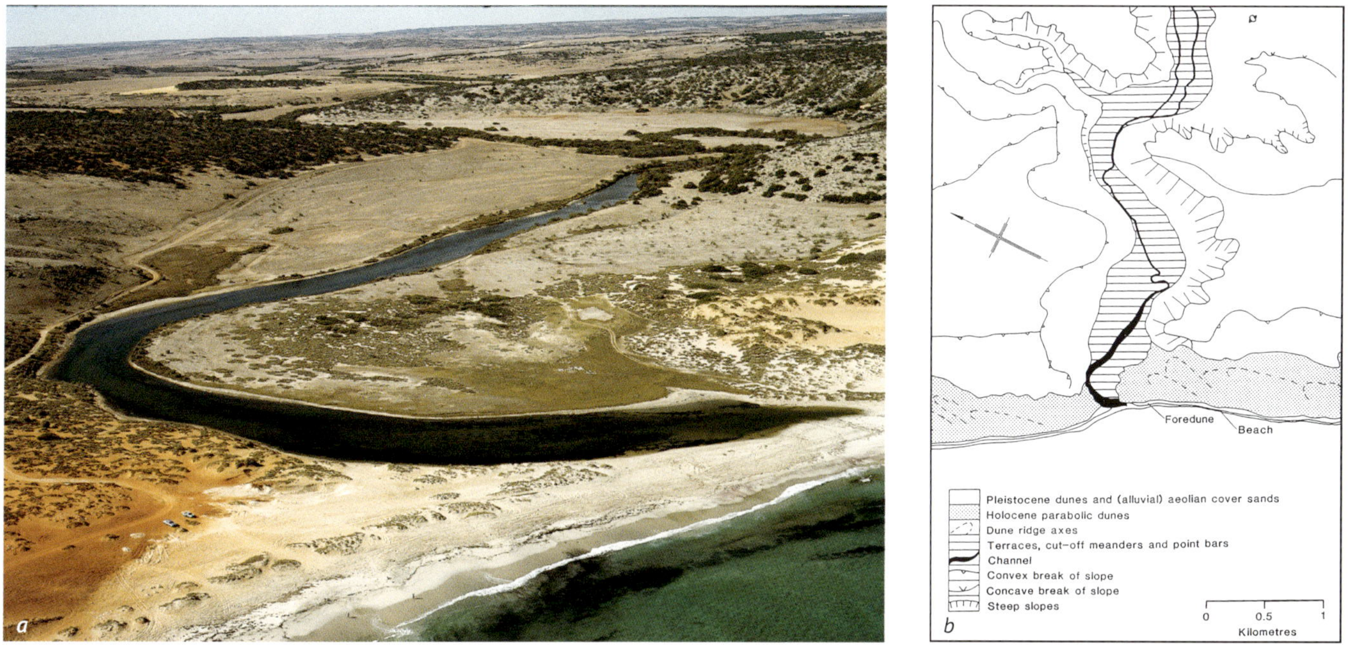

FIG. 5.14 *Bowes River Estuary: (a) aerial view, April 1982. The river flows through a break in the Tumblagooda Sandstone, and springs replenish the river pools and support paperbarks upstream year round. The water is often fouled by sheep droppings and rubbish. Aboriginal paintings on the rock outcrop in this area (Wolla Wolla or Willigulli), about a kilometre from the mouth, are a reminder that this area was once home to large numbers of Aboriginal people.* E. P. Hodgkin Collection. *(b) The Bowes River landscape showing the typical features of a riverine estuary with alluvial point bars and terraces surrounded by high dunes of Pleistocene age.* Hesp, 1984 from aerial photograph WA 1756 (c, No. 5126).

New plans — Oakajee to Buller

Plans for a new port, steel mill and industrial park between the Oakajee and Buller Rivers during the 1990s heralded another phase in the history of the landscape (Figs 5.16 and 5.17). Although land was rezoned for industry and the development was approved, the plans were 'mothballed' for some years. However, in 2004 interest in construction of the port was renewed despite expansion and redevelopment of the Geraldton Port. Situated on sparsely populated former cropping and grazing land, development will bring more people to the whole Mid-West Region: a boost to the region's economy but with implications for the environment, not just near the site but also from people-pressure in the surrounding city and hinterland (Map 5.4). New noises and odours in the immediate area and further afield along the railway and access roads will be issues for people in the vicinity. The proposed port structure extends into the bay about 2 kilometres, past the 20 metres depth contour. Changes to water and sediment flows along the coast are likely: the question for now remains as to how the pluses and minuses are balanced when other ports are planned.

FIG. 5.15 *Oakabella Creek, aerial view, April 1982, showing the sediment-filled channel that extends upstream for about 5 km. In places the red alluvium is ~10 m high, capped in part with rubble of similar form to deposits found in the Oakajee and Buller. On old Admiralty charts the creek is marked as the Yellow Gully. The creek flows to the sea during most winters.* E. P. Hodgkin Collection.

FIG. 5.16 *The Buller River Estuary looking eastwards, aerial view, April 1982. Although the catchment covers 74 km², most of the rivers flows through dunes along the coast west of the flat-topped mesas of the Moresby Ranges in the background. The river flows to the sea most winters but usually remains open for only a few days. In the right foreground the river channel is lined with river gum* Eucalyptus camaldulensis, *sheoak* Casuarina obesa, *tea-tree* Melaleuca *sp., wattles* Acacia *sp. and Tamarix. The left side shows farmland now in the buffer area of the proposed industrial park.* E. P. Hodgkin Collection.

FIG. 5.17 *Oakajee River, aerial view, April 1982. Most winters, the Oakajee River discharges through a narrow gap in the stable dunes to a wide area of mobile foredune sand with a 3 m-high cliff on the north side. In the past the river probably flowed southwards along a path now blocked by dunes. Inland of the dunes, the river winds through flats of beach sand with a dense tree cover. The planned new port area will probably cross the current opening of the Oakajee with a new entrance created further north.* E. P. Hodgkin Collection.

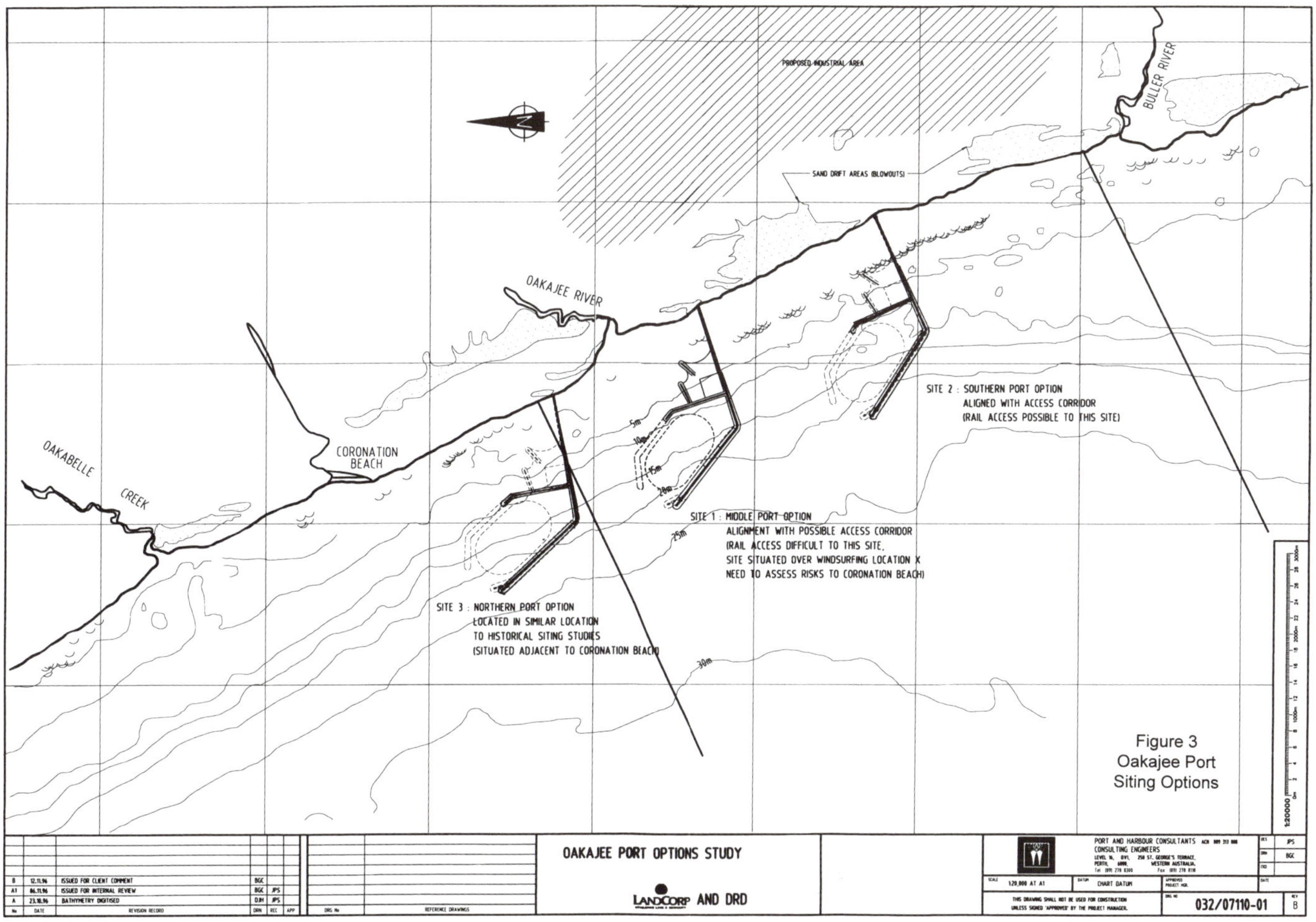

MAP 5.4 *Concept plans for a new deepwater port at Oakajee. Published EPA Bulletin 849, 1997.* Alan Tingay & Associates (ATA Environmental) 1996, 1997. Courtesy DIR.

CHAPMAN RIVER

The Chapman River has a relatively large catchment (1,160 square kilometres) compared to other rivers nearby that flow from the Moresby Flat Topped Range. Rising about 60 kilometres to the north-north-east near Yuna on the Victoria Plateau, the Chapman flows in a southerly direction east of the Moresby Flat Topped Range. After entering the coastal plain just north of Moonyoonooka (15 kilometres east of Geraldton), it changes direction and flows north-westwards for about 11 kilometres to Bluff Point at Geraldton in Champion Bay (Fig. 5.18). Filled with red alluvial sediments derived from the upper rocks of the upper catchment, the rich soils of the Chapman Soil Association attracted the first settlers in the 1840s. As a result cleared farmland covers most of the catchment, where little of the original vegetation remains. Although the deep valleys and river sediments are evidence of natural weathering and erosion over thousands of years, steeply sloped cleared farmland is particularly prone to erosion. As a result gully formation, loss of soil and expensive fertiliser is a problem throughout the area. (See **Box 5.5 River running red**.)

FIG. 5.18 *The Chapman River at Geraldton looking eastwards towards the Moresby Flat Topped Range, aerial view, April 1982. The estuary is now surrounded by urban areas, little riparian vegetation remains, and most is degraded. The estuary waters are often brown due to eroded sediment, the river now little more than an urban drain. Although the mouth can open directly to the sea (August 1981), it is often deflected to the south. On close examination, sand around the mouth is pink in colour due to the tiny garnets washed downstream. Nazareth House ca 1941, on the south bank of the Chapman River, is typical of buildings in Byzantine style designed by Monseigneur Father John Dawes.* E. P. Hodgkin.

Box 5.5 RIVER RUNNING RED

In an area receiving about 400 millimetres of rain annually, unusually heavy rain in June 1986 (185 millimetres of rain in six days) saturated the catchment resulting in erosion affecting approximately 50,000 hectares of land. About 22,000 tonnes of suspended sediment was lost to the ocean at a rate of one tonne of sediment every seven seconds at peak flow (119 cubic metres per second). If the losses in July were similar to those of June, an additional 17,000 tonnes entered the ocean, bringing the yearly loss for 1986 to 39,000 tonnes at an estimated cost of $73,000 for nitrogen based on $520 a tonne.

Phosphorus losses were not considered as costly as most was bound to soil and not available for plant growth, but based on a loss of 10%, phosphorus replacement value was estimated at $900 (based on $130 a tonne). The amount of organic material washed away remains unknown. When the more coarse sand is included in the calculations, 16,000 to 25,000 tonnes of soil was lost or approximately 3 to 5 tonnes per hectacre.

Most losses were from uncontoured, traditionally cultivated cropland with little loss from land under pasture even on steeply sloping paddocks. Areas of contour banking showed reduced erosion, and reduction in cultivation allows rainfall to penetrate while reducing soil loss and maintaining or increasing crop yields on similar soils.

Source: Bligh, 1987.

The Chapman River flows to the sea most winters but does not generally flow from summer rain, although some water generally remains over summer in the lower 1.5 kilometres near the bar. When the river is flowing, the floodwaters erode the lower banks. The water level in 1972 was about 3.5 metres. Although water quality measurements have only been taken infrequently, they indicate the estuary is eutrophic with high concentrations of nitrogen, phosphorus and chlorophyll *a* (Graph 1.2). While the bar is open, salt water intrudes about 1.5 kilometres upstream. Salinity in this estuarine area, however, is usually quite low. Generally sand and accumulations of seaweed and seagrass from the offshore reefs block the estuary mouth. The sandbars near the estuary also vary with the frequency and direction of storm waves and erosion of the adjacent dunes. In 2002–2003 these shoreline changes appeared to have been compounded by a marina development and dredging during port modifications.

A small strip of water world

Although somewhat degraded and considered in moderate to poor condition due to a long history of disturbance and proximity to urban areas, areas of natural vegetation along the Chapman still retain conservation merit as a refuge for wildlife representing the original natural environment. The value of parkland was recognised as long ago as 1849, when 500 acres (200 hectares) of bushland (Spalding Park) was reserved by the Geraldton Municipal Council. In 1989, 364 hectares of remnant bush, historic structures and recreational facilities were proposed as a Regional Park. This has now been expanded to include The Chapman Regional Wildlife Corridor Project, funded by grants from the National Heritage Trust, to implement management plans covering the riparian area of the river and the floodplain from the City–Shire boundary at Strathalbyn to the Cutubury Reserve north of Moonyoonooka, and incorporating public reserves and freehold farmland.

The estuarine reaches of the Chapman support waterbirds, including pelicans, swans, spoonbills, five species of duck, two species of grebe, three species of herons, four egret and five cormorant species as well as the occasional gulls, terns and waders including eleven species of trans-equatorial migrant. Further upstream the lining vegetation provides refuge for over sixty species of bush birds.

Little is known of the invertebrates within estuarine waters but they appear to be fresh to brackish water species. Similarly the fish, although not very abundant or diverse, are characteristic of brackish to fresh water conditions; the presence of introduced mosquito fish and *Tilapia* reflecting the disturbed nature of the system (Fig. 5.19 and Table 5.2). The presence of *Tilapia* is a great concern, a threat to the native fish, but also a potential source from which other waterways can be invaded. (See **Box 5.6 *Tilapia*—the Mozambique Mouth Brooder**.)

TABLE 5.2 *Fish species in the small riverine estuaries caught by Rod Lenanton using a seine net in the Bowes, Chapman, Irwin and Hill River Estuaries, 21 September 1977. Salinity: Bowes 4.7‰, Chapman 22.4‰, Irwin 4.1‰, Hill 7.2‰ at surface and 9.1‰ at bottom. P indicates presence, * from records of McCotter, 1993 and Fisheries, 2000. Int. sp. = Introduced species. Life cycle categories; O, marine estuarine–opportunist species; and E, species confined to estuaries.*

Fish name	Species name	Bowes	Chapman	Irwin	Hill	Life cycle and habitat
Black bream	*Acanthopagrus butcheri*		P	P	P	E
Yellow-eye mullet	*Aldrichetta forsteri*		P*		P	O
Sea mullet	*Mugil cephalus*	P	P	P	P	O
Swan River goby	*Pseudogobius olorum*	P	P	P		E
Mosquito fish*—Int. sp.	*Gambusia affinis*		P*			Fresh–brackish
Tilapia Mozambique Mouth Brooder—Int. sp.	*Tilapia mossambicus*		P			Fresh–brackish

Box 5.6 *TILAPIA* THE MOZAMBIQUE MOUTH BROODER

Tilapia mossambicus or *Oreochromis mossambicus*

As the name Mozambique Mouth Brooder implies this fish is an import to Australia; its habit of mouth brooding is a protection against conditions usually stressful to immature fish, allowing it to invade new territory.

Tilapia can grow to 400 millimetres and up to 2 kilograms in weight, although most are smaller. They are generally dark in colour, with large fleshy lips and have a single long dorsal and a large anal fin. They often swim in groups or small schools.

Probably brought to Australia as ornamental fish, they are now found in many rivers throughout Queensland and in Western Australia. First recorded in the Gascoyne at Carnarvon in 1981, *Tilapia* are now found in the Minilya and Lyndon Rivers to the north in the Pilbara area and the Chapman. They are considered a major threat to native fish, competing for food and altering the environment. *Tilapia* eat a wide variety of food, mostly plants and detritus but also some small insects, competing with native fishes, and they prey on eggs of other fish. The nest building by males, often covering large areas, increases the turbidity of the water and changes the waterway. It would appear that in the Gascoyne where *Tilapia* is abundant some native fish have disappeared; in contrast these species are still present in the Murchison where the invader has not been caught.

Because they are farmed in some areas of the world, there has been some interest in farming them in dams throughout Australia. In Western Australia however, imports of the fish are banned as they are considered a great threat to native fish. Invasions of tropical waterways are a particular concern as *Tilapia* could threaten our prized and unique barramundi.

Source: Fisheries Western Australia, 2000; Morgan et al., 1998.

FIG. 5.19 Tilapia—*a threat to native fishes and river habitats.*
Courtesy D. Morgan, Murdoch.

GREENOUGH RIVER

Rising 240 kilometres north-east of Geraldton in the Tallering–Wandina area on the Yilgarn Plateau, the Greenough River flows in deep valleys in a south-westerly direction across the Waterloo Range on the southern edge of the Victoria Plateau (Fig. 5.20). After emerging from the ranges, the river flows to the north-west along a flood plain behind the coastal dunes through the fertile alluvial soils of the Greenough Flats, entering the sea at Cape Burney.

It has been suggested that thousands of years ago the Greenough probably flowed north-westwards behind the limestone ridge (the Back Flats), along the path where the Northampton Railway Line was built to the area of Rudds Gully from where both rivers flowed northwards to the

FIG. 5.20 *The Greenough River Estuary looking eastwards, aerial view, April 1982. The mouth of the estuary is usually closed by a sandbar and accumulations of seagrass. Seagrasses growing around sand and patchy reef offshore (dark areas) are important nursery grounds for rock lobster. The rich alluvial soils in the background were cleared for agriculture in the 1850s and the river channels are now full of sediment mobilised during floods. A narrow band of remnant vegetation growing on a limestone ridge lines the eastern bank. On the right is Cape Burney where dune vegetation although reserved is dissected by tracks.*
E. P. Hodgkin Collection.

TABLE 5.3 *Floods in the Greenough River Flats. River flows have only been recorded since 1971; earlier information based on historical records and anecdotal evidence,* Bureau of Meteorology, 1929. *Flow figures 1971–1999,* Courtesy R. Bretnall, WRC.

Year	Average Recurrence Interval (ARI) (years)	Time of flooding	Peak flow (m^3/s)	Notes
1888	170	summer rain in upper catchment	NA	South of Greenough, water was 16–17 feet (5–5.5 m) deep, for over a mile, the tops of the telegraph line only just visible. (Treadgold, 1941)
1927	40		NA	
1934	15		NA	
1953	60		NA	
1961	15		NA	
1963	20		NA	
1970	15		NA	
1971	30		410	
1988	15	May (closed Brand Hwy) *	320	
1994		March upper catchment*	240	
1999	15		435	

Average Recurrence Interval (ARI) is the statistical estimate of the average period in years between the occurrence of a flood of a given size or larger. This is not a projection of when a flood of a given size may occur again. A flood having an ARI of 100 years has a 1% chance of occurring in any one year and a 50% chance of being experienced at least once in an average person's lifetime. The 100-year ARI flood has been generally adopted as the basis for floodplain management. Adapted from *Water Facts 13*. * Notes Kim Griffin, Water Corp. Geraldton.

Chapman River. Later Rudds Gully probably formed a shorter course to the sea, probably beheading the lower Greenough. Later still a short stream cut back through the limestone hill around Walkaway to capture the Greenough, and diverted the main flow to the present course west of the limestone ridge (Front Flats), the old course now marked by Rudds Gully.

The alluvial soils that supported the Aboriginal people and attracted the first farmers in the 1850s are an indication of a long history of erosion and flooding from the hinterland. Floods have continued to be a feature in the development of the area (Table 5.3). One of the most severe, the flood of 1888, initiated the gradual demise of the Greenough Hamlet (Fig. 5.21).

FIG. 5.21 *Gray's Store, built on the flood plain of the Greenough in 1861 by Henry Gray, a merchant. The reports of George Grey, who named the river after George Ballas Greenough, President of the Royal Geographical Society at the time of his expedition, were the impetus for further exploration and eventual settlement of the area in the 1850s. A few buildings of limestone, mortared with mud cement and roofed with timber shingles are the only remains of the thriving community of about 2,000 people, once supported by the rich alluvial soils. Major floods, particularly the flood of 1888 that killed four people and inundated land for 50 km, brought financial ruin leading to the amalgamation of small holdings and a reduction in population. Today, little of the original vegetation, probably a low forest of* Acacia–Melaleuca *with scattered river gums* Eucalyptus camaldulensis, *remains; the surviving trees growing bent away from the prevailing south-westerly winds that scald the growing buds.* Courtesy National Trust of Australia (WA).

Features of the wind

Mobile dunes are a prominent feature along the coast, migrating northwards under the influence of strong south to south-westerly winds. Historically the migrating dunes probably influenced the northwards movement of the river mouth from the south near Lucy Beach. More recently, massive sand movement initiated or amplified by disturbance and fire is a potential threat to roads, urban developments and agricultural land.

In 1985 the Cape Burney drift on the southern side of the river was moving towards the Greenough River at a rate of about 7 metres a year (Table 5.4). Calculations based on a continued supply of sand and similar rate of drift suggest that the drift could reach the estuary in 65 years (2045) when it could affect siltation, increase flooding and possibly change the course of the river. At the same time, the Southgate Dune on the northern side of the estuary was advancing about 10 metres per year with the potential of encroaching on subdivisions at Tarcoola in about 80 years and reaching the Brand Highway in 35 years (2015). The Southgate Dune was also advancing onto the shoreline fuelling the littoral drifts and accretion of the beaches (Fig. 5.22).

Much of the dune area between the Greenough River northwards to the City of Geraldton is likely to be developed for housing. South of the estuary the dune area around Cape Burney, although part of a shire reserve, is dissected with tracks and invasive weeds are common. In 1998 the EPA noted the lack of an adequate foreshore reserve to protect the remnant vegetation, and recommended a mapping program to facilitate a protection and restoration program. The EPA also noted that proposed urban developments around the Greenough township on the northern side were within areas originally defined as buffers around a waste water treatment plant.

TABLE 5.4 *Movement of parabolic sand dunes near the Greenough River Estuary. The dunes are variously referred to as South Gates and Southgate; and were named after Sir James Southgate who arrived in the colony in 1858. Southgate owned and cultivated land in the Greenough area from 1866 using 'ticket of leave' labour and returned to London in 1875.* Table from Clayton & Elliott, 1985.

	Period	Rate (m/yr)
Southgate Dune		
	1942–56	11.9
	1956–62	10.7
	1962–68	8.6
Mean	1942–68	10.4
Greenough Dune		
	1942–56	16.4 (east)
		3.0 (west)
	1956–70	7.1 (east)
		7.1 (west)

The Greenough Estuary

The Greenough is a narrow inter-barrier estuary, similar in form to the Harvey and Leschenault Estuaries, lying between the dune lines. The estuary is 1.5–4 metres deep and estuarine conditions probably extend for about 7 kilometres upstream to near Bootenall Springs, however the extent of the salt wedge when the bar is open has never been examined. Surrounded by farmed land and with an every growing township to the north, the estuary still retains a natural charm. To the west lie the vegetated, although somewhat disturbed, dunes of Cape Burney and to the east some native shrub vegetation still covers the limestone ridge. Vegetation surveys in April 2000 rated the foreshore vegetation and verge vegetation as very good, with only occasional slumping of banks in the

FIG. 5.22 *The Greenough River Estuary looking north-west, aerial view, April 1982. The shallow waters are often turbid due to suspended sediment brought by floodwaters. The green colour of the estuary waters in the photograph show the abundance of phytoplankton fuelled by fertiliser in run-off from agricultural land in close proximity to the right. The sharp bend in the river is Devlins Pool, and Rudds Gully can be seen entering on the right. Cape Burney is to the south (left) and Southgate Dune to the north (top right)* E. P. Hodgkin Collection.

estuarine reaches. Vegetation and bank stability, however, declined in quality further upstream where much of the fringing vegetation had been removed, weed invasions were common, and banks unstable and eroded by stock.

The bar does not break every year and it may break in winter or in summer after thunderstorms in the upper catchment. Water levels often remain high due to ground water flow from the dunes, seepage of sea water through the bar or waves overtopping the bar. The bar usually breaks naturally but the Shire broke it when there were floods in 1983 and 1991. The bar, once open, often remains open for several months. In May 1999 the bar opened across the whole width of the channel and remained open for 3–4 months. It reopened in March 2000 on the south side, the section that had closed most recently, and remained open for 1–2 months. The northern part of the bar has been stabilised with brushwood to prevent sand from it blowing into the estuary and to force the bar to open against the more stable southern dune.

Water world habitat of the Greenough River Estuary

The waters of the Greenough River and Estuary are turbid, laden with sediment and eutrophic (Fig. 5.22 and Graph 1.2). Some reports indicate that large numbers of migrating wading birds no longer visit the sand flats, possibly because populations of worms and prawns appear to have also declined in recent years, although recreational anglers are still able to catch fish and prawns.

The infrequent opening of the bar limits the number of fish entering. Isolation of the estuary waters and evaporation also raise the salinity as indicated by Ernest Hodgkin's records of 1977, placing physical constraints on the species living in the waters. Black bream, mullet, cobbler and mulloway are among the most common with other species entering when the bar opens. Crustaceans such as prawns and blue manna crabs as well as mud crabs also live in the estuary, although prawn catches now appear to be much lower than they were in the 1970s. Anecdotal evidence suggests that large mullet are caught in river pools upstream, at least as far as Ellendale about 60 kilometres upstream from the mouth. A list of species caught in a 1977 survey by Ernest Hodgkin with Rod Lenanton, a former student now a member of Department of Fisheries, Government of Western Australia, illustrates the type of fish most frequently found in the Greenough, that is, species that enter when the bar is open and are quite tolerant of elevated salinities (Fig. 5.23 and Tables 5.5 and 5.6). (See **Box 5.7 Impression of the Greenough—handwritten notes of Ernest Hodgkin**.) Surveys of fish communities in estuaries such as the Greenough were important for Rod Lenanton's research on the overlap of fish communities of estuaries and near shore waters. This research formed the basis for fisheries management policy.

TABLE 5.5 *Fish species and crabs caught by seine net in the Greenough River Estuary in 1977 by Ernest Hodgkin and Rod Lenanton. U indicates upper estuary, M near mouth of estuary. Life cycles defined following Potter & Hyndes, 1999. Field notes made during sampling indicate that waves were washing onto the bar, but only occasional strong waves washed into the Estuary. Life cycle categories; O, marine estuarine–opportunist species; E, species confined to estuaries; E&M, species represented by estuarine and marine populations; and A, semi-anadromous.*

Fish name	Species name	Area of estuary where fish were caught	Life cycle
Black bream	*Acanthopagrus butcheri*	U, M	E
Yellow-eye mullet	*Aldrichetta forsteri*	M	O
Yellowtail grunter	*Amniataba caudavittatus*	M	E
Elongate hardyhead	*Atherinosoma elongata*	M	E
Giant herring	*Elops machnata*	U	O
Long-finned goby	*Favonigobius lateralis*	M	E&M
Flat-tail mullet	*Liza argentea*	U, M	E
Sea mullet	*Mugil cephalus*	U, M	O
Perth herring	*Nematalosa vlaminghi*	U	A
Swan River goby	*Pseudogobius olorum*	M	E
Tarwhine	*Rhabdosargus sarba*	U, M	O
Mulloway	*Argyrosomus japonicus*	U	O
Blue manna crab	*Portunus pelagicus*	U, M	
Smooth shore crab	*Cyclograpsus audouinii*		

Box 5.7 IMPRESSION OF THE GREENOUGH—HANDWRITTEN NOTES OF ERNEST HODGKIN

These notes made by 'Hodge' during his visits to the Greenough covering a period of 21 years illustrate his enthusiasm for his subject. In this era of striving for scientific rigour, of detailed measurements and replication we often lose sight of the *big picture*. In some ways the notes are simple, *but* they are the only historic ecological records of the Greenough Estuary. They are also important because they were made by a scientist familiar with other estuaries along the west coast, travelling with other scientists with whom he was formulating ideas and theories about estuarine processes and health.

30.3.1973 Greenough River Probably about 1 mile holding water behind the bar. Near the bar shore is fine sand and mud with some filamentous algae. Much detritus in suspension. Some rock along the shore. Many small fish in shallows, ? mullet. Surface salinity at mouth bar 34.5‰ 24.7°C.

Algae Gracilaria confervoides (L.) Grev. (Red) + Chaetomorpha sp. (Green), Centroceras sp. (Red), Cladophora sp. and pennate diatoms (Green).

Australonereis sp. (ehlersi) +++ and Boccardia sp. (Family Spionidae) +

20.9. 1977 (notes made when sampling fish with Dr Rod Lenanton, Fisheries WA). Bar ≈200 m wide of fine-mud, mainly calcareous sand; 1.5 m above S/L. Sea and estuary level not much different. Sea washing onto bar at visit. Probably only strong waves wash into the estuary.

Except at the mouth & head, the estuary runs between limestone ridges with rock outcropping along the banks, especially on east bank. Ridges are parallel to the shore. Scrub vegetation on west bank—to the mobile dune. Steep west bank with trees—Casuarina, Eucalyptus. Picnic spots here.

Small parallel flowing tributary from N enters halfway along east bank—this swampy at visit but carries free water.

Head of the estuary in paddocks. Upstream of this there is no discrete river channel. When flowing, the river spreads out flooding the Greenough river flats. Whole area subject to flooding with heavy rain in the catchment. Big floods reported for 1862, 1872 (Bain, M.A.), 1888 (locals), July 1963, 1907.

Bar said to have last broken 1974, probably not since. Rain in April 1977, raised water level ≈1 m—this was from 2 nights of thunderstorms. Would have brought down a lot of fine sediment.

Marginal vegetation: Salicornia, Juncus—would indicate W/L remains fairly stable most of the time. Ectocarpus on Sporobolus rhizomes.

29.5.1983 bar is closed but waves have been washing over it, probably overnight, carrying seaweed into the estuary. Salinity at mouth 41‰.

Interviews Colin Desmond and son Anthony 3.5.1994. Surface salinity at first bend 39 ppt (3.5.94), near south end of river track 35 ppt (6.5.94). Waves have recently washed over the south part of the bar which is only about 1 m above sea level.

Had discussion with Kim Griffin: There have been 4 floods in the last 6 years. May 1988, June 1991, March 1994 from 'cyclonic' rain in the upper catchment. Before 1988 there had been no floods since March 1971 (confirmed from the flow record). The 1988 flood closed Brand Highway, it was not closed in 1994. Suggests that the high water levels maintained in the estuary result from a) waves over the bar and b) ground water from the dunes.

Source: Hodgkin, E. P. Collected letters, diaries, notes and photographs pertaining to Estuaries of South-Western Australia. To be archived at The University of Western Australia Library.

TABLE 5.6 *Salinity, temperature and oxygen in the Greenough River Estuary. (Estuary said to be 2–4 m deep. Water carries much sediment.)*

20.09.1977		S (‰)	T (°C)	O_2 (%)
4 km upstream (2100 h)	Surface	36.7	21.9	132
	1 m			94
Mouth (2120 h)	Surface	36.3	21.2	120
	≈1 m	36.3	19.9	96

FIG. 5.23 *Preparing to sample fish in the upper Greenough River Estuary, September 1977. The alluvial soil on the eastern bank in the foreground is in sharp contrast to the white sand of the younger dune system along the coast.* E. P. Hodgkin.

IRWIN RIVER AND ESTUARY

Rivers of sand

The river channels of the Irwin as well as the estuary contain extensive alluvial deposits, reflecting the types of rocks upstream. Rising between Mullewa and Mingenew on the Yilgarn Block, the main channel of the Irwin flows in a south-westerly direction to the west of Mingenew where it is joined by the Lockier and Green Brook Rivers (Map 5.3). In the upper reaches the Irwin and Lockier flow through easily eroded sedimentary rocks of Permian age (Nangetty Formation, Holmwood Shale and Irwin Coal Measures), a feature of the Coalseam National Park (Fig. 5.24). Further downstream the river flows through an area of laterite with occasional outcrops of Jurassic shales and siltstones surrounded by areas of easily eroded quartz sand. Green Brook flows through a large area of older Proterozoic (gneiss with quartz, feldspar) covered by sands. Near the coast the Irwin flows through ridges of Tamala or coastal limestone that mark older dune lines. Dunes bordering the coast contribute to the river fill and bar, particularly where the plant cover is sparse. Sediment from the dunes is mostly carbonate derived from shells and algae.

The composition of sediments in the river indicates that most of the bed load in the Lower Irwin comes from the Upper Irwin and Lockier Rivers, derived from the shales, siltstone and sandstones of the Permian rocks. In contrast, Green Brook bringing feldspars, quartz and muscovite appears to contribute a smaller amount of sediment. At Yarragadee on the Irwin the depth of sediment indicates that erosion from the Upper Irwin is much greater than that of the Lockier River or Green Brook channels. In the estuarine reaches some quartz and carbonates are derived from the coastal dunes.

Changes to the landscape

Like the other areas around Champion Bay and Geraldton, the rich fertile soils of the Irwin River attracted early agricultural development and a prosperous town developed along the banks at Dongara served by Port Denison. Originally named 'Dongarra', the name is either a corruption of the Aboriginal name 'Thungarra' meaning the place of the fur seal or the

Aboriginal people's name for river 'arranoo murraja' or 'yarranoo murraja' meaning river flows only in flood (yarranoo = river in flood, murraja = river) reflecting the character of the river and estuary where floodwaters race down the dry river bed spilling out across the flat landscape.

Generally winter flows are light and the river is dry over summer except for a few pools. Heavy falls in summer or winter, however, bring floods, erosion and reworking of sediments throughout the length of the river (Figs 5.25 and 5.26). Floods in the Irwin may occur in the same years as in other systems nearby, such as the Greenough, or may be quite different depending on which part of the catchment receives rain. Although the alluvial deposits indicate that cycles of erosion are characteristic of the region, sedimentation appears to have increased in historic time.

In the lower reaches, over bank deposits extend for more than 30 metres away from the main channel around Pell Bridge and in the Dongara township. Near Pell Bridge about 2.5 metres of sediment has accumulated in the 30 years since the bridge was constructed in 1967. In other areas evidence of reworking of sediments includes the burial and re-exposure of wells probably built between 1850 and 1880, car bodies from the late 1960s buried under about 1.4 metres and buried fence lines. In 1999 winter floods overtopped the Irwin Bridge in Dongara and spread across the riverbanks to the Priory grounds and to historic Russ Cottage where the heights of all floods are recorded. The 1999 flood was classed with an Average Recurrence Interval of 10 years. The height of the largest floods in the summer of 1871–1872 and 1971 were classed as 100-year floods: interestingly they were just that, one hundred years apart.

The bar breaks most winters and with summer floods, but remains open only for a short time while there is river flow. The mouth opening has shifted several times as the result of flood regimes and human intervention, and there appears to have been a northward migration of the bar opening, the

FIG. 5.24 *View of the Coalseam National Park in the Upper Irwin Valley about 30 km NNW of Mingenew, July 1997. The wide valleys indicate an ancient history of great river flow, where different layers of sedimentary rocks representing glacial, marine, river and lake environments in the Permian era are exposed. The coal seams contain fossils of 'seed ferns'* Glossopteris. *Discovered by the Gregorys in 1846, the coal deposits have been mined in the past but are not considered economic. In the foreground the wide alluvial plain between the breakaways and the current sediment channel illustrates cycles of erosion and deposition.* Courtesy S. Grimes.

FIG. 5.25 *The height of floodwaters (probably 1971) indicated by the tyre lodged in a tree about 1.5 m above the main river channel in the Upper Irwin at Yarragadee, August 1997. Nearby a tyre and star pickets buried 1.5 m below the bank illustrate the amount of sediment brought down river in recent times.* Courtesy S. Grimes.

FIG. 5.26 *Flood damage on the Irwin River, July 1999. Trees were uprooted and soil eroded along the banks of the river flowing through the township of Dongara.* A. Brearley.

swampy area to the south perhaps marking an earlier channel. Generally the river has broken to the south of the bend at a low point in the bar, although in more recent floods it has been closer to the bend. In summer the only remaining water is generally close to the beach.

The estuarine reaches also appear to have shallowed since the first maps were drawn in the 1850s and the first photographs taken in the early 1900s (Figs 5.27 and 5.28).

Although in summer the only remaining water is generally near the beach, the estuary is however a haven for bird life. Fish such as sea mullet enter during the brief time that the bar is open. Black bream also live in the river and estuary along with small gobies (Table 5.2). Mud crabs have become quite plentiful in recent years in the Irwin, as they have in other estuaries along the west coast.

Searching for explanations

The large changes in erosion appear to be linked to the clearing of native vegetation in the catchment (Table 5.7). Today over 95 per cent of the catchment has been cleared for agriculture with only isolated pockets of native

FIG. 5.27 *Irwin River Estuary, ca 1900. An island lies in the middle of the channel and a property on the southern bank has been cleared and fenced, the house (not visible) built before this time. The limestone outcrop is the site of one of Dongara's first quarries; on the left the fenced area, Location 688, is the site of the first tillage lease. The palm tree shown in the 1977 photograph probably dates from this time.* Courtesy G. Grundy, Irwin District Historical Society Inc., Acc. No. 1993/0464, IRME 116.

FIG. 5.28 *Irwin River Estuary, September 1977: comparison with early photographs and maps indicates that a large amount of sediment was deposited in the 2 km-long estuarine channel by floods following the passage of cyclone Mavis in 1971. The island in middle of the river appears to be less defined and smothered with sediment that also fills the side channel. In the background lies Port Denison in Arurine Bay; originally 'Denison' named in 1851 after the Governor of Tasmania, an adviser on convict transportation. Leander Point named after the wreck of the* Leander *in 1853 shelters a fishing boat harbour in Arurine Bay.* E. P. Hodgkin.

vegetation remaining in rocky areas of the northern catchment. The story of development and clearing along the Irwin is characteristic of the whole of the Champion Bay area. Removal of native vegetation by clearing or grazing and replacement with annual species that die over summer exposed the soil to winds and rains, particularly the flooding rains of summer cyclones.

Clearly erosion and redistribution of sediments has increased in the catchment since the 1850s when the first clearing started. Photographs reveal clear differences in the state of the river between 1939–1954 and 1955–1996. Major floods and erosive events such as the 1971 flood have followed the large-scale changes since the mechanisation of clearing after the 1950s. Rich cropping lands are now dissected with gullies, topsoil has moved downstream filling the riverbeds or out to sea. The removal of deep-rooted species that tapped ground water has also allowed the watertable to rise and some springs now flow year round.

IRWIN RIVER TO HILL RIVER

Between the Irwin and Moore Rivers, a number of short creeks flow across the coastal plain without entering the sea (Map 5.5). The waters of some such as Cockleshell and Stockyard Gullys in the Gairdner Range area north-east of Jurien, the Arrowsmith and Nambung (south-east of of Cervantes) Rivers as well as Eneabba Creek (Weelawadji Cave) percolate through the dune limestone in caves, while other terminate in fresh and saline lakes. The Hill

TABLE 5.7 *Changes in the Irwin catchment after European settlement: these changes also apply to other parts of the Champion Bay District. Note river flows in the Irwin have only been recorded since 1976.* Source: Grimes, 1997; Bain, 1975.

Era	Change	Impact
early 1850s	Introduction of hard-footed grazers	Removal of vegetation and compaction of soil
late 1850s	Farming particularly on alluvial soils	Gradual clearing, generally some chop and burn
late 1860s – early 1870s	Droughts, fires destabilised soil followed by flood 1871 & 1872	Erosion
late 1860s – early 1870s	Economic downturn, closure of The Cattle Company, crop losses due to rust and locusts. Search for alternative incomes—sandalwood cutting	Clearing
1871	Occupation Licence—legislation for fencing and clearing	More clearing; major flood 1/100 years ARI
1873	Poison Land Act—allocation of 1,000 acres at £1 ($2) per year if cleared of poisonous native vegetation and fenced	More clearing
	Rabbits	Remove vegetation, destabilise soil
1890s	Increase in cattle and sheep; more land released for cropping	Compaction and erosion on stream banks
1890–1900	Clearing alluvial flats	'Heavy country' cleared by gangs of 5–6 men employed to chop and 'light country' burnt or flattened by horse-drawn roller
1930s	Clearing of heavy country	
1950s	**50% CLEARED**	
1950s	Introduction of trace elements; mechanisation—introduction of plough, chains, bulldozers	Clearing of poorer sand plain country
1960s	**MAJORITY OF THE CATCHMENT CLEARED**	
1971	Major flood, tropical cyclone Mavis (estimated peak flow 1,900 m^3/s)	Loss of houses, bridges, estimated 1/100 years ARI
1988	Major flood (peak flow 1,350 m^3/s)	Estimated 1/10 years ARI
1996	Flood (peak flow 350 m^3/s)	
1999	Major flood (peak flow 2,000 m^3/s)	Estimated 1/10 years ARI

River drains a much larger area and flows to the sea. The river flows from the Arrowsmith Region, an area of hills of Cretaceous and Jurassic rocks capped with laterite, about 50 kilometres inland between Eneabba in the north and Badgingarra in the south. The river descends the Gingin Scarp and then crosses low swampy areas of the Bassendean Dunes on the coastal plain. Tributaries join south of Mt Lesueur and flow across the coastal plain to a small estuary about 10 kilometres south of Jurien Bay.

The lower reaches and estuary of the Hill River are considered to be of state and regional significance as they are in a relatively natural condition compared to most of the rivers in the northern coastal plain (Figs 5.29 and 5.30). The offshore coast is also considered pristine due to the relatively low population and limited industrial development. A chain of islands and reefs offshore break the oceanic swells, creating a mosaic of habitats that accounts for the area's high biodiversity representative of the central west coast of Western Australia's marine waters. In recognition of these features, the Jurien Bay Marine Park was proclaimed in August 2003. The Jurien Bay Marine Park extends from Green Head to Wedge Island and is about 811 square kilometres in area.

The town site of Jurien Bay, currently a focal point for a rock lobster fishery and holiday-makers, has a population of about 600. The area, however, is expected to grow particularly since it was identified by the Western Australian Planning Commission in 1996 as the most suitable district centre for the entire central coast. Expansion of the town is likely to be to the south, to the Hill River, largely over land previously used for grazing. Development plans include an expansion of existing reserves to create a buffer around the

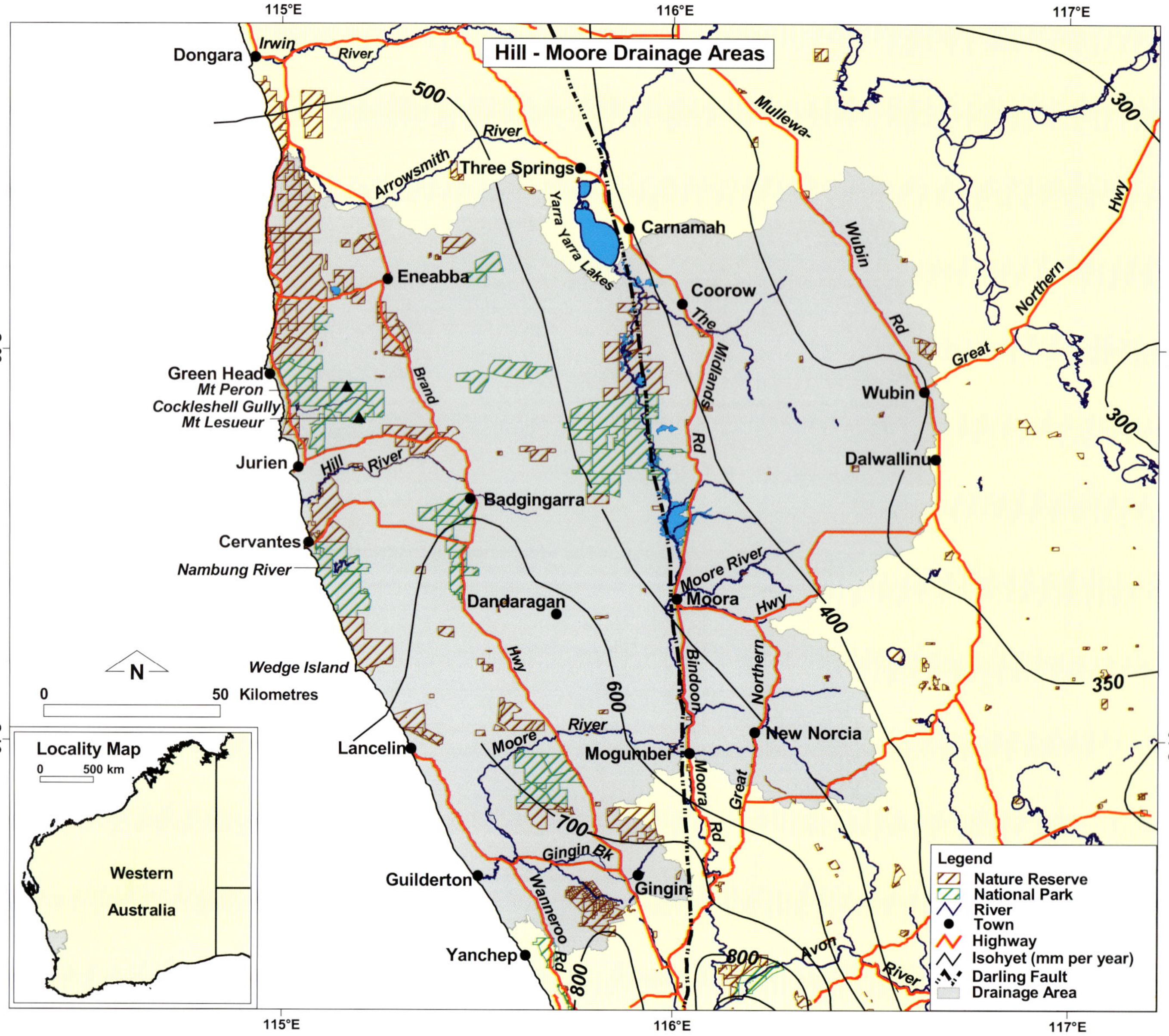

river and estuary. These will include setbacks based on models for a '100-year' flood event. Similarly, development along the coast will include reserves with setbacks in accord with current coastal planning guidelines. Although largely protected from oceanic swells, the rate of sand erosion and accretion varies along the coast and from year to year. Around the foreshore of the Jurien Bay town site, 150 metres of sand accumulated between 1942 and 1999 (2–3 metres per year). South of Island Point sand is eroding at a rate of 1 metre per year, while further south near Booka Valley accretion is 1.5 metres per year. The entrance to the Hill River also varies from year to year but appears to be migrating northwards under the influence of the northward migration of sand (Fig. 5.30).

The bar across the Hill River usually breaks every year but the timing depends on the first heavy and consistent rains. The estuary mouth stays open for only a short time, a few days or weeks, depending on river flow and the longshore movement of sand along the coast. The bar usually closes when strong waves bring sand onshore. Ernest Hodgkin made notes following a conversation with J. Williams, Fisheries Inspector at Jurien Bay, 1976: 'Mullet and black bream have been collected in the estuary and are taken there and there could be other species' (see Table 5.2).

MAP 5.5 *Drainage basins Hill River to Moore River. Many of the short rivers south of the Irwin never reach the sea terminating instead in lakes or caves. The Arrowsmith River drains farmland at the southern end of the Victoria Range, Cockleshell Gully drains the Gairdner Range including the main peaks of Mt Lesueur and Mt Peron, while the longer Hill River drains the Dandaragan Plateau through the Arrowsmith Region and the Swan Coastal Plain.* Courtesy DoE.

FIG. 5.29 *Hill River on the 'Turquoise Coast', the view from the mouth, September 1977. In the 1980s the water was mostly brackish but was thought to have been fresh prior to clearing. Main flow is in winter, but springs also contribute water over summer.*
E. P. Hodgkin.

FIG. 5.30 *The mouth and estuary of the Hill River, aerial views: May 1995 with the channel in the north position and (inset) September 1996 with development of the channel to the north and then south where it has opened to the sea.*
See Alan Tingay & Associates (2000b). Figures reproduced with kind permission from Ardross Estates Pty Ltd.

MOORE RIVER

It's a very scenic place where people can escape from their normal environment and feel like they're in a wilderness environment—but they are only an hour away from Perth...For me it is an inspiring place. It gives me a sense of the natural processes and the importance of the ecology. If people are educated about the fragility of this river, they will take that on board.

Writer Rod Moran talking to Phil Smeeton of the Local Community Association at Guilderton; The *West Australian*, 14 September 1999.

History and the changes to come

Changes have come slowly to the Moore despite its proximity to the Swan River and Perth (Figs 5.31 and 5.32). Apparently it was overlooked by the early Dutch or French explorers, perhaps it was hidden by the sandbar. After European settlement however, colonial travellers and explorers such as Bannister (1830), Norcott (1836), Moore (1836), Grey (1838–1839), Roe (1839), the Gregory brothers (1846, 1848) and von Sumner (1847) all crossed the Moore. Records of the state of the estuary are scant, although Bannister and Roe may have visited the mouth. George Fletcher Moore in 1835, however, recorded a description given by an Aboriginal person named Weeip of the Gingin Brook flowing west to the sea (the 'Garban') after joining a river from the north-east. The ephemeral nature of the river flow revealed by the statement, 'No, they are dead but when the rain comes they will be alive and walk into the sea' (de Burgh, 1976). The name used by the Aboriginal people referring to the mouth was 'Gabbadah' or 'mouthful of water'.

Farming on the fertile alluvial soils around the lower Moore River and Gingin Brook started during the 1850s and 1860s, with the town of Gingin established in 1871. Cattle and sheep were grazed through the coastal areas by the Brockman family, although there was no permanent dwelling downstream of 'Neergabby' on the confluence of the Moore and Gingin Brook, about 20 kilometres from the coast, until 1900.

The estuary was used as a holiday area from the late 1890s by the Brockman and de Burgh families and more generally after 1907 when it was declared a picnic ground. Public access was restricted from 1940–1943 when the area

FIG. 5.31 *The Moore River Estuary, aerial view, June 2002. The large dune on the southern side of the estuary is known descriptively as 'Sand Sliders'. Although the area was bare during the 1930s, the area of exposed sand and beach increased greatly between the mid-1950s and mid-1960s. Between 1973 and 1983 the dune vegetation disappeared exposing more sand, and by 1988 sand was spilling into the estuary. A path to the 'slide' is clearly visible in a photograph taken in 1994. In 1995 the area was fenced and the pathway covered with brush to reduce sand drift into the estuary.* Courtesy D. Tracey, WRC.

was occupied by Australian Defence Forces, but after the war pressure for a permanent camping ground increased. The townsite was declared during the 1940s with the town of Guilderton gazetted in 1951. The name Guilderton was suggested by the historian and writer Henrietta Drake-Brockman to link the town with the nearby wreck of the 'Gilt Dragon'. The name derived from 'guilders', Dutch coins, carried in the cargo that had been found near the estuary mouth. Although the coins were later found not to be guilders, the name of Guilderton however remained.

Life blood of the estuary — source of problems

The estuary of the Moore River is influenced by water from two very different catchments. The Moore River with a catchment of over 12,000 square kilometres brings saline water to the estuary, along with sediment and some nutrients. While the smaller Gingin Brook and its tributary Quin Brook, with a catchment of only 826 square kilometres, flowing from an area of higher rainfall (720 millimetres), more intensively used for agriculture and horticulture, brings fresh water and much higher concentrations of nutrients. Water also enters the estuary from ground water seeps, as a result of which the water is fresher than that of the nearby Swan.

The Moore Catchment

The headwaters of the Moore River lie on the ancient Yilgarn Block, north-east of Dalwallinu, more than 100 kilometres inland, an area of broad shallow valleys 300–350 metres above sea level. In ancient times the Coonderoo River, the main tributary of the Moore, probably connected to a chain of lakes, the Monger palaeochannel in the Yarra Yarra Basin between the Darling and Dandaragan Plateaus (Fig. 5.33). This old river system is reactivated by heavy rain inland (Maps 5.5, 5.6 and 1.10). (See also **Chapter 1 Understanding Swanland: The rivers of Swanland—back to front rivers**.)

Emerging from the Darling Scarp, the Moore River flows south through Moora and Mogumber then westwards for about 30 kilometres through a well-developed valley on the southern edge of the Dandaragan Plateau. Descending the Gingin Scarp near Regans Ford on the Brand Highway, the braided courses of the Moore wind southwards across the coastal plain through the Karakatta Sands of the Spearwood Dunes and a small area of the Bassendean Dunes, where it is joined by Gingin Brook, and through to the estuary at Guilderton.

FIG. 5.32 *Moore River, 2001. The dark waters of the estuary and river are surrounded by vegetation typical of the Swan Coastal Plain. In many respects the Moore is a 'scaled-down' reminder of the essence of the Swan when it was visited by the first European explorers, Vlamingh (1696) and Fraser (1828). In the lower reaches of the estuary, Tuart trees* Eucalyptus gomphocephala *and flooded gums* E. rudis *extending down to the water's edge form a spectacular backdrop surrounded by Zamia and grass trees* Xanthorrhoea, *along with* Allocasuarina *and dune plants. The Moore marks the northern extent of large areas of tuart and marri* Corymbia *(formerly* Eucalyptus) calophylla *woodlands. The banks are also lined with freshwater or swamp paperbark* Melaleuca rhaphiophylla *that form 'islands' in the main channel. Further upstream the more developed soils are covered with Banksia communities.*
A. Brearley.

FIG. 5.33 *The Coonderoo River at Namban West Road about 30 km north of Moora in the upper Moore River catchment showing floodwaters that linked the salt lakes after heavy rain, aerial view, 23 March 1999. The road crossing was cut for all of winter 1999.* Courtesy P. Muirden, WRC.

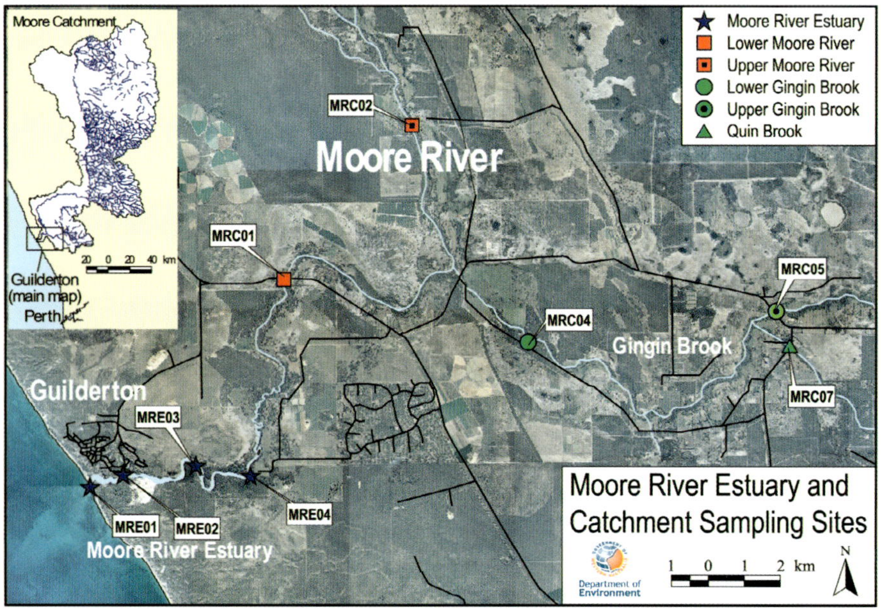

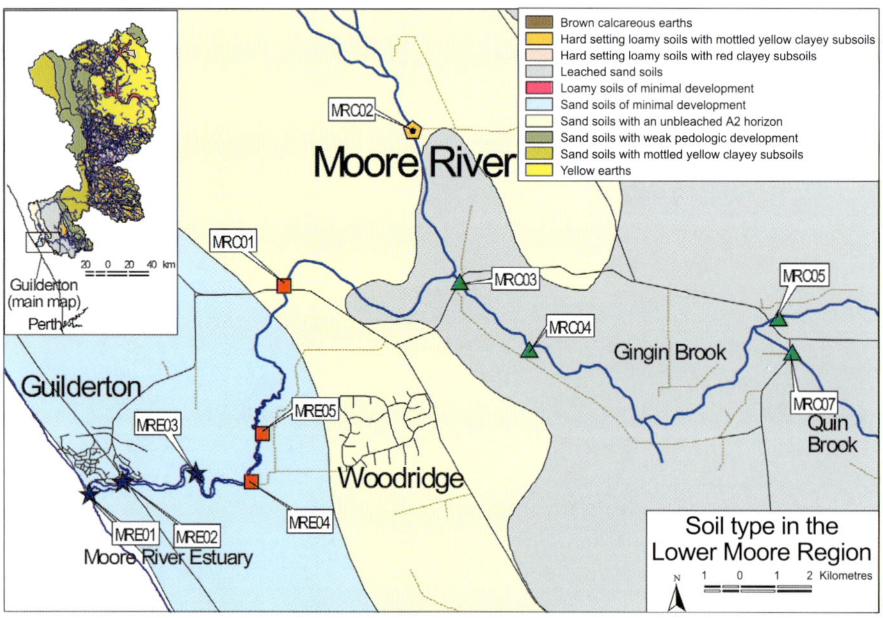

MAP 5.6 *The Moore River Estuary and locations of water sampling, MRE01 Moore River Bar; MRE02 Silver Creek; MRE03 Diamond Island; MRE04 Caraban Rocks; MRC01 Lancelin Road Bridge; MRC02 Waterville Road; MRC04 Bookine Bookine; MRC05 Nolans Bridge; MRC07 Military Road: (a) the path of the river, cleared land and locations of intensive horticulture; and (b) the soils and landscape surrounding the river and estuary. Grey colour indicates Bassendean Dune System and yellow and blue Spearwood Dune System.* See Cousins, 2003; Cousins & Tracey, 2003a & b; Courtesy WRC.

Gingin catchment — a different landscape

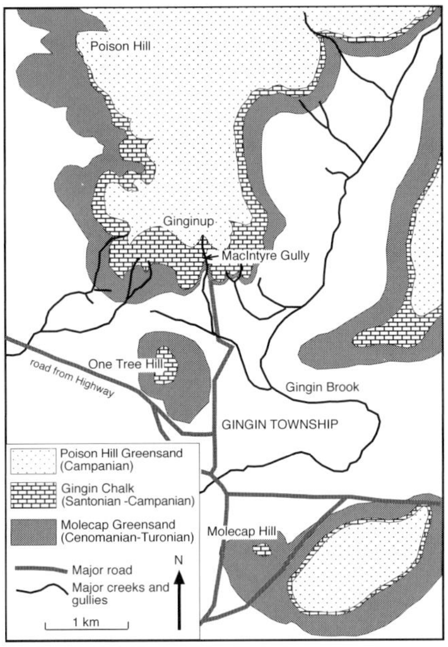

FIG. 5.34 *Area around Gingin showing the outcrops of cretaceous sediments.*
Adapted from Feldtmann, 1963; Courtesy D. Haig, UWA.

Gingin Brook rises in freshwater springs on the Dandaragan Plateau. Unlike the hard Precambrian rocks of the Darling Plateau from where the Moore flows, the rocks of the Dandaragan Plateau are sedimentary and younger in age, laid down during the Jurassic and Cretaceous periods. (See also Maps 1.1–1.3, Figs 1.1 and 2.2). The rounded hills are a surface expression of the same strata that underlie the Perth Basin. The valley of Gingin Brook and the town of Gingin lie on the non-marine rocks of the South Perth Formation of Early Cretaceous Age, iron-rich sandstone formed of coarse to fine shales (Fig. 5.34 and Table 5.8). Above this a gap in the time sequence (an unconformity) is followed by marine sediments of Late Cretaceous age, the Molecap Greensand, the Gingin Chalk and the Poison Hill Greensand.

The Molecap Greensand on the lower slopes is fine grained and dark greyish-green in colour due to the mineral glauconite. Glauconite (a hydrated silicate of iron, potassium, aluminium and magnesium ($K_2(Mg,Fe)_2Al_6(Si_4O_{10})_3(OH)_{12}$), related to mica, was mined between 1932 and 1960 for use as a phosphate fertiliser or water softener. Fossils include microscopic marine animals, phosphatised wood, and bones of large marine reptiles such as icthyosaurs and plesiosaurs. Plesiosaurs up to 20 metres in length with paddle-shaped flippers and long necks are characteristic of the Cretaceous Period.

The highest hills (about 200 metres) are topped by harder rock of the Poison Hill Greensand (Moorgup, Ginginup, Poison Hill), while the softer Gingin Chalk tops the lower hills of Molecap and One Tree Hill. The best outcrops of Gingin Chalk however, lie between the upper and lower greensands on the southern slopes of Ginginup. The chalk is rich in fossils such as sharks teeth, bivalves, gastropods, crustaceans, solitary (non-reef-building) corals, sea-urchin spines, plates of feather stars (crinoids), lampshells (Brachiopods) and ammonoids (related to modern *Nautilus*). Obviously the seas that once covered the land here abounded with marine life. The Gingin Chalk is remarkably similar to the European Chalk in rock type and fossil content.

The diversity of environments that laid down the sediments is reflected in the soils around Gingin that vary from yellow sands to 'rendzina', a fine, almost black soil that forms over chalk. The Gingin rendzina with a high phosphorus content is particularly interesting and unusual as Australian soils are generally deficient in phosphorus.

TABLE 5.8 *Rock strata outcropping and underlying the Gingin area on the Dandaragan Plateau.*
Adapted from Feldtmann, 1963, See also Seddon, 1972 and Playford et al., 1975.

Period	Formations	Age (m. yr)	Thickness (m)	
Upper Cretaceous		65		Marine
	Poison Hill Greensand		50	
	Gingin Chalk		0–20	
	Molecap Greensand		10–30	
	UNCONFORMITY			
Lower Cretaceous		< 90		Non marine
	South Perth Formation		50–60 exposed in Gingin area	
Upper Jurassic		< 160		
	Yarragadee		2,000	

The Cretaceous spanned a period from 65 to 142 million years and the Jurassic 142 to 205 million years. Rocks of Cretaceous Age were first described in Europe, the name derived from the Latin cret-a-ceus or chalklike, although not all rocks of Cretaceous age are chalk or greensands.

On the coastal plain the Gingin River flows through the nutrient poor sands of the Pleistocene Bassendean Dunes: these are easily leached so that additional nutrients added to agricultural land move through into the streams and ground water.

Rains, floods and seepage

In the catchment of the Moore River rain falls predominantly in the winter months of June and August, but high rainfall events in summer also occur about every 25 years, usually associated with tropical cyclones. Run-off in most years is very low, about 1 per cent of the rainfall, with peak flows generally considered low for a catchment the size of the Moore (13,557 square kilometres) when compared to peak discharges of other rivers in the South-West. Generally peak flow occurs in July and August, but summer flows are significant about every 25 years. Although parts of the river dry out in summer, the estuary generally contains water from springs.

High rainfall events in the upper catchment clearly relate to floods in the lower river, but flooding also appears to be more extensive when the lower catchment around Gingin is also flooded. Winter flooding usually follows prolonged periods of rain, but summer flooding is usually due to intense falls over a short period (Fig. 5.35). In some areas levee banks have been built but floodwaters often break though these into old channels and spread out across the flat landscape. (See **Box 5.8 Levee banks**.)

Except for the large river flows of 1974 (over 197,000 megalitres), more than 218,000 megalitres in 1995 and nearly 395,000 megalitres in March 1999, total annual discharge from the Moore River rarely exceeds 150,000 megalitres (Graph 5.2). Following the Moora Floods in March 1999, peak discharge at Quinns Ford exceeded 25,000 megalitres a day, averaging 0.29 megalitres per second on 22 March, and affecting flows until May 1999. Heavy flows also continued through winter eventually declining in November to 100 megalitres per day.

Gains and losses to the ground water also have a significant influence on the water of the lower Moore River and the estuary. In some areas water moves from the main watercourse into the ground water remerging as springs. Ground water flows are generally fresh, often with high concentrations of nutrients, and are particularly significant during periods of low rainfall.

Box 5.8

LEVEE BANKS

Levee banks protect landowners from the more frequent nuisance flooding but allow a false sense of security to develop. During major flooding, levee banks have considerable effects—floodwaters can no longer spread out across the natural flood plain but are channelled through the banks. As flood levels within the banks rise, pressure on the banks increases and breaching occurs. When breached, the higher velocity of water has higher erosive force and floodwaters take longer to subside. In the 1995 flood there were four major levee breaches and an overtopped levee. Two breaches and overtopping occurred upstream of Bennies Road and directed flow to the east before returning to the main channel.

Source: Pen, 1999; WRC, 1997.

Shallowing rapidly

Since the 1960s sediment appears to have accumulated in the estuary, making it shallower. The causes are many. Most of the catchment has been cleared, and rabbits, a problem from the 1920s and in plague proportions in the 1950s until numbers were reduced in 1954, removed vegetation and destabilised riverbanks. Streamlines have been altered, new channels created, and floods have eroded riverbanks. In particular, floods following cyclonic rains in summer are likely to be major contributors of sediment as they occur when the bare catchment is vulnerable to erosion.

Floodwaters have brought soil from the largely cleared catchment where evidence of erosion is quite clear (Fig. 5.36). Sediments from upstream have filled deeper channels and extended sandbanks and bars around Diamond Island and Silver Creek, while near the active dunes (Desert and Sand Sliders) sand has spilled into the estuary. Premature opening of the bar has reduced the head of water leaving the estuary, reducing the strength of flushing water to remove sediment. Near the mouth development of the foreshore has also extended the foreshore area into the main channel.

The area on the northern side near the bar was possibly 6–10 metres in depth. Photographs taken in the 1930s show exposed rocks used in the 1960s

FIG. 5.35 *Moore River in flood, August 1975. On the Swan Coastal Plain, the Moore River divides into many shallow channels where the flow is generally sluggish due to low gradients. During periods of high flow, the water spreads out across the flat plain flowing though a number of channels where levees have been constructed to control floodwaters and direct flow. Particularly strong flows, however, often break levee banks and the river reverts to old natural channels, the highly erosive water damages property, deposits sediment on the land and in the river. The waters, turbid with valuable soil and nutrients, is carried through to the sea at Guilderton.* E. P. Hodgkin.

GRAPH 5.2 *River discharge from the Moore River catchment 1968–2000. Years of particularly high flow occurred in 1974, 1995 and March 1999. Floods in 1900, 1915, 1917, 1923, 1926, 1932, 1934*, 1939, 1945, 1946, 1955*, 1956 (Nov)*, 1958, 1968, 1974, 1995, 1999* are documented as major flood years. Summer floods marked with *.* Adapted from Cousins, 2000; Courtesy WRC.

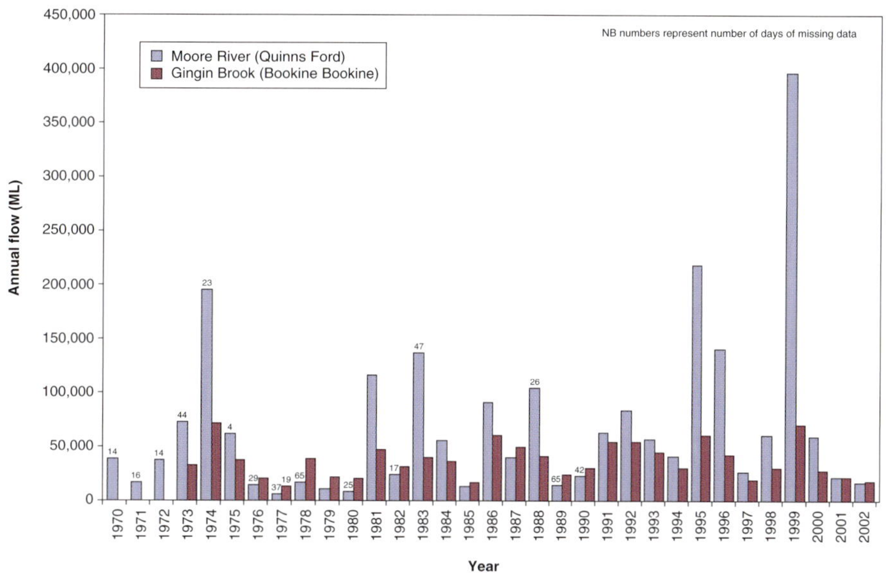

as the base of a diving platform, where the water must have deep enough for diving. Anecdotal evidence supports the contention that the depth was sufficient to cause pressure on the eardrums of 'duck-divers'.

Water in the estuary — opening the bar a contentious issue

Water level in the estuary varies with catchment rainfall and river flow, the rate of ground water recharge, frequency of bar opening, barometric pressure and the height of the bar. Little is known about the effect of floodwaters on the bar. For example, the bar opened following high rainfall and flooding in the lower river in July 1995, however it is unknown if this opening was natural or deliberate. The bar also appears to have opened eight times between 1996 and 1997 when rainfall was relatively low (less than 100 millimetres). This frequent opening and closing could indicate that the bar had not build up to the usual height and that low rainfall and water levels were sufficient to initiate breaching.

FIG. 5.36 *The banks of the Moore River are badly eroded by stock grazing and trampling vegetation on the riverbanks in the lower reaches. (a) The erosion in this photograph taken in 1978 during sampling for fish remains a problem. (b) Old trees whose roots are a major protection against flood damage are now undermined, July 2001. In other areas, vehicles and pedestrians have also contributed to the loss of riparian vegetation along the river and estuary banks.* (a) E. P. Hodgkin; (b) A. Brearley.

Natural breaching of the bar usually occurs when water levels are high after heavy rain in the catchment. The dynamics of closing or remaining open are, however, also dependent on the sea level (wave and wind climates, tides and barometric pressure), nearshore sediment movement (longshore and oscillatory) as well as estuarine water levels (Figs 5.37 and 5.38).

Natural breaching generally creates a deep channel that is likely to remain open for some time; conditions conducive to effective flushing of fine sediments from the lower estuary. In contrast, deliberate and premature openings lower the head of water in the estuary, producing a shallow channel that usually closes quickly. As a result hydrology and sedimentation in the estuary are affected by the less rapid and shorter flushing periods, that in turn affect water quality and the biological communities and processes.

The bar has probably been opened artificially during the main holiday periods since the late 1890s, except perhaps for the period during the war when public access was restricted. Generally the bar would have been opened to allow fish to enter, although it has also been done to release floodwaters. Bar breaching has always been controversial and remains a problem today. Openings are regarded as fun, creating better surf once the river water forms a wave in the nearshore zone, and released fish increase the opportunity for fishing along the coast while allowing others to enter. On the down side is the lack of proper flushing that follows the less vigorous outflow. The fall in water level also restricts boat access upstream, and appears to allow light to penetrate the dark coloured waters providing ideal conditions for algal growth.

In May 1998, after remaining closed for six months, the bar reached a height of 1.5 metres above sea level allowing the estuary waters to rise to a record height flooding estuarine vegetation and structures along the foreshore. After the opening on 1 June, water levels fell rapidly scouring a channel at least a metre below the height of the bar. Three days later however, the channel was wider but only 10 centimetres deep.

Salinity

Like all estuaries the water salinity in the Moore Estuary reflects the connection to the sea and flow from the catchment. When the bar is open and river flow is low, marine water of 35‰ enters with the tides. The dense saline water moves upstream along the bottom to near Diamond Island, although at times it can reach beyond Caraban Rocks (Graph 5.3). In many estuaries, salinity rises as river flow decreases and the bar closes. However when the bar of the Moore is closed, the trapped marine water is diluted by fresh ground water inflows. Further upstream flows of the saline water from the upper Moore River are also an influence. For example, at Waterville Road (approximately 24 kilometres upstream) salinity was less than 1.5‰ in October, but after the flood in March 1999 salinity increased to 7.6‰ due to saline water coming from the catchment (Graph 5.3).

FIG. 5.37 *The Moore River Estuary, July 2001. The estuary is riverine in nature with only a small lagoon near the mouth. The brown colour of the water due to tannins gives the Moore a particular character, more similar to estuaries along the south coast. Generally the estuary is isolated from the sea by a sandbar that reforms quite quickly after opening under the influence of wave action moving sand onshore. The high dunes are covered with typical sand dune vegetation, and the area south of the estuary has been zoned for a housing development.* A. Brearley.

FIG. 5.38 *Water levels in the Moore River Estuary: (a) when the bar was closed; (b) after the bar opened. The fish is the only remaining artwork from the 1999 Artists-in-Residence project to encourage awareness of the area's beauty. The others, made of natural materials, were dismantled or decomposed. The fish however remains and provides a measure of water depth related to the state of the bar. At times it is completely submerged.* M. Cousins, WRC.

Oxygen in the estuary is influenced by salinity and algal growth. At the surface, oxygen concentrations are usually fairly high due to diffusion from the atmosphere or algal photosynthesis. Near the bottom, however, low oxygen concentrations reflect isolation of the water due to the dense layer of salty water. In the deep pools, such as those near Caraban Rocks, decomposition of organic matter also reduces oxygen concentrations. These low oxygen conditions favour the release of nutrients from the sediment, indicated by the peaks in nutrient concentrations that coincide with periods of deoxygenation. As a result, the occurrence of low oxygen can be considered a threat to the health of the Moore.

Years of warning signs

As long ago as 1989 the Moore River was regarded as seasonally eutrophic with high concentrations of nitrogen; the lack of problem algal blooms reflecting the poor light conditions in the dark tannin-stained waters rather than lack of nutrients.

Although leaching rates of nutrients from the catchment of the Moore River Estuary were lower than those from the Harvey River in 1987, adjustment for flow (division by water yield) indicated that the loads were very similar. The leaching was probably due to the poor phosphorus retention properties of the Bassendean Sands found around both estuaries. High concentrations of nitrogen and phosphorus in Gingin Brook were thought to

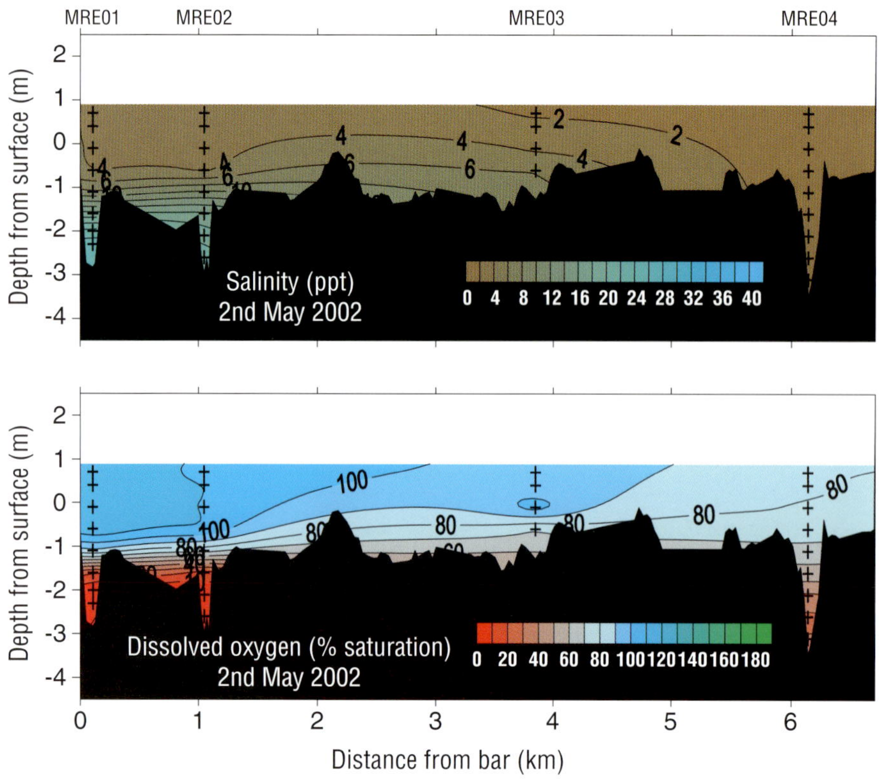

GRAPH 5.3 *Salinity and oxygen profiles along the Moore River Estuary, May 2002. MRE01 Bar, MRE02 Silver Creek, MRE03 Diamond Island, MRE04 Caraban Rocks. The marked salinity stratification between the surface and the bottom is obvious at the lower estuary locations near the mouth. Surface water is generally 2–10‰ within the whole estuary. Bottom waters generally contain less oxygen than surface waters as indicated by the red colour.* Source: Cousins & Tracey, 2003a & b; Courtesy WRC.

reflect the relatively large number of nutrient point sources in the catchment. 'The phosphorus export rate can be expected to increase over the next few years (5–10 years?) as the soils within the Moore River Estuary catchment become saturated with phosphorus.' (Bott & Maling, 1989).

Quality of the water 1998–2002

In response to ongoing concerns for the health of the Moore River and its estuary, a four-year program of water sampling started in December 1998 as part of a Natural Heritage Trust project to better understand the Moore River catchment.

The distinctive brown colour of the water in the Moore River Estuary originates from dissolved organic matter 'tannins' (Gilvin) leached from the native vegetation, particularly the tuarts, tea-trees or paperbarks. The Gilvin scatters and absorbs light, and hence the amount of light available for plant growth through photosynthesis. The Gilvin colour was also lower in the bottom waters due to the layer of salty marine water. The highest colour was in Gingin Brook with much lower colour at Waterville Road on the Moore River. The colour of the estuary water is quite striking near the bar where the contrast with the lightly coloured ocean can be seen. (See **Box 5.9 Tea-coloured water**.)

Arriving in the water

Nutrient concentrations in the Moore tend to follow a seasonal pattern with peaks in winter. However nutrient concentrations are also high in summer with storm flow and in areas where ground water enters. Total nitrogen in the Moore is higher than in the Leschenault and the now more marine Peel–Harvey and the Swan, and total phosphorus and chlorophyll *a* are similar to the Peel–Harvey but less than the Swan (Graphs 1.2 and 1.3).

Total nitrogen concentrations were highest during periods of high flow, and generally higher and more consistent in water from Gingin Brook. In the estuary there were higher concentrations at the surface. The lower

Box 5.9 TEA-COLOURED WATER

The brown-coloured tea-like waters of lakes, rivers and estuaries with high concentrations of humic (decomposed plant and animal material) organic matter are termed dystrophic. Characteristically, in these dark waters light penetration is reduced and lower wavelengths are absorbed at the surface resulting in low and altered light spectras. Consequently temperatures are often lower at depth, setting up a thermal gradient that restricts circulation. The low and altered light coupled with low temperatures limits photosynthesis reducing productivity even when nutrient concentrations are high. Combined with other stressful conditions, such as low pH and low concentrations of dissolved oxygen, the flora and fauna in coloured water bodies are often limited and productivity is low.

In addition to the effects on light, compounds leaching from the leaves of tuart trees *Eucalyptus gomphocephala* have been shown to inhibit or to stimulate algal growth. The compounds with high molecular weight (humic and fulvic acid with about 10,000 daltons) have high chelating or bonding attractions to iron and copper and to a lesser extent zinc but not manganese, possibly indicating that these trace metals were not available for algal growth. Inhibition of growth is termed allelopathy. In contrast, compounds from the leaves with low molecular weights (1,000 daltons) appear to stimulate algal growth.

Source: Pen, 1999; Wilson, 1991.

concentrations in the bottom waters reflect the nutrient poor marine waters creeping upstream, with the surface concentrations reflecting nutrient enriched water from the catchment (Graph 5.4a).

The oxidised forms of nitrogen used by growing plants, nitrate (NO_3^-) and nitrite (NO_2^-), were generally lower and less variable in the catchment, and higher in the upper estuarine areas (Graph 5.4b). Generally, however, concentrations were low in Gingin Brook. Concentrations of total nitrogen and oxidised nitrogen exceeded ANZECC, 2000 guidelines.

Concentrations of ammonia, the breakdown product of animal waste and plants, were also higher and more variable in the estuarine areas (Graph 5.5). Concentrations of ammonia were particularly high, nearing toxic concentrations, in the saltier low oxygen bottom water in the upper

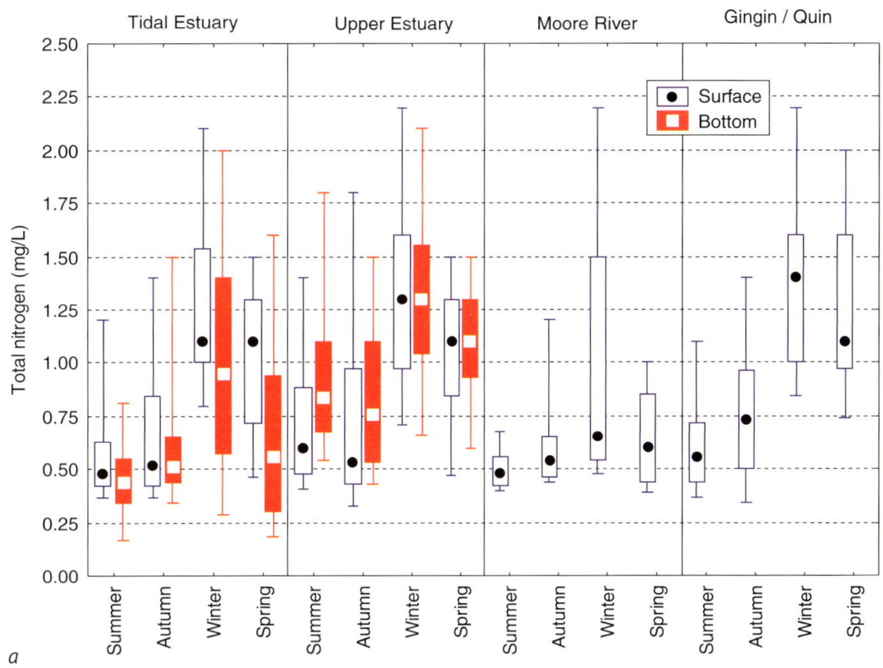

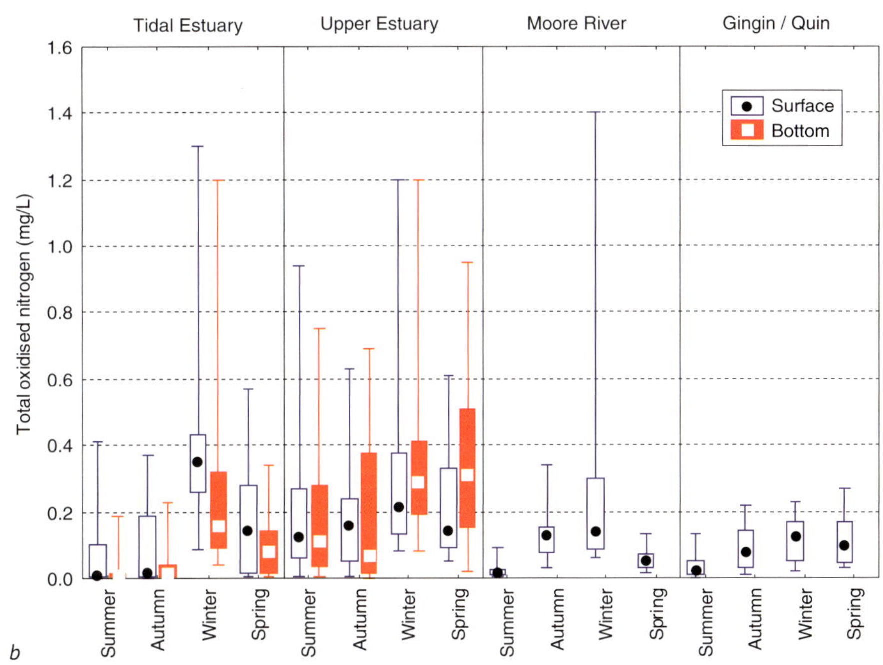

GRAPH 5.4 *(a) Total nitrogen concentrations and (b) total oxidised nitrogen concentrations in the Moore River Estuary, December 1998–November 2002.* See Cousins, 2003; Cousins & Tracey, 2003a & b; Courtesy WRC.

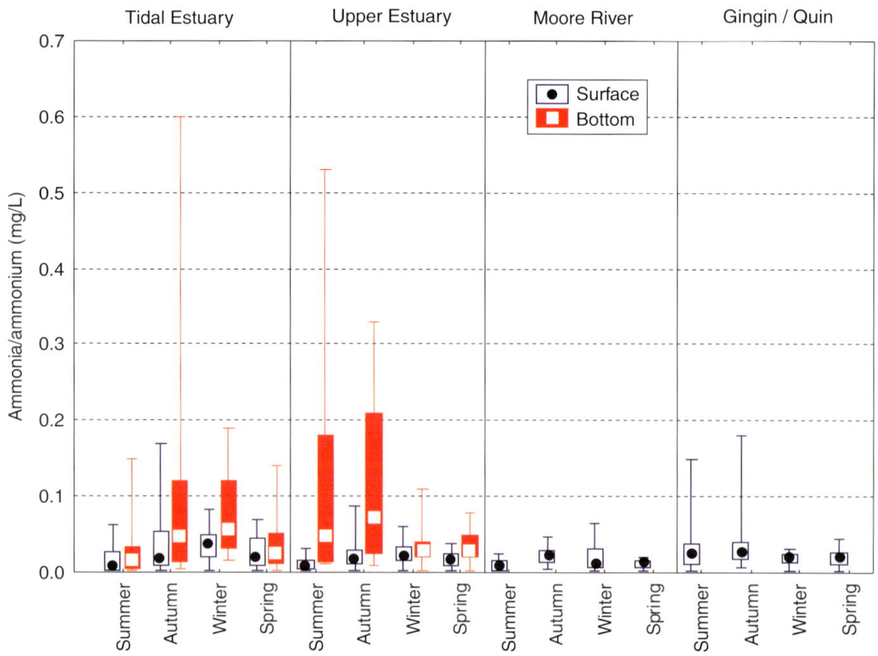

GRAPH 5.5 *Ammonium concentrations in the Moore River Estuary, December 1998–November 2002. Note the higher concentrations in the bottom waters upstream particularly around Caraban Rocks.* See Cousins, 2003; Cousins & Tracey, 2003a, b & c; Courtesy WRC.

estuary at Caraban Rocks, probably indicating release from the sediment. The presence of foul-smelling hydrogen sulfide gas was an indication of these processes occuring. The dissolved inorganic nitrogen (DIN), the sum of the total oxidised nitrogen (nitrate and nitrite) and ammonia, the forms used most readily by plants, was often high in the upper estuary. The high concentrations were probably due to ground water and surface flows, and decay of organic material in the system.

There were distinct patterns in total phosphorus in different parts of the estuary. Gingin Brook consistently showed higher concentrations than the lower Moore, while concentrations in the estuary were midway between the two sources. However concentrations increased in the bottom waters of the estuary when oxygen concentrations were low (Graph 5.6a). Similarly, filterable reactive phosphorus (FRP), the soluble form, was greatest in Gingin Brook and lowest in the Moore River except for the flood in 1999, and also high during times of low oxygen (Graph 5.6b). Soluble phosphorus concentrations consistently exceeded guideline values for healthy estuaries.

The flowing water also brings sediment that not only shallows the channels but also reduces light for primary production, interferes with filter feeders and blocks the gills of fish. A measure of sediment in the water column (total suspended solids) used as an indication of evidence of erosion indicates that, although these are generally low, the amount coming out of the catchment is higher than in the estuary and, perhaps not surprisingly, increases with river flow. Sediment in the estuary water varies and is highest when the bar is open, possibly due to flocculation where salt and fresh water bodies mix.

Dissolved silica also plays an important role in the growth of phytoplankton as it is incorporated into the cell walls. Silica concentrations in 1999 were highest at Gingin Brook and in bottom waters of the Moore River, while the estuarine locations were much lower.

Big floods

Floods generated by the dissipating tropical cyclone Elaine in March 1999 that devastated the town of Moora also brought large volumes of water to the lower Moore River and Estuary, creating a plume of discoloured water

GRAPH 5.6 *(a) Total phosphorus and (b) soluble phosphorus in the Moore River, December 1998–November 2002. Concentrations of total phosphorus in the water flowing from Gingin Brook were very high.* See Cousins, 2003; Cousins & Tracey, 2003a, b & c; Courtesy WRC.

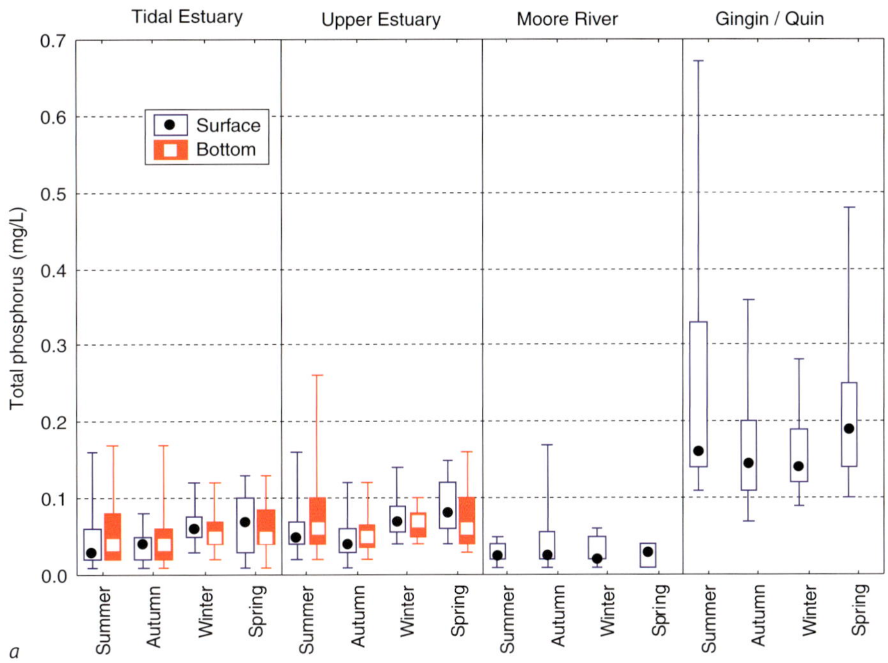

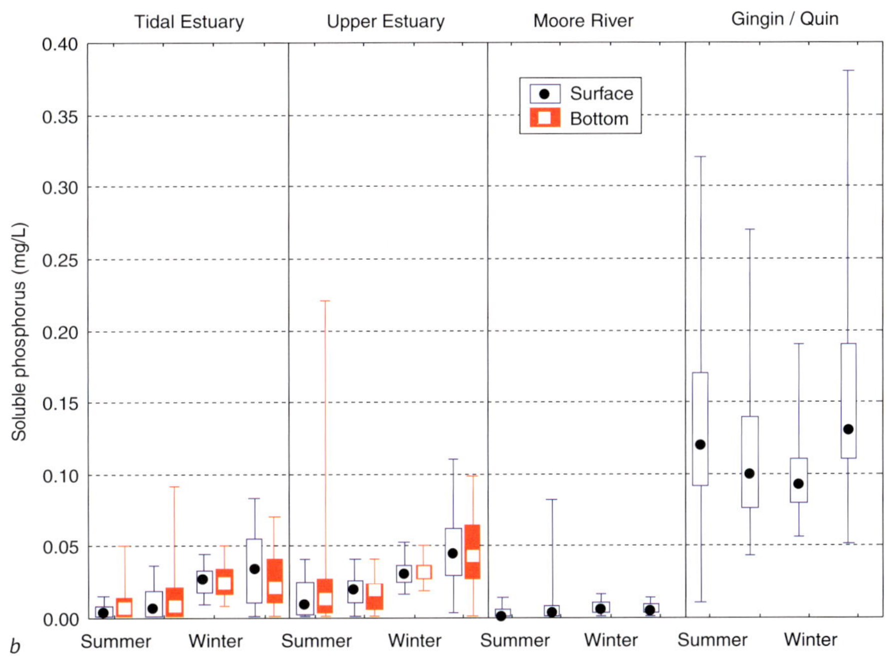

that extended much further out to sea than the usual plume from winter flow (Fig. 5.39). The floodwaters also contained high concentrations of nutrients in excess of ANZECC guidelines. Nitrogen concentrations in samples collected at Mogumber were particularly high (Graph 5.7). Concentrations of total nitrogen at Waterville Road and Guilderton were only slightly higher than concentrations in winter flows, however the high concentrations were more obvious in the estuary than the catchment and remained higher for longer. In contrast to nitrogen, phosphorus concentrations in the floodwaters were higher than usual except at Mogumber (Graph 5.8). Suspended solids were also lower at Mogumber but silica concentrations were higher than at other sites.

Seeking the sources – the ground water springs

Samples of ground water from bores in Guilderton generally contain high concentrations of total oxidised nitrogen (> 3.5 milligrams per litre), and the concentrations appear to have risen since 1988. The potential for ground water springs to contribute to nutrients in the river and estuary has been confirmed by analysis of spring water at Caraban Rocks and Waterville Road, where concentrations of total nitrogen, total oxidised nitrogen, and phosphorus were higher than samples of estuarine water (Table 5.9). These rising concentrations are a concern for the health of the estuary, but also for the use of ground water for human consumption.

FIG. 5.39 *The plume of river water from the Moore River flood following cyclone Elaine in March 1999 discoloured the water offshore and extended 15 km northwards up the coast from Guilderton. Photograph taken at the lighthouse 2 km north of the mouth on 26 March 1999, one week after passage of the cyclone.* Courtesy V. Hosja, WRC.

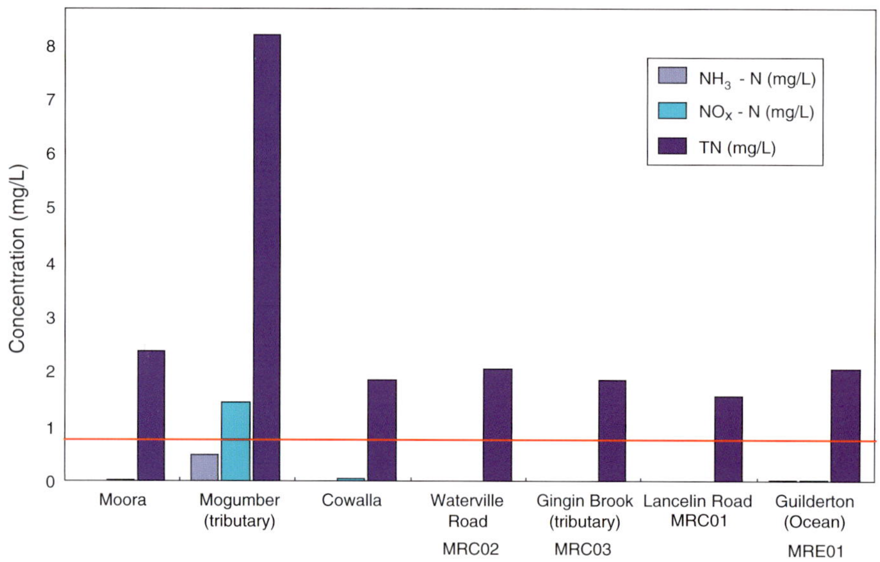

GRAPH 5.7 *Concentrations of total nitrogen and soluble inorganic oxidised nitrogen (NOx) and ammonia (NH_3) in floodwaters, March 1999. Red line indicates 1992 ANZECC guidelines for total nitrogen.* See Cousins, 2000; Courtesy WRC.

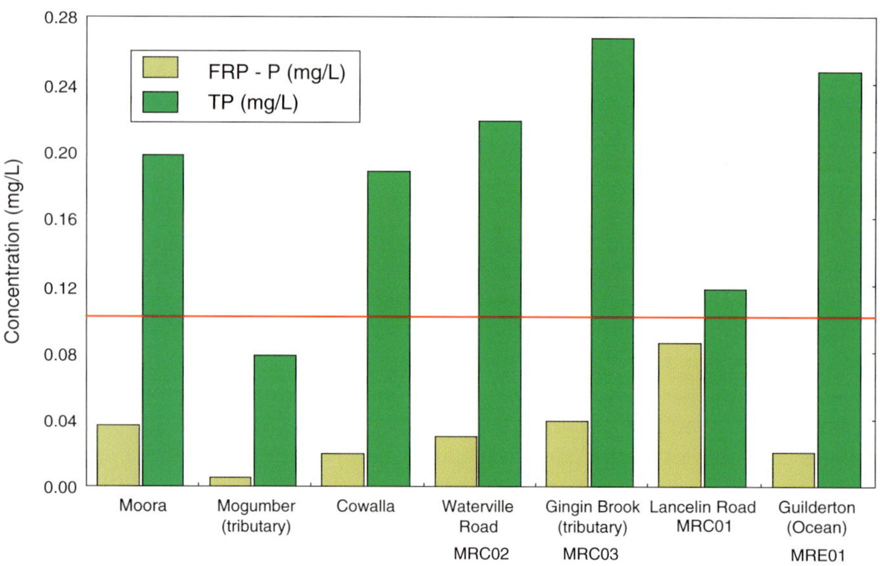

GRAPH 5.8 *Concentrations of total phosphorus and soluble or filterable reactive phosphorus in floodwaters, March 1999. Red line indicates ANZECC 1992 guideline for total phosphorus.* See Cousins, 2000; Courtesy WRC.

TABLE 5.9 *Comparison of nutrient leaching rates from ground water springs and adjacent estuarine area, 10 December 1999. Ground water springs along the river are exposed when water levels are low, generally after the bar has opened. Guideline concentrations for nitrate are 0.1 mg/L for protection of aquatic life in estuarine waters and 10 mg/L for drinking water.* Adapted from Cousins, 2000.

Units (mg/L)	Caraban Rocks		Waterville Road	
	Ground water spring	Estuary surface	Ground water spring	River
TN	3.3	0.91	0.78	0.62
NO_x–N	2.5	0.2	0.57	0.026
NH_3–N	0.007	< 0.005	0.084	< 0.005
TP	0.08	0.06	0.03	0.03
FRP–P	0.011	0.005	0.005	< 0.003

Living in the dark waters of the Moore River Estuary

Microscopic phytoplankton and larger algae

The amount of chlorophyll *a*, a measure of plant growth in waters of the Moore, was generally low at all sites indicating that there were few plant cells despite the high nutrient concentrations. Chlorophyll was highest in the upper estuarine area and the Moore River, and generally highest in the warmer months of spring and summer.

Diatoms, chlorophytes and cryptophytes dominated the phytoplankton community of the Moore. Chlorophytes were more common in the upper reaches than the lower reaches probably due to the lower salinity of the area. To date, algal blooms have been due to harmless species. Diatoms formed a bloom in late 1999 with another in February 2000, principally due to the chlorophyte *Tetraselmis*. Blue-green algae including the undesirable species *Microcystis*, *Anabaena* and *Oscillatoria* have occasionally been found in water samples in relatively low numbers. Although not a major threat at present, there is concern that they could increase if conditions were favourable.

More obvious were the mats of filamentous green algae, such as *Spirogyra*, *Enteromorpha* and *Cladophora* particularly in shallow, slow moving stagnant water where they grew quickly before collapsing. In summer of 1999, mats of the green algae *Enteromorpha* were later replaced by other species upstream of Diamond Island and at Waterville Road. Green filamentous algae were also found near freshwater springs. Although not toxic, anoxic low oxygen conditions under large algal mats favour release of sediment-bound nutrients that perpetuate cycles of algal growth creating stressful conditions for many animals. (See **Chapter 3 The Peel–Harvey Estuarine System: Stinking masses in the Peel**.) The presence of macroalgal growth needs to be viewed with some concern as similar events in other estuaries throughout Swanland have, in the past, often preceded more serious phytoplankton blooms.

Zooplankton — tiny animals of the water column

Compared to other estuaries in south-western Australia there were relatively few types of animals in the zooplankton of the Moore River. The most abundant were larvae of bivalve molluscs (85 per cent) and copepod crustaceans (9 per cent). The low species diversity was thought to be due to the relatively limited habitats and the fluctuations in salinity due to flooding. The zooplankton assemblage did not appear to have changed between late 1988 and 1999. The most abundant of the copepods were species only found in estuaries in south-western Australia. The abundance of bivalve larvae (although unidentified) is in accord with the abundance of bivalves, such as the large sunset shell *Soletellina* (previously *Sanguinolaria*) *biradiata* and the tiny false mussel *Fluviolanatus subtorta*. In the Moore these bivalves are the major dietary component of black bream. Juvenile black bream feed on the smaller bivalves and consume the larger species and individuals as they grow.

Fish

The Moore is a favourite spot for fishing, the quiet waters offering an opportunity to fish out of the wind and surf. Popular targets include tailor, mulloway, whiting and other opportunistic marine species that enter the estuary when the bar is open, while the estuarine black bream shelter among the tree roots that line the estuary edges.

A study of the fishes in 1994 and 1995 recorded twenty-seven species, but only eight were abundant (Table 5.10). Most were small, the yellow-eye mullet *Aldrichetta forsteri*, one of the most common species, the only fish larger than 100 millimetres. The most abundant were the small hardyheads (atherinids) and gobies, species commonly eaten by cormorants, pelican and herons. Wallace's hardyhead *Leptatherina wallacei* accounted for almost 75 per cent of the fish sampled and over 20 per cent of the biomass. Yellow-eye mullet, sea mullet *Mugil cephalus* and black bream, although fewer in number, accounted for most of the biomass.

TABLE 5.10 *Fish caught by seine net in the Moore River, February 1994–February 1995;* Young et al., 1997. *Samples were collected near the mouth and at distances from the mouth (1.5 km lower, 5 km middle and 5.5 km upper). Life cycle categories; S, marine straggler species; O, marine estuarine–opportunist species; E, species confined to estuaries; E&M, species represented by estuarine and marine populations; A, semi-anadromous; F, freshwater species.*

Fish name	Species name	Life cycle	Number of fish		Biomass of Fish	
			Number	%	kg	%
Wallace's hardyhead	*Leptatherina wallacei*	E	78,890	74.8	29.41	21.8
Swan River goby	*Pseudogobius olorum*	E	9,346	8.8	1.57	1.2
Black bream	*Acanthopagrus butcheri*	E	5,705	5.4	20.07	14.9
Yellow-eye mullet	*Aldrichetta forsteri*	O	2,440	2.3	46.17	34.2
Yellow-tail trumpeter	*Amniataba caudavittata*	E	2,408	2.3	8.02	5.9
Long-finned goby	*Favonigobius lateralis*	E&M	2,184	2.1	0.81	0.6
Sea mullet	*Mugil cephalus*	O	1,444	1.4	21.43	15.9
Goby	*Afurcagobius suppositus*	E	1,439	1.4	1.25	0.9
Yellowfin whiting	*Sillago schomburgkii*	O	810	0.8	1.61	1.2
Trumpeter whiting	*Sillago burrus*	O	375	0.3	1.44	1.1
Western school whiting	*Sillago vittatus*	S	297	0.3	2.16	1.6
Striped (six-lined) trumpeter	*Pelates sexlineatus*	O	33	< 0.1	0.11	< 0.1
Blue sprat	*Spratelloides robustus*	O	30	< 0.1	0.03	< 0.1
Mosquito fish	*Gambusia affinis*	F	19	< 0.1	< 0.01	< 0.1
Roach	*Gerres subfasciatus*	O	14	< 0.1	0.19	0.1
Sandy sprat	*Hyperlophus vittatus*	O	14	< 0.1	0.03	< 0.1
Perth herring	*Nematalosa vlaminghi*	A	11	< 0.1	0.03	< 0.1
Common carp or goldfish	*Carrassius auratus*	F	8	< 0.1	0.05	< 0.1
Tarwhine	*Rhabdosargus sarba*	O	6	< 0.1	0.11	< 0.1
Bridled goby	*Amoya bifrenatus*	E&M	5	< 0.1	< 0.01	< 0.1
Tailor	*Pomatomus saltatrix*	O	3	< 0.1	0.08	< 0.1
Ogilby's hardyhead	*Atherinomorus ogilbyi*	O	3	< 0.1	< 0.01	< 0.1
King George Whiting	*Sillaginodes punctata*	O	2	< 0.1	< 0.01	< 0.1
Small-toothed flounder	*Pseudorhombus jenynsii*	O	2	< 0.1	< 0.01	< 0.1
Swan River hardyhead	*Leptatherina presbyteroides*	E&M	1	< 0.1	< 0.01	< 0.1
Southern anchovy	*Engraulis australis*	E&M	1	< 0.1	< 0.01	< 0.1
Prickly pufferfish	*Contusus brevicaudus*	S	1	< 0.1	< 0.04	< 0.1
TOTAL NUMBER OF FISH			105,491		134.9	
TOTAL NUMBER OF SPECIES			27			

The unpredictable and limited openings of the bar at Moore River certainly restrict access for other ocean dwelling fish, the marine stragglers. In contrast to the Swan, only 75 kilometres to the south, with a permanent opening to the sea where thirty-seven fish species classed as marine straggler are found, only two species of fish in the Moore fall into this category, probably reflecting the lack of opportunity to enter. The less saline waters also place a physical stress on fish living in estuaries such as the Moore.

The largest group of fish (thirteen species), classed as marine opportunists, were most common near the mouth but accounted for less than 5 per cent of fish caught. In contrast, the less diverse estuarine group (nine species) accounted for almost 95 per cent of fish caught.

Of the two species of mullet and whiting, most were juveniles exploiting the productive estuarine environment. The most abundant of the whiting, the yellowfin whiting, declined after the bar opened probably indicating migration from the estuary. In contrast, during the same opening of the bar more juvenile mullet entered, and resident mullet appeared to move into the lower salinity areas of the middle and upper estuary while sea mullet moved further upstream into the riverine areas. Perth herring, a species generally tolerant of changing salinity and breeding in the upper less saline reaches of the Swan, appeared to be less common in the Moore probably due to the limited opportunity for fish to enter through the bar.

The effect of low salinity in the Moore River Estuary is also reflected in a comparison of hardyheads and gobies with those in the Swan–Canning Estuary. The most abundant species in the Moore are mostly found in the upper reaches of the Swan, where salinities are low. Other hardyheads common throughout the Swan and other estuaries along the coast appear to be absent from the Moore.

The low salinity of the Moore River Estuary is also reflected in the presence of two introduced species, the mosquito fish *Gambusia affinis* and the common carp or goldfish *Carrassius auratus* in the upper estuary. These are generally more common in wetlands and drains in urban areas.

The low salinity waters also appear to influence the biology of some species such as black bream. Black bream shows a quite remarkable salinity tolerance, living and breeding in most estuaries in conditions varying from fresh to hypersaline. (See **Box 2.13 A popular fish—the black bream** and **Box 10.2 Black bream—a real estuary fish**). In the Moore, black bream spawn in mid-spring to early summer. They tend to mature at an older age but at a smaller size compared to those in the Swan River, probably indicating that they grow more slowly; a pattern similar to black bream in Walpole–Nornalup on the south coast. In contrast, in the variable and harsh hypersaline conditions of Wellstead Estuary, black bream mature at a younger age and smaller size, a strategy thought to ensure a successful breeding event before conditions become too stressful.

Black bream, although feeding on a wide range of plant and animal matter, appear to feed slightly differently in other estuaries depending to a large extent on the abundance of suitable items. In the Walpole–Nornalup they consume large amounts of swan grass *Ruppia* and fish, in the Wellstead the alga *Chaetomorpha* and tube dwelling amphipods *Corophium minor*, while in the Moore they been known to consume the filamentous green alga *Cladophora* and large numbers of the sunset shell *Soletellina biradiata*, a bivalve that commonly lives in sand along the coast and in estuaries.

Of interest is the recent appearance of mudcrabs *Scylla serrata* in the estuary. Mud crabs are usually found further north where mangroves are common, but are occasionally found in estuaries further south including the Swan, Leschenault and even Wilson Inlet, probably arriving as larvae brought south by the Leeuwin Current. (See **Box 1.4 Leeuwin Current, El Niño and ENSO years**.) The muddy sediment-filled Moore River lined with *Melaleuca* provides a different but suitable habitat where they appear to be thriving.

What of the future — the Moore and other estuaries of the mid-west coast?

The area surrounding the lower Moore River and Estuary is the last enclave of this type of native vegetation on the Swan Coastal Plain. The township of Guilderton, predominantly occupied by retirees and weekend visitors and popular for fishing and canoeing, is marked for expansion fuelled by the number of people wanting to live by the coast. Further development in the townsite and rural catchment will, if current land use practices continue, contribute more nutrients to a system already showing signs of stress. Some of these might be avoided if we take heed of examples in the Swan–Canning and the Peel–Harvey; protecting the ground water by reducing nutrient additions to the sandy soils where horticulture is rapidly expanding, and leaving adequate buffers along streamlines in both residential areas and in rural areas where erosion due to trampling by stock is already obvious. Elevated nutrients not only affect estuary health, but the presence of nutrients raises the issue of the potential sources and the likelihood of other contaminants being present. This also has implications for use of ground water for potable supplies.

The future for Guilderton is also echoed further north at Jurien Bay through to Geraldton and Kalbarri. As the population in each area grows, nutrients in the river and estuarine waters are likely to increase, and water flows change. Port developments will be required for the fishing and lucrative rock lobster industries, and to improve access to the hinterland for agricultural and mining products. These developments in turn will further promote population growth, and modifications to the small rivers and estuaries will follow. There is no doubt, however, that they will remain foci of interest; and maintenance of healthy water bodies is in the interest of the local and wider community.

Chapter 6 Hardy Inlet and estuaries of the Naturaliste–Leeuwin Ridge

Along the Swanland coast between Cape Naturaliste and Cape Leeuwin the coastal plain is insubstantial. Most freshwater habitats are short rivers or creeks rising as seeps between old granites and the overlying limestone. The only true river system, Margaret River, drains the north-west corner of the Blackwood Plateau and enters the sea through a small estuary (Maps 6.1, 6.2 and 4.8).

In lee of Cape Leeuwin, the Blackwood River and the smaller Scott River flow to the Southern Ocean through the Hardy Inlet. The Blackwood River, about 330 kilometres in length extending from the Lake Grace area on the Precambrian Yilgarn Block in the east, is one of the longest rivers in Swanland, only exceeded by the Frankland and Murchison. The inlet and estuary basin cover about 9 square kilometres and extend from the ocean bar about 5 kilometres upstream to the mouths of the rivers.

MARGARET RIVER: LET'S NOT FORGET THERE IS ALSO AN ESTUARY

Margaret River is perhaps one of the best-known names on the west coast. Images of rolling surf, beaches, fine wines, fresh fruits and cheeses—the list could go on and on. Little noticed, but important for our story, the Margaret River has a small estuary, probably known by most as a place to swim and canoe when the sea breeze blows and the surf pounds outside the bar. Known to our early settlers, the area was first farmed in the mid 1800s when Ellen and Alfred Bussell settled at Wallcliffe on the banks of the river estuary. Early settlers also used the areas near the smaller rivers such as Ellens Brook and Turners Brook at Deepdene, as no doubt the Aboriginal people of the Bibbulmun Tribe did before European settlement. The interesting nature of the Margaret River and other small watercourses is now recognised by inclusion of some of the surrounding area in the Leeuwin–Naturaliste National Park.

Margaret River has a catchment area of only 39.8 square kilometres, and the estuary too is small by any standards. Only 1 kilometre long and approximately 200 metres at its widest point, it narrows further upstream to 10 metres wide extending inland for 2 kilometres (Fig. 6.1). The central basin of about 20 hectares is connected to the ocean via a short narrow channel. The bar opens intermittently depending on the strength and duration of river flow. In this respect, despite its small size, it mirrors the behaviour of many estuaries in Swanland that have progressively filled with sediment over the last 5,000 years, only opening now when rainfall in the catchment is heavy.

At Margaret River flow restrictors in the lower reaches of the river and a weir above the town reduce flow to the lower reaches. With increasing urbanisation there have been indications of change in the water quality, but this tiny estuary has some factors in its favour. On the north-west it is bounded by Kilcarnup Reserve and a buffer of wetlands, and riparian vegetation remains along much of its length.

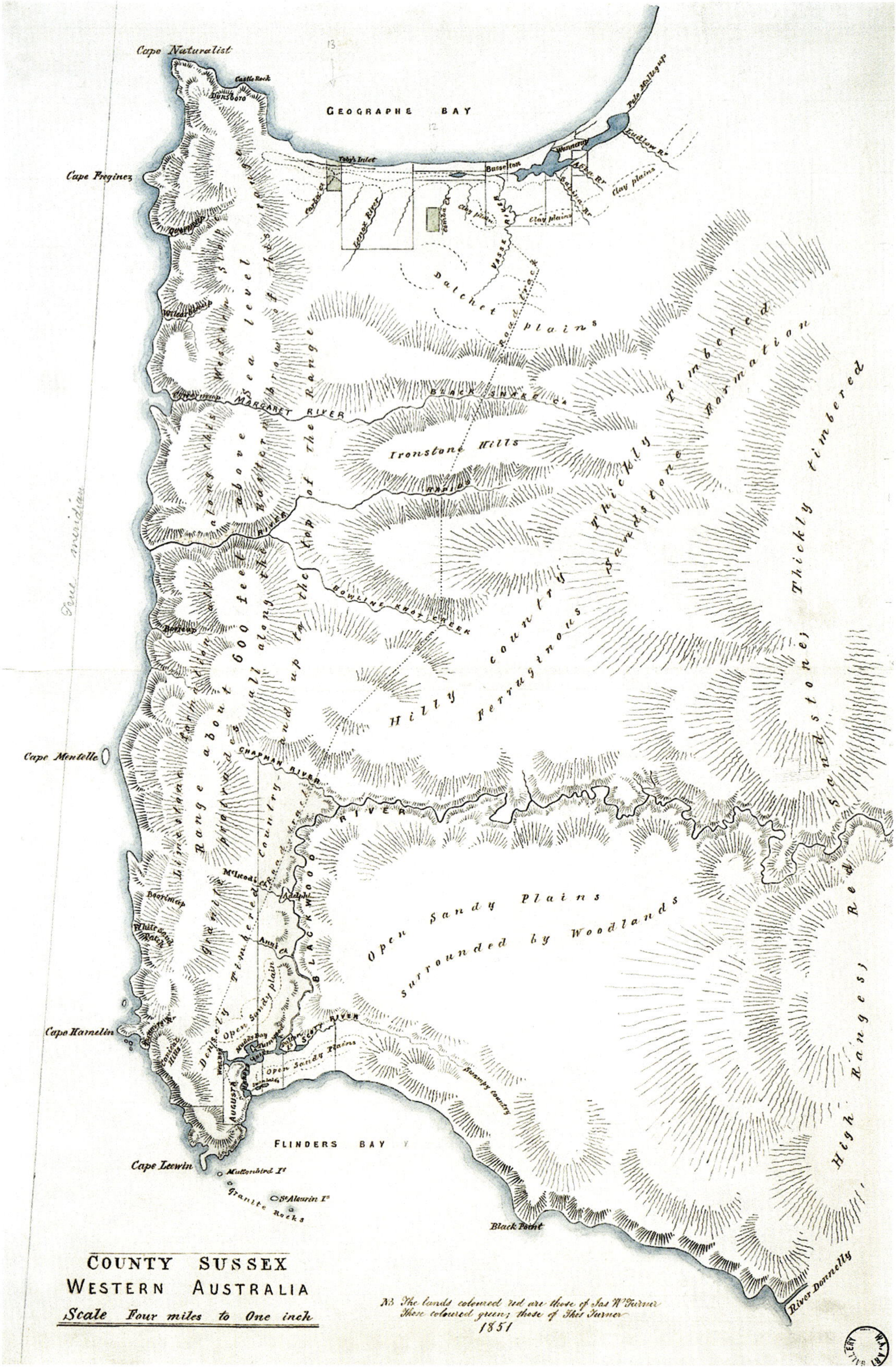

MAP 6.1 *Topographical map drawn by Thomas Turner in 1851, one of the first settlers at Augusta. The map clearly shows the flat coastal plain to the north where the Vasse–Wonnerup forms a great wetland, the short rivers flowing from the Naturaliste Ridge, and the mighty Blackwood River that drains a great area to the east before entering the sea at Augusta in Flinders Bay. Note the route to the Busselton Vasse–Wonnerup is now part of a Heritage Trail.* County Sussex, Western Australia, 1851 1953/00D6, Courtesy Art Gallery of Western Australia.

FIG. 6.1 *Margaret River looking upstream, aerial view, July 1975. Swell lines marking the famous surfing site at the mouth of river also move sand onto the bar as river flow decreases. In the distance, farmland originally cleared for grazing is now cloaked with vineyards.* E. P. Hodgkin.

HARDY INLET— BLACKWOOD RIVER ESTUARY

What a sight as the visitor arrives at the small coastal town of Augusta: a grand vista of houses nestled between gum trees along the banks of a peaceful river in the lee of Cape Leeuwin, the grey-green treescape (peppermints) to the east and in the distance the vast expanse of the Southern Ocean where waves roll onto the long sandy beach (Cover image and Fig. 6.2). This lovely spot sheltered some of our earliest pioneers, and the river served as a gateway to other areas of the South-West. Originally settled in 1830, Augusta more recently developed as a retirement and holiday town renowned for its peaceful lifestyle. The scene of tranquillity we view today, however, could have proceeded quite differently since the 1970s. Mining companies pegged claims to mine the river sands and coastal dunes for heavy minerals (ilmenite, magnetite, zircon and red garnet). Objections to the plans were made by citizen groups, government departments and the National Parks Board (now CALM) on social, aesthetic and environmental grounds, with particular concerns raised about the potential damage to fish and bird populations.

It became obvious in the 1970s that little was known about the ecology of the estuary and the possible effects of mining. Only limited information was available about similar places to allow comparisons. Although there was some information about individual organisms in estuaries of the South-West, there were no studies of the overall structure and functioning of any estuary. On this basis the response of the Environmental Protection Authority (EPA) was to 'recommend that no mining be allowed on these claims, and they not be listed for hearing in the Mining Warden's Court until further research has been undertaken'. These recommendations were accepted by the Government, and the EPA asked the Estuarine and Marine Advisory Committee (EMAC) to undertake a detailed investigation of the area. The study had two primary objectives: in the short term 'To attempt to predict the probable effects of mining and dredging in the estuary and environs', and in the long term 'To understand the working of the Blackwood estuary ecosystem as the basis for making decisions about the management of this and other estuaries of south west Australia'. Ernest P. Hodgkin was appointed Research Coordinator for the study undertaken in 1974 and early 1975, and fourteen technical reports were published in this major collaborative study.

In the 1970s a study like this was relatively unusual, and the Blackwood study has provided a background for understanding other estuarine systems in south-western Australia. After publication of the report, the EPA recommended that the dredging and mining applications should not be approved. The mining company advised the Department of Mines that it did not wish to contest the various objections and, on its recommendation, the Mining Warden refused the application on 12 January 1977. The township could go back to sleep—or could it?

From vantage points above the inlet we notice houses nestled under the trees on the eastern bank (Fig. 6.2). There were only a few shown on the 1970's map. In the distance we see the Swan Lakes. Before the 1920s these lakes contained fresh water, but an opening of the channel to the estuary allowed salt water to enter and the water is now brackish. Looking seaward, the inlet near Seine Bay turns eastwards, runs behind a dune and sandbar and enters the sea through a narrow channel beyond the entrance to Swan Lakes. Something is different! For previous visitors, our memory says there is a large lagoon, the Deadwater, east of the entrance channel and our map confirms the difference. The river mouth has changed, moved eastwards and swallowed the Deadwater (Fig. 6.9b). Human lifetimes seem so transitory in comparison with geological processes, the power of nature over landscapes and the relentless forces of the Southern Ocean.

FIG. 6.2 *The Hardy Estuary at Augusta, February 1976. Although the inlet still retains great charm some changes are evident. The townsite has now grown and on the eastern bank the now much larger settlement of East Augusta lies among the peppermint trees. Water now flows through the Deadwater and the bar opens to the sea further east.* E. P. Hodgkin.

An ancient setting

As with all the estuaries of Swanland, the Hardy Estuary is a very recent feature in a very old landscape (Maps 6.2 and 4.8). The catchment of the Blackwood River covers an area of 28,000 square kilometres extending from Augusta, 330 kilometres eastwards almost to Lake Grace. The upper part of the Blackwood River catchment lies on the ancient Precambrian Shield, also termed the Yilgarn Block, a plateau 300 metres above sea level. Here the small creeks and river tributaries run through gently undulating country with broad sandy (quartz) valleys and extensive salt lakes. Gradients are low, drainage is poor, evaporation high and the rivers flow infrequently (Maps 1.10 and 1.11, Fig. 6.3).

In the middle part of the catchment, west of the Meckering Fault, the rivers flow through U-shaped valleys with low gradients where the underlying laterite is exposed. Rainfall is higher at 450–700 millimetres, and the rivers are seasonal, flowing during winter but drying to strings of brackish pools in summer. Further west in the Darling Range Drainage Zone, the rivers flow in V-shaped valleys cut to 60–200 metres below the laterite uplands, exposing granitic rock along the valley sides.

West of the Darling Fault the rivers run through the Blackwood Plateau, previously called the Donnybrook Sunklands or the Bunbury Trough. Here, about 60 metres above sea level, between the two fault lines, sediments have accumulated over million of years. At Nannup, the Blackwood River and its tributaries are steep-banked, cutting meandering courses through gently undulating country where drainage is generally poor (Figs 6.4 and 6.5). Rainfall in the Sunklands area of the catchment is above 1,000 millimetres.

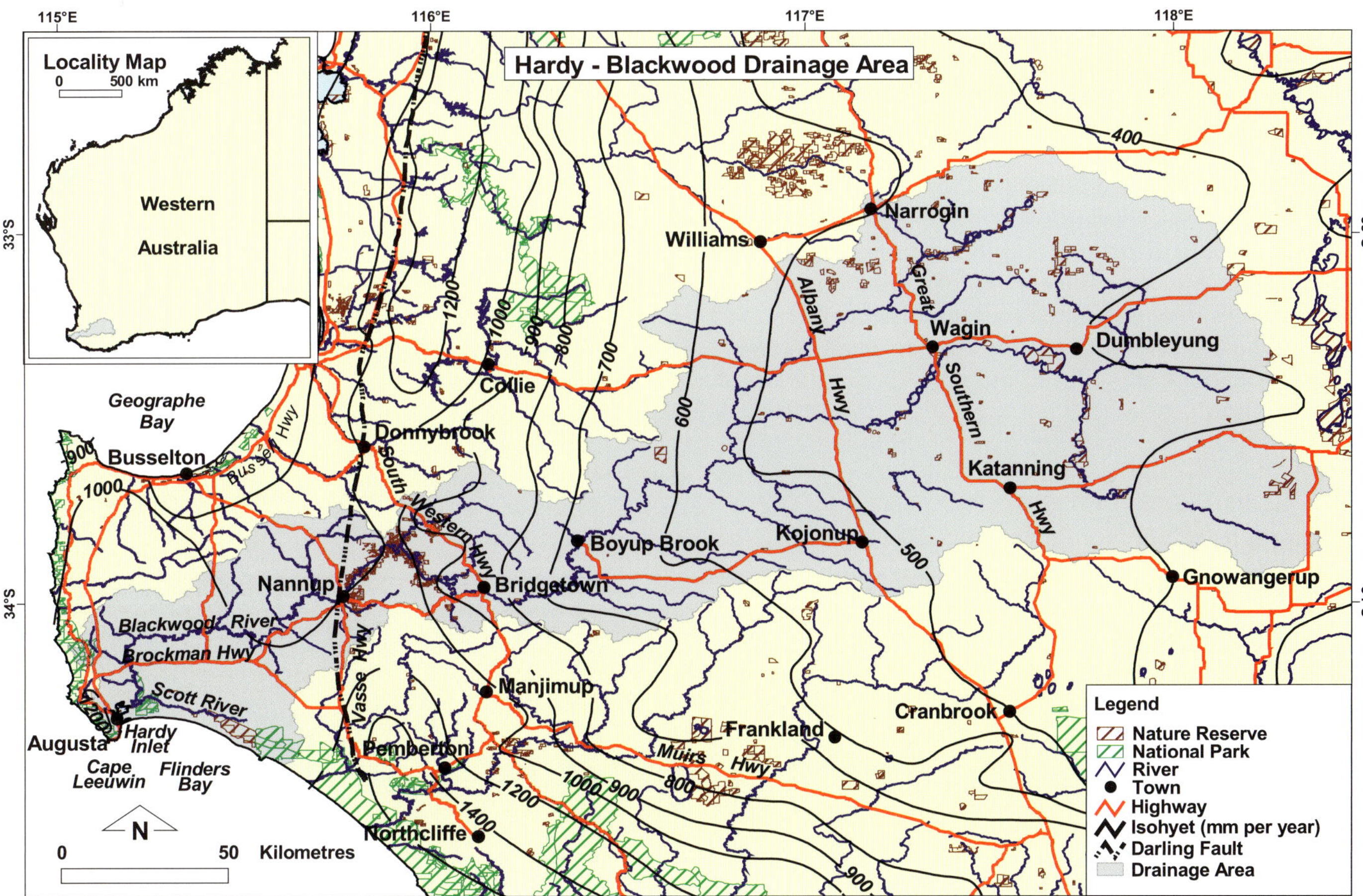

MAP 6.2 *Catchment of the Blackwood and Scott Rivers. Although the catchment of the Blackwood is about sixty-six times larger than the Scott, it generally only contributes five to six times the amount of water to the inlet as the Scott drains an area of much higher rainfall.* Courtesy DoE.

FIG. 6.3 *Lake Dumbleyung from the north, Zone of Old Drainage, upper catchment of the Blackwood River, October 1974. Note the low topography of the surrounding country.* E. P. Hodgkin.

FIG. 6.4 *Landscape around Nannup, about 60 m above sea level, in the Donnybrook Sunklands of the Blackwood River, September 1974.* E. P. Hodgkin.

South of the Brockman Highway drainage is to the Scott River, where the streams are ill-defined and there is often sheet drainage in winter. Four old (Pleistocene) shorelines are recognised. North of the Scott River the two oldest dunes have weathered to flat acid peaty quartz soil. South of the Scott River, which flows parallel to the coast through an interdunal depression, the dune formations are still recognisable. In this region heavy minerals appear to have been deposited along the old shorelines from sediments transported by the Blackwood. Different parts of these deposits led to the controversial mining applications in the 1970s and their rejection; and development of a mine at Beenup in 1997 that only operated for a few years before closing, leaving a legacy of large ponds of acidic waste that required rehabilitation (see Fig. 6.15). (See **Chapter 9 Albany Harbours and the smaller estuaries of the Albany region: Torbay Inlet**.)

Another fault line, the Dunsborough Fault, separates the catchment from the Precambrian rocks along the west coast called the Naturaliste–Leeuwin Block. Near Augusta the fault line passes through Thomas Island, Point Irwin and Dukes Head (Map 6.3 and Fig. 6.6a).

Flooding seas and melting icecaps

About 15,000 years ago during the last ice age in the Pleistocene era, the sea level was about 100 metres lower than present. The coastline would have been considerably further south beyond the islands off Flinders Bay and

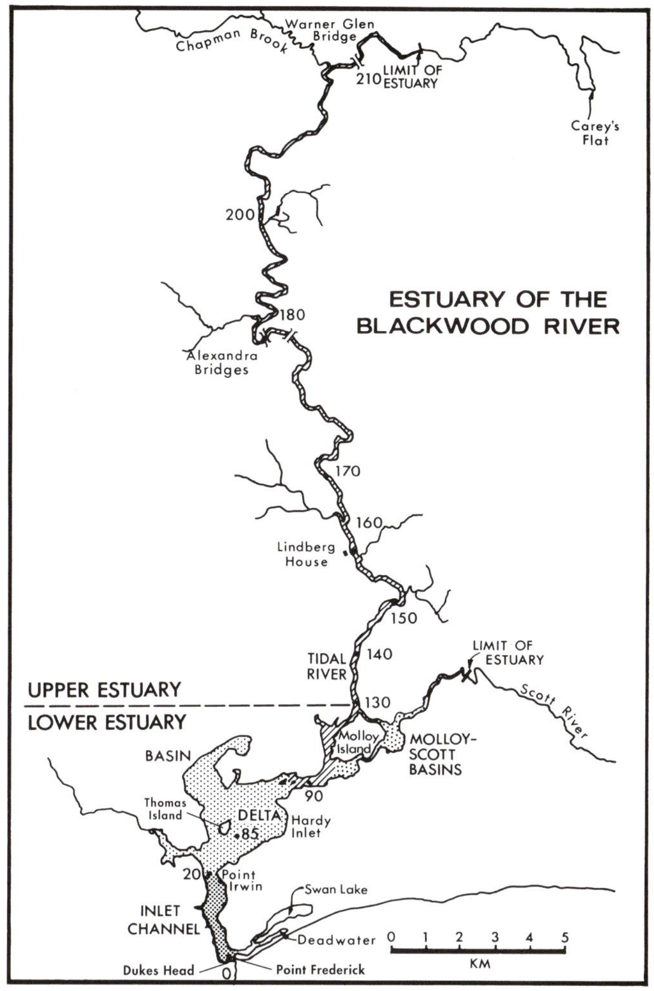

FIG. 6.5 *Blackwood River: (a) Sues Road Bridge, September 1974; (b) Great North Road Crossing, March 1978.* E. P. Hodgkin.

MAP 6.3 *Hardy Inlet showing the main divisions of the estuary system: numbers show the main sampling stations used in the 1974–1975 surveys. Note the position of the Deadwater and entrance from the sea.* Adapted from Hodgkin, 1978.

Cape Leeuwin (Map 6.1), and a river probably ran through a valley where the estuary is now. The estuary as we know it only developed more recently in the Holocene period, less than 10,000 years ago when the sea level rose and the river valley was flooded. Sea level has been fairly stationary for the last 5,000 years with some minor fluctuations up to 3 metres.

All the estuarine sediments have accumulated during this brief period of geological time. Moreover the present is not static; the same sedimentary processes are still at work although now they may be modified by human activities.

Hidden history

If we could look down through the sediment in the estuary we would trace the history of the estuary's development. The types of sediment and fossils allow us to follow the changing habitats for animals and plants, and compare these to those living in the estuary today. In fact geologists during the 1970's study did this to assess how the sediments were likely to be affected by any mining activities (Fig. 6.6).

The oldest basement sediments in the estuary are Pleistocene soils formed from the ancient granites. To the west these are orange to red material with a layer of pure clay or kaolin, while to the east the soil is composed of leached sands. As the estuary gradually flooded, soils in the inlet and basins were moved or reworked and remains of terrestrial plants and estuarine marine fossils were deposited in a thin sheet (Fig. 6.7). Later still a thick layer of shells, mostly mud oysters, was deposited in a black mud through which worms burrowed. Some of the shells have been dated at 4,000 years old. This type of deposit is typical of a sheltered situation, probably more marine than the present estuary, and represents an early phase of development when sea water penetrated more freely. Oysters probably lived in the Hardy for about 600 years, but today they are only present in the more marine Oyster Harbour at Albany.

To seaward the Oyster Unit is covered by a layer of sand from the beach and the nearby Pleistocene dunes that contains fossils of marine shells and sea urchin spines. Upstream in the basin the Oyster Unit is covered by mud derived from fine organic and inorganic material brought down by the river, and from plants and animals of the estuary. There are shells of species still living in the basin and of others no longer present. The fossil fauna in this unit changes with depth; fewer species and especially fewer marine species occur in the upper level indicating that the flow of marine water into the estuary gradually decreased. This layer would have accumulated more rapidly as the mouth of the estuary was gradually obstructed by build up of the bar.

In the upstream part of the basin, a delta unit lies over the basin mud. This deposit illustrates variation on both the discharge of the river and in wave action, with medium to coarse river sand interspersed with fine mud,

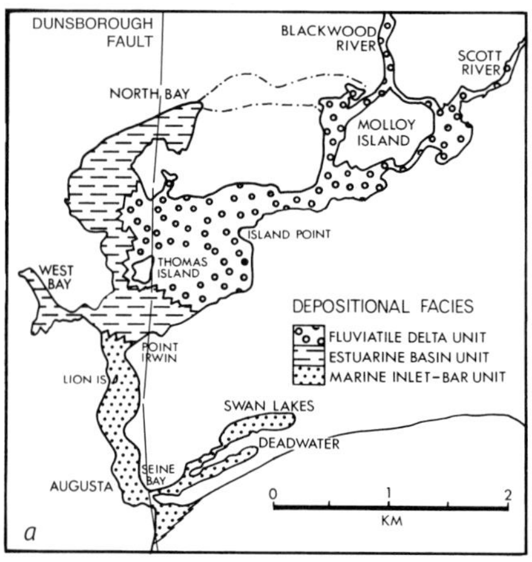

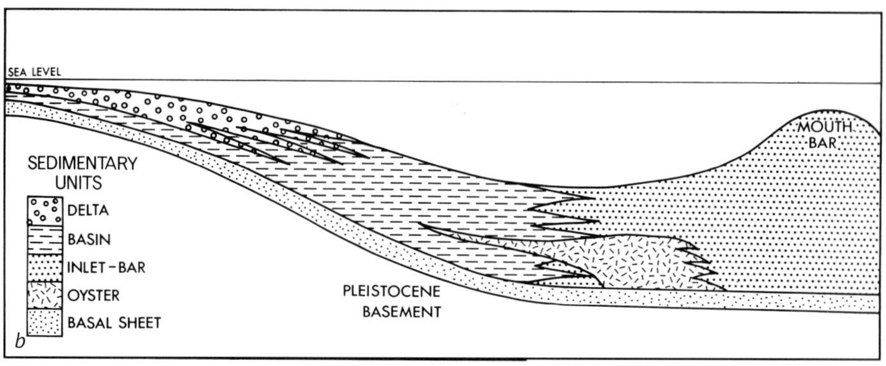

FIG. 6.6 *Sediments of the Blackwood River estuary: (a) surface sediments and (b) idealised stratigraphic section.* Adapted from Hodgkin, 1978.

FIG. 6.7 *(a) Drilling rig used to investigate the underlying sediments in the Hardy Inlet, October 1974. (b) Cores extracted from the Hardy Inlet: note the heavy shells of the Oyster Unit in the top core and the change in sediment type between the Oyster Unit and the old Pleistocene soil (basal sheet) in the bottom core.* E. P. Hodgkin Collection.

plant debris and faecal pellets formed by animals living in the estuary. In the 1970s the river carried little bed load, and the delta sands were thought to be reworked material scoured by winter floods from the Molloy Island region of the tidal river. Winter rains and summer floods like those associated with cyclones such as Bruno and Errol in 1982 brought mud to the estuary. The amount of sediment brought to the estuary has probably increased with clearing of the catchment and erosion of river banks. It appears that the lower estuary area has not increased or decreased very much in area during the last 10,000 years of the Holocene. There has been change with the infilling of the old basin with more than 8 metres of sediments, and accumulation of marine sand to infill the inlet and form the bar. This basin area is small compared to other Swanland estuaries, which generally have large basins and smaller rivers.

These same sedimentary processes continue at the present day. The estuarine mud continues to accumulate in the deeper parts of the basin, and the delta sands continue to prograde over it. Today, most of the basin area is less than 2 metres deep, about half is less than 0.5 of a metre such that extensive shoals are uncovered at low tide. Sediment deposition in any location can be quite rapid, and the northern channel has silted up and become almost impassable during the last one hundred years. Because of this infilling the boating channel was dredged in 1956, and re-dredged in 1973 in a slightly different alignment to a depth of 6 metres. Infilling continued at the downstream end of the channel and the requirements of larger boats necessitated some re-dredging to 2.5 metres in 1997. The formation and subsequent disappearance of the Deadwater is another example of recent sedimentary change, illustrating the transitory nature of estuaries and habitats.

Shifting sands

Conspicuous changes have occurred at the mouth of the estuary in the years since the first European settlement in 1830. The bar has moved eastwards and the rich area known as the Deadwater has formed and disappeared again.

A survey of the tiny Augusta settlement in 1832 by Hillman, Turner and Edwards showed the entrance through the bar quite close to Dukes Head. Swan Lake appeared as two lakes, Rushy Creek entered the estuary near its mouth, and there was no Deadwater although there was probably a depression between the coastal dunes and dunes separating Swan Lake (Fig. 6.8 and Map 6.4). The 1878 Admiralty Chart and the 1899 Townsite Plan both indicated a similar entrance but showed erosion of the dunes along Point Frederick. Greater change was apparent in the 1925 hydrographic survey, which showed an easterly movement of Dukes Head and the mouth at the expense of Point Frederick, which had retreated. Swan Lake no longer entered the estuary through Rushy Creek.

FIG. 6.8 *View of Augusta, Western Australia by Thomas Turner, October 1833. In 1830 Governor Stirling and some fifty settlers arrived in Flinders Bay at the mouth of the Hardy Inlet aboard the* Emily Taylor. *The Turners, Molloys and Bussells were among the early settlers; their names are immortalised today in place names throughout the area. In the early days, sheep and cattle were grazed through open forest, and clearing appears to have been limited to areas for houses and gardens. The estuary was important for transport and fish supplemented the diet. The small settlement gradually declined because of difficulties with clearing the massive trees, poor soils and isolation. By the 1850s the shores of the inlet were silent again until the 1870s when the timber industry commenced. The inlet was probably named for a comrade of James Stirling and John Molloy, Thomas Hardy, who was with Admiral Nelson on the* Victory *at Trafalgar. The view is from the east towards the settlement and clearly shows the western opening of the inlet to the sea at that time.* Courtesy Hon. David and Mrs Marie Louise Wordsworth.

Between 1925 and 1935 the mouth migrated eastwards about 2 kilometres. According to letters on Fisheries Department files: 'the bar has so silted up that getting into or out of it from the Bay is a matter of impossibility'. In 1930 the spit from Dukes Head was firm enough for vehicles to be driven along it, and a photograph dated January 1931 shows a Tiger Moth aeroplane on the spit close to the river mouth. An aerial photograph taken in 1943 (Fig. 6.9a) shows a dune 200–300 metres wide extending east from Dukes Head, and the river flowing behind the dune. Point Frederick was eroding in spite of efforts to stabilise it with marram grass in 1927 or 1928. There is no evidence that the bar ever closed completely, although the channel has at times been very shallow.

The winter of 1945 was critical with the greatest river flow for seventeen years. The river mouth had silted up completely, the camping area was flooded and residents were concerned. Local residents and government engineers agreed to open the bar and decided on the best location. Frustrated by delays, a band of men started digging. After five days of battling, and rumour has it with the help of dynamite, the river suddenly broke and in a few hours floodwaters dropped over a metre. The floodwaters spread as far as St Alouarn Island 9 kilometres from the mouth, and carried away fishing nets set in the estuary.

Photographs taken in 1955 show the mouth as a deep channel near Dukes Head; the 1943 river channel as the Deadwater with a broad shallow bar of mobile sand at Point Frederick and the western end of the Deadwater. This was also the case during the study in 1974–1975 when the Deadwater was a rich habitat for seagrass, fish and bird life. During floods in summer 1982 the river mouth was diverted against the western bank in the lee of Dukes Head, and the seaward part of the channel deflected to the east initiating erosion of Point Frederick. The eroded sand was deposited in the old channel. By 1983 the spit from Dukes Head had grown eastwards, and the dune line along the Deadwater had eroded. There was a bar between the Deadwater and estuary but it was located near the entrance to Rushy Creek.

Notes made by Ernest Hodgkin in August 1997 and again in May 1998 only months before his death in September are testament to his great interest

FIG. 6.9 *The mouth of the Blackwood estuary, aerial views: (a) 1943;* official RAAF photograph # 27738 Mao 1136 Busselton–Karridale Run 2. *(b) In September 1974 the entrance was close to Dukes Head with a large Deadwater to the east;* E. P. Hodgkin Collection. *(c) In 2000 the sand bar at Dukes Head had migrated eastwards as it had done in the 1940s, and water flowed through the Deadwater to the sea to the east of the Swan Lakes opening. Note the 2003 entrance was about a further 100 m east.* S. Neville/Ecotones, Courtesy WRC. See also Cover photograph taken in February 2005.

in the Hardy and the processes affecting the bars of all the estuaries of Swanland, the key to understanding how estuaries function.

> 23 August 1997 The spit has built out from Dukes Head in a curve towards the entrance to Swan Lake leaving only a narrow channel into the Deadwater, but Point Frederick appears to be little different from when last seen (photos). The sea bar is now only about 0.5 m at low tide and fishing boats can only get in and out on extreme range tides.

> 12 May 1998 to Augusta took photos from the usual location. The Point Gray [correctly Dukes Head] spit is further north east than I have ever seen it and from the end of it one looks straight into the mouth of the Swan Lake channel.

Later still the sandbar between the Deadwater and estuary eroded and the Deadwater became an arm of the inlet. By 2003 the estuary flowed to the sea through the old river channel of the Deadwater, and opened to the sea through a shallow passage on the eastern side of Rushy Creek almost in the position where it silted up in 1943 (Map 6.4d and Fig. 6.9a). The flow eroded the dune face along the channel and the entrance of Rushy Creek.

We can only wonder at the forces that have led to these large changes twice in 175 or so years. It is tempting to talk about the original river mouth as it was at the time of settlement in 1830, but the sequence of bar growth could indicate that there is no 'proper channel'. In the early days, the first changes to the mouth appear to have occurred before large-scale clearing of the catchments and before the channel was artificially deepened, when river flow and ocean currents combined to build the western spit out from Dukes Point pushing the mouth eastwards. However there were other changes to the vegetation and drainage on the eastern shore that may have precipitated the sandbar movements after the 1920s.

Most of the eastern shore was used for cattle grazing but the low areas seaward of Swan Lake through which Rushy Creek flowed were particularly valuable. Generally the creek entrance was blocked by sand, but this was removed in spring to release winter water and extend grazing land for summer. In the 1920s it would appear, however, that sand did not accumulate and sea water flowed upstream changing the fresh water of the lake to salt and killing the surrounding vegetation. Attempts to block this flow are probably visible in the 1943 air photo (Fig. 6.9a). Thus the combination of grazing and changes to drainage, both leading to a reduction of vegetation and culminating in further loss due to fire, individually quite small scale human interferences, changed the patterns of erosion on the eastern bank and at Point Frederick. Subsequently natural forces took over with the movement of sand (longshore drift) along the coast. This led to the development of a spit from Dukes Head that diverted the river and caused it to gouge a new channel which carried the mouth eastwards. Human intervention to protect property by breaching the bar in 1945 probably pre-empted a natural breaching that would have eventually occurred when

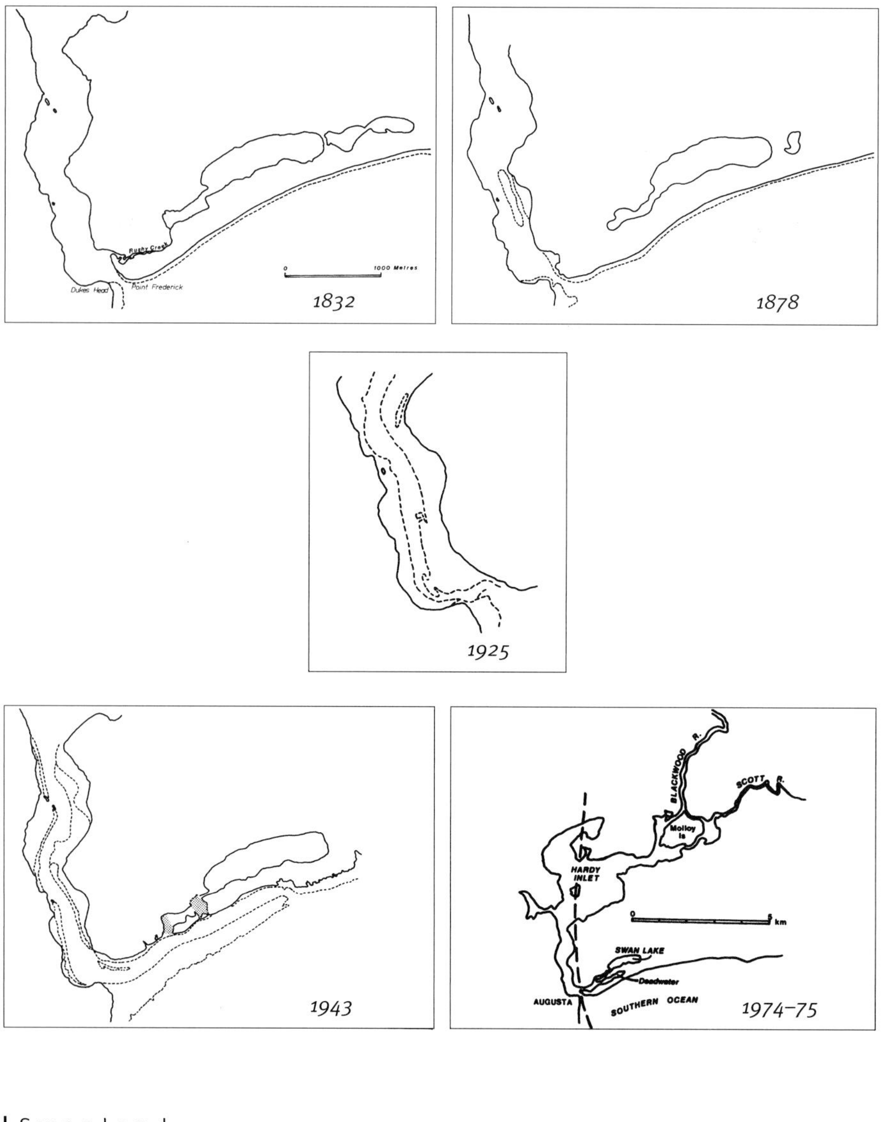

MAP 6.4 *Various configurations of the Hardy Inlet mouth: (a) Map dated 1832–4–8 Hillman, Turner & Edwards; (b) 1878 Admiralty Chart; (c) 1925 Hydrographic Survey (PWD 23962); (d) tracing of air photo 1943 with stippled areas indicating bare sand thought to be the result of attempts to block the creek; and (e) 1974–1975. Broken line indicates the line of the Dunsborough Fault. Small dashed lines on all maps indicate sandbars.* Adapted from Hodgkin 1976.

rising floodwaters forced the shortest route to the sea. We can guess that a breaching may have occurred in the years immediately before European settlement, hence the 1830s channel.

The patterns of bar migration between 1925–1945 and 1983–2005 are very similar. Years of low river flow allowing sand to accumulate followed by years of above average flow, and storm waves moving sand along the shore have been suggested as other mechanisms leading to the eastwards migration of Dukes Head. The causes are likely to be manifold and intertwined, but nevertheless a reminder of the variability of river flow and storms, of the mighty forces of nature and the transitory nature of habitats.

It is interesting to note that in his 1976 paper Hodgkin observed, 'The Deadwater is silting up slowly and, left to itself, it might eventually return to the 1830 condition'. Clearly it did not do this, and the 1982 diversion of the river eastwards led to a prediction that the Deadwater might eventually again form the entrance channel; and this has now occurred. In 1985 when concerns were raised about the shallow sandbars in the entrance, Hodgkin's notes record the response, 'that while you could push a channel through the bar it would close again rapidly—unless there (was) a considerable head of water to scour a deep channel & I can't see any sign of that'. The ending statement of his 1976 paper is also of interest, 'However, it is clear that there is a situation of uneasy balance in which quite small changes to the natural environment may precipitate major movements of the poorly consolidated sands near the mouth of the river'.

Winter flood—summer drought

Of all the Swanland estuaries, the Hardy with an annual mean flow of approximately 860,000 megalitres receives the largest volume of water although it does not have the largest catchment (Appendix 1 and Graph 6.1). Although the combined catchment of the Blackwood and Scott Rivers is only about 21,4000 square kilometres, much of the lower catchment lies in an area of relatively high rainfall. In contrast, despite the Swan–Canning estuary's much larger catchment of about 140,000 square kilometres, the mean flow is only about 600,000 megalitres as much of the catchment lies in areas of lower rainfall. The Blackwood River is the main source of water inflow to the Hardy, while the Scott River accounts for only 5–10 per cent of the annual flow. Most of the Blackwood catchment except for the forested areas downstream of Nannup has been cleared and is affected by salt. Clearing has not been as extensive in the Scott catchment, about 47 per cent, but the heavy application of agricultural fertilisers results in high nutrient flows to the estuary.

Water from the Blackwood River has a brown muddy appearance caused by suspended material, while the Scott River is darkly coloured by tannins. The water bodies maintain their integrity for some distance into the estuary;

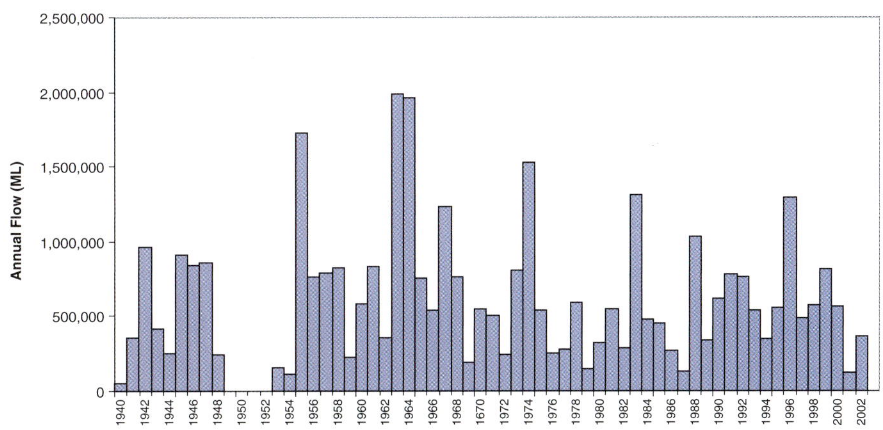

GRAPH 6.1 *River flow of the Blackwood River, 1940–2002, measured at the gauging station at Hut Pool approximately 45 km from the estuary mouth and 10 km upstream from the Warner Glen Bridge near the Great North Road crossing.* Courtesy WRC.

the contact between them only becomes blurred by turbulent mixing over the 3 kilometres to the basin, but even there the Scott water moves as a separate body down the eastern margin while the Blackwood water flows over the northern and central part of the basin. In some years, when river flow from the Scott is relatively high as in September 1999, the differences in salinity and turbidity are striking and the mixing zone resembles an 'oily scum' from a pollution event. (See **Box 5.9 Tea-coloured water.**)

Like all estuaries in Swanland, the hydrology of the Hardy is controlled by the climatic pattern of wet winters and dry summers. Estuary water is fresh throughout winter, and in summer the inlet and basin are marine (Graph 6.2). The transition to the winter fresh phase is usually abrupt but the change from fresh to salty may take several months. The Blackwood is estuarine for about 30 kilometres (near Warner Glen Bridge) and the Scott for 8 kilometres upstream. Tidal changes in water level may extend 55 kilometres upstream to Careys Ford (Map 6.3 Careys Flat).

As the river flow of winter decreases, the outgoing current is not strong enough to exclude sea water. On the rising tide sea water flows into the estuary beneath the out flowing fresh water and, because of its greater density, forms a salt wedge in the inlet channel. On the falling tide the upper part of the saline layer is caught up with the outgoing surface flow, making the halocline between surface and deepwater masses more abrupt.

Outside the mouth mixing takes place, and water of intermediate salinity may return to the inlet part of the estuary with a rising tide. Because the basin

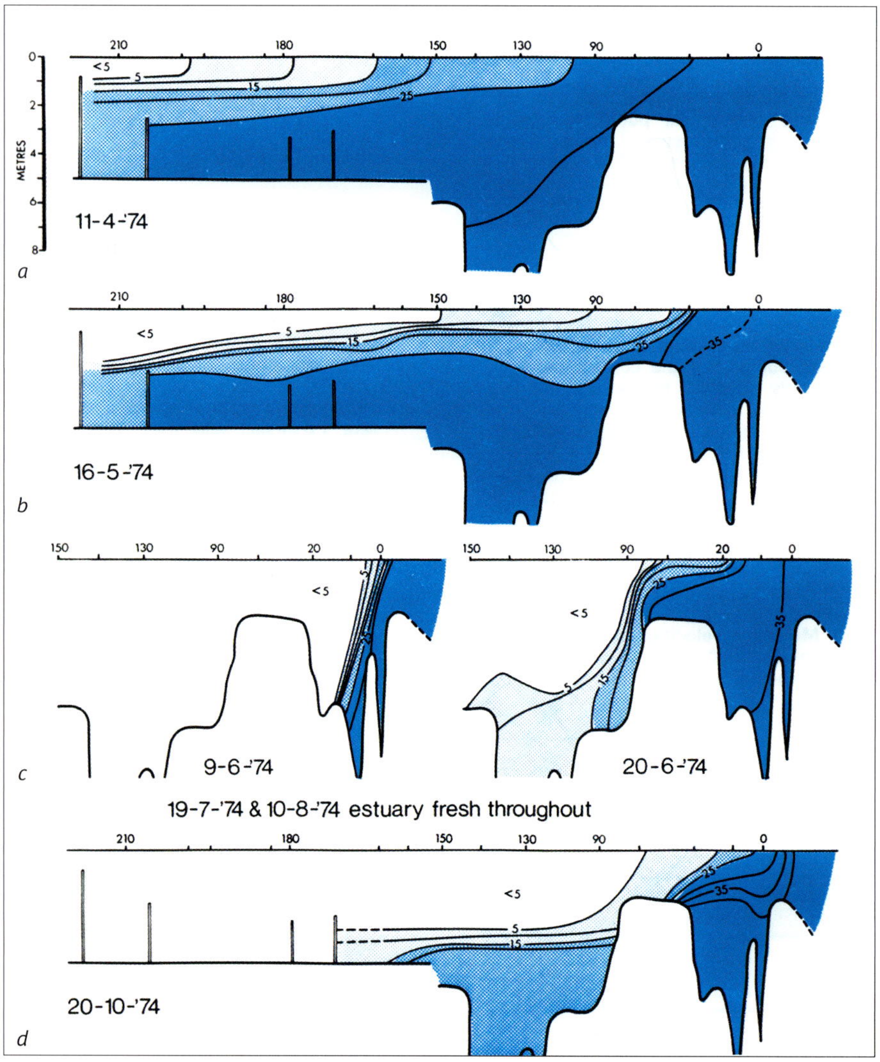

GRAPH 6.2 *Salinity profiles to show seasonal hydrological change in Hardy Inlet: (a) (11/4/74) shows late summer conditions; marine water has penetrated upriver almost as far as Warner Glen Bridge with some fresh water at the surface. (b) In May (16/5/74) increased flow has moved fresh water downstream. (c) In June (20/6/74) fresh water had penetrated most of the inlet, marine waters are excluded and isohalines at the mouth are almost at vertical. (d) In October (20/10/74) flow has decreased and saline marine water is again penetrating along the bottom of the inlet. Different densities of fresh and salt water produce a two-layered structure in which the two water bodies tend to retain their identity, i.e. it is stratified. The bottom waters below the halocline may become deoxygenated in the tidal river in summer.* Source: Hodgkin, 1978. Courtesy DoE.

and inlet are of relatively small volume, there is a considerable exchange of water with the sea on each tide. In the absence of any strong ocean longshore currents or waves however, much of the estuary water that goes out on a falling tide will return on the next rising tide.

When high salinity water has accumulated in the lower estuary, some of it flows up the estuary beneath the out-flowing fresh water. With a further decrease in river flow, high salinity water from the inlet traverses the basin with the rising tides and cascades into deeper upstream parts of the profile to form a salt wedge in the tidal river. The replenishing jet of saline water can travel up to 10 kilometres upstream on each rising tide.

Soils in the catchment also contain salts that have accumulated from rainfall over thousands of years. (See **Chapter 1 Understanding Swanland: Hidden in the flowing waters—salt and nutrients**.) When most of the catchment was covered by deep-rooted vegetation, little water and salt escaped to the rivers. After clearing (over 85 per cent of the catchment is cleared) and replacement of the natural vegetation with annual shallow rooted crops and pastures, salt flow to streams increased. While this is of great concern in some sections of the Blackwood, most of the water entering the estuary comes from the relatively less disturbed forested part of the catchment downstream of Nannup. Extractions from the underlying aquifers in the South-West including the Yarragadee for water supplies, however, have the potential to reduce flows of fresh water to the Blackwood and other rivers.

Nutrients for plant growth

The winter river carries dissolved nutrients such as nitrogen, phosphorus and silica into the estuary. Nitrogen may be in the form of ammonium (NH_4), nitrate (NO_3^-), nitrite (NO_2^-) as well as dissolved organic nitrogen and particulate nitrogen. Phosphorus is usually found as a constituent of organic compounds or adsorbed onto minerals. Generally carbonate soils such as those found in coastal areas of Swanland do not bind phosphorus, so much of the superphosphate that is spread on farmland moves into rivers and ground water. Phosphorus suitable for plant growth can also be released from sediments under the deoxygenated water found in the stratified tidal part of the upper estuary and river.

Changes to catchment vegetation have increased river run-off and inputs from domestic animals and fertiliser, particularly from the Scott Basin, are much higher now than in former times. Waters of the Scott River, although accounting for only 5–10 per cent of water inflow to the estuary, account for almost 40 per cent of the total phosphorus and 6 per cent of the total nitrogen. As a result, total phosphorus in the lower reaches of the Scott and channel near Molloy Island have at times exceeded ANZECC guidelines. Nitrogen levels in excess of guidelines have been recorded during winter in West Bay and upstream areas (Graph 6.4). (See also Graphs 1.2 and 1.3.)

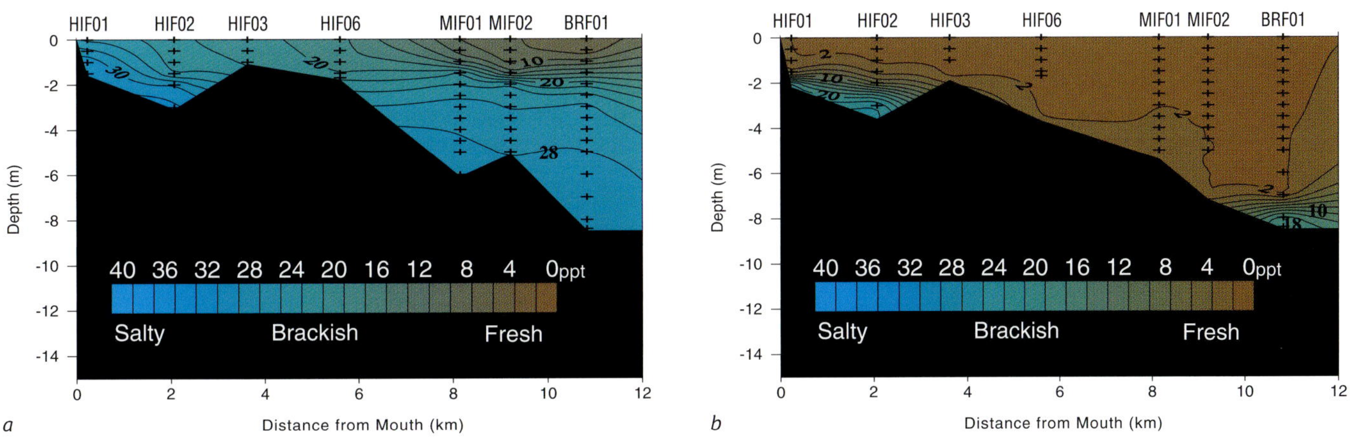

GRAPH 6.3 *Salinity profile: (a) summer December 1998 and (b) winter June 1999. Note the difference in depth between the shallow estuary, 90% of which is less than 2 m deep, and the deeper riverine reaches. In December marine water (blue) has entered the estuary and moved upstream along the bottom, only a small layer of fresh water remains near Molloy Island and the riverine reaches of the Blackwood at top right. In June the saline marine water lies beneath out-flowing fresh water in the estuary, and a pool of saline water lies trapped in deeper pools. In upstream areas, the trapped more saline waters often become deoxygenated falling to below 3% saturation. In most of the shallow areas oxygen levels are high due to plant growth or wind mixing. Locations are Hardy Inlet (HIF01, HIF02, HIF03, HIF06), vicinity of Molloy Island (MIF01, MIF02) and Blackwood River 0.5 km upstream of Molloy Island (BRF01).*
See Hardcastle & Cousins, 2000. Courtesy WRC.

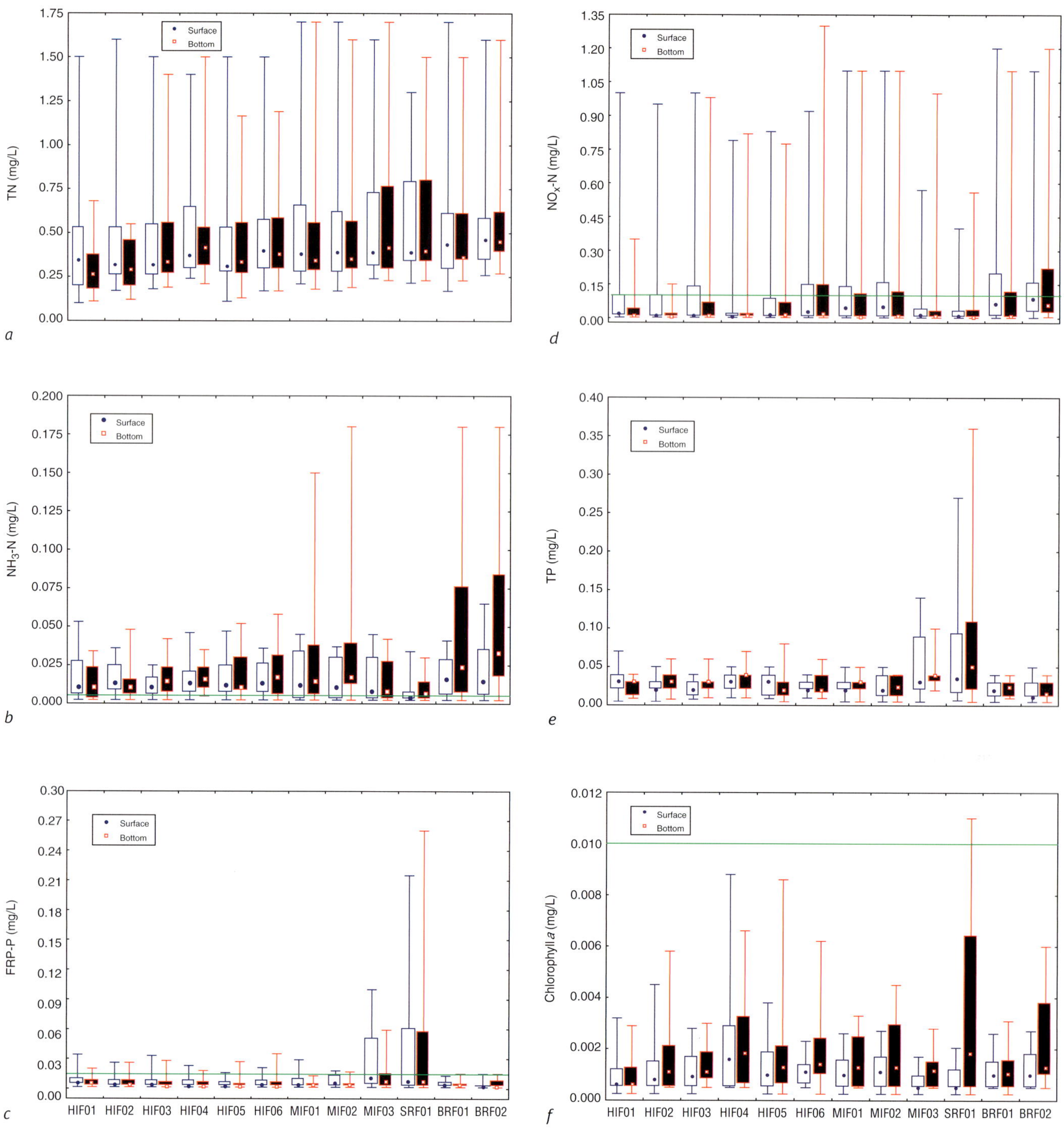

GRAPH 6.4 *Nutrients in the Blackwood River and Hardy Inlet, January 1999 to January 2000. Sampling locations indicated on the horizontal axis from the sea upstream to the main basin HIF01–6; Molloy Island area MIF01–3; Scott River SR; and Blackwood River BR01–2. Graphs (a) TN (total nitrogen), (b) NH_3 (ammonia), (c) FRP (filterable reactive phosphorus), (d) NOx (oxidised nitrogen), (e) TP (total phosphorus) and (f) Chl-a (chlorophyll* a*). Green horizontal lines indicate ANZECC 1992 guidelines. The large error terms in some graphs indicate the great variability in concentrations. Generally this was due to seasonal differences in flow, for example, strong run-off and flow after a dry period flushing nutrients from the soil profile or river pools.* See Hardcastle & Cousins, 2000. Courtesy WRC.

The higher nutrients in the upstream areas are reflected in the concentrations of chlorophyll *a*, a measure of the amount of microscopic algal growth. Chlorophyll *a* concentrations are much higher in West Bay, the lower reaches of the Blackwood and the area to the north of Molloy Island.

Scenic views: homes for many

Plants of the land

This is the country that early European settlers battled to clear before moving onwards to the Vasse, and where Georgiana Molloy started to love the flora of her new home (Fig. 6.8). (See **Box 6.1 Georgiana Molloy—lady of flowers: a change in perception**.)

Our bird's-eye view of the estuary landscape reveals different areas of the water body and a variety of habitats for many plants and animals: to the sea the sandbar, the shallow tidal flats along the inlet widening into the larger basin upstream with more complex vegetation along the banks, and the fringing or riparian vegetation extending down to the water's edge (Fig. 6.10).

A closer view of the fringing vegetation reveals an intermingling of different shapes, colours and textures. Along the banks, growing into the water, rushes such as the salt tolerant *Juncus krausii* and *Baumea juncea*

Box 6.1 GEORGIANA MOLLOY—LADY OF FLOWERS: A CHANGE IN PERCEPTION

At the time of European settlement 'botanising' was a favoured pastime of the more privileged classes and, while there was a desire to remember and recreate the gardens of their homeland, interest in the native plants of Western Australia was developing. The story of Georgiana Molloy, an early settler in Augusta and Vasse, illustrates this gradual change in perception. Soon after arriving in 1830 Georgiana started a flower garden, because as she wrote: 'I cannot be without flowers'. Yearning for the violets and primroses of her homeland she wrote in an early letter (1832) to her sister: 'This is certainly a very beautiful place—but were it not for domestic charms the eye of the emigrant would soon weary of the unbounded limits of thickly clothed dark green forests where nothing can be descried [sic] to feast the imagination...'.

Georgiana's love of gardens and growing interest in native flowers obviously became well known for in 1836 she received a package of seeds from a well-regarded amateur botanist, Captain Mangles, together with a request for an exchange of local seeds. This started a friendship that lasted until her death in 1843, and from the tone of her letters must have been of great importance to this lonely, isolated woman. Georgiana's enthusiasm for collecting seeds is shown by her idea of enclosing a leaf and description of the flower, and a request for Mangles to send names for her duplicate set. While noting that some from the Vasse were 'exquisite floral beauties', the less spectacular and common plants were also included in the boxes to Mangles. One letter in 1838 records that if she accompanied the box to England:

> *I should have a very extensive conservatory, or conservatories, of no plants but from Augusta. I do not say this vauntingly, but to inspire you with the ardour and interest with which the collection leaves me, and cordially thank you for being the cause of my more immediate acquaintance with the nature and varieties of those plants that we have exchanged for the productions of our own country...but from your request, I should never have bestowed on the flowers of this Wilderness any other idea than that of passing admiration.*

It is difficult to estimate Georgiana's importance to botany: Mangles and his correspondents were full of praise for the quality of her collections, and she also assisted the German botanist Ludwig Preiss who stayed with her for over a month at Vasse leaving with over 2,700 specimens. Perhaps it is a reflection of the times, but sad that neither Preiss nor Mangles acknowledged her in their publications. Fellow countrymen were no doubt influenced by her interest in the flora and found her, as the clergyman Wollaston noted, 'a perfect botanical dictionary'. Possibly her main contribution was to the progress of horticulture, as noted by George Hailes in 1844 on hearing of her death, 'not one in ten thousand who go out into distant lands has done what she did for the Gardens of her Native Country'. Ironically some of the weeds that invade our bushland today probably originated as escapees from her gardens.

Source: Hailes, 1844; Hasluck, 1955; Lines, 1994; Molloy, 1832; 1838.

grow along the whole length of the estuary while others, such as the sedges *Lepidosperma tetraquetrum*, *L. effusum*, *Cyperus* sp. and the Tall Kangaroo Paw *Anigozanthos flavidus* are confined to the upper reaches. Rushes can withstand winter flooding and may be dry for long periods in summer. Their importance is great and many faceted: they trap sediment and stabilise the shoreline, and create habitats. They are feeding grounds for bird species and, when water levels are high, for small invertebrates and fish. Although few animals eat them, when they decay they contribute to the mass of detrital plant material through which a large part of the energy and nutrients of the system are channelled.

FIG. 6.10 *Fringing vegetation on the western shore of Hardy Inlet, August 1974. Rushes line the water's edge with grass-trees,* Macrozamia *and eucalypts on the slightly higher ground.* See Congdon, 1981; Congdon & McComb, 1980; 1981). E. P. Hodgkin.

Observations in the 1970s showed eroding *Juncus* in some places with the rhizome mass exposed (Fig. 6.11). In other places it was growing or prograding into the estuary and trapping sediment. Active invasion by *Juncus* into shallow water is principally by rhizomes from which shoots grow up. Over time sands are trapped, a ridge develops and a swamp can form which favours either the samphire *Salicornia* or the paperbarks *Melaleuca* sp. rather than *Juncus*. In places the ridges have built up sufficiently for cycads *Macrozamia*, grass trees *Kingia australis* and peppermints *Agonis* to be established (Fig. 6.12).

Characteristic shore zonation is seen in many places, though the composition of the flora varies. To some extent this zonation represents a plant succession: as peat accumulates in fringing *Juncus* marsh, the salt tolerant paperbark tree *Melaleuca cuticularis* becomes established and itself builds up more peat further from the water's edge. Gradually this may be replaced by less salt tolerant trees, *Melaleuca hamulosa* and *M. raphiophylla*; as the ground level rises these are succeeded by flooded gums *Eucalyptus rudis*, and finally by jarrah *E. marginata* and marri *Corymbia calophylla*. This zonation is related not only to the salinity gradient, but also to soil water content, pH and dissolved oxygen.

FIG. 6.11 *(a) Small rush island with salt tolerant paperbark* Melaleuca cuticularis. *Note the shoots growing from rhizomes across the shallow sand where sand is accreting, November 1974. (b) Eroding rushes showing exposed rhizomes. Once the mat is damaged, the steep face at the water's edge is vulnerable to water movement during storms or boat-wash, and erosion is likely to increase making re-establishment of the rush community difficult.* E. P. Hodgkin.

Algae and aquatic plants

Tiny single celled algae or phytoplankton float in the water where they capture light for photosynthesis and absorb nutrients from the water. When phytoplankton are abundant they may colour the water, accumulating in 'blooms' or forming slimy mats. In the surveys conducted in 1974–1975 there were no such blooms. Counts of cells and concentrations of chlorophyll (the light harvesting pigments) extracted from water were also low, even in spring, when a bloom could have been expected because of nutrients washed in with winter rains. More recently, there have been reports of algal blooms in the upper basin particularly around Molloy Island and West Bay. There is no doubt that the higher concentrations of nutrients entering from the catchment areas of the Blackwood and Scott Rivers, and the gradual leaching of waste from Molloy Island and the Augusta townsite, have contributed to growth of these blooms. (See above **Nutrients for plant growth**.)

The phytoplankton generally contain diatoms, cryptophytes and occasionally non-toxic blue-green algae and dinoflagellates. In 1999 diatoms dominated more estuarine areas throughout the year reaching bloom numbers in West Bay (130,000 cells per millilitre) in September 1999, a time when salinity was still low, nutrient levels were high and temperatures were rising. The lower reaches of the Blackwood and north of Molloy Island were dominated by cryptophytes in summer and autumn (January–April 1999). Blue-green algae were found in bloom numbers in April 2000 (3,500 cells per millilitre) although the bloom was not visible in the water column. Blooms of another blue-green alga *Trichodesmium erythaeum*, commonly known as 'sea sawdust' or 'red tide', have also occurred in the inlet, usually in late summer. This blue-green is more usually found in the ocean but proliferates in the inlet when it is salty.

Compared to other estuaries in Swanland, there are relatively few plants growing in the Hardy. In the 1974–1975 survey this was attributed to the very short saline phase and the absence of rocks and other solid substrate. Larger water plants were only abundant in spring and summer when blooms of the stonewort *Lamprothamnium papulosum* (Charales) were found near Molloy Island and *Rhizoclonium riparium* in Swan Lake and the Deadwater. The sudden growth was not surprising in shallow water that heats up in summer, conditions that are particularly conducive to release of nutrients from the sediment. It is interesting that the nuisance algal species *Cladophora* and *Chaetomorpha*, which grow excessively in parts of the Swan, Peel and Walpole estuaries, were not abundant in the Hardy.

Five types of aquatic plant are found in the Hardy with distribution patterns that reflect hydrological conditions. In the lower part of the estuary the eelgrass *Zostera mucronata* and the narrow paddle weed *Halophila decipiens* grow where there is only a brief period of low salinity. This type of paddle

FIG. 6.12 Kingia australis *on Molloy Island, February 1978.* E. P. Hodgkin.

weed is only known from the lower reaches of the Swan, Nornalup–Walpole and Oyster Harbour and offshore from deeper marine waters (Fig. 6.13). Although recorded in 1966 it was not found in the 1974–1975 survey, but was found in deeper water through the 1980s. Seed of this species of *Halophila* appear to germinate in spring, the plants grow through summer and set seed in the autumn. At the opposite end of the estuary, the freshwater ribbon weed *Potamogeton pectinatus*, abundant at Molloy Island, extends downstream almost to the basin in spring and disappears with the advance of brackish water, being intolerant of salinities in excess of 8‰. Between these two species and overlapping them, swan grass (sea tassel) *Ruppia megacarpa* (referred to as *R. maritima* in older publications) is most abundant in the inlet, basin and marginal shallows of the tidal river in summer, and in the 1970s was the dominant macrophyte species in the Deadwater and Swan Lake. Another macrophyte of similar appearance, *Lepilaena cylindocarpa*, occurs sparsely in the basin. Most of the *Ruppia* growing in the estuary is found in shallow water where there is plenty of light. Its roots extend down into the black muddy anoxic layer where nutrient levels are high. Plant growth in the Hardy varies considerably according to the length of the fresh and saline phases. (See **Box 8.3 This thing called *Ruppia*.**)

Meadows of *Ruppia* and *Potamogeton* provide rich feeding for swans, which at times remove extensive areas by eating the foliage and disturbing the rhizomes. Both *Ruppia* and *Zostera* are often covered by growths of diatoms and other algae. This epiphytic cover can vary greatly depending on the extent of grazing by the tiny invertebrate fauna. Thick epiphytic cover can affect the photosynthesis of plants through the shading effect. In other parts of the coast and estuaries, huge increases in epiphyte cover have contributed to a decrease in seagrass growth with follow-on effects on associated animals. (See **Box 9.3 Shrinking seagrasses versus growing algae**.)

Inhabitants

Despite the extreme hydrological conditions in the Hardy Estuary, there is a successful fauna of resident estuarine (euryhaline) invertebrate animals and a smaller group of (stenohaline) animals that live in the estuary only during the summer marine phase. The resident species are also the most abundant, accounting for a great part of the biomass.

FIG. 6.13 *The tiny paddle weed* Halophila decipiens *is only found in a few of the southern estuaries of Swanland. The leaves, bright green in colour, are small, about 25 mm long and 5 mm wide. Flowers and fruits develop progressively as the plant grows. Fruits are located at the base of the short erect stems, and developing flowers and fruits on younger leaf nodes at branches.* See Kuo & Kirkman, 1995. Courtesy J. Kuo, UWA.

Plankton and tiny animals

In general, estuaries of Swanland have very small populations of planktonic animals (zooplankton) although some copepod crustaceans may be abundant in the marine and brackish waters. In the 1970s, except for the area of the Deadwater, they appeared to be particularly sparse in the Hardy. The four dominant copepod species were also typical of the other estuaries. *Gladioferens imparipes*, the most common, can breed in the wide range of salinities that occur here. At night the adults are in the plankton where they feed on microscopic algae, but by day they occur on the bottom, probably feeding on benthic detritus. This copepod is an important food item for many fish such as gobies, silver bream and yellow-eye mullet. Two other species of copepod, *Acartiura* sp. (sometimes referred to as *A. clausi*) and *Oithona nana*, are cosmopolitan and are also found in sheltered marine environments such as Cockburn Sound. The fourth species, *Sulcanus conflictus*, is normally found in low salinities (2–20‰ and in the Blackwood it is found in pools in middle reaches of the river near Bridgetown and Duranillin), but it is never common. Some other larval worms and fish may also spend time in the water column. (See **Chapter 2 The Swan–Canning Estuary: The zooplankton of the Swan**).

Other small crustacea, such as ostracods, were also abundant in the fresher waters of Swan Lake where they formed a major component of the sediment, but they appeared to be rare in the basin, delta and tidal river areas, particularly the more marine parts of Seine Bay and Deadwater. In contrast, the many types of single celled foraminferans were typically marine species.

Larger animals we can see

In total, the 1974–1975 survey found fifty-five species of invertebrate with forty-two of these probably confined to the estuary environment. Seven species made up 90 per cent of the estuarine population: three worms (*Capitella*, *Scoloplos* and *Ceratonereis*), two bivalves (*Arthritica* and *Fluviolanatus*) and two snails (*Potamopyrgus* and *Hydrococcus*). The large bivalves, such as cockles (*Katelysia*) and sunset shells (*Soletellina* formerly *Sanguinolaria*), amphipods (*Melita* and *Paracorophium*), the shrimp (*Palaemonetes*), and the larvae of the midge (*Pontomyia*) were also common and an important part of the benthic fauna.

The greatest number of species was found at the marine end of the estuary (Table 6.1). The more common species can be grouped according to their distribution, but these do not always represent their physiological potential or their distribution in other estuaries. The number of species in the upstream parts was generally less during the fresh phase than during the marine phase with some species confined to the marine end of the estuary in winter, extending their range into the basin during the saline phase over summer. We see this type of response in *Katelysia* and *Soletellina*, placed in Group 3.

Many of the animals were present throughout the area surveyed (Group 2), although most were found only as far upstream as Molloy Island. The less diverse fauna in the tidal reaches of the river may have been due to the absence of suitable habitats rather than a response to salinity. Generally animals were also less abundant during the fresh than the saline phase. Animals confined to the upper reaches of the estuary (Group 1) include the mussel *Xenostrobus securis*, false mussel *Fluviolantus* (previously called *Anticorbula*), and the tiny gastropod snail (*Potamopyrgus*). It is not known why they were absent from the marine part of the estuary as both the bivalves can tolerate sea water; *Xenostrobus* closes the valves and becomes inactive in low salinity while *Fluviolantus* remains active at low salinity.

Habits and homes

The habits of many estuary animals are not known nor is their role in the estuary ecosystem very well understood. The study conducted in 1974–1975 remains one of the most comprehensive.

TABLE 6.1 *Ecological groupings of the principal benthic invertebrate fauna in Hardy Inlet.* Adapted from Hodgkin, 1978.

	Popular name
Group 1 Species confined to the tidal river and basin	
Xenostrobus securis	black mussel
Fluviolantus (previously called *Anticorbula*)	false mussel
Potamopyrgus	tiny black snail
Group 2 Species found throughout the estuary	
Capitella capitata	polychaete bristle worm
Scoloplos simplex	polychaete bristle worm
Ceratonereis erythraeensis	polychaete bristle worm
Melita zeylanica kauerti	amphipod
Corophium sp.	amphipod
Paracorophium sp.	amphipod
Palaemonetes australis	shrimp
Pontomyia natans	midge
Group 3 Species not found upstream of lower basin (Stn 61)	
Hydrococcus graniformis	
Tellina deltoidalis	
Soletellina (previously called *Sanguinolaria*) *biradiata*	sunset shell
Katelysia scalarina	ribbed cockle
Rush fauna	
Tatea preissii	tiny high spired snail
Potamopyrgus sp.	tiny black snail
Orchestia sp.	amphipod
Cyclograpsus audouinii	crab
Leptograpsodes octodentatus	crab

Many animals burrow into the sediment where they play an important part in the physical breakdown of detrital plant material and their faecal pellets are an important part of the organic mud. During the course of a year, the volume of sediment that they 'rework' is substantial (many times greater than their body volume). The burrowing activity (bioturbation) varies with the grain size and interstitial space of sediments, but produces a lightly coloured deeply oxidised surface layer that controls the release of stored nutrients. In turn, the depth of the surface oxygenated layer may determine the type of animals found in the sediment. The worm *Ceratonereis* is generally in surface sediment, when it leaves the sediment it may be eaten by fish such as hardyheads. In contrast, the worms *Capitella* and *Scoloplos* may burrow into the darkly coloured anoxic sediments to a depth of 10 centimetres. These worms are less likely to be eaten by fish.

Bivalves also burrow in the sediments; *Soletellina* burrows into the anoxic layer and extends its siphons to the surface while the cockle *Katelysia* (Fig. 6.14) is confined to the top 5 centimetres. As both *Katelysia* and *Soletellina* have been collected for bait by fishers for many years, it is hard to know how abundant they really were. Certainly there were fewer in the 1990s than in the quieter days of the 1970s. Among the rushes, the tiny snail *Tatea* or *Hydrococcus* is found on the mud surface and crawling up the leaves, and the small crabs *Cyclograpsus* and *Leptograpsodes* can be abundant under logs or burrowing in damp mud. These are eaten by wading birds, and at high tide shrimps, hardy heads and other small fish feed among reeds.

The aquatic plants *Ruppia*, *Potamogeton* and *Lamprothamnium* create important habitat for many invertebrates and fish. Abundance of the shrimp *Palaemonetes* seems to be directly related to the density of *Ruppia*, although it is probably an omnivore feeding on dead and living animal and plant tissue. In the Deadwater and Swan Lake, the *Ruppia* has been often seen supporting large populations of hydrobiid gastropods that browse on epiphytic algae as does the amphipod *Melita* and the larvae of the midge *Pontomyia*.

FIG. 6.14 *The venerid bivalve* Katelysia scalarina. *Shell width approximately 4.5 cm. This species, now only found in estuaries along the south coast, also occurred in estuaries and bays along the west coast 3,500–6,500 years ago.* E. P. Hodgkin.

The dog whelk *Nassarius pauperatus* crawls over the mud and is a scavenger feeding on dead animal tissue. *Nassarius* appear to grow more slowly in the Hardy Inlet in comparison with other estuaries. Here it breeds only once a year and snails will not reproduce until two years old. Under favourable conditions it can be expected to complete development from egg to adult in less than a year.

Rocks, logs and jetty piles provide attachment for the mussels *Xenostrobus* and *Fluviolanatus* and the barnacle *Balanus*. This habitat is not as common as in other estuaries and low abundance of attached animals is clearly related to the limited availability of such sites. Other animals, such as isopods, find shelter among the mussels.

Resident invertebrate fauna are predominantly suspension feeders and deposit feeders. Most of the bivalves such as *Xenostrobus*, *Fluviolanatus*, *Katelysia* and the barnacle *Balanus* are suspension feeders, eating phytoplankton and suspended detritus. The large bivalve *Soletellina* (*Sanguinolaria*), which burrows deeply into the anoxic sediment, and many of the worms are deposit feeders, eating detritus and a diversity of small organisms within the sediment, including the bacteria that are involved in decomposition processes. The amphipods *Corophium* and *Paracorophium* also feed on the sediment surface deposits while another species *Melita* and isopods feed on larger particles.

Fish and fisheries

As with other animals within the estuary, the fish fauna consists of true estuarine species, species confined to estuaries and other brackish waters, and marine species with varying degrees of ability to contend with lower salinities. Composition of the fish fauna therefore varies with the seasonal changes in hydrology. Unlike the invertebrates, which are represented by a suite of resident species, there are only five or six resident fishes. These residents are nevertheless an important part of the estuary fauna.

Although there have been some more recent surveys, the 1974–1975 study organised by Dr Ernest Hodgkin and Dr Rod Lenanton has been the most comprehensive, examining fish throughout the estuary and tidal rivers. Fifty-seven species were recorded at that time, and the Hardy was considered as rich as other south coast estuaries. Most of the fish species are known in other estuaries of Swanland, particularly those along the south coast. To some extent the species present and their abundance will vary with the availability of larvae in the offshore areas, which in turn is controlled by larger scale oceanic effects. One species, the flathead (*Platycephalus* sp.), common in the 1940s was not abundant in the 1974–1975 survey nor was it common during the 1990s.

Of the resident estuarine species, only the black bream was more abundant in the upper reaches of the estuary (tidal river and Molloy basin) than in the lower estuary. During high winter flows, black bream moved down to the lower estuary and were abundant in Swan Lake. The two common resident gobies had slightly different distributions: the south-west goby *Favonigobius tamerensis* was most abundant in the tidal river and Molloy basin, while the long finned goby *F. lateralis* was most abundant in the inlet, Deadwater and Swan Lake. Hardyheads (*Atherinosoma*, two spp.) were abundant in the lower estuary at all times. In contrast, freshwater catfish or cobbler *Tandanus bostocki* normally confined to the riverine reaches entered the estuary with river run-off.

Of the non-resident fish, the sea mullet *Mugil cephalus* and yellow-eye mullet *Aldrichetta forsteri* were abundant in the estuary throughout the year though numbers decreased in the tidal river during the fresh phase. The western sand whiting (yellowfin whiting) *Sillago schomburgkii* and King George whiting *Sillaginodes punctata* showed a more marked downstream movement at this time, as did silver bream *Rhabdosargus sarba* and striped trumpeter *Pelates sexlineatus* and these were always more abundant in the lower estuary.

Another group such as the southern garfish *Hyporamphus melanochir*, sandy sprat *Hyperlophus vittatus*, blue sprat *Spratelloides robustus*, old wife *Enoplosus armatus*, flathead *Platycephalus* sp., mulloway *Argyrosomus japonicus* and blue rock whiting *Haletta semifasciata* were absent during the fresh phase. Some species left in May with the first flush of fresh water, others as the flush increased and reduced salinity, only returning in January with the more marine water.

The estuary is a nursery ground for some of the common species, although few actually spawn there. The estuary provides a relatively rich feeding ground compared with the open sea, and a greater measure of protection from large active predators that are generally not found in lower salinity estuarine waters. Juvenile fish seek the shelter of plants and are taken in greatest abundance in the shallow areas where seagrasses grow. In the 1974–1975 study, 98 per cent of the juvenile sea mullet, King George whiting and silver bream were found in the *Ruppia* beds in the then extensive Deadwater. Blue swimmer crabs *Portunus pelagicus* also enter the estuary after larval and early juvenile development in the ocean, and are probably more abundant during the years in which the marine phase begins early. They may persist in the lower reaches of the estuary in years when the winter flows are weak and some salty water remains in deeper areas of the inlet. (See **Box 4.2 Crabbing—a summer pastime**.)

In the past the estuary supported a productive fishery but, in the pattern of most other estuarine fisheries, commercial fishing has declined as recreational fishing has grown. Between 1979 and 1987, six licences were issued but this was reduced to four by June 1998, and to two by October 1998 with only one remaining in 2003, but under great pressure from recreational fishers for this to be withdrawn. The 1974–1975 survey estimated the recreational catch as black bream (20 tonnes), yellowfin whiting (12.8 tonnes), Australian herring (9.2 tonnes), King George whiting (2.5 tonnes), silver bream (2.3 tonnes), tailor (2.2 tonnes) and skipjack trevally (2.2 tonnes). These catches were based on the estimated 16,500 days of fishing effort per year and a catch rate of 3.8 fish per angler per hour. Since that time the recreational fishing pressure has increased substantially, however the current catch is unknown. Generally the commercial catch is sea mullet, yellow-eye mullet and whiting (mainly yellowfin) and in years when the whiting catch is low by higher catches of yellow-eye mullet. Between 1978 and 2002 the annual commercial catch varied between 10 and 34 tonnes. Black bream were a major part of the fishery until the mid-1980s, with a maximum of 3.6 tonnes in 1983. After this the catch declined due to fishing and other factors related to the environment, such as sprays to control blackberry, or heavy sediment loads following the 1982 floods. Black bream appear to be increasing again. Assessment of the black bream population and investigations into the restocking with hatchery raised fish commenced in 2002. Restocking is, however, problematic as each estuary has a genetically distinct population, and the likelihood of long-term improvements to fish abundancies by this method is minimal. (See **Box 2.13 A popular fish—the black bream** and **Box 10.2 Black bream—a real estuary fish**.)

Food for fish

The seagrass *Ruppia* is eaten by garfish, probably the only species in the estuary that can be regarded as an herbivore, but it is not known if the seagrass is the most important source of nutrition. Algae and seagrass are also found in the stomachs of black bream, silver bream and trevally. For garfish it is not known if these are actually digested, and they are probably omnivores. A number of fish studied contained detritus, but the only species known to use the microbiota as a source of food is the sea mullet.

Most other fish including the resident black bream, hardy heads and gobies are predators feeding on invertebrates. Prey is usually selected according to size, depth to which a fish can probe the sediment, and abundance at a particular time rather than selection of a particular type of animal. Prey taken by black bream is perhaps the most varied, and food found in their stomachs was the most abundant prey available in the area in which they were taken. Molluscs are a minor part of fish diets. The smaller *Arthritica*, *Hydrococcus* and *Potamopyrgus* are those most frequently eaten, partly because they are also the most abundant but also because the large species have heavier shells or are not as accessible as they burrow deeper into the sediment. Small and large yellow-eye mullet and silver bream select similar kinds of food, but bigger fish select bigger prey and therefore some different species. Some marine fish such as yellowfin whiting and yellow-eye mullet enter the estuary on a rising tide to feed, remaining in the estuary for only a short time.

Feathered friends

Like many of the plants, invertebrates and fish that live within the waters of the estuary and respond to the seasonal changes, birds also make different use of the estuary throughout the year. Seasonal usage by birds is different however, and related to the availability of other aquatic habitats rather than conditions within the estuary itself. Many waterfowl come to the Hardy and other south-west estuaries when inland waters dry up over summer. Estuaries are therefore important for maintaining bird populations from a much greater area. Although some species nest nearby, estuaries are important for most birds because they provide rich feeding grounds. In fact a great deal can be learnt about estuarine food chains by a study of birds because they are more easily observed than fish, the other main predators. In addition they have a great deal of emotive appeal. People enjoy seeing them, and changes in their abundance can be appreciated far more than the disappearance of fish or invertebrate species. Few people are not touched by the sight of pelicans (*Pelecanus conspicillatus*) perched on jetties, parading for fish handouts on the sand flats or performing acrobatics on the branches of dead trees.

The estuary and surrounding bushland offer many different habitats for birds. In the 1974–1975 study, fifty-seven bird species were recorded on the estuary and an additional thirty-nine in the peppermint thickets and rushes on the eastern bank. There are probably many others. Birds can be grouped into summer migrants from overseas, summer migrants from other parts of Western Australia, winter visitors, and residents.

The estuary provides rich feeding grounds for eighteen migrant species that breed in the Northern Hemisphere and fly south to feed here during the northern winter. Generally the Hardy received fewer of these visiting species than the Swan and Leschenault, probably because shallow feeding grounds are less extensive. Some, such as the Sanderling *Calidris alba*, were common in the 1970s in flocks of sixty or more. Migrant species started to arrive in September and left quite abruptly in March–April, but at the height of summer over 1,000 of these wading birds were feeding on the sand and samphire flats. (See **Box 1.10 Global travellers: living a double summer life—transequatorial migrant birds**.)

Local migrant species arrived later than the international visitors, with numbers building up only after inland waters dried up in summer, some staying only a short time. Black Duck *Anas superciliosa* arrived in December leaving again in February, while Black Swans *Cygnus atratus* arrived in January but stayed until May. Pelicans returned to the Hardy earlier, after nesting at Shark Bay or further north between March and August. White-faced Herons *Ardea novaehollandiae* were present from December to May. The numbers of birds fluctuated greatly between years depending on conditions in other areas. A few birds such as the White Egret *Egretta alba* are regular autumn–winter visitors while the little black cormorant *Phalacrocorax sulcirostris*, Little Pied Cormorant *Phalacrocorax melanoleucos* and Silver Gull *Larus novaehollandiae* had a more uniform abundance throughout the year.

Bird food and habitats

The distinct habitat preferences of many birds are usually related to availability of the type of food consumed. Birds require a great deal of food because of their high metabolic rate and this requires a healthy ecosystem. In turn, bird feeding behaviour has considerable effects on the ecosystem (Graph 6.5).

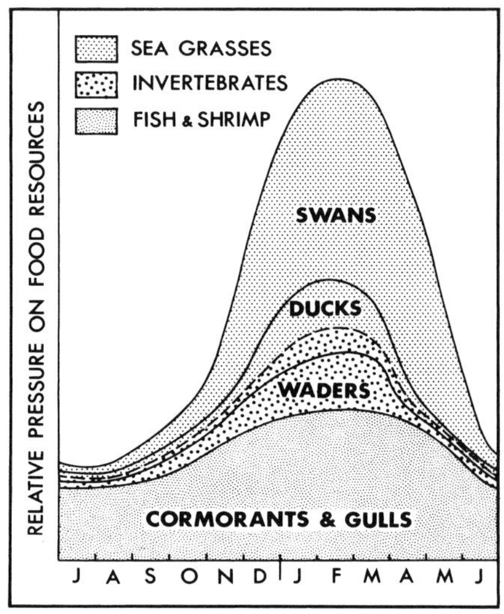

GRAPH 6.5 *Relative seasonal pressure on the estuarine food resources by the main bird types.* Hodgkin, 1978. Courtesy DoE.

Bird hides

Rush beds provide good cover for birds and are favoured by the White Egrets, Swamp Hens *Gallinula porphyrio* and Little Grass Birds *Megalurus gramineus*. White Egrets feed mainly on invertebrates such as crabs and small fish at high water, but Swamp Hens also eat plant material. In a similar way the samphire marsh, while not as extensive as in other estuaries, provides shelter for small semi-aquatic invertebrates and is a favoured hunting ground for the small migrant waders, the White Egret and White Ibis. White-faced Heron are also found here, but they and Silver Gulls also feed in the shallow water of the tidal flats. Tidal flats cover a large area of the basin, the inlet, Molloy Basin, Swan Lake and the Deadwater. They may be uncovered or covered only by a few centimetres at low tides but are completely covered by normal summer tides. Many migrant waders are opportunistic feeders taking the most abundant and accessible invertebrates in the sediments, so there is heavy predation during summer months. The worm *Ceratonereis* appears to be an important part of the diet of these birds.

Slightly deeper areas that remain permanently submerged, where the seagrasses *Ruppia*, *Potamogeton* and *Lamprothamnium* grow, are important feeding grounds for Black Duck, Musk Duck *Biziura lobata* and Grey Teal *Anas gibberifrons*. Black Swans can feed in slightly deeper water, about one metre, provided they can reach the bottom with their long neck. In fact, swan numbers seem to reflect seagrass abundance as they are most abundant where *Ruppia* is most dense, for example, in Swan Lake. Each swan consumes about 2.5 kilograms wet weight of *Ruppia* per day, and it was calculated that in Swan Lake they consumed 230 tonnes of seagrass during the year, a significant part the yearly production of 1,000 tonnes. Swans have two methods of feeding, a lawn mower technique in which foliage is chopped into short lengths of 2–10 millimetres and another in which birds dig circular pits. In this case they eat the rhizomes, which are rich in food stores. They actually dig to a greater depth than the rhizomes. In both feeding methods there is considerable wastage. Not all that is pulled up is consumed and much of the foliage passes through the gut undigested. In the basin area Black Swans destroyed most of the *Potamogeton* in spring, but plants regenerated from rhizomes when salinity was low again. It has been suggested that feeding Black Swans could have a cultivating effect and make worms available for fish predation. (See **Chapter 2 The Swan–Canning Estuary: Recreating Swanland—bringing back the Black Swan**.)

Deeper open water is favoured by the fish eaters: cormorants, eagles and gulls—birds that swim or dive for their prey. These include some of the most abundant birds of the estuary such as cormorants, although they are not as abundant on the Hardy as on other estuaries such as Leschenault. It is interesting to note that cormorants probably consume a similar amount of fish to that caught by humans, but the types of fish caught by cormorants were principally the small gobies, non-commercial species and the shrimp *Palaemonetes*. (See **Chapter 8 Wilson Inlet: Birds; Box 2.19 Black cormorant feeding frenzy**.)

Where do we go from here?

The story of an estuary is one of change. The Hardy we see today is not the estuary of times gone by. Processes that have shaped the estuary range from geological events such as faulting to climate change affecting sea levels and rainfall. The strength and duration of river flow affects the sediments and salinity; with salinity the master factor controlling the types of plants and animals that can live within and near the estuary. Biological patterns in the Hardy depend on how long the winter fresh water phase persists before changing to the summer salty phase when marine animals can invade. The layers of fossils preserved in the sediments indicate that these same stages have been occurring throughout the estuary's history.

The large birds and fish that visit for short periods to feed are also affected by the character of the estuary and variability of the resident flora and fauna. Likewise many of the people fishing are interested in marine species that visit the estuary for only a short time, their numbers varying greatly from year to year depending on marine stocks available for recruitment. Interestingly, a favoured fish habitat in the 1970s, the Deadwater, has now virtually disappeared as part of a natural cycle. One fact remains certain: all who use the estuary depend on its health. So how healthy is the Hardy?

River waters have changed since European settlement. The upper reaches of the Blackwood are now saltier and species such as the marron are now only found in fresher parts of the river in forested areas. The estuary is saved

FIG. 6.15 *Beenup Mine, aerial view, 2003. After two years of operation, the mine was closed in 1999 following problems with acid sulfate soils and the clay content of the ore body. The site is being rehabilitated.* Courtesy BHP Billiton.

to some extent by the vegetation in the lower catchment, as water emerging from here is less salty. Decline of the black bream in the 1980s, while possibly affected by heavy fishing, has also been attributed to changes in the river catchment due to a loss of food items resulting from sprays used to control blackberry or much heavier sediment loads as the river banks erode. Black bream now appear to be increasing again. Waters from the Scott, flowing from the low-lying sand to the east bring only a small inflow but gather large amounts of nutrients from farm and pasture lands, passing the now defunct Beenup Mine where considerable effort has been made to contain the acidic waste (Fig. 6.15).

While the number of animals using the estuary varies from year to year, the number of people using the estuary has increased and will continue to do so. Fish are continually under pressure, but many of the favoured types are probably little threatened by overfishing as they are restocked from the sea. The major problem from the human point of view is that there are less fish to go round. The greatest impact of people, however, is on water quality. It is worth considering that subtle changes that are hardly noticeable at first may have far reaching effects on a whole pattern. The influence of people and their waste extends from Molloy Island to the bar, and houses now lie on both sides of the estuary. Waste from septic tanks and garden fertiliser gradually leaches into ground water, which finds its way to the estuary where nutrients increase the risk of nuisance and toxic algal blooms. Probably the greatest help we can give the estuary is to limit these inflows and protect the remaining vegetation along its borders. Well-vegetated surrounding areas would also have the added effect of maintaining the much-loved vistas and the wildlife that depend on a healthy estuarine system (Fig. 6.16).

FIG. 6.16 *Pelicans and others at rest Hardy Inlet, 1974.* E. P. Hodgkin.

Chapter 7 Estuaries of the wet Western South Coast

This chapter describes the estuaries of the western south coast extending from Hardy Inlet to Wilson Inlet. These estuaries are fed by large rivers, which drain the drier ancient heartland of Western Australia, and by numerous smaller rivers, creeks and seeps around their margins. Their drainage areas determine the type of water that the estuaries receive, while the amount and timing of water flow is determined by the weather patterns—for this is the wettest part of the south coast, often referred to as the Rainbow Coast. The location and exposure of the mouth, and the sand supply in combination with river flow, determine the type of bar and the exchange of water with the sea, affecting the animals and plants that live within them. Some, such as Broke, are surrounded by uncleared wooded catchments. Others, such as Wilson, Irwin and Parry, are surrounded by cleared and cultivated land from which salt and nutrients are leached affecting township water supplies or causing eutrophic conditions with rich growths of aquatic plants or algal blooms. Wilson Inlet is described in a separate chapter, as the problems of nutrient enrichment there are the best known. (See **Chapter 8 Wilson Inlet**.) The larger estuaries, Broke and Walpole–Nornalup, are described in the greatest detail. Irwin and Parry Inlets, smaller in size, shallow and hypersaline or dry over summer, perhaps at the end of their life as estuaries, are described together. The smaller riverine systems in the west (Maps 7.1, 1.1, 1.2 and 1.10), although only loosely fitting the definition of estuary, are described below in Stretching the definition—not really estuaries.

PLACE IN THE LANDSCAPE

The upper reaches of some of the rivers (see Appendix) extend onto the ancient Yilgarn Block or Craton, some 2,900–2,500 million years old, where numerous salt lakes overflow to the rivers during periods of excessive rain. Part or all of the catchments of larger rivers such as the Warren and Donnelly, Frankland–Gordon and Kent, lie on the southern edge of the Western Shield plateau of granitic rocks in the Albany Fraser Orogen where the rocks are younger, about 1,300–1,100 million years old. Here the land is gently undulating with broad valleys about 300 metres above sea level. During the Paleogene and Neogene, previously termed the Tertiary (65–2 million years ago), the edge of the Yilgarn plateau was tilted upwards along the Jarrahwood Axis that passes roughly parallel to the coast through Manjimup. Southwards the Ravensthorpe Ramp, a strip 40–80 kilometres in width, slopes gently to the sea. As a result the rivers deepened and widened in a process known as rejuvenation.

The 1,000 kilometre-long Darling Fault lies along the western margin of the Yilgarn Block, first formed during rifting in the Jurassic and early Cretaceous periods (160–120 million years ago). The fault lies along the western side of the Donnelly River catchment, and the lower reaches of the river and its western tributary, Barlee Brook, are on younger (less than

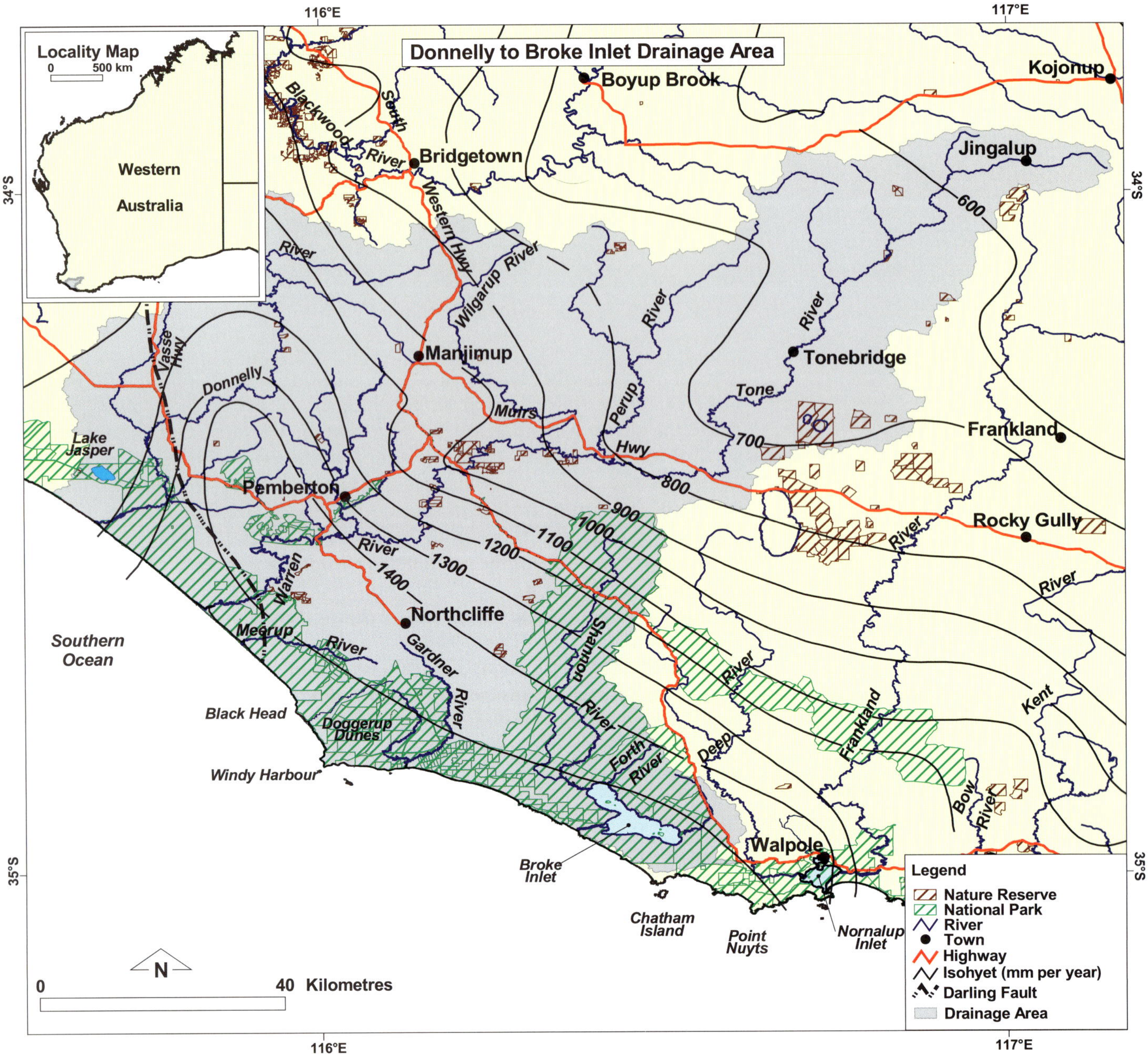

MAP 7.1 *Location of estuaries on the western south coast: between the Donnelly River and Broke Inlet the coast faces south-west. The upper reaches of the rivers flow from the north and north-east and drain areas of low rainfall. The lower river reaches and estuaries lie within the higher rainfall area in the D'Entrecasteaux National Park.* Courtesy DoE.

65 million years old) rocks. Spectacular basalt cliffs at Black Point (Cape Beaufort) and rocks in the Donnelly River (Fig. 7.1) mark the line of the Darling Fault.

Like the west coast, the rise and fall of sea level has played a major role in forming the near coastal landscape where the estuaries lie. In the western section of the south coast old shorelines form the Scott Coastal Plain. Underlying the coastal plain but visually more prominent, east of Albany are layers of clay (siltstone) and sponge spicules together with shells (spongolite), the Pallinup Siltstone Formation deposited during the Eocene, 40 million years ago. (See **Box 10.1 Sponge garden to rock formation—the Pallinup Siltstone or spongolite**.)

At times during the early Pleistocene (2 million to 10,000 years BP), the granite headlands we see today would have been islands, and the estuaries and inlets would have connected more directly with the ocean. During the last major ice age, which peaked only about 20,000 years ago, sea level was more than 100 metres lower than it is now and there was dry land as much as 200 kilometres beyond the present coastline. Today's estuaries were river

FIG. 7.1 *(a) Cliffs at Black Point (Cape Beaufort) mark the line of the Darling Fault, February 1975.* E. P. Hodgkin. *(b) Basalt outcrops in the Donnelly River, February 1987.* Courtesy R. Clark.

valleys, and those underlain by easily eroded rocks were many metres deeper than their present beds.

As the sea levels rose at the end of the last ice age in the Holocene period 10,000 years ago, the sea flooded the valleys and the sea may have risen about 2 metres higher than it is now. The estuaries and inlets would have been open to the sea and the nearby swamps would have been part of the inlets. They would have experienced the full ocean range of tides. Sea level fluctuated several times during the Holocene, leaving lines of dunes in the Scott and Swan Coastal Plains marking old shorelines as the sea retreated to about its present level around 6,000 years ago.

The estuaries are gradually filling in from both ends, quite a natural process although in some cases it has accelerated in the last hundred years. During the last 6,000 years, sediments from the rivers and a large amount of organic material have gradually made the estuaries and swamps shallower. Sandy shoals around the margins and beach ridges, which have formed from the erosion of shorelines and redistribution of sandy sediments, are also contributing to the infilling of the estuaries and swamps. In places the coastal dunes have moved and are continually moving inland, spilling sand down their steep slopes to the inlet waters.

Beach sand is also contributing to the filling of the estuaries. While the finer sand was blown into dunes and stabilised by plants, wave actions moved coarser sand onshore, offshore and alongshore, gradually building bars and narrowing the estuary entrances. Tidal currents carried sand into the estuaries further obstructing the channels and forming flood tide deltas. Under the influence of waves and wind, the bars grew and the channels narrowed and shallowed to such an extent that neither the seasonal river floods nor the small tides could prevent the seasonal closing of the entrances.

This natural rate of change is slow in human terms, but in geological time scales is relatively rapid. Parry Inlet is now so shallow that it may dry out completely in summer, and in Irwin Inlet, areas that were estuarine 6,000 years ago are now swamplands. The tidal delta at Wilson Inlet is now more extensive, possibly due to the premature opening of the bar and subsequent reduction in scouring of the channel.

Like all the estuaries of Swanland where there is only small river flow in summer, the entrance channels through the bars can only remain open continuously where sand is in short supply (Oyster Harbour), or where there is shelter from waves (Nornalup). Where sand is abundant and wave action strong, the bars close during periods when river flow is reduced and only river floods can break them (Wilson). This process has been going on for about 3,000 years and has resulted in the brackish water environments with the extreme fluctuations in salinity that we see today.

Facing the southern ocean

In the western part of the south coast, the rivers and estuaries (the Donnelly, Warren and Gardner Rivers, the large Broke Inlet, the Walpole–Nornalup Inlet System and the much smaller Meerup River and Doggerup Creek) open onto long sandy beaches except for some coastal limestone on either side of the mouth of Broke Inlet and granite in the Gardner River bar.

To the west, the orientation of the dunes dictates the predominantly WSW–ENE orientation of the lower sections of the river channels and the channel of Broke Inlet. The beaches face south-west or south into the prevailing winds and swell. The beaches are backed by high coastal dunes from Black Point (Cape Beaufort) to Point D'Entrecasteaux (Windy Harbour) and along much of the coast through to Nornalup Inlet. In many places the dunes have consolidated into limestone forming high cliffs at Point D'Entrecasteaux and near Broke Inlet.

At Nornalup Inlet and at Rocky Head, steep granite cliffs shield the entrance from the south-westerly winds and swell. Sand movement along the coast however varies greatly. Although winds are from west to south-west in winter and sand is transported away from the entrance to the east, the winds also promote a countercurrent that brings sand westwards. In summer, winds are from east to west bringing sand across the entrance channel shallowing the bar when there is no river flow. The south-facing Bellanger Beach is backed by 20–40 metre-high dunes before the next rocky headland at Conspicuous Cliffs. The dune system encloses the western margin of the inlet and the lower reaches of the Frankland River. To the west of Rocky Head (Map 7.4) at Circus Beach, the dunes cover an older entrance to the estuary system.

Irwin, Parry and Wilson Inlets open to the ocean towards the western end of bays, in the shelter of rocky headlands affording some protection from the winter south-westerly winds and swells. Irwin Inlet opens via a narrow, 1.5-kilometre channel through low, poorly consolidated dunes onto the exposed Quarram Beach. The mouth is sheltered from the south-west by Point Irwin, but is 3 kilometres from the nearest rock at Peaceful Bay (Fig. 7.15). There are no rocky outcrops along this stretch of coast, and the entrance channel can change position and may have been much further west within the last 100 years. Parry Inlet also opens via a narrow 2-kilometre channel through low unconsolidated dunes in the shelter of Point Hillier, although the nearest exposed rock is 1 kilometre away at Parry Beach. Eastwards there are low dunes along William Bay. Wilson Inlet opens in the shelter of Wilson Head peninsula where the wide beach to the east is backed by densely vegetated unconsolidated dunes with some large blow-outs. Further east the high limestone cliffs of the Nullaki Peninsula back the narrow shore of Ratcliffe Bay (Maps 8.1 and 8.2, Fig. 8.2). Generally the dunes have been shaped by the south-westerly winds, except at Peaceful and William Bays where coastal eddies and summer south-easterlies have had an influence.

RAINBOW COAST — DRIZZLE ON THE SOUTH COAST — FOREST COUNTRY

Like the more northern areas of Swanland, the climate on the south coast is determined by the eastwards passage of the high and low pressure systems. The south coast however is slightly cooler and receives higher rainfall over a longer period. In summer, the high pressure systems move slowly over the lower west coast, blocking the low pressure systems and cold fronts so that they pass to the south of the continent. In winter, the pressure systems move north and the low pressure systems bring rain-bearing fronts to the coast of Swanland. When the frontal systems are not very strong, their influence is often only felt on the south coast. Hence the terms so favoured by weather-presenters of fronts 'slipping away to the south' or 'drizzle on the south coast' and the promotional title of Rainbow Coast for this region. Most rain falls between May and September (with peaks in June and July), although the south coast also receives light rainfall during summer and autumn. Thunderstorms in summer also bring isolated heavy falls to inland areas that may result in river flows to the estuaries. (See **Chapter 1 Understanding Swanland: climate present**.)

Rainfall near the coast may be as high as 1,350 millimetres annually but varies from year to year from as little as 1,000 millimetres to 2,000 millimetres (Maps 7.1, 7.2 and 1.8). Annual rainfall decreases inland with less than 800 millimetres in the Tone and Perup river catchments, 600–700 millimetres in the cleared agricultural lands of the upper Kent and Hay river catchments,

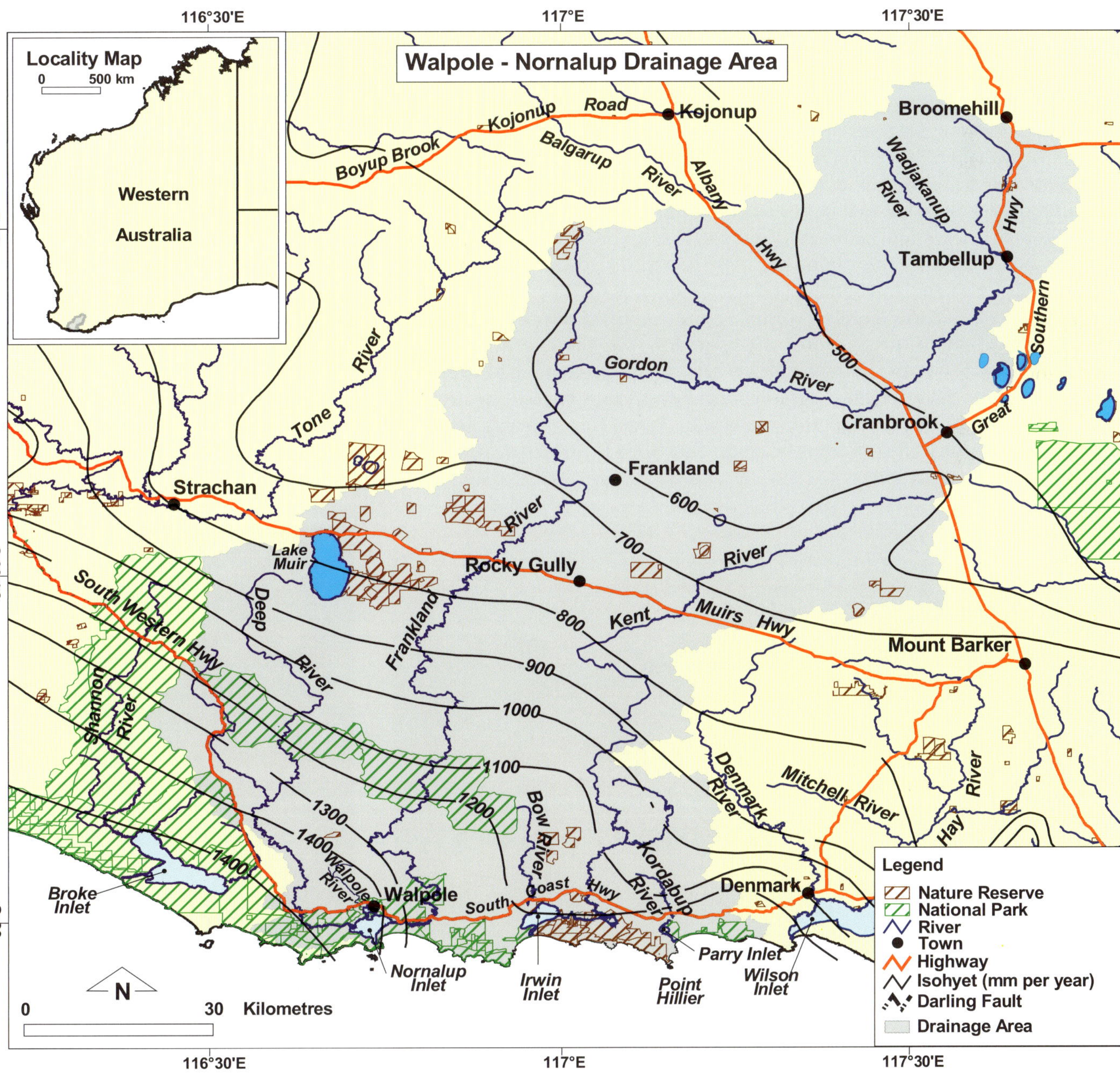

MAP 7.2 *The catchment of the Walpole–Nornalup Estuary showing the decrease in rainfall inland.* Courtesy DoE.

600 millimetres at Cranbrook near the upper Hay, and about 750 millimetres at Mount Barker and in the upper Kent catchment. In the last 30 years the rainfall near the coast has decreased by about 200 millimetres.

The higher and more reliable rainfall compared to the west coast has influenced the soils and vegetation. Karri forest covers the loamy soils of the hilly country in the southern part of the forest, giving way to jarrah and marri forest to the north following the decrease in rainfall (Fig. 7.2).

Jarrah reaches its best development in the hilly lateritic areas where mean annual rainfall is above 900 millimetres, the forest becoming more open as the rainfall decreases. Blackbutt *Eucalyptus todtiana* and bullich *E. megacarpa* grow in the damper valleys and along creeks. The broad swampy drainage lines and open sandy flats that may be inundated with water carry low open woodland of *Melaleuca* and *Banksia* with an understorey of heaths and sedges. Other eucalypt forests of tingle trees also occur around Walpole and Nornalup. The coastal dunes are covered with peppermint trees *Agonis flexuosa* together with shrubs and sedges typical of coastal dune communities throughout Swanland.

The forests have been heavily logged for many years. Under the native forest erosion is low but it increases where native vegetation is removed by activities such clear felling. Guidelines have been drawn up defining the width of remnant vegetation that should be left intact along streamlines.

The river waters are stained a dark colour by dissolved tannins. During winter the river water also carries silt and clay-sized particles, mostly clay minerals, organic matter and fine siliceous sand. The sediments accumulate in river deltas where river flow slows. The fine sediments flocculate or settle to the bottom when the river waters meet the saltier waters of the estuary, although they may be flushed out to sea when the river flow is strong. There is little information about sediment loads carried by the rivers during normal flows or floods, and it is not known exactly how much these have increased due to clearing in the catchments. Heavier coarse sands that accumulate in the river pools and the largely forested lower catchments probably reduce the amount of sediment reaching the estuaries, but these can be moved downstream with floods. The river deltas can change from year to year, and lower reaches of rivers such as the Frankland and Deep have been dredged periodically to maintain navigable channels. The need for dredging, however, may be due to the redistribution of sediments already present rather than additional sediment inflow.

FIG. 7.2 *(a) Seasonal swamps with reeds and the red summer flowering* Beaufortia sparsa *(swamp bottlebrush), surrounded on higher ground by jarrah and marri trees, near Walpole, 2001.* Courtesy T. Alford, UWA. *(b) Karri forest where mature trees reach over 80 m in height, Warren River National Park, 1982.* Courtesy S. Hopper, BGPA.

The effective river flow occurs in five months (June to October), and most of the flow to estuaries is from the southern high rainfall areas near the coast. Although the major river flow is in the winter and spring months, some flow continues through summer and autumn. After the shallow entrance channels close in summer, evaporation may reduce the water level in the estuaries by a metre or more. Low water levels expose large areas of the shallows, and some estuaries such as Parry Inlet may dry out almost completely. Some of the river tributaries are dammed for water supplies for townships or farms.

STRETCHING THE DEFINITION — NOT REALLY ESTUARIES

Three small rivers (the Donnelly, Warren and Gardner) defined as riverine estuaries drain the western part of the Scott Coastal Plain (Map 7.1). Bordered by high dunes that dictate their river-like character, they have no basin or lagoon although the Donnelly has a small 'broadwater' lagoon area. In some respects they do not really fit the criteria of estuaries as the marine influence is limited, but the 'scaled down' and transitory estuarine state helps our understanding of the larger systems. They are also a focus of interest in a drought prone land.

The bar at the mouth of the Donnelly River opens and closes several times each year, and sea water flows back in under the out-flowing river water. The Warren River flows out over the high beach most of the time and is rarely if ever tidal or saline. A rock bar at the mouth of the Gardner River keeps it open, and sea water flows in under the fresh water when river flow slackens and tides are high, although it may be closed briefly by sand and seaweed.

The Donnelly River

The Donnelly River rises between Bridgetown and Manjimup about 60 kilometres inland and 200 metres above sea level (Map 7.1). The catchment is mainly forested hilly country. The river has cut a deep valley where it merges from the Darling Scarp onto the Scott Coastal Plain about 80 metres above sea level. Barlee Creek joins the main river about 23 kilometres from the mouth.

At the boat ramp 12 kilometres upstream, the banks are sandy and 1–2 metres high. Mouthwards the river winds through a freshwater swamp. Nearer to the mouth the steep slopes of dunes rise to about 40 metres especially on the south side where the banks are narrow and densely vegetated. Small outcrops of basalt between 1 and 2.5 kilometres from the bar mark the line of the Darling Fault. Less than a kilometre from the sea, the estuary widens. Low dunes border the north shore and limestone cliffs and rocks border the southern shore, and there are wide sand flats through which the river has formed a number of channels (Fig. 7.3).

The bar is part of the beach composed of well-sorted fine mainly siliceous sand with 10 per cent carbonate derived from shells. The small beach, only 2.5 kilometres long between two limestone outcrops, faces south-west in line with the prevailing swell, and a sandbar about 250 metres offshore runs parallel to the shore. The bar only builds to about 2 metres above sea level and, during high seas, waves wash over the bar into the estuary carrying sand into the mouth. The bar breaks with the first heavy rain sometimes as early as March, and may close and open for short periods before closing for a longer period in December or January.

When the bar is closed the water level may rise to 2 metres above sea level, but drops rapidly when the bar breaks. The estuary is generally shallow about 2 metres below MSL particularly upstream of the broadwater area, although some areas are about 4 metres deep and there is a 9-metre deep hole near the basalt outcrop. The estuary is tidal for about 12 kilometres to Boat Landing when the bar is open, but tides are very attenuated by the bar.

Hydrologically and biologically the Donnelly is an atypical estuary. When the bar is open and river flow slackens sea water flows into the estuary for several kilometres before intruding below the fresh water. The two water bodies do not mix except within 3–4 kilometres of the mouth, and the brackish water may penetrate along the bottom the whole length of the estuary. Presumably the brackish water is flushed out and replaced by fresh water when the river flows strongly. Visibility is only about a metre in areas where tannin-stained river water dominates, but where sea water is present the water is much clearer.

The fringing vegetation reflects the generally fresh condition of the estuary; salt tolerant species such as samphire only grow in a small area near

FIG. 7.3 *Donnelly River, aerial view, June 2000. The bar may break at either end of the beach against the south-eastern cliff or the north-western foredune; old flow channels are visible at both locations. When the bar is closed, vegetation establishes on the bar and wash-over delta.* S. Neville/Ecotones, Courtesy WRC.

the mouth and freshwater plants such as lake club-rush *Schoenoplectus validus* dominate the broadwater area. Similarly, although some estuarine water plants and animals live in the lower reaches they are replaced by freshwater biota upstream.

The seagrass *Ruppia megacarpa* and stonewort or charophyte *Lamprothamnium papulosum* have been found in the lower reaches. The green alga *Cladophora* and mats of blue-green algae were also observed near the mouth in the 1970s. A few marine fish enter the estuary, and it is especially popular with recreational anglers although netting is not permitted. Tarwhine (silver bream), black bream, cobbler, sea mullet, yellow-eye mullet, Australian herring, flathead, and rainbow trout are taken in the estuary and rivers. The inlet was occasionally fished by commercial fishers but in July 2002 was closed to further commercial fishing (Table 1.15).

The Warren River

The Warren River meanders through forest about 150 kilometres from the junction of the Tone and Perup Rivers (Map 7.1). It occupies a deep valley through the hills where it flows to the coastal plain. The Tone River rises 110 kilometres inland on the ancient plateau near Kojonup, where land salinisation and soil erosion are major problems in this extensively cleared agricultural land. Land along the Perup River is also cleared, but most of the catchment is forested.

In the last 8 kilometres the river is only 50 metres wide and meanders through high steep-sided dunes with narrow alluvial flood plains before emerging at the coast. The banks are densely vegetated except where the steep banks have eroded or sand banks have formed. Sedges also occur as a narrow band in some places. A large blow-out approaches within 100 metres of the northern shore near the mouth but does not reach the river. The dunes and river valleys are oriented WSW and ENE and the coastline NW–SE. Close to the coast the dunes are hardened to limestone and this is exposed along the riverbanks near the mouth.

The river discharges onto a narrow coastal strip 300–600 metres wide with a foredune and low hummock dunes between the beach and the steep slope of the dunes behind, rising to 150 metres (Fig. 7.4). The river widens where it flows to the beach and when the river flow slackens the beach builds up, confining the flow to the back of the beach near to the foredune. The flow changes to the NW or SE with channels up to 2 kilometres long before entering the sea. In July 1977 there were channels in both directions with an opening channel on the NW end. In March 1984 the river cut a channel almost straight through from the river mouth. The SE channel had closed and the NW channel persisted as a pool against the foredune. Aerial photos in November 1985 and October 1988 show the river flowing in a more northerly course behind the bar opposite the mouth. The position of the mouth and formation of the bar depend on the combination of river flow and onshore transport of sediment. These events also affect salinity within the river and, although a little salt water enters, the water is always fresh (less than 0.5‰).

Defining the Warren as an estuary is an extension of the definition, as it is really the mouth of a river that is probably never tidal. Parts of the riverbed are above sea level and when flow slackens sandy shallows separate the deeper pools. The river is narrow and shallow to where it flows out over the beach apparently without ever cutting a tidal channel through it. There is some river flow in most months of the year and little sea water enters. When river flow slackens, the beach may build up and close the mouth. The sea rush *Juncus kraussii* grows in a swamp on the beach. Further inland, the rushes around the margins are predominantly freshwater species and the bullrush *Typha* grows within half a kilometre of the beach.

FIG. 7.4 *Warren River, aerial views: (a) July 1977 and (b) November 1985.* DLI. *(c) June 2000.* S. Neville/Ecotones, Courtesy WRC.

Little is known about water quality in this riverine estuary, but in 1989 there were low concentrations of phosphate in samples despite the high concentrations in the Tone River originating from farmland and the trout hatcheries on Lefroy and Treen Brooks. The estuarine waters are diluted by the greater run-off from the higher rainfall forested lower catchment. There has been a progressive increase in salinity of the Warren River water from less than 400 milligrams per litre in the 1940s to 700–800 milligrams per litre in the 1970s. The increase in salinity has been correlated with land alienation and clearing in the 1920s, 1960s and 1970s. River water becomes progressively saltier upstream with more than 6,000 milligrams per litre in the headwaters of the Tone River, but as with the nutrients, this high salinity water is diluted by the greater volume of freshwater run-off from the southern high rainfall area.

The Gardner River

The Gardner River, about 35 kilometres long, rises in low (less than 100 metres above sea level) undulating country near Northcliffe (Map 7.1); much of it is farmed. Its eastern tributary, the Canterbury River (27 kilometres) traverses higher, steeper, mainly forested country joining the Gardner 20 kilometres above the mouth. Nearer to the mouth the river meanders across the coastal plain and is joined by Blackwater Creek that drains swampland to the west.

The river meanders through low-lying land subject to inundation for much of its length, and is probably tidal for about 5 kilometres. In the lower reaches it may be 3 metres in depth. Granite bars at the mouth and at 1.5 kilometres and 2.5 kilometres upstream obstruct the river. It flows between steep sandy banks through scrub covered flat land less than 5 metres above sea level. Small dunes and patches of higher ground carry denser vegetation with peppermints and eucalypts. Just beyond the first bar the river is joined on the west by Blackwater Creek. The Gardner is only 10 metres wide in the upper reaches but widens to 150 metres within 300 metres of the mouth (Fig. 7.5). A small blow-out on the west near the bar is reported to have developed less than 50 years ago along a cattle track. The blow-out is partially stabilised by marram grass, a species introduced from Europe for dune reclamation but now regarded as an environmental weed.

At the sea, rounded granite rocks restrict the flow to narrow channels. Sand from the beach builds up onto the granite that extends for some distance on either side forming a small bay. The bar is open most of the time, closing only briefly when blocked by masses of seaweed and soft sand. The river flows for much of the year and fresh tannin-stained water covers the surface. When river flow slackens and sea level is high due either to tides or swell, sea water flows in under the out-flowing fresh water. The river is then sharply stratified with a halocline (an abrupt change in salinity) between the surface and bottom.

FIG. 7.5 *Gardner River, aerial views: (a) October 1980 showing the winding river.* E. P. Hodgkin. *(b) Detail of the river mouth showing regrowth of vegetation to the west, June 2000.* S. Neville/Ecotones, Courtesy WRC.

The marginal vegetation and benthic fauna are predominantly freshwater species, except for a short distance from the mouth where estuarine species are found. Little is known about water quality in the Gardner, however nutrient concentrations are likely to be low as the greater part of the catchment is uncleared.

The Meerup River and Doggerup Creek

The Meerup River and Doggerup Creek also flow to the coast between the Warren River and Point D'Entrecasteaux. Neither is estuarine and both flow directly to the ocean. The Meerup, a small version of the Warren River, rises in swampy land, some of it cleared, west of Northcliffe. It passes through the dunes in a narrow valley before spreading out in braided channels over the narrow coastal plain to the high beach (Fig. 7.6).

Doggerup Creek rises in swampy ground between Northcliffe and Windy Harbour, an area little affected by human activity. It flows to the sea past a mobile dune between the granite shore of Black Head and the limestone cliffs of Point D'Entrecasteaux. At times both the Meerup and Doggerup flow behind the foredune before breaking into the sea.

FIG. 7.6 *Meerup River, aerial view: (a) June 2000,* S. Neville/Ecotones, Courtesy WRC. *(b) Detail showing the tannin-stained water, March 1984.* E. P Hodgkin.

BROKE INLET

Impressions

The approach to Broke Inlet is very special. The turn-off from the South Western Highway takes the traveller along a gravel road winding through forest country of alternating jarrah, marri and tall karris. The road ends among the squatter's huts at Camfield with its eerie ghost town atmosphere. The estuary is glimpsed through the thinner tree cover and suddenly the estuary spreads out before the visitor. The size of it is quite staggering, some 48 square kilometres. To the west the hills show the typical coastal shapes of sand dune country, very smooth outlines due to the low vegetation (Fig. 7.7). At the boat-launching beach, the sand is dotted with granite boulders, gnarled paperbarks and banksias (Fig. 7.8). Cast up bivalve shells (*Soletellina*) looking like stranded purple butterflies show the hidden richness of the area, and the changing conditions leading to massive kills of the estuary animals.

Because of the wide sandy shallows a boat has to be dragged for hundreds of metres through ankle deep, clear ginger coloured water. In deeper water

FIG. 7.7 *Broke Inlet, aerial view, December 1988. Broke Inlet lies parallel to the coast in a NW–SE direction, surrounded by relatively undisturbed forest and coastal dunes. The dark tannin-coloured waters visible in the narrow channel are typical of southern rivers, a sharp contrast to the clear blue of the Southern Ocean.* Courtesy DLI.

FIG. 7.8 *A beach on the north shore of Broke Inlet with granite rock and paperbark trees, June 1987.* E. P. Hodgkin.

we can start the motor and skim across the shallow sandy areas. We see little in the way of plant life growing on the bottom. The trip can be quite bumpy as the prevailing wind along the long axis of the estuary creates small waves in this shallow area. As we near the inlet channel to the sea, the flapping door of a small hut is the only sound above the murmur of the wind. Back across the inlet we head to the small granite islands, continually altering course across the ever-changing sandy shallows. Westward of the islands nearer to the entrance of the Shannon, we can no longer see the bottom, but the water is still only 4 metres deep. Looking at the hills we are suddenly aware that the impact of humans is not obvious, this is almost wilderness.

The rounded coastal hills are not scarred by the tracks of vehicles; only the occasional stake marks the edge of a sandbank. There are no other boats, and the quiet is broken only by the sound of waves lapping against the boat and the cry of cockatoos as they sweep across the waters. Broke Inlet is the only estuary in Western Australia regarded as pristine; its ecological processes unchanged by human activity, although the estuary waters are fished and people camp along its shores.

A large body of water

The major flow to Broke Inlet (Map 7.3) is from the Shannon River, 47 kilometres long, that rises in forested hills 200 metres above sea level and flows mostly through national park. Two small rivers and creeks also flow to the inlet from swampy uncleared country on the Scott Coastal Plain. The Forth River (10 kilometres) flows from the Pingerup Plains to the north-west end of the inlet, and Inlet River (14 kilometres) enters to the east.

MAP 7.3 *Broke Inlet: topography of the surrounding areas.* From Hodgkin & Clark, ESS # 6, 1989. Courtesy DoE.

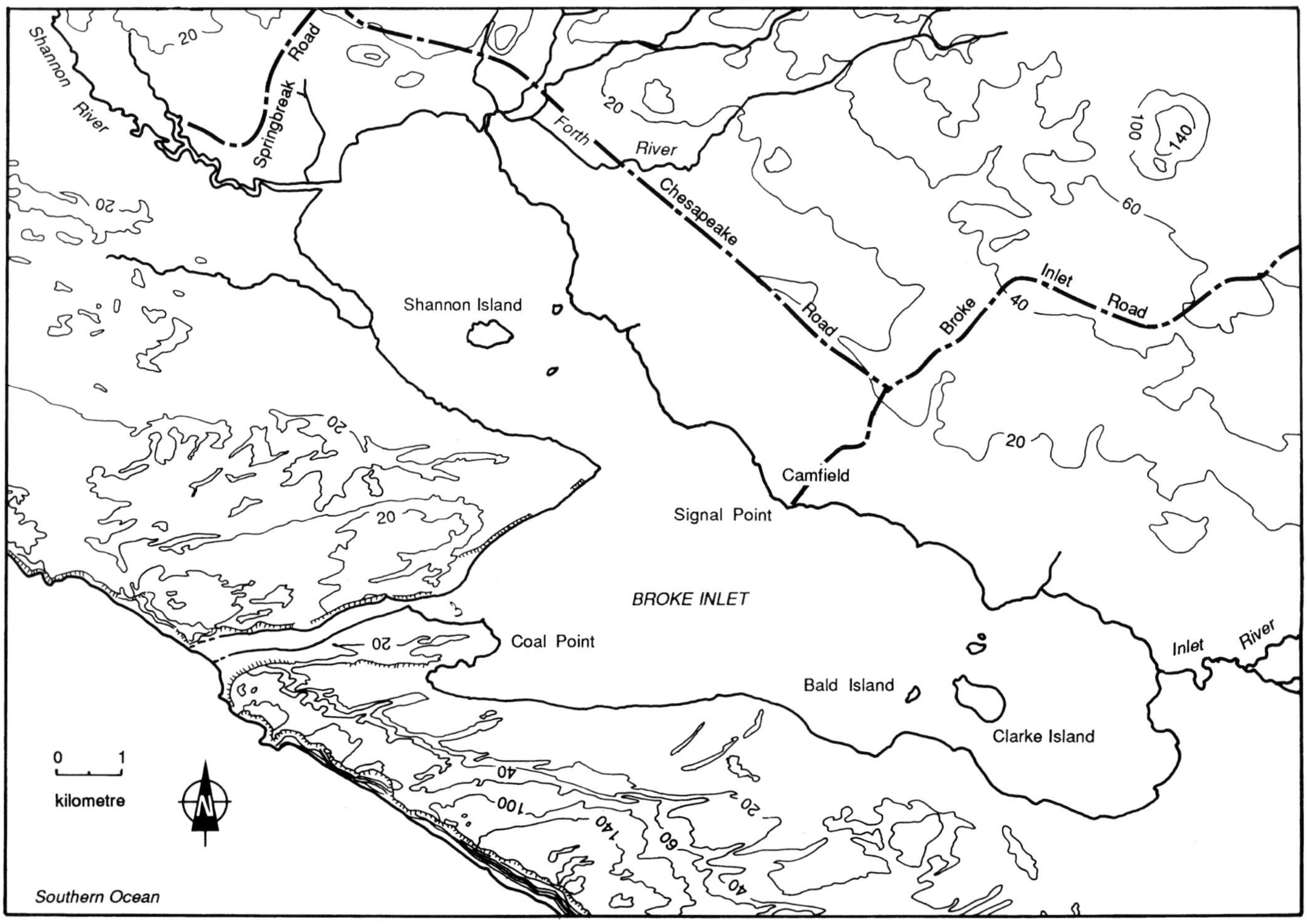

Broke has a large lagoon, 48 square kilometres in area; the same size as Wilson Inlet. These estuaries are the largest on the south coast (Maps 7.1 and 7.2). The lagoon is connected to the sea by a long narrow channel through the coastal dunes, and the lower river reaches are also estuarine for some distance. The water in Broke Inlet is nearly fresh in winter before the bar opens, but in summer mixing and evaporation makes it brackish. The waters are seldom more than half the salinity of sea water except in deep water of the inlet channel.

The Shannon is estuarine (tidal when the bar is open) at Springbreak Road, 3 kilometres from the river mouth, and perhaps beyond. The Shannon is about 3 metres deep and deepens to 10 metres where it is scoured on a bend. It meanders through swampy lowlands and discharges through a triangular delta of beach ridges and swamps. The Forth River is estuarine for 300 metres, and the Inlet River has a meandering estuarine channel of about 2 kilometres through swamp and dune.

Broke Inlet, 15 kilometres long, lies in a NW–SE direction parallel to the coast and is 2–3 kilometres wide except in the middle where it widens towards the inlet channel. Assuming an average depth of 1.5 metres below sea level (MSL), the volume when full is approximately 72×10^6 cubic metres. Sandbanks border much of the shoreline and extend out from the shores dividing the lagoon into three main basins of 2–3 metres depth. Sometimes at the end of summer it is possible to drive across the inlet on these banks. A tongue of dune projects into the lagoon, narrowing the shallow channel between the western and middle basins. Two large islands, Clarke Island and Shannon Island, and a number of small, low islands are granitic, indicating the presence of Precambrian rock not far beneath the surface.

The north-eastern shoreline of Broke is divided by low rocky outcrops into a series of sandy bays where the beaches rise to narrow, densely vegetated dunes. There are sandy lignites and clays near the surface along the shoreline, and also in the banks of the Shannon River, and at Coal Point. Tree stumps about 7,000 years old buried in the sediments serve as a testament to the landscape before the estuaries filled with water. Aboriginal fish traps, probably not used since the bars formed and tidal exchange decreased about 4,000 years ago, lie between Camfield and Inlet River. (See **Box 9.1 Stone circles—echoes of the past**.) The sheltered south-western shore is of different character with a wide or narrow fringe of paperbark swamp along much of its length, or a narrow sandy beach generally with a fringe of vegetation at the water's edge. There is only one outcrop of granite along the southern shore, opposite Clarke Island.

The inlet channel, 3.5 kilometres long, runs nearly due west through the parabolic coastal dunes. Less than 250 metres wide, it is the deepest part of the estuary with 5–6 metres of water; and water flow can be quite strong when the bar is open. The northern shore of the channel is steep with Pleistocene dune rock outcropping along the shoreline and at the mouth. In contrast, a wide sandy beach borders the southern shore. Thus the northern shore is eroding and the southern shore accreting so that the whole channel appears to be migrating slowly northwards. The limestone 'island' in the bare sand at the mouth separates the present channel from what may have been an earlier southern channel. The bar now breaks north of this island, but is reported to have broken through south of it in the 1920s. The mouth is 300 metres wide between the island and the northern rock wall. The active channel is constricted to less than 100 metres by deltaic and wash over sand, but its form varies greatly from season to season.

The bar at the mouth of Broke Inlet may build up to a width of 500 metres and 1.8 metres above MSL, but even at this height waves can still break across it when seas are high. It is composed of clean white fine to coarse-grained sand with small amounts of shell grit. The bar is usually closed for the first half of the year and breaks or is broken between June and September when the inlet water reaches the top. It generally breaches in the middle but often breaks across its whole width. The bar opened, or was breached by fishermen, every year from 1964 to 1989 except 1969 and 1986, years of very low rainfall. Records of more recent openings are sporadic due to the area's isolation and have not been collated since Ernest Hodgkin and Ruth Clark's 1989 report. However the opening and the length of time the channel stays open would depend on the time and volume of river flow. In the past it never stayed open for more than six months and it may have closed after only one month. Water level in the estuary can vary by nearly 3 metres, from 2 metres above MSL before the bar breaks to 1 metre below this level at the end of summer.

When the bar is open the estuary is tidal, but tidal exchange is so restricted by the bar that the daily astronomic tides in the inlet are probably less than 10 centimetres (10 per cent) of the ocean tide. When river flow slackens, the greatest exchange of water with the ocean is due to high sea levels associated with low barometric pressure. Passage of high and low pressure systems brings sea level changes of about 30 centimetres, and the low pressure allows sea level to rise pushing water into the inlet.

Where river meets sea

In general, when the rivers are flowing in winter the salinity in Broke Inlet is about 5‰ rising to 30‰ or more at the end of summer. There is great variation from year to year depending on the volume and timing of river flow, and the time and duration of the bar opening. Mixing is rapid in the open water of the lagoon and salinity is generally uniform throughout. There is usually little difference in salinity across the inlet or from top to bottom, except when the river flow is strong. For example, in October 1976 surface salinity ranged from 1‰ in the western basin to 10–13‰ near the inlet channel. High salinity water also flows along the bottom from the lagoon into the Shannon River under the river water, causing strong stratification.

Salinity in the inlet channel is very similar to that in the lagoon. When river flow slackens and the bar is open, sea water flows back into the estuary under the out-flowing fresh water creating a marked stratification between the surface and deep water that can persist after the bar closes.

The clarity of the water varies greatly. River water is tannin stained but clear and, in calm weather, the inlet waters may be clear to the bottom. Wind stirring however can increase turbidity markedly and a secchi disc, used to measure light penetration and water clarity, is not visible from the surface at 0.5 metre.

Generally the whole inlet is well oxygenated. Some deoxygenation has been observed when the lagoon is stratified as it was in August 1974, although this condition is unlikely to persist because of wind mixing. The deep water in the inlet channel also becomes deoxygenated when it is stratified. In January 1989 it was anoxic (0.5 milligrams per litre) at 5.5 metres depth and oxygen was low in the deeper water of the Shannon River.

Inhabitants

Broke has a typical estuarine aquatic and fringing flora, but it has not been studied extensively. Within the estuary the swan grass *Ruppia megacarpa*, the stonewort *Lamprothamnium papulosum*, the filamentous red alga *Polysiphonia* and the green alga *Cladophora* have been recorded. They appear to have been more abundant in 1976 than in 1988, especially in the inlet channel. Epiphytic algae were also found on the fringing reeds and *Ruppia* in the channel in 1976. Detrital marine algae and seagrasses washed in from the sea accumulate in the deeper water near the mouth and probably contribute to the productivity.

Marine copepods *Acartia* sp., *Gladioferens inermis* and *Oithona* were present in February 1972, and the common estuarine copepod *Gladioferens imparipes* was abundant in plankton samples taken at Broke Inlet in October 1976. The estuarine copepods were caught in the lagoon at the surface and at a depth of 2.5 metres, and in the channel at the surface and at 3.5 metres depth.

Animals living in the sediment of the lagoon and inlet channel are predominantly common estuarine species, although the assemblages vary according to the type of sediment and presence or absence of aquatic plants (Table 1.9). Populations of invertebrates also change greatly, often dying suddenly with changes in salinity. King prawns have been abundant in Broke Inlet in some years, and a flood from the Shannon River is reported to have killed many prawns. Similarly, an influx of small cockles *Katelysia* that proliferated after the 1968 opening died, probably from fresh water, when the bar did not open the following year.

Broke Inlet is an important habitat for many species of birds because of its largely pristine surroundings. The many animals such as small molluscs, crustaceans and worms are important in the diet of many waterbirds, and fish eating species are also abundant.

Thirty-five species of fish have been recorded from Broke and seventeen of these are commercial species. The area attracts a number of recreational fishers but supports only a small commercial fishery in contrast to the much richer fishery in similar sized Wilson Inlet. Broke Inlet, by contrast, is nutrient poor or oligotrophic with little aquatic plant growth compared to Wilson Inlet, which may be considered eutrophic supporting a rich growth of *Ruppia*. The catch is predominantly yellow-eye mullet *Aldrichetta forsteri*, together with sea mullet *Mugil cephalus*, King George whiting *Sillaginoides punctata* and Australian herring *Arripis georgianus*. It has been suggested that catches of these have decreased since 1960s. All the fish species spawn in the sea and the young are recruited to estuarine populations. Black bream are not often caught because populations are mainly confined to riverine habitats that are seldom fished. Possibly the low salinities of the rivers are not favourable habitat for them. The Shannon is one of the rivers in Swanland through which the pouched lamprey *Geotria australis* passes on its way to the inland waters. (See **Box 7.1 Swimming through**.)

Broke in the scheme of things

Broke Inlet is of similar size and form to Wilson Inlet, although it is shallower with more extensive sandy shallows. The major difference lies in its largely undisturbed state compared to Wilson Inlet, which shows signs of nutrient enrichment. Broke Inlet is therefore important as a benchmark for other systems allowing comparison in our changing world. (See also Graphs 1.2 and 1.3.) The surrounding wetlands are considered of National Significance and the inlet has some protection as it lies within a national park, but there are discrepancies. There is the increasing pressure of tourism with associated problems: off-road vehicles, rubbish, dieback, erosion, fire hazards, erosion of sandbanks by powerboats, and nutrients from human waste. There is a small squatters' settlement at Camfield and in the inlet channel. The bar is broken (when it does not break naturally) to allow recruitment of fish, and the area is fished quite heavily. The practice of opening the bar may indeed be favourable in the short term for seasonal recruitment of fishes. However opening the bar is also associated with decreased scouring that leads to early reforming of the bar and a reduction in time for fish to recolonise. In addition, the open channel does not always coincide with large numbers of fish in the waters offshore.

Box 7.1 **SWIMMING THROUGH**

The pouched lamprey *Geotria australis* lives the early part of its life in fresh water where it breeds but spends part of its adult life in the ocean, passing through the estuaries and lower reaches of the rivers at the beginning and end of its life (Fig. 7.9).

The eel-like pouched lamprey is one of the few remaining representatives of the jawless (agnathan) fishes thought to have changed little in the last 280 million years. Lampreys, unlike the bony fishes, lack paired fins, have a skeleton of cartilage and a soft skin. They have two dorsal fins and one tail fin, seven small round gill slits on each side, one nostril and a pineal gland (an eye-like structure thought to regulate day–night rhythms of the body) on the back of the head. The males develop a large sac or pouch (hence their name) of unknown function that hangs down just behind the mouth. Marine adults are bright blue and silver with greenish stripes along the top and sides. As they migrate upstream, they become greyish blue changing to brown. Juveniles are brownish-black in colour and the young adults (downstream migrants) resemble the parents.

During its time in the ocean the lamprey feeds by attaching to a larger fish with its toothed sucker, eating the muscle tissue and blood that it converts to a fat store. The length of this marine feeding stage is unknown, but it lasts for at least a year. When about 650 millimetres in length and with sufficient fat reserves, it stops feeding, re-enters the river and migrates upstream to spawn, using the fat reserves in this non-feeding last stage of its life.

Along the south coast adult lampreys enter the rivers in late winter or spring when water levels are high, temperatures low and rain is falling. They move at night when the sky is dark due to cloud cover or the absence of the moon, sheltering during the day in crevices. When their passage is blocked, they can climb rock faces with the suction disc or even move overland and enter the river again above the blockage. They may journey 80–100 kilometres in three months.

In the following late spring, 15–16 months after entering the fresh water, they spawn and die. The next generation of eyeless juveniles (ammocoetes) burrow into the stream bed of permanently flowing cool fresh water creeks and rivers where they filter feed on algae and bacteria that live in the organic material. This stage lasts for about three to four years until they are about 80–95 millimetres in length. They gradually change to the adult form between January and July passing through seven distinct stages. The young adults then migrate downstream to the ocean to start the feeding stage. Again they move during darkness and when the river flows are highest.

Lampreys are found in the relatively undisturbed rivers of the cool moist karri forest region along the south coast such as the Donnelly, Warren, Gardner, Shannon and Deep Rivers. They are less common further east but have been found in the Waychinicup River. They are also found in other streams south of the Swan River and occasionally as far north as the Moore River. Lampreys are found in colder water rivers in eastern Australia, New Zealand, Chile and Argentina.

The story of the lamprey's epic journey is remarkable. They have to cope with the dramatic change in salinity. They also swim against the river flow, negotiating obstacles of rapids, waterfalls and, more recently, the hazards of weirs and dams such as Pemberton Weir and the Rainbow Trail gauging weir on the Lefroy Brook, until they reach their final spawning grounds. The presence of lampreys may be a good indicator of river quality. A river that is suitable for lampreys will probably be suitable for most other river inhabitants.

Source: Morgan et al., 1998; Potter et al., 1986.

FIG. 7.9 *Pouched lamprey migrating around Pemberton Weir, Warren River. The seven gill slits at the side of the head can be seen on some individuals.* Courtesy D. Morgan, Murdoch.

WALPOLE–NORNALUP ESTUARINE SYSTEMS

The Walpole–Nornalup estuary is dominated by trees; giant ones that border the roadways along all approaches (Fig. 7.10). Together with the rich green colour of the surrounding hills and the dark waters of a large body of water softened by a gentler light, this area is very different. This is our forest country, and the annual rainfall here is the highest in Swanland so the rivers discharge a large volume of fresh water in winter.

A journey seawards

Impressions

To travel the inlet is a wonderful treat. At the boat ramp near the town of Walpole sounds seem muted on the dark still waters surrounded by high hills and tall green trees. Along the inlet edge, high swamp cedars *Agonis juniperina* tower over the fringing vegetation. Passing out of the narrow channel we pass by the small shallow basin of Walpole Inlet overlooked by the small town of Walpole. Further along we pass by the Rest Point Caravan Park overshadowed by the hills that border the narrow channel to the larger Nornalup Inlet. Karri trees dominate the skyline, their dead crowns reminding us of ravaging fires. Looking eastwards, the eroded cliffs at Coalmine Beach are a stark contrast to the eucalypt forest. To the west we notice the unusual colours on Newdegate Island, an outcrop of granite. The vegetation has changed here: has it been cleared or is the soil shallow so only small scrubby species can grow? We notice many small boats anchored in the lee of the high hills, for the days are often windy and fishing is a major attraction; in fact Nornalup Inlet has the highest diversity of fish species in a Western Australian estuary. As we cross the inky dark waters that look purple in the soft sunlight penetrating the light drizzle we see the coastal dunes. The shapes attract our attention, steep sided, some conical, all covered in harsher green low vegetation. At the estuary edge we can see the dune face wearing away, the cause of the erosion unknown, but now an unstoppable process in the persistent wind. Further towards the entrance channel we cross shallow shoals where the waters are a red-purple colour. We pass a rocky limestone outcrop and a hillside covered with drooping peppermint trees *Agonis flexuosa* where we notice a staircase under the trees: where does

FIG. 7.10 *Walpole–Nornalup Inlets, aerial view, June 2000. The town of Walpole and the Rest Point development lie on the edge of Walpole Inlet, and a narrow channel connects the two inlets. Except for a caravan park at Coalmine Beach, centre left, the land surrounding Nornalup Inlet is largely uncleared. Despite the extensive sandbanks and shoals that surround the inlet, the narrow entrance to the sea in the lee of Rocky Head remains open all year.*
S. Neville/Ecotones, Courtesy WRC.

it lead? Moving through a narrow channel, across the dark brown water, we pass between a high hill edged with granites and the white sands of the bar. We round a corner, and there is the Southern Ocean where incredibly sea green waves roll onto the shore. Unlike most other estuaries on the south coast, this gateway to the sea is always open. The sandbanks and channels are always moving, constantly changing as sea water flows in with the tide or seawards with the winter rush of fresh water.

A wetter area — larger rivers

Rainfall on the coast at Walpole–Nornalup is a high 1,300 millimetres a year, but decreases inland to less than 500 millimetres in the northern part of the catchment. Rain falls predominantly in winter but there is also some rain in summer. Annual rainfall has been decreasing in the South-West, however the magnitude of the decline is uncertain partly due to the absence of long and reliable records. The reasons for the decline in rainfall are much more uncertain, as the factors affecting climate change and variability are not fully understood and interactions between the different influences are complex. (See **Chapter 1 Understanding Swanland: Climate future—less rainfall?**) The records from Walpole are characteristic of the difficulties in the interpretation of trends. Annual rainfall decreased from 1956 to 2003 (Table 7.1) but rainfall varied greatly from year to year. For example, in 1996 rainfall was 1,477.7 millimetres but in 2001 only 1,081 millimetres.

TABLE 7.1 *Annual rainfall at Walpole (Station 009611); 10-year periods except for 1996–2003 (8 years).* Source: BOM, WA.

Years	Annual rainfall (mm)
1956–1965	1,465
1966–1975	1,319
1976–1985	1,265
1986–1995	1,215
1996–2003	1,278

Nornalup Inlet receives water from the Frankland–Gordon River, about 400 kilometres long, and the Deep River, 120 kilometres long (Maps 7.1, 7.2 and 7.4). The Walpole River, 15 kilometres long, and several small creeks flow to Walpole Inlet from the north and east. Together the rivers bring to the estuary an estimated mean annual flow of 390,000 megalitres. Although flow varies greatly from year to year, the annual flow is about ten times the volume of the estuaries (about 40,000 megalitres). Eighty per cent of river flow occurs between June and October with negligible flow during summer, except for infrequent floods in the upper Frankland. In most years 40 per cent of the flow comes from the Deep River and 45 per cent from the Frankland. Rainfall and run-off decrease away from the coast, with 80 per cent of river flow coming from the high rainfall forested area south of the Muir Highway. Rainfall during the winter months can vary greatly from year to year and within a year. In July 1964, 387 millimetres in July resulted in an August river flow of 146 × 10^6 cubic metres that was almost half the total for the year and about twice the volume of the estuary. The summer downpour in January 1982 produced a flow of 118 × 10^6 cubic metres mostly from the upper catchment.

The waters of the Deep and Walpole Rivers are fresh and the Frankland River is saline. As with other rivers in Swanland, saline waters in the upper catchment (Gordon 10‰) are diluted by fresh water in the higher rainfall area of the lower catchment.

Most of the land in the north of the Frankland catchment has been cleared and is used for cereal cropping, sheep farming and cultivation of vegetables. Part of the Deep River and the Frankland between the Muir Highway and Beardmore Road is forest reserve. Settlement around Walpole–Nornalup began in the 1920s, and the natural beauty of the area has been appreciated since the earliest days. The Walpole–Nornalup National Park covers most of the land south of the estuarine reaches of the Deep and Frankland Rivers, and the area to the north and west of the estuarine part of the Frankland. (See **Box 7.2 One of the most beautiful single sights—history of a national park.**) Forest fires have greatly changed the landscape; fires in 1986–1987 burnt areas south of the inlet and north of the Frankland, and most other areas of the park have been burnt quite recently.

Box 7.2 **ONE OF THE MOST BEAUTIFUL SINGLE SIGHTS— HISTORY OF A NATIONAL PARK**

The area around Nornalup has held a place in folklore from the beginning of European settlement on this coast. While settlement at Albany was in its infancy in the 1820s, the largely unexplored area to the west was referred to as the 'Deep River of the Sealers'. The lawlessness of these earliest settlers added to the mystique of the area. The inlet was probably first sighted by Dr Thomas Braidwood Wilson with the Aboriginal guide Mokkare in 1829. Then in 1831 a group led by Captain Thomas Bannister on the way from the Swan River Settlement to Albany passed through when they strayed off course and become lost due to faulty instruments and the inexperience of the surveyor Mr. Smythe. Ascending a 'two headed mountain' to obtain direction Bannister recorded, 'that nothing satisfactory could be seen; as far as the eye could reach, was one great forest'. Making for the coast through the dense bush he also recorded the trees so famous today. 'The Trees were principally the Blue Gum*; and if others had not seen them I should be afraid to speak of their magnitude; I measured one. It was breast hight (sic), 42 feet in circumference. In height, before a branch, 140 or 150 feet we thought at least, and straight and smooth as the barrel of a gun.' On reaching the sea near Point Nuyts to the west, Bannister and his party passed around the inlet taking nineteen days to reach Albany.

The estuary waters were first officially explored by Lieutenant Preston in 1831. Preston entered in a whaleboat launched from the *Sulphur* passing over the bar in 10 feet (3 metres) of water. In July 1833 surveyor Alfred Hillman and a party arrived at the banks of the Frankland River recording that the river (about 3 kilometres from the estuary) was 200 yards (60 metres) wide and the water was salty. Hillman also noted: 'The scenery on the banks of this river is certainly beautiful, far more so than anything I have yet seen in the Colony'. Hillman formally ascribed Nornalup to the area. The Aboriginal people originally occupying the area, the Minang or literally 'Southerners' or South people, referred to the area as No-Nor-Nup, the place of the Norne (black snake).

The next recorded visitor was the lawyer and entrepreneur William Nairne Clark in 1841. Clark noted that despite the abundance of names on maps there was a lack of detailed information on the area—a lack he set out to address with the publication of his diary, letters to politicians and newspaper articles detailing an enthusiastic assessment of the grazing and agricultural potential. However when searching for safe haven, his diary of 21 February records an enchantment.

> *The swell of the ocean was heavy outside, but we easily swept over the bar, got into deep water as smooth as a mill-pond, and moored the boat to a tree by the river bank (left side going up). I cannot describe my sensations on entering this beautiful river during the gloom of night. A sense of awe mingled with pleasure was chiefly predominant. We were all much fatigued and wet, but a brisk fire was soon made and we procured plenty of fine spring water at hand.*

On his trip up the Deep River, Clark also commented: 'The sail up was truly delightful. The river actually appeared in places to be embosomed (sic) amongst lofty wooded hills, thus casting a deep gloom over the waters and making the scenery the most romantic that I ever witnessed in other quarters of the globe'.

The area was first settled in early 1845 by the Landor brothers, and later in the year by Charles Burns. The group fished and grazed livestock on Newdegate Island and near the Deep River for only about one year, but interest in the area was already fuelled. After exploring the area around Kojonup and the upper reaches of the Gordon River, a party led by A. C. Gregory in 1846 followed the Frankland River southwards

FIG. 7.11 *Red flowering marri* Corymbia *(previously* Eucalyptus*)* ficifolia*: the species is unique to the south coast and particularly abundant near Nornalup.* T. Alford, UWA.

reaching the inlet and sea before returning to Perth with reports of stands of large trees. Later a small timber industry operated in the 1850s for a period but floundered due to lack of a harbour and the distance from markets: factors that were to plague later settlers but afford some protection from hasty development. Other early pioneers included the Hassells with leases at Quarrimup on the coast east of Nornalup, and the Muirs further to the west. Generally these families had lands in the upper river catchments and used the coastal areas for summer grazing.

The interest in the hardwood timbers, however, was growing, prompting a visit by the Melbourne-based botanist Baron von Mueller. Mueller's report included details of the different types of timbers, including the unique red flowering gum *Corymbia* (formerly *Eucalyptus) ficifolia* (Fig. 7.11), karri and tingle. In his enthusiastic report on the timber resources, but also advocating development of plantations of tropical species and Northern Hemisphere softwoods, von Mueller noted:

> *But as nowhere, not even in the most extensive woodlands, can the supply of timber from natural forests be considered inexhaustible, a rational far-seeing provision for the maintenance (if not for the enrichment) of its forest-treasures is needful for Western Australia, however indiminishable (sic) these may appear to be at present.*

FIG. 7.12 *Giant Tingle trees and the Tree Top walk, 2002.* Courtesy S. Jacobs.

The timber industry throughout the South-West continued to grow and by the 1880s accounted for about 10 per cent of the state's export income. Timber cutting in the Nornalup area was originally centred on Rest Point with sawn logs shipped across the inlet in shallow-drafted boats built in Albany. There are also reports of oysters being gathered from the rocks and exported.

In response to concern about the growing scale of timber exploitation, the Premier Sir John Forrest appointed James Ednie Brown the first Conservator of Forests to head the Woods and Forests Department, working under the jurisdiction of the Commissioner for Crown Lands. Lobbying for long-term leases by the timber developers versus the agriculturalist and others wary of over-exploitation, and the issue of huge scale leases for timber extraction delayed the passage of the Bill to 1899. Many of the companies eventually forfeited their rights and a 1903 Royal Commission recommended that the leasing of forestland for timber cutting should be abolished, and timber should be disposed of on a royalty system based on the measurement of logs in the round.

In anticipation of future development in the area, two areas to the west of Walpole were declared Reserve as Town and Landing Place in 1872 and 1886 (later to become reserves). A number of potential investors were granted leases that they subsequently dropped. Frank Skinner Thompson settled on the Deep River at Tinglewood, and Pierre and Andre Bellanger moved to the Frankland to farm in 1908. The Bellangers established the first tourist attraction, the Nornalup Park Homestead, renowned for its fine table and entertainment of picnics and fishing trips down the river onto the Inlet. Among the many prominent guests who regularly holidayed there was the Premier Sir James Mitchell. Mitchell, already captivated by the area but known for his championing of development, was instrumental in the declaration of a national park. He may have been influenced by his friend Professor E. H. Wilson, Assistant Director of the Arnold Arboretum, Harvard University, who told a reporter from the *West Australian* on 16 November 1920:

> *You have got everything there, wonderful forest scenery, mountains, landscapes, seascapes, boating, fishing. It is one of the most beautiful single sights I ever saw in all my life. It is a fascinating place, I don't know of any other that appealed to me in the same way. It was so unexpected. It would be a great shame to parcel this land out for dairying purposes as had been done with the forest land at Denmark.*

The prevailing view in the 1920s and 1930s highlighting the competing demands of industry is captured in the remark of C. J. Moran, a Trustee of the Agricultural Bank, in an article in the *West Australian* 17 October 1930. 'It is standard forestry country, mostly karri and tingle. There is no better country in Western Australia for dairying and its sidelines.'

Against this background the national park was declared in 1924, with the first park ranger Tom Swarbrick of the Rest Point Guest House appointed in 1927. Since that time debate and lobbying of different interest groups has continued. The status of the park and the boundaries have been changed: the area known as Giants Block was excised and reclassified as State forest in 1927; the Walpole–Nornalup National Park was created in 1972 by the transfer of existing C Class National Park Reserves to A Class status (Fig. 7.12; Maps 7.1 and 7.2). It was intended that the inlet waters would be included, however this was not possible under legislation at that time. The issue remained unresolved despite the recommendation in 1994 by the Marine Parks and Reserves Selection Working Group that 'declaration of Walpole and Nornalup Inlets and tidal parts of the Deep, Frankland and Walpole Rivers as marine park be implemented as a matter of high priority, and its management integrated with that of the surrounding national park'. The planning process involving the local community, stakeholders and government agencies for a proposed marine conservation reserve commenced in 2003.

In the words of Professor Wilson: 'The natural wealth belongs not only to the present but to posterity, and all time. It is ours only in trust'.

* A general term for Eucalyptus, not *E. globulus* currently planted for timber.

Source: Bannister, 1831; Bellanger, 1980; Clark, 1841; CALM, 1991b, 1992; Fernie & Fernie, 1989; Hillman 1833; Moran, 1930; Mueller, 1882; Wilson, 1920

A natural setting

Along the steep forested hillsides fringing the Walpole–Nornalup channel, the south-western shore of the Nornalup Inlet, and along the Deep and Frankland Rivers, fine stands of karri *Eucalyptus diversicolor*, red tingle *E. jacksonii* and yellow tingle *E. guilfoylei* dominate the tree line. The lower sandy ridges are covered with *Banksia* woodland, while the low ground is occupied by paperbark (*Melaleuca*) woodlands, and swamps with sedges and rushes.

In low-lying areas near the mouths of the rivers and near Walpole, a low shrub community is dominated by the yellow-flowered pea shrubs *Oxylobium heterophyllum* and *Jacksonia horrida*, bottle brush *Beaufortia* sp., wattle *Acacia pulchella*, *Astartea fascularis*, kangaroo paw *Anigozanthus* sp., zamia palm *Macrozamia riedlei*, grass trees *Xanthorrhea preissii*, sword-sedge *Lepidosperma gladiatum* and *Euphorbia* sp., and near the boat ramp swamp cedars *Agonis juniperina* tower above the vegetation. Near Walpole this community has been invaded by many weed species such as bracken and grasses, and is covered with dodder *Cassytha* sp.

Where the sand is stabilised in the lee side of the coastal dunes, peppermint trees *Agonis flexuosa* together with shrubs *Spyridium globulosum*, *Olearia axillaris*, *Rhagodia baccata*, *Billardiera variifolia* and the sedge *L. gladiatum* form typical coastal dune communities.

Around the estuary, the rush plant community dominated by sea rush *Juncus kraussii* is often narrow due to the steep slopes around the inlets.

Where the ground is flatter, species of sedge and grass such as bare twig rush *Baumea juncea*, marram grass, sword-sedge *L. gladiatum*, salt water couch or seashore paspalum *Paspalum vaginatum* (introduced) and *Isolepis nodosa* grow behind the *Juncus* community.

The sand spit at the bar is colonised by introduced weeds such as the Mediterranean marram grass *Ammophila arenaria*, dune cabbage *Arctotheca populifolia*, and the native club rush *Isolepis nodosa*. Shrubs *Leucopogon parviflorus*, *O. axillaris*, *Acacia littorea* and the sedge *L. gladiatum* dominate the more stable dune vegetation.

Where river meets sea

The Nornalup and Walpole Inlets are aligned west to east parallel to the coast, and together cover an area of 14.8 square kilometres. Our eagle high view shows the small Walpole Inlet and the narrow channel, 200 metres wide, to the larger Nornalup Inlet. Beyond the extensive marginal sandbanks the inlet opens to the Southern Ocean (Fig. 7.10).

Walpole Inlet, about 2 kilometres from west to east, and a kilometre from north to south, is very shallow, generally less than a metre except for the dredged boat channels. The southern and north-eastern shores are higher and rocky, while the western and eastern shores where the Walpole and Collier Rivers enter are low-lying with sand or swampy land. The kilometre-long channel with a depth of 2 metres leading to the Nornalup Inlet is bordered by steep granite hills with rocky shores (Map 7.4).

Nornalup Inlet is about 5 kilometres in length from east to west, 3 kilometres from north to south and about 6 metres deep in the central basin. When river flow is low, the Deep River flowing from the west is

MAP 7.4 *Walpole and Nornalup Inlets showing landforms and vegetation.*
Adapted from draft by I. Eliot, from Hodgkin & Clark, ESS No. 2, 1988 revised 1999.

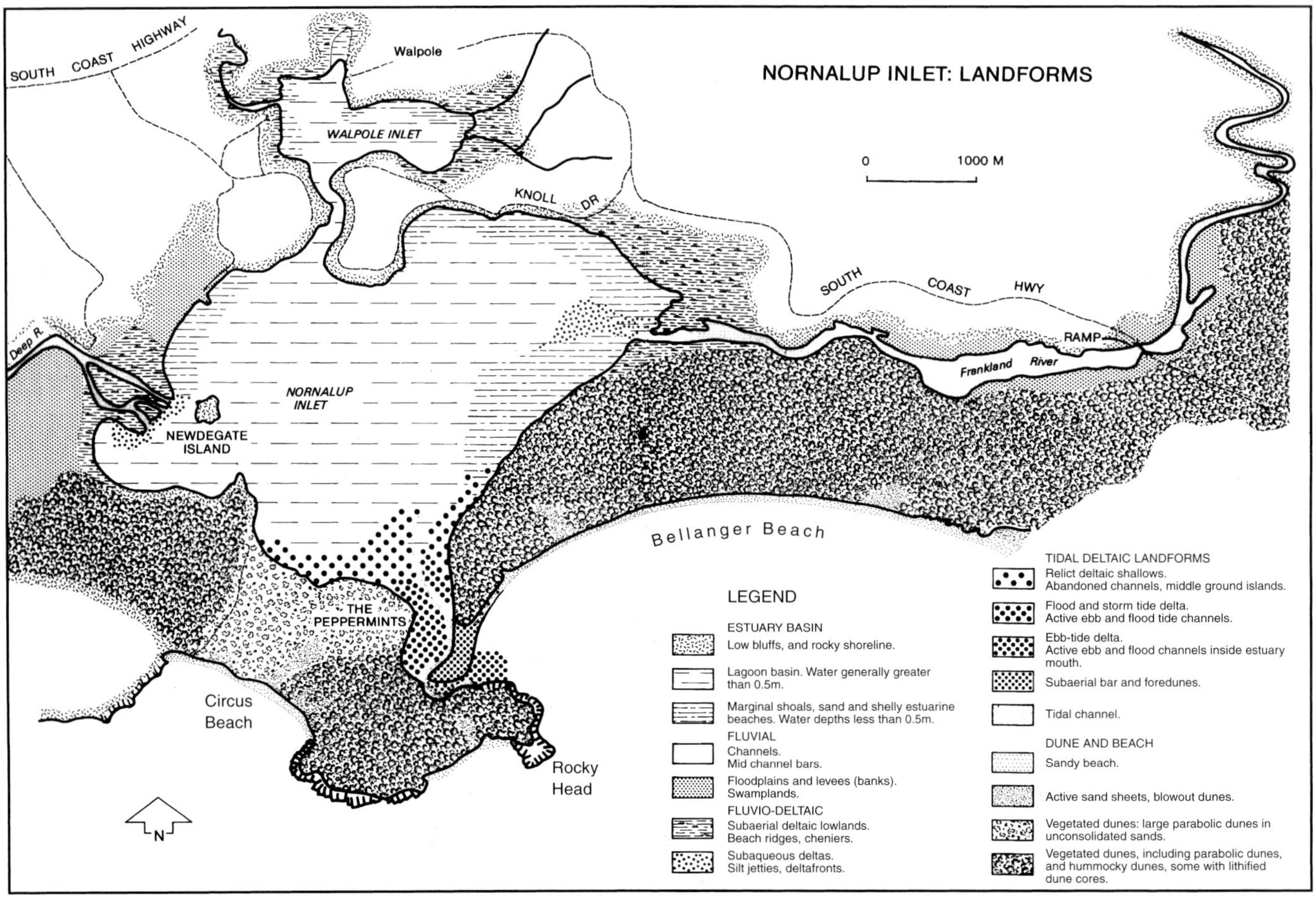

estuarine for 6 kilometres and the Frankland River flowing in from the east is estuarine for 12 kilometres.

The Frankland River has about 6 kilometres of well-defined channel scoured in places to a depth of 5 metres. Further downstream it widens and shallows to less than a metre, flanked by riverine flats, swamps and rock. Water levels in the riverine areas can increase greatly during floods: the Frankland River bridge was covered during the floods of 1982. Over the years, floods have eroded the banks and filled in the river channels and today the river flows in a well-defined (dredged) channel through the flood delta. On the northern side of the Nornalup Inlet between the Walpole Inlet channel and the Frankland delta, the narrow beach is backed by the cliffs cut into the softer tertiary siltstone of Coalmine Beach (Fig. 7.13). Along the south-eastern shore the dunes are truncated, and sands eroded from the shore have formed wide marginal shoals that slope steeply to the 3–5 metre deep central basin.

The Deep River also flows in a well-defined channel between steep hills on the north bank and sandy flats and swamps on the south bank. The river is 5 metres deep in the upper reaches shallowing to 1–2 metres in the lower reaches. Like the Frankland, logs block the upper reaches. At the inlet there are three river channels through the delta, only one navigable. The western shore around the Deep River delta is bordered by swamps and beach ridges. Sand is gradually enclosing a small bay, and a shoal links the delta to Newdegate Island (Snake Island) that rises to 19 metres. In the south-west, rock and poorly consolidated sandy peat or lignite outcrop on the sandy floor of the inlet.

Nornalup Inlet opens to the sea in the shelter of Rocky Head, which protects the estuary mouth from the south-westerly winds and swell. Although there are extensive sandbars in the entrance channel, unlike most other south coast estuaries the entrance remains open year round due to its sheltered position and the scouring river flows in winter (Fig. 7.14). The channel takes different routes through the mobile sands and may be several metres deep in some places, but shallows to less than a metre deep where it crosses the beach to the sea. It is interesting to note that when the inlet was first explored by William Nairne Clarke in 1841 the channel appears to have been much deeper.

> March 3rd…I sounded the passage over the bar, when the tide was only beginning to pour itself slowly in. The depth near a black rock in the centre of the river, and on the *left* side of which vessels of small burden can only pass, averaged from 10 to 21 feet. We found 12 feet of water, but, to be accurate, I made an allowance for swell of the sea, making 10 feets at low water the true mark. What the depth may be at high tide, or in the winter season, when the three rivers pour their freshes into the sea, I cannot pretend to say.
>
> William Nairne Clark (1841)

FIG. 7.13 *Siltstone cliffs at Coalmine Beach, January 1988.* E. P. Hodgkin.

FIG. 7.14 *Sandbars and shoals at the mouth of Nornalup Inlet, aerial view, June 2000.* S. Neville/Ecotones, Courtesy WRC.

The bar, entrance channel and beach consist of well-sorted fine sand, predominantly quartz with some shell material. Along the margins the sands are well-sorted becoming muddier with some shells at their deeper edge. In the deepest parts of the Nornalup Inlet the sediment consists of sand and clay silts, and the bottom of the Walpole Inlet is covered with fine organic mud.

The water world

Like most estuaries in Swanland, timing, volume and persistence of river flow influence the salinity and nutrients of the estuary. Although water levels in the river reaches may rise considerably, changes in the inlet are usually smaller because of the large surface area and the open ocean channel. The tides follow those in the ocean but are reduced by 40 per cent through damping by the narrow entrance.

Unlike many of the other estuaries, the permanent channel to the ocean through the bar is the major factor in determining salinity within the estuary, and marine salinity water (36‰) fills the inlets for about six months of the year. In winter the surface water of Nornalup Inlet may be nearly fresh, 10‰, although the deep bottom waters are seldom less than 30‰. Winter salinities in Walpole Inlet are usually lower than in Nornalup Inlet.

In summer, saline water gradually extends up the rivers. Salinity in the deeper parts of the Frankland River is usually about 30‰ to the head of the estuary (about 12 kilometres) with a layer of lower salinity water flowing over the surface. The layer of salt water may persist in the deeper parts of the river over winter or the river may be flushed completely. The marine waters also penetrate the Deep River for a distance of about 6 kilometres in summer, although there may be an abrupt change or halocline between the bottom saltier water and the surface layer of fresh water.

In summer, water temperature in Nornalup Inlet is 20–22°C and in winter 12–17°C at the surface, but the deeper waters may be several degrees warmer. The river waters show a greater range of temperature, from 11–26°C at the surface and up to 28°C in the deeper water at the end of summer.

The inflowing fresh water affects water clarity. In winter, the darkly tannin stained waters with suspended loads of silt reduce light penetration. In summer, the clearer sea water invasion gradually increases water clarity. The change in colour may, however, not only be due to dilution with sea water but to the light mediated breakdown of the organic compounds or tannins. The dark colour may be an important factor in reducing algal growth.

Where waves stir the open inlet, the waters are generally well oxygenated. When the waters are stratified in winter however, low oxygen concentrations have been observed near the bottom.

Water quality and assessing health of the inlets

The river waters also bring dissolved nutrients to the estuary. Early records of nutrients collected during the 1970s Fisheries Department surveys indicate that nutrients were low in river and estuary waters. Samples of river water collected in 1995–1996 showed similar low nutrient concentrations, although the amount of nitrate was higher in the Frankland River than the Deep River. At the beginning of 1999, the Water and Rivers Commission funded by the Natural Heritage Trust initiated a series of water quality surveys in summer, late summer, autumn and spring through to 2002, at eight monitoring sites. The monitoring sites included two sites in Walpole Inlet, the channel between the two water bodies and five areas in Nornalup, near the mouths of the Deep and Frankland Rivers, central basin, and near and in the entrance channel to the ocean (see also Graph 1.2).

The late 1990s surveys confirmed the earlier observations of seasonal changes in salinity, and related these to nutrient concentrations and algal growth. Summer conditions in 1999 indicated that there was no effect of river flow, although the water in Walpole Inlet was slightly less salty (about 4‰) than Nornalup Inlet, which was just below that of sea water. The inlets were well mixed from top to bottom, although temperature varies slightly with 28°C at the surface of Walpole Inlet and shallow sandbanks near the channel to the ocean.

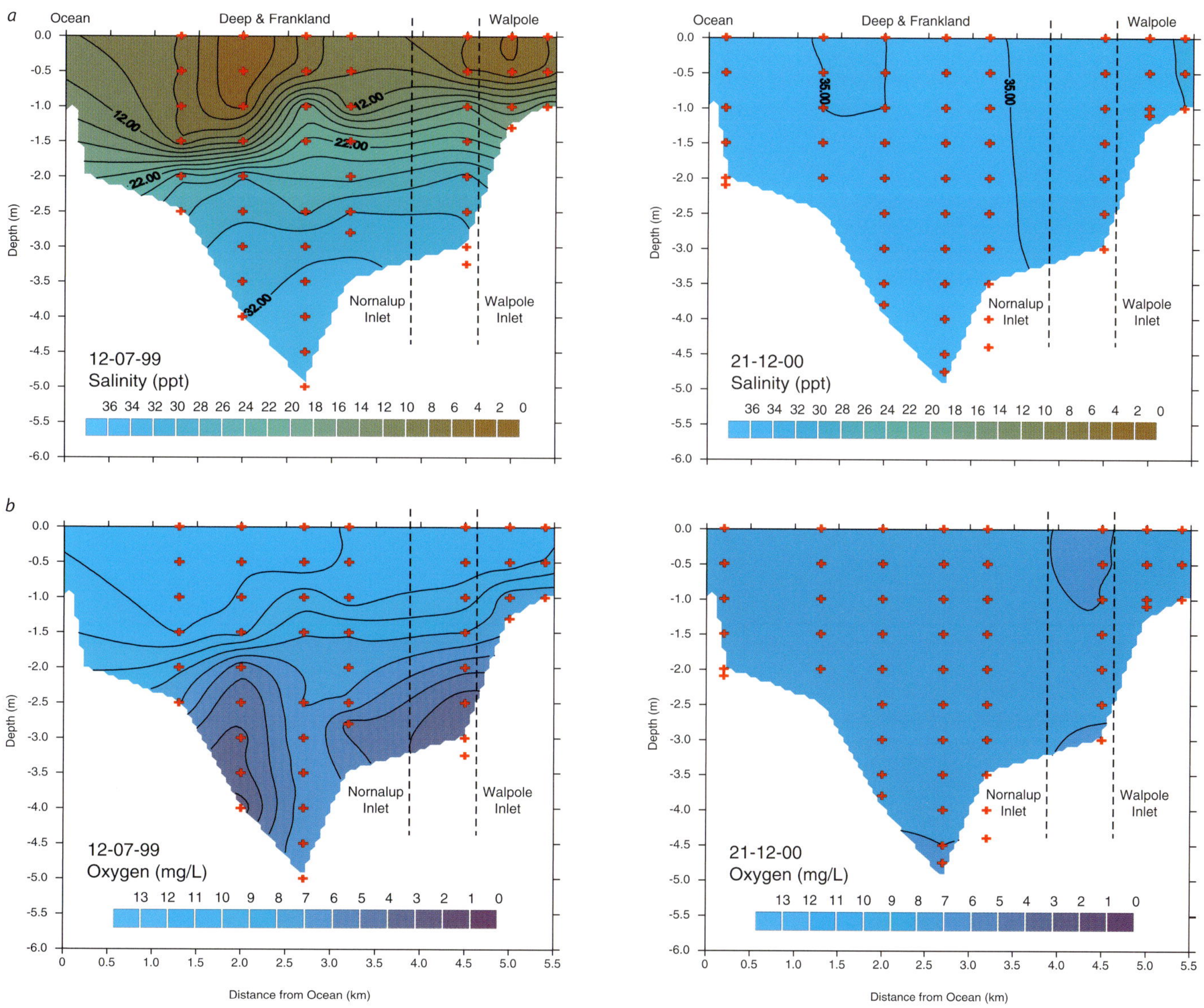

GRAPH 7.1 *(a) Salinity (ppt or ‰) and (b) oxygen (mg/L) in Walpole and Nornalup Inlets, mid-winter 1999 and summer 2000. The ocean is on the left-hand side of the graphs, Nornalup and Walpole Inlets are marked by dotted lines, and the entry of Deep and Frankland Rivers to Nornalup Inlet and the Walpole River to Walpole Inlet are indicated.* Courtesy WRC.

The 1999 winter river flow brought a plume of fresh water to the surface of the inlets that produced a strong stratification (Graph 7.1). Walpole Inlet almost completely flushed with fresh water. In Nornalup the surface layer was 1–1.5 metres deep and the bottom salty layer about 2 metres deep. The slightly lower oxygen concentrations in the bottom waters indicated that the saline layer was isolated, probably due to the depth of the surface fresh water preventing salt water intrusion over the flood tide delta. Although not anoxic, the oxygen concentrations were low enough for nutrient release from sediments to commence and denitrification, the release of N_2 gas, to cease. Although low oxygen conditions are a concern and conditions in Nornalup could have deteriorated, it was considered that due to the depth of the saline layer and extent of the inlet this would probably take longer than in Wilson Inlet where the saline layer is generally much thinner. In addition, at this time salinity was 32‰ rather than 36‰ of marine water, indicating that some mixing was occurring probably due to winds. Generally there were higher oxygen concentrations at the surface, where temperatures were also lower.

With the decrease in river flow in spring and early summer the stratification, although persistent, weakened and oxygen concentrations in the bottom waters of Nornalup Inlet increased. An increase in salinity of the surface water indicated that the water body was mixing, although Walpole Inlet was generally fresher than Nornalup. The return to more marine conditions, however, in summer 1999–2000 was earlier than in the previous summer due to the smaller volume and earlier slowing of flow. In the winter of 2001, the low winter rainfall resulted in low river flows and noticeably more marine conditions due to a strong tidal influence.

Most of the nutrients with the exception of total phosphorus showed marked seasonal changes, and varied from year to year depending on flows. Total phosphorus concentrations however were fairly constant throughout the year and low (50 micrograms per litre) compared to most other south coast inlets (Graph 7.2). Interestingly there was only a small amount of phosphorus with the particulate material washed into the inlets during winter.

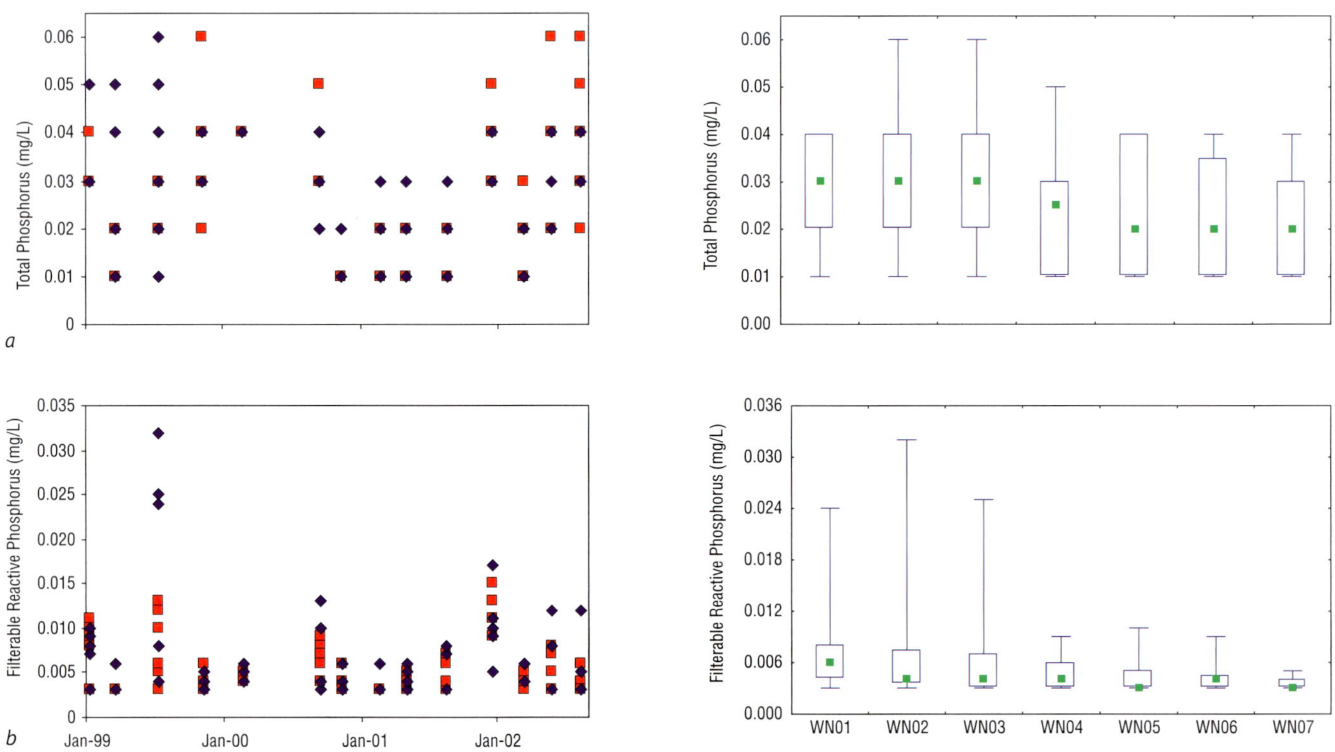

GRAPH 7.2 *Phosphorus concentrations in Walpole and Nornalup Inlets, 1999–2002: (a) total phosphorus and (b) filterable reactive phosphorus. The site numbers are Walpole Inlet near river WN01, east of centre WN02, channel between inlets W03; Nornalup Inlet middle WN04, near Deep River WN05, near Frankland River WN06, near channel to the sea WN07, and channel WN08. Red and blue squares indicate surface and bottom on seasonal graphs. Green = median, blue = 75% and 25%, and blue whiskers = maximum and minimum on site box plots.* Courtesy WRC.

Concentrations of total phosphorus were higher in Walpole Inlet possibly due to less flushing and more re-suspension of sediments by wind in the shallow water body. Concentrations of filterable reactive phosphorus were low most of the year, between detectable limits and 10 micrograms per litre, but increased in surface waters with winter flows particularly in Walpole Inlet.

The dissolved forms of inorganic nitrogen (ammonium and nitrate) were highest in winter, particularly in the fresher surface waters due to river flows (Graph 7.3). In mid-winter to early summer most of the nitrogen was due to nitrate. In summer both ammonium and nitrate were present although concentrations were very low. In late summer ammonium concentrations increased, possibly derived from sediments. In winter 1999 concentrations of nitrate were high (100 micrograms per litre to 750 micrograms per litre) but most flushed through to the sea. In contrast, in the winter of 2001, a year of lower rainfall, there was less nitrate and more ammonium possibly due to differences in flow.

Total nitrogen also showed strong seasonal responses. High concentrations were present in winter with the freshwater inflow. In winter dissolved inorganic nitrogen accounted for 60 per cent of the total nitrogen but only 5 per cent in summer. The increase in algal biomass in summer probably indicates that the inorganic nitrogen had been taken up by algae.

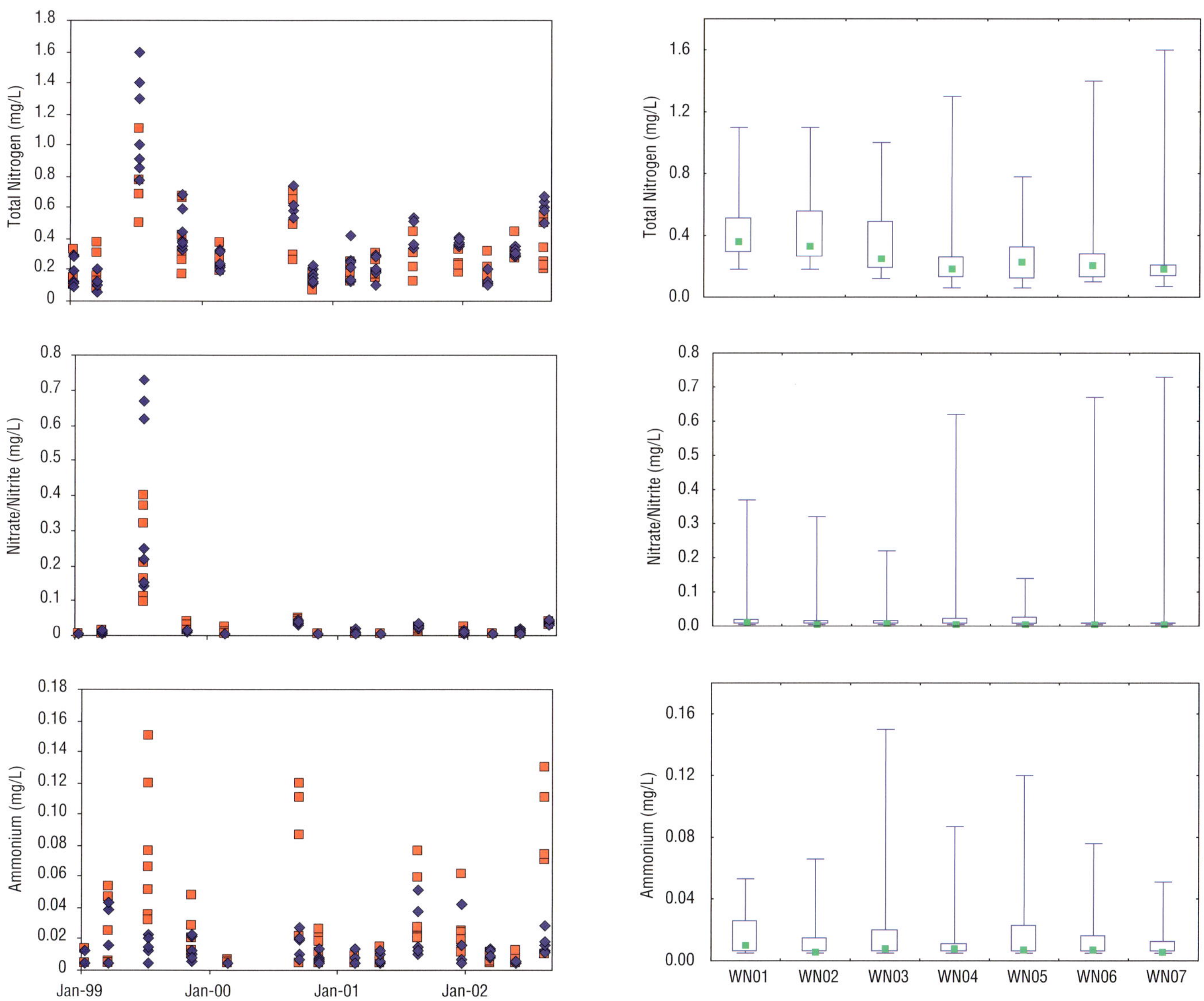

GRAPH 7.3 *Nitrogen concentrations in Walpole and Nornalup Inlets, 1999–2002: (a) total nitrogen, (b) nitrate and (c) ammonium. The site numbers are Walpole Inlet near river WN01, east of centre WN02, channel between inlets W03; Nornalup Inlet middle WN04, near Deep River WN05, near Frankland River WN06, near channel to the sea WN07, and channel WN08. Red and blue squares indicate surface and bottom on seasonal graphs. Green = median, blue = 75% and 25%, and blue whiskers = maximum and minimum on site box plots.* Courtesy WRC.

The ratio of dissolved inorganic nitrogen to dissolved inorganic phosphorus (N:P), the forms used by plants, was 5:1 in summer and 100:1 in winter. This suggests that in summer the low nitrogen may be a limitation to algal growth (primary production). In contrast, in winter, light, temperature and water flow may be more important in controlling algal growth than the availability of nutrients.

Where the nutrients go — fuelling the food chain and into the sediment

Microscopic plants or phytoplankton often respond quickly to increased nutrients in the water column. The types of algae and their abundance often assist in determining if nutrient concentrations may be high enough to initiate nuisance or toxic algal blooms. Between 1999 and 2002 there were no obvious patterns in the chlorophyll *a* concentrations indicating algal growth (Graph 7.4). They appear to be minimum in winter. The slightly lower concentrations in midsummer compared to early and late summer probably reflect the lower concentrations of inorganic nitrogen at that time. Walpole Inlet and channel between inlets had much higher chlorophyll, although these did not indicate bloom conditions. The microscopic flora generally contained estuarine diatoms, cryptophytes, chlorophytes and less often dinoflagellates.

Sediments in the estuary fall into two main groups. There are fine sediments with high concentrations of organic material and nutrients near Newdegate Island, the channel between Walpole and Nornalup Inlets, and in Walpole Inlet. There are some indications of pollutants above background concentrations in the Walpole Inlet and channel and near Newdegate Island, however the source of these is unknown. Sediments near the channel to the sea and near the Frankland River inflow are coarser with low concentrations of organic matter and nutrients, indicating that they may be reworked by the stronger water movements in these areas.

Inhabitants

Water plants

There are not many types of aquatic plants growing in the Walpole–Nornalup estuaries. The swan grass *Ruppia megacarpa* grows on the shallow sandbanks near the Deep River and in places along the eastern shore, and is often covered with algal epiphytes such as *Polysiphonia*. The *Ruppia* however is not as prolific as it is in Wilson Inlet. The more marine eel grass *Zostera* sp. and the paddle weeds *Halophila ovalis* and *H. decipiens* are also found along the eastern shore. *Polyphysa*, also known as *Acetabularia calyculus*, commonly grows attached to dead shells and rocks. The deeper part of the central basin contains black organic mud, and the dark tannin stained waters are too dark for plant growth.

GRAPH 7.4 *Chlorophyll* a *concentrations in Walpole and Nornalup Inlets, 1999–2002: based on integrated samples taken through the water column. The site numbers are Walpole Inlet near river WN01, east of centre WN02, channel between inlets W03; Nornalup Inlet middle WN04, near Deep River WN05, near Frankland River WN06, near channel to the sea WN07, and channel WN08. Red and blue squares indicate surface and bottom on seasonal graphs. Green = median, blue = 75% and 25%, and blue whiskers = maximum and minimum on site box plots.* Courtesy WRC.

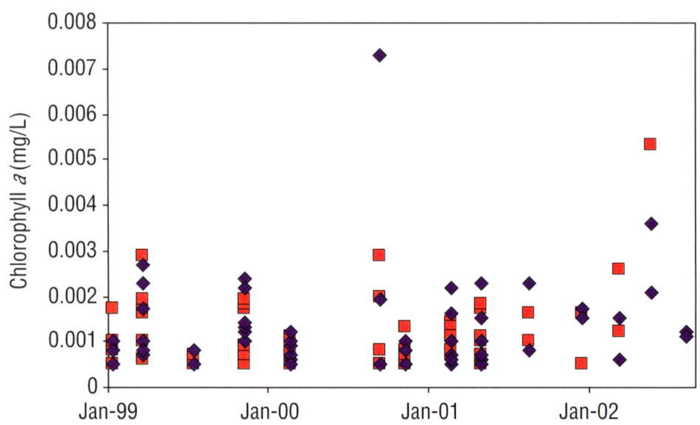

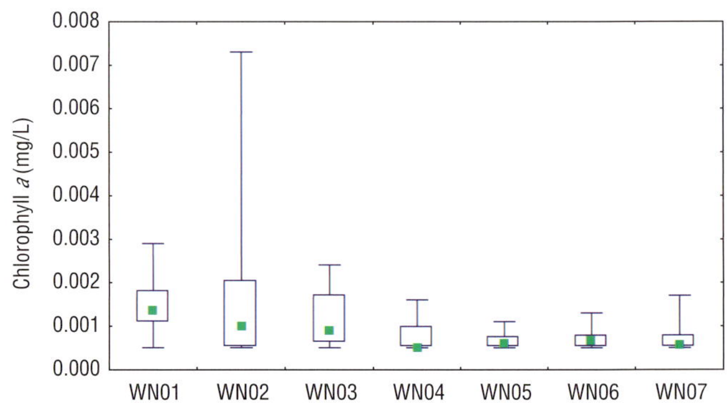

Only a few algal species are recorded from the inlets; the brown alga *Cystoseira trinodis* growing attached to rocks in the channel between the inlets and green algae 'nuisance' types in the shallow, higher nutrient waters of Walpole Inlet. *Chaetomorpha linum* and *C. aurea* were abundant in 1973–1974 and *Cladophora* sp. were abundant on the muddy sands in 1976. In 1976 living algae covered a layer of dying and dead algae 10–20 centimetres thick over black deoxygenated ooze in 1–2 metres of water. In 1987 floating mats of *Enteromorpha intestinalis* and *Chaetomorpha billardierii* were noted in the shallow areas on the eastern side of the inlet.

Floaters: the zooplankton

In studies of the 1970s, the common estuarine copepod *Gladioferens imparipes* dominated the zooplankton of the Deep and Frankland Rivers, and was also found in both the Walpole and Nornalup Inlets when salinity was low. The more marine species *Acartia tonsa* dominated when salinities were higher in both the Walpole and Nornalup. The larval stages of bottom dwelling animals were also abundant at times, particularly the veliger stages of gastropod molluscs. More recently, in 1995 a similar suite of zooplankton was collected and the biomass was highest in Walpole Inlet where the microscopic phytoplankton are most abundant.

Animals of sand and mud

The permanently open channel through the bar that allows good oceanic exchange results in the establishment and survival of a greater diversity of invertebrate species in Walpole–Nornalup than in other estuaries of the south coast except for Oyster Harbour. The large number of Foraminifera recorded from the inlets also reflects this marine condition.

In the late 1980s, the greatest number of invertebrate species was found on the shallow sandbank areas around the main basin. The small polychaete *Capitella capitata* was abundant throughout the estuary. A number of small gastropod snail species such as *Tatea preissii*, *Assiminea* and *Hydrococcus brazieri* were found in the fringing rushes. The typical estuarine bivalves the mussel *Xenostrobus pulex*, the trough shell *Spisula trigonella* and *Arthritica semen* were found near the Frankland River entrance. Larger prawns *Metapenaeus dalli*, *Palaemonetes serenus*, and *Leander* sp., the blue swimmer crab *Portunus pelagicus* and mud burrowing crab *Cyclograpsus audouinii* are also found in the inlets.

Fish and fisheries

The fish of Walpole–Nornalup and Wilson Inlets were quite extensively studied in the 1970s–1990s because differences in fishing pressure and marine influences allowed interpretation of driving factors in the environment. Recreational and commercial fishers both utilise Wilson Inlet, but only recreational fishing occurs in the Walpole and Nornalup Inlets. However recreational fishing pressure in Walpole–Nornalup has increased substantially in recent years, and current information on fish communities has not been documented. The permanent connection to the sea at Nornalup is reflected in the diversity of fish species found within the estuary, a greater number than in Wilson Inlet where the bar closes for some part of the year. Most are species that spawn in the sea and enter the estuary after they have completed larval development. In some years when the water in the entrance is very turbulent because of the ocean swells or strong outflow, entry of fish may be hindered. (See **Chapter 1 Understanding Swanland: Fish and fisheries**.)

Despite the open entrance channel, the fish communities of shallow areas were dominated by small estuarine-spawning species. Hardyheads or atherinids *Leptatherina wallaceii*, *Atherinosoma elongata* and *L. presbyteroides* together with the goby *Gavonigobius lateralis* were the most abundant species in both Walpole–Nornalup and Wilson Inlets, accounting for over 90 per cent of the individuals.

The studies have revealed that the deeper waters of Nornalup Inlet harboured many more marine species than Wilson Inlet. In Nornalup, marine species account for 64 per cent of the fish species compared to Wilson Inlet where only 37 per cent are marine species. King George whiting *Sillaginodes punctata*, the pink snapper *Pagrus auratus*, five species of elasmobranchs (sharks and rays) and marine turtles have been recorded in Nornalup. The gummy shark *Mustelus antarcticus* and the eagle ray *Myliobatis australis* were relatively common. Gummy sharks were confined to the outer basin of the Nornalup Inlet where salinity was highest, although catches were greater in August when freshwater discharge was high; salinities even at this time still exceeded 26‰ in the deeper areas. The eagle ray was more widely distributed within Nornalup Inlet, but was only recorded in the Walpole Inlet or the Frankland River during summer when salinity was greater than 23‰. The absence of sharks and rays in Wilson Inlet was thought to be due to lower salinities that rarely exceed 30‰ even when the mouth is open.

A number of studies have indicated a similarity between the fish faunas of Walpole Inlet, the main river tributaries leading to Nornalup Inlet and Wilson Inlet. This similarity is probably due to the preference of estuarine black bream *Acanthopagrus butcheri*, sea mullet *Mugil cephalus* and the yellow-eye mullet *Aldrichetta forsteri* for reduced salinities and/or features associated with riverine environments. In the estuarine reaches black bream is probably the most important fish caught by recreational fishers.

In a study of fish larvae conducted in the 1980s and 1990s, thirty-six species of larval fish were caught throughout the estuary, twenty-two species were classified as marine stragglers, three species as marine estuarine–opportunists, ten estuarine species and one freshwater species (*Galaxias occidentalis*). Of the total numbers of larval fish caught, marine estuarine opportunists and freshwater species each accounted for less than 1 per cent of the larvae caught, and estuarine species accounted for 99 per cent. Not unexpectedly, this varied across the estuary; near the entrance marine species were more common decreasing within the inlet, while the reverse was found in the inner areas with estuarine species being more common.

Of the more marine species, twenty-three were never found as juveniles or adults. The species that dominated the larval assemblage in the two basins of Walpole–Nornalup Inlet were also abundant in Wilson Inlet. The southern anchovy *Engraulus australis*, the blue spotted goby *Pseudogobius olorum* and long-finned goby *Favonigobius lateralis* collectively accounted for more than 95 per cent of the total in the two inlets. Larvae of the hairy pipefish *Urocampus carinirostris* and the blenny *Parablennius tasmanianus* were also abundant in both systems.

Most of the larvae, in early stages of development, were typically inshore marine species common in the surf zone outside the inlets and appeared to be passively transported into the inlets with the flood tides. The striped trumpeter *Pelates sexlineatus* classed as a marine opportunist is one species that may actually spawn within the inlet. Their larvae were found in the entrance channel and in all three basins within the Walpole–Nornalup Inlet system suggesting that the tidal flow was sufficiently strong to transport them there. Alternatively, as adults were also present in the inlet, the marine-like conditions may have been suitable for breeding.

The fish faunas in the shallow and deeper waters of Walpole–Nornalup Inlet and Wilson Inlet are less diverse than those in comparable waters of estuaries such as the Peel–Harvey and Swan on the lower west coast. This is because several species are not found this far south or are present only in small numbers. The tarwhine *Rhabdosargus sarba* was the only subtropical–warm-temperate marine species that was even moderately common in the Walpole–Nornalup Inlet and Wilson Inlet. The marine King George whiting *Sillaginoides punctata* and the Australian herring *Arripis georgianus*, commonly found in the two estuaries and in the Blackwood–Hardy, are temperate species living in this region.

King George whiting, mullet, striped trumpeter and herring were usually found at a larger size in Walpole–Nornalup and Wilson Inlet on the south coast compared to the size of fish recruited to estuaries on the lower west coast. Based on the difference in size at which fish recruit to estuaries, it has been suggested that the spawning areas are at a greater distance from the south coast estuaries and that spawning may occur in the large marine embayments and fringing limestone reefs of the lower west coast.

The future

There is fairly wide agreement that Walpole–Nornalup is beautiful, but it is also special among the estuaries of Swanland as it is surrounded by national park that is largely forested and has a permanent opening to the ocean. Thus it is largely unspoilt with a free exchange of water to the ocean outside, features that are not found in many other places. The high diversity of fish species and their abundance make this a very pleasant and productive place for residents and holiday-makers.

The waters of Nornalup Inlet are not eutrophic and are likely to remain in good condition while the catchment is forested, development nearby is limited and the bar remains open. There are, however, problems of eutrophication in Walpole Inlet where growth of algae has occurred. The waters in here are shallow, there are unsewered residential areas nearby, and water flow to Nornalup is limited. Boating channels dredged in Walpole Inlet appear to accumulate algae that rot, releasing stored nutrients which promote more algal growth.

The channel through the bar is constantly changing and often shallow with swell waves creating a potentially hazardous entry on this unprotected coastline. The maintenance of the open channel is probably due to the comparatively high river flow and the relatively protected location against the rocky headland. The permanent opening is a major influence in the ecosystem: a reduction of river flow could change the open status of the channel with the bar closing in summer, although winter flows would probably still breach the bar unless the rivers were dammed. Any reduction in the time the bar is open would alter the dynamics of the estuary leading to a loss of some of the marine fauna so that it would become more similar to Broke or Wilson Inlets. Additional nutrients would have a greater effect on the health of the system if flows from the catchments were reduced.

In common with other estuarine areas, more tourist interest is leading to change. Fishing pressure is increasing every year. The tranquillity of the area is threatened with an increase of noise from powerboats. Areas once isolated are now visited more frequently by adventurers in canoes, who make overnight camps in the riparian vegetation, increasing the risk of fire outbreaks.

Continuation of this healthy estuarine state will depend on protection of the surrounding natural habitats for the maintenance of ample, fresh, low nutrient waters from the catchment.

IRWIN AND PARRY INLETS

Irwin Inlet of 10 square kilometres and Parry Inlet of 1.4 square kilometres (Map 7.2) are smaller than the Walpole–Nornalup (14.8 square kilometres) and Wilson Inlet (48 square kilometres). Like Wilson Inlet they are also affected by nutrient inflows from the catchment and the length of time the bar is open. Although surrounded by cleared lands, the coastal dunes are relatively well vegetated. Part of the Bibbulmun Track (extending from the Perth metropolitan area to Albany) passes Parry Inlet and Irwin Inlet where there is a canoe crossing on the channel leading to the ocean.

Irwin Inlet

Irwin Inlet, originally referred to as Quarram Inlet, receives waters from the Kent and Bow Rivers and small creeks draining the surrounding low-lying lands (Fig. 7.15). The Kent River, over 100 kilometres long, rises from salt lakes and saline seeps in cleared agricultural land to the north-east near Albany Highway. In the lower reaches the Kent passes through the large Owingup Swamp. Below the swamp, the river forms a 2-metre deep channel and passes through densely vegetated sandy lowlands before entering the inlet on the east (Fig. 7.16). The last 5 kilometres are estuarine. The Bow River, 20 kilometres long, enters from the north-west. The Bow River, about 5 metres deep, is estuarine for about 2 kilometres passing through sandy lowlands. Both of the river mouths are barred by deltas that are part of the wide (up to 500 metres) marginal shallows that dry out in most summers.

The inlet is 5.5 kilometres long and 2.5 kilometres wide, oriented north-east to south-west behind a narrow strip of low coastal dunes. Most of the inlet is shallow with only the central basin reaching 2 metres deep. The marginal shallows are muddier in the north, and a boat channel has been dredged though them. There is a small granite island in the south-western corner, just offshore from the boat launching area.

The ocean channel from the central basin is about 2 kilometres long, 200–300 metres wide and 1.5–2 metres deep with a maximum depth of 3.5 metres at the northern end, gradually shallowing near the sea. In the

FIG. 7.15 *Irwin Inlet, aerial view of opening to Quarram Beach in Foul Bay. Peaceful Bay lies just out of view in the southern sweep of the coast in the foreground, June 2000.* S. Neville/Ecotones, Courtesy WRC.

middle reaches of the channel, a low island 400 metres long is covered with swamp vegetation. Flow channels run either side of the island, but the main channel is generally close to the western bank. Some rocks outcrop on the western bank but the banks are generally sandy and bordered by low rushes, although there is some erosion in places.

The channel discharges to shore at Quarram Beach in Foul Bay near Peaceful Bay (Fig. 7.17). The opening appears to be migrating westwards. On the western side, the high (10 metre) well-vegetated dunes are collapsing where the channel has cut into them. A low scarp on the western dunes, 550 metres from the beach, and a series of low beach ridges suggest that the mouth may have been considerably west of the present position quite recently. To the east, there is a wide area of bare sand through which the channel sometimes flows, backed by low vegetated dunes.

The bar is 150 metres wide and only about a metre above sea level when closed. The flood tide delta extends 500 metres into the channel. The bar contains pale yellow moderately well-sorted medium to fine sand, 90 per cent quartz and 10 per cent shelly fragments. The sand behind the bar is 'quicksand'. The bar opens after heavy rain in the catchment, usually between June and August, although it is often breached by fishers or landowners to reduce flooding of the low-lying surrounds. Sometimes it remains open all

FIG. 7.16 *Fringing vegetation of* Juncus, Melaleuca incana *and* Agonis juniperina *along the banks of the Kent River, January 1988.* E. P Hodgkin.

FIG. 7.17 *Western shore of Irwin Inlet with salt marsh and eroding bank, January 1988.* E. P. Hodgkin.

summer. In 1988 the bar breached about 40 metres from the western shore, but the channel developed towards the west cutting to about 110 metres wide. The estuary drained to the sea and tidal influence was obvious after 20 hours. While the channel remains open, the flow channels meander from west to east behind the bar often running parallel to the beach.

Salinity varies from fresh in winter to close to sea water when the bar is open and there is no river flow. Towards the end of summer it becomes hypersaline reaching up to 40‰. The inlet waters are generally well mixed, but when the rivers first flow it may be stratified in the deeper parts with river water lying over the more saline inlet water, or when the bar breaks and the denser sea water moves in as a wedge under the out-flowing fresher inlet water. Flow between the inlet and the Kent River is good, and the river waters are similar to those of the inlet when river flow ceases. Exchange between the inlet and water in the Bow River appears to be more restricted, as the waters are less well mixed and may be stratified.

Much of the land surrounding the inlet has been cleared. Remnant vegetation in Owingup Swamp, which is no longer part of the estuary, is a renowned bird habitat and probably an important trap for nutrients from the catchment. Total phosphorus concentrations in Irwin Inlet are higher than those in Wilson Inlet and exceed guideline concentrations. Total nitrogen and chlorophyll *a* concentrations are, however, lower than those in Wilson Inlet. (See Graphs 1.2, 1.3, 8.4–8.6). The inlet experiences great extremes of water level, varying from high when the rivers are flowing and the bar is closed to dry and very shallow at the end of summer. Changes in water level may be affecting the bordering swamp cedars *Agonis juniperina* and banksia trees that are dying along the lower slopes.

Swan grass *Ruppia megacarpa* is found in a continuous band on the sandy shallows, but the amount of growth varies from year to year. Similarly, the green alga *Enteromorpha paradoxa* and *Chaetomorpha* sp. were recorded in the shallows in 1987 but vary in abundance from year to year. The green alga *Polyphysa peniculus* also grows on the northern shore of the inlet attached to shells.

There are fewer species of fish in Irwin Inlet compared to Wilson Inlet, but the invertebrate fauna appears to be as diverse. The fauna is probably more prone to periodic mass mortalities as a result of lower salinities and exposure of the shallow areas over summer. Despite the extreme conditions, Irwin Inlet was on occasions fished commercially but was closed to this activity in June 2002. Two marine species, a type of stingray and the common blowfish or banded toadfish, are present in Irwin Inlet but do not appear to have been reported from Wilson Inlet and may be favoured by the higher salinity in Irwin Inlet. There was a black bream fishery in Irwin Inlet but by the 1980s the number of fish had become too small to be economic. In 1987 local fishers with permission from the Department of Fisheries 'seeded' the Kent and Bow Rivers with juvenile black bream; however the success of the experiment is unknown.

Parry Inlet

Parry Inlet receives water from one river, the small 12 kilometre-long Kordabup, which enters through swamp lands on the north-east. Several drains also enter the inlet from swampy land to the west. The catchment is about 120 square kilometres and most of the land around the inlet has been cleared for cattle farming.

The inlet is small, only 1.4 square kilometres in area, with a short estuarine stretch on the Kordabup River (Fig. 7.18). Swamplands on the eastern shore are almost as large as the inlet itself. Much of the shallow basin, 0.5 metre average depth, dries out in summer except for a slightly deeper channel about 2.5 metres deep on the eastern shore. The inlet is surrounded

FIG. 7.18 *Parry Inlet, aerial views: (a) October 1980 and (b) January 1988.*
E. P. Hodgkin.

by low sandy rush-covered shores, and has a low rush island. There are small granite outcrops at the entrance to the inlet channel.

The inlet channel, over 2 kilometres in length, follows a north-west to south-east direction. Only 20 metres wide when it leaves the inlet, it widens to over 100 metres halfway along narrowing again near the bar. In the northern half there are low sandy shores with small rush and samphire covered islands and swamps. Towards the seaward end of the channel the eastern shore has a series of low beach ridges for some distance between the channel and the dunes, while the western shore is cut into dunes about 10 metres high with steep, sparsely vegetated slopes to the channel. The channel has eroded the western dunes and the mouth has progressed south-west in recent times. The location of the flow channel varies and may approach either bank; it is up to 2 metres deep with shallow banks on either side.

The bar is low, about a metre above sea level, forming part of the beach where it flows into William Bay on the eastern side of Point Hillier. There is no rock near the bar. When the channel is open it flows close to the south-western shore but gradually migrates north-east cutting a channel 100 metres wide and 2 metres deep. The bar opens naturally or is opened to prevent flooding of roads and paddocks along the western shore, and may close and open again several times during winter. The bar normally closes in summer, but may have remained open for about five years in the 1940s.

Water within the inlet varies from fresh to hypersaline in summer. *Ruppia* grows in the shallows of the inlet and channel, but may be less abundant than previously. *Enteromorpha intestinalis* and another species of *Enteromorpha* were present near the mouth and inlet basin in a 1980s survey, together with *Chaetomorpha* and the blue-green alga *Lyngbya*. Only a few species of fish and invertebrates have been recorded from the inlet and there are often large fish kills as the waters dry up in summer. Parry Inlet has been fished commercially in the past, although this has not been done every year or heavily. The inlet was closed to commercial fishing from 30 June 2002, but will still be open to recreational anglers.

State of Parry Inlet

In 1999 the Water and Rivers Commission initiated a sampling program to assess water quality in Parry Inlet. Monitoring commenced in January 1999, and the inlet was sampled ten times from January 1999 in summer, autumn, winter and through to spring 2001. The investigation was conducted at five sites: three within the lagoon and two in the entrance channel, but some of these were not always accessible due to low water levels.

The study found that despite the very shallow depths, salinity stratifications still developed and these were sometimes accompanied by a deoxygenation and release of nutrient from the sediments. Oxygen concentrations were high in winter due to the lower water temperatures, and occasionally high in summer when growing algae produced oxygen. Parry Inlet was slightly more acidic than the larger and deeper south coast estuaries, possibly due to the types of soils in the catchment or the wetting and drying of estuary sediments over winter and summer. The estuary waters were also coloured with tannins from the river water.

Total nitrogen concentrations in Parry Inlet were high for a south coast estuary; the concentrations more similar to inlets further east or north. This was possibly due to release of nitrogen from the sediments, concentration through evaporation or high concentrations entering from the catchments. Nitrate nitrogen concentrations were high when fresh water was entering from the catchment, and low in summer. The low summer values may indicate that nitrate was being used by the algae and plants or converted to N_2 gas by bacteria during denitrification. Ammonium concentrations were generally around 0.01 milligrams per litre; concentrations fairly typical of estuaries on the south coast where dissolved inorganic nitrogen is generally low. High concentrations of ammonium (0.3 milligrams per litre) occurred when the inlet was stratified and oxygen was low.

Total phosphorus concentrations in Parry Inlet were also higher than in most other south coast estuaries, similar to the Vasse–Wonnerup and exceeded only by those in Torbay Inlet (Graphs 1.2 and 1.3). Like the nitrogen concentrations this could be due to catchment loads, sediment release as well as concentration through evaporation. Filterable reactive phosphorus concentrations varied but were neither particularly high nor low. Sporadic measurements of chlorophyll *a* indicate that phytoplankton in Parry Inlet is probably higher than in Irwin and Wilson Inlets.

Chapter 8

Wilson Inlet

The pretty town of Denmark lies on land sloping to the river that shares the same name, surrounded by pockets of mighty forest, giant karri trees, interspersed with 'Oh so green' paddocks dotted with cattle. A country atmosphere continues in the midst of the increasing tourist traffic. The countryside is changing, more houses are scattered through the trees and vineyards cloak the slopes with different colours.

The dark waters of the river edge the town park, and a walkway leads through karri and paperbark trees down past the old railway line that skirts the north-eastern side of the estuary. We can stand on the old log bridge and watch people fishing for bream along the banks, and admire the ducks and herons. At the boat ramp, through the caravan park we see the estuary spread before us.

Driving around the estuary we catch glimpses of water through the trees. Eventually our trip takes us past low country lined with paperbarks at Poddy Point, where the sand flats are often covered with huge flocks of Black Swans. We drive up a rise, a glance northwards shows a view back across the estuary and southwards to the wild Southern Ocean (Figs 8.1 and 8.2).

Standing here above the bar at the lookout guarded by sentinels of sea and forest, a wooden dolphin and eagle, we are at the junction of two worlds: a spot unique for contemplation of estuaries, where exchange between ocean and estuary controls processes that affect all life around the inlet. On one side the dark still waters backed by hilly treed slopes with the pure white sand below, and on the other a bright blue-green ocean where swell lines move landwards to crash onto the beach.

If our visit is in winter the channel may be open, with dark tannin-stained waters from forests and farmlands flowing into the sea, staining the bay before merging with the ocean waters. If the tide is rising we may see the clear blue

FIG. 8.1 *Wilson Inlet meets the Southern Ocean, March 2003.* A. Brearley.

ocean waters moving inwards through the channel pushing back the brown inlet waters, a line of foam marking their meeting. Along the channel we may see little whirlpools and flurries of sand, where sand is held in suspension by the inward moving heavy salt water and the outward moving lighter fresher inlet water.

If we visit at the end of the dry season when the rivers have not been flowing and the water level in the inlet is low, we may look down on a wide sandy beach where the ocean has moved sand onto the beach blocking the entrance. The great expanse of sand is crisscrossed with tyre tracks made by the vehicles parked next to parties of fishers or surfers, or simply driven along the beach; so many that we have to take care to obey road safety rules. Such changes in a year: two very different states for the plants and animals.

Beyond our view, hidden below us there is a different story. Wilson Inlet is at a crossroad. Clearing for cultivation in the catchments of the major tributary rivers has led to them becoming increasingly saline; a potential threat to Denmark's water supply. Increased nutrients from fertiliser have produced potentially eutrophic conditions and a rich growth of macroalgae and aquatic plants such as the swan grass *Ruppia megacarpa*. This growth is a double-edged sword: it fouls fishing nets and banks up in rotting piles along the shore, but catches of some fish such as cobblers increase when there are more plants.

Because the estuary must remain healthy, surrounded as it is by growing residential and tourist populations and supporting a fishing industry, Wilson Inlet has been the subject of more studies than any other estuary on the south coast. To preserve the estuary and waterways, the clearing of remnant vegetation has been banned and revegetation programs initiated to aid retention of salt and nutrients in the catchment, and reduce inputs to the inlet. The outflow of nutrient-rich waters from the inlet however, depends on the time the bar is open and the volume of water exchanged with the ocean. As the entry of marine water, subsequent stratification of the water column and fish recruitment to the estuary are also affected by the state of the bar, it is hardly surprising that there is great controversy over the location, timing and water level at which the bar is breached.

The Shire of Denmark established the Wilson Inlet Management Authority in 1994 to monitor the inlet and recommend management options. This committee combines local knowledge and experience with advice from

FIG. 8.2 *Wilson Inlet, aerial view, June 2000. Wilson Inlet is surrounded by cleared farmland except for the largely uncleared coastal dunes of the Nullaki Peninsula that form the seaward barrier to the Southern Ocean in Ratcliffe Bay. The peninsula was subdivided into large housing blocks in the late 1990s. Extensive shoals bar the mouth of the inlet.* S. Neville/Ecotones, Courtesy WRC.

state government departments including the Department of Conservation and Land Management, Department of Fisheries, Department of Planning and Infrastructure, Department of Environment (previously the Water and Rivers Commission) and the Environmental Protection Authority. The Shire of Albany is also represented because the eastern shores of the inlet and the eastern catchments are within their boundaries. Recent studies under the National Eutrophication Management Program (NEMP) have been unravelling the important nutrient cycling processes in the inlet, including the role of *Ruppia*, phytoplankton and sediments.

Changing scenes

In 1829 Dr T. B. Wilson named the river after Dr Alexander Denmark RN, a physician. The Aboriginal name for the area 'Koorrabup' was recorded by Alfred Hillman in 1833. Settlement by Europeans commenced in the mid-1880s by the Youngs, near Youngs to the east of the inlet, and the Randall family who farmed near the bar at Ocean Beach. Denmark on the northern shore of the inlet was originally established as a mill town in 1895 and declared a townsite in 1909. Connection to the outside world became more reliable when Millar's Karri and Jarrah Forest Limited built a railway to Albany to take railway sleepers to the Great Southern Railway. The railway also provided a route for fish from the estuary to reach markets. At the height of the timber industry there were three mills around Denmark cutting about 90,000 super feet (200 cubic metres) of timber per day for export worldwide and throughout Australia. Most of the timber was cut out by 1905, the railway closed, the population declined and Denmark was almost a ghost town. The original mill site near the estuary is now a reserve and the track line a popular tourist walk along the shoreline to Crusoe Beach. The railway reopened in 1907, and market gardening, potato growing and dairy farming developed near the town. The coastal lands were extensively cleared during the 1920s and 1930s by settlers in The Group Settlement Schemes, and by Work Groups in the depression years of the 1930s. The inexperience of the settlers and the problems of mineral deficiencies in the soils resulted in many farmers abandoning their holdings. Denmark was a centre for gradual development to the west, and the railway line to Nornalup opened in 1926.

The soil and rainfall have determined the patterns of land use. In the laterite country, sheep are grazed for wool and meat production with some cropping for barley and oats. Sandy areas nearer to the coast are used for beef cattle production, and some are now planted with blue gums for timber. The rich loamy soils associated with granite outcrops are used for potato, fruit and vegetable growing including the new crop of grape vines, with some dairying where grasses continue to grow over summer. The population declined after World War II but since the 1970s it has developed again as a centre for retirement and tourism. The estuary waters are fished commercially, and black mussel spat are collected for farming in King George Sound.

PLACE IN THE LANDSCAPE

Wilson Inlet lies on the narrow coastal plain between the coastal dunes of the Nullaki Peninsula and the granite hills at the southern end of the ancient plateau. Two million years ago in the early Pleistocene when sea level was much higher, there would have been an open seaway from Irwin Inlet to the hills west of Wilson Inlet probably extending through to Princess Royal Harbour. The granite headlands of today would have been islands offshore. When the sea level was much lower 20,000 years ago, the ancestral estuaries were much deeper river valleys. After the last ice age in the Holocene period about 8,000–6,000 years ago, sea levels rose flooding the valleys and the nearby swamps. Since that time Wilson Inlet, the low-lying areas and other inlets have been gradually filling with sediment. (See **Box 8.1 Sediments revealing times past.**)

River sources

Wilson Inlet has two major rivers, the Hay and the Denmark Rivers (Maps 8.1 and 8.2). The Hay River, 80 kilometres in length, enters the estuary at the eastern end and the Denmark River, 60 kilometres long, enters on the northern shore. These two rivers drain 89 per cent of the catchment with about 70 per cent of the Hay and 30 per cent of the Denmark cleared for cropping, pasture and horticulture. Water flows from both rivers are increasingly saline. Most of the Denmark River catchment and middle reaches of the Hay River are in dissected granite country, and much of the area is still forested. A dam to provide water for the town of Denmark was built on the Denmark River 10 kilometres upstream from the inlet in 1961, and another on the Quickup in 1988 to ensure the growing population had continuity of supply. Use of the Denmark River as a potable water supply ceased for some years due to increased salinities, however current revegetation programs along watercourses and in the catchment appear to have reversed this trend.

The Sleeman River, 22 kilometres long, drains low hills and swamps to the east of the inlet, and together with other small tributaries brings fresh water to the inlet. The Lake Saide Drain and the small Cuppup River drain coastal swamps to the east, and the Little River flows from hilly country to the west emerging near Poddy Point. The lower reaches of the Denmark and Hay Rivers and the smaller Bow (Irwin Inlet), Kordabup (Parry Inlet) and Sleeman Rivers flow through tertiary sandstone and Pallinup siltstone. This formation also underlies the sediments on the coastal plain where the soils are sandy and acidic. The waters are highly coloured with tannins and carry organic materials.

The Denmark River is estuarine for about 2.5 kilometres and up to 4 metres deep, discharging to the inlet across the northern shallows through which there is a dredged boat channel. The Hay River is estuarine for about 5 kilometres and is 3–4 metres deep in places, entering the inlet over the wide eastern shallows of Morley Beach (Fig. 8.3). In summer when water levels are low, the shallow sand flats bar the mouths of the Hay and Sleeman Rivers. In the estuarine reaches, the rivers with steep banks over 3 metres high cut through sandy alluvial and estuarine deposits revealing the extent of the estuarine area in earlier times. Because the land was low-lying and prone to flooding, gates were installed where the channels cross the Eden Road to prevent water entering Nenamup Inlet and the Lake Saide potato fields.

Box 8.1 SEDIMENTS REVEALING TIMES PAST

The sediments of the marginal shallows are sandy, varying in coarseness and the amount of silt and shells. There are coarse shell deposits south of Pelican Island dominated by the cockles *Katelysia*. The sediments of the deep basin are dark grey, sandy, silt with some fossils to a depth of 50 centimetres, and below this mottled pale to dark muddy sand with some shell fragments. This type of layering is characteristic of transgressive phase deposits where the sea was retreating, leaving behind coarser material with progressively finer material lain on top. The fine materials have the highest nutrient contents, and hence nutrient concentrations are highest in the deeper areas where sediments are finest.

Cores collected from the flood tide delta near Poddy Point in the 1970s also showed changes in marine influence. Those from the bottom showed a high initial influx of marine sediment (with marine skeletons of sponges, bryozoans) deposited during formation of the delta. A layer of coarser fragments of estuarine animals above suggested that the marine influence was then reduced and the estuary barred. An overlying mixed layer indicated that the sediments had been moved or reworked by the changing channels within the delta.

Subsequently pumps were installed upstream from Nenamup Inlet to pump water from Lake Saide, and the Eden Road gates are now generally open.

The sea rush *Juncus kraussii* and the salt tolerant paperbark tree *Melaleuca cuticularis* fringe the inlet, the channels and a short distance up the rivers. *Juncus* forms single species stands or grows mixed with samphire *Sarcocornia quinqueflora* and the herb *Samolus repens* near the mouth of Wilson and Irwin Inlets. In disturbed areas of the salt marshes there are introduced grasses: salt water couch *Sporobolus virginicus*, *Polypogon monspeliensis* and 'buffalo grass' *Stenotaphrum secundatum*. A second community can be found in flatter areas behind the *Juncus* and *Melaleuca*; this is dominated by the sedges *Isolepis nodosa* and *Baumea juncea*, and *S. repens*.

Where river meets sea

Wilson Inlet has an area of 48 square kilometres, similar to Broke Inlet; the lagoon, 14 kilometres long and 4 kilometres wide, lies parallel to the coast in an east-west direction. Assuming an average depth of 1.8 metres, the volume

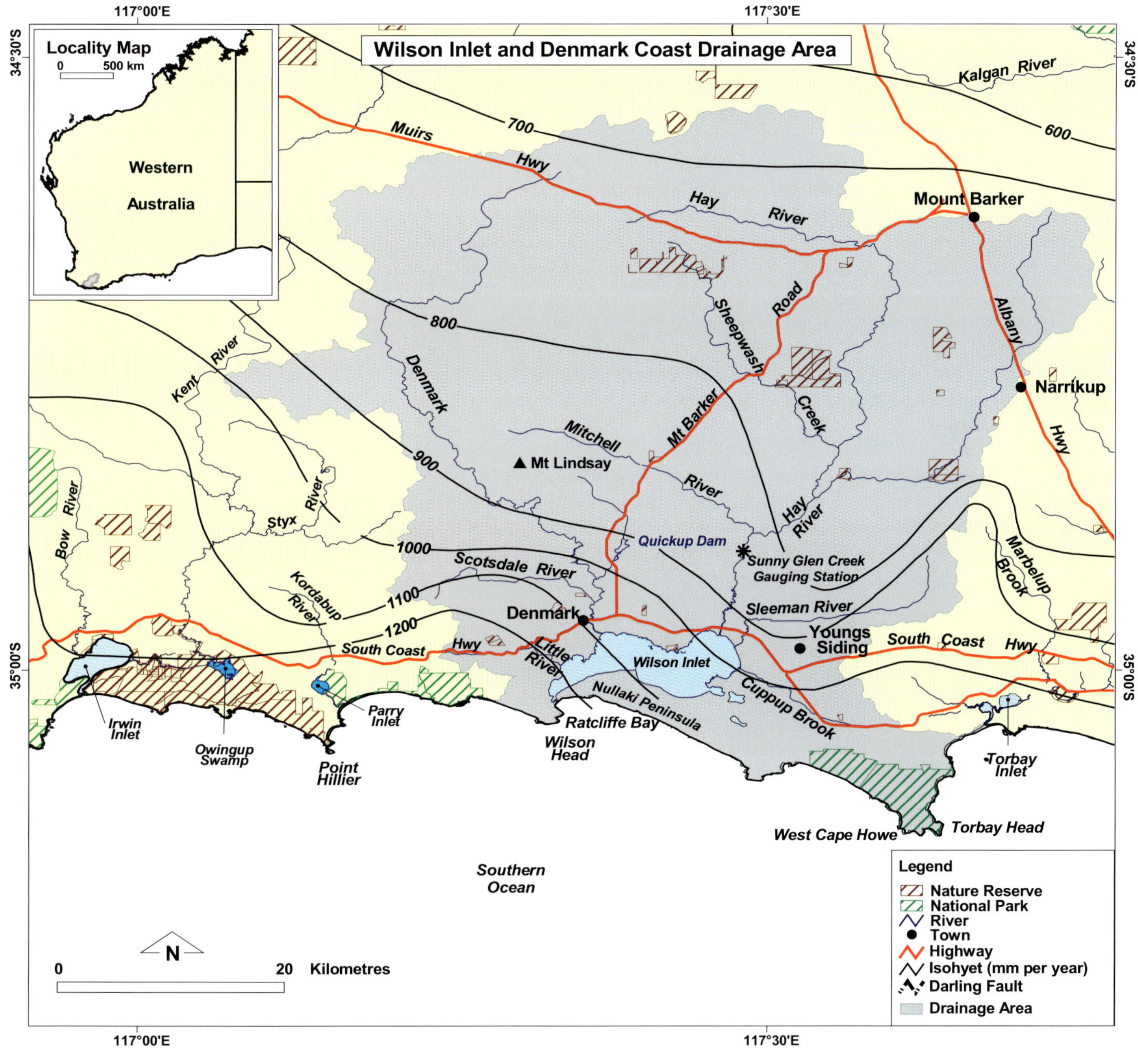

MAP 8.1 *Catchments of Wilson Inlet. The upper reaches of the Denmark River extend into areas of lower rainfall where there has been extensive clearing. Note the Kent River flowing to Irwin Inlet also extends to the north of the Wilson catchment. Near the coast water also enters the inlet and sea from ground water.* Courtesy DoE.

at mean sea level is approximately 90 gigalitres at 0 metre AHD and when 1 metre above AHD about 130 gigalitres. The inlet is partially divided into eastern and western basins by the shallow sill between Pelican Island and the southern shore (Map 8.2). The extent of the estuary varies greatly with the seasons as the extensive shallow areas around the margin in the east and west often dry out in summer. The central basin is more than 2 metres deep with some pools up to 5 metres.

The western and northern shores of the inlet are mainly rocky (granite) with small swamps and sandy bays of quartz river sands. Several small rocky islands lie close to shore, and there are small granite outcrops in the inlet. At the eastern end of the inlet, Morley Beach is bordered by low beach ridges and backed by a swamp and a small beach ridge plain east of the Denmark River mouth (Fig. 8.3). At the western end of the inlet near Jacks Island and in the east near Marbalup Farm, arrangements of stone along the shoreline mark the location of Aboriginal fish traps (Fig. 8.4). (See **Box 9.1 Stone circles—echoes of the past**.)

The channel to the sea is located at the western end of Wilson Inlet. The channel, a kilometre wide at Poddy Point, narrows to 500 metres at the mouth in the shelter of Wilson Head where it is almost completely blocked by a large flood tide delta (sandbar) about 2 kilometres long. The sands of the delta are unstable with different areas filled in or scoured out according to the pattern of flow when the bar is open, usually between August and January. The location of the channel opening has been a subject of great debate, and the bar is usually opened on the western side near the limestone cliff where flow through Prawn Rock Channel is favoured. Openings less than 200 metres from the cliffs are termed a 'western opening', while a breach 300 metres or more away is termed an 'eastern opening'.

The bar is composed of moderately well sorted, medium to coarse sand (quartz 65 per cent and shelly fragments 35 per cent). The bar is about 150 metres wide and slopes gradually from the crest across the flood tide delta. Bar height depends on the strength and duration of the south-easterly winds that build it up to about 1.5 metres above MSL.

FIG. 8.3 *Hay River entering Wilson Inlet at Morley Beach, aerial view, June 2000. Except for a narrow band of salt-tolerant vegetation, most of the low-lying country bordering the river and eastern margin of the inlet has been extensively cleared for agriculture.* S. Neville/Ecotones, Courtesy WRC.

THE WATER WORLD

River flow to Wilson Inlet is highly seasonal (Graph 8.1). Water levels in the inlet now vary about 1.5 metres between the summer–dry condition and the winter–wet condition. Tidal differences (when the bar is open) are small, about 10 centimetres or 10 per cent of oceanic tides due to the narrow

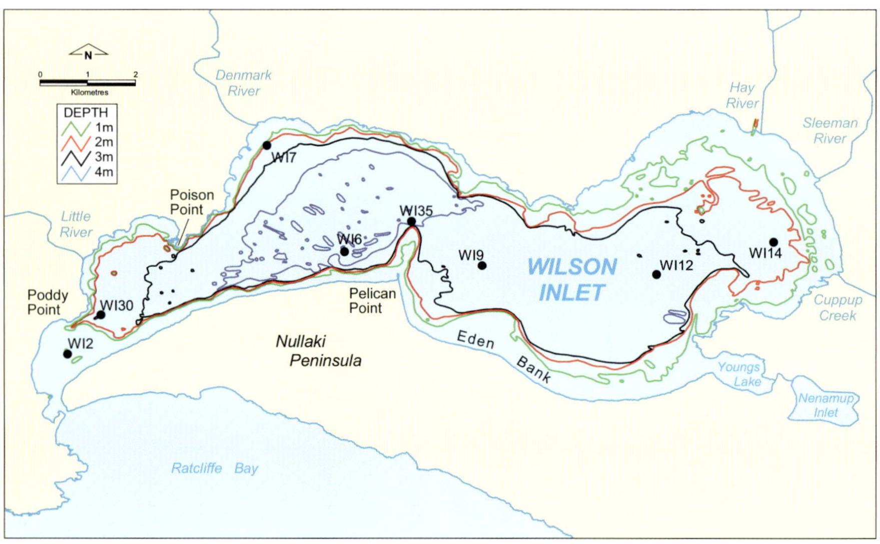

MAP 8.2 *Bathymetric map of Wilson Inlet showing locations of Water and Rivers Commission water quality sampling sites: WI2 shallow delta area; WI30 Poddy Point; WI17 near Denmark River; WI6 and WI35 Western Basin, west and east of Pelican Point; WI19, WI12 and WI14 Eastern Bank.* Courtesy WRC.

entrance channel. The greatest difference in water level affecting exchange with the ocean is due to the barometric pressure (meteorological tides). With the passage of a low-pressure system and the associated rise in sea level, the water level in the inlet may rise 0.3 of a metre or more over a period of days and fall again when the pressure rises. Heavy rain over a short period also causes a rise in water level. Strong winds along the long fetch of the inlet cause the water level to rise and fall up to 0.2 of a metre at the opposite ends of the inlet, setting up a seiching effect with periods of 55, 90 and 250 minutes. Waves generated by wind can reach 1.2 metres in height.

The bar is opened when the water level in the inlet reaches about 1.01 metres above AHD (designated originally as 5.4 feet (1.65 metres) above datum on the old railway bridge), but has varied from 0.71 metres to 1.25 metres AHD. In the mid to late 1920s, the bar was broken to prevent flooding of the Elleker–Nornalup railway line. In earlier times water levels rose to the top of the bar and the inlet waters broke through or it was breached when the inlet waters were lapping over the top. Early records indicate that the Aboriginal people often breached the bar and the early settlers continued

FIG. 8.4 *Stone fish traps, near Marbalup Farm, eastern end of Wilson Inlet, April 1988. Aerial photograph taken at an altitude of 300 m; size is indicated by the crosses with arms 10 m long. These are similar to those in Oyster Harbour (Fig. 9.6) and were probably used in a similar manner.* Courtesy C. Dortch, WAM.

GRAPH 8.1 *Total flow to Wilson Inlet, measured at gauging stations.* Courtesy WRC.

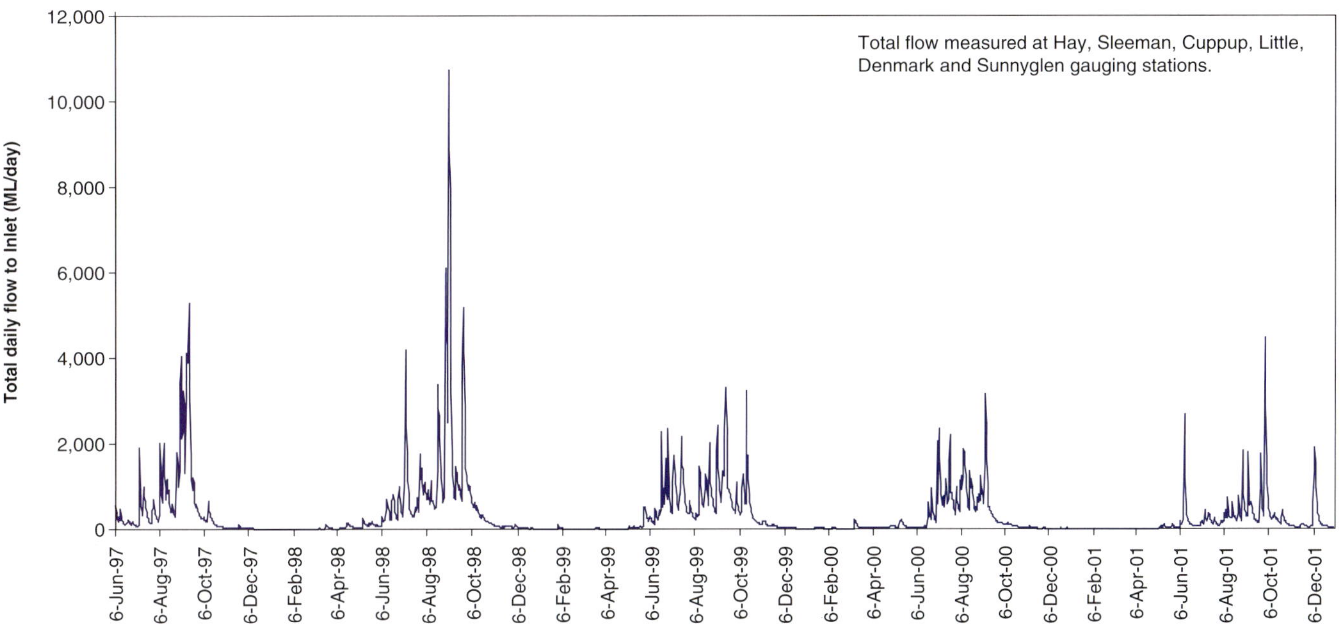

the tradition. Today the bar is breached to prevent flooding of roads and developments in low areas along the shoreline, and ease drainage from the farmlands in the Sleeman and Cuppup River catchments and the Lake Saide potato fields.

The most efficient breaking of the bar occurs when the estuary waters are high and the ocean level is low. For example, in 1975 the inlet water was 1.11 metres above AHD and the ocean level 0.10 metres below AHD, creating a head difference across the bar of 1.21 metres. When the bar broke, the maximum rate of discharge was 464 cubic metres per second, made up of 420 cubic metres per second from the lowering of the water level and an estimated 44 cubic metres per second input of river flow. The water level fell at an average of 0.03 metres per hour, and it took about eight hours before the level fell at the maximum rate. During this time the energy of flow was used in cutting the channel through the bar. Over the next three to four days the rate decreased regularly, as expected with flow through a constant outlet. Thus it took three to four days for the inlet water to drain to the sea and the level to fall to sea level.

Closure of the bar results from a combination of factors that reduce flow in the bar channel; particularly growth of the subtidal estuary sill, and build-up of the wave-built ocean bar. The bar generally starts to close in November. Other factors contributing to its closure are the slowing of river flow, evaporation, and the generally lower sea levels of summer when the dominant long period waves bring sand back onto the beaches after the scouring effect of the short period waves of winter storms. The longshore movement of sand to the east and west also changes from time to time, largely dependent on the prevailing winds. Net longshore sand movement is about 2,000 cubic metres a year to the west with a gross movement of 10,000 cubic metres along the beach; this is however much less than areas such as Dawesville on the west coast. Sand is transported inwards when the estuary water level is low and tidal flow through the channel dominates. Sand also blows in from the beach with the prevailing strong onshore winds. Blown sand also adds to the above-ground sandbar, and the tip of the Nullaki Point appears to be growing westwards as the sand is stabilised by vegetation (Figs 8.5 and 8.6).

It has been claimed that the bar now closes sooner than when the water built up high enough to break it naturally. Natural breaching produces stronger flows scouring a deep wide channel and moving accumulated sand to the sea. In contrast, the absence of strong outward flows under artificial opening creates conditions favouring sand accumulation and bar closure. As there are no records of opening and closing dates before 1955 these claims are difficult to assess. Clearly the damming of the Denmark River in 1959–1960 reduced the flow of water into the inlet, which in turn probably contributed to a reduced scouring effect through the channel. The magnitude of outward flows (flushing) depends on run-off from the catchment and river flow, which in turn depend on the amount and timing of rainfall. Sea level at the time of opening, height of the inlet water and the extent of the sandbar, however, also affect the volume and rate of outflow. Outflows calculated for 1996 and 1998

TABLE 8.1 *Estimated outflow from Wilson Inlet 1995–2001.* Source: Boughton, 1999, 2000*; WRC, 2002. *Note dates of bar openings vary slightly in the three reports. Data in Boughton presented as million cubic metres converted to gigalitres (GL). NA = not available. Outflow is the sum of river flow and rainfall minus evaporation plus the exit of marine water flushed into the inlet. Marine exchange indicates the entry of sea water when sea level is high, which displaces water in the inlet as the ocean level falls again.*

Year	Breached	Shoaled	Open days	Rainfall (mm)	River flow (GL)	Marine exchange = inflow (GL)	Outflow* (GL)
1995	19 August 95	26 February 96	191	950	105	105	190
1996	5 August 96	23 February 97	202	970	185	165	329
1997	18 August 97	19 January 98	154	830	115	60	183
1998	7 August 98	23 May 99	289	1050	185	250	398
1999	30 June 99	6 March 00	250	1000	135	110 (to 25 Oct)	282 (to 25 Oct)
2000	21 July 00	30 November 00	132	1030	90	NA	NA
2001	3 October 01	15 February 02	135	930	NA	NA	NA

FIG. 8.5 *Wilson Inlet with the bar closed, aerial view, January 1988. Prawn Rock Channel is to the west of the sandbar and the flood tide delta is to the left of the photograph.* E. P. Hodgkin.

FIG. 8.6 *Wilson Inlet shortly after a western breaching, 0930 hours, 15 June 1988.* E. P. Hodgkin.

were almost double those recorded in 1995 and 1997 (Table 8.1). In contrast, tides bring marine water into the inlet and influence marine exchange. When the outflow has been low, the bar opening is less effective and less marine water enters the inlet. In recent years flushing has removed about 40 per cent more water than marine exchange. This in turn affects salinity, the nutrients flushed to sea, and the entry and exit of aquatic life.

Changes of the water world

The waters of Wilson Inlet are always brackish with strongly seasonal differences in salinity as well as temperature and oxygenation. (See **Box 8.2 The four states of Wilson Inlet**.) In spring and early summer, salinity ranges from 12 to 20‰ (sea water 35‰). In late summer and autumn before the first rains, salinity rises to around 30‰ when the bar is closed, so there is no ocean inflow, no river flow and high evaporation (Graph 8.2). If the rainfall is late, salinities may stay high for much longer. Although strong winds keep the waters well mixed, there are small differences in salinity that are worth considering as they may affect the plant and animal life as well as the nutrient balance. Higher salinities are particularly obvious near the bar where denser ocean water penetrates below the lighter fresh water, the so-called salt wedge (Graph 8.3a). Lower salinities are most obvious in the surface waters near the river mouths and in the eastern basin; and the rivers may be stratified when the freshwater flow decreases and saltier estuarine water moves up river.

Temperature also varies seasonally ranging from 10°C in winter to 24°C in summer, with the most extreme fluctuations in the shallow areas and rivers. When the water is stratified, there may be a temperature difference between the top and bottom waters.

The waters are usually well oxygenated although there may be some deoxygenation when the waters are stratified, but these events do not persist for long (Graph 8.3b). Between August and February 1995–1997, dissolved oxygen concentrations were below 40 per cent saturation in all areas except for the inlet channel and near the Denmark River mouth. The deoxygenated layer was deeper nearer to the bar (except for the channel) but mostly as a thin layer about 0.5 of a metre off the bottom. The dense salty deoxygenated layers

Box 8.2 THE FOUR STATES OF WILSON INLET

The hydrodynamics of Wilson Inlet have been characterised into four states.

1 The Closed Dry State (typically January/February – May/June): The inlet water column is well mixed with low concentrations of nutrients and oxygenated waters. There is little river flow and the salinity increases over this period due to evaporation.

2 The Closed Wet State (typically June/July – October/November): The start of the winter rainfall changes the state of the Inlet slightly. Inflows from the rivers form freshwater plumes on the surface, but these are destroyed by wind mixing resulting in an inlet-wide reduction in salinity. Freshwater plumes in the eastern basin are only able to form after there is sufficient run-off and move downstream as a buoyant overflow. The run-off is tannin stained with significant concentrations of inorganic oxidised nitrogen (NOx) and total nitrogen (TN).

3 The Open Wet State (typically August – October/November): After the mechanical opening of the sandbar, marine water intrudes into the Inlet within four days as outflow slackens. The intrusion moves into the western basin forming a saline bottom layer up to a metre in depth. Up to two weeks later the salt water penetrates into the eastern basin, where the saline layer reaches a thickness of 20 centimetres. Due to continued freshwater discharge, the Inlet behaves as a salt wedge estuary over this time. Wind mixing and convective cooling are generally insufficient to completely destroy the salinity stratification. Only strong and continuous winds (moderate gales) are able to completely break up the stratification. The well-mixed water column does not persist for long after strong winds as the stratification is re-established by successive tidal inflows bringing new marine water into the Inlet.

The oxygen concentrations in the Inlet show strong dependence on the salinity structure. Deoxygenation occurs below the halocline several days after salinity stratification becomes established. Persistent stratification leads to severe deoxygenation and anoxic conditions.

Sulfate reduction of organic matter in the sediments may also lead to anoxic conditions. When anoxic conditions occur, nutrients such as ammonium are released from the sediments. Elevated concentrations of filterable reactive phosphorus and silica are also often associated with anoxic conditions.

4 The Open Dry State (typically October/November – December/January): With the reduction in stream flow, the channel to the ocean shoals and tidal exchange is restricted. Marine water intrusions do, however, continue. The stratification is less severe than in the Open Wet State, and wind mixing is more effective at mixing the water column. The salinity of the Inlet usually increases over this period. The oxygen concentrations in the water column are influenced by the strength and persistence of the salinity stratification.

Phytoplankton blooms dominated by dinoflagellates and diatoms generally occur in October and November. The nutrient concentrations decline at this time, reaching baseline concentrations that remain steady until the start of the winter rains and the Closed Dry State.

Source: Kalnejais & Robb, 1999.

GRAPH 8.2 *Surface and bottom salinities in Wilson Inlet at site WI6 near Pelican Point in 1996. When the bar is closed the inlet behaves like a lake. The salinity rises with evaporation over summer and is similar from top to bottom due to wind mixing. Salinity declines with freshwater inflow in winter. When the bar opens, saline marine water enters along the bottom and there is a large difference in salinity between the surface and bottom.* WRC, 1999. Courtesy WRC.

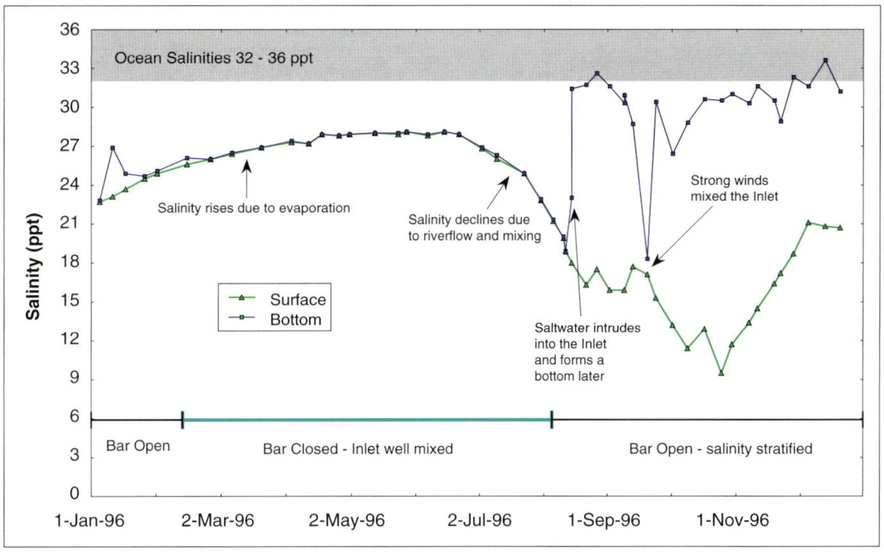

GRAPH 8.3 *Salinity and oxygen profiles in Wilson Inlet, September 1996. (a) Note the salinity differences between the surface and bottom waters; saline water lies along the bottom of the inlet. (b) The area of low oxygen corresponds with the saline bottom water. Supersaturated conditions also occur when phytoplankton concentrations are high.* Modified from WRC, 1999. Courtesy WRC.

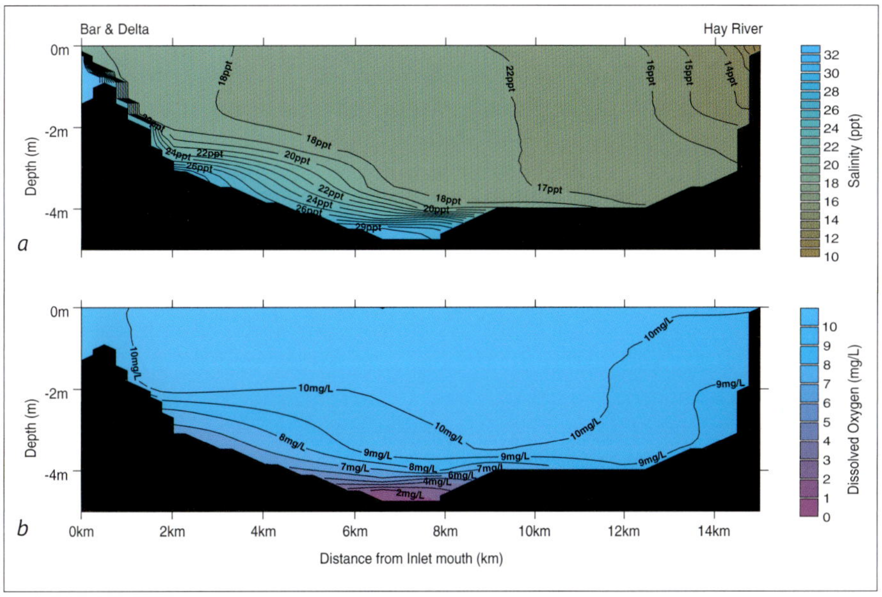

are not only unfavourable for most animal life but also facilitate the release of nutrients in the sediments, and set in train a series of events increasing the nutrients in the estuarine waters. The linking of salinity stratification, deoxygenation and nutrient release illustrates that more frequent occurrence of these conditions would be disastrous for water quality in the inlet.

Nutrients enter the inlet in river and ground water from the land and from the sediments. Nutrient concentrations in the Mitchell, Denmark, Hay, Scotsdale and Little Rivers are generally below the accepted guideline concentrations. The Sleeman River and Cuppup and Sunny Glenn Creeks however, are nutrient enriched due to a combination of factors including fertiliser run-off, the presence of stock and lack of fringing vegetation. Fertilisers, manure, septic tank waste and soil all contribute to the inflows of nutrients, organic matter and sediments when water flow is high.

The nutrient concentrations in the inlet vary greatly over the year depending on river flow and on the conditions within the inlet. Generally total phosphorus and total nitrogen concentrations are below the guideline values used to initiate concern for water quality in slightly disturbed estuaries (Graph 8.4). (See also Graphs 1.2 and 1.3.)

Most of the nitrogen and particulate phosphorus (sediment-bound) in the rivers and drains is organic, derived from plants and animals and not

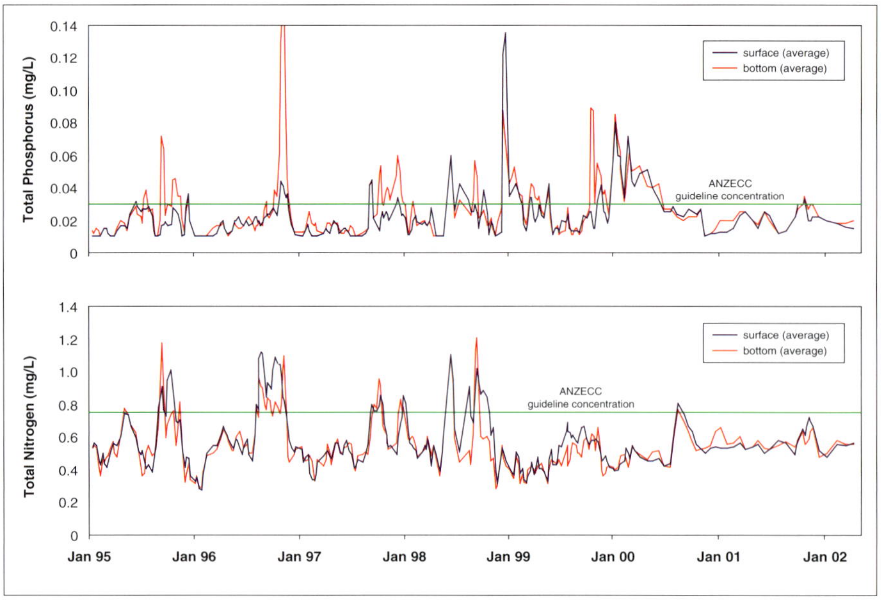

GRAPH 8.4 *Total phosphorus and total nitrogen in Wilson Inlet, 1995–2002. Average concentrations at sites WI6, WI35, WI12, WI14 and WI9. Green line indicates ANZECC guideline value. See Map 8.2 for localities.* See WRC, 1999; 2000; 2000a. Courtesy WRC.

directly available for plant growth. Dissolved and inorganic nitrogen or oxidised nitrogen (NOx, principally nitrate) and phosphorus orthophosphate, the forms used by plants, are highest in the early run-off due to the flushing of fertilisers from agricultural areas, and concentration peaks often reflect the river flow (Graph 8.5). Oxidised nitrogen usually peaks in June or July, and inorganic phosphorus in July or August. Peaks of organic nitrogen and phosphorus usually occur later as they take longer to move from areas higher in the catchment. Oxidised nitrogen, ammonium and orthophosphate are usually low within the inlet, near the limits of detection due to rapid uptake by algae and plants.

Concentrations of ammonium, particularly those in the surface waters, are generally much lower than nitrate concentrations. Ammonium concentrations in the bottom waters are also low for most of the year, except for a few weeks when the bar is open and there is a layer of salty water along the bottom. The dense salty water restricts mixing, and bottom water becomes deoxygenated favouring the release of ammonium and orthophosphate from organic material in the sediment. When the waters are oxygenated most of the ammonia released is removed by bacteria, which produce nitrogen gas through the processes of nitrification and denitrification. Ammonium also enters the inlet with winter run-off and from the Sleeman River and Cuppup Creek in summer.

Nutrient concentrations decline after the winter peaks and remain low for most of the year. However, dissolved forms such as nitrate and ammonium are readily available to phytoplankton, *Ruppia* and bacteria with the main growth period occurring as temperatures rise in spring. Chlorophyll *a* (Graph 8.6) concentrations indicate spring blooms and summer blooms in 1995 to 1998 but these were less intense in 1999 and 2000. There were threefold differences in chlorophyll *a* concentrations and twenty-fold differences in cell numbers between years. Spring blooms generally followed the bar opening and subsequent salinity stratification, deoxygenation and sediment nutrient release (principally ammonium). The smaller summer blooms in some years are possibly due to the release of nutrients with the decay of *Ruppia*. Nutrients not taken up by plants are deposited in the sediments or flushed to sea. Probably about half the nutrients entering the inlet remain within the inlet. They may be buried in the sediment from which they can be released during deoxygenation events or taken up by plants and released when the plants die.

GRAPH 8.5 *Ammonium, oxidised nitrogen (NOx, principally nitrate) and orthophosphate in Wilson Inlet, 1995–2002. Average at sites WI6, WI35, WI12, WI14 and WI9.* Courtesy WRC.

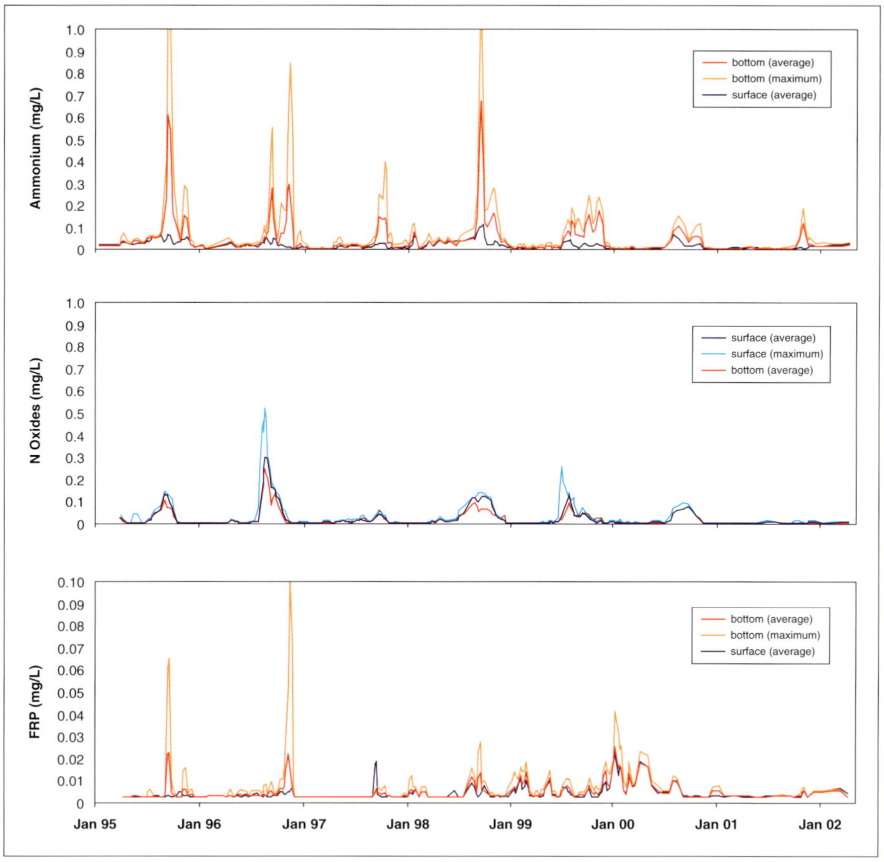

GRAPH 8.6 *Water chlorophyll* a *concentrations in Wilson Inlet, 1995–2002 at sites WI6, WI35, WI12, WI14 and WI9. Phytoplankton increase when nutrient concentrations are high and temperatures are rising in spring—the spring bloom.* Courtesy WRC.

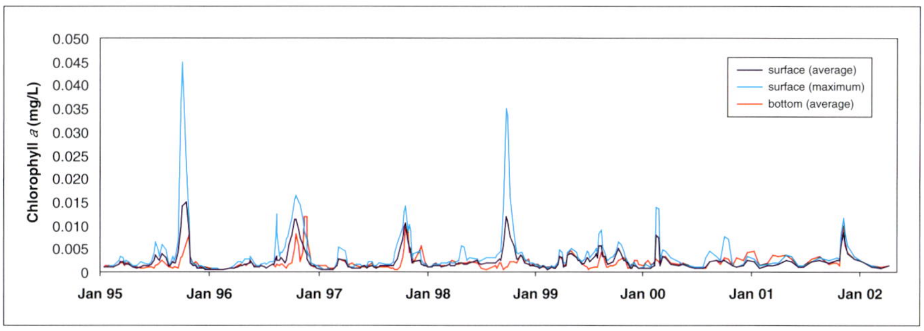

Inhabitants

Water plants

Phytoplankton samples collected over two years in the late 1990s showed the number of cells in the water column varied with the seasons, but did not vary greatly across the inlet. Between January and September the number of cells was low, mainly cryptophytes and a few diatoms, dinoflagellates and chlorophytes. The number of cells increased in October and reached bloom proportions in November with diatoms, usually *Chaetoceros* sp. and dinoflagellates *Prorocentrum minimum*, dominating the blooms. In contrast to *Ruppia*, the biomass of phytoplankton probably varies with the availability of nitrogen rather than phosphorus. Late summer blooms in some years appear to be associated with nutrients released during the senescence and decay of *Ruppia* and its epiphytes as well as macroalgae. Other summer blooms appear to be due to nutrients released by winds stirring the sediments.

Swan grass, *Ruppia megacarpa* is the dominant plant in the estuary. *Ruppia* extends over most of the shallow areas of the inlet flourishing in the brackish water. It does however require fresh water to germinate, so that delays in the onset of rain and persistence of high salinity may affect germination. Estimations of how much *Ruppia* occurs in Wilson Inlet have been very difficult (Fig 8.7 and Graph 8.7). A study in the mid-1990s found the number of shoots in a given area varied greatly across the inlet, and there were different growth forms and canopy heights. The longest plants were generally found near the Hay River (about 1,500 millimetres) and the shortest plants on Poddy Delta (about 300 millimetres). Growth was found to decline with high salinity, increased turbidity (light penetration) and depth, but effects of these occurred at different time scales. At all study sites above-ground biomass varied by an order of magnitude between seasons. At Eden Bank above-ground production varied from 75 to 138 grams dry weight per square metre per year. In some parts of the inlet *Ruppia* appeared to grow as an annual, while in other areas it was perennial: the different growth forms appeared to depend on the stresses and disturbances such as water depth, winds, currents, desiccation and salinity. *Ruppia* growth also appears to depend on the availability of phosphorus rather than nitrogen, unlike phytoplankton growth that is limited by nitrogen. (See **Box 8.3 This thing called *Ruppia*.**)

The stonewort or charophyte (probably *Lamprothamnium*) was found growing at the eastern end of the inlet in 1976, near the mouth of the Hay River and in the shallow bay adjacent to Nenamup Inlet. *Ruppia* is usually covered with a heavy load of algae such as *Cladophora*, *Polysiphonia* and *Audouinella* and other epiphytic species, including microscopic diatoms that exude a mucous slime. This slime also traps fungi, bacteria and particles of organic material. Only a few types of algae such as brown algae *Cystoseira trinodis* and *Ectocarpus*, red alga *Gracilaria verrucosa* and green algae *Chaetomorpha*, *Cladophora* and *Enteromorpha* colonise the rocky littoral areas. (See **Chapter 1 Understanding Swanland: Macroalgae.**)

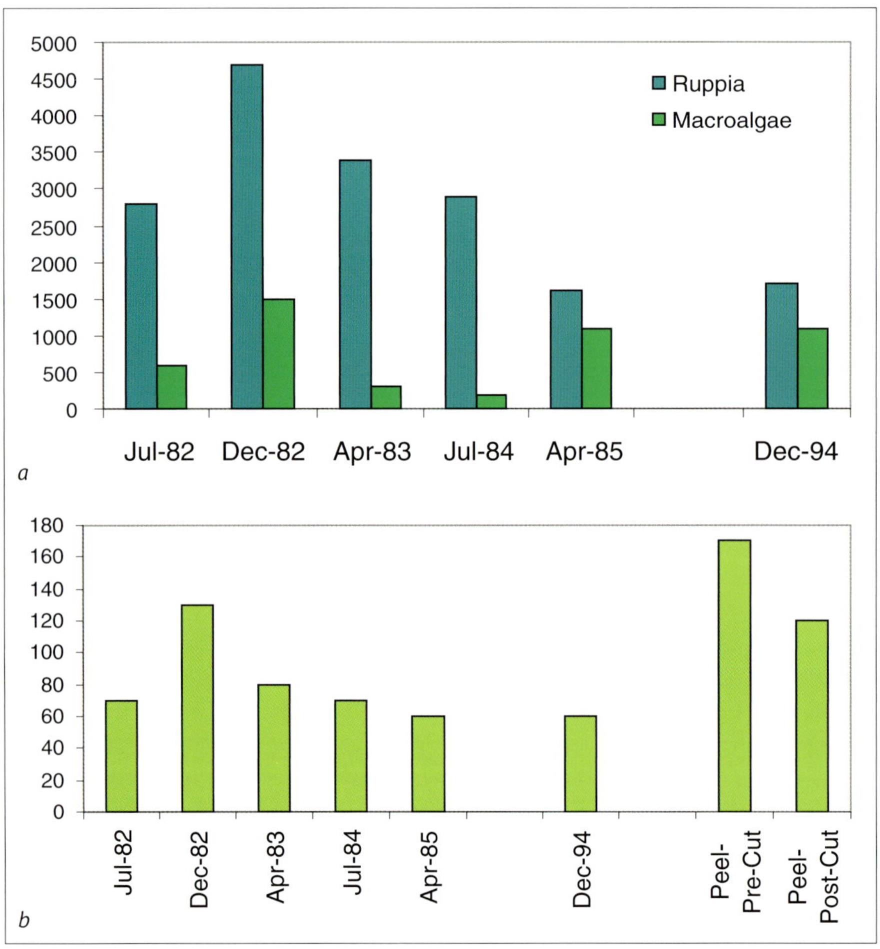

GRAPH 8.7 *Macrophytes* Ruppia *and algae in Wilson Inlet: (a) estimated biomass (dry weight in tonnes); (b) estimated areal load of combined macrophytes (dry weight in tonnes per km²) compared to the Peel–Harvey prior to and post construction of the Dawesville Channel. Note these are probably overestimates due to the patchy distribution of* Ruppia *within the inlet. Previously published estimates of biomass in July 1982 and April 1985 were higher and the values here have been recalculated.* See WRC 2003. Courtesy WRC.

Waterlife

Wilson Inlet with its brackish water (10–30‰) and relatively large volume has a rich fauna when compared with other seasonally closed estuaries of the south coast. It is impoverished, however, in comparison to permanently open Nornalup–Walpole where the connection to the sea allows a greater inflow of larvae, and maintains higher salinities suitable for a greater range of marine species. Fauna within the inlets can be grouped according to the types of habitat created by the water column, sand flats and deeper water where organic material accumulates.

Samples of plankton collected from the water column in the 1970s contained three common estuarine copepods, *Acartia*, *Gladioferens imparipes*, and *Oithona*. In the late 1980s, it was suggested that the high densities probably reflected the relatively high nutrient loading to the system and the limited exchange with the sea. Larval stages of copepods are important for larval fish that feed on plankton.

More recent studies in 1989–1990 and in 2000–2002 recorded about forty species of invertebrate, of which fourteen were species of polychaete worms and eighteen species of molluscs. The most abundant organisms were polychaete worms and the small gastropod snail *Hydrococcus brazieri*. The studies found their distribution related to water depth rather than location within the estuary. The highest diversity and abundance corresponded to the amount of *Ruppia* seagrass, which grew in shallow sandy areas where light penetration was good. This suggested that the habitat of the grass and shell debris created more favourable conditions for a greater number of species. In 1989–1990 six species, the bivalves *Spisula trigonella*, *Tellina deltoidalis*, *Katelysia scalarina*, the dog whelk *Nassarius burchardi* and the worms *Heteromastus filiformis* and *Ceratonereis aequisetes* accounted for 75 per cent of the biomass. Most of the species were present throughout both basins except for a few more marine species, such as two species of bristle worm and the

FIG. 8.7 *Shallow* Ruppia megacarpa *along Beach Road covered with* Cladophora *sp., December 1995.* Courtesy T. Carruthers.

Box 8.3 THIS THING CALLED *RUPPIA*

The seagrass *Ruppia megacarpa*, often called 'Swan grass', grows in estuaries, coastal salt lakes, inland brackish to hypersaline lakes and dams in southern Australia. Other members of the family are found throughout the world in similar situations. Although *Ruppia* can look more like an alga, as with all seagrasses it has true plant features such as roots, underground stems or rhizomes, leaves and flowers (Fig. 8.8). Compared to other seagrasses that grow in more exposed marine conditions along the southern Australian coast and in estuaries such as Oyster Harbour, Princess Royal Harbour and Waychinicup, *Ruppia* is a finer less robust seagrass.

The thin stems can grow to over 2 metres long in a zig-zag branching pattern. The narrow (< 1 millimetre) fine, flat leaves are expanded around the base and sheath the stem. The flowers, made up of four male pollen sacs and two to six female carpels are borne between the leaf bases. With maturity, the flower stem or peduncle elongates and grows to the water surface where the pollen is expelled and fertilisation occurs. The peduncle then coils, withdrawing the fertilised carpels back into the leaf base where the seeds develop. The seeds lie on the bottom until the freshwater flush of winter, when they start to germinate. The seedlings grow rapidly as the waters become warmer, more saline and the light increases with spring. Like all plants *Ruppia* also needs nutrients for growth, and in Wilson Inlet the increased growth of *Ruppia* since the 1920s has been linked to the increased nutrient concentrations.

Like other seagrasses and to some extent algae, *Ruppia* creates habitat for many small invertebrates and fish. *Ruppia* is also eaten by fish such as mullet and bream, swans and other water birds and is an important part of the estuary's ecology.

Source: Carruthers et al., 1997; 1999; WRC, 2003.

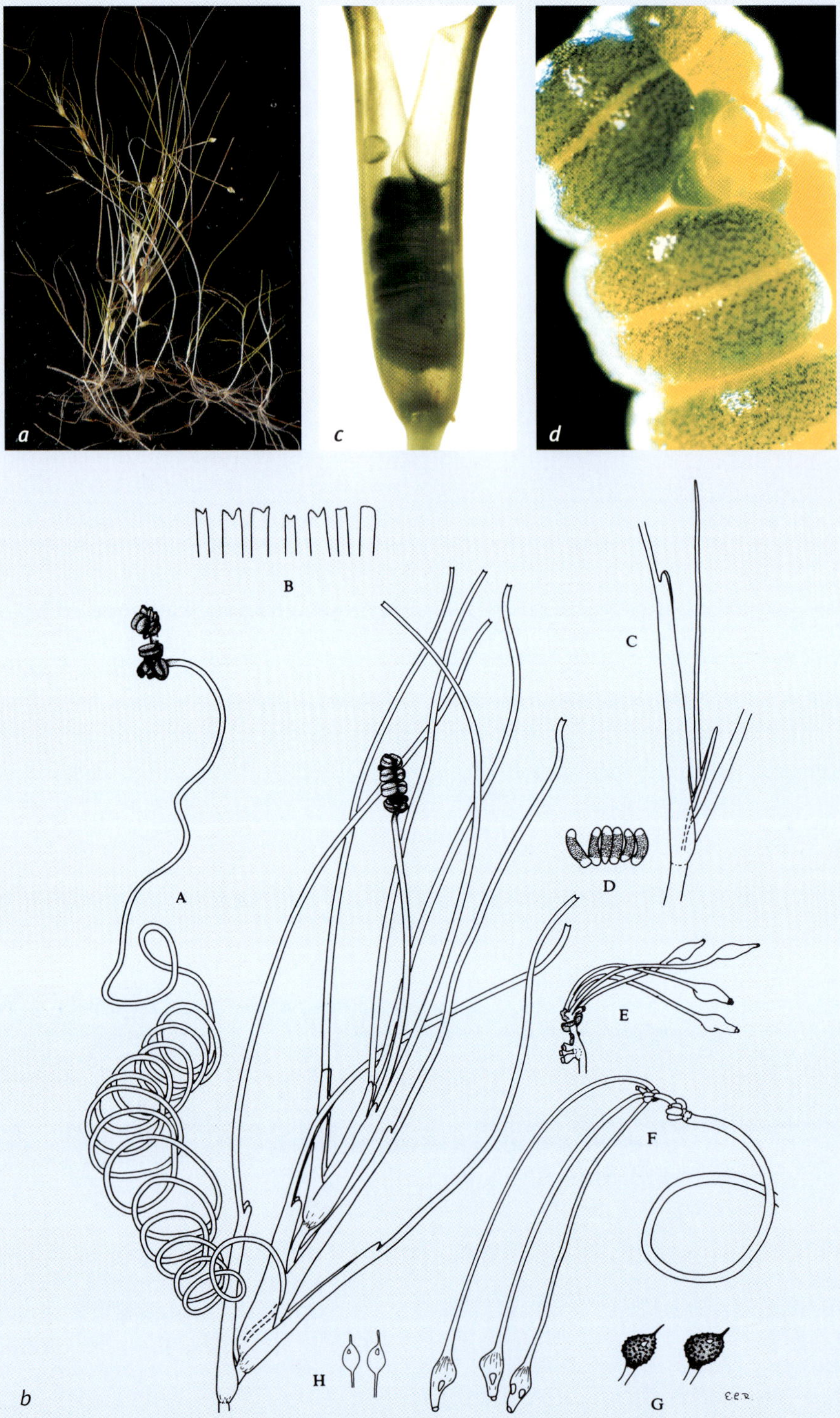

FIG. 8.8 Ruppia megacarpa *plant structure. (a) Whole plant: the flower rises on a stalk or peduncle to the surface where fertilisation occurs. After fertilisation the peduncle recoils bringing the fertilised carpels back to the leaf base. (b) Drawings:* A *plant with straight young and coiled older inflorescence;* B *leaf apices;* C *leaf base and axillary shoot;* D *pollen grain;* E *young fruits;* F *older fruits;* G *endocarp of ripe fruit;* H *fruits.* Adapted and renumbered from Black, 1986 from drawings by E. L. Robertson. *(c) Coloured images of developing inflorescence within a leaf sheath (5 mm). (d) Pollen sacs and carpel (1.5 mm diameter).* Courtesy T. Carruthers.

bivalve molluscs, the pipi *Donax columbella* and the venus shell *Dosinia sculpta* that were only found near the ocean entrance. In 2000–2001 another large bivalve *Soletellina biradiata* was found to be abundant in areas of bare sand in shallow water.

The deeper areas had relatively impoverished faunas with only one abundant worm, the detritivore *Prionospio cirrifera*, presumably favoured by the greater proportion of particulate organic material in the fine sediments.

Rocks around the inlet are sometimes colonised by tubeworms *Ficopomatus*, the mussels *Mytilus* and *Xenostrobus*, and barnacles *Balanus*. The black mussel *Mytilus edulis* can be very abundant, and spat collected in the inlet is transferred to King George Sound for the grow-out phase in aquaculture leases. The fauna in the estuarine reaches has not been studied, but the small mussel-like bivalve *Fluviolanatus* probably attaches to solid structures.

Low salinities often result in high mortality of some marine species such as mussels *Mytilus* and venus shells *Irus*. In the 1990's study, it was suggested that the decline in some worms and small gastropod snails could be also be related to low salinity in spring. The populations then need to re-establish from larvae produced by the few survivors or from the ocean beyond the estuary. Mud oysters *Ostrea angasi* were once also abundant in Wilson and Irwin Inlets. Their low numbers in Wilson Inlet are probably due to over-collecting in the 1880s, but they are thought to have died out in Irwin Inlet in the 1970s after floods and periods of low salinity.

The most recent study in 2000–2002 highlighted the importance of *Ruppi*a not only as a habitat for animals but also as a food source for communities within the inlet. The small gastropod *Hydrococcus* was shown to be a proficient grazer of epiphytic algae growing on *Ruppia*, while the shrimp *Palaemonetes australis* utilised decaying *Ruppia* along with the associated microbial community (detritus) and algae living on the *Ruppia*. In turn, small invertebrates are important sources of food for larger animals: crabs, king prawns and fish such as gobies, mullet, black bream, flathead and cobbler. *Palaemonetes* is hunted by many species of fish including the blue-spotted flat head *Platycephalus speculator* that also eat smaller fish: calculations based on summer and spring populations of flathead indicate that 1,000 kilograms of shrimp and 450 kilograms of fish are eaten each day.

Fish and fisheries

Wilson Inlet has the largest estuarine fishery on the south coast, both recreational and professional. The success of the fishery is probably attributable to its present character as a seasonally closed estuary, where there is an abundance of food and a limited range of salinity (10–30‰) favouring a few estuarine species and excluding most predatory species. The abundant seagrass *Ruppia* creates habitat for micro-organisms and larger invertebrates provide rich feeding grounds for many fish. Seagrass also provides shelter, particularly for juvenile fishes from predatory birds and fish. The permanently open, more marine Nornalup Inlet may have a more diverse fish fauna, with marine 'straggler' species such as gummy shark, trevally and salmon, but some common species in Wilson Inlet, such as King George whiting, appear to be less abundant.

Marine fish species enter from the sea when bars are open, so recruitment is greatly influenced by the time the channel is open and the depth of the channel. Populations of fish vary from year to year particularly in the lower basin compared to the upper and middle regions, which have slightly different species. Recruitment also depends on larvae and juveniles available in the nearby coastal waters. Anchovy *Engraulis australis*, black bream *Acanthopagrus butcheri*, blue-spotted flathead *Platycephalus speculator*, sea garfish *Hyporhampus melanochir* and estuary catfish or cobbler *Cnidoglanis macrocephalus* (Fig. 8.9) can complete their life cycle within estuaries if conditions are not too extreme, as do many gobies, pipefish and hardyheads.

Black bream is popular with amateur anglers in riverine areas. Black bream were reported as common in estuarine reaches of the Hay River in the 1980s, but appeared to have declined. Lampreys also pass through the estuaries on their way to rivers, where they spawn and where larvae develop until they are ready to return to the sea. (See **Box 7.1 Swimming through**.) Rainbow trout are present in the Denmark River above and below the dam and sometimes enter the estuarine reaches of the river. Recreational fishers target King George whiting *Sillaginodes punctata*, pink snapper *Chrysophrys auratus*, Australian herring *Arripes georgianus* and a variety of marine species that are most abundant near the bar. Some, such as the snapper entering as juveniles, are trapped when the bar closes—they grow well in the productive protected inlet resulting in an unusual bounty for people fishing.

King prawns *Penaeus latisulcatus* are sometimes fished commercially, the number of prawns varying greatly, probably because of erratic recruitment to the estuaries. Blue swimmer crabs *Portunus pelagicus* are also present in summer, and in recent years mud crabs *Scylla serrata* have also been caught. Mud crabs have become common in many west coast estuaries such as the Greenough, Moore, Swan, Leschenault and Hardy. The two-spot sand crab

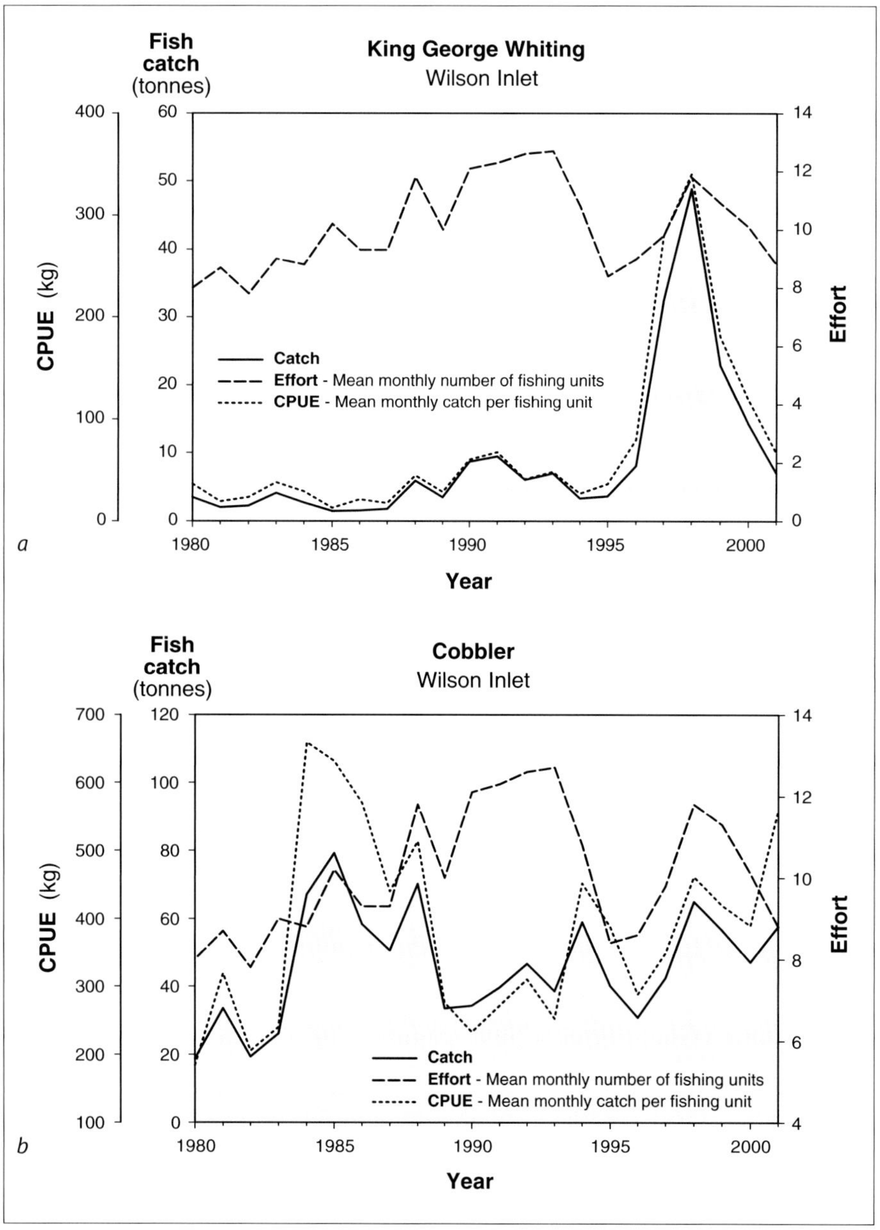

GRAPH 8.8 *The annual catch, effort and catch per unit effort (CPUE) for (a) King George whiting and (b) cobbler in Wilson Inlet 1980–2001. King George whiting was a major part of the catch in Wilson and Irwin Inlets. In 2001 the catch of King George whiting in Wilson Inlet continued to decline from the high level of 1998, but was still above the early 1990s catch. The very high catch in 1998 resulted from high juvenile recruitment into Wilson Inlet several years earlier, a trend evidenced across all south coast estuaries. In 2001, cobbler accounted for 77 t of the fish caught in estuaries on the south coast, 75% of which were caught in Wilson Inlet. The 2001 catch was 10 t higher than the 2000 catch.* Source: Ayvazian & Nowara, 2002a. Courtesy DoF.

Ovalipes australiensis is sometimes so abundant in Wilson Inlet that it fouls fishing nets.

Thirty-three commercial species have been recorded from the inlet but the bulk of the catch consists of only a few species: cobbler, King George whiting, yellow-eye mullet *Aldrichetta forsteri*, sea mullet *Mugil cephalus*, Australian herring, blue-spotted flathead, long-snouted flounder *Ammotretis rostratus*, pink snapper and sea garfish. (See **Box 8.4 Cobbler: spines to fear—great to eat** and **Box 8.5 Blue-spotted flathead**.)

Wilson Inlet is part of the restricted entry south coast estuarine fishery: new licences only being issued when existing licences expire with approval from the Minister of Fisheries. The fishery of six to ten persons using five to eight boats mainly fish for cobbler. Only one or two professional fishers commonly use Irwin Inlet, and Parry Inlet was closed to commercial fishers in 2002.

Box 8.4 COBBLER: SPINES TO FEAR—GREAT TO EAT

The cobbler or estuary catfish *Cnidoglanis macrocephalus* is an eel-tailed catfish of the family Plotosidae (Fig. 8.9). It hides in banks of seaweed in shallow water, feared by waders and prawners who dread a sting from the large spines on the head. The cobbler is also a favoured eating fish caught by commercial and recreational fishers. Cobbler may reach a size of 910 millimetres and 2.5 kilograms in weight. In an effort to restore declining cobbler populations in Western Australia, regulations in place since 1994 restrict the recreational catch to fish greater than 430 millimetres in length and a daily bag limit. With continued concern for cobbler populations the previous bag limits were reduced in 2002 and the status of the stocks is reviewed annually.

Cobblers live in nearshore and estuarine waters from the Abrolhos to Kingston in South Australia and from Jervis Bay to Kirra in New South Wales. Recent studies in Western Australia have shown that they breed either in the sea or in estuaries, and that the populations are discrete. The populations in different estuaries are quite distinct and they probably do not interbreed. The divisions between the populations have arisen because of the breeding habits.

Adults become sexually mature when about 42.5 centimetres. In Wilson Inlet most do not reach sexual maturity until the fourth year, and spawning occurs between October and January when the temperature is 18°C and the salinity is 20‰. In the Swan River, water temperature is usually 22°C and the salinity 22‰, and cobbler appear to grow faster and breed earlier than those in Wilson Inlet on the south coast. Some of the cobbler in the Swan may move into the ocean to spawn where the salinity is 35‰.

Cobblers lay large eggs in burrows excavated in fine sand under large rocks, where the male shelters them under his pelvic fins until they hatch. At hatching, the larvae are 2.9 centimetres in length growing quickly over 17–18 days to 4.3 centimetres, by which time the egg sac is resorbed.

Cobblers are generally larger and more highly developed at hatching than other members of the catfish family, and take a shorter time to transform to the juvenile stage.

The advanced stage of development at hatching and the fast growth probably limit dispersal, which in turn contributes to the genetic isolation of populations, a feature that makes cobbler particularly vulnerable to overfishing. These factors and the parental guarding of eggs and newly hatched larvae probably help survival within the fluctuating conditions that characterise estuaries.

Source: Ayvazian et al., 1994; Cliff & Lenanton, 1983; Hutchins & Thompson, 1983; Laurenson et al., 1993.

FIG. 8.9 *Cobbler* Cnidoglanis macrocephalus *vary from a greyish-black colour to a mottled pattern or pale and dark blotches. The mouth is surrounded by tentacles that are used for detecting buried prey. Venomous spines are present at the front of the pectoral fins and first dorsal fin.* Courtesy B. Hutchins, WAM.

Fishing in Wilson Inlet is mainly by gill nets. To protect juvenile fish, stocks like cobbler and set aside an area for recreational line fishing, netting is not permitted near the bar, in the rivers or near their mouths. Catches are sold locally or freighted to Perth mainly for human consumption.

Annual catches of commercial species have varied greatly from year to year (Graph 8.8). Numbers given for catches need careful interpretation as they do not necessarily represent abundance, because catch can also reflect fishing strategies and market demand. (See also Graph 1.4).

Catches of cobbler and whiting are exceptions, and generally reflect abundances because these fish are always in demand. Environmental factors affecting abundance are complex and affect species in different ways. The time and duration of bar opening must affect recruitment of species that spawn at sea. Low winter salinities can cause mortality of fish and their prey, as does the summer drying of the shallows. The abundance of seagrass appears to favour cobbler especially, and probably other species although less directly. The great increase in cobbler catches in the 1970s and 1980s may be due to the reported increase in seagrass. As cobbler has only been considered a good table fish in recent times however, the numbers caught may simply reflect a greater demand. Mud oysters appear to have been scarce ever since overfishing in the 1880s.

Birds

Seventy-nine species of birds have been recorded from Wilson, Irwin and Parry Inlets and in Owingup Swamp, but bird numbers are debatable as censuses protocols have been inconsistent. The most common species at Wilson Inlet are the Black Swan (Fig. 8.10), Australian Shelduck, Pacific Black Duck and Grey Teal. They are seasonal visitors, most abundant in winter and seldom seen in Irwin and Parry Inlets in summer. Banded Stilt and Red-necked Avocet were abundant in Wilson Inlet in the mid-1980s. The Fairy Tern is a summer visitor to Wilson Inlet and regularly attempts to breed at Ocean Beach, but often fails because of disturbance from vehicles, people and dogs when the bar is closed. Small numbers of Fairy Tern feed at Morley Beach. In the 1980s there were also very large numbers of Little Black Cormorants (1,000), Little Pied Cormorants (1,000), Pied Cormorants (50) and Black Cormorants (50) in Wilson Inlet. In a study comparing the

Box 8.5 **BLUE-SPOTTED FLATHEAD**

The blue-spotted flathead *Platycephalus speculator* breeds in Wilson Inlet where it can grow to over 3 kilograms and 90 centimetres in length. Like the cobbler they also have venomous spines, but these are on the dorsal fin. Flathead can reach 12 years of age, a much longer life span than most other species that spawn in Swanland estuaries. The species occurs along the south coast of Australia from Tasmania, Victoria, South Australia and on the west coast as far north as Kalbarri.

Sexual maturity occurs in males at the end of the first year but is delayed a further year in females. Breeding occurs predominantly during summer and early autumn (December to March). Spawning takes place when daylight, temperature and salinity are rising, although there is some indication that spawning may be delayed when the rise in salinity is delayed. Flathead spawn at sea but in Wilson Inlet they spawn at salinities less than 30‰, that is below full strength sea water when the estuary mouth is barred. They can spawn more than once during the breeding season. Flathead eggs and larvae are pelagic (floating), in contrast to benthic eggs produced by most other species that spawn in estuaries.

Source: Hyndes et al., 1992.

diet of cormorants to that of the blue-spotted flathead fish, calculations suggest that they collectively consume 10 kilograms of *Palaemonetes* shrimp and 150 kilograms of fish per day. The smaller cormorants feed on shrimp and similar fish species to the flathead, and the large cormorants on some of the larger fish including cobbler. The presence of such large numbers of birds and the similarity of their diet to an abundant predatory fish serves as an indication of the richness of the estuary habitat. (See also **Box 2.19 Black cormorant feeding frenzy**.)

Better or worse

Given the concern over the health of the estuary and the need to develop management strategies to control or reduce the input of nutrients, comparisons need to be made between past and present water quality.

Nutrients within the inlet waters originate externally from the catchment and internally from the sediment. Not all nutrients are flushed from the inlet indicating that the inlet is retaining some of the nutrients. The low concentrations of nutrients during the Open Dry and Closed Dry states probably indicate that nutrients are taken up by the *Ruppia* seagrass and its epiphytes, benthic algae and phytoplankton. Nutrients stored in these may be released again when they die and decompose.

While there is little doubt that there has been a proliferation of macrophytes, *Ruppia* and algae since the 1950s, changes in water quality are not obvious and blooms of nuisance phytoplankton have been infrequent. As most of the nutrients are taken up by *Ruppia* and algae, we could say that they are buffering the system.

Recycling of nutrients from the sediments is clearly linked to deoxygenation of the sediments (anoxic conditions) that occurs when a layer of salty marine water enters through the open bar. This recycling accounts for about fifteen times the amount of nitrogen and four times the amount

FIG. 8.10 *Black Swans grazing on* Ruppia megacarpa *growing in shallow water between Poison and Poddy Points, Wilson Inlet, April 1996.* Courtesy T. Carruthers.

of phosphorus as the inputs from the catchment, and is in forms suitable for plant growth. In contrast, when the water column is well oxygenated about twenty-five times as much nitrogen and a significant amount of phosphorus, more than the amount transported out to sea through the bar, are converted to forms unsuitable for plant growth.

If stratification and anoxic events increase in frequency or duration, and in other parts of the inlet or estuarine reaches of the rivers, it is likely that the sediment-stored nutrients will be released in greater quantities. Persistence of nutrients in the water could then lead to a change in the species composition from *Ruppia* seagrass to macroalgae, and more intense and frequent algal blooms that persist for a longer time. The inlet is thus in a delicate balance: to some people the occurrence of anoxic events suggests the inlet is already mildly eutrophic, and further changes to the biotic community will lead to severe eutrophication.

Chapter 9 Albany Harbours and the smaller estuaries of the Albany region

The Albany region borders the Southern Ocean for 280 kilometres from Wilson Inlet in the west to the Pallinup River and Beaufort Inlet in the east. Around the town of Albany the large embayments of King George Sound, Princess Royal Harbour and Oyster Harbour create a safe haven in this spectacular but rugged coastline. Further east the bays are more open. As rainfall decreases and becomes more sporadic, the rivers are smaller and the estuaries isolated from the sea by a sandbar that breaks only briefly and infrequently.

In 1791 Vancouver had mapped the area and named the three large harbours: King George Sound in honour of the reigning British monarch George III, Princess Royal Harbour commemorating the birthday of the king's eldest daughter, and Oyster Harbour for its abundant oysters; a recognition from the beginning of the area's rich biology (Map 9.1).

Maps of the area were also made by Freycinet on the *Casuarina* and Baudin on the *Géographe* during the Baudin Expedition in 1801–1803 (Fig. 9.1). Specimens collected by one of the naturalists François Peron on that expedition, particularly from King George Sound (*au port du Roi Georges*), and another French expedition led by M. J. Dumont d'Urville on the *Astrolabe* in 1826–1829, were taken back to France and later described by leading scientists of the time. This interest in the area by the French was the main impetus for settlement by the British.

In a prelude to British settlement, Australia's greatest navigator Commander Matthew Flinders RN visited the area in 1801 on the *Investigator*. Using Vancouver's map he chose King Georges Sound 'as the proper place in which to prepare ourselves for examination of the south coast of Terra Australis'. Flinders spent a month in the area around Albany and Esperance

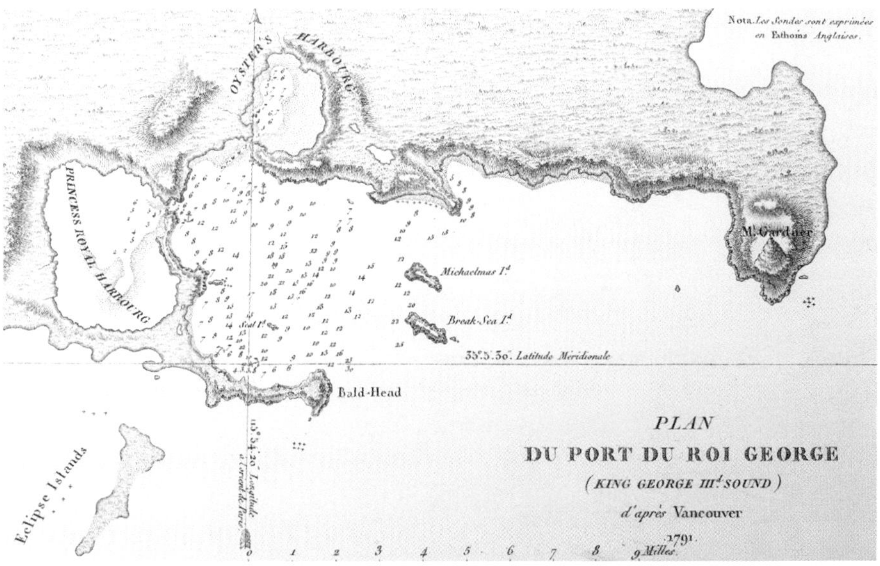

MAP 9.1 *Vancouver's map of King George Sound and Oyster Harbour (d'après Vancouver 1791). This is an insert in the corner of a large map made following the Baudin Expedition, titled Plan du Port Roi Georges (King George III Sound), (A la Terre de Nyuts, Nouvelle Hollande par M. M. L. Freycinet, Faure & Ransonnet (Février 1803)). Baudin carried a copy of Vancouver's map and acknowledged that the fine details were of great importance to the success of his own expedition.* See Baudin, 1974; Marchant, 1982. Courtesy T. Burrows, Scholars' Centre, UWA.

FIG. 9.1 *A view of a pond near King George Sound (Vue d'un Etang près de la Baie du Roi Georges) by Louis de Sainson, 1826 Baudin Expedition. The lake in the foreground is undoubtably Lake Seppings with Middleton Beach on the right. Oyster Harbour lies in the background where Green Island and the high ground of Bayonet Head can be seen. The dune lines between Lake Seppings and Middleton Beach mark the gradual isolation of the lake from the sea.* Baudin, 1974; Marchant, 1982. Courtesy Hon. David and Mrs Marie Louise Wordsworth.

before proceeding eastwards, chancing upon the Baudin expedition in the area now known as Encounter Bay in South Australia. Flinders mapped the coast in great detail and concluded that the land he saw was not suitable for settlement and that Terra Australis was one landmass (Map 9.2). Collections made during the expedition of the terrestrial plants and algae formed the basis for the first extensive work on the flora by botanists Robert Brown, Ferdinand Bauer and Peter Good.

Captain Phillip King RN also explored the area in 1818, noting in particular the plentiful stocks of fish and the fish traps on the rivers used by the Aboriginal people. Like his predecessors King planted a garden, but this had long since died when he revisited the area three years later (Fig. 9.2).

The Albany region was the first area on the western margin of the continent to be settled by Europeans. Arriving in the brig *Amity* in late 1826, Major Edmund Lockyer proclaimed the settlement of Frederickstown (later to be renamed Albany) three years before settlement at Perth and Fremantle on the Swan River.

In the early days of settlement there was great enthusiasm for the area because of the wonderful natural harbours, the bounteous supply of fish and shellfish, and the relatively rich hinterland. Aboriginal people also enjoyed the bounty of these sheltered waters, as evidenced today by the fish traps around Oyster Harbour and up the Kalgan River (See **Box 9.1 Stone circles—echoes of the past**.) But change to the harbours and hinterland is also a story of misuse (Fig. 9.3).

In the 1980s the rich seagrass habitats of Princess Royal and Oyster Harbours were dying, smothered below dense seaweeds that grew quickly because of high nutrient loads and sediment washing in from the catchment, probably in flood events. The story is sorrier as we seem to learn slowly from our mistakes. Also in the Albany region, the lands surrounding Torbay, once the source of vegetables for the goldfields, became acidic in the early 1900s when the soils were disturbed by drains and the peat swamps burned. A parallel exists today at Beenup Mine near the Hardy Inlet where mining has ceased because of problems with acid sulfate soils. In addition, nutrient rich waters currently drain to Torbay where they cause toxic algal blooms.

Describing this area, its natural setting and ancient history will highlight how a once rich resource was almost lost. The smaller estuaries to the east

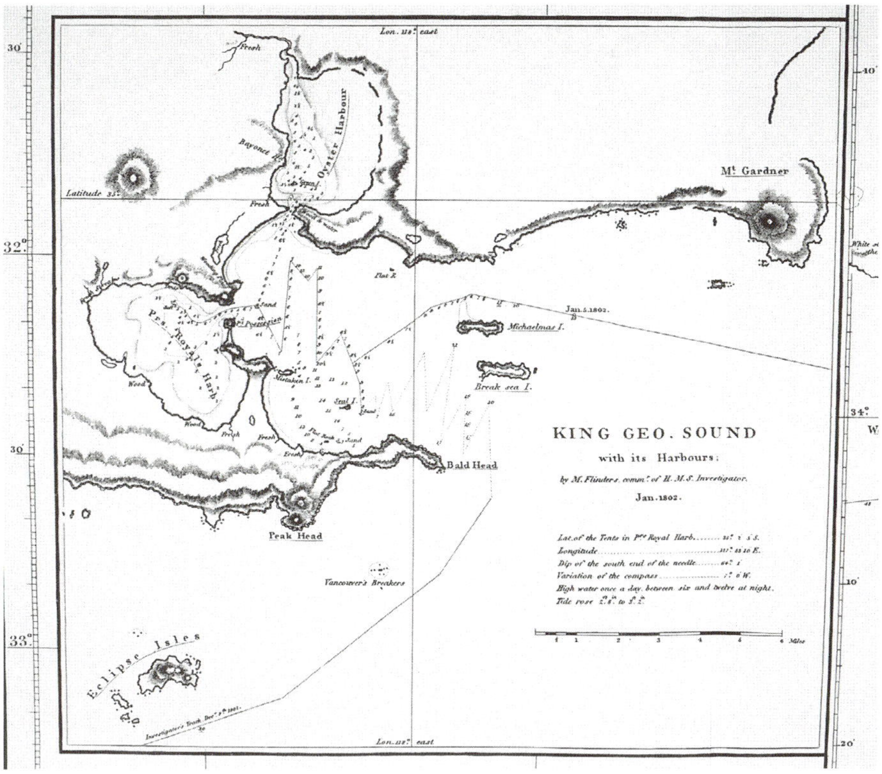

MAP 9.2 *Matthew Flinders' chart of King George Sound drawn during his exploration on the* Investigator *January 1801 – May 1802. Note this section 'King Geo. Sound with its Harbour' is one of three, the others labelled 'Part of New Holland discovered by Peter Nuyts 1627' and 'Archipelago of the Recherche by M. Flinders from* Chart of Terra Australis' *by M. Flinders Commr. His Majesty's Sloop* Investigator *South Coast, Sheet 1 1801.2.3. Published as the Act directed by G. & W. Nicol, 29 Jan. 1814.* Courtesy T. Burrows, Scholars' Centre, UWA. C704/A69922.

FIG. 9.2 *Entrance to Oyster Harbour: a pen and ink and watercolour sketch and painting dated 31 December 1821, with reference to P. P. King. An etching and aquatint of this work entitled 'Entrance of Oyster Harbour, King George III Sound. Interview with the natives' was published as the frontispiece to King's journal. The incident depicted took place on King's fourth voyage on the* Bathurst *during his visit to the area between 23 December 1821 and 6 January 1822. The native in the foreground is spearing fish: probably attracted to 'burley' made of crushed shellfish. The view shows Green Island in the middle distance. The inscription 'Mermaid' on the drawing sometimes quoted is incorrect. Although King visited the area in the* Mermaid *in 1818, he did not see any natives at that time.* Courtesy Art Gallery of Western Australia, AGWA, 1921/00W2.

could so easily go the same way. Some are surrounded by reserves, others by farmland originally cleared for sheep grazing and cropping, now wooded again but with a sea of blue gums.

FORMATION OF THE ESTUARIES

The estuaries are young, only about 6,000 years old, but they overlie sediments and rocks that are much older. To the north the ancient Yilgarn Plateau covers the inland heart of the continent. About 100 kilometres inland the younger hills of the Stirling and Porongurup Ranges, around 1,000 million years old, remain after cycles of deposition and erosion have sculptured the broad gently sloping land (Map 9.3).

In the early Eocene, 40 million years ago, the sea level rose and covered the coast as far inland as the Stirling Ranges. The sediments deposited during this time form a fine-grained rock, the Pallinup Siltstone (spongolite) that

FIG. 9.3 *Oyster Harbour, aerial view, June 2000. On the left is a creek flowing from Lake Seppings and in the middle the King and Kalgan Rivers enter the harbour across wide sandflats. Note the deep channel through to King George Sound with Mt Martin to the east and the stone breakwaters built to reduce erosion of the low dunes at Emu Point on the western side of the channel. The white sands of Middleton Beach lie in the foreground. The Porongorup and Stirling Ranges are visible in the background surrounded by extensively cleared hinterland.* S. Neville/Ecotones, Courtesy DoE.

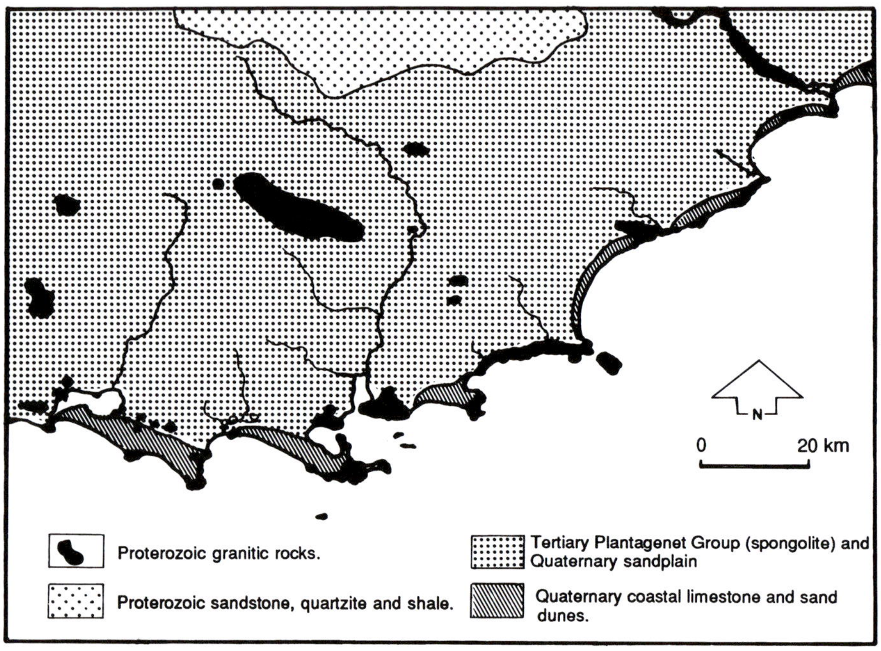

MAP 9.3 *Orientation of the coast and main geological formations of the Albany area. The Proterozoic granites (shown in black), high points above the gently sloping land, form the Porongurup Range and coastal headlands. From the left Wilson Head and Inlet, Knapp Head, West Cape Howe/Torbay Head, Flinders Peninsula with Bald Head protecting Princess Royal Harbour, King George Sound and Oyster Harbour, Mount Martin, Mount Gardner and Cape Vancouver, followed by Two Peoples Bay, Mermaid Point with Bald Island, Green Range, Cape Riche and Groper Bluff with the Pallinup River and Beaufort Inlet.* Hodgkin & Clark, 1990, Courtesy DoE.

lies in a gently sloping plain about 200 metres above sea level. (See **Box 10.1 Sponge garden to rock formation—the Pallinup Siltstone or spongolite**.) Later in the Eocene, about the time Australia and Antarctica separated, the southern edge of the ancient plateau rose along the Jarrahwood Axis, from which the area known as the Ravensthorpe Ramp slopes southwards to the coast and out to the edge of the continental shelf. The uplift resulted in rivers carving deep valleys through the spongolite, which are particularly obvious east of Albany.

At the end of the Pleistocene (in the last two million years), when the sea level was similar to the present, dune sands formed along the coast between the granitic hills. Where there were large amounts of shell material, the dune sand cemented into limestone rocks, the Tamala Limestone. During the last ice age when the sea level fell again to about the edge of the continental shelf, rivers such as the King and Kalgan were tributaries of an ancient river that formed a deep valley through Oyster Harbour out to the 60-metre contour line seaward of King George Sound. At the end of the last ice age (20,000 years ago) the sea again covered much of this coastline, reaching its present level about 6,000 years ago when the valleys were flooded forming the estuaries and coastal lagoons of today.

At first the estuaries would have been more open to the sea. The fossil shells lying beneath areas such as Lake Seppings and Yakamia Creek behind Middleton Beach, the Grasmere Valley between Torbay Inlet and Princess Royal Harbour, and Gull Rock Lake and the lakes behind Two Peoples Bay remind us that these areas were parts of larger more marine estuaries. Later the dune building started again, sand accumulated in flood tide deltas and in bars, filling and narrowing the mouths of the estuaries in the same way that the dunes of Middleton Beach and Emu Point Peninsula closed off Oyster Harbour.

In some estuaries the bars are now only open for short periods so that exchange with the sea is limited and salinity is quite different to earlier times, and some previously estuarine areas such as Lake Gardner are now completely cut off from the sea. This progression of increasing isolation from the sea becomes more evident in the estuaries further to the east.

The estuaries have also become shallower, filled with sediment brought by the rivers and by coastal processes, a pattern of infilling that continues today. Dune sands are gradually encroaching on the water bodies. Plants such as samphires that can grow in harsh saline conditions around the edges trap sediments and add organic material, further decreasing the size and depth of the open water body.

Facing the southern coast

The character of the coast is determined by the different rock types and the orientation of the coastline to the prevailing south-westerly winds, the ocean swell and the summer south-easterlies. The coastline of hard rock (granitic) headlands alternates with limestone cliffs and sandy beaches where the estuaries open to the sea (Maps 9.4–9.10).

Wilson Inlet to Mount Gardner

The long line of high dunes on the Nullaki Peninsula shelters Wilson Inlet and the small lakes draining to its eastern shore (see also Maps 8.1 and 9.11). The granite cliffs of Knapp Head, Torbay Head/West Cape Howe, and Bald Head on the Flinders Peninsula are each at the eastern end of limestone cliffs over 100 metres high. Torbay Head just west of Cape Howe is the southernmost point of Western Australia.

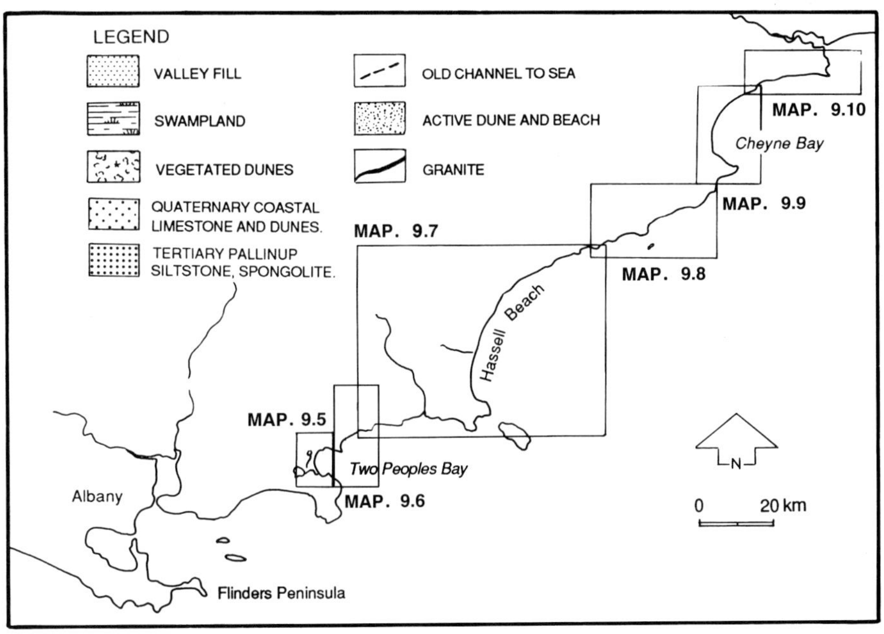

MAP 9.4 *Location of and legend for Maps 9.5–9.10.* Modified from Hodgkin & Clark, 1990.

The Flinders Peninsula (Torndirrup National Park) and Bald Head protect King George Sound from the south-west and south. The large embayment (30 square kilometres) of Princess Royal Harbour in the shelter of Vancouver Peninsula has no major rivers, and only receives fresh water from rainfall, run-off, and ground water seepage from the surrounding land and the Grassmere Valley. Low swampy areas along the western shore mark an older estuarine system through the Grasmere Valley and Torbay Inlet to Wilson Inlet. The entrance to Oyster Harbour (20 square kilometres) off King George Sound is the most protected mouth of any estuary in Swanland.

From Oyster Harbour the rugged granite shores of Mount Martin and Mount Taylor alternate with sandy beaches. Gull Rock Lake, behind the dunes of Ledge Bay, drains to a swamp at the western end of the bay where it is blocked by dune sand. Taylor Inlet lies to the east of Mount Taylor at the western end of Nanarup Beach. At the eastern end of Nanarup Beach, west of Mount Gardner, limestone onshore reefs and cliffs border the shoreline and a large blow-out of mobile sand cascades into Moates Lake.

Two Peoples Bay

Two Peoples Bay faces due east between the granite headlands of Mount Gardner and Boulder Hill (Map 9.5 and Fig. 9.4). The shoreline of Two Peoples Bay appears to have advanced from the surrounding hills, trapping swamps and Gardner and Angove Lakes behind the low dunes. Two rivers flow towards the lakes and low-lying swamps. Goodga River, with a 45 square kilometre catchment, flows to Moates Lake from which there is a small creek through dunes and swamps to Gardner Lake. Moates Lake is deep but Gardner Lake is shallow and contains shell beds. The creek from Gardner Lake turns behind the foredunes to the south end of the bay where it flows across the beach in most years, or the beach is cut to drain water from the swamps. Sea water and stranded seagrass debris wash back into the creek. At the north end of the bay Angove River, with a catchment area of 29 square kilometres, flows to Angove Lake and swamp. The swamp also drains via Gardner Lake Creek to the south end of the bay, but once it may have drained to the northern swampy area near the beach.

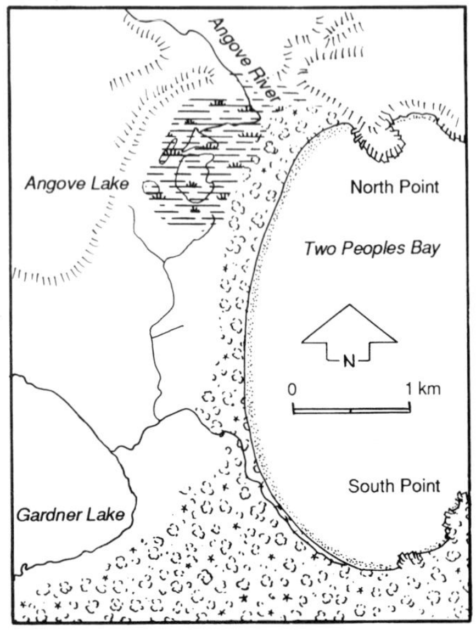

MAP 9.5 *Lakes of the Two Peoples Bay area have been isolated as the shoreline has advanced seawards. Gardner Lake is in Two Peoples Bay Nature Reserve.*
Hodgkin & Clark, 1990, Courtesy DoE.

Normans Inlet and Mount Manypeaks Range Area

At the north end of Two Peoples Bay, Boulder Hill towers above Bettys Beach and Normans Beach where King Creek and Normans Inlet flow into the south-eastern facing bay (Map 9.6). King Creek drains a 17 square kilometre catchment flowing in a narrow valley in the granite before turning southwards behind the foredune at the western end of the bay. The creek flows almost continuously over the beach. Normans Inlet lies in a narrow valley of the Mount Manypeaks Range. The granite cliffs of the range face south, broken only by the rocky gorge of the Waychinicup River estuary.

The 20-kilometre long curve of Hassell Beach faces south-east then east at its southern end where it is sheltered by the Mount Manypeaks Range and Bald Island (Map 9.7). A number of creeks flow from the spongolite hinterland, however the belt of blow-outs, stable dunes and limestone between the spongolite and the beach block normal flows, and there are small lakes and swamps in the narrow valleys.

At the south end of Hassell Beach a small lake drains through the dunes in a 250-metre long creek behind the foredune. Further along, Little Bluff Creek and Bluff River have narrow pools where they wind to sand-filled gullies in the stable dunes and flow across the narrow beach. Coal Creek flows through the vegetated dunes across a narrow blow-out. Wongerup Creek finishes at a dune, but can overflow when in flood. McKennas Lake at the end of a small creek blocked by limestone and dunes contains fresh water and does not appear to flow to the sea.

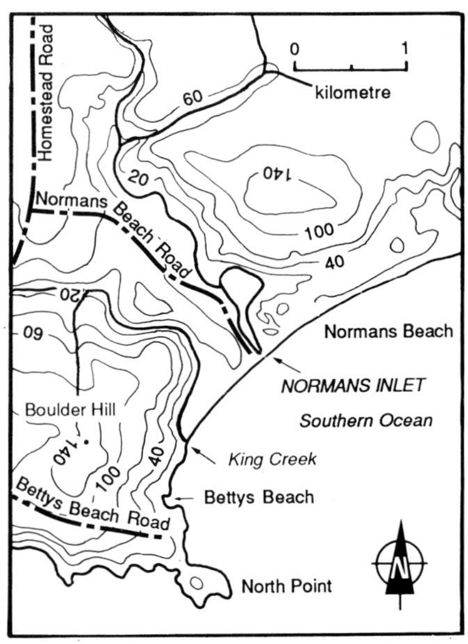

MAP 9.6 *Normans Inlet and King Creek flow into a south-eastern facing bay.*
Hodgkin & Clark, 1990, Courtesy DoE.

FIG. 9.4 *Lake Gardner, Two Peoples Bay, Albany, aerial view, June 2000.* S. Neville/Ecotones, Courtesy WRC.

At the north end of Hassell Beach two lakes have formed where creeks from Green Range end against stable dunes and a large blow-out. The larger, Mirrambeen Lake (on Mullocullop Creek), is 1.5 kilometres from the beach and about 8 metres above sea level. Swan Lake is 4 metres higher. The lakes have no connection to the sea but after exceptionally heavy rain, water flows through dense dune vegetation, a small swamp and across the blow-out to the sea. Flow from the swamp is normally absorbed by the mobile dune. Well Creek is a narrow trough on a small creek from Green Range; it lies between the dune and a spongolite hill and may fill to a depth of 3 metres. The catchment area of these lakes is 150 square kilometres. The lake water is slightly saline (5‰) from creek water flowing into them.

Green Range to Cape Riche

Alternating headlands of gneiss and limestone line the coast to the east and Wilyunup Creek catchment, with a relatively large (112 square kilometres) sandplain, forms a large pool behind the dunes (Map 9.8). Near Giants Causeway, Cowla Creek flows from limestone cliffs to a bay in a half-kilometre break in the cliffs where there is a sandy beach and blow-out. Between Giants Causeway and Cape Riche, the south-east facing gneiss cliffs are interrupted by two small bays to which creeks flow in narrow valleys.

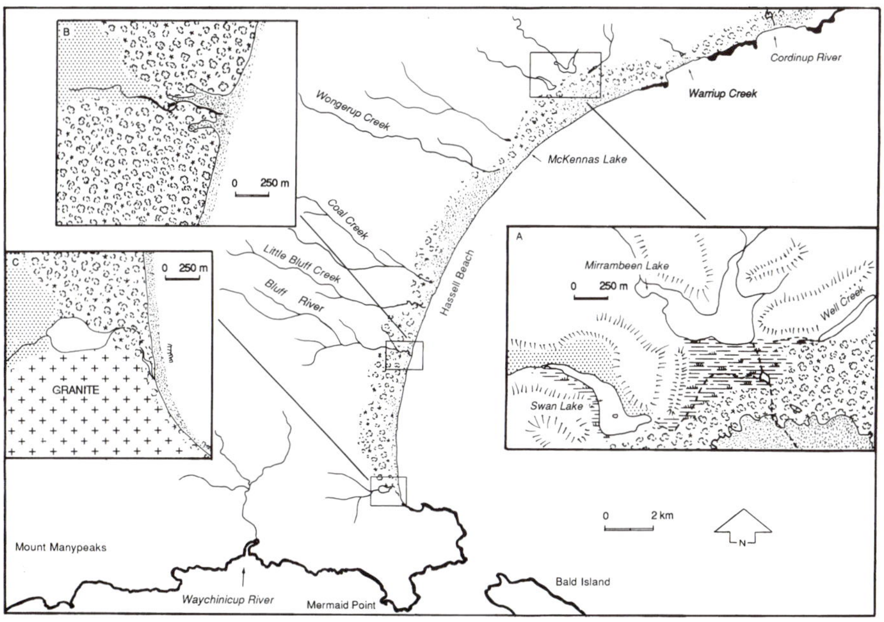

MAP 9.7 *Small rivers and estuaries from Mount Manypeaks Range to Hassell Beach.* Hodgkin & Clark, 1990, Courtesy DoE.

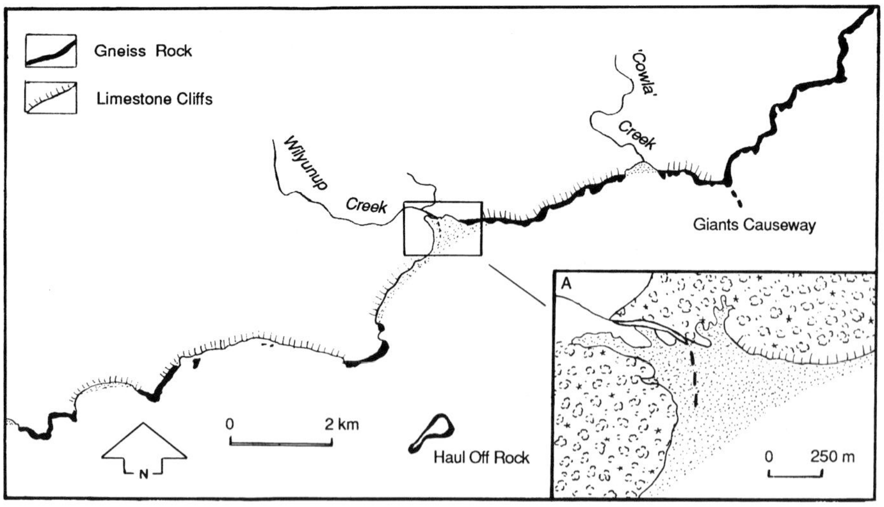

MAP 9.8 *Estuaries and creeks between Green Range and Cape Riche.* Hodgkin & Clark, 1990, Courtesy DoE.

The Cordinup River (Map 9.7) flows in a valley on the eastern side of Green Range entering the sea in a small bay between Hassell Beach and Cheyne Bay. Here granite cliffs and headlands alternate with shallow bays backed by steep dune slopes with occasional onshore limestone reefs. West of Cordinup, Warriup Creek flows from Green Range to a V-shaped blow-out between the cliff and the small Oven Lake, which has been trapped behind stable dunes over which the creek occasionally floods.

Cheyne Bay—Cape Riche, Boat Harbour to Groper Bluff

The coast of Cheyne Bay from Cape Riche to Groper Bluff runs east and south-east. Ten creeks drain to the beach from the spongolite hills. The Eyre River at the south of the bay forms the estuary of Cheyne Inlet (Map 9.9). In the northern part of the bay, stable dunes and limestone front the beach but there are spongolite cliffs along much of the coast. In the middle of the bay Swan Gully with a narrow estuarine reach flows through sandstone rock and stable dunes to the beach. Several other small creeks hold temporary pools where they have cut narrow gullies through the spongolite rock to the beach and cliffs (Fig. 9.5).

A small creek enters the embayment of Boat Harbour, sheltered to the west by Black Head. Between Boat Harbour and Groper Head a number of creeks with very small catchments carve through the coastal dune and limestone. One slightly larger (27 square kilometres) catchment drains from the spongolite hills to a small tree-fringed lake that curves behind the dune to the sea. At the eastern end of the Albany region, Mount Groper and Groper Bluff shelter the Beaufort Inlet from the south-westerly winds and swell (Map 9.10).

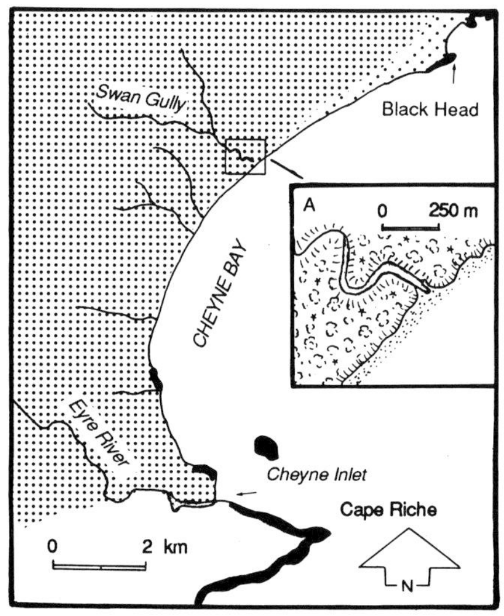

MAP 9.9 *Creeks flowing to Cheyne Bay between Cape Riche and Boat Harbour.* Hodgkin & Clark, 1990, Courtesy DoE.

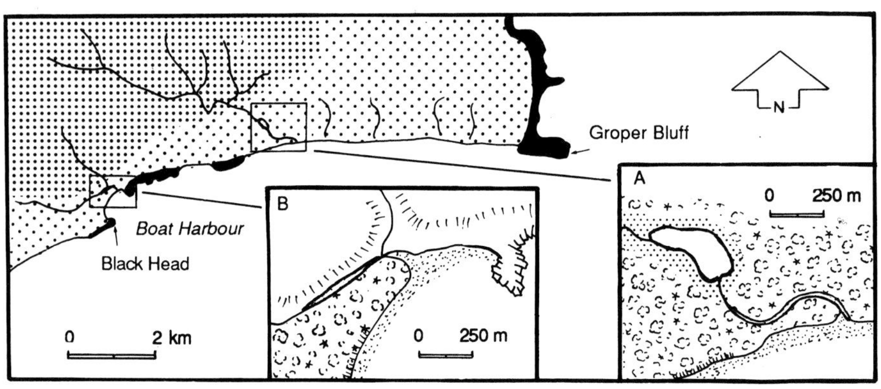

MAP 9.10 *Boat Harbour to Groper Bluff.* Hodgkin & Clark, 1990, Courtesy DoE.

FIG. 9.5 *(a) Swan Gully, aerial view, and (b) a cliff 2 km south of Swan Gully, Cheyne Bay, September 1990.* E. P. Hodgkin Collection.

Becoming drier

Rainfall in the Albany region ranges from the wetter rainfall area in the west to the more arid area to the east (Map 9.11). The mean annual rainfall of 880–1,000 millimetres near the coast west of Albany decreases inland to 500 millimetres near Cranbrook and Tenterden, and eastwards along the coast to about 570 millimetres at Cape Riche (Map 10.1). Annual extremes may be much higher or lower than this and, although most rain falls in winter, infrequent summer cyclonic storms can also bring very high falls that can cause extensive flooding. Run-off from the catchments varies greatly depending not only on the size of the catchment but also the location. The Kalgan River drains the largest area (Appendix 1) but its discharge is relatively small, reflecting the large part of the catchment that lies in the low rainfall area away from the coast. In comparison, Torbay Inlet receives a higher run-off from the smaller Marbellup Brook catchment.

The salinity of river water increases as rainfall decreases both inland and eastwards along the coast. The Kalgan River is saline with an average concentration of 3,100 milligrams per litre but less than 500 milligrams per litre during extremely high flows, and is highly coloured with tannins. (See **Box 5.9 Tea-coloured water**.) The waters of the King River ranged from 1,000 to 2,000 milligrams per litre when sampled in 1988. Marbellup Brook is fresh, averaging 360 milligrams per litre, and is highly coloured. Waychinicup River is moderately coloured and averages 720 milligrams per litre.

ALBANY HARBOURS

From the eagle's-eye view from Mount Melville and Mount Clarence above the city, the wide calm waters of Oyster Harbour, Princess Royal Harbour and King George Sound lie behind this wildest of coasts with a backdrop of patchwork farmlands stretching to the distant blue hills of Mount Barker, the Porongurups and the Mount Manypeaks Range. Oyster Harbour is just through a short narrow channel in the sand dunes from the Southern Ocean, collecting water imprinted with all the past and present stories of the hinterland.

Oyster Harbour

> In our way out of this harbour, the boats grounded on a bank we had not before perceived; this was covered with oysters of a most delicious flavour, on which we sumptuously regaled; and, loading in about half an hour, the boats for our friends on board, we commemorated the discovery by calling it Oyster Harbour…
>
> Commander George Vancouver, 29 August 1791

At beach level near Emu Point, long-dead shells crunch underfoot on the wide flats of muddy sand; the pheasant and spindle shells the same as those thrown up by the surf on the gleaming white sands of Middleton Beach, along with oysters and cockles from the protected rich waters of the Harbour. The variety of shells shows the combined influences of the wild sea coast and the calmer estuary waters.

At the north of the Harbour between the two rivers, wading birds forage on the wide sand shallows among the shells and circles of brown necklace seaweed. Areas of swampy ground alternate with gum trees growing almost to the waterline, where at the end of a tidy street small bikes propped against the bus shelter wait for the end of a school day. With the children's homecoming, the area becomes a meeting place and playground as it has done for generations, for a few steps away along the water's edge circlets of stones lie as memorials to early times when Aboriginal people gathered here particularly over summer and autumn to trap fish and feast on the Estuary's rich bounty (Fig. 9.6 also Fig. 8.4). (See **Box 9.1 Stone circles—echoes of the past**.)

Oyster Harbour's place in the landscape, rivers and catchment

Oyster Harbour receives water from the Kalgan and King Rivers (Map 9.11). The Kalgan River rises near Cranbrook 130 kilometres from the coast where its tributary, the Young River, drains the southern side of the Stirling Range. In the upper reaches only 500 millimetres of rain falls in a year, but in the lower reaches the river flows through sandplain country where the average rainfall is almost double. The catchment of the King River and its tributary Mill Brook is smaller, but it drains the higher rainfall area south of Mount Baker.

The Kalgan River is tidal for 9 kilometres to Upper Kalgan Bridge. Downstream from the bridge, a rock bar separates the upper and lower estuary. Upstream the river is only 20–30 metres wide but widens downstream to 100 metres about 3 kilometres from the mouth. Further downstream near the Lower Kalgan Bridge, it widens still further to 250 metres before passing through swampland and sandy shallows to the mouth. At the Estuary, the river has deposited a large delta through which it flows in a well-defined channel for about 2 kilometres into the main estuary basin.

In the upper reaches the Kalgan River is 2–3 metres deep with some holes on the river bends up to 4 metres, but the main channel gradually shallows in the lower reaches to a depth of only a metre. There are three small islands in the river: Elbow Island lies in the river at 4 kilometres, smaller White Island near the Lower Kalgan Bridge, and a small island in the rock bar. The rock bar was largely removed by blasting in the 1890s and 1920s to allow passage of boats carrying fruit and wool from farms upstream.

The King River is estuarine for 7 kilometres to the Upper King Bridge where it is joined by Mill Brook. Upstream the river is narrow, winding through 1–2-metre high sandy banks. The rock bar near the Upper King Bridge is exposed across the width of the river at low tide, and granite rocks outcrop downstream for about 1.5 kilometres. Downstream the river flows through low-lying pastoral land fringed with sedges and paperbarks. Two kilometres upstream of the Lower King Bridge the river winds through salt marsh and sand flats before narrowing and deepening again near the bridge. The river is gradually shallowing with sediments washed in from cleared land, and the lower reaches of the King River are not navigable at low tide.

Fresh water also enters the Harbour from Johnston Creek, which flows to the northern shore of the estuary, and Yakamia Creek that enters the Harbour between Emu Point and Bayonet Head.

Box 9.1 STONE CIRCLES—ECHOES OF THE PAST

While recognising traditional oral history, there are few written records as to how the Aboriginal people used the south-west estuaries. The material evidence we have is circles or alignments of stones in the estuaries, and some middens containing discarded shells. Early explorers and settlers provided written accounts: the first, Vancouver in 1791 commented on the stone fish traps or tidal weirs found along the shorelines of Oyster Harbour and the Kalgan River. It would appear that fish swam over the stones at high tide and were trapped behind them when the tide fell. Fish may also have been herded into traps or 'burleyed' up with 'cockles' (Fig. 9.2). Other traps were constructed of brushwood (Fig. 3.7). If the traps depended on tidal changes, similar structures at Wilson (Fig. 8.4) and Broke Inlets were probably constructed when the sea level was higher and the estuaries were larger and permanently open to the sea. Their later use was probably affected by the lower sea levels in the Holocene, and more frequent closure of the entrance bars. Many traps are now difficult to see as they are covered with sediments brought to the estuaries following erosion in the catchments.

Aboriginal use of aquatic life in the estuaries appears to have been centred on fish stocks which characteristically fluctuate in abundance, declining with high salinity and being replenished from marine populations when the bars open. Fishing activities would have followed changes in abundance and movements of the different types of fish.

Some estuarine molluscs were eaten but apparently in smaller quantities compared to other places in Australia. For example, the Sydney cockle (also known as the blood clam *Anadara trapezia*) found in Oyster Harbour is common in middens on the eastern seaboard, but there are no similar middens here on the south-west estuaries. Shells of the ridged Venus shell *Katelysia scalarina* in the Katelysia Rock Shelter near the bar at Wilson Inlet provide the only direct evidence of estuarine molluscs eaten by groups camping at the mouth of an estuary. Fish and other wildlife were abundant enough to sustain the relatively small population, but it is not clear if estuarine shellfish also formed a major part of the diet. It is of interest that the large abundant oysters from which Oyster Harbour received its name were not observed directly as being eaten by Aboriginal people before Europeans arrived, although early records indicate that they did start to cook and eat the shellfish fairly soon after settlement. Perhaps this is an indication of the rapidity of social changes after European settlement, or of the decline in other food sources.

FIG. 9.6 *Oyster Harbour fish traps, 1981. The age of the fish traps is not known and they have not been dated by radiocarbon or other methods, but they cannot be older than the Estuary's formation about 7,000 years BP and were still in use in December 1831. The stone works have been disturbed in recent times, but are currently being restored and interpretive signs prepared by custodians of the area.* Photographs Moya Smith, Anthropology Department, WAM. Copyright WAM.

During the 19th century Aboriginal people had a number of food alternatives, the people moving from place to place within defined areas dictated by the availability of food and the seasons. People moved inland during winter, and other groups from inland moved to the coast during summer and autumn when fishing was an especially important activity. Evidence of their presence, scattered shells and worked stones, lie along the shores of many estuaries but many more are probably buried under the vegetation or moving dunes. At the time of European settlement, fishing activities were often the time for interaction with other groups and ceremonies; a time when the sheltered shores of Oyster Harbour with abundant schools of fish, other wildlife and plants would have echoed with the sounds of large groups of people feasting and celebrating.

Source: Dix & Meagher, 1976; Dortch, 1997a, 1997b; Dortch et al., 1984. Meagher, 1974. See also Hallam, 1984, 1989; O'Connor et al., 1995.

Oyster Harbour is a large lagoon, 15.6 square kilometres in area, roughly rectangular in shape and extending 5.5 kilometres in a north-south direction. Extensive sand and shell beds border much of the Harbour. Along the eastern margin the shallow sandy flats shelve to deeper (5 metres) water along their western edge. The western shore of the harbour mouth is bordered by the unstable sand dune of Emu Point and by the wide swampy area of Lake Seppings (Fig. 9.7) and Yakamia Creek. Further north, a steep laterite-topped ridge extends from Bayonet Head to the King River. The northern shore is also steep except for the area around Johnston Creek. The north-western shore is bordered by beach ridges around Swan Point and Mount Mason. In the south-east the shore rises steeply to Mount Martin. Green Island rises from the deeper southern part of the Harbour.

The entrance from the sea between the granite rocks of Mount Martin and the low sand dune at Emu Point is less than 200 metres wide and 12.5 metres deep. Today the Emu Point side of the channel is walled and a number of groynes have been constructed to reduce the damage of storm waves. The steep-sided dunes along the seaward side of Middleton Beach

MAP 9.11 *Catchment and drainage to the Albany Harbours and Torbay Inlet. Taylor Inlet opens to Nanarup Beach to the east of King George Sound. The Kalgan River flows from an area of lower rainfall near the Stirling Ranges, the smaller King River catchment drains an area of higher rainfall. The estuaries further east have much smaller coastal catchments. East of Mount Manypeaks the orientation of the coast changes and the bays are more open and exposed.* Courtesy DoE.

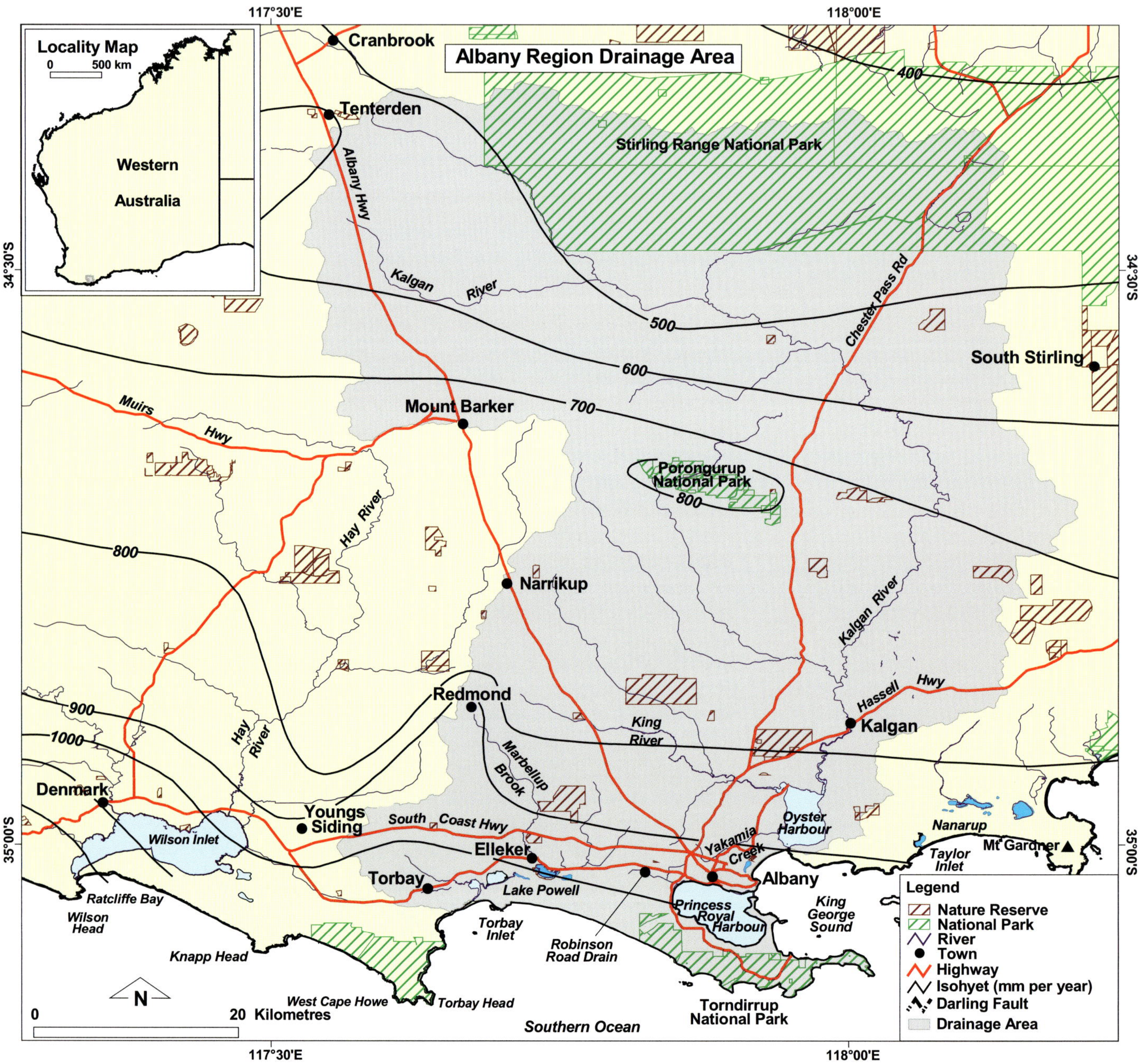

illustrate the instability of this coastline where, as recently as the 1880s, Lake Seppings is reported to have flowed through to Middleton Beach.

Unlike most estuaries on the south coast, the ocean tide is not reduced by the presence of a bar. The tides are usually semi-diurnal but may be diurnal. Water level varies between 0.3 and 1.2 metres daily, with extreme tides up to 2 metres. Large areas of sand in the east, north and south are exposed at low tides. Shoals extending from the eastern shore divide the harbour into a northern and southern basin, connected by a 10-metre deep channel that follows the ancient river course.

Sediments in the Harbour range from fine to coarse and are generally finer along the gently sloping eastern shore and at depth. Around the margins, outcropping shell beds up to 3 metres deep that extend 100 metres inland near Emu Point, and as far as a kilometre and 6 kilometres inland around the Kalgan and King Rivers respectively, indicate that Oyster Harbour was once much larger and marine animals flourished.

A large body of water—like the sea?

The good tidal exchange between Oyster Harbour and King George Sound is unique among the estuaries of Swanland. Flushing times vary from 10 to 20 days during periods of low water flow, in contrast to other south coast estuaries with flushing times of months or years. Salinity in Oyster Harbour is generally close to sea water (35–37‰), creating a more marine environment for a diverse flora and fauna. In contrast, the estuarine reaches of the Kalgan and King Rivers, although tidal, experience marked seasonal changes in salinity and have fewer species living in them.

In summer the small inflow of fresh water from the rivers may lower salinity in the surface waters of the northern part of Oyster Harbour, but mixing due to wind usually breaks down the stratification in 5–10 hours. Within 12–48 hours of heavy rain in the catchments the flow from the rivers spreads in a layer less than a metre thick across the Harbour. If conditions are calm, this buoyant low salinity surface water, high in nutrients, may move through into King George Sound. Strong winds of over 24 knots blowing continuously for six hours are needed to break down the stratification. Although a freshwater layer can persist in the Harbour briefly after heavy flows, the surface waters are seldom less than 30‰ for more than a week due to wind mixing and the good tidal exchange.

The estuary is 'salt-wedge' in nature, particularly in summer, and although there can be marked stratification of salinity, temperature and density, these gradients can break down particularly in the central area under windy conditions. In the shallow marginal areas, extremes of temperature can create density currents.

FIG. 9.7 *Creek entering Oyster Harbour from Lake Seppings, November 1999, showing samphires and rushes underlain by a thick layer of shells.* A. Brearley.

Princess Royal Harbour

Princess Royal Harbour is not strictly an estuarine system, and is generally regarded as a marine embayment. Although freshwater inflows are not obvious they are however considerable, largely entering from ground water, and they do affect the biota. In addition, the similarities to Oyster Harbour including the same environmental problems that resulted in wide ranging studies make it a useful comparison.

Princess Royal Harbour is roughly oval in shape, about 8 kilometres long and 4 kilometres wide oriented in a north-west to south-east direction with an area of about 30 square kilometres. The basin is 5–10 metres deep bordered by shallow sand flats; approximately half the area is less than 2 metres deep. The sand flats are particularly extensive along the western and southern shores and the Vancouver Peninsula which forms the eastern barrier to King George Sound (Fig. 9.8 and Map 9.12).

Although no major rivers enter Princess Royal Harbour, a network of drains collects water from industrial and agricultural land in the Robinson Estate and the Marbelup–Elleker areas. The largest surface source of water, Limeburners Creek, enters the harbour in the south. Total catchment is 8,351 hectares, about 35 per cent of which has been cleared. A large area of the southern catchment lies in the Torndirrup National Park.

Like Oyster Harbour the salinity of Princess Royal Harbour is similar to sea water, although slightly higher in summer due to evaporation and lower in winter due to fresh water seepage. The seasonal fluctuations are, however, not as pronounced. Princess Royal Harbour is generally well mixed and only stratifies weakly due to wind circulation and tidal currents through the narrow entrance channel, which is dredged for shipping access. Flushing time of about 20 days is slightly longer than Oyster Harbour (10–20 days), but much less than other estuaries on the south coast that may not flush for months or years.

FIG. 9.8 *Princess Royal Harbour showing the extensive sand flats on the north-western side, and the city of Albany with the port buildings, aerial view, June 2000.* S. Neville/Ecotones, Courtesy WRC.

As the site of the first European settlement, the area around the City of Albany has been an agricultural, urban and industrial centre for many years. It was an important naval base during the world wars and the departure point for Australian troops to the Middle East and Europe. Expansion of the city during the 1950s led to construction of a domestic waste water treatment plant and discharge of effluent from King Point near the ocean entrance, while development of industries brought industrial discharges (Map 9.12).

Algal growth is more pronounced than in Oyster Harbour as nutrient sources have been more direct and continuous throughout the year. Additionally, the long residence time of water means that nutrients remain in the system long enough for algal growth. In contrast, in Oyster Harbour, although nutrient additions are very high, the nutrients enter in pulses with sporadic strong river flow and some flush to sea in the plume of river water.

Water world of the Albany Harbours

Flora and fauna in the Albany Harbours are influenced by the salinity range and by the availability of suitable substrates such as rocks, water plants, sand and mud. The area of Oyster Harbour can be considered in four major zones: the shoreline from near the high water level to about half a metre below the high water level that includes some rocky areas and beaches; the sandbanks extending down to 1.5 metres; the bank slopes colonised by seagrasses; and the deep central channel area that extends up the rivers. In many respects the habitats in Princess Royal Harbour are very similar, but areas of hard substrate are dominated by structures such as wharves and road works.

The combination of low riverine influences and the permanent opening to the sea means that in many ways Oyster Harbour and Princess Royal Harbour can be considered sheltered marine embayments with a flora and fauna more diverse and marine in character than other estuaries of Swanland. For example, the large seagrasses, the wide leafed ribbon weed *Posidonia australis*, the narrow leafed ribbon weed *P. sinuosa*, and the wire weed *Amphibolis antarctica* are generally found in coastal marine waters.

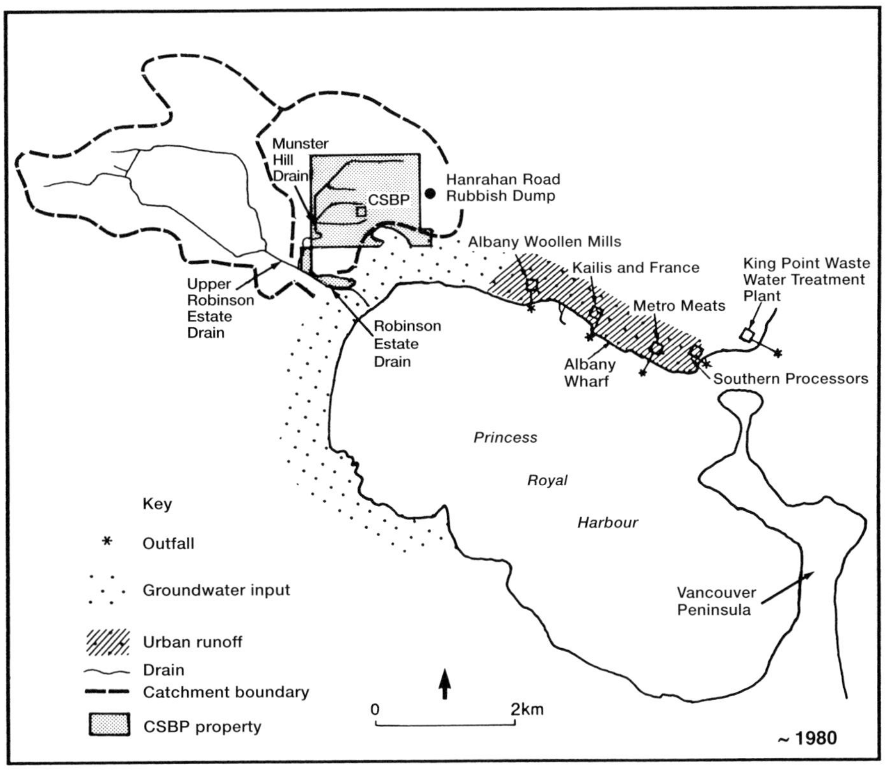

MAP 9.12 *Location of outfalls contributing nutrients and other pollutants to Princess Royal Harbour in the 1980s.* Adapted from EPA, 1990.

THE SHELLS The mollusc fauna is better known than most other groups of animals and is of particular interest; their hard shells preserved in fossil beds can be used to understand the development and changes that estuaries undergo through geological time. In total ninety species of molluscs have been reported from Princess Royal Harbour and 109 species from Oyster Harbour. Most of the molluscs are marine species that only enter estuaries for short periods or at irregular intervals, and are most common on the seaward margin. The next group includes those of marine affinity with essentially marine distributions that are frequently found in other estuaries throughout Swanland. Despite the marine influence, large changes of salinity occur during floods when populations of cockles and other shallow sand-dwelling species are killed. The populations however generally recover quickly because larvae can recruit from other areas inside the harbour and from the sea.

A HISTORY IN SHELLS Molluscs in the Albany Harbours are also of special historical interest because collections made by the earliest explorers became important references in museum collections. The first were made by the French expedition led by Baudin in 1801–1803 when the naturalist François Peron collected many bivalves and other molluscs in the area, particularly from King George Sound (*au port du Roi Georges*). These were deposited in the Musée National d'Histoire Naturelle in Paris and many were used by Lamarck when he described and named species in 1822. Another French expedition led by M. J. Dumont d'Urville on the *Astrolabe* in 1826–1829 also collected shells in the area that were later described by Quoy and Gaimard in 1830–1834 (Fig. 9.9; see also Figs 2.29 and 2.30).

SHELLS FROM RIVER TO SEA The estuarine reaches of the King and Kalgan that flow to Oyster Harbour are similar to other estuaries of the area where salinity may change seasonally from fresh to marine, and they have a similar fauna dominated by a few common estuarine species. In addition, they also have some more marine species because of the influence from Oyster Harbour. Common species include the conniwink *Bembicium melanostomum*, the boring venus shell *Irus crenatus*, the black mussel *Xenostrobus securis*, the barnacles *Balanus* (Fig. 2.25) and the tube worms *Ficopomatus* that are abundant on rocks and logs downstream of the rock bar. Other black mussels are found throughout the Harbour, their location depending on the salinity. While *X. securis* is found in the fresher parts of the Harbour and in the rivers, a similar species *X. pulex* grows on rocks and stakes near the entrance, and a third species *X. inconstans* in the intermediate areas. The large black mussel *Mytilus edulis* is also found in the most marine areas of the Harbour.

The mud oysters *Ostrea angasi* for which the harbour is named are today found mostly in the lower reaches of the rivers and the northern part of the Harbour (Fig. 9.10). The identity of some of the surviving oysters is currently uncertain as a study of oyster DNA indicates that the European flat oyster *Ostrea edulis* is also present. One of the clues to the introduction of European flat oysters is the presence of the parasite *Bonamia*, which was implicated in oyster disease throughout Europe and America.

The shallow margins are criss-crossed by tracks of many creeping snails such as the large mudwhelk *Zeacumantus diemenensis*, small mudwhelk *Batillariella estuarina*, fragile air breather *Salinator fragilis*, and estuarine snail *Hydrococcus brazieri*. The common periwinkle *Austrocochlea constricta* crawls on the rocks and the brown 'beads' of the algae known as Neptune's necklace *Hormosira banksii* grows in clumps along with the ribbed mussel *Brachidontes erosus*; a combination quite characteristic of the south coast (Fig. 9.11). Venus cockles *Katelysia scalarina* and *K. rhytiphora* are common in the sand and are much sought after as a seafood delicacy and as fishing bait (Fig. 6.14).

FIG. 9.9 *Many of the molluscs found in Oyster Harbour and Princess Royal Harbour are marine species reflecting the permanent connection to the sea. Some such as the spindle shell* Fusinus australis *were collected by the first explorers who took specimens to Europe where they were named. (a)* Brachiodontes erosa ≈ *60 mm, (b)* Vulsella spongarium *120 mm, (c)* Pinna bicolor ≈ *400 mm, (d)* Phasianotrochus irisodontes ≈ *10 mm, (e)* Bittium granarium ≈ *10 mm, (f)* Batillaria estuarina ≈ *10 mm, (g)* Eubittium lawleyanum ≈ *20 mm, (h)* Zeacumatus diemenensis ≈ *30 mm, (i)* Bembicium melanostomum ≈ *15 mm, (j)* Polinices conicus ≈ *40 mm, (k)* Bedeva paivae ≈ *25 mm, (l)* Cominella tasmanica ≈ *25 mm, (m)* Fusinus australis ≈ *110 mm, (n)* Bulla quoyii (previously botanica*)* ≈ *40 mm, (o)* Salinator fragilis ≈ *12 mm.* McKenzie, 1964; Roberts & Wells, 1980; Wells et al., 1991; Petersen et al., 1994. C. Bryce, Courtesy WAM.

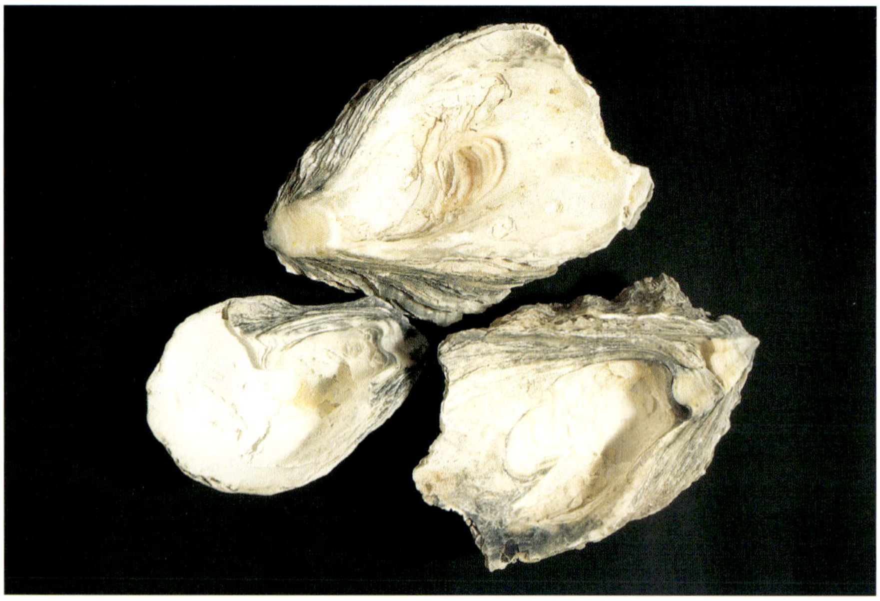

FIG. 9.10 *Mud oysters (*Ostrea angasi*) after which Oyster Harbour was named. The oysters grow to about 150 mm or larger. The young attach themselves to stones or other shells but adults usually lie unattached on the bottom. Adults take about 6 years to mature; eggs are fertilised within the parent and are brooded before being cast out. Time to maturity and brooding are features that make them vulnerable to overfishing. Populations have probably been affected by changes in water quality, and large individuals are now quite rare.*

FIG. 9.11 *The brown algae* Hormosira banksii *or Neptune's necklace grows in protected intertidal areas, November 1999.* A. Brearley.

The beds of *Posidonia* seagrass, which characterise the subtidal area and extend into the intertidal, have the same mollusc species as the flats, as well as species such as the Tasmanian Buccinum whelk *Cominella tasmanica* and dog whelk *Nassarius pauperatus* (Fig. 2.29). The razor shell *Pinna bicolor* is also found here in large numbers throughout the harbours, but the large number of dead shells suggests that these may be wiped out under unfavourable conditions, such as periods of low salinity.

One species, the ark shell *Anadara trapezia*, is of special interest (Fig. 10.35). This large bivalve is found near the seagrasses partly buried in the sediment, often with a tuft of green alga growing on the exposed point. Oyster Harbour is now the only place in Western Australia where it is found alive, although it is a common bivalve in eastern Australian estuaries where there are seagrass meadows. *Anadara* shells are common in Pleistocene shell deposits around many south coast estuaries, including the Blackwood Hardy at Augusta. They are not present, however, in the younger extensive Holocene shell beds of estuaries east of Albany, indicating that they died out before this time. (See **Box 10.6 A shell and understanding climate change**.)

A variety of other characteristically marine fauna, such as sea stars, are recorded from the harbour, and the sea urchin *Temnopleurus michaelseni* has been found grazing on seagrasses.

The Albany Harbours are rich feeding grounds for a wide variety of fish. Except for black bream, the commercial species are common coastal marine fish that spawn at sea and use estuaries as nursery and feeding areas. The harbour supports considerable commercial and recreational fisheries based on leather jacket, cobbler, crab, sea mullet, herring, garfish and cockles. Recently the number of commercial licences has been reduced and restrictions placed on netting to protect the resource and maintain fish populations. Cockle collecting is only permitted by hand to limit the disturbance of sand and there is a limit to the volume that can be collected by each person per day.

In Oyster Harbour a 125-hectacre area covering three leases for aquaculture under the jurisdiction of the Department of Transport (now DPI) were established in ca 1992 (subtidal areas) and 1995 (intertidal areas) with current activity based on rock oysters and mussels. The industry was controversial because of concerns for seagrass health, loss of traditional fishing grounds, disturbance of seabird rookeries, detraction from overall view of the Harbour, and increased noise and boat traffic. There has, however, been a greater acceptance of the industry due to the economic returns to the Albany community.

The surrounding wetlands and fringing vegetation are important roosting and feeding areas for waders and waterfowl. Groups of waders are present over summer, and migratory birds such as sand pipers, stilts and stints use the rich shallow bands for feeding before returning to the Northern Hemisphere to breed. Pelicans nest intermittently on Green Island, where they were caught by the French explorers in the 19th century. (See **Box 3.7 The Pelican**.)

Step by step—a slow decline

The hinterlands of Oyster Harbour and Princess Royal Harbour have been farmed or settled for nearly 170 years. Over the years patterns of water flow have changed, the nutrient content of the soils has increased, erosion has accelerated, and the two great harbours are now considered eutrophic to the extent that most of the seagrasses in Princess Royal Harbour and Oyster Harbour have died. We could question why the problem was so bad before it was noticed! It is now widely recognised that decline of ecosystems is usually rapid and extreme, the culmination of many small changes that are hard to detect. This appears to have been the case in Oyster Harbour when a large flood brought an accumulation of sediments and nutrients that triggered algal growth. In Princess Royal the long slow accumulation of nutrients from ground water and industry brought a similar result.

Hindsight is a wonderful thing; a comment often made cynically, but the lessons learnt from the Albany region as well as the Peel–Harvey have taught us many things. Studies in both areas have shown how water quality, plants and animals interact, and illustrate that constant surveillance of the water quality in the catchments can show us where problems are originating, hopefully allowing the problems to be stopped before they become irreversible.

FIG. 9.12 *Signs of change: Princess Royal Harbour, 1984. The fishing ban in the western half of the harbour was in place 1984–1992.* EPA Bulletin 412, 1990; Francesconi et al., 1997. Courtesy R. Masini, DoE.

In the 1960s a survey of biological communities in Princess Royal and Oyster Harbours considered the area 'pristine', but by the 1970s there was concern for the quality of water entering the Albany Harbours. A number of studies were initiated by the Department of Conservation and Environment (later the EPA) 'to assess the extent of the contamination or deterioration of the main body of water in relation to the various trade wastes and other inputs' (Western Australia Government Chemical Laboratories, 1978).

In 1976 some sewage contamination was reported, but the overall nutrient levels in the waters were low and Princess Royal Harbour was considered sufficiently well flushed to cope with the nutrient loads at the time. Alarm bells sounded in 1983 when mercury was detected in fish. Subsequent investigations showed the western end of Princess Royal Harbour was extensively contaminated by mercury and lead, and fishing was banned for about eight years (1984–1992) (Fig. 9.12). The CSBP fertiliser factory, which discharged effluent directly to the Harbour, was identified as the source of the contamination, a practice that ceased in 1986. Other industries, including vegetable and fish processors, an abattoir, a woollen mill, a power plant, domestic waste and urban run-off also discharged nutrients, organic matter, pesticide residues, zinc and chromium to the Harbour, and spillage from unloading fertiliser on the wharf added to the problem.

In 1984 surveys conducted in response to concerns for the health of the marine habitats detected a loss of seagrasses and an increase of algae (Fig. 9.13, Maps 9.13 and 9.14). By 1988 almost 90 per cent of the seagrass meadows in Princess Royal Harbour and 80 per cent in Oyster Harbour had been lost. Initially seagrasses in areas deeper than 2 metres were lost, but as the filamentous macroalga *Cladophora prolifera* increased, the seagrasses in shallower water were also smothered. The amount of algae was so large that an estimated 1,200 tonnes dry weight grew per year in Oyster Harbour and 12,000 tonnes in Princess Royal Harbour. In Oyster Harbour over 90 per cent of the algae was concentrated in the south-east corner. The report concluded:

> If current pollutant loadings into the harbours continue, most of the remaining seagrasses will be lost within five years in Princess Royal Harbour, and within five to ten years in Oyster Harbour, and the general ecology of the harbours will continue to deteriorate.
> EPA, 1990

FIG. 9.13 *(a) Ribbon weed* Posidonia *and (b) wireweed* Amphibolis *smothered by the algae; (c) eroded face of a healthy* Posidonia *meadow in King George Sound with leaves clear of the dense algal growth. Note the upward growth of the meadow above dense rhizomes that have trapped sediment.* See Bastyan, 1986; Bastyan et al., 1996. Courtesy G. Bastyan.

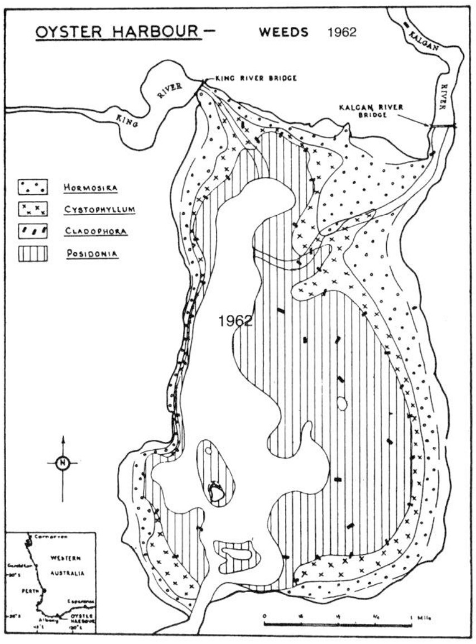

MAP 9.13 *Seagrass distribution Oyster Harbour, 1962 before the major losses. Note the unshaded areas represent deeper channels, the legend indicates areas dominated by brown algae* Hormosira *and* Cystophyllum *and the ribbon weed seagrass* Posidonia. *The green alga* Cladophora *that later grew prolifically was present in small patches.* Modified from McKenzie, 1964.

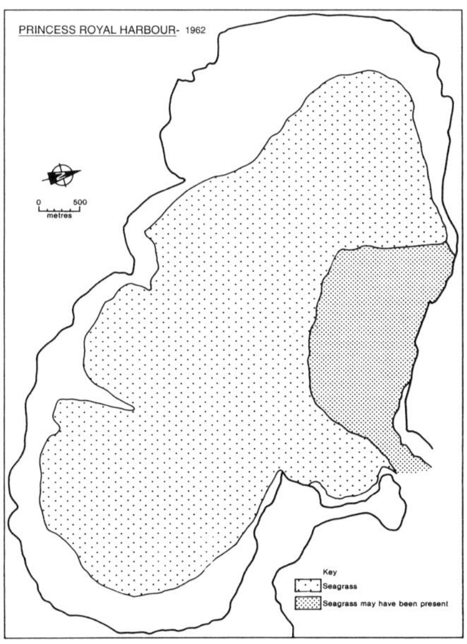

MAP 9.14 *Seagrass distribution Princess Royal Harbour, 1962.* Modified from McKenzie, 1964.

Too many nutrients—too much algal growth

Unnaturally large accumulations of seaweeds (macroalgae) in Princess Royal Harbour and past discharges of wastes containing large quantities of plant nutrients suggested that the decline of the seagrass meadows in Princess Royal Harbour was linked indirectly to nutrient enrichment. Prolonged reductions in light reaching the seagrass meadows due to shading and smothering by these algae were implicated as the direct cause of the seagrass decline.

Studies of water quality showed nutrient concentrations were high in the sediments with the next largest pool of nutrients in the macroalgae (Table 9.1). In oligotrophic (nutrient poor) ecosystems, the nutrient pool in the macroalgae and epiphytes combined is usually smaller than the nutrient pool in the seagrasses. The nutrient concentrations in seagrasses, epiphytes and periphyton indicated that Oyster Harbour was more nutrient enriched than Princess Royal Harbour and King George Sound.

The sources of nutrients in the two harbours were found to be quite different. Most of the nutrients in Princess Royal Harbour came from local sources (industry) around the Harbour with a constant input for most of the year. The high nutrients, good mixing and long residence time of the harbour waters created an environment for year-round algal growth.

In contrast, many of the nutrients to Oyster Harbour came from the river catchments in episodic flow events, much of which flowed straight to sea before mixing with the main water body where the nutrients could be used for algal growth. The large river flows also brought sediment that increased turbidity, reduced light and coated seagrasses, adding to the algal problem.

The input to Oyster Harbour varies greatly from year to year depending on the volume and pattern of river flow, both of which are influenced by climatic and catchment conditions. Usually the amount of nutrients in the river water is low but under flood conditions, often the first heavy winter rains, the concentrations can be very high. Most of the phosphate comes indirectly from the broad scale application of superphosphate in the catchment, and nitrogen from the legume pastures that fix atmospheric nitrogen in the soil along with some input from urban development and intensive agriculture (piggeries, dairies and horticulture). (See **Box 9.2 The P problem—can it be controlled?**)

Solving the problem— catchments, harvesting and regrowing

Understanding the role nutrients play in promoting algal growth, which shades and smothers seagrasses, has led to the suggestion that reducing nutrients should help solve the problem. New regulations for urban run-off and effluent management, together with reduced fertiliser application and revegetation of streamlines, have reduced some nutrients entering the harbours although nutrient rich ground waters still pose a problem. The effect of these strategies on the algal mass is unclear.

Harvesting of algae to reduce smothering of seagrasses, remove nutrients and reduce nutrient release from sediments is another option. Algae were harvested from Princess Royal Harbour, where drift algae were the major problem, and used for soil improvers and garden fertilisers. Equipment and weather constraints, however, meant the area where the harvesters could be used was limited and only a small proportion of the algae was removed. In addition, there were concerns about the effect of harvesters on bottom dwelling animals and fish such as King George whiting and cobbler. Given the cost of harvesting, its questionable value and unknown effects on animals, removal of algae through harvesting has ceased. (See **Chapter 3 The Peel–Harvey Estuarine System: Managing the problems, Harvesting**.)

	Princess Royal Harbour		Oyster Harbour	
	TN(%)	TP(%)	TN(%)	TP(%)
Water column	0.8	2.1	2.4	2.2
Sediment	78.9	90.4	88.7	94.4
Seagrass	1.1	0.8	1.9	1.3
Epiphytes	0.1	0.1	0.2	0.2
Macroalgae	19.1	6.6	6.8	1.9
TOTAL	100.0	100.0	100.0	100.0

TABLE 9.1 *Size and distribution of nutrient pools in Princess Royal and Oyster Harbours, 1988.* Source: Simpson & Masini, 1990.

If the water quality improves and the smothering algae declines, the question remains: will seagrasses regrow? We do know that some seagrasses have regrown but their progress is slow, the sediments may have changed and the animals that lived among them may have disappeared. (See **Box 9.3 Shrinking seagrasses versus growing algae**.) Transplanting areas is another option, and is being tested in the Albany Harbours with considerable success. One fact is certain however, whatever method is used to restore seagrasses, reducing nutrient inputs, harvesting or transplanting, it will be quite a slow process, not without expenses. (See Figs 1.11 and 1.12.)

Why seagrasses? What's wrong with algae?

Seagrasses create a whole world for animals, providing food, shelter and reducing sediment and water movement. Seagrasses form the base of many food webs in coastal areas through the capture of light energy or primary production. Although only a few animals eat living seagrass, the dead leaves are broken down by bacteria, fungi and small animals that in turn are eaten by larger animals.

Seagrasses provide many places for animals to live. This may be a firm base for attachment on an otherwise unstable sandy bottom, or shelter from predators and water movement among the leaves and buried stems. This creates a paradox: for the seagrasses also provide a place for algae to grow, and these too shelter small animals. Yet it is the algae that can smother the seagrasses—a very fine balance that can be thrown into a spin.

Algae, particularly the mat-like types that grow rapidly in nutrient-rich waters, not only smother seagrasses but their less complicated structure does not provide the same three-dimensional habitat for animals. They are unstable, accumulating in downwind areas where they smother plant and animal communities and then decompose, causing unpleasant noxious odours.

Seagrass meadows can be important nursery areas for numerous species of fish, providing shelter and harbouring the small animals on which juvenile fish feed. When seagrasses are lost, so too are the animal populations that depend directly upon them, with effects flowing on to other animals in the wider ecosystem.

Seagrass meadows reduce water movement, hence they stabilise sediments and reduce erosion. The leaves create baffles that can reduce water movement by more than 50 per cent depending on the type of seagrass, how thickly the leaves are grouped and the depth of water above the leaves. This reduction in water movement can be observed where seagrasses grow in very shallow water—the water is often smooth and calm even when conditions are windy. In contrast, the water surface above the nearby areas of bare sand is disturbed. Sand and silt settle out where water movement is reduced under the leaf canopy and gradually build up. In fact, seagrass meadows can become shallower over time because of sand accumulating so that in very shallow areas, such as the Albany Harbours, some seagrasses are now exposed at low tides and the leaves can be sunburnt. In addition to the leaves that reduce water movement, the underground stems also bind the sediment.

Box 9.2 THE P PROBLEM—CAN IT BE CONTROLLED?

With recognition that the nutrients to Oyster Harbour come from the catchment, it is worth considering if the catchment can be managed to meet dual objectives of sustainable agriculture and healthier waterways. Generally farming practices are similar to those developed when the land was first cleared, with an annual fertiliser application regardless of soil nutrient status and little consideration of potential off-site impacts.

When the phosphorus problem was first examined, attention focused on the silica-rich sandy soils around the harbour, like those of the Peel–Harvey, which lacked phosphorus-adsorbing substances such as iron and aluminium oxides, and clay minerals. These soils are prone to nutrient leaching. More recently it has been recognised that run-off in the Oyster Harbour catchment has the potential to carry higher concentrations of phosphorus where the soils contain high amounts of phosphorus and reactive iron (more than 1,600 ppm). Estimates in the early 1990s suggested that although fertiliser was applied to most paddocks, 90 per cent already had more than adequate amounts of phosphorus for good clover growth. In addition, the amount of phosphorus in soils was highest where conventional farming had been established for the longest time. These findings suggest that nutrient inputs (applications) had been exceeding nutrient outputs (products) and the difference was accumulating in soils or being lost.

For soils where excess phosphate can accumulate (high reactive iron), intense rainfall events combined with erosion result in nutrient loss proportional to the phosphorus stored in the soil. During intense rainfall, major run-off events lasting a few days can account for most of the annual flow to Oyster Harbour, and most of the nutrient input occurs during a few storm events each year. For example, over three days in July 1991 the Kalgan River delivered more than 90 per cent or 36 tonnes of its annual load of phosphorus (39 tonnes) to Oyster Harbour, an event often missed if water quality sampling is infrequent. Currently in excess of 2,000 tonnes of phosphorus is applied to pastures in the catchment each year, mostly at the break of season in April or May. Most of the river flow occurs in July and August, although there are unseasonal flows as in January of 1982 when the highest flow ever was recorded.

Besides dissolved phosphorus, 70 per cent of the P-load has been measured in particulate material. To give a sense of scale, in 1992 (a year of relatively mild flow conditions) more than 4,500 tonnes of suspended sediments were discharged from the Kalgan River to Oyster Harbour. While up to 70 per cent of the P-load has been measured as particulate phosphorus, other years of measurement have shown equally high proportions of phosphorus in dissolved form. This suggests a wide range of factors influence nutrient processes in this catchment, ranging from leaching to run-off and erosion depending on climatic and other factors.

The recognition that phosphorus lost from farmland varies in form highlights the value of strategically located vegetated buffers, particularly along creek lines where most run-off is across the surface. These streamside buffers can act physically on eroded material and dissipate the energy of floodwaters, trapping material and acting as a filter. The loss of streamline vegetation through clearing and livestock grazing reduces the natural filtering and dissipating capacity of streamline buffers (Fig. 9.14). In instances where the subsurface flow is dominant however, the physical filtering capacity of streamside vegetation can be bypassed, allowing nutrients into streams unimpeded and altering nutrient forms.

Initially only easily accessed and traversed streams (low order depressions in paddocks) in the catchment headwaters were cleared. Those in the deeper, closely vegetated valleys downstream were left or fenced to reduce stock losses because they were not cleared as easily. As a result, most of the buffering capacity of the catchment headwaters has been lost and only the lower main channels retain this buffering function. The main channel of the Kalgan is 140 kilometres long with about 2,000 kilometres of tributaries. While most of the main channel and about 30 kilometres of tributaries are in reasonable condition, this represents a small percentage of the buffering capacity of the original system.

Today the loss of streamline vegetation continues, but mostly passively through livestock grazing, trampling and fires. The native perennial species are gradually being replaced by annual grasses, which are less able to support the stream banks. In 1992 a survey of 94 kilometres in the lower reaches classed only 25 per cent of the river banks as in an undisturbed pristine state, 46 per cent degraded and 29 per cent as a ditch or drain.

Despite the degradation the remaining vegetation provides one explanation for differing nutrient and sediment transport to Oyster Harbour in different years (1991 and 1992) (Table 9.2) for similar flows. During 1992 flow was spread over more days resulting in a greater time for nutrients to be taken up by vegetation or deposited with sediment because of longer travel time, reduced velocity and lower energy. The large size of the catchment also means there is a greater chance of nutrients being trapped or assimilated.

How can we change the inputs?

A number of management actions are available to reduce nutrient flow to streams. Some actions will simply delay nutrient loss or move the problem from one location to another rather than reducing inputs,

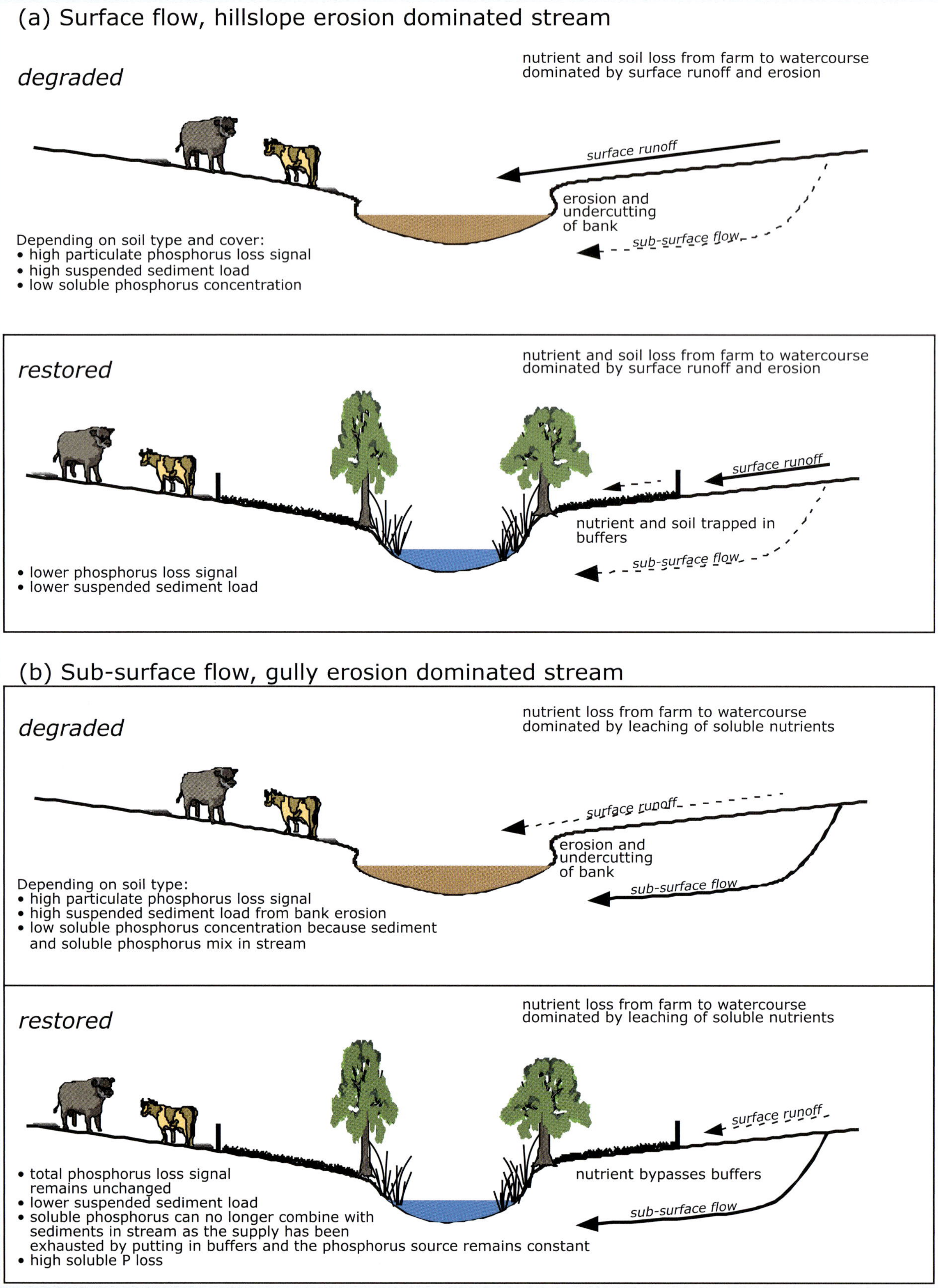

FIGURE 9.14 *Effect of vegetation on the movement of nutrients and sediment into watercourses through (a) surface and (b) subsurface flows. Note that even with streamside vegetation, nutrients can still be high in a watercourse due to subsurface leaching of ground water, illustrating the importance of understanding the major hydrological pathways and nutrient sources, and reducing nutrient additions to the surrounding land and catchment.* Courtesy D. Weaver, DAWA.

TABLE 9.2 *Water flow and nutrient loads in the Kalgan River.* Source: Weaver et al., 1994.

Year	Flow (L × 10^6)	P load (t)	Flow days	Flow events	Maximum daily flow (L × 10^6)	Maximum instantaneous P concentration (mg/L)
1991	70	39	3	1	11.8	3.0
1992	75	9.2	28	4	4.8	0.8

or increasing outputs or biomass stores, depleting stores of nutrients or offering opportunities for nutrient source replacement. Ultimately inputs should match outputs, but we need to remember that the current stores of nutrients are already too high. Matching inputs and outputs for newly developed land should not lead to increases in nutrient store, and therefore not lead to nutrient loss problems. This is not, however, the starting point as we need to consider the current nutrient store as part of the nutrient balance, and aim for actions that can deplete these nutrient stores while reducing inputs and increasing outputs.

On the input side, most soils have high phosphorus concentrations and additional application is of questionable benefit. Soil testing would allow judgement of situations where it is required, in what amount, and if alternative nutrients are in fact deficient. Reduced phosphorus applications would reduce loss in run-off and also decrease costs to the farmers. In the longer term, money saved could be used to combat other nutrient problems such as sulfur deficiency, which is common on sandy soils. As superphosphate contains significant amounts of sulfur, it is not unusual to see pastures respond to applications because of sulfur rather than phosphorus deficiency.

The effectiveness of vegetated streams in slowing down and trapping nutrients is quite clear, but the vegetation needs to be located where it can intercept contaminated run-off rather than being seen as a cure-all for nutrient problems. It should be noted that vegetated streams also provide other benefits, such as ecosystem function, as additional reasons for their establishment. Eventually some trapped nutrients may be exported from buffering vegetation, although the rate of loss should be less than where buffers are absent. It could also be argued that buffers are lost farmland. Grazing in buffer strips that also provide shelter is a particular problem because animals destroy the vegetation and add high concentrations of nutrients directly to the watercourse. Clearly the fencing of streams is a priority for a range of reasons including nutrient control. There is also merit in the creation of upslope vegetation buffers to extend the effect of streamside buffers. This requires a landscape approach to management or an integrated system in which protection of existing streamside vegetation is one component. Such a system would include the following elements:

- reductions in nutrient inputs to farmland through effective fertiliser management;
- reductions in nutrient exports from the farmland through effluent control and animal management near streams;
- improved utilisation of nutrients through the use of perennial pastures;
- improved retention of nutrients leaving the farm through streamside vegetation; and
- increased assimilation of nutrients in the receiving waterways through streamside vegetation.

Source: Weaver & Prout, 1993; Weaver et al., 1994.

Can we bring them back again?

Seagrasses grow from seeds and by growth of runners or rhizomes. In the past it was assumed that once disturbed, seagrasses would not regrow because the sediment would no longer be suitable, there would be no seeds available and lateral spread would be very slow. More recent evidence suggests that if the water quality is improved and the amount of algae reduced, some seagrasses may slowly regrow. Experiments in Princess Royal Harbour have shown the ability to regrow depends on to what extent the light reaching the leaves was reduced and how long they were stressed.

Mapping in the Albany Harbours has shown that in some areas seagrasses are regrowing naturally. In Oyster Harbour this has occurred in shallow water, particularly along the eastern margin above the 2-metre contour along the northern end of the harbour and in deeper areas near Emu Point and Green Island (Fig. 9.15). The wide ribbon weed *Posidonia australis* appears to be recolonising areas where seagrasses died or were removed by dredging. In Princess Royal Harbour regrowth has occurred in the deep and shallow areas surrounding South Spit. In this area and along the western margin, the dense accumulations of macroalgae that have covered the seagrass have declined. It is now obvious that some of the seagrasses had not died but had defoliated and are now resprouting. There also appears to have been a change in the dominant species with wire weed *Amphibolis antarctica* growing in areas once covered with *P. australis*, while in other areas of 1.5–3 metres depth the *P. australis* appears to be replacing the narrow ribbon weed *Posidonia sinuosa*.

In this early stage of recovery the meadows are not dense. Projecting how fast the seagrass area will recover is extremely complicated. We know that recovery is likely to be slow and dependent on the continuation of favourable conditions, however the rate could increase if the seagrasses can again produce seeds. How long it will take for the accompanying seagrass animals to re-establish is another unknown.

Seagrass loss—what else have we lost?

Due to the importance of seagrasses that provide most of the primary production in the harbours and form the basis of a detrital food web, it is interesting to consider how the loss of seagrasses affects the biological communities. The seagrass habitats have been more extensively studied in Princess Royal Harbour than in Oyster Harbour, but as the faunas are probably very similar these studies can be used to assess the overall consequences of seagrass decline.

Studies in many areas including Albany clearly demonstrate the importance of seagrasses compared to nearby sand areas. The different seagrasses had slightly different forms creating different habitats, so the amount of animal matter varies. It is also worthwhile remembering that many of the animals found in the sand areas also depend on seagrass indirectly because they feed on fragments of dead seagrass leaves. In Princess Royal Harbour the number of animals, the amount of animal material (biomass) and the amount of matter produced each year (secondary production) was much greater: 19.39 grams dry weight per square metre per year in seagrass compared to unvegetated areas with 0.47 grams dry weight per square metre per year (Table 9.3). Molluscs dominated the production in unvegetated sediments and accounted for half the production in seagrass beds; filter feeding ascidians were another important group in the seagrasses.

Box 9.3 **SHRINKING SEAGRASSES VERSUS GROWING ALGAE**

All plants require sunlight for photosynthesis, the process that produces energy for use in respiration and growth. The decline of seagrass meadows has been linked to a reduction in light reaching the meadows. Light can be reduced by decreased water clarity (due to microscopic plants, phytoplankton or sediment re-suspension through dredging), shading due to seaweeds growing on leaves (epiphytic algae), or unattached algae. Seagrass leaves that are covered with large amounts of epiphytes therefore receive less light than leaves with less epiphyte cover.

In an annual cycle, seagrass growth occurs in summer months when day length is long and light levels are high; energy is stored as starch in the underground stems or rhizomes. In winter, some of the stored starch is used to meet respiratory demands for energy.

Experiments in Princess Royal Harbour showed that the annual growth of macroalgae could be twenty-five times that of seagrasses, and an area can change from a seagrass community to an algal community. This is because algae are by comparison quite simple and can absorb carbon and nutrients more quickly, so energy can be used for growth not in the support of structures like roots, shoots and leaves.

Smothering by macroalgae appeared to be the major cause of seagrass decline in Princess Royal Harbour. In contrast, in Oyster Harbour except for the south-east corner where macroalgae accumulated, algal epiphytes were implicated as the main cause of seagrass loss. This difference may have be due to the better water clarity in Princess Royal Harbour favouring proliferation of macroalgae, while the higher nutrient loading and relatively poor light conditions in Oyster Harbour may favour the growth of epiphytes.

Source: Cambridge et al., 2002; Simpson & Masini, 1990.

FIG. 9.15 *Seagrass regrowth in Oyster Harbour in an area where all the seagrass died, aerial view, 2003. Small round patches indicate recent seedling growth; larger patches have been established for longer periods and some patches have coalesced. 'Grid-like' patterns are experimental trials using shoot and rhizome cuttings. The sharp lines across the meadow were cut by boat propellers.* Cambridge et al., 2002; Coupland & Crawley, 1997. Courtesy G. Bastyan.

TABLE 9.3 *Secondary production rates of benthic marine invertebrates in seagrass beds and unvegetated sediment in Princess Royal Harbour, Albany, WA. Data for seagrass is mean of production rates in the seagrasses* Posidonia australis, P. sinuosa *and* Amphibolis antarctica. *Standard errors (SE ±) are shown where sufficient data are available.* Table adapted from Wells et al., 1991.

Faunal group	Unvegetated sediment (g dry wt m^{-2} yr^{-1})	Seagrass (g dry wt m^{-2} yr^{-1})
Molluscs	0.37 ± 0.01	9.69 ± 0.29
Ascidians	0	6.92
Polychaetes	0.03 ± 0.00	1.08 ± 0.03
Echinoderms	0	1.55
Crustaceans	0.04	0.10
Nemerteans	0	0.01
Sipunculans	0	0.04
TOTAL	0.47 ± 0.00	19.39 ± 0.14

Although some caution is needed because seagrasses often grow in patches, and determining how much of each type of seagrass is present is difficult, an estimate of lost biological activity can be made by coupling information on the production of invertebrates and the knowledge of seagrasses losses (about 90 per cent in Princess Royal Harbour and 80 per cent in Oyster Harbour) (Table 9.4).

In Princess Royal Harbour benthic invertebrate biomass within the seagrass beds fell from 1937 tonnes in 1960 to 340 tonnes in 1988, and invertebrate production fell from 279 tonnes to 49 tonnes in the same period, clearly demonstrating the importance of seagrasses. Similar calculations for Oyster Harbour where 80 per cent of the seagrass was lost indicate that the biomass of invertebrates probably decreased from 817 tonnes in the 1960s to 174 tonnes to 1988, while secondary production fell from 115 tonnes to 25 tonnes.

Of course when the seagrasses declined, the sand areas increased and these would not have been completely devoid of life. To some extent the losses of primary production of seagrasses may also be offset by some increase in microscopic plant cells (phytoplankton) and macroalgae, however there is also evidence that drift algae (which was so abundant) might inhibit settlement of invertebrates, further adding to the problem.

Undoubtedly invertebrates formed the food supply of secondary consumers, including commercially valuable fish species. Changes to invertebrate communities due either to loss of species or changes in abundance are usually followed by changes in fish production or the types of fish, however the extent of these cascading effects in the Albany Harbours has not been quantified.

TABLE 9.4 *Comparison of dry weight of standing stock and production of seagrass and benthic marine invertebrates in Princess Royal Harbour.* Adapted from Wells et al., 1991. See also Hutchings et al., 1991; Walker et al., 1991.

	Area of seagrass beds (km^2)	Biomass seagrass (tonnes ± SE)	Seagrass primary production (tonnes ± SE)	Invertebrate biomass (tonnes ± SE)	Invertebrate secondary production (tonnes ± SE)
Princess Royal Harbour					
1960 seagrass	14.4	4,074 ± 284	8,380 ± 1493	1,937 ± 235	279 ± 2.0
1988 seagrass	2.5	716 ± 50	1,473 ± 262	340 ± 41	49 ± 0.4
Bare sand	11.9	0	0	43 ± 12	6 ± 0.0
TOTAL 1988		716 ± 50	1,473 ± 262	383 ± 53	55 ± 0.4
Oyster Harbour					
1960 seagrass	6.1	1,769	3,538	817	115
1988 seagrass	1.3	377	754	174	25

What type of future?

The major management problem for the Albany Harbours is the control of nutrient inputs and eutrophication; a problem that also applies to the other small estuaries with cleared catchments such as Torbay and Taylor Inlets. In Princess Royal Harbour foreshore industries that accounted for 75 per cent of phosphorus and nitrogen in 1988 only contributed 25 per cent of the discharge of those nutrients in 2000. More diffuse sources still contribute more nutrients than is desirable for maintaining a healthy ecosystem.

In Oyster Harbour superphosphate from pastures in the King and Kalgan catchments is the principal source of phosphorus, while legume pastures are probably the main source of nitrogen. There are other sources of nutrients such as those from septic systems, urban run-off and industries that deliver bacteria as well as nutrients. While many of the industrial sources have been significantly reduced, the recent although temporary bans on fishing and oyster harvesting because of bacterial *E. coli* contamination is particularly illustrative and timely. This is a reminder of how easily a wonderful environment can be destroyed, and how eventually it affects all of us (Fig. 9.16).

TORBAY INLET

Torbay Inlet enters the sea at Perkins Beach between the granite headland of Torbay Head on West Cape Howe and the Flinders Peninsula (Fig. 9.17 and Map 9.3). Torbay Inlet, Lake Powell (formerly Lake Grasmere) and the nearby swamps lie only just above sea level. Marine fossils found during the excavation of Marbelup Drain indicate that the area was relatively recently (about 1 million years ago in the Late Pleistocene) part of the sea connecting to Princess Royal Harbour through the Grasmere Valley. (See **Box 10.6 A shell and understanding climate change**.) Today, the inlet and the section of Marbelup Brook to the High Level Drain are the estuarine part of a highly modified drainage system.

The fertile soils have been farmed since the early days of European settlement. Vegetables have been grown since the 1890s when supplies from Albany were sent to the goldfields, and the area is also used for cattle and sheep farming. The land has always been prone to flooding in winter and invasion by salt water. To combat the problem, as in nearby Wilson Inlet, drains were dug diverting water from the swamps to the inlet but they failed to keep water off the land. Worse than this, there was the unexpected effect of making Seven Mile Swamp (now Ewarts Swamp) acidic and infertile.

To quote Woodward (1917): 'The poisonous mineral solutions in the soil of the "Seven Mile Swamp" are due to the oxidation of the iron pyrites in

FIG. **9.16** *Ernest Hodgkin and Ruth Clark, authors of the* Estuarine Studies Series *on the banks of Oyster Harbour, ca 1989.*

FIG. 9.17 *Torbay Inlet and Lake Manarup, aerial view, June 2000. The surrounding country is low lying and has been cleared extensively for agriculture. Water flow to the system is regulated to prevent flooding when rainfall is high, and to prevent sea water entering when rainfall is low. The inlet channel passes through the dunes to Perkins Beach where the position of the opening varies depending on the prevailing swell.*
S. Neville/Ecotones, Courtesy WRC.

the soil itself owing to the mineral being brought in contact with the air by drainage and cultivation'. Similar acid sulfate soil problems have arisen at Beenup near the Hardy Inlet, around the Peel–Harvey and Swan River, and throughout Australia where the soils are rich in organic material—it would seem that over 85 years later we are still slow to learn from our mistakes.

In 1912 a floodgate barrage was built half a kilometre upstream from the bar to prevent salt water flowing onto farmland. The system worked quite well for a few years, but by 1920 the sea was winning because sand build-up on the downstream side and gaps between the timbers allowed sea water to enter. Further modifications using sheets of iron prevented the seawater flow but also slowed the floodwaters. After numerous episodes of burning and rebuilding it was finally demolished in 1985, the remaining line of posts a reminder of the difficulties of trying to control natural water flows.

Not what it used to be

The present drainage system was developed to prevent salt water flooding agricultural land east of the Elleker–Grasmere Road and to maintain a fairly constant water level on the low-lying swamps (Map 9.15). Maintenance of the system also necessitates breaking the bar.

Water in Torbay Inlet comes from four sources:

1 Low-lying land to the west drains to Torbay Main Drain and then through a creek to Torbay Inlet.

2 The hilly sand plain country to the north is drained by Marbelup Brook that flows via a constructed channel, the Marbelup High Level Drain, bypassing Lake Manarup and rejoining its old natural channel about a kilometre from the inlet. The natural channel into Lake Powell is blocked by a bund (earthen bank), the 'Plug'. A limited amount of water can leak through a pipe used to regulate flow from the lake.

3 In the east, the Grasmere Drain takes water from Five Mile and Seven Mile Creeks to Lake Powell. Water from the lake flows via the Marbelup Sub B Drain (Penstock Drain) to the Marbelup High Level Drain. Floodgates (penstocks) at the Elleker–Grasmere Road retain higher ground-water levels for potato farming and reduce salt water flooding. Some salt water does enter as the gates are old and leak.

4 Seven Mile Swamp (Ewart Swamp) and the paddocks south of Lake Powell drain via North Creek to the old Marbelup Brook, passing in a siphon under the Marbelup High Level Drain to Lake Manarup. At the outlet from Lake Manarup floodgates prevent salt water entering from Torbay Inlet. These are opened to allow inlet water in when the lake dries up or to hold excess water so that opening the bar can be delayed during the salmon season.

A little body of water

The Torbay Inlet lagoon is shallow, less than a metre, with about 1 square kilometre of open water. Torbay Inlet is surrounded by a margin of dense saltwater tolerant species, the paperbark tree *Melaleuca cuticularis* and the sea rush *Juncus kraussii*. Behind this the taller swamp sheoak *Casuarina obesa* occurs with an occasional *Banksia attenuata* present in the flats.

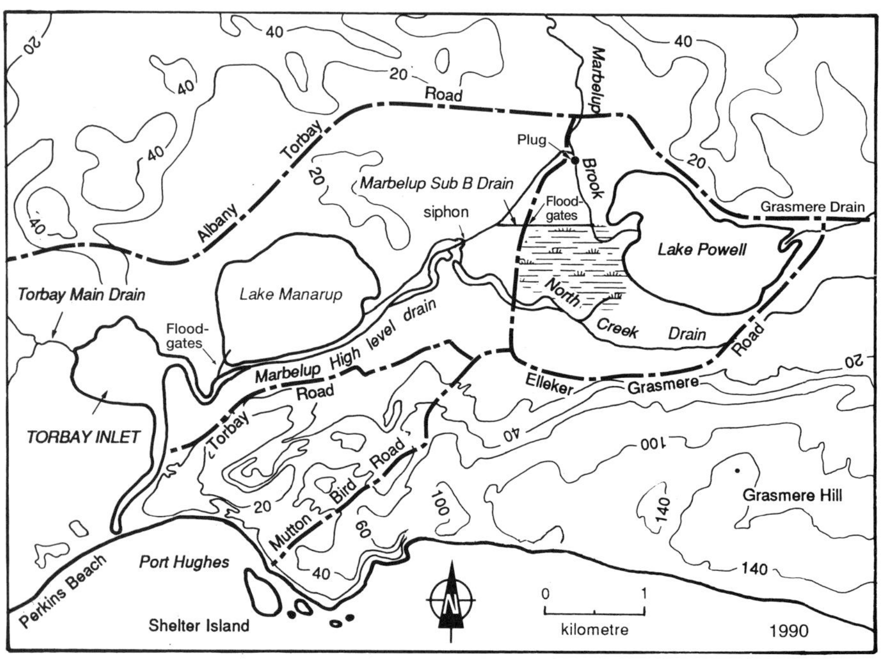

MAP 9.15 *Torbay Inlet showing topography of surrounding areas and modifications to the drainage system; compiled 1990. See Fig. 9.17. Note Marbelup is sometimes spelt Marbellup.*
Adapted from Hodgkin & Clark, 1990. See also Adeney, 2001.

Between Torbay Inlet and Lake Manarup, the river has a fringe of coastal sword-sedge *Lepidosperma gladiatum* with couch *Cynodon dactylon* and occasionally *J. kraussii.* This is backed by swamp cedar *Agonis juniperina*, peppermint tree *A. flexuosa*, pea shrub *Oxylobium lanceolatum*, the coast saw-sedge *Gahnia trifida* and the occasional *M. cuticularis*. Further upstream on the bank of the Marbelup Sub B Drain near Elleker–Grasmere Road, freshwater species such as the paperbark *Melaleuca rhaphiophylla* and jointed twigrush *Baumea articulata* become more dominant along with the yate *Eucalyptus cornuta* and *Agonis flexuosa*.

The 1.2 kilometre-long inlet channel narrows from 150 metres wide near the inlet to 50 metres near the beach. In the upstream part near the old floodgates the channel is only 3 metres deep and gradually deepens downstream to 5 metres. The channel ends on Perkins Beach 1.5 kilometres west of Mutton Bird Island (Shelter Island), the position varying about 100 metres to the east or west depending upon the direction of swell and longshore drift at the time of opening. The flood tide delta is poorly developed and probably washes out when the bar breaks.

The bar is part of the beach composed of well-sorted medium to fine sand 200 metres wide and built up to 1.2–1.5 metres above sea level. The bar is usually broken by the Water Corporation when water level rises 0.8–1.3 metres above mean sea level (AHD) to prevent flooding of properties on the western side of Torbay Inlet. It may open and close a number of times during the year. The bar generally stays open longer when opened in winter, and usually stays closed over summer. Waters within the inlet can vary from fresh when the river is flowing to the sea, or have a gradient from sea water near the bar to fresh water in the Marbelup Drain. Sea water penetrates upstream as far as a bund on the original river causing a saltwater stratification in the deeper section of the drain for much of the year.

What a state

Torbay Inlet and Lake Powell are eutrophic, containing high concentrations of nutrients (see Graphs 1.2 and 1.3). Until the late 1990s algal blooms were not a great problem in Torbay Inlet although there were some brief but dense blooms of blue-green algae before the floodgates were blown up in 1985, and some more recent growths of the wire-like red alga *Gracilaria*. More recently there have been blooms of toxic blue-green algae such as *Anabaena*, *Microcystis*, *Nodularia*, *Anabaenopsis*, straight chain *Anabaena* and *Pseudanabaena*. Rotting algae have also caused a serious midge problem. There are also regular algal blooms in Lake Powell, which was once very productive with freshwater mussels and Murray Cod introduced in the 1880s, but these disappeared when the waters became quite acidic and the algal blooms increased. The fish decline has also been attributed to weed spraying along the drains. Water in Lake Powell is now neutral to alkaline particularly during algal blooms.

The reasons for differences in algal blooms in the two water bodies are not clear. They may be due to the tannin-stained waters and the seasonal flushing with saltwater in Torbay compared to the lack of flow, higher nutrients and temperatures in Lake Powell.

The problems in Lake Powell highlight the long legacy of past land use. Waste water, now treated near the airport and used to irrigate blue gum plantations, once flowed directly to Five Mile Creek. Nutrients from a piggery, now closed, and used as a blue gum plantation are probably still moving from the surrounding low-lying areas through to the lake. Although some streamlines are in good condition, many are degraded by livestock and no longer provide a buffer to the larger water bodies downstream. Despite water quality problems Lake Powell is a nature reserve and the whole area is an important feeding and breeding habitat for waterbirds including migratory species.

THE SMALLER ESTUARIES — WATERY REFUGES

Taylor Inlet

Taylor Inlet lies between the hard gneiss rock of Mount Taylor and the coastal dunes of Nanarup Beach (Map 9.11, also Map 10.1). This is a tiny estuary, less than a square kilometre, of similar size to Torbay, consisting of a lagoon, inlet channel and a few metres of creek. One small creek drains the mallee–heath area of Mount Taylor–Mount Richards to the west, while another two creeks bring water from the now largely cleared lateritic soils and jarrah woodland to the north. It is really a saline lake that has retained its estuarine state because of the relatively sheltered location of the entrance. Despite its pretty setting, the estuary is eutrophic. There have been reports of the green alga *Chaetomorpha* growing near the creeks, and prominent signs caution against swimming because of toxic blue-green alga and amoebic meningitis.

The surrounding area, although allocated for farming in the 1870s, was not cleared for cattle grazing until the 1960s. Since then change has come fast, and today most of the catchment is cleared with a narrow Recreation Reserve around the edge of the inlet. High above an old house commands a view of the inlet and ocean, and a few fisherman's huts on the eastern shore and one on the western shore are built on very low-lying land next to the inlet (Fig. 9.18).

The inlet is fringed by the salt tolerant paperbark *Melaleuca cuticularis* with an understorey of sea rush *Juncus kraussii*, creeping brookweed *Samolus repens*, club rush *Isolepis nodosa*, and saltbush *Atriplex cinerea* backed by coastal saw-sedge *Gahnia trifida* and bare twigrush *Baumea juncea*. Further back there are freshwater paperbarks *Melaleuca rhaphiophylla* and pea shrub *Oxylobium lanceolatum*, which also line the river banks together with peppermints *Agonis flexuosa* on higher ground behind.

The lagoon lies behind the coastal dunes with a steeply wooded slope on the north-western shore rising to 30 metres, and a gentle slope on the western shore that then rises more steeply to Mount Taylor. Shallow banks

FIG. 9.18 *Taylor Inlet, aerial view, June 2000. The inlet, although pretty, is surrounded by farming land and the waters are eutrophic. The catchment is only 10 km² in area. The inlet channel area provides access to Nanarup Beach where salmon are fished.* S. Neville/Ecotones, Courtesy WRC.

surround the main basin but most of the lagoon is 2–3 metres deep, with holes up to 5 metres deep near the northern creek and the inlet channel. The water level can vary nearly 2 metres from a metre above sea level when full to a metre below sea level at the end of summer. The water is deeply coloured by tannins, and salinity varies with the seasons from near sea water (35‰) to less than half this during winter due to freshwater inflow of about 1,000 megalitres per annum.

The inlet channel, only 300 metres long and 100 metres wide, is very shallow and the lagoon is gradually filling with sand. The bar is part of the gently sloping beach at Nanarup and is only 1 metre above sea level. Nanarup Beach is sheltered from the south-west by Islet Point and a small island (both granite–gneiss) 600 metres to the south. A broken limestone reef about 200 metres offshore also protects the beach line. The beach is composed of poorly sorted coarse to fine sand.

The bar is closed most of the time but has been broken both naturally or artificially in most years since the 1890s, usually staying open for a few weeks. As with most bars, there is a conflict over why and when the bar should be opened. The inlet was originally opened in the 1890s to create an area suitable for the cultivation of Sydney Rock Oysters on the north side of the lagoon. Currently it opens naturally or is opened by the City of Albany to lower the water level to prevent salt water intruding into shallow bores, and to prevent flooding of the road. Fishermen sometimes break the bar to allow fish to enter. The length of time the bar stays open depends on when the bar is opened, the height of water in the estuary, and conditions along the coast including the longshore movement of sand, sea level height and swells. Growth of the inlet delta is aggravated by erosion along fishing tracks to Nanarup Beach. This may lessen, however, as vehicle access except for professional fishers during the salmon run has been stopped.

Because of the concern for water quality, the Water and Rivers Commission sampled the inlet during 1999 in April and August. At these times the bar was closed but opened sometime in August. In April the inlet water was about 24‰ and the water column was well mixed. The salt was thought to have been due to the previous period of marine intrusion, diluted by freshwater run-off. The water was supersaturated with oxygen, indicating that algae in the phytoplankton were active.

In August the water level was a metre higher, and slightly stratified with surface water about 25‰ and about 28‰ below 2 metres in the slightly deeper part of the inlet near the rivers and away from the bar. The slight increase in salinity was thought to be due to a small amount of sea water washing over the bar. Generally the well-mixed condition was due to winds. The surface water was still supersaturated with oxygen, but the saltier water below the halocline was only 20 per cent saturated.

Nutrient concentrations were generally low in April 1999 but in August they were higher in deeper low oxygen areas away from the bar. In April ammonium and nitrate were below these guideline values, but dissolved phosphorus was substantially higher. In August, except for anaerobic areas, ammonium had risen above guidelines while phosphate had declined and nitrate remained low. In April the ratio of inorganic nitrogen to dissolved inorganic phosphorus (0.25:1) indicated that phosphorus was in excess and algal growth was probably limited by nitrogen. Nitrogen is possibly increasing in the sediments and not being released through denitrification. Also it appears that phosphorus is being released from the sediments. Concentrations of total nitrogen and total phosphorus are higher than in Wilson Inlet and Oyster Harbour. (See Graphs 1.2 and 1.3.)

Chlorophyll *a* concentrations were low to moderate in April 1999 (0.0006 milligrams per litre) composed mainly of chlorophytes and cryptophytes along with a few dinoflagellates and diatom. Chlorophyll *a* concentrations are similar to those of Parrys Inlet.

Normans Inlet

The river and lagoon of Normans Inlet lie in a small valley between high granite hills just behind the high coastal dunes (Figs 9.19–9.21). The inlet opens to the sea at Normans Beach in the shelter of North Point to the east of Two Peoples Bay (Map 9.6). The tiny water haven is located on the border of private land where a futuristic dome house overlooks the open ocean, Two Peoples Bay and Bettys Beach.

The catchment is only 18 square kilometres and drains the spongolite plain to the north and western side of the Mount Manypeaks Range. In the upper reaches the river descends through a rocky gorge before entering a tangle of paperbark trees where it is about 3 metres wide, becoming navigable over the last kilometre to the lagoon.

The lagoon covers half a kilometre in area and is only about 50 centimetres deep, except for a narrow channel along the eastern shore. The lagoon base probably lies at about mean sea level, and when the bar breaks most of the bottom is covered by only a few centimetres of water. On the eastern side the ground rises steeply to the granite hills of the Mount Manypeaks Range. The western side is low lying with some paperbarks *Melaleuca cuticularis*, Christmas trees *Nuytsia floribunda* and sea rush *Juncus kraussii* with a small cleared area at the base of the hills.

The inlet channel is about 200 metres long and 20 metres wide, and is slightly deeper than the lagoon at about 1–1.5 metres deep. On the eastern side the steep dune is covered with dense vegetation to the water's edge and on the western shore the cliff of soft dune limestone is falling into the channel.

The bar generally opens every year in about October but only stays open for a few days generally closing before the inlet has drained completely, probably because there is not enough flow to keep it open. It is reported to have stayed closed for about four years in the 1960s. The bar is part of the beach about 1.5 metres above sea level and 2 metres above the limestone platforms that edge the beach. Storms sometimes move most of the sand offshore, exposing limestone to the base of the dunes. The beach sands are predominantly quartz of medium to fine grain.

The river water is deeply stained with tannins, and the lagoon appears to be about one-quarter the salinity of sea water with a gradient to fresher water near the river. Only a small amount of sea water enters the lagoon when the seas are high or during the short time that the bar is open. The flora and fauna reflect the low salinity with a gradation to fresh water in the

FIG. 9.19 *Normans Inlet and King Creek, aerial view, January 1990.* Courtesy DLI.

FIG. 9.20 *Normans Inlet, aerial view, June 2000. The inlet is surrounded by high wooded hills of the Mount Manypeaks National Park.* S. Neville/ Ecotones, Courtesy WRC.

FIG. 9.21 *Normans Inlet, tranquil waters, a view upstream from near the bar, March 1987.* E. P. Hodgkin Collection.

upstream part. In the lower part, the seagrass *Ruppia* and fine pond weed *Lamprothamnium* sometimes grow quite densely. The fauna is predominantly estuarine with false mussels *Fluviolanatus subtorta*, small black snails *Hydrobia buccinoides*, shrimps *Palaemonetes*, and isopods (pill bugs). The low salinity water is also suitable for larvae of caddis flies, damsel flies and dragonflies, the adults of which are a spectacular sight as they dart among the rushes.

Waychinicup River and estuary

Waychinicup River is quite different to other south coast estuaries that are surrounded by low-lying land (Fig. 9.22; Maps 9.7, 9.11 and 10.1). Waychinicup lies in a gorge carved through granite. Rocky headlands on either side of the mouth maintain a permanent opening to the ocean allowing free exchange of water through tides and swell. The gorge, 1,300 metres in length and 2 metres deep, is essentially the arm of the sea with no sand or rock bar at the entrance; more a fiord than an estuary. The water is only diluted by fresh water briefly when there is strong river flow.

FIG. 9.22 *Waychinicup River, aerial view, June 2000. Waychinicup opens to the sea between high granite headlands.* S. Neville/Ecotones, Courtesy WRC.

FIG. 9.23 *Waychinicup Inlet, December 1989.* E. P. Hodgkin Collection.

The only riverine source, the Waychinicup River, has a catchment of 145 square kilometres, 40 per cent of which is cleared. The lower reaches and estuary lie within the Waychinicup National Park, an area noted for its diversity of flora. The 17-kilometre long river has a number of rock pools along its length. Upstream, the valley has a flat swampy floor in the spongolite and sandplain with headwaters on grazing land north of the South Coast Highway, 100 metres above sea level. The lower 3 kilometres of the river passes through a valley 40–60 metres deep, incised into granite with smooth slopes and occasional rock outcrops (Fig. 9.23).

The unusual beauty of the inlet, where ocean swells send pulses of water up the calm reaches, has attracted visitors for many years but its history is largely undocumented. The coastline was explored by the earliest European whalers, sealers and settlers. A domed structure marked on maps as a 'sealer's oven' on the eastern side may date from this time (Fig. 9.24). Alternatively, it may have been built during the early years of settlement by shepherdesses who used the area as a watering point for their flocks. More recently, until 1987, the 'hermit' Frank Cooper lived for 30 years in a shack on the western side from where he overlooked the inlet and supervised the visitors. The stonework of his camp is now guarded by large lizards. Although the inlet lies in a national park tourism is increasing, and the demand for potable water for urban developments at nearby Cheyne Beach and Albany could pose a threat to the inlet's health.

FIG. 9.24 *Sealer's or shepherdesses' oven on the eastern shore, April 2000: a reminder that Europeans have been visiting this special area for many years.* A. Brearley.

The inlet was largely unstudied until September 1995 when water quality, plants and animals were investigated. Records of water flow indicate that monthly and yearly flows are highly variable with most flow occurring in winter (June–September), a mean annual flow of 9.15×10^6 cubic metres (1970–1995) and about 870 milligrams per litre salinity.

Salinity and temperature at the surface were slightly higher at the seaward end, 33.9‰ and 16°C compared to 21‰ and 14.7°C at the head of the estuary, with slightly warmer saltier water at the bottom. Light penetration was also greater nearer to the sea, but oxygen concentrations were consistently high throughout the estuary. Concentrations of dissolved inorganic nitrogen and phosphorus changed with river flow. Phosphorus concentrations were very high near the river mouth, particularly in the fresher surface layer (0.11 milligrams per litre) that was almost four times the concentration of the bottom layer. Inorganic nitrogen, principally nitrate, was 0.07 milligrams per litre at the surface and 0.03 milligrams per litre at the bottom. As a result there was a gradient in N:P ratios of the water column. The upper and middle estuary had ratios less than 16:1, possibly indicating that nitrogen might be a limit to algal growth. Although these records were only a single 'snapshot' of water quality, they are comparable to those in highly eutrophic estuaries of Swanland indicting that despite its pristine appearance the waters of Waychinicup show signs of eutrophication. The study in 1995 was conducted during a period of low river flow, and tidal exchange was probably reducing the retention of nutrients.

The sediment also changed along the estuary, the amount of organic material decreasing away from the river mouth towards the sea. Fine material followed a similar pattern with the finest sediment found near the river mouth and coarse sand and shell fragments near the ocean. In the middle of the estuary the seagrass meadow had the greatest mixture of sediment types (fine and coarse sand, organic material and shells) and organic material.

The water world

Forty species of macroalgae and six species of seagrasses have been recorded in the estuary. This is a very rich flora compared to most other south coast estuaries, which are dominated by only three species *Ruppia megacarpa*, *Polyphysa peniculus* and *Lamprothamnium papulosum*. The algae and seagrasses in Waychinicup, like the Albany Harbours, are more typical of southern coastal marine areas, not estuaries.

The macrophytes can be considered in two main groups: those on the estuary floor and those on the vertical rock walls. Along the estuary floor, fine green alga *Chaetomorpha* and red algae occurred in the upper estuary. The green alga in particular is a species adapted to changing salinity and low light typical of dark tannin-stained fresh water in this area. A mixed assemblage of algae dominated by the brown seaweed *Cystoseira trinodis* was restricted to an area of hard substratum between 150 and 300 metres from the river mouth. Elsewhere, the substratum was muddy to sandy and supported seagrasses. The seagrasses *Posidonia australis* and *Zostera* (formerly *Heterozostera*) *tasmanica* were widespread, although they were generally limited to areas more than 200 metres from the river mouth. *Posidonia sinuosa*, a seagrass usually found in higher energy environments, was found only near the mouth of the estuary. Two other species, *Halophila australis* and *Amphibolis antarctica*, occurred in small patches throughout the estuary.

The vertical rock walls in the lower reaches of the estuary supported a mixed algal assemblage dominated by brown and red algae. *Cystophora brownii* and *C. retorta* dominated along the north-western shore. Along the channel area there was a vertical distribution of algae with the *Cystophora* species near the surface, *Halimeda cuneata* and *Metamastophora* sp. slightly lower and *Zonaria turneriana*, *Hypnea* and *Jania micrarthrodia* near the bottom below 6 metres.

The fauna can be classified into upper, middle and lower estuary groups. The upper estuary group were most different. Small crustaceans dominated sites at the mouth of the estuary while polychaetes were dominant in the upper reaches. The change corresponded to the different sediment types with polychaetes favouring the finer, more organic-rich sediments of the upper estuary. Most of the fish are marine species found in the nearby bays, and Waychinicup is fished commercially.

Despite the small amount of river inflow and the good exchange with the sea, the riverine flow is an important biological determinant affecting salinity and turbidity. The gradation from estuarine to marine flora has not been recorded from any other system along the south coast. It could be argued that Waychinicup is a simple river mouth with granite headlands, but overall it has the character of a seasonally stratified estuary.

Cordinup River

The Cordinup River with a 104 square kilometre catchment area has carved a narrow valley through the granite rock to the east of Green Range and Warriup Hill. The small estuary opens to a 2-kilometre long bay in the rocky coastline between Hassell Beach and Cheyne Bay (Fig. 9.25).

FIG. 9.25 *Cordinup River, aerial view, May 1989. The western shore was planted with blue gums in the late 1980s and the first harvest completed in 2002.*
E. P. Hodgkin Collection.

A small stone house built by H. W. Hassell recalls a period when the country was first settled in the 1920s and 1930s. The land was cleared for grazing in the 1970s.

The river, which is only 10 metres wide and about 2 metres deep, is estuarine for about a kilometre (Map 9.7). In the riverine part it winds through a delta and alluvial flats that surround the northern part of the lagoon. The saltwater paperbark *Melaleuca cuticularis* forms a fringe along the lagoon and the lower riverine part of the estuary, and along the two spits in the lagoon with generally sparse sea rush *Juncus kraussii* except in the more open parts. The flats are covered with samphires and dead paperbarks trees (reported to have been dead for many years). The fringing vegetation has been and is still disturbed by cattle and sheep, and there are weeds, predominantly pasture species, in some areas.

The lagoon can reach 2 metres deep before the bar breaks, but is only about 0.5 of a metre when open to the sea. The lagoon is gradually filling with sand from a dune encroaching from the beach.

The inlet channel, 100 metres long and 20 metres wide, passes through steep well-vegetated dunes for a short distance before the beach. The bar is reputed to break when the water level in the inlet is high, about every two or three years, although it can break more often and winter storm waves may wash sea water inwards. The water is turbid when heavy rain falls in the catchment but otherwise the water is very dark due to the high tannin content, or clear when the bar is open and sea water washes inwards. Salinity is seldom more than half sea water.

Little is known about the animals living in the estuary but they appear to be 'estuarine species', those found only in estuaries and not in adjacent marine and freshwater areas. The inlet is a popular fishing spot for bream. Turtles are reported to be quite common; their presence possibly indicating that a freshwater phase may dominate at times.

Cheyne Inlet

Cheyne Inlet is the estuary of the Eyre River that opens into a small bay on the north side of Cape Riche, a granite rock headland. The mouth faces north-east near the south end of the 12-kilometre long Cheyne Bay (Fig. 9.26; Maps 9.9 and 10.1).

The upper part of the catchment extends onto grazing land at about 80 metres above sea level, and 61 per cent of the 77 square kilometre catchment is cleared. The area south of the inlet has been farmed since 1839, first by George Cheyne and later by the Moir family who still graze sheep and cattle on the alluvial soils. The small settlement was used as a depot by many of the early explorers and travellers to the Mount Barren area. (See **Chapter 10 Estuaries to the east of Albany: Crisscrossed paths—travellers, collectors, explorers and miners**.) To the north and west the terraced spongolite slopes are still bush, and Crown Land north of the river is proposed for vesting as a conservation park. The small headland immediately north of the lagoon is on privately owned land and is mostly still bush. Parts of the riverbanks are well vegetated with paperbarks and sedges, but there have been concerns that these are becoming degraded.

The lagoon is a kilometre long and only 100–200 metres wide, except where a small swampy bay forms a loop at the upstream end. Although shallow (about 1 metre) it never dries out completely, and bristle worm tubes high on the rocks indicate that at times before the bar breaks it may be much deeper. It is reported to have been much deeper in the past when the main basin was about 2.5 metres and as deep as 6 metres near the bar. Large floods can scour a channel to bedrock (about 2.3 metres) and shell beds along the banks are evidence that the lagoon was once more open to the sea.

FIG. 9.26 *Cheyne Inlet, aerial view, June 2000. The Eyre River has cut a narrow valley through soft spongolite rock, and about 4.5 km is estuarine including the narrow lagoon. Upstream there are small pools in the river and these are reported to hold fresh water. It widens to 20 m where it opens into the lagoon.* S. Neville/Ecotones, Courtesy WRC.

The bar, only 1.5 metres above sea level, is composed of fine white quartz sand that becomes quicksand when wet. The bar often contains a large amount of decaying seagrass. The bar breaks naturally most winters following heavy rain, and occasionally in summer following cyclonic rains. It generally stays open for a few weeks, but once in the 1940s and again in the 1950s it remained open for about 18 months. In 1930s it is reported to have stayed closed for nine years with black bream the only surviving fish. The bar was broken artificially for the first time in 1975. Salinity varies from about 10‰ in winter to about 50‰ in summer.

The fringing vegetation is dominated by salt marsh with the saltwater paperbark *Melaleuca cuticularis*, sea rush *Juncus kraussii*, seablite *Suaeda australis*, saltbush *Atriplex cinerea* and samphire *Sarcocornia blackiana* in the lower estuary. Further upstream, about 2 kilometres from the bar, the banks carry a fringe of *M. cuticularis* and *J. kraussii* with coastal sword-sedge *Gahnia trifida* behind, and bullrushes *Typha* near the head of the estuary.

In 1988 infrequent but considerable growth of filamentous green algae were reported to have occurred in the lower reaches of Cheyne Inlet, possibly indicating the nutrient concentrations were high at times. The inlet, however, was not considered to be seriously eutrophic. In response to growing concerns about the health of the inlet in the late 1990s, the Shire of Albany and the Water and Rivers Commission funded four quality surveys. River flow was very low throughout the survey period and was not sufficient to break the bar. Salinity was highest (25‰) in March 2000 and lowest (20‰) in June 2000 (Graph 1.2). Comparison with the earlier data indicates that although nutrients were slightly higher than in 1988, these varied over the recent survey period with the highest concentrations of total phosphorus, ammonium and nitrate in June 2000 after the onset of winter flows. Phosphorus was highest in March 2000.

Swan grass *Ruppia* grows in the lower estuary and the green alga *Ulva* is also quite common further upstream. *Ulva* is uncommon in other south coast estuaries but on the west coast it has often been associated with nutrient enrichment and freshwater seepage. Compared to Normans Inlet and Cordinup River, the other two small estuaries nearby, the higher salinity of Cheyne Inlet favours a more marine fauna including cockles *Katelysia* and tellin shells *Soletellina* (formerly *Sanguinolaria*) that sometimes grow to maturity in the lagoon close to the bar.

Little water havens—lessons to be learnt

Most of the small estuaries lie in small nature reserves where their picturesque settings are a focus for visitors and havens for water life in a largely cleared landscape. Because the surrounding land use has changed, the major threats are eutrophication due to increased nutrients, and increased sedimentation due to increased erosion. Insecticides and herbicides from crops are also potential threats to animal and plant life in the rivers and estuaries.

The immediate surroundings are also changing. Around many of the inlets eroding dunes cascade into the entrance channels. As a result of increased visitor access, feet and tyres trample the fringing vegetation. Finally, the premature breaking of the bars has reduced scouring, allowing sediment to accumulate. Together these human impacts are changing the character of Swanland's smaller southern estuaries.

The ongoing health of the small inlets, like the larger water bodies, will depend to a large extent on management in the catchment, although their small size increases their vulnerability. Some are near the end of their estuarine life as they become shallower and more isolated from the sea, but their health as coastal lakes should still be valued. We need to take care that we do not love them to death through overuse, as some of the greatest threats lie in damage to the immediate surroundings.

Chapter 10

Estuaries to the east of Albany

East of Albany, the lower rainfall and the underlying rock formations transform the colours and the contours of the landscape. The trees become smaller, changing to the mallee habit as we move from the Menzies Jarrah–Marri Forest Region to the mallee–heath of the Esperance Plains Region. Infrequent and unseasonable heavy rains and floods in summer can be as important to the estuaries as the seasonal winter rainfall. Evaporation is high from the shallow water of the lagoons and they often dry out in summer. When there is water, salinity varies from one-third to about twice that of sea water. The waters of the eastern estuaries are often murky due to fine silts, unlike those further west where the waters are tannin coloured but generally clear.

The estuaries no longer fit the conventional definition, a claim they probably forfeited about 3,000 years ago (Prelude Map 3; Tables 1.7 and 1.8). Some (Beaufort, Wellstead, Gordon, Fitzgerald, Dempster, Hamersley, Oldfield and Stokes) have been termed 'Normally Closed Estuaries' for they are seldom open to the sea, the bars opening only following heavy rain. Others that have opened even less often have been termed 'Permanently Closed' (Culham Inlet and Jerdacuttup Lakes). 'Permanently closed' is a term that needs to be used with care as it suggests a degree of stasis and finality, whereas it reflects our short-sighted view of natural cycles, for these estuaries still open to the sea after particularly heavy rainfall. We have little real idea of how often this may be because of our short history (less than 150 years) of recording weather patterns. (See **Chapter 1 Understanding Swanland: Classifying Estuaries** and **Climate future—less rainfall?**)

As well as these larger estuaries, there are also small rivers and lakes such as Hunter River, Kellys Creek, Boondadup River, Saint Mary River, Lake Nameless, Torradup River and the lakes around Esperance, where the earlier estuarine state is recorded in the topography and fossil shell basement but the sea barrier is covered with a series of dunes. Changes in sea level and sedimentation together with low volume and sporadic flow mean that these small estuaries have progressed further along the path to extinction.

Most of the estuaries in this eastern part of the coast are relatively small and valley-like. The Beaufort, Hamersley, Stokes and Oldfield have high bars (2–3 metres) and only open briefly. The water depth can vary by as much as 3 metres when the bars are closed. Fitzgerald and Saint Mary also break infrequently and stay open briefly, but they are much shallower.

The Beaufort entrance at the western end of a long south-facing bay is sheltered by a headland from the south-west, while the entrances of the Hamersley, Oldfield and Stokes are exposed to the south although they have some protection from bars or reefs. The Wellstead and Gordon are in relatively sheltered locations: Wellstead in a deep bay and the Gordon on an east-facing coast with low dunes, and their bars stay open for several seasons. Stokes Inlet with deep water is more similar to Oyster Harbour, but the high sea bar only breaches briefly every few years so that the water level and salinity vary greatly with river flow and evaporation.

Box 10.1 SPONGE GARDEN TO ROCK FORMATION— THE PALLINUP SILTSTONE OR SPONGOLITE

The spongolite of the Pallinup Siltstone is a pale coloured lightweight rock, composed largely of tiny glassy silica spicules, skeletons of sponges that carpeted the relatively deep quiet water ocean floor 40 million years ago. The sponge community of many different forms, about thirty-two species, is recognised as one of the most diverse in the world. Around Albany this formation can be up to 60 metres thick.

The Pallinup Siltstone overlies the older Werrilup Formation. Around the Nornalup and Denmark areas, the Werrilup consists of sandstone, clay and lignite (a low-grade coal). East of Albany between Nanarup and Manypeaks, the Werrilup Formation contains the Nanarup Limestone and many marine fossils. The Nanarup Limestone is never more than 12 metres thick, and was deposited in shallow water around the high areas of much older granites and gneisses (Stirlings, Porongurups, and the high coastal hills like Mount Melville). The warm waters teemed with life: bryozoans that make up most of the deposit along with bivalve shells, sea-urchins, starfish, brachiopods, corals, crabs and more rarely sea lilies. Sponges, so common in the Pallinup Spongolite, are less common in the Nanarup Limestone.

Some of the fossil material is broken and the fragments of calcite form the limestone rock. Some larger fossils are often represented by only a few spines, fragments or moulds of the interior or exterior, so recognising their origin requires great skill and experience. There would have been many animals in these ancient seas, and fossils indicate only part of the community that lived in an area. Many animals are not represented in the fossil record because they did not have hard skeletons so they did not preserve, the form of calcium carbonate (aragonite) skeleton or shell leached away, or they were eaten, decayed or broken by waves.

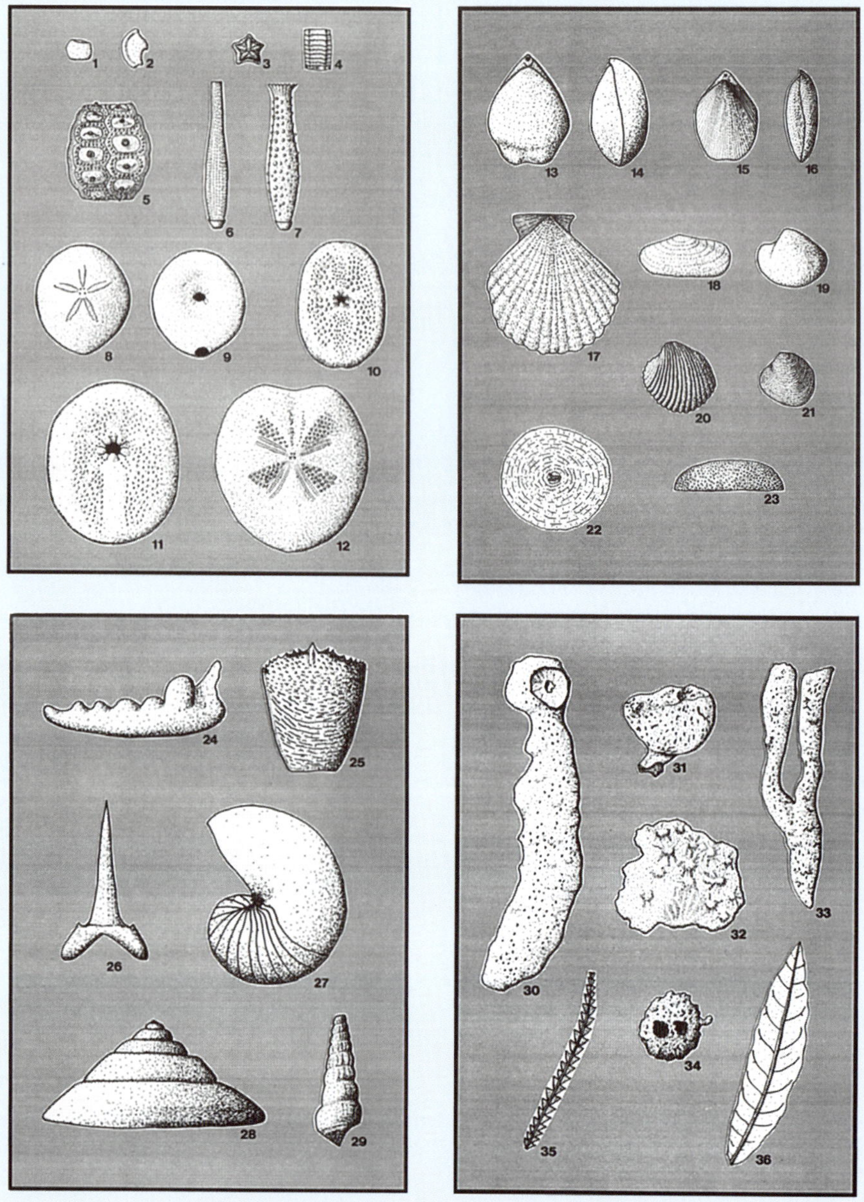

FIG. 10.1 *Fossils in Pallinup Siltstone (spongolite) and Nanarup Limestone: Echinoderms 1–12; Brachiopods 13–16; Bivalves 17–21; Bryozoans 22 & 23; Crustaceans (crabs) 24 & 25; Fish 26; Nautilus 27; Gastropods 28 & 29; Sponges 30–33; Terrestrial plant leaves 35 (banksias) & 36.* Drawings drafted by Dr Ken McNamara, WAM.

In the Nanarup Limestone the most common fossils, and perhaps the most easily recognised, are echinoderms (sea urchins), although much of the rock is made up of the broken skeletons of coral-like bryozoans. A few echinoderms are also found in the Pallinup Siltstone. There are two types, regular and irregular. The regular urchins are rounded in shape with thick heavy spines (often the only part preserved) similar in form to urchins found today on rocky shores where they graze on algae. The irregular or egg-shaped urchins have very fine spines and a five-petal pattern on the topside, and on the bottom side a central mouth and posterior anus. Some are pear-shaped and others flattened discs. Most irregular urchins bury in sand and mud.

Bivalve molluscs are also common in the Pallinup Siltstone with twenty-three types so far recognised. The shell of bivalves is in two parts of similar size and shape, and they are not bilaterally symmetrical. One of the most common, the scallop *Chlamys*, can be up to 60 millimetres in size. There are also brachiopods or lamp-shells. These are similar to bivalve molluscs, but their valves are of different sizes and they are bilaterally symmetrical down the mid line with a hole in the pointed end where a fleshy stalk attaches to a rock or shell.

Many other fossilised animals have been found in one or both rock types: crab claws are found in the Nanarup, shark teeth and nautilus-like shells (related to squids and cuttlefish) show that these carnivorous species were present in both formations, and a great variety of marine snail shells are abundant in the Pallinup Siltstone but relatively less common in the Nanarup Limestone.

The Pallinup Siltstone is notable because it also contains land plants: leaves, wood and fruit. Large chunks of fossilised wood of plants like banksia that washed into the sea from the Porongurup 'islands' have been found around the base of the Porongurups. Because we know the age of the marine animals in the same area, we can determine the age of the plant material. This helps with understanding how the Australian flora evolved. The fossils and history they reveal tell us that Australia was much further south, separated from Antarctica by a narrow shallow sea and that the seas around Antarctica were much warmer. This helps us to interpret the ocean currents and climate, and helps with understanding climate change on a global scale. The diversity of plants such as banksia, casuarinas and the southern beech *Nothofagus* indicates that the surrounding area was a moist temperate forest. The Nanarup Limestone does not contain fossil plants, although parts of the Werrilup Formation, of which it is a part, around Walpole contain wood and some banksia cones. These along with others from the Kennedy Ranges of the Gascoyne region are among the earliest evidence of banksia in Australia.

Rocks containing fossils are often buried under layers of sediment and bush, exposed only in quarries or road cuttings. Such places are often areas on private property, and collectors need to seek permission of landowners before collecting fossils. As with any specimen collecting, there are three mottoes to bear in mind: **R**estraint, w**R**apping, and **R**ecording. Restraint in extracting a fossil from its bedrock base, in taking only what is really needed and leaving some for others. Fossils can be very fragile so wrapping to prevent damage is most important. Recording details of where the fossils were collected with a description of the area need to be included with the packaging, as our memories are unreliable and specimens are only valuable in telling a story when we know their origin.

Source: McNamara, 1992.

The less frequent and shorter period of connection to the sea has large effects on the ecology of the water bodies. They dry out more often and the waters become more saline creating conditions tolerated by few plants and animals. The shorter period of inflow from the sea also reduces the chance of recruitment of animals when water conditions are favourable. Only five or six species of mollusc live permanently in these estuaries, although additional species typical of sheltered bays sometimes enter and flourish for a short time after the bars open. In comparison, Permanently Open Oyster Harbour and Princess Royal Harbour only a few hundred kilometres west have more than 100 species living in them. The fish fauna is also poor, with 102 species in Permanently Open estuaries, sixty-four in Seasonally Open, forty in Normally Closed, and only four in Permanently Closed. However a few species can be very abundant, providing a rich source of food for birds such as pelicans and cormorants as well as the basis for recreational and commercial fisheries.

THE ANCIENT LAND—ANCIENT SHORELINE

Some of the larger rivers rise from the northern watersheds in the salt lake country on the ancient Yilgarn Plateau (Maps 1.1, 1.2, 1.6 and 1.10). Here the basement rocks are Proterozoic granitic rocks (migmatites more the 1,200 million years old), but near the coast they are covered by the Eocene Pallinup Siltstone that takes the form of siltstone or spongolite (40 million years old) (Figs 10.1 and 10.2). The spongolite rock consists of flat-bedded marine

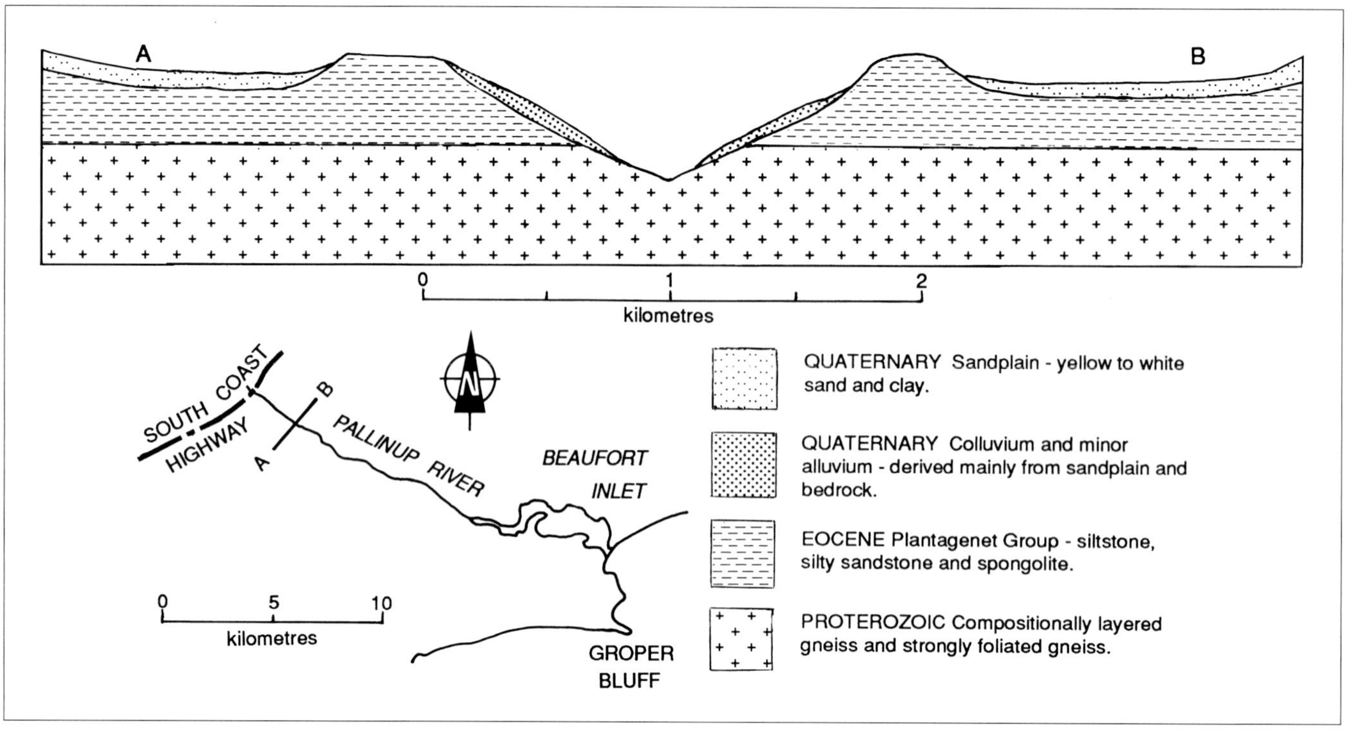

FIG. 10.2 *Geological cross-section of the Pallinup River valley showing the relationship of strata; vertical scale is exaggerated.* Adapted from Hodgkin & Clark, 1988 (ESS #4) originally from Bill Moir. *The Plantagenet Group indicates rocks laid down in an ancient sea when Australia was much further south near Antarctica.*

sediments with many sponge spicules; it is relatively soft and the rivers have cut spectacular gorges though to the basement granites. (**Box 10.1 Sponge garden to rock formation—the Pallinup Siltstone or spongolite**.)

The landscape of the eastern south coast is overlain by Quaternary deposits formed between two million and 10,000 years ago, sand plain on the higher land and eroded colluvial and alluvial deposits in the valleys. At the coast the higher dunes have a core of Pleistocene dune limestone covered by more recent Holocene dunes formed in the last 10,000 years. Dune sand can extend up to 5 kilometres inland from the coastal dunes, either anchored by vegetation or mobile in blow-outs.

Along the coast the Mount Barren Group form a range of hills from Middle Mount Barren to East Mount Barren near Culham Inlet. Schist and quartzite rocks exposed along the Phillips, Hamersley and Dempster Rivers possibly follow old fault lines. Schist also outcrops at Point Charles and Point Ann on a coastline otherwise dominated by dunes. Inland from Point Ann, the Mount Bland and West Mount Barren hills are quartzite. The Mount Barren Group is similar in composition to rocks in Antarctica, marking the point at which the two continental masses separated 50 million years ago in the Eocene. Prominent 'wavecut' benches at about 80 metres on the seaward face probably mark earlier sea levels of Eocene age when the Pallinup Siltstone was deposited.

FACING THE SOUTHERN OCEAN

Jerramungup Shire

In the western part of the region managed by the Shire of Jerramungup, the three estuaries and the estuarine reaches of the rivers have a roughly north-west to south-east alignment (Map 10.1). The mouths are partially protected by headlands from the prevailing winter south-westerly winds and swell.

The mouth of Beaufort Inlet faces south-east near the western end of 25-kilometre long Wray Bay, protected from the south-westerly winds and

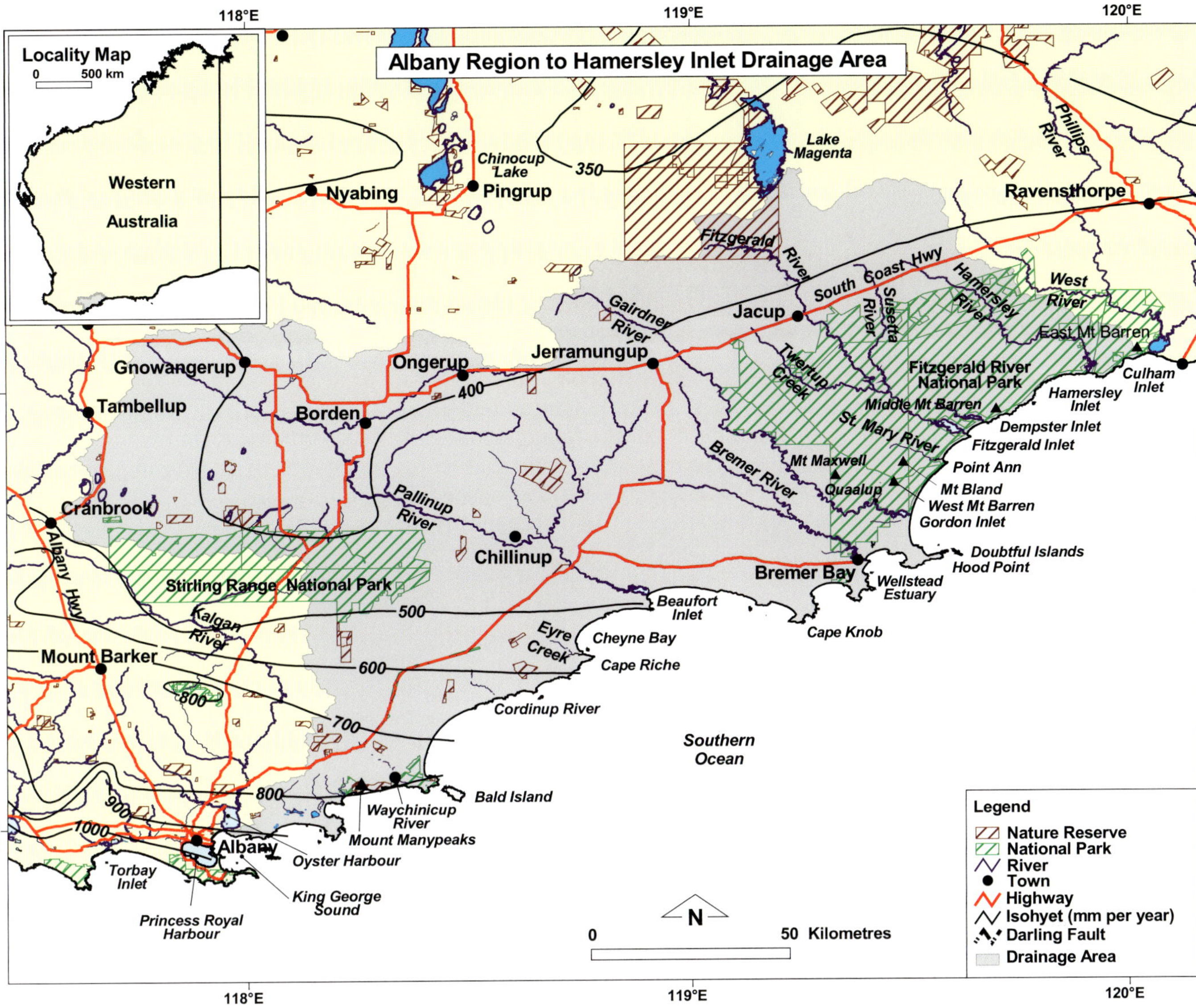

MAP 10.1 *Catchment and drainage of the Albany region to Culham Inlet: the rainfall is much lower than the area around Albany to the west and decreases rapidly with distance from the coast. The coastal areas of some catchments and estuaries lie in the Fitzgerald River National Park.* Courtesy DoE.

swell by Mount Groper and Groper Bluff (Maps 9.4 and 9.10). Wellstead Estuary faces east at the extreme south-western end of Bremer Beach, deep in the shelter of Bremer Bay. In the middle of Bremer Bay, the Hunter River flows to a freshwater lake impounded by a marl bar covered with sand. North of the prominent Hood Point, Gordon Inlet opens directly onto the long, east-facing 10-kilometre long Tooregullup Beach.

The Fitzgerald River National Park

Eastwards from Gordon Inlet to Dempster Inlet, the 40 kilometres of sandy coastline is backed by Pleistocene and Recent dunes that face first south-east and then east. Two rocky (schist) headlands at Point Ann and Point Charles break the sweep of the coast and shelter the mouths of Saint Mary River and Fitzgerald Inlet respectively from south-westerly winds and swell. Inland from the dunes the inlets are bordered by flat-topped hills with deep valleys and cliffs cut in the spongolite rock. Quartzite is exposed in the Mount Bland—West Mount Barren range of hills through which the Boondadup River drains, and waterfalls mark the quartzite ridges.

Between Dempster Inlet and Culham Inlet jagged cliffs and narrow wave-cut benches in the quartzites and schists of the Mount Barren Beds form a very different coastline to the rounded rocks of granite coasts (Fig. 10.3).

The rocky shore is only interrupted by short stretches of dune sand and rock, the largest of which forms the small bay east of Hamersley Inlet, the largest estuary along this stretch of coast.

Eastwards to Esperance

The 200 kilometres of coast from East Mount Barren near Culham Inlet and Hopetoun through to Esperance faces south in a series of dune-backed shallow bays between granite headlands (Map 10.2).

The winds are strong and blow consistently, although the direction varies forming and shaping the dunes. The dunes have trapped a series of saline coastal lakes that have no connection with the sea: Jerdacuttup Lakes, Lake Shaster, Lake Gore and the Esperance Lakes as well as a number of smaller lakes and swamps. Most of the lakes have small catchments and are not likely to have been estuarine in the last 10,000 years, while others may simply be low points in the landscape that intersect the watertable. In many places the calcareous beach sand has hardened into rock, forming onshore and nearshore reefs that probably mark old beach lines. The reefs appear to protect the mouth of the estuaries, which unlike most further west lack the protection of nearby headlands.

Esperance and beyond

The coastline to the east of Esperance as far as Cape Arid follows a similar form to that seen west of Esperance, that is a series of shallow south facing bays between rocky headlands. The bays are however smaller, the headlands more numerous, and offshore lie the many islands of the Archipelago of the Recherche. East of Cape Arid from Cape Pasley the coast veers north-east and is of totally different character devoid of watercourses.

Drier and drier—unpredictable and erratic

The mean annual rainfall is about 600 millimetres at the coast decreasing inland to about 400 millimetres. Winter rainfall is relatively reliable with 60–70 per cent of rain falling between May and October but cyclonic storms in summer may bring more than 200 millimetres in a few days, and extreme rainfall events have been recorded in all months (Maps 10.1, 10.2 and 1.8).

FIG. 10.3 *Rocks near the creek mouth at Quoin Head with Thumb Peak and wave-cut bench in the distance, May 1989.* E. P. Hodgkin Collection.

Annual rainfall can vary greatly even on the coast. As an example, at Cape Riche where rainfall records span nearly a hundred years (1897–1985), over 900 millimetres was recorded in 1917 but less than 300 millimetres in 1936. Records for other places do not go back so far, with those in the eastern part around the Young and Lort Rivers (Stokes Inlet) dating only from 1969.

Extremes of rainfall on a yearly basis affect how much water the estuaries hold, with many drying to shallow saltpans over summer or for longer periods when winter rainfall is light. Conversely, heavy falls in the catchment can fill the estuaries, break the bars and change the whole estuarine environment. Although the change between a dry closed state and deep open state may take decades, we need to remember that the potential for estuaries to fill rapidly remains, depending on the amount of rainfall, the run-off due to the catchment area and the soil moisture. Culham Inlet was dry for decades, but rainfall records for Ravensthorpe and Hopetoun that cover the years from 1901 to the present day indicate that heavy falls are not uncommon at any time of the year. Rainfall in excess of 50 millimetres, as in October 1992, can recur every 7 years and similar falls to those in May 1993 (50 millimetres) can occur every 2.5 years.

A salty land washing away

The rivers in this eastern part of Swanland have been naturally salt for hundreds if not thousands of years, while others have become saltier more recently. For example, the Gairdner was always salt and the Pallinup was

MAP 10.2 *Catchment and drainage Culham Inlet to Cape Arid.* Courtesy DoE.

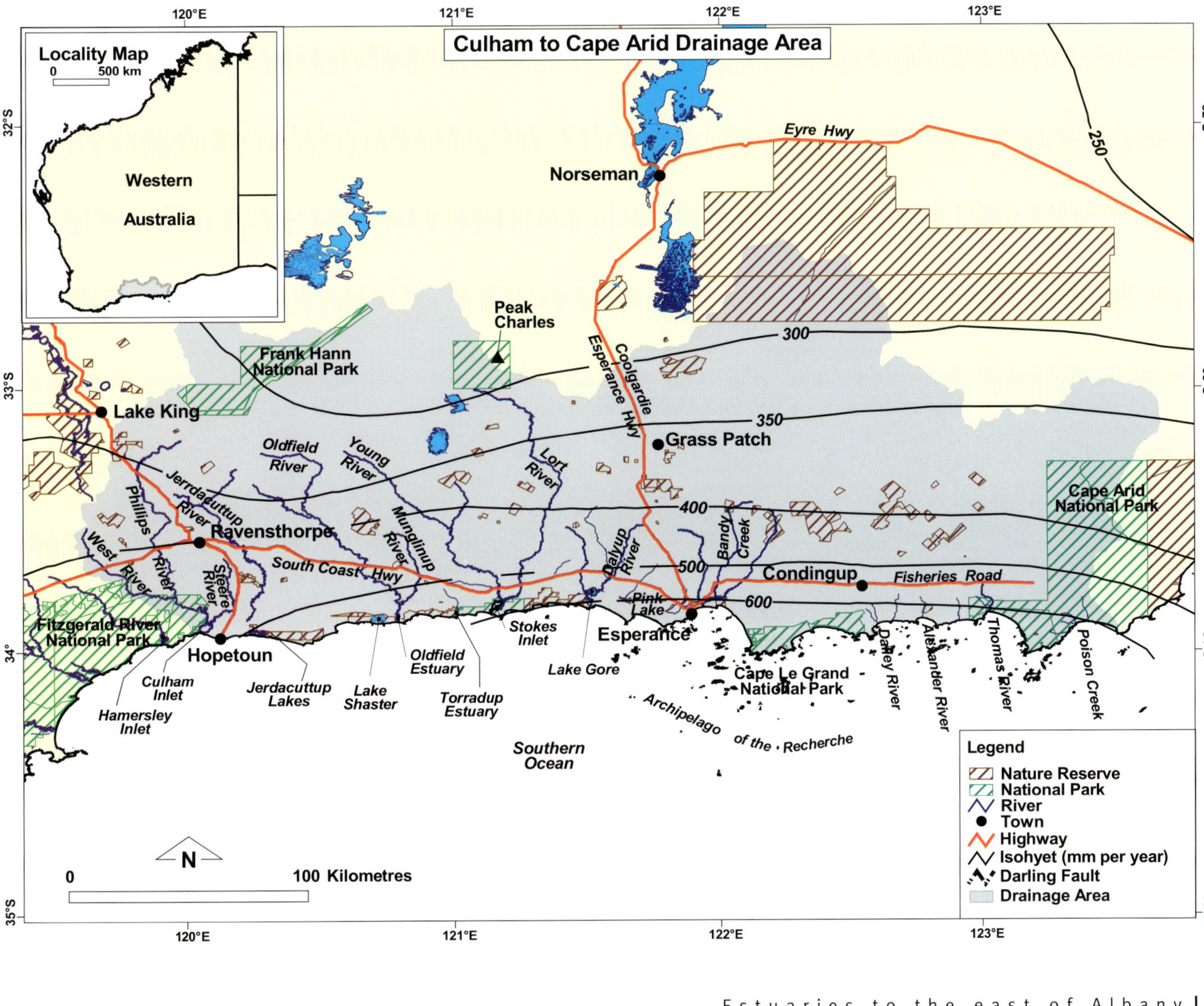

originally called the Salt River. The inland fish fauna consists of typically saline species not freshwater species.

River water from the cleared catchments has become increasingly salty. Salinity now varies from about 10 per cent of sea water (3–4‰) when the rivers are flowing to 50‰, more salty than sea water, in the stagnant river pools in summer. The salt originates from two sources: connate salt that seeps from the ancient (Eocene) marine spongolite rock, and cyclic salt that is sea salt deposited by rainfall within the last 100,000 years. This salt from rainfall is usually trapped in the subsoil and rises to the surface with the rising watertable following clearing. Eventually after many hundreds of years this salt source should decrease.

Run-off to the rivers and estuaries is also affected by vegetation, which in turn affects sediment transport and the infilling of the estuaries. Extrapolating from gauging stations at Jacup on the Fitzgerald River (1974–1986) and on the Pallinup River, run-off to the estuaries averages about 6 millimetres per year from cleared catchments but probably less than half that from uncleared land.

There is little doubt that there has been a considerable increase in sediment mobilised in the catchments due to clearing and the resulting sheet erosion and gully wash. With heavy falls of rain, fine topsoil and coarser material are eroded from riverbeds, and trees are torn up, dropped into lagoons and carried to the sea (Fig. 10.4).

This is especially true of the Pallinup River catchment, 84 per cent of which is cleared. In January 1982 it is estimated that 100,000 tonnes of sediment was eroded from the Pallinup River catchment and deposited in Beaufort Inlet, representing a layer of sediment 25 millimetres in depth over the bottom of the lagoon. Floods in the Gairdner River, that flows to Gordon Inlet, in 1955 cut across a meander upstream from Quaalup depositing a large sandbank. In May 1988 floodwaters rose 6.5 metres at the Quaalup Bend and swept across the property on the northern end of the peninsula.

Run-off can increase following fires, so that the timing and frequency of burning can affect erosion. In December 1989 lightning strikes started fires that burnt about 1,500 square kilometres of the Fitzgerald River National Park; in January 1990 heavy rain caused floods that carried much ash and other debris and broke the bars of all the park estuaries.

A new ingredient

Along with the changes of salt and sediment in the river waters, the amount of nutrients from fertilisers has also risen. The Pallinup and Bremer Rivers and their estuaries, the Beaufort and the Wellstead, have contained

FIG. 10.4 *Gairdner River in flood, May 1988. Much of the gently undulating mallee country has been cleared and increased run-off after heavy rain erodes riverbanks, washing away roads.* Courtesy D. Lullfitz, Print E. P. Hodgkin Collection.

relatively high concentrations of nitrogen and phosphorus probably linked to fertiliser application in the catchment, and the estuaries are now quite eutrophic. (See also **Chapter 1 Understanding Swanland**; Graphs 1.2 and 1.3.) The Gordon Inlet lying within the national park also has high concentrations of nutrients. The alternating wet and dry cycles mean that the estuaries are very productive for a time when conditions are favourable, but when conditions change the dead plants add to the organic sediment, and decaying plant material favours the release of nutrients that promote nuisance algal blooms.

THE HAND OF HUMANS

Occupied by an ancient people

Aboriginal use of the estuaries has been relatively poorly documented. As in other areas of Swanland, communities probably visited the estuaries and coastal areas at particular times of the year when fish were abundant. (See **Box 9.1 Stone circles—echoes of the past**.) This is a tradition that continues today between Hunter River and Wellstead Estuary where Aboriginal groups from inland areas have met since the 1940s.

Many of the place names in use today are based on Aboriginal names: Jerramungup was derived from the Aboriginal name of 'Yarra-mo-up' or 'place of the tall yate trees'. Early records of Aboriginal words and customs were recorded by Ethel Hassell in an account of her life on Jarramungup Station in the late 1870s and 1880s. The Aboriginal people burnt tracts of land to encourage new growth that would attract kangaroos and other game, a practice followed by the early pastoralists in the 1800s.

Mapped early—cleared so recently

The coastal areas were visited by many of the early European explorers, including the Dutch in the 1600s. Captain George Vancouver named the Doubtful Islands in 1791 and noted the 'high detached clusters of craggy mountains, on a base of low and to all appearances level land', probably the earliest record of the Barrens' coastline. The following year, two French ships the *L'Esperance* and *La Recherche* under the command of Bruny D'Entrecasteaux passed by King George Sound and explored the south-eastern coast, commemorating the discoveries and safe anchorage with the naming of Esperance Bay and the Archipelago of the Recherche. Both Vancouver and D'Entrecasteaux carried scientific collectors and, although specimens from the Great South Land were taken back to Europe, it is interesting to note that D'Entrecasteaux was not impressed with what he saw along the south coast: 'The same barrenness was everywhere; only rocks and sand dunes could be sighted, and there was little hope of finding water' (Duyker & Duyker, 2001, p. 111).

Matthew Flinders in the *Investigator* during his circumnavigation of the continent in 1802 sailed between the Doubtful Islands, noting the seals. He sighted 'some masses of tolerably high land which appeared to be granitic; and for distinctions in the survey, they are called West, Middle and East Mt. Barren' (Flinders, 1801–1803, p. 76). The term 'barren' is apparently a reference to Flinders's view of their appearance, and they are noted on his charts as 'barren hills'. On many maps the term 'Mid' has been used for Middle Mount Barren, however Flinders's original name has now recognised officially.

Sealers and whalers from Britain and North America followed, and as early as the 1830s a decline in seal numbers was noted. Fur seals were probably hunted to a greater extent than sea lions as the pelts were more sought after. From all accounts:

> All the common sense rules of harvesting a renewable resource were broken: the young as well as females were taken just as readily as males, and where possible, all individuals on a single island were collected.
>
> Ian Abbott, 1979.

It is perhaps ironic that one of the earliest botanical collectors, William Baxter, who recognised the rich diversity and unique character of the area and whose name is immortalised in a number of plant names, travelled with the sealers who wrought so much havoc on the seal populations.

Pathfinders

Although the sealers were active along the coast little is known of how far inland they ventured, but they probably did as the early wave of pastoralists had apparently heard stories about the country. Sandalwood cutters such as George Maxwell, who roamed inland in search of the valued timber, were also responsible for the early reports about the land. Among the earliest travellers through the area were James Newell and James Manning, who walked to Albany after they were marooned on the coast east of Esperance in 1835 and added to the growing interest and knowledge about the area.

Six years later Edward John Eyre and his Aboriginal companion Wylie during their journey from South Australia to Albany followed a similar path. They crossed the Nullarbor in the hottest months of late summer and autumn when food and water were very scarce. The part from Esperance to Albany was in winter and they walked through many miles of waterlogged country, noting that in many places the water lying on the ground after rain quickly became too salty to drink. They passed by Jerdacuttup Lakes and proceeded towards the Barrens, struggling through the dense mallee around Culham Inlet. Eyre's notes show his naming of the hills was different to that of Flinders, confusing East Mount Barren with the area known today as Eyre Range. Eyre and Wylie travelled near Hamersley Inlet then headed inland to avoid the dense coastal scrub, although they still had to contend with rough 'broken country' and had trouble finding water. Although his journey was heralded as a major achievement, Eyre himself was disappointed with the results, 'I have no important rivers to enumerate, no fertile regions to point out…on the contrary, all has been arid and barren in the extreme' (Eyre, 1845).

With the interest in collecting the flora and finding suitable pasture land, official explorations increased. Among the earliest were the botanist and collector James Drummond in 1847, the First Surveyor General John Septimus Roe in 1848, and the Gregory brothers in 1849. Following close behind were the first pastoralists such as John Hassell, who took up land around Jerramungup in 1849 and developed a large pastoral business in this lightly wooded area of natural grassland.

Further east the Dunn brothers took up a lease at Cocanarup west of Ravensthorpe on the Phillips River in 1868. James Dunn discovered gold in 1899 and a number of mines were opened soon after. The population of Ravensthorpe grew from 500 in 1900 to 3,000 in 1909. Copper gradually replaced gold as the principal mineral mined in the Phillips River Mineral Field, with the ore exported from Hopetoun. A fall in the price of copper in 1913, however, saw many mines close. Small-scale mining continued in the late 1950s, and with the mineral boom of the 1970s this appeared likely to escalate. Plans for large-scale operations to mine montan wax near the Fitzgerald River prompted the declaration of the Fitzgerald River National Park.

The Overland Telegraph Line was constructed east–west through the area in 1875–1877. After the line was abandoned the track continued to give access for road vehicles to many isolated areas. Many others tracks were opened by beach, rock and estuary fishermen. The No. 2 Rabbit Proof Fence that extends on the south coast to Point Ann was erected in 1904–1905 and

FIG. 10.5 *Rabbitee or Rabitto—the longest fence in the world, June 2000. The fence was built to halt the invasion of rabbits from the east, but by the time it was finished there were as many rabbits inside as out. Today it protects crops and pastures from emus.*
Courtesy West Australian Newspapers Ltd.

only abandoned in the 1950s and 1960s after the introduction of the viral disease myxomatosis. The No. 1 Fence, begun in 1901 and completed in 1906, ran from Eighty Mile Beach between Point Hedland and Broome on the north-west coast and ended in Starvation Boat Harbour 40 kilometres east of Hopetoun, a distance of 1,822 kilometres and deserving the status of the Longest Fence in the World (Fig. 10.5).

Clearing on a large scale

Following World War II and the extension of farming with new land releases for both ex-servicemen and civilians, the scale and rate of clearing in the low mallee scrub on the sand plain country was unprecedented. Stories abound of bulldozer drivers instructed to work along compass bearings for hours before turning. For some years good rainfall, and the addition of superphosphate and trace elements saw successful crops grown, but the situation changed quickly. The ground water rose with the removal of the deep-rooted perennial vegetation and the water was salt, as it had always been in this ancient area. Years of unreliable rain diminished the cropping potential. Then the impact of winds began to show when the topsoil blew away and banked up against fences, spilling over onto roads (Fig. 10.6). Later torrential rains washed the unbound sand with the unused fertiliser into creek lines and rivers from where it moved down to the estuaries, hastening the natural infilling process and raising the nutrient concentrations, changing forever some of the estuaries.

BEAUFORT INLET

A place in the landscape

Beaufort Inlet receives water from the Pallinup River that rises about 250 kilometres to the north-west near Broomehill (Map 10.1 and Fig. 10.7). Downstream it is joined by several small tributaries draining from the northern side of the Stirling Range. The main tributary, Corackerup Creek, rises near Jerramungup to the north. In the upper reaches much of the riverbed is sandy but in the lower reaches the river has cut to the granite base where some pools hold water through summer.

Land in the upper Beaufort catchment around Jerramungup was settled by John Hassell in 1849. At first sheep were grazed on the native grasses and the shrubs were burnt to promote grass growth. Extensive clearing began in 1910–1912 on the heavier soils in the Gnowangerup–Borden–Ongerup area, and in the 1920s with soldier settlement west of Gnowangerup.

FIG. 10.6 *Sand drifts near Jerramungup, Gairdner River, June 1981. The light soils of the mallee country are prone to erosion. At times sand from paddocks has drifted across the road, banking up against fences; however this happens less regularly with changes in land practice.* Courtesy D. Carter, DAWA.

The discovery of the trace element deficiency led to widespread clearing on sandy soils north of Chillinup and in the upper catchment around Ongerup and Borden in the 1950s and south-west of Chillinup in the 1960s. Today 84 per cent of the Pallinup River catchment is cleared for cereal cropping and sheep grazing. There is some uncleared bush along Corackerup Creek and in the Corackerup Nature Reserve (45 square kilometres) to the north-west. Vegetation around the estuary is largely uncleared and is vested in the Jerramungup Shire and the National Parks and Nature Conservation Authority.

The wide range of habitats in the Beaufort Inlet area, including low woodlands, mallee shrub lands, samphire flats, permanent water bodies, river and estuary, beach and sea, provide habitat for a great variety of birds. Some migratory birds stopover at the estuary during summer, although the number of species and individuals are thought to be lower than at nearby Bremer Bay due to the lack of suitable mudflats and low sand patches. The inlet is a home for huge flocks of black swan.

The rare Newbey's Mallet *Eucalyptus newbeyi* is confined to Millers Point (along the main access road to the inlet) where it forms a dense shady grove. The understorey has suffered trampling however, and there is unlikely to be regeneration.

A small shallow body of water

The estuary is 14 kilometres long and runs south-east to north-west perpendicular to the coast. The upper 8 kilometres are river-like with a deep valley incised through the spongolite rock exposing the granite bedrock. Most of the channel is only 1–2 metres deep with sandbanks in the wider part, but the riverbed scours to 5–7 metres on the bend following floods.

Small swamps border the middle estuary at 3–6 kilometres from the bar. Here sediments have accumulated in the river delta and sandbanks enclose the small bays. A slightly higher sand ridge with paperbark and she-oak trees borders the swamp opposite Millers Point.

The lagoon, 3 kilometres long and 1.5 kilometres wide, is less than 1.5 metres deep throughout. On the northern shore, a 40-metre high cliff cut into the spongolite rock forms a spectacular background (Fig. 10.8). Except for a small section of limestone cliff, alluvial flats border most of the southern shore. About a kilometre from the bar, a sandy spit partially encloses a small bay. Shell fossils, mostly *Katelysia* spp. embedded in the sand spit and bar area, show that the inlet was at times more open to the sea 3,000–4,000 years ago.

The estuary mouth is closed by a high sandbar 3–3.5 metres above sea level, about 500 metres long, 100–150 metres across, and oriented NE–SW. The bar, built of poorly sorted coarse quartz sand with some calcareous shell material, is backed by a small delta of sand washed over the bar during

FIG. 10.7 *Beaufort Inlet, aerial view, June 2000. The Stirling Ranges, surrounded by cleared farmland, can be seen in the background. The area surrounding the inlet is largely uncleared. Spongolite cliffs border the eastern side of the inlet and steep dunes border the ocean on either side of the bar. Millers Point is on the north-east (right side) of the Inlet, where the main basin narrows opposite the island.* S. Neville/Ecotones, Courtesy WRC.

FIG. 10.8 *Spongolite cliffs bordering the Beaufort Inlet near Millers Point, May 1988. The eucalypt trees with their golden tops growing over the cliffs are a spectacular sight—the mallee form quite different to the trees lining the estuaries in the wetter west.* E. P. Hodgkin Collection.

storms. On the northern side a tongue of beach and dune sand is covered with vegetation. On the southern side an outcrop of limestone and fossil shells (often covered by sand) reduces scouring when the bar breaks. The bar usually breaks at the south side cutting a channel about 50 metres wide; however a big flood can cause a break more than 200 metres across, as in 1955 and 1982. Waves and swell rapidly rebuild the bar so that it seldom stays open for more than a few weeks. When the bar closes and the water level in the lagoon falls below sea level, sea water seeps through the bar into the estuary.

The bar broke in 1945, 1949 and 1953. In 1955 February floods opened the inlet and the bar stayed open for 18 months. Following this long opening it did not open again for 12 years until 1968 when it was open for less than a month in June. The bar also opened in 1971, and was broken by the Shire in 1974. It opened again in 1978, 1982 and briefly in 1983 when it was disturbed (Fig. 10.9). Ernest Hodgkin's notes made while talking to local residents indicate that the bar also opened in 1988, 1989, 1990, 1992, 1993, 1994 and 1998. Records of the Water and Rivers Commission funded by the Natural Heritage Trust collected after Ernest Hodgkin's death in 1998 indicate that the bar opened in December 2001, and remained open for about two weeks. Documentation of bar openings is important as they dictate the degree of connection with the sea.

FIG. 10.9 *Floodwaters flowing from the Beaufort Inlet, aerial view, January 1982. The inlet remained open until May.*
E. P. Hodgkin Collection.

Like the sea but saltier

The shallow lagoon water is very turbid, due mainly to plant plankton and fine sediment resuspended by wave action. A thick layer of plants covers the estuary bottom. Anecdotal evidence suggests the plant material has increased due to high nutrient levels from the catchment. In response to concerns for the water quality and as a baseline for understanding estuaries, the Water and Rivers Commission (funded originally under the Natural Heritage Trust) started a regular monitoring program in 1997. The sampling locations cover the length of the estuary and record salinity, temperature, oxygen, nutrients and phytoplankton at different depths.

The estuary waters are always quite salty (Graph 10.1). Although river flow reduces salinity it is seldom less than 18‰, about half that of sea water, and increases with evaporation over summer. This rise is shown by the change from October 1997 when the inlet was 28.5‰ and 17°C to 56‰ and 22°C in March 1998. Generally the shallow inlet water was well mixed except for an area about 8 metres deep upstream of Millers Point where the inlet is narrow. In this area the waters below 5 metres were very salty, about 68‰, and the temperature several degrees lower. Although the difference between the surface and the bottom became less as the surface salinity increased over summer, the bottom layer probably did not mix and it is likely that it is not disturbed by the river flow.

In October 1997 the oxygen concentrations were high with about 80 per cent saturation in the upper waters of the inlet, although at lower depths the oxygen concentrations were lower. In the deeper saline water upstream the water was anoxic, completely lacking in oxygen, and smelt of hydrogen sulfide. By the end of summer in March 1998, oxygen concentrations were also low throughout most of the shallower parts of the inlet. These low oxygen concentrations are of concern. As salinity and temperature were similar from top to bottom, it could be expected that oxygen concentrations would also be high. The low oxygen concentrations suggest that the oxygen demand in the water column and/or these sediments is high.

The nutrient concentrations were high at all times throughout the inlet. (See Graphs 1.2 and 1.3.) These nutrients are believed to come from fertilisers applied to catchment soils. A deep hole where the water was anoxic had particularly high concentrations of total nitrogen (18 milligrams per litre)

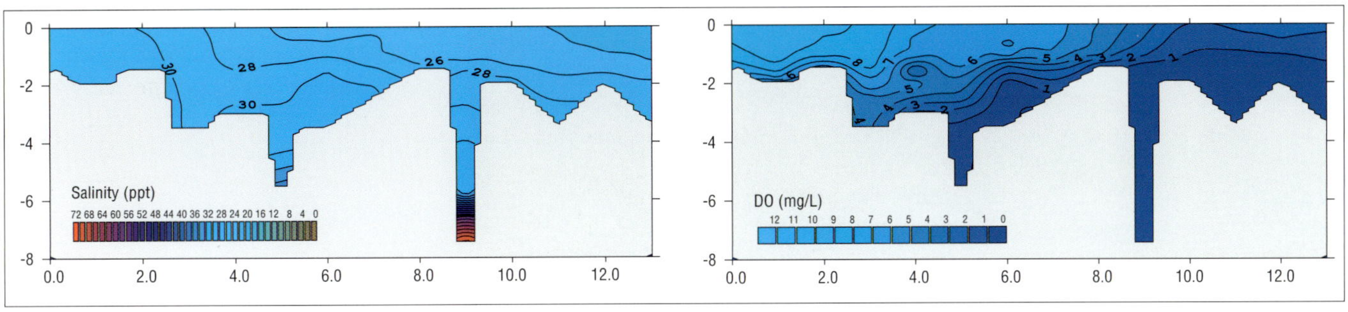

GRAPH 10.1 *Salinity and oxygen profiles of Beaufort Inlet, February 2000. The bottom axis shows distance from the ocean bar. Note the strong salinity stratification in the deep hole 9 km upstream where oxygen concentrations are very low. The deoxygenated water is also affecting the estuary downstream.* Courtesy WRC.

and total phosphorus (2.7 milligrams per litre). These high concentrations were mostly due to dissolved fractions: ammonium (14 milligrams per litre) and Filterable Reactive Phosphorus (2.4 milligrams per litre). Silica concentrations were also high, above 14 milligrams per litre.

In October–November 1997 the amount of larger phytoplankton was very low but there were high concentrations of very small plankton. By March 1998 there was a significant bloom of phytoplankton, especially dinoflagellates and diatoms, and there was a small amount of *Prymnesium*, a species that can produce a fish poison.

Homes for many

Although the shallow waters are quite turbid, there are dense growths of seagrass *Ruppia megacarpa* and the stonewort *Lamprothamnium papulosum*. The faunal species known from the inlet are also restricted to those few that can cope with extremes of salinity. The animals that can cope with the conditions, such as the tiny gastropod snail *Hydrococcus brazieri* that lives on the aquatic plants, are however often very abundant. Mounds of tubes made by bristle worms *Ficopomatus enigmaticus* on the rock and trees stumps that fringe the inlet are also a feature (Fig. 10.10). Although these worms are found in the western wetter estuaries, they are more common and obvious in the estuaries as we move eastwards into a drier, saltier area where the bars open less frequently.

The dead shells along the strand lines show that other invertebrates invade when the bar is open, dying when conditions become unfavourable. This is a cycle of invasion and demise mirrored in the fossil shell beds along the margins of the estuary. Certain marine fish also enter the estuary from the sea when the bar is open, and the inlet becomes a popular fishing area for recreational and commercial fishers with the most abundant species caught in nets. The more abundant species such sea mullet *Mugil cephalus* and yellow-eye mullet *Aldrichetta forsteri* can live in the estuary for some years after the bar opens. King George whiting *Sillaginodes punctata* are less tolerant of the harsh estuarine conditions and are most abundant soon after the bar has been open to the sea. Black bream *Acanthopagrus butcheri* live permanently in the estuary and river, and the Beaufort is an important part of the South Coast Estuarine Fishery (Fig. 10.11). (See **Box 10.2 Black bream—a real estuary fish**.)

Too harsh a place to live

Massive 'fish kills' in Beaufort Inlet have been reported in 1936, 1939, 1942, 1953, 1961–1962?, 1972, 1988, 1989 and 2000, and there may have been others that have gone unnoticed because of the relative isolation of the inlet. The kills have usually been preceded by reports of salt burns on the fish, of fish swimming at the surface with mouth gaping, and the deaths of shrimp and other invertebrates. The cause of fish kills can be difficult to detect because many factors, such as toxic algae, deoxygenation and temperature, may all contribute. Often there are delays before a kill can be

FIG. 10.10 *Shoreline at Beaufort Inlet following a dry summer, May 1983. The rocks in the foreground are covered by masses of worm tubes. The shoreline is bordered by trees sheoak* Casuarina obesa *near the estuary and mallee eucalypts on higher ground.*
E. P. Hodgkin Collection.

FIG. 10.11 *Black bream are caught by professional fishers in Beaufort Inlet but Wellstead Estuary to the east can only be fished by recreational fishers. June 2000.*
Courtesy West Australian Newspapers Ltd.

investigated and the fish are often decayed. Generally the kills have been attributed to the combined effects of high salinity and temperature, but there may be complicating factors; the eutrophic condition of Beaufort Inlet suggests that the water may be fouled by decomposing organic matter and become deoxygenated.

In June 1998 there was a fish kill and a bloom of dinoflagellates, predominantly *Heterocapsa*. The inlet was very green in colour due to tiny cyanobacteria, probably *Synechococcus*. There were however no cells of *Prymnesium*, a known toxic species that had been recorded in the inlet earlier in the year in March. Following the fish kill, run-off broke the bar in June 1998 and the phytoplankton around Millers Point fell to the concentrations expected in flowing waters.

Portrait of a fish kill

In February 2000 another fish kill was reported. Although the fish had been dead for some time when water samples were taken, water colour, smell and oxygen concentrations suggest that it was probably due to an algal bloom

Box 10.2 **BLACK BREAM—A REAL ESTUARY FISH**

The black bream *Acanthopagrus butcheri* is found in southern Australian estuaries from Murchison River to Myall Lakes on the north coast of NSW. It is one of the most important commercial and recreational fish species caught, growing to 2 kilograms in weight and considered excellent eating. In Western Australia the commercial catch is regulated as part of the larger estuarine fishery in the regions: Swan–Canning, Peel–Harvey, Leschenault Inlet, Hardy Inlet and South Coast Estuarine Fisheries. The industry is currently under review to ensure the future of the industry and equitable sharing of the resource. Management options cover limitations on gear, season and time closures as well as a reduction in the number of fishing units. The fishers within a region target a number of estuaries depending on where fish stocks are high. In the mid-1990s catches in the South Coast Estuarine Fishery were concentrated in Stokes, Beaufort and Oldfield Inlets, although other inlets such as Culham produced good yields in the early 1990s before the bar was opened. The estuaries open to commercial catches are under review and access to some south coast estuaries is likely to be withdrawn. Recreational fishers have a bag and size limit, and for netting a licence is required. Regulations are likely to change in the future and those limiting net mesh size may be a more important consideration than bag limits for maintaining recreational stocks. Although fishers are required to attend nets so as to remove undersized and unneeded fish alive, many fish are dead or damaged when retrieved.

Unlike most other fish in estuaries, the black bream completes its whole life cycle within the harsh changeable estuarine environment. They generally prefer brackish conditions in the middle and upper reaches, but can cope with salinities from virtually fresh to marine (35‰). When waters become hypersaline in the closed estuaries along the south coast, many bream move upstream and fish kills are not uncommon.

Although some fish are flushed into the sea when the bars open, there is little migration along the coast as most remain outside the estuary and move back in when the flow slackens. As a result each estuary has a distinct genetic population, and depleted populations through overfishing are not replenished by fish from other estuaries.

Black bream feed on a wide variety of plant and animal matter depending on the particular estuary and abundance of food. Older fish in all of the estuaries appear to eat more algae and larger prey than younger fish. Although some feed on animals such as burrowing polychaete bristle worms, most feed on animals above the sediment. In the Swan River they feed on the black estuarine mussel *Xenostrobus securis*, crabs, amphipods, fish, polychaete worms buried in the sediment and some plant material. In the Moore River they feed on the sunset shell *Sanguinolaria biradiata* as well as the macroalgae *Cladophora* sp., and in landlocked Lake Clifton they mostly feed on polychaete worms. In Walpole–Nornalup Inlet they eat large volumes of the seagrass *Ruppia megacarpa*, small gobiid fish and worms; while in the Wellstead Estuary they consume large amounts of macroalgae *Chaetomorpha* and tube-dwelling amphipods *Corophium*.

Fish in the Swan River grow faster and mature earlier than those in the Moore River and estuaries along the south coast. In the Swan River black bream usually reach sexual maturity at 2.2 years (≈ 215 millimetres) but this may take 4.3 years (≈ 179 millimetres) in Walpole–Nornalup. In Wellstead Estuary the fish mature at about 2 years but at a much smaller size (145 millimetres), a feature that could indicate selection to breed early because of the harsh conditions or that fishing pressure has been high. The particularly slow growth is probably due to the high salinity and the diet of low quality algae.

Spawning varies from year to year depending on conditions, with very low and high salinities both affecting success. In the extreme conditions of the south-eastern estuaries this can result in the loss of whole year classes. In a study conducted in the mid-1990s, fish older than 15 years spanning a number of age classes were caught in the Moore River, Swan River and at Walpole–Nornalup, indicting that fish probably spawned each year. In the Wellstead there were no fish older than 7 years (spawned 1984–1985), indicating that the fish did not spawn or had limited success in the strong flow after the bar opened between September 1984 and May 1985 when adults were flushed to sea and caught along the coast.

To ensure that some fish have bred before they are caught, regulations set a Minimum Legal Capture Length of 250 millimetres. In the Swan River with heavy fishing activity from recreational fishers through the year, and commercial fisheries during winter and early spring, only 5 per cent of the fish caught were more than 5 years old. In comparison, 30 per cent of the fish in the lightly fished Moore River were more than 5 years old. In the Walpole–Nornalup, closed to all netting and considered one of best inlets for black bream angling, 45 per cent were more than 5 years old. Although the decline in black bream numbers in many estuaries like the Hardy has been attributed to commercial and recreational fishing, the loss of streamline vegetation, salinisation and herbicides may have affected spawning in the river reaches. Care of the catchments and estuary systems thus remains an important objective if fish are to remain for the future.

Source: Fisheries Department of Western Australia, 1993a; Young et al., 1997; Pearn & Cappelluti, 1999; Sarre & Potter, 1999, 2000; Sarre et al., 2000.

that deoxygenated the water column rather than a toxic bloom that poisoned the fish.

Several days after the dead fish were found, the waters were turbid with a gold-brown colour in the lower estuary and gold-green further upstream. Oxygen concentrations were relatively high in the lower reaches, decreasing upstream. In the middle estuary the surface water was only 44 per cent oxygenated, and this decreased with depth becoming anoxic below 2 metres. This anoxic layer smelt of hydrogen sulfide (H_2S) and was strong enough to discomfort the survey team. Further upstream the anoxic layer was nearer the surface. The lower oxygen concentrations appeared to be associated with the gold-green plume of water. Fish were active in the middle estuary but this activity decreased further upstream except where a plume (10 metres in width) of darker, 'cleaner', slightly oxygenated water was being driven upstream along the eastern bank for a distance of about 1.75 kilometres. The observations suggest the fish kill was due to deoxygenation of the water column that may have occurred over a distance of 5 kilometres. The H_2S may have also contributed. (See also **Chapter 4 Leschenault Inlet and Estuary and smaller estuaries of the Geographe Bay area: Vasse–Wonnerup wetland system: Green soup and dead fish**.)

WELLSTEAD ESTUARY

Wellstead Estuary lies at the end of the 70-kilometre long Bremer River, which rises near Jerramungup (Map 10.1 and Fig. 10.12). The catchment of only 716 square kilometres is much smaller than the Beaufort catchment (4,970 square kilometres) only a short distance to the west. Like the Beaufort, some of the catchment land was cleared quite early after European settlement with about half the catchment cleared by 1968, but land clearing continued until quite recently and by 1984 over 80 per cent was cleared. A narrow strip of Shire Reserve borders the lower reaches of the river and the upper part of the estuary, and this is surrounded by the Fitzgerald River National Park. The town of Bremer Bay and a caravan park lie along the southern shore.

FIG. 10.12 *Wellstead Estuary, aerial view, June 2000. The cleared farmland that surrounds much of the catchment can be seen on the southern side of the estuary to the left. The northern side of the estuary is uncleared land vested in shire and state authorities. The small town of Bremer Bay lies along the banks of the estuary.* S. Neville/Ecotones, Courtesy WRC.

River and lagoon

The estuary runs in a north-west to south-east direction for about 13 kilometres perpendicular to the coast. In the upper reaches the river has cut a narrow channel through the soft, flat-bedded spongolite rock, and the fine sediments that make the water turbid are gradually filling the small bays. In the middle reaches 4–5 kilometres from the bar, the inlet widens and there are shallow bays backed by swampland. The banks are lined with low outcrops of spongolite rock bordered by the saltwater paperbark *Melaleuca cuticularis* (Figs 10.13 and 10.14). In the lower reaches the estuary widens still further to about 600 metres with a lagoon 2.5 square kilometres in area. The lagoon is very shallow, generally less than a metre, except for a channel along the northern shore so that the volume of the estuary is estimated to be only 2.4×10^6 cubic metres.

The bottom of Wellstead Estuary is covered with fine dark grey to black organic mud overlain by quartz sand. Interlayering of sand and mud probably reflects past flood events and periods of lower flow. There are indications that sedimentation has increased since the hinterland was cleared, but the organic matter probably originates from the rich growth of aquatic plants in the estuary itself. (See below Stokes Inlet.) The high organic content of the mud and the presence of siliceous diatom skeletons from phytoplankton suggest the estuary has been productive for a long period of time.

FIG. 10.13 *Wellstead Estuary, August 1978. The north shore at 3 km from the bar showing the exposed spongolite rocks and the mallee eucalypts with* Melaleuca cuticularis *fringing the water's edge.* E. P. Hodgkin.

The mouth of the estuary at the extreme south of Bremer Bay is sheltered from the south-west by a headland of granite. The bar, of mostly quartz sand, is low (1–2 metres) with a wide tidal delta. When the bar opens, the water scours a winding channel 20–30 metres wide and 1–1.5 metres deep, generally along the northern side of the tidal delta before meandering to the southern shore where it may divide into two channels (Fig. 10.15).

The bar breaches more often and probably stays open for a longer time than many of the other estuaries along this coastline. There are also long periods when the bar does not open, and it is said to have remained closed for 20 years between 1860 and 1880. It was also closed between 1972 and 1977. Like the other estuaries on this coast, the estuary may fill with winter rains or after summer storms.

The estuary remains tidal while the bar is open and sea water can penetrate up to 2 kilometres. Sand, seaweeds and seagrass carried inwards can block the channel for a short time before they are dispersed during heavy seas. When the bar is closed the water level varies with river flow and evaporation; it may rise to a metre above sea level but fall to about a metre below sea level, exposing the rocky shoreline during dry periods.

FIG. 10.14 *Paperbark tree growing along the edge of the Wellstead Estuary, November 1999.* A. Brearley.

From fresh to salt

The waters of the Wellstead vary from nearly fresh at the head of the estuary following rain to more than twice seawater salinity in the basin when the bar has been closed for a long period over summer, reaching over 111‰ in March 1997. Surveys of water quality have recently been conducted because of concern for the estuary's health. Like those in the Beaufort Inlet, these surveys complement some early reports in the 1980s and will form a baseline for the future.

In October 1997 the lower 3 kilometres of the estuary near to the bar was close to marine salinity (35‰) and well mixed (Graph 10.2). Further upstream the water became less saline at about 22‰, but the deeper pools were similar to sea water. By March of 1998 the salinity in the lower estuary had increased to more than 70‰. The water was well oxygenated throughout the length of the estuary and at all depths.

Concentrations of total nitrogen (0.4–1.1 milligrams per litre) and total phosphorus (0.08–0.11 milligrams per litre) in October 1997 were moderate with slightly higher values in upstream areas (Graphs 1.2 and 1.3). These became higher in summer (March 1998) with total nitrogen 2.5 milligrams per litre and total phosphorus 0.22 milligrams per litre, probably because of the concentrating effect of evaporation when there was no river flow, although there could have been some input from ground water or the internal breakdown of organic matter. At this time Wellstead had lower concentrations of total phosphorus and total nitrogen than the Beaufort and Gordon Inlets.

In late 1997 there were low numbers of phytoplankton in the water, although some blue-green algae were present in November. By March 1998 however, there was a large 'red tide' (ca 85,000 cells per millilitre) mostly of the small dinoflagellate provisionally identified as *Heterocapsa* sp. Diatoms were also present but only in low densities (100 cells per millilitre).

FIG. 10.15 *Wellstead Estuary Bar, October 1978. In the distance lie West Mount Barren and the Fitzgerald River National Park.* E. P. Hodgkin Collection.

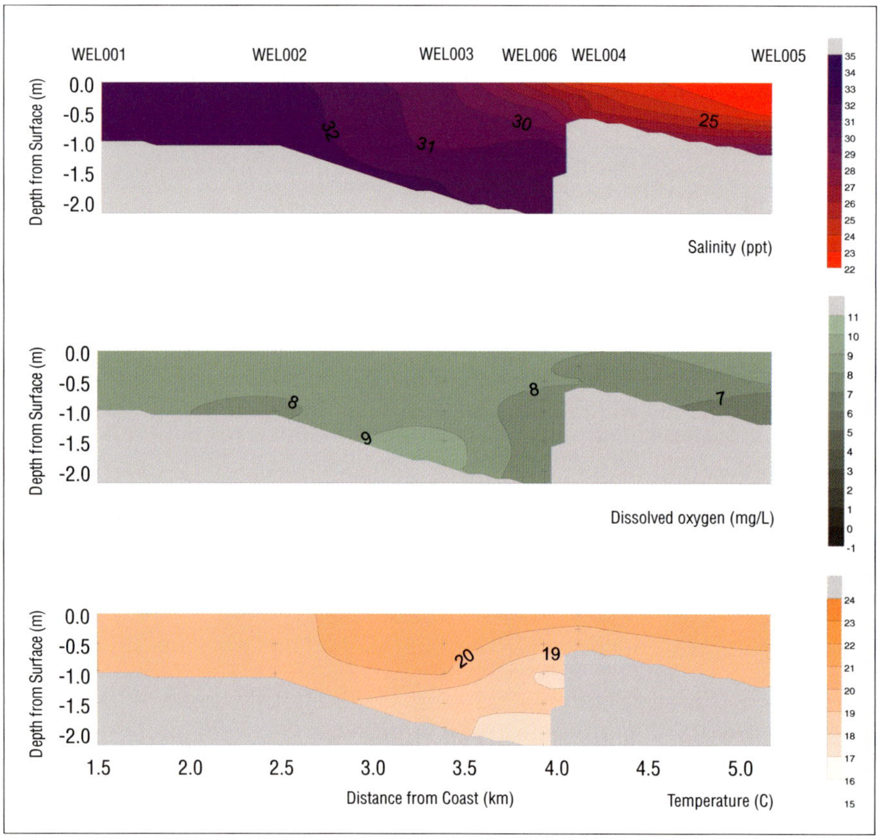

GRAPH 10.2 *Salinity, dissolved oxygen and temperature in Wellstead Estuary, October 1997. Bottom axis indicates distance upstream from the ocean bar. Saline water is present near the bar and in the upper reaches. Note except for a deeper area about 3–4 km upstream from the bar, most of the estuary is very shallow.* Courtesy WRC.

A 'weedy' place—underwater garden

Compared to the estuaries further west, one is struck by the amount of plant material growing in the estuary—an underwater garden. Tracks of boats, where the propellers have cut through the strands, can be seen quite clearly. When the waters are low the margins are covered with dying water plants. Beds of the fine swan weed or seagrass *Ruppia megacarpa* grow throughout the estuary, especially in the area of the flood delta and in the shallow areas of the middle and upper estuary. (See also **Box 8.3 This thing called *Ruppia*.**) In the basin the charophyte, stonewort *Lamprothamnium papulosum*, is dominant forming a continuous cover over the mud. Often covered by filamentous green algae both the plants grow well, except in the most extreme hypersaline conditions, supporting large flocks of Black Swans. In the lower estuary the tiny alga *Polyphysa peniculus* grows attached to dead shells and rocks (Fig. 10.16). Other algae such the green *Enteromorpha* and *Vaucheria* as well as red filamentous forms have been reported over the years.

A hard place to live

The invertebrates living in the inlet are those species able to cope with the great changes in salinity. When the bar has been closed for a long time only a few estuarine species survive, but when it is open a number of species enter and flourish until conditions become too extreme. Cockle shells (*Katelysia*) are one of the animals that settle, grow and then die when the water becomes too salty. This pattern of change is recorded in the shell beds. Shells from the riverbanks have been dated at 3,930 ± 265 years BP, and the deeper shells as old as 4,900 ± 375 years. These dates suggest the Wellstead has changed from a sheltered marine embayment or open estuary system, a system with regular flushing like Oyster Harbour and Princess Royal Harbour, to the more recent period when the bar closed, and the fine river sediments and organic material started to accumulate.

The tiny black snail *Hydrococcus brazieri* can be very abundant and is often eaten by fish. The black bream *Acanthopagrus butcheri* feed on algae, particularly the less abundant green algae *Chaetomorpha*, which they appear to select in preference to the more common *Lamprothamnium*. Like the

a

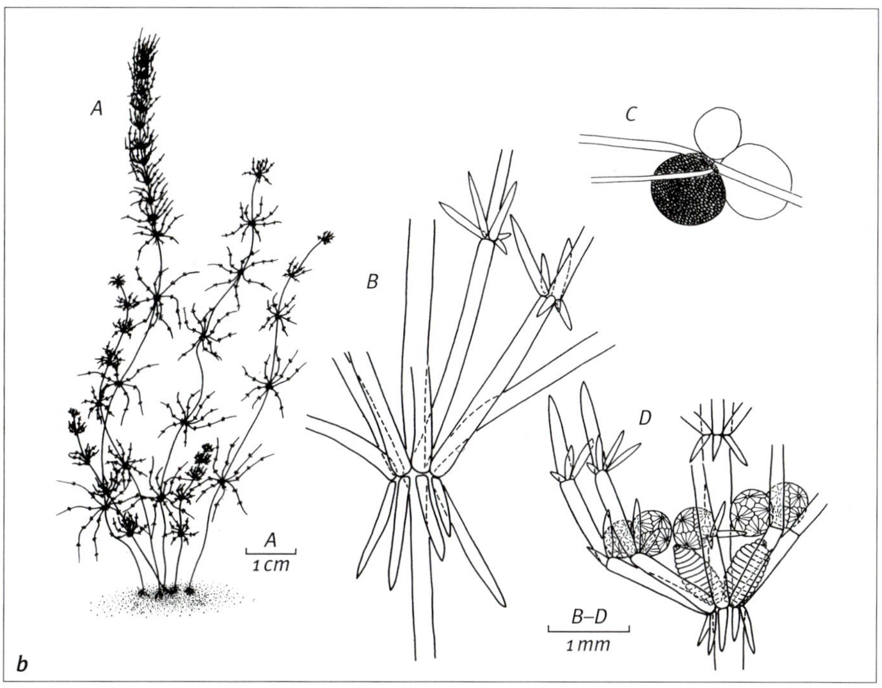

FIG. 10.16 *(a)* Polyphysa peniculus *and (b)* Lamprothamnium papulosum *make Wellstead Estuary a 'weedy' water body, a real water garden.* L. papulosum *A. Plant form; B. branching at nodes showing opposite stipulodes and bracts in two lateral branches; C. starch bulbils on rhizoids from muddy substrate; D. part of fertile spike with antheridia (male) on first node of branchlets and oogonia (female two only shown) at base of whorl.* Photograph of *P. peniculus*, E. P. Hodgkin Collection. *L. papulosum*, Courtesy H. B. S. Womersley.

invertebrates the fish species change with the condition of the bar, although there are some species such as mullet, black bream, hardy heads, gobies and spotted minnows that persist in the estuary. Black bream that breed in this estuary appear to reach sexual maturity at a smaller size than black bream in other estuaries, which may be an adaptation to the extreme salinities. (See **Box 10.2 Black bream—a real estuary fish**.)

Marine species enter when the bar is open and die as conditions become too extreme (salty). Leatherjackets are said to be the first to die, within two weeks of closure. Fish kills also occur during floods, such as those in May 2005, as the waters often become anoxic. With the exception of black bream most of the fish caught for recreation are marine stragglers and marine–estuarine opportunist species, so their presence depends on the bar being open. A winter opening (March–September) favours the recruitment of mullet that spawn in winter, while a summer opening allows newly spawned cobbler to enter.

GORDON INLET

The Gairdner River rises 25 kilometres north of Jerramungup and flows 130 kilometres to Gordon Inlet (Map 10.1 and Fig. 10.17). South of the South Coast Highway the river meanders along a narrow valley cut through sand plain, first into granite and then spongolite below Devils Creek Road. Water also seeps from the spongolite, forming pools along the riverbed.

The area around Gordon Inlet was visited by many of the early explorers with pastoralists soon following. John Septimus Roe, the first Surveyor General of Western Australia, when camped on the river flats at Quaalup in early 1848 described an abundance of grass suitable for hay, a report noted no doubt by John Wellstead of Bremer Bay and other early settlers when they established properties in the area. Leases around the Gordon were taken up by the Hassells of Jerramungup in the 1890s as part of their extensive station network. Sheep were shorn and wool shipped from Fishery Cove House Beach, a sheltered area near Whalebone Point in the Doubtful Islands area that had earlier been the site of a whaling camp.

Clearing in the northern Gairdner River catchment for cereal crops and sheep grazing removed about 60 per cent of the native vegetation. Most of the clearing occurred before the 1970s, unlike the catchments of the Beaufort and Wellstead where clearing continued into the 1980s. In the lower catchment the river is bordered by bush; to the east some of it is in the Fitzgerald River

FIG. 10.17 *Gordon Inlet, aerial view, June 2000. Gordon Inlet lies within the Fitzgerald River National Park, acclaimed for its diversity of flora. The wide shallow bays with extensive sand flats that dry out in summer are rich feeding grounds for great flocks of waterbirds. The mouth of the Gordon Inlet opens onto a sandy shore that extends for about 8 km to the south, terminating at Point Hood and Doubtful Island. About 2 km south of the mouth the small Kellys Creek winds through the dunes for 1.5 km to a small perched lake that is often dry. Sea water washes over the beach into the mouth of the creek.* S. Neville/Ecotones, Courtesy WRC.

National Park (FRNP). A small area within the park at Quaalup is privately owned; some of this has been cleared and is cropped, while other parts are regenerating or remain as bushland.

River and lagoon

The riverine reaches of the inlet meander through a shallow valley cut into the spongolite rock where cliffs or steep slopes border the outer bends, and gentle slopes and sand deltas the inner bends. A low rock bar marks the head of the estuary about 13 kilometres upstream near the river flats at Quaalup, where pools up to 4–5 metres deep are scoured in the riverbed. About 5 kilometres from the mouth, the river opens to the lagoon bordered by a high spongolite cliff on the north and a samphire flat on the south. Here the lagoon is about 60 metres wide and 0.5–1 metre deep (Fig. 10.18).

Further downstream the wide lagoon is shallower still, probably no more than 0.5 of a metre below sea level and seldom holds more than a metre of water. During the 3-year long drought 1994–1996, the inlet dried out completely. Shallow spits border small bays to the north and south of the main channel. At the eastern end of the lagoon, a sand flat and swamp have filled a former bay. The shores are sandy with many dead shells.

The 200-metre wide channel to the sea passes through a kilometre of well-vegetated coastal dunes. On the south side, half a kilometre from the bar, sub-fossil marine shells (*Katelysia rhytiphora*, *K. scalarina* and *Nassarius* sp.) are found in beds up to about 2 metres above sea level (Fig. 10.19). The channel opens to the sea opposite the centre of the lagoon. This is an unusual location for a bar opening on the south coast and probably reflects the effect of an east-facing beach where there is little net longshore drift.

The low bar of well-sorted fine quartz sand and a small amount of shell is 1–1.5 metres above sea level. The height varies with the strength of the easterly winds, and from its crest slopes gradually to the sand of the flood tide delta. On the seaward side, heavy seas sometimes expose a rock platform. The rock of silcrete and lateritised material derived mainly from the spongolite rock upstream is quite different to the limestone rock that borders many other estuaries of the south coast.

The bar breaks every 3–5 years following heavy rain in the upper catchment. Generally the channel stays open for only a few weeks or months while the river is flowing, although it probably stayed open for two years following

FIG. 10.18 *Gordon Inlet showing the spongolite cliff and samphire swamp, October 1988.* E. P. Hodgkin Collection.

the 1955 flood and three years in the 1980s. In 1973 and 1975 floods broke across the whole width of the bar, scouring the banks and carrying away trees on the north bank of the inlet channel. The bar of the Gordon opened during the December 2001 floods and was still open in June 2002.

Saltier again

Like Beaufort and Wellstead estuaries, the Gordon undergoes large changes of water level and salinity that affect water quality and estuarine ecology. When the river is flowing, the water level rises and water in the riverine reaches may be less than 10‰. After the bar breaks and river flow slackens, water levels decrease and sea water can enter the lagoon and move up to the riverine reaches. As summer progresses, the bar closes and the water remaining becomes hypersaline with salinities more than four times that of sea water recorded in the basin and more than 100‰ in riverine reaches.

Although the major changes of salinity are due to river flow, or the concentration of marine water due to evaporation, there are also small pockets of fresh water due to the inflow of ground water. Freshwater seeps from the spongolite cliffs and dunes may dilute the saline water and create refugia for less tolerant plants and watering areas for birds. Seawards, some salt water can seep through the bar into the estuary when the bar is closed and the water level in the lagoon is low.

As a baseline for understanding these cycles, Gordon Inlet was one of the south coast estuaries monitored by the Water and Rivers Commission between October 1997 and March 1998. At the beginning of the survey in October 1997, sea water had entered the lagoon and salinity graded from near seawater salinity (31‰) in the more coastal areas to 29‰ about 5 kilometres inland. At this time, the inlet was generally well mixed from top to bottom and well oxygenated. As the summer progressed temperatures rose from 17°C in October to 23°C in March, evaporation increased salinity, and the inlet became hypersaline and slightly stratified with about 50‰ at the surface but only 5‰ higher at the bottom. Further upstream in water over 2 metres deep, bottom waters were up to 70‰ with oxygen only 48 per cent of saturation. Water more than 2.5 metres deep was completely devoid of oxygen (anoxic) and slightly warmer. These low oxygen concentrations were probably due to the salinity difference from top to bottom that reduced exchange with the atmosphere. In contrast, deeper saline water in the Beaufort was cooler than the surface. In the Gordon this deep saline water probably originated from another source such as the ground water or solar heating in the shallows. The low oxygen concentrations suggest that it had been isolated from the surface for some time.

In 1997 concentrations of total nitrogen (≈1.4 milligrams per litre) and total phosphorus (0.12 milligrams per litre) were moderate and varied little between sites across the inlet. In November 1997 the total phosphorus appeared to decrease relative to nitrogen, possibly due to biological processes or chemical adsorption, with the highest concentrations in the deepest water where there was less oxygen. As in Beaufort and Wellstead, nutrient concentrations (2.3 and 0.07 milligrams per litre for TN and TP respectively) were slightly higher in March 1998 due to the concentrating effect of evaporation and/or release from the sediment. (See Graphs 1.2 and 1.3.)

Larger diatoms and dinoflagellates in the phytoplankton were moderately low in October and November 1997, but there were high concentrations of small phytoplankton and traces of cyanobacteria over summer. Chlorophyll *a* concentrations in the Gordon are among the highest in Western Australia reflecting the high nutrient concentrations, high temperatures and shallow calm waters.

FIG. 10.19 *Shell bed on the banks of the Gordon Inlet bar, January 1979.*
E. P. Hodgkin Collection.

Wildlife habitat

The fauna of the Gordon Inlet is impoverished with only a few invertebrates and fish species able to tolerate the range of conditions, but these are often present in very high numbers when conditions are less extreme. Polychaete worms dominate the fauna in a number of species (eight) and may be very abundant. Large clumps of the tubeworm *Ficopomatus* are very conspicuous on the trunks of *Melaleuca* trees lining the river banks (Fig. 10.10).

Fish enter from the sea when the bar is open; the species and their numbers depending on the larval and juvenile fish in the nearby coastal waters. Although the inlet is bordered by the Fitzgerald River National Park to the north and east, the inlet has been fished commercially when fish populations are high; a practice currently under review. Fish populations are often killed when conditions are extreme with mass mortalities reported in 1939, 1944, 1953 and 1976 and more recently.

The inlet is also a refuge for many waterbirds with Black Swans feeding on the shallow aquatic plants and pelicans feeding on the fish. In 1988 Gordon Inlet was found to be the most important wetland in the south-west region for the Australian Shelduck, and the second and third most important wetland for the Chestnut Teal in 1987 and 1988 respectively. (See **Box 10.3 Chestnut Teal, *Anas castanea*** and **Box 10.4 Chestnut-breasted Australian Shelduck, *Tadorna tadornoides*.**)

ESTUARIES OF ANOTHER TIME

To the east of Gordon Inlet a number of other small rivers and estuaries lie within the Fitzgerald River National Park: Boondadup River, Saint Mary River, Lake Nameless, Fitzgerald Inlet, Dempster Inlet, Quoin Head, and Hamersley Inlet. The rivers extend in a north-westerly direction, most of the catchments and estuaries are surrounded by the undisturbed bush of the national park; only the upper watersheds of the Fitzgerald and Hamersley lie in cleared land (Maps 10.1, 10.2 and Prelude Map 3).

These estuaries of another time are quite ephemeral, changing from water bodies full of water bustling with water life and birds to shallow muddy puddles or dry glimmering salty pans, meeting the sea only rarely as they near their demise as an estuary—a place where river meets sea.

The craggy peaks of the Barrens, prominent features in the flat southern Australian landscape, had been noted as early as 1791 by Vancouver, and in 1802 by Flinders. Despite the whaling and sealing activity along the coast and settlement at Albany in 1827, little was known about the area. Interestingly, the very factors that interested the early visitors who explored the area, the plants and the geology, were the same features that led to the confrontation of different interests culminating in the eventual declaration of a now highly prized national park. (See **Box 10.5 Fitzgerald River National Park—wilderness versus development**.)

Crisscrossed paths—travellers, collectors, explorers and miners

FIG. 10.20 *The distinctive* Hakea victoria *occurs in patches throughout Fitzgerald River National Park but is rare outside it, October 1989.* E. P. Hodgkin Collection.

In 1847 the botanist James Drummond travelled around the Stirling Range and as far east as West Mount Barren where he collected the Royal Hakea *Hakea victoria*, which is found almost exclusively in the area (Fig. 10.20). The bracts were impossible to press but Drummond was so impressed by their magnificence that he bundled them up and strung them across his horse's back (poor horse!), and carried a 14-foot (4.2-metre) specimen by hand back to Cape Riche. Excited about the botanical diversity, Drummond made another trip in 1848. At the same time John Septimus Roe, the first Surveyor General, was also exploring the area. Roe had seen the Doubtful Islands area from the sea in 1839, but during this land exploration Roe named the Fitzgerald River and Inlet after the Governor Charles Fitzgerald RN (1848–1855) and checked on the locations of coal that had been reported.

In 1849 the Gregory brothers, one of whom had been with Roe, sailed to Doubtful Island Bay to further investigate the reports of coal 'on the east Fitzgerald'. They found a small outcrop of low quality coal, and were so optimistic of the importance of their find that they even proposed a harbour near Point Ann. The Gregorys also travelled past Dempster Inlet and Hamersley (Wonjarup) Inlet, through to Culham Inlet and the Phillips River but considered the country 'altogether worthless and covered

Box 10.3 **CHESTNUT TEAL, *ANAS CASTANEA***

To the Aboriginal people, Chestnut Teal are known as Gnwool-ye-nug-ger-rang, Ngwol-yi-naggirang in the Perth area and Calyung near the Pallinup River.

The Chestnut Teal is found throughout south-eastern and south-western Western Australia, mostly in coastal areas (Fig. 10.21). Common in Tasmania and the islands of Bass Strait, they are less common from Sydney across the southern coast. In Western Australia they are mainly found in the south-west, but there are records of its occurrence and breeding at North West Cape. Chestnut Teal are most common in the brackish estuaries and lakes eastwards of Cape Leeuwin, including Lake Muir. Further east in the barred estuaries between Albany and Esperance they can be more common than Grey Teal. Flocks of thirty were reported by Serventy in the Hamersley River in the 1950s and 1960s, and pairs and small groups in the pools of the Phillips, Young and other rivers of the region.

At the end of summer, waterbirds are usually concentrated in wetlands near the coast where they are easily counted. Following the first rains in April or May waterbirds disperse to temporary seasonal wetlands where they remain until the summer drying. Counts of waterfowl in March 1986 and 1987 found the most abundant species were the Grey Teal and Australian Shelduck; while Freckled Duck, Chestnut Teal and Hardhead were least abundant.

FIG. 10.21 *Chestnut Teal.* Source: Babs and Bert Wells Collection, Courtesy CALM.

Elegant small ducks, Chestnut Teal are about 38–44 centimetres in length weighing 500–850 grams, although males can be slightly heavier. The male Chestnut Teal is among the most brightly coloured of the local ducks with the head and neck blackish-brown with a brilliant greenish sheen, rich chestnut body, white flank mark and black rump, a dark blue-grey beak, and grey-brown or dark grey legs. Females are generally less colourful and similar to Grey Teal but with a richer brown with a pronounced rufous colour on the underside. In flight, males and females show a narrow white wing-stripe and narrow white wedge on the 'armpit'.

Chestnut Teal breed in spring, nesting on the ground, in tree hollows, and even artificial nest boxes located near water producing six to ten eggs. They feed on the aquatic plants *Ruppia* and *Lamprothamnium*, seed and other plant material as well as insects.

Source: Jaensch & Vervest, 1988; Pizzey, 1985; Saunders & Ingram, 1995; Serventy & Whittell, 1962.

in scrub' (Gregory, A. C., 1849). Their hopes of wealth were further deflated when on their return to Cape Riche they met the geologist Dr F. von Sumner and Mr Bland who had visited the Barrens area twice and considered the existence of extensive mineral fields unlikely. More than a century later this scenario would be repeated in the 1970s when the Geological Survey established there was not sufficient coal to support an industry; although by this time the proposal to mine had been rejected in favour of the park.

Despite the early interest in the area, there was little clearing of the hinterland until the 1950s except around Ravensthorpe and Hopetoun. More recently extensive areas of the upper catchments of the Fitzgerald, Hamersley, West and Phillips Rivers have been cleared, and there was further clearing in the catchments of the Jerdacuttup and Phillips Rivers until 1987.

Box 10.4 CHESTNUT-BREASTED AUSTRALIAN SHELDUCK, *TADORNA TADORNOIDES*

The Chestnut-breasted Australian Shelduck is also called the Chestnut Sheldrake, Grunter, and Mountain Duck. The popular name is thought to have originated from the old English word 'sheld' meaning pied. The family to which Australian Shelduck belongs is widely distributed in Europe, Africa, Asia, the Australasian region and New Zealand. There are two species of Shelduck in Australia and one, the Australian Shelduck, is endemic. In Western Australia, Australian Shelducks occur from Shark Bay to Esperance and are sometimes seen in the north around the Kimberley. Shelduck numbers vary greatly from year to year. For example, surveys recorded 19,395 in March 1986 and 65,552 in 1987. The lower number in 1986 may have been due to poor recruitment in the previous dry year or summer rains in 1986 that allowed ducks to disperse to other areas during the survey period. The increase was, however, mostly attributed to the high survival of waterfowl in 1986–1987 due to good rainfall throughout most of the state in 1986.

The Australian Shelduck is richly coloured with a fine head and bill. The male has a white neck ring and buff breast, a white shoulder patch and is deep chestnut on the wing feathers held near the body (outer secondaries), with a large green patch on the wing. Females are smaller in size with a chestnut breast, white eye ring and white around the base of the bill; males sometimes have white marking around the bill but never around the eye (Fig. 10.22).

Australian Shelducks are found in pairs, family groups or flocks often with paired birds. They graze on short grass and in shallow water that may be fresh to brackish, water typical of estuaries. They also visit grasslands, young crops, stubble, irrigation areas, open woodland and are occasionally see in the sea. Around Rottnest Island they nest in caves on the offshore islets. When only a few hours old, the chicks are shepherded by the parents to swim across to the main island where they walk overland to a family territory, a lakeside area of freshwater seepage. In other areas, the nests of grass lined with down are made in a hollow tree some distance from water, on ground under vegetation, or in rabbit burrows. Ten to fourteen cream coloured oval eggs are laid in July to December. Shelducks migrate to more permanent water in summer to moult, often in company with other birds, during which time they are unable to fly.

Source: Jaensch & Vervest, 1988; Pizzey, 1985.

FIG. 10.22 *Australian Shelduck.* Source: Babs and Bert Wells Collection, Courtesy CALM.

Box 10.5 **FITZGERALD RIVER NATIONAL PARK—WILDERNESS VERSUS DEVELOPMENT**

The Fitzgerald River National Park (FRNP) of 329,039 hectares was declared in 1973 and enlarged along the northern boundary in 1989 to cover an area from the coast to 30 to 40 kilometres inland along the old Ongerup Road, and from Gordon Inlet in the west to Culham Inlet in the east. Four management zones are designated: special conservation, wilderness, natural environment, and recreation. The park forms the core of the International Biosphere Reserve surrounded by the agricultural area 'Zone of Co-operation'. The only other Biosphere Reserve in Western Australia is the Prince Regent Nature Reserve in the Kimberley.

The park is one of the richest areas for plants in Western Australia with over 1,750 known species, about 20 per cent of the State flora, a rich diversity that was recognised as early as the 1800s by collectors such as Baxter and Drummond. (See above Crisscrossed paths—travellers, collectors, explorers and miners.) More than seventy species are found only in the park and a further 250 species are found only in a restricted geographic area or are rare.

The park also contains a richer fauna than any other conservation area in Western Australia. Many habitats are represented in the park: uplands, plains, gorges, ranges and coastal dunes, each with their particular plants and animals. The rivers and the estuaries to which they flow have carved spectacular cliffs and gullies in the many coloured layers of spongolite rock. The catchments of the Dempster and Saint Mary Rivers lie entirely within the park while those of the Fitzgerald, Hamersley and Phillips extend northwards into cleared land, although there are some reserves and other uncleared land along most of the major rivers.

The original impetus for creating a reserve to preserve the diverse flora came from the Western Australian Naturalists' Club supported by the Government Botanist C. A. Gardner. More generally the area was viewed as extremely poor and suitable for land settlement. The 1950 'C' Class Reserve status meant it had some legal protection, although the boundaries were not fixed by survey, rather they reflected the amount of vacant Crown land available at the time. The area was totally unmanaged but its status as a C Class Reserve ensured that the area was not included in the large-scale clearing in southern Australia.

The reserve however was not popular with many farmers, some of whom regarded it as a haven for vermin (dingoes, kangaroos and emus), while others considered it a wasteful use of an area suitable for good farms. In 1957 and again in 1961 it was suggested that consideration should be given to reducing the area to a strip five kilometres wide along the coast.

In contrast, in 1962 the Australian Academy of Science suggested that it should be raised to 'A' Class Reserve status and include an additional area between the Phillips River and the Hopetoun Road north of Culham Inlet. The Academy also recommended the establishment of a central body with adequate finance and administrative power to manage national parks in Western Australia. Although the National Parks Board was formed from the State Gardens Board in 1958, it had neither funding nor power at this time.

Despite recommendations to increase the status of the reserve there were other plans to excise 39,000 hectares from the reserve, but these were defeated because of opposition from government departments, the shire and public concern that Western Australia was losing too much of its wildflower heritage. Champions of conservation at the time including Dr Dom Serventy suggested that the time had come to move away from the idea that reserves should only be of land unsuitable for agriculture, and that reserves should contain representative flora, fauna and scenery.

There had always been interest in the area's mining potential for magnesite, gold and building stone, but the scale of interest increased with the mineral boom of 1970 when leases were pegged over 11 per cent of the reserve. Leases covered most of the Fitzgerald Valley for kyanite and coal, the West River for copper, the spongolite cliffs around Twertup and Mount Maxwell for kaolinite and diatomite, and a dredging claim for mineral sands near Gordon Inlet. Concern about the cost of some developments, pollution and environmental issues was triggered by plans to mine coal for montan wax, and culminated in the first coordinated biological survey. A period of legal wrangling followed with an increasing public interest in the conservation debate petitioning against mining, matched by interest in developing an industry in an area of agricultural recession. To the conservation movement this was viewed as a test case for mining in other parks and reserves.

On advice of the Environmental Protection Council, the Environmental Protection Authority, established in late 1970, endorsed the elevation of the reserve to an 'A' class National Park and recommended that the extent of the coal deposit should be examined and botanical surveys made of the prospecting areas. Following exploratory drilling that revealed insufficient montan wax (a component of lignite or brown coal) of good quality to support an industry, the applications to mine were rejected by cabinet. The decision to create a national park was, however, made before the drilling results were publicised with the reserve gazetted as an 'A' class reserve on 19/1/1973 as two National Parks, the Main Inland Park and a narrow coastal strip over which the Government retained 'power to lease'. Today the status of the park is ensured and its value more widely appreciated. However there are continuing threats from mining, salinisation, feral animals, fires, plant diseases such as dieback, *Phytophthora*, and the need to provide access to areas of interest and beauty while preserving and maintaining conservation values. The Department of Conservation and Land Management is responsible for the overall management of the park, a huge and complex duty, balanced by the Fitzgerald River National Park Association (re-formed in 1980) that provides a 'grassroots' community approach to management and a reminder that the original impetus for the park's declaration was born in the community. The Association has been a model for other Friends of National Parks groups in Western Australia, although they are continually faced with problems of funding and limits to their time and energy.

Source: Bradby, 1989; CALM, 1991, Chapman & Newby, 1995.

Boondadup River

East of the Gordon River two small streams hold long pools where they flow to the sea. The larger, Boondadup, flows from the north slope of Mount Bland where it forms cascades as it crosses a quartzite ridge. The river then flows across the spongolite plain before winding through dunes along the coast. The catchment covers 36.5 square kilometres of bush within the FRNP. The river is estuarine for 700 metres and is up to 10 metres wide. The banks are low where the river has cut a broad valley in the spongolite, but high where it winds through the dunes. Paperbarks make a dense growth along the banks through the dunes but more sparsely in the open valley. The tannin-stained water is quite fresh, about 5‰.

The smaller river, 2.5 kilometres south-west of Point Ann, lies in a gully in the coastal strip of schist. It holds a narrow pool, 250 metres long, which discharges across a sandy beach in the mouth of the valley.

Saint Mary River

The catchment of the smallest estuary, Saint Mary Inlet, lies entirely within the FRNP. The greater part of the catchment lies on spongolite plain, although small tributaries flow from the areas of harder quartzite rocks of the Mount Barren Group near the coast. The river flows to the inlet in a steep rocky valley with several pools, the last nearly a kilometre long, and discharges to the inlet through a densely vegetated swamp.

The lagoon, only a kilometre long and 250 metres wide, has filled with sediment to about sea level and the estuary is often dry in summer (Figs 10.23 and 10.24). The southern shore of the lagoon has a narrow sand flat with samphire and other marsh plants backed by paperbarks trees. The northern shore is much steeper without a bordering sand flat except where the schist gives way to limestone near the inlet channel. There is a small sand flat with beach ridges, and a swamp fills the mouth of a gully in the contact between limestone and schist.

The inlet channel, 700 metres long, is bordered by spongolite on the south side near the lagoon and limestone on the seaward side. On the north shore, the dune is collapsing below a hard cap along much of the exposed section. The bar, only 1.5 metres above sea level, is really only part of the beach where waves wash over and carry sand into the channel. It probably behaves similarly to the Fitzgerald Inlet bar, breaking infrequently and only staying open briefly; although it opened in the spring of 1998 and again in January 2000.

The inlet water alternates from fresh to brine. The riverine input is almost fresh at 4‰, and sea water enters when the bar is open or seeps in through the bar when the water level is low. The mollusc shells lying around

FIG. 10.23 *Saint Mary River and Point Ann, aerial view, October 1978. The long narrow lagoon is very shallow where sand has accumulated and closed the narrow channel to the sea. Point Ann is a popular location for viewing whales that calve along the south coast in late winter and spring.* E. P. Hodgkin Collection.

FIG. 10.24 *Saint Mary River, October 1989, showing the west bank 2 km from the mouth bordered by* Wilsonia *sp., samphire* Halosarcia, *wattle* Acacia *and paperbark* Melaleuca. E. P. Hodgkin Collection.

the margin indicate that moderate salinities, about 25–50‰, may persist for a year or more; enough time for some species to flourish before conditions become too harsh. Fish populations are probably small and the estuary is not fished commercially.

Water quality in Saint Mary River has recently been part of a study that also examined the Beaufort, Wellstead and Gordon Inlets. (See Graphs 1.2 and 1.3.) In October 1997 the waters were stratified with a salinity of 26‰ in the top 1.5 metres, whereas at 2 metres the water was hypersaline (43‰), slightly cooler and devoid of oxygen (Graph 10.3). Salinity, temperature and oxygen increased over the next month with no indication of increased salinity or anoxia in deeper areas. By March 1998 salinity had doubled to 70‰ and temperature increased to 23°C.

The concentrations of total nitrogen and total phosphorus were low to moderate in each part of the inlet with total nitrogen of 1 milligram per litre and total phosphorus of 0.07 milligram per litre in October 1997. Bottom waters had higher concentrations of total nitrogen (3.9 milligrams

GRAPH 10.3 *The salinity profiles and oxygen profiles in Saint Mary Inlet, October 1997. The bottom axis indicates distance from the ocean bar. Saline water is present in the bottom waters where oxygen concentrations are low.* Courtesy WRC.

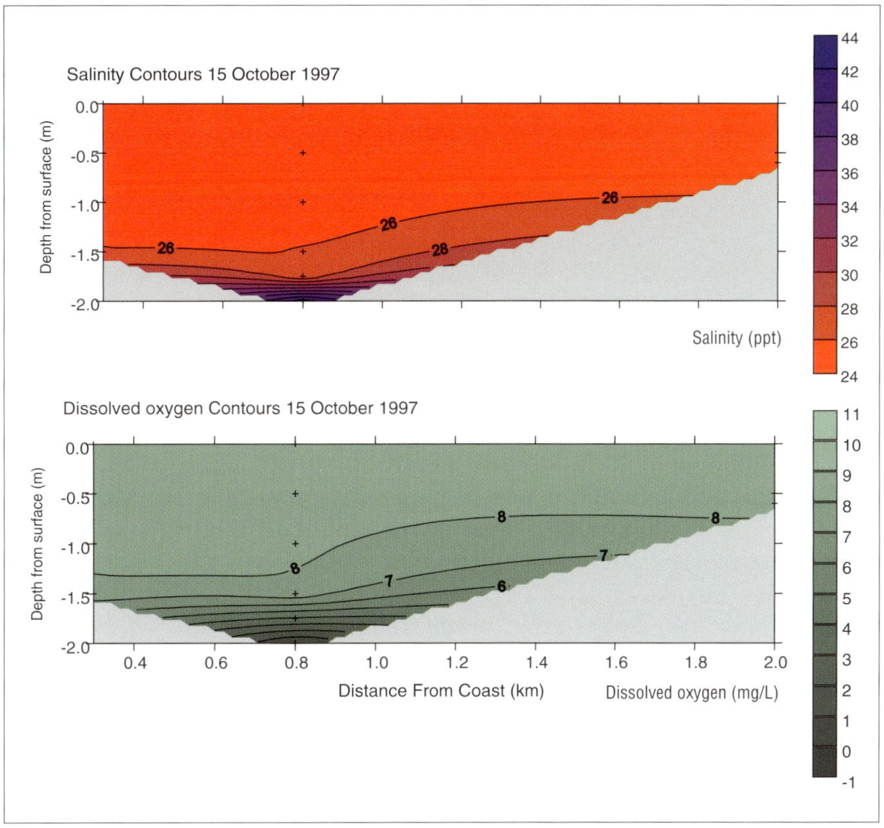

per litre) and total phosphorus (0.13 milligram per litre) under salinity-stratified anoxic conditions. These values are comparable to those in Gordon Inlet and Wellstead Estuary, where the catchments are cleared, possibly reflecting cycles of inflow and drying that concentrate the dissolved nutrients, and low oxygen concentrations that favour release of nutrients from the sediment.

Diatoms and dinoflagellates dominate the phytoplankton community in the Saint Mary River, and a bloom of *Chaetoceros* was observed in October 1987. In March 1998 phytoplankton numbers were particularly high, dominated by the dinoflagellate *Oxyrrhis*. To date there have not been any blue-green algae in any of the Saint Mary samples.

FIG. 10.25 *Lake Nameless, sometimes salt, sometimes fresh; the changing water levels evidenced by dead paperbarks* Melaleuca cuticularis, *February 1989.* E. P. Hodgkin Collection.

Lake Nameless

The name conjures up an image of inconsequence but Lake Nameless, now a salt lake, was in time past an estuary, the former channel to the sea blocked by a kilometre-wide dune. Apart from the lack of an outlet to the sea, the lake is similar in most respects to the larger Fitzgerald Inlet and smaller Saint Mary River, lying in a branching valley in the spongolite plain surrounded by cliffs and steep slopes. The former wide inlet channel is now filled with longitudinal dunes and bordered on either side by a higher cemented (lithified) dune landscape (Fig. 10. 25).

Fitzgerald Inlet

The Fitzgerald River flows in a south-east direction from its headwaters 80 kilometres from the coast and 300 metres above sea level in Lake Magenta Nature Reserve (Map 10.1). Between the Nature Reserve and FRNP, the river flows in a narrow strip of uncleared reserve and Crown Land between cleared farmland. Two small tributaries, Twertup Creek (15 kilometres) and Sussetta River (35 kilometres), join the main stream 25 kilometres from the sea. The Twertup rises in cleared land between the South Coast Highway and Fitzgerald Road, and the Sussetta to the north of the highway.

Fitzgerald Inlet is registered as a wetland of national significance (Figs 10.26 and 10.27). The Fitzgerald has been a 'focus' catchment in a three-year Natural Heritage Trust Program that aims to re-establish, conserve and manage native vegetation along the river and tributaries between the Lake Magenta Nature Reserve and the Fitzgerald River National Park.

The Fitzgerald is the largest inlet in the FRNP with an area of 7 square kilometres. The estuarine part of the river winds for about 4 kilometres through deltaic sediments and samphire swamps, some of which mark former river channels while others were part of the inlet when sea level was higher. The river widens from 10 metres to over 100 metres before discharging into the lagoon. For the last 2 kilometres the river flows against a cliff on the south side and a curving spit on its north bank that partially cuts off the western segment of the lagoon. The 10-metre high spit is well vegetated at the river end but descends to a samphire swamp where it ends in the lagoon.

Most of the inlet lies in a valley in the Pallinup Siltstone, which forms spectacular red and yellow 40-metre high cliffs around the margins. The beaches are narrow and there are only small patches of swamp. The flat bottom of the inlet lies at about mean sea level so that even when full the maximum depth is only about 1.5 metres deep, although the channel to the sea has scoured to about 5 metres deep. Following a flood (in about 1963) many tree stumps lay scattered over the lagoon floors for years, but most of these have now disappeared.

FIG. 10.26 *Fitzgerald Inlet, aerial view, October 1978. The large lagoon is very shallow and a slightly deeper channel leads to the sea in the shelter of Point Charles.* E. P. Hodgkin Collection.

FIG. 10.27 *Fitzgerald Inlet. A quiet moment on a spongolite outcrop at the pool below the crossing: Ernest Hodgkin with members of an expedition examining the south coast estuaries, May 1987. From top left to right: Greg Davis, John Kelly, Ernest Hodgkin, Paul Pavlenko, Ruth Clark, Jenny Shaw, and Francis Beronie.*

The narrow inlet channel winds for 1.5 kilometres through the coastal dunes to the ocean at the south-western end of a 6-kilometre long beach, where the mouth is sheltered from south-westerly winds and swell by Point Charles. The bar is generally low, about 1.5 metres, but can build up higher. The bar is built of fine white silica sand, which becomes saturated forming 'quicksand' when the estuary waters are high.

The bar breaks infrequently after heavy rain in the catchment, and generally only stays open for a few weeks. It is said once to have remained open for 18 months and to have had abundant fish and shellfish at the turn of the century. A flood broke the bar in about 1963, and again in 1971. It broke in July 1978 and was still open in January 1979. It broke in late winter 1988 after record rains in early winter and again in January 1990 after heavy rain that damped the December 1989 fires, but on neither occasion did it stay open for very long. Recent bar openings are not officially recorded.

There are few salinity records from the inlet. Presumably the water may be quite fresh with river flow, about 5‰, when first filled by winter rains but becomes progressively more saline as the water dries up in summer. Salinities in excess of 200‰ have been recorded. In summer sea water seeps through the bar into the channel and waves wash over the bar, but the volume is only sufficient to retain water in the channel.

Only the most hardy of estuarine invertebrates and fish live within the estuary. In the past when black bream were abundant the inlet was fished commercially, but this ceased in July 2002.

Dempster Inlet

Dempster Inlet is only about 4 kilometres from the Fitzgerald Inlet and its mouth opens at the north-eastern end of the same bay (Fig. 10.28). The main tributary Coppermine Creek flows to the western end, and several small gullies flow from the bare slopes of Middle Mount Barren to the lagoon. The inlet and catchment lie within the wilderness area of the FRNP.

The 7-kilometre long estuary lies east-west in a narrow valley in the Mount Barren Schists and the bare slope of Middle Mount Barren rises steeply from its northern shore (Fig. 10.29). In this respect it is similar in character to Hamersley Inlet and different to the Fitzgerald, which is surrounded by spongolite cliffs. Like the Fitzgerald, the lagoon has filled with sediment to about sea level and is dry in summer. The Gregorys noted in 1849 that it was covered with 6 inches (15 centimetres) of salt.

FIG. 10.28 *Dempster Inlet in the foreground (right) with Fitzgerald Inlet on the left, aerial view, June 2000. A number of small creeks flow to the lagoon from the lower slopes of Middle Mount Barren. The large sand sheet west of the Dempster mouth approaches to within 300 m of the lagoon.*
S. Neville/Ecotones, Courtesy WRC.

FIG. 10.29 *Dempster Inlet, February 1989. (a) The hills of Middle Mount Barren tower above the flat sediment-filled estuarine landscape. (b) Freshwater swamp in Dempster Inlet with* Melaleuca cuticularis *backed by a calcarenite hill.* E. P. Hodgkin Collection.

The lagoon, less than a kilometre wide and about 2.3 square kilometres in area, has a flat muddy bottom and narrow sandy shores from which the land rises steeply. Schist is exposed along the northern shore and patchily along the southern shore. Dune limestone and mobile sand reach the southern shore near the mouth, and about 2 kilometres from the mouth fresh water drains from the limestone to a small swamp. This is probably the Taylor Spring marked on some maps.

The narrow 25–50 metre-wide inlet channel curves for 700 metres from the lagoon to the beach. On the northern side there is a wide beach ridge at about 2 metres above sea level, which bars the samphire swamp in the mouth of a small creek. The southern shore is a steep well-vegetated sand slope with a steep rock face with schist to the waterline further seaward.

The inlet mouth faces south with no protection from the south-west. The bar builds up to 2 metres above sea level and retains this depth of water when the inlet fills (Fig. 10.30). The bar broke following heavy rain in 1971, 1988, January 1990 and, although there are no records, probably at other times as well. Like the Fitzgerald the sand can behave as 'quicksand' when saturated.

Shell deposits, possibly Holocene (<10,000 years), containing the cockles *Katelysia scalarina*, *K. rhytiphora*, heart cockles *Fulvia tenuicostata*, oysters *Ostrea angasi*, and scallops *Chlamys asperrimus* are evidence of more marine conditions when the inlet was more open to the sea.

FIG. 10.30 *Dempster Inlet, 1989. Ruth Clark examining algae marking the height of flood waters. Ruth Clark worked with Ernest Hodgkin for a number of years collating much of the material in the Estuarine Study Series as well as accompanying him on many field trips to the estuaries.* Used in Hodgkin & Clark, 1990 (ESS#7) and in E. P. Hodgkin collection.

Currently the lagoon only holds water for a few months each year, changing rapidly from fresh to brine. Invertebrates such as the bivalve *Spisula trigonella* (Fig. 2.29f) sometimes live in the lagoon, and sea mullet have been caught in the pool in Coppermine Creek. The inlet is closed to commercial fishing but is netted by recreational fishers.

Quoin Head

Quoin Head is one of a number of gullies, sometimes referred to as a ria, cut into schist and quartzite along the precipitous coast between East Mount Barren and West Mount Barren (Fig. 10.31). The mouth, blocked by a small sandbar, holds water in a 300-metre long pool.

Hamersley Inlet

Hamersley Inlet, about 7 kilometres long, lies in a valley carved through the hard quartzite and schists of the Mount Barren Group (Figs 10.32 and 10.33). Fed by the Hamersley River that rises on the edge of the plateau 50 kilometres from the inlet and 300 metres above sea level, most of the catchment lies within the FRNP with only 20 per cent of the headwaters in cleared agricultural land.

FIG. 10.31 *(a) Quoin Head Creek, aerial view, June 2000.* S. Neville/Ecotones, Courtesy WRC. *(b) View of the quiet estuary waters, May 1989.* E. P. Hodgkin Collection.

FIG. 10.32 *Hamersley Inlet, aerial view, June 2000. The lagoon is divided by a long spit into two arms unlike many of the other eastern estuaries that are linear in form. The dune erosion on the eastern side of the mouth is said to have started after a fire in the early 1900s.* S. Neville/Ectones, Courtesy WRC.

The lagoon, about 4 kilometres long and about a kilometre wide at widest point, covers an area of 2.3 square kilometres. Near the mouth it is 2 metres below sea level and shallows gradually upstream to the wide marginal shoals. The upper arm of the lagoon is narrow in a south-easterly direction from where the river first widens out into it. When the water level is low, shoals isolate small salt lakes in the mouths of two creeks that enter from the north. The peninsula on the western shore is rock with a beach of moderately rounded pebbles.

The 3-kilometre long lower arm of the lagoon turns abruptly to the south-west. The tortuous northern shore and part of the southern shore are composed of quartzite and schist (Fig. 10.34). Nearer the mouth the southern shore cuts into an old dune that overlies the hard rock. About a kilometre from the mouth, shell beds of Pleistocene age are exposed. (See **Box 10.6 A shell and understanding climate change**.) Here the shore is sandy and slopes gently from the dune to the deeper part of the basin. Closer to the mouth a dense stand of paperbarks fringes the shoreline.

Near the mouth, the basin narrows to a short channel 150 metres wide and 500 metres long. A flood tide delta fills the seaward end and slopes down

FIG. 10.33 *Hamersley Inlet: the expanse of the inlet set in the Fitzgerald River National Park, August 1978.* E. P. Hodgkin Collection.

FIG. 10.34 *Hamersley Inlet Gorge showing the schist of the Mount Barren Ranges, April 1978.* E. P. Hodgkin Collection.

Box 10.6 **A SHELL AND UNDERSTANDING CLIMATE CHANGE**

Beds of long-dead shells are found in the estuaries and salt lakes along our southern and western coasts. The shells show us that the areas were once connected to the sea, that exchange with the ocean was good, and that it was a quite different aquatic environment. This is a story about how studies of these deposits have led to a greater understanding of past conditions and climate.

The story focuses on one shell and has many parts: chronicling observations of living populations, enzyme analysis for relating living populations, collections of dead 'fossils', mapping fossil deposits with other deposits of known age and against the modern coastlines, examining collections housed in museums to establish relationships with other shells (taxonomy) and, finally, modern techniques for ageing the shells using radiocarbon, uranium-series disequilibrium, amino acid racemisation and electron spin resonance dating.

So what is this shell? The Ark shell or 'Sydney Cockle' *Anadara trapezia* is a large, heavy ribbed bivalve growing to about 9 centimetres (Fig. 10.35).

FIG. 10.35 *In Western Australia the bivalve shell* Anadara trapezia *now lives only in Oyster Harbour, but fossils indicate that it was more widespread during the Last Interglacial Maxima when sea level was higher, when rain may have fallen throughout the year, and temperatures were higher. Adult shells are about 90 mm in length.* Photograph C. Bryce, Courtesy WAM.

It lives in the shallow water of marine bays and estuaries, partially buried in stable soft silt, mud or fine sand, the exposed part of the shell often bearing a little algal growth.

The shell burrows and moves by pushing the foot outwards, enlarging the foot and then rapidly contracting it. It feeds by filtering phytoplankton from the water, and probably breeds when phytoplankton is abundant in late summer. Living in estuaries from subtropical Townsville in north Queensland, where the water temperature ranges from 20–32°C, to cool temperate Port Phillip Bay in Victoria (9–22°C) and warm temperate Oyster Harbour, Western Australia (13–25°C), its presence in Oyster Harbour but not Princess Royal is thought to be linked to the freshwater inflow to Oyster Harbour. However there are other estuaries on the southern coast where environments could be suitable but where it is not found. The record in fossils tells us that it was once more widespread: so why did it die out? The enzyme evidence comparing living populations tells us that they are all quite similar, indicating that the Oyster Harbour population has probably not been isolated from the east coast populations for very long, even though the shells have a few more ribs. However, we cannot rule out the explanation that the species has a very stable genotype because conditions across its range are quite stable.

Shells in museums show that the *A. trapezia* had a much larger distribution in the past, indicating that estuarine environments were once more widespread, and a close relative and possible ancestor of *A. trapezia* lived in Argentina. In the Middle Pleistocene (730,000–130,000 years BP) the shell was found in Australia and New Zealand. In Western Australia it was found from as far north as Urala Station near the Ashburton River, 4.9 kilometres

FIG. 10.36 *Hamersley Inlet, May 1982. George Kendrick taking notes and Patrick Hesp collecting from shell beds. George Kendrick's extensive collecting and collation of information of molluscs from many estuaries and coastal areas during more than 50 years association with the Western Australian Museum has contributed to the understanding of estuarine evolution. Patrick Hesp co-authored Ernest Hodgkin's final paper (1998) on estuarine evolution.* E. P. Hodgkin Collection.

inland from the modern coast, at Hamelin Pool, Shark Bay and along the west coast at Hill River, Peppermint Grove in the lower Swan Estuary and near Baldivis east of the present Peel–Harvey Estuary.

Later in the Late Pleistocene (125,000 years BP) it was found from Lake Macleod and Cardabia near Carnarvon through to Israelite Bay including parts of Shark Bay, Mimim Cove on the Swan River, Augusta, Wilson Inlet, Torbay–Elleker, Bremer Bay and Hamersley Inlet, Steere River (Culham Inlet), Stokes Inlet, Esperance and salt lakes such those NNE of Israelite Bay. The sediments and shells tell us that many of these places were near freshwater outflows (Fig. 10.36).

In the most recent part of the Holocene period (6,000 BP), the range of the species shrank and there are no fossils from this time known from Western Australia. In support of the idea that the shell was no longer present and that we simply have not found them, is the lack of these shells in Aboriginal middens from this time.

The wider distribution up the west coast in the late Pleistocene indicates that conditions were wetter and warmer and the seasons less marked, possibly due to a stronger Leeuwin Current. Shell beds in places now distant from the coast show that sea levels were higher between the ice ages and the estuaries were larger with more freshwater inflow. So the combination of greater river flow and more protected embayments favoured *Anadara*. The shells from around the Lyndon River in Lake Macleod north of Carnarvon are of particular interest to our estuary story. Once the northern arm of Shark Bay, the Lake is today a dry saltpan illustrating the transitory nature of estuaries—a story we are observing in our most eastern estuaries that open to the sea very infrequently.

Source: Murray-Wallace et al., 2000.

into the basin. Rock, sometimes covered with sand, borders the western shore of the channel and extends to the beach. Fire and erosion in the early 1900s is thought to have influenced the sparse vegetation on the eastern shore, however this damage may be more recent.

The estuary mouth is closed by a sandbar about 300 metres wide composed of off-white, moderately well-sorted fine grained sand, 70 per cent quartz and 30 per cent calcareous shell. The bar builds up to 2 metres or more above sea level. On the seawards side sand often covers quartzite rocks that are continuous with the rocky shore west of the mouth: when the bar breaks the outflow exposes the jagged rock that is about a metre above sea level.

The bar breaks infrequently and only remains open for a few weeks. Reports indicate that is was closed for a long period between 1915 and 1955, although it may have opened in 1923. It was broken in 1971, November 1977, August 1978, August 1986, May 1988, January 1990 and January or February 2000. The break in 1955 may have followed February and May rains, but late winter or spring breaks appear to be the more usual pattern. The occasional heavy summer rains are generally absorbed by the dry catchment soils and are unlikely to break the bar unless the estuary water level is already high.

Hamersley Inlet is deeper than other estuaries with a lagoon bottom about 2 metres below sea level, and sometimes holds water for longer periods than other inlets in the area. Some areas of the inlet dry out however, leaving only patches of shallow hypersaline water along the lower arm of the lagoon (as in January 1968, April 1977 and July 1983). Pools in the river section of the estuary up to 4 metres deep also hold water when the lagoon dries. There is probably little stratification in the shallow open water of the lagoon, but deep water in the riverine part may be stratified with fresher surface water of 5‰ flowing from the river over hypersaline bottom water of 50‰. Nutrients in Hamersley Inlet are similar to those of the other eastern estuaries (Graphs 1.2 and 1.3).

As Hamersley Inlet sometimes holds water for longer than other inlets in the area, it provides a favourable environment for estuarine and marine animals. These sometimes flourish for several years, however at other times the invertebrate fauna has been impoverished. At times there has also been a greater variety of fish in the inlet compared to the nearby Fitzgerald and Culham Inlets, and recreational and commercial fishing are allowed.

CULHAM INLET

A sleeping giant

At the eastern end of the Fitzgerald River National Park the roads winds spectacularly around East Mount Barren. Here is a reminder of ancient times and a once mightier continent, for the scattered white rocks are similar to those on the Antarctic Continent now separated by thousands of kilometres of wide blue Southern Ocean. This is where the continents divided in the Eocene Epoch about 50 million years ago. About 80 metres above sea level there is a marine terrace left from when the sea covered the land, either in the Eocene or more recently in the Pleistocene. But before entering or leaving the park there is the crossing of Culham Inlet to contend with. For many years a track followed the dune lines to Hopetoun, but when the inlet filled this was impassable and the FRNP was cut off from the town. The bar broke in the 1920s when the inlet was very high, but for the following 70 years the inlet was mostly shallow or dry, and even when the inlet was full there was no connection to the sea. In the last decade or so the floodwaters have increased in size and frequency, probably exaggerated by greater run-off from the catchment after clearing. The roads have been cut by floodwaters more often, and roadwork repairs have become more substantial. This cycle culminated in 1993 when despite earthworks to control the area of a break, the inlet broke to the sea washing away dunes and roads. A new road was built with culverts to reduce flooding and 'sacrificial areas' to be broken when floods were high, but so great were the floods in 2000 that the inlet again broke free gouging a great passage to the sea (Fig. 10.37).

The dilemma arises as the inlet clearly follows natural extremes from flooding and bursting to the sea, as reported by the explorer Gregory in April 1849, or being covered with only a film of water that dries to a saltpan in summer. How often do these extreme conditions occur? Can we estimate its extremes? Can we tame these extremes? Or should we learn from our mistakes, admit our interference with its rhythm and learn to live with the extremes? The processes that created the nearby mountains, the wild Southern Ocean that moves beaches and sand dunes, the above 'average' rainfall events all remind us of nature's power and changeable states. The shell beds at the inlet base also provide evidence of marine invasions, a timely reminder of how our coastline is still evolving on a scale we cannot hope to tame.

Where the waters come from

Culham Inlet receives water from the Phillips and Steere Rivers. The Phillips River and its tributary West River rise in sand plain north of the South Coast Highway, but much of their catchments and that of the Steere River are in mineral-rich hilly country of ancient volcanic and sedimentary rocks south-west of the Ravensthorpe Range (Map 10.2 and Fig. 10.37). Most of the catchments south of the South Coast Highway (a distance of about 40 kilometres) are in natural bush, part of the FRNP. The catchment area north of the highway, and east and south of the Steere River as well as the east bank of the inlet, however, has been cleared. Overall about 40 per cent of the 2,300 square kilometre catchment has been cleared.

The Phillips River lies along a fault line in the Mount Barren Group, and in the upper reaches cliffs 20 metres or more high line the banks. The river is estuarine for about 7 kilometres (slightly more when flooded) from the inlet to a sharp bend called Pitchie Ritchie. The river is 4 metres, possibly 8 metres, deep and possibly extending below mean sea level against the cliffs, but has sandy shallows in the upper reaches. The Steere River has a

FIG. 10.37 *Culham Inlet, aerial view, June 2000. The western side of the inlet lies within the FRNP. The eastern side of the inlet and the catchments of the Steere River as well as the upper catchment of the Phillips River are surrounded by cleared land.* S. Neville/Ecotones, Courtesy WRC.

meandering course across a low-lying alluvial plain, and about 2 kilometres of the river is estuarine.

The Phillips and Steere Rivers enter the inlet from the north through a wide sand and gravel delta, about 2 metres above the level of the lagoon bottom. The delta contains marine and estuarine fossil shells, probably of Pleistocene age. There are also wide marginal terraces of predominantly cemented, fossiliferous Pleistocene sands in the lagoon with a thin veneer of Holocene sediment again with abundant mollusc shells.

Like other areas along this part of the coast, most of the rain falls in the winter between May and October. Falls can be heavy exceeding 100 millimetres a day, and similar heavy falls also occur in summer after the southwards passage of a tropical cyclone. The catchment is not gauged, but flows have been estimated to vary from zero to more than 50×10^6 cubic metres annually with an average of 3.5×10^6 cubic metres. Although there is no record of river flow to Culham prior to clearing, estimations suggest it was 1.3×10^6 cubic metres; however A. C. Gregory recorded a deluge in 1849 when the river rose 10 feet (3 metres). Although flow varies with soil moisture of the catchment due to rainfall, the increase in flow since clearing has been caused by the removal of deep-rooted vegetation, and the accompanying rise in the watertable together with the reduced penetration of rain because the soil has been compacted by stock. The present much higher flows are expected to increase still further.

The river waters are naturally saline, as early explorers such as John Forrest in 1870 found to their anguish during their search for fresh drinking water. Salinity in the Phillips River is about 10‰ when flowing and in the Steere River about 5‰, becoming more saline due to evaporation when flow slows. Although there are few records of water quality, high concentrations of total nitrogen and total phosphorus have been recorded in stagnant hypersaline water in Culham Inlet and the Phillips River. Other records indicate that nutrient concentrations were low in the West River in the mid-1980s, but there were brief periods when phosphorus concentrations in the Phillips River were high following superphosphate applications.

A large basin

The lagoon, oval in shape (approximately 3 kilometres × 4 kilometres) with an area of 11.3 square kilometres, narrows to about 500 metres close to the dune line bordering the Southern Ocean. A shallow pan about 1–1.5 metres below mean sea level with gentle slopes to the shorelines and the river delta area, the lagoon is so shallow that it often dries up completely although it may hold water for several years as in 1988 and 1989. At other times the lagoon may be deeper than 2 metres because of the high bar: in August 1989 there was about 4.5 metres of water in the deepest part in the centre.

The bar, a barrier

The inlet and ocean are separated by a kilometre-long, dune-topped bar (Fig. 10.38). In the west the dune is 10–15 metres high, 100 metres wide and covered with trees and shrubs, while the eastern section, 400 metres long, is only 4–5 metres high, 40–60 metres wide and sparsely vegetated. The dune line widens on both sides of the mouth with a belt of dunes 20 metres high and a kilometre or more wide in the east. The dunes also widen to the west, but a low swampy area in the south-west corner of the inlet near the FRNP Ranger's cottage suggests that this is possibly the location of an old flow channel to the sea at Barrens Beach. The bar is subject to erosion, and storm waves expose patches of hard limestone reef rock in the sea beach at the eastern end of the bar in winter, to be covered again by sand in summer.

Lake, lagoon or inlet?—floods and flooding

Shells reveal time past

Abundant shells of Holocene marine molluscs testify to the inlet having been open to the sea within the last 4,000 years. Cockle shells *Katelysia scalarina* from the eastern shore of the inlet have been aged at 3,660 ± 185 years BP. This cockle is no longer present in Culham Inlet but lives today in estuaries like Hardy Inlet, Oyster Harbour, Princess Royal Harbour and other estuaries after they have opened to the sea (Fig. 6.14). (See **Box 10.6 A shell and understanding climate change**.)

In the last 4,000 years, lower river flow as the climate changed meant that the bar was able to build up, and it probably opened infrequently like Hamersley Inlet today. Further drying and sand movements allowed the dune line to develop and stabilise, so the inlet has opened so infrequently that it has at times been described as a 'permanently closed estuary', that is a coastal salt lake. This is perhaps an unfortunate title because it implies stability and predictability, whereas long-term weather cycles and the changes wrought by humans on the catchment make the situation anything but predictable.

FIG. 10.38 *Culham Inlet, view from East Mt Barren, August 1972. The road between Hopetoun and the FRNP lies just behind the dune line.* E. P. Hodgkin Collection.

Opening and closing

We do know that the bar broke naturally in the past because of the report by Gregory in 1849, and it is also reported to have been open in the 1870s. After that time human interference became an influence, but exactly when is unclear. Some reports suggest it was broken by miners in 1906, and there was pressure from a proposer of a fishing and fish curing business for the Public Works Department to open it in February 1918. Then in 1918 or 1920, following years of heavy rainfall, the local farmers with eight horses and a scoop dug a channel from the inlet to the sea at the eastern end, the narrowest point. The bar broke through and the inlet is reported to have emptied leaving 'stinking black mud'. When sea water washed back in, it covered only a small area on either side of the narrow channel from the break in the lagoon. The bar is variously reported to have remained open for a few weeks up to three months. Another report says that the bar broke in 1918 at the eastern end and the water flowed out over rock so that the inlet did not drain completely. Inlet water is also reported to have trickled out in 1955.

In 1986 the inlet water almost reached the level of the road at the lowest point, then following the heavy rains of May 1988 water covered the road to a depth of 0.9 metre, and in August 1989 there was 1.6 metres of water over the road. The inlet water was then 3.2 metres above MSL and water was seeping through the dune to the beach.

In September 1988 run-off from above-average rain raised the inlet level above 3 metres with 0.8 metre over the road at the lowest point. The inlet rose again in 1989 to 3.3 metres, roads and farmland were flooded, and paperbark and coastal moorts along the eastern shore of the inlet died. In May 1990 a pipe was installed at the 2.2-metre level in the eastern end of the bar onto a reef line. With a drought the following year, the inlet level fell to about 0.4 metre in July 1992. Rains then saturated the catchment and heavy falls between August and October raised the inlet to 3.9 metres. Water seeped from the inlet along the whole bar and dune line, and flowed out through the 1990 pipe.

The inlet was still 2.2 metres high in March 1993 when rain saturated the catchment and raised the inlet to 3.3 metres in early May. A cut was commenced at the western end of the dune, but before it was completed further heavy falls raised the inlet to 3.7 metres. The waters tore through the cut, opened a 70-metre wide and 3-metre deep channel, then scoured a channel parallel to the dune (Fig. 10.39). The water level fell 3 metres in three days, but it was a week before sea water entered. Within six weeks the beach had rebuilt to about 2.3 metres above sea level and closed the inlet.

Since the 1993 break out, the inlet has only briefly risen above sea level. In February 1997 the inlet was a saltpan, and in April 1997 sea water was seeping in through the eastern bar and in the area of the breach in the western bar. Following the 1993 floods that opened the bar and broke the road access to the FNRP, options to maintain the road access while maintaining reasonable water levels in the inlet were considered. Eventually after controversial discussions as to the likelihood of similar floods and the height to which the estuary could and should be allowed to rise, a plan was implemented to:

> rebuild the road in an arc about 500 m inland from the break in the dune at 4 m, with a 100 m long relief floodway at 3.7 m, eleven 1.8 m diameter culverts through the road embankment at an invert level of 1 m to release flood water, and a smaller culvert at –1 m to equalise water levels between the Inlet and the pool between the roadway and the bar. If, as anticipated, the bar rebuilt to 3.5 m it would have to be broken when flood water reached that level.
>
> Hodgkin, 1997 quoting Anon 1995.
> Proposed structure at Culham Inlet. Main Roads, Perth, WA.

The new roadway was opened in April 1997.

A new road—nature's revenge?

In January 2000 cyclonic rains brought heavy flows from the catchment, and by 23 January the level was 4.03 metres and still rising despite the culverts. The road was cut in the planned sacrificial section west of the culverts to cause a break out, but the inlet waters flowed through the old 1993 channel east of the culverts with a 300-metre wide gap in the road (Fig. 10.40). The Culham was no longer a salt lake but an estuary again, isolating the FRNP from Hopetoun and the cleared lands to the east. The roadway was rebuilt in 2002.

Home for many

The plant and animal life of Culham Inlet follow the vastly different water states of this large lagoon. The inlet is surrounded by salt tolerant species such as the saltwater paperbark *Melaleuca cuticularis*, the samphire *Sarcocornia quinqueflora* and the saltbush *Atriplex isatidea*. Trunks of dead paperbarks mark the changing status of the inlet; many died back after the high water

FIG. 10.39 *Culham Inlet bar–dune area showing the recently completed road washed away, May 1993.* Courtesy P. Blond, print in E. P. Hodgkin Collection.

FIG. 10.40 *Culham Inlet showing the break through the dune barrier destroying the new road to the FRNP, aerial view, June 2000. The 900 mm pipe installed in 1990 is located in the small bay on the right of the photograph, and a section of the old road behind the dunes lies to the bottom left. The new road built in 1997 and again in 2002 follows a route 500 m inland through swampland. The culverts, 1 m above AHD, are located left of the break through the road.* S. Neville/Ecotones, Courtesy WRC.

levels in 1989–1990 and in 1993, but by 1994 appeared to be re-sprouting. Behind this paperbark fringe, the eastern banks are backed by coastal moort *Eucalyptus platypus*, many of which were killed by the 1989–1990 flood. On the western side, Yate *Eucalyptus occidentalis* grow behind the paperbarks. The estuarine reaches of the Phillips River are bordered by the same species that also died following the high water levels, although the extent of the die-off was less because the steeper river banks reduced the effect of water inundation.

In 1993 when the inlet waters were high, filaments of the green algae *Chaetomorpha indica* and some *Cladophora* sp. were found growing on submerged dead branches, and there were wind-driven accumulations of floating algae in the north-west. Nine months after the bar opened the seagrass *Ruppia*, a common species in Wellstead and Wilson Inlets further west, was found growing throughout the inlet and the red alga *Polysiphonia* was found near the bar. The growth of *Ruppia* marks the changing cycle in the inlet: although it can tolerate and grow well in quite saline water, it needs fresh water for the seeds to germinate, which they obviously did after the floods, flourishing as the waters became saltier.

Clumps of calcareous tubes made by the filter feeding bristle worms *Ficopomatus enigmaticus* remain high on the dead trees marking the flood levels when the waters recede. Other species that can tolerate salty conditions such as bristle worms and the false mussel *Fluviolanatus subtorta*, two amphipods and midge larvae, were also found in quite high numbers among the rock and mud. For a brief period when the inlet was tidal in 1993, the fauna also included other invertebrates such as the bristle worm *Ceratonereis* sp. and the trough shell *Spisula trigonella*. These are all species found in inlets that open regularly.

Periodically when the inlet is relatively high and salinity is less than 45‰, the inlet is highly productive and commercial fishers have taken large catches of the black bream *Acanthopagrus butcheri*. Prior to 1980 catches were low, but particularly large catches were recorded in 1990–1991 (32 tonnes), 1991–1992 (61 tonnes) and 1992–1993 (≈77 tonnes). Recreational fishers also took large numbers. Other small fish in the inlet include the blue spotted goby *Pseudogobius olorum* and two hardyheads *Antherinosoma elongata* and *Leptatherina wallacei*. These are all species that can tolerate a wide range of salinity and breed in the estuary, but retreat to the semi-permanent river pools in the upper reaches of the Phillips River where they can still breed when conditions in the inlet become hypersaline or dry up. The large catches of fish can be attributed to the favourable conditions for small invertebrates that flourish in less extreme conditions. As well as the estuarine species, opportunistic fish species such as sea mullet *Mugil cephalus* and herring *Arripis georgianus* as well as prawns and crabs were caught after the bar opened. The sea mullet in particular is a species that can tolerate a wide range of salinities and is found in estuaries that regularly open to the sea.

The inlet also supports large waterbird populations with twenty-five species recorded from the inlet. Most of these are opportunistic, visiting different areas depending on where conditions are most favourable.

A different environment

Sometimes regarded as a dead or permanently closed estuary, Culham Inlet really behaves similarly to Beaufort and Hamersley Inlets where the bars break every few years following floods. The cycling of opening and drying is impossible to predict, and any structures placed near the bar or within the area of the fossil estuary need to be considered dispensable. There are likely to be more types of plants and animals while conditions are not too harsh. Given favourable conditions, sea mullet should live and grow for several years until the next drying cycle and black bream are likely to flourish while water levels are high until they need to retreat again to the deeper river reaches.

EASTWARDS TO ESPERANCE

Along the coast from Hopetoun to Esperance there are considerable areas of internal drainage to salt lakes behind coastal dunes, such as Lake Shaster west of the Oldfield Estuary. Barker Inlet would probably also be a salt lake but for the position of the mouth close to a western headland, which protects it and allows the water to flow to the sea.

The Oldfield Estuary and Stokes Inlet have large catchments that extend 100 kilometres inland from the coast to the southern edge of the plateau, beyond which is an area of internal drainage and salt lakes where the northern watersheds are poorly defined. The Jerdacuttup Lakes catchment, while not as long, extends 50 kilometres inland to the north of Ravensthorpe (Map 10.2 and Prelude Map 3). Torradup Estuary and Barker Inlet catchments lie within 15 kilometres of the coast.

Jerdacuttup Lakes—the end of the line

The Jerdacuttup Lakes have probably not been open to the sea for about 6,000 years, but they are a coastal lagoon system illustrating the progression from estuary to coastal lagoon as riverine and coastal processes change. The lakes receive water from the Jerdacuttup River that rises in granite country 300 metres above sea level, about 50 kilometres inland north of Ravensthorpe. The river follows a tortuous course about 60 kilometres to the east of the Ravensthorpe Range and then across the coastal sand plain to where it enters the lakes. There are many pools in the river, some over a kilometre long and 5 metres deep. Over 75 per cent of the catchment is cleared but the lower reaches of the river and the lakes are bordered by reserves.

The open water and paperbark swamps of the Jerdacuttup Lake system extend for 15 kilometres along a shallow depression behind the coastal dunes (Fig. 10.41). The Jerdacuttup River enters the western lake through a swamp, which restricts flow to the lake, resulting in flooding north of Springdale Road. Two small streams drain to the smaller eastern swamps. The curving

FIG. 10.41 *Lake Jerdacuttup, aerial view, December 1988. The lakes lie parallel to the coast behind a narrow (60 m wide) dune line. Although the fringing vegetation is gradually encroaching reducing the amount of open water, the wetlands are very extensive spreading out across cleared cropping land after heavy rain.* Slide in E. P. Hodgkin Collection, Courtesy DLI, 13.12.1988, WA 26699 (c) 5178.

lines of paperbark trees show where the swamps have progressively grown out into the lakes greatly reducing the area of open water. There are small outcrops of granite (migmatite) on the northern shore of the swamp. The narrow, low area of less consolidated dune sand bordering the seaward side of the lakes probably marks the original outlet to the sea.

When flooded the lake is about 5 metres in depth with 2 metres of water over the samphire fringe. In January 1990 the lake was 3.3 metres above MSL. All this water evaporates after years of low rainfall leaving the lake dry, although the river always holds water. The lake waters vary from fresh when full to brine as it dries out. It is reported that local residents introduced the black bream to the river. At times these can be very abundant and they are fished commercially.

Oldfield Estuary

The Oldfield River is 95 kilometres long and flows from the north-west where its headwaters drain the eroded granite hills north of the Ravensthorpe Range (Map 10.2). Nearer to the coast the Oldfield and its smaller tributary the Munglinup River have cut shallow valleys through the sand plain. South of the South Coast Highway, where the river has cut through spongolite rock, the valleys are steeper. River pools up to 55 metres long are probably fed by springs from the spongolite.

In 1998 the Oldfield Catchment (Fig. 10.42) was declared one of the 'focus catchments' of the South Coast Region, a scheme providing extra assistance to land-holders for fencing and rehabilitation in the catchment though Landcare agencies. The Oldfield and Munglinup Rivers were also identified as priority waterways by the Water and Rivers Commission, as they had broad vegetated corridors along the main channels where the fringing vegetation provided rich bird habitat (Fig. 10.43). There are reserves 500 metres to 4 kilometres wide bordering both the Oldfield and Munglinup Rivers along most of their length, although these are subject to grazing at times. The Lake Shaster Nature Reserve extends along the coast on either side of the recreation reserves that border the estuary itself. Much of the lower half of the catchment has been cleared for agriculture, but only 35 per cent of the total catchment has been cleared.

The Oldfield River is estuarine for 8 kilometres from the coast almost to Springdale Road. It lies north-south perpendicular to the coast. The upper part of the estuary is a winding, riverine channel through a sandy plain, which then widens where there is higher ground and granite rock on the eastern shore (Fig. 10.44). Finally the estuary opens to a 3-kilometre long meandering basin, less than 600 metres wide, with an overall area of about a square kilometre. There is little evidence of active fluvial deposition, and the river delta is small with an elongate silt 'jetty' extending well into the basin. Near the mouth the eastern dunes rise to 10–15 metres with a core of coastal limestone that outcrops on the estuary shore about 200 metres from the mouth. The western dunes are low and only sparsely vegetated for some distance from the mouth. A shallow arm of the basin extends westwards behind the fore dunes to the west, probably marking an older channel to the sea.

The bar is high and narrow, a part of the steep ocean beach, only 100 metres across and 200 metres long from east to west (Fig. 10.45). Early reports suggest the bar used to open briefly every three to four years. It remained closed for seven years until 1986 but opened again in 1987, in May 1988, in January 1999, and December 2001 following heavy rain. The high beach and exposed coastline suggest that the bar is unlikely to remain open for very long. Strandlines observed along the estuary shore indicate that the bar opened and closed several times after the May 1988 opening before finally closing.

FIG. 10.42 *Oldfield Estuary, aerial view, June 2000. The lagoon is surrounded by uncleared land vested in the shires of Ravensthorpe to the west and Esperance to the east. These areas are bordered by the Lake Shaster Reserve that connects along the river reaches to the upper catchment.* S. Neville/Ecotones, Courtesy WRC.

FIG. 10.43 *Oldfield Estuary: vegetation 5 km upstream from the bar, October 1989.* E. P. Hodgkin Collection.

FIG. 10.44 *A granite outcrop at the edge of the Oldfield Estuary—a nice place to rest, October 1988. The vegetation around this favourite picnic and boat launching place is deteriorating, crisscrossed by tracks, invaded by weeds and scattered with rubbish. Hopefully we are starting to appreciate such special places.* E. P. Hodgkin Collection.

FIG. 10.45 *Oldfield Estuary ocean bar, October 1988. The bar rises to 3.5 m above sea level and is built of off-white, well-sorted medium to coarse-grained quartz with 20% calcareous skeletal remains.* E. P. Hodgkin Collection.

After the bar breaks the water level falls emptying the estuary, leaving water in only a small central channel at about sea level. When the bar rebuilds, the river flow brings nearly fresh (about 5‰) darkly tannin-stained water to the inlet. There can be a brief period when sea water enters and mixes with the river water increasing the salinity, but because the bar is open only briefly, the water is not usually as saline as some of the other estuaries on the south coast. The salinity in the lower estuary rises with decreased river flow and evaporation. In the upper estuary salinity appears to remain lower, possibly due to ground water seepage. Nutrients in the Oldfield are generally lower than in the other eastern estuaries. (See Graphs 1.2 and 1.3.)

Although the seagrass *Ruppia megacarpa* grows in the Oldfield, generally its distribution appears to be restricted by low light in the deeply stained waters. In contrast, exceptionally clear water in early 2001 appears to have promoted luxuriant growth. After the inlet breaks to the sea it is fished commercially for two or three years while the water level stays high and salinity is favourable, but with poor rains the remaining water becomes too saline for fish other than black bream. At such times the small salt lake snail *Coxiella* is very abundant on mud in the shallow water.

Torradup River Estuary

Torradup Estuary (Fig. 10.46) lies east of the Oldfield between the Lake Shaster Nature Reserve and Stokes National Park. The small catchment, only 15 kilometres long, drains an undulating area of sandy, colluvial deposits with granite outcrops in the valleys. Most of the catchment (71 per cent) is cleared, though the estuary itself is surrounded by uncleared land.

The estuary is small, only a 3.5-kilometre long stretch of open water about 200 metres wide. Most of the estuary is shallow and the marginal sand flats are exposed in summer, but the narrower parts are more than 2 metres below sea level. The river always holds water, which may be almost fresh when the river flows, and is probably seldom more salt than sea water. However the presence of salt lake snail *Coxiella* suggests that the water may sometimes become highly saline, but perhaps only in isolated pools on the sand flats. The sea bar breaks every year and stays open briefly, for about a week. The mouth opens towards the western end of a 3-kilometre long south-facing beach in the lee of Margaret Cove.

Despite its small size the inlet can hold quite large numbers of fish. In the past these were fished commercially, however this ceased with the new fisheries management plan that came into effect in July 2002.

FIG. 10.46 *Torradup Estuary, aerial view, June 2000. The estuary mouth, 100 m wide, opens regularly but only stays open for a short time.* S. Neville/ Ecotones, Courtesy WRC.

Stokes Inlet

Stokes Inlet is one of the most attractive estuaries of the south coast being set in a deep valley between high densely wooded dunes with shady paperbark trees along the beaches bordering relatively deep permanent water (Fig. 10.47). It is one of the largest estuaries (14 square kilometres) in the area and unusual among south coast estuaries (except for Oyster Harbour) because of the deep water. The high sea bar breaks briefly at intervals of several years, and in consequence water level and salinity vary greatly with river flow and evaporation. Nevertheless it is a productive estuary and popular with amateur and commercial fishers.

The Young River, 120 kilometres long, and Lort River, 100 kilometres long, flow to Stokes Inlet from salt country to the north with a combined catchment of about 4,500 square kilometres. The river channels have cut into the sand plain and exposed granitic rock at intervals along their length (Fig. 10.48). Spongolite rock outcrops in the valleys within 20 kilometres of the coast, but elsewhere colluvial and alluvial deposits fill the valley bottoms. South of the highway the sand plain gives way to a belt of dune sands 5 kilometres wide.

The rivers are estuarine for a few kilometres upstream, but are cut off from the lagoon by a massive river delta when the water level is low. Lake Cobinup, north of the estuary, floods in winter and overflows into the inlet, but holds little or no water in summer.

Changing roles: pastoral, cropping and national park

Like much of the eastern part of the south coast, exploration was followed by pastoral activities in the late 1800s although most of the land was cleared only quite recently. In 1863 Charles and William Dempster took up a 41,000 acre (16,600 hectare) pastoral lease west of Esperance that included land on either side of Stokes Inlet. In 1873 Alexander and John Moir were also granted a lease of 14,000 acres (5,700 hectares) around the inlet with an additional 57,000 acres (23,000 hectares) granted in 1888. The Moirs established a homestead near the eastern shore of the inlet and grazed sheep through the coastal bush, which they burnt periodically to provide fresh feed. Later a small area near the homestead was cleared for barley and then other grain crops as feed for the stock. The limestone walls of the Moir homestead survived until they were razed by fire in 1993; the remains lie

FIG. 10.47 *Stokes Inlet, aerial view, December 1988. The estuary is surrounded by the uncleared land of the Stokes National Park, although much of the upper catchment is cleared. Lake Cobinup lies behind the main basin on the left of the photograph. Close to the coast and bordering the lower estuary, dunes rise 60–80 m and are cut into steep gullies on both sides of the estuary. The estuary mouth is protected by a chain of reefs offshore shown in the foreground.* S. Neville/Ecotones, Courtesy WRC.

FIG. 10.48 *A pool on the Young River at the Jonegatup Road crossing, April 1987. The mallee and acacia trees form a backdrop to the shallow pools and granite outcrops.* E. P. Hodgkin Collection.

in a small reserve for the Preservation of Historical Buildings within the Stokes National Park. Wool together with sandalwood from the surrounding country was shipped from Fanny Cove (to the east of Stokes Inlet); and in the 1890s the cove was used as a port for miners going to the area around Norseman in the Dundas Goldfields.

There was little further development of the Stokes Inlet area until Noel White, a pastoralist from the goldfields, established Young River Station at the head of the inlet in 1950 with a 70,000 acre (28,300 hectare) pastoral lease. At this time the road from Ravensthorpe was unsealed. In the 1960s a smaller area south of the road was cleared for sheep and cattle pasture. Land was released for farming progressively from the late 1950s to the late 1960s. At first this occurred near the coast, with the upper parts of the Stokes and Oldfield catchments released in the 1970s and 1980s. The lower reaches of the Young and Lort Rivers are both in narrow river and foreshore reserves, although 60 per cent of the catchment has been cleared.

The inlet and Lake Cobinup are surrounded by Stokes National Park, a 3–5 kilometre-wide (10,667 hectare) strip of densely vegetated high dunes along the coast. Many bush tracks were badly eroded by heavy rains in 1986 and have been closed.

A pretty shoreline

The inlet, 10 kilometres long and 2 kilometres wide in the northern part with an area of about 14 square kilometres, lies in a north-east to south-west direction. The elongate shape of the estuary and deep river valley may be due to an underlying fault structure. The inlet is very shallow in the north gradually deepening towards the mouth where it is over 4 metres deep. There are holes up to 10 metres deep opposite the mobile dune on the western shore, although near the bar the deepest areas are by the eastern shore.

The beaches are shelly sand with extensive deposits of shells. In the southern part, limestone outcrops along the shore and scattered outcrops of granitic rocks are exposed near the western shore when the estuary is low. Paperbark thickets and dense scrub border much of the shoreline (Figs 10.49 and 10.50).

The inlet narrows to a channel 2 kilometres long and 500 metres wide that winds between the steep well vegetated dunes, where much of the

FIG. 10.49 *Stokes Inlet, October 1988. Granite rocks and paperbark* Melaleuca cuticularis *line the shores of the inlet that lies within the national park. Managed by a resident ranger, camping is permitted in some areas along the western shore.* E. P. Hodgkin Collection.

FIG. 10.50 *Stokes Inlet, January 1987.* Melaleuca *seedlings colonising a saltpan on the eastern shore.* Courtesy Ruth Clark.

shoreline is beach limestone embedded with shells. A wide sandy bay borders the eastern shore, while on the west a steep active dune cascades sand into the estuary.

The mouth of the estuary is closed by a sandbar about 200 metres long and 1.5–2 metres in height above sea level. At the eastern end, a well-vegetated dune 7–10 metres high extends onto the bar and at times extends westwards, only to be carried away with a major flood. The western shore of the bar is hummocky bar sand with scattered dune plants and wattle on the higher ground. The bar is built of pale grey, poorly sorted, medium to fine-grained sand, mainly quartz with 25 per cent calcareous shell material.

The bar may break across its full width cutting a channel along the eastern shore and eroding the dune on its inlet face, but as flow decreases the channel swings across to the western shore. The channel does not scour very deeply, possibly due to underlying rock. Offshore, two lines of limestone reefs protect the shoreline and bar from waves and swell. However waves wash over the bar carrying sand into the inlet, and sea water seeps through the bar into the estuary when the water level is below that of the sea.

Mostly closed

The bar breaks infrequently and remains open briefly (less than six weeks). Before the 1970s and the extensive clearing in the catchment, the bar opened in 1919, 1927 (or 1932) and then not again until 1967. Since then the bar has broken more frequently, although there was a period in the late 1970s and early 1980s when it remained closed for seven years. In general, the pattern appears to be of spring breaks following wet winters and late heavy rains in October–November. Occasional heavy rains in summer, like those in 2000, have also produced sufficient flow to break the bar.

Although the rainfall records cover only a short period, and records from flow gauging stations are even shorter, calculations suggest that 10×10^6 cubic metres of flow is required to break the bar naturally on this high energy coast. Falls of rain of 100 millimetres inland or about 200 millimetres near the coast within a short period of one or two months in winter would probably be needed to create this type of flow.

The water level varies by nearly 3 metres, from 1.5 metres above to 1.5 metres below sea level. The level rises rapidly following heavy rain and river flow, and falls slowly with evaporation or rapidly at first if the bar breaks. In floods, the roots and lower trunks of the paperbark trees are submerged, while low water levels expose wide sandy beaches to the west and areas of muddy sand to the north and east.

Seepage of low salinity water from the coastal dunes and of sea water through the bar contribute a small but unknown input, however these are unlikely to make much contribution to the volume of estuary water.

As with many estuaries there has been a long-held belief that the bar should be breached to maintain a healthy estuary that would then continue to open naturally. But although it was broken artificially in July 1972 when the water nearly topped the bar, the opening of the bar has generally been left to nature. The rationale for opening the bar is questionable as premature openings reduce scouring so that sediments accumulate.

The water world—river flow and salty sea water

Salinity in the inlet is seldom less than that of seawater (35‰), but may be more than twice that at the end of summer. The salt content of estuary water derives from seawater inflow during the brief period while the bar is open, from water seeping through the bar, and from saline river water. Floodwater briefly lowers the surface salinity causing quite an intense stratification until wind and waves mix the water body. The 1979 floods brought surface salinity down from 42‰ to 4‰, but two weeks after the rains had stopped it had risen to 30‰ with the deep water slightly higher at 36‰.

There are few records of nutrient concentrations but the figures suggest a minor degree of nutrient enrichment, and the presence of blue-green algae in the water on some occasions would support this view. Nutrient concentrations may be higher in the riverine reaches of the estuary when cut off from the lagoon (Graphs 1.2 and 1.3).

The aquatic flora is dominated by salt tolerant species: the small green alga *Polyphysa peniculus*, the seagrass *Ruppia megacarpa*, and the stonewort *Lamprothamnium papulosum* (Fig. 10.16). *Polyphysa* grows extensively in shallow water throughout the inlet, sometimes forming a continuous cover on both sand and rock. *Ruppia* is sometimes abundant in the lagoon and riverine reaches where it may be heavily grazed by swans. *Lamprothamnium* is also recorded from shallow water, and the filamentous green alga *Cladophora coelothrix* was along the inlet shore in 1987.

Although many of the shells and worms are estuarine species, some marine species are also found in the estuary following the opening of the bar (Fig. 10.51). Juvenile prawns *Penaeus latisulcatus* and mussels *Mytilus edulis* were found in a survey in May 1987 nine months after the bar had been open, and they had matured by October 1988. A few cockles *Katelysia scalarina* and *Fulvia tenuicostata*, blue manna crabs *Portunus pelagicus*, shore prawns *Leander littoreus*, and small jellyfish have also been found. These marine species have mainly been found in the southern part of the estuary near the bar, while the estuarine species are more widely distributed throughout the inlet: a pattern typical of the eastern estuaries.

Because the inlet opens infrequently black bream flourish, and other species particularly sea mullet enter when the bar is open. There are reports of particularly large numbers of Australian salmon, whiting and silver bream entering when the bar opened in 1968. When salinity rises the number of fish species declines, and there are reports of mass mortalities of fish in 1932, 1938 and 1983 similar to those experienced in Beaufort and Gordon Inlets.

Stokes Inlet is open to commercial fishers, and is also popular with recreational fishers who principally target black bream and sea mullet. In 2003 net fishing was prohibited from 1 December to 30 April inclusive each year in all waters of Stokes Inlet and all rivers and streams flowing into it. There are currently no restrictions on recreational net fishing in the Oldfield Estuary or Torradup and Barker Inlets. The growing body of recreational fishers and the emergence of tourism as a major industry has increased pressure for the inlet to be closed to all forms of netting.

FIG. 10.51 *Ernest Hodgkin at work examining the invertebrate fauna in Stokes Inlet: a familiar sight to many students and other estuarine workers, May 1987.* Courtesy Ruth Clark.

Sediments reveal time past

Cores taken from the inlet bottom in deep water in 1987 and dated by the ^{137}caesium technique show that 50–60 centimetres of wet sediment (20–25 centimetres dry) has accumulated in the 30 years following clearing in the catchment (B. Campbell & K. Heinz-Wyrwoll, Geography Department, UWA). It must be supposed that a considerably greater depth of river sediment has accumulated less rapidly since the estuary was first flooded, but proof of that awaits deep coring. Dune sand is pouring in from the western shore and beach sand washes in from the bar.

The shells taken from the sediments, principally cockles (*Katelysia*), were dated by the ^{14}carbon technique at 4,380 ± 355 years old, showing that the estuary was permanently open to the sea until that time. Another interesting finding is from the wood of a *Melaleuca* stump found rooted about a metre below the present level of living paperbark trees: this was dated at 7,300 ± 200 years old indicating perhaps that the trees were growing there while the sea was still rising to its present level (G. Kendrick, pers. comm.).

With the growth of the bar and its closure, the character of the estuary changed from that of a marine bay to its present condition of being only briefly open to the sea, with very variable water level and salinity, and a very restricted flora and fauna.

FIG. 10.52 *Barker Inlet, June 2000. (a) The inlet is very shallow and probably contains little water over summer except for a deep channel near the bar, aerial view. (b) High dunes on the eastern side of the bar, aerial view.* Photographs S. Neville/Ecotones, Courtesy WRC. *(c) Dunes to the east of Barker Inlet, October 1988.* E. P. Hodgkin Collection.

Barker Inlet

Barker Inlet (1.8 square kilometres) is the remnant of a much larger water body whose extent is marked by the surrounding swamps, salt lakes and dune ridges (Fig. 10.52). Almost all of the 259 square kilometre catchment lies within low sand plain, about 80 per cent of which has been cleared. South of the highway the stream has cut a valley in the spongolite to the swampy alluvial area within the nature reserve. Two branches of a small river drain from the northern swamp, which is fed by Coomalbidgup Creek.

Unlike Oldfield, Torradup and Stokes Estuaries that open into open coast, the mouth of Barker Inlet is at the extreme western end of a bay against a rocky headland protected from the south-west winds and swell. The headland has probably prevented the surrounding dunes from completely blocking the entrance. The low bar, 1–1.5 metres above sea level, is part of the beach line. Although there is little information on how often the bar breaks or remains open, the absence of vegetation on the bar indicates that it probably breaks across its full width fairly frequently. The presence of the gutter suggests that tidal scour keeps it open on the eastern side for weeks before it finally closes.

The inlet probably seldom holds water through summer except in a deep, narrow channel to the mouth. The water must vary from nearly fresh when the basin first fills with river water to almost brine when the basin dries. The marginal vegetation includes both salt tolerant plants, samphire and paperbark trees *Melaleuca cuticularis*, and others with only limited salt tolerance. Shells of the salt lake snail *Coxiella* are abundant on the beaches showing that the water is hypersaline at times (Fig. 10.53).

Esperance and beyond

In this most eastern part of Swanland the yearly rainfall is about 600 millimetres near the coast but rapidly decreases inland (Maps 1.8, 10.2 and Prelude Map 3). The rivers are short and the flows erratic. Sandy soils mantle the spongolite Eocene sediments that formed the spectacular gorges to the west, but in this eastern area the topography is flat broken only by a few large granitic domes.

FIG. 10.53 *The salt lake snail* Coxiella *(≈9 mm) found in some of the ephemeral estuaries along the south coast also lives in salt lakes much further inland. Thick deposits of dead shells left when lakes and estuaries dry up are a reminder of how productive these systems can be, October 1988.* E. P. Hodgkin Collection.

The coastline from Esperance to Cape Arid is similar in character to the section eastwards of Culham Inlet, lying west-east with a series of shallow bays between rocky headlands. However eastwards of Esperance the bays are generally smaller, the headlands more numerous and offshore lie the islands of the Archipelago of the Recherche. East of Cape Arid from Cape Paisley the coastline veers to the north-east, and is of a totally different character, devoid of watercourses.

The Esperance Plains are recognised as a discrete biogeographic area, and the area around the Esperance lakes can be subdivided into the Gore and Torregullup land systems (Fig. 10.54). The Torregullup system is a series of large parabolic dunes running in a south-west to north-east direction along the coast with numerous sand sheets and blow-outs. The Gore system is a narrow discontinuous coastal plain overlain in parts by the Torregullup sand dunes, where internal drainage lines flow to wetlands that include Pink, Warden, Windabout, Wheatfield, Station, Woody, and Mullet Lakes. Shark Lake is, however, regarded as a separate system.

Many of the lakes dry out in summer but may reach about 2 metres depth in winter. They are generally saline due to poor flushing and concentration of salts with evaporation; some may be influenced by sea water entering the ground water. Shark Lake, however, contains fresh water. Expansion and development in the area has brought an accompanying demand for ground water, although this has been offset to some extent by increased run-off from the catchment due to rising watertables following clearing.

Some of the lakes have Ramsar Convention listing as water bird habitat, for example Lake Warden regularly supports 20,000 waterbirds, which includes over 10 per cent of the Australian population of Hooded Plover.

FIG. 10.54 *Lake Gore, aerial view, June 2000. Lake Gore receives water from the Dalyup River, part of the large area of internal drainage, and is an important refuge for wildlife.* S. Neville/Ecotones, Courtesy WRC.

The Esperance Region served as a gateway from Adelaide to the west via the Overland Telegraph erected in 1897. The town of Esperance was gazetted in 1893 following the discovery of gold at Dundas, Coolgardie and Kalgoorlie, but a delay in a rail link to the goldfields brought a decline in the population. Despite the early interest, development of the mallee country for wheat farming was halted by drought in 1914–1915. At the time salt was also considered a potential problem, with half of the land regarded as containing too much salt for profitable farming. The droughts and anticipated salinity problems delayed work on the rail link through Salmon Gums that eventually opened in 1925. Another push for development of the mallee country was halted with the Great Depression in the 1930s when 75 per cent of the farms failed. The opening of the Gibson Research Station in 1949 and experimentation with superphosphate and trace elements such as zinc and copper heralded a new push for development. A new stage in 1956 initiated by Alan Chase and US syndicates with the State Government, eventually gave rise to the Esperance Land Council and the emergence of a major agricultural region. The industry has not been without its problems, with massive areas affected by soil erosion and salinisation. Today Esperance is a major port and centre for tourism and fishing with infrastructure for mining and agriculture. Bandy Creek is now the site of a small boat harbour for recreational vessels with support facilities for commercial vessels engaged in the abalone, table and bait fish industries (Fig.10.55).

Small estuaries before the desert

In Esperance Bay the 30-kilometre long Bandy Creek drains salt lake country lying about 160 metres above sea level, and passes through the Lake Mullet Nature Reserve. The lower reaches formed a small estuary about 3 kilometres in length until 1982 when the small boat harbour was constructed at the mouth, and the upstream movement of sea water was blocked by a dam wall.

Like other normally closed estuaries on the south coast, the fauna in Bandy Creek would have consisted of only a few hardy species. The harbour development now provides a relatively well flushed but protected bay on an otherwise exposed sandy coast, an additional habitat for a greater variety of marine organisms. However one species, the sea squirts or ascidians *Ciona intestinalis*, originally from the north Atlantic that has spread to North

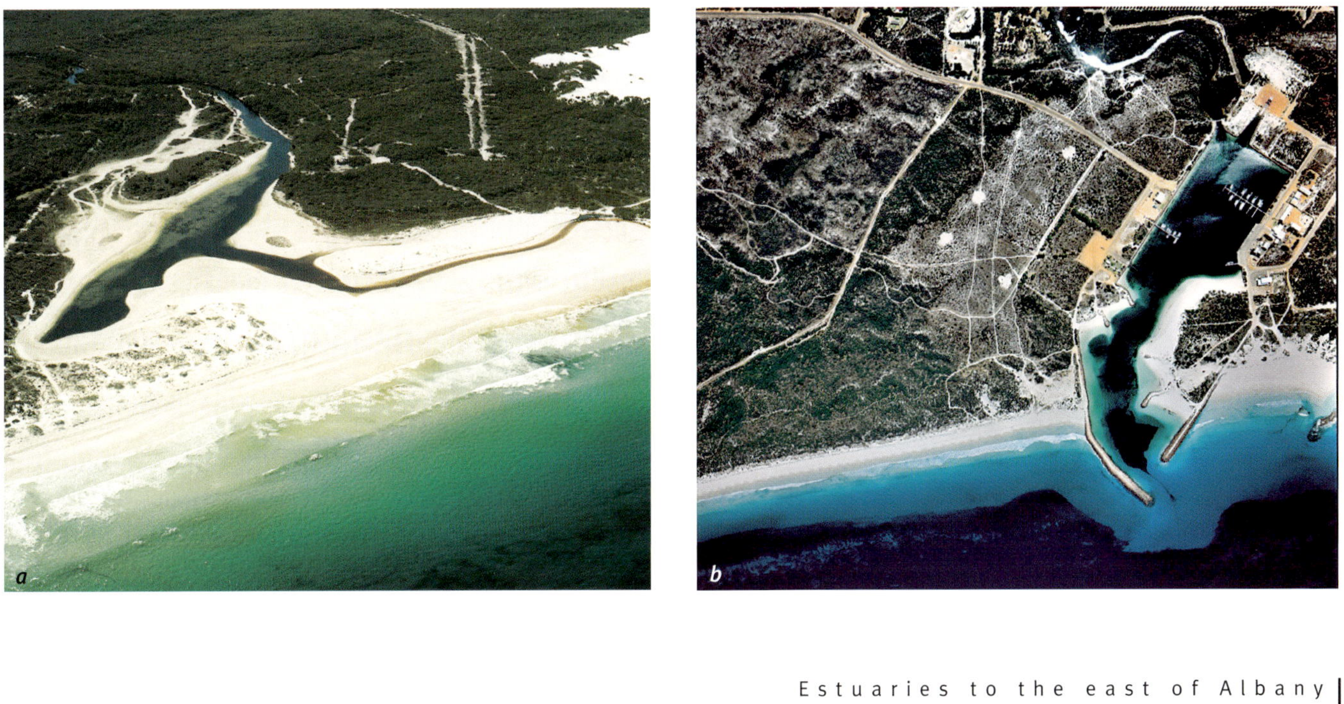

FIG. 10.55 *Bandy Creek, aerial views: (a) September, 1981.* E. P. Hodgkin. *(b) 8 March 2002 (11.14 hours).* Photograph number 5131, WA number 4730, Project 010151, Courtesy DPI, Asset Information Branch.

America, Chile, the Mediterranean, South Africa, Korea and southern Australia—where they are regarded as a nuisance to aquaculture—is of interest and concern.

Towards Cape Le Grand a series of creeks drain through the wide belt of Quaternary dune sands behind the long, west-facing, high-energy beach. Between Cape Le Grand and Cape Arid, the land rises steeply to a sand plain plateau about 80 metres above sea level. The short creeks have cut narrow valleys through the soft spongolite or Pallinup Siltstone that borders the coast and overlies the older and harder granitic bedrock. The heads of the creeks are only about 10 kilometres from the coast on the edge of the plateau, which is still largely uncleared although some of the creek lines have been dammed where they extend into farmland cleared in the 1950s and 1960s. These are the Dailey River, Mungliginup Creek, Alexander River, Blackboy Creek, and Thomas River (Map 10.3 and Fig. 10.56). The estuaries, 12 kilometres in length, lie in reserves of uncleared coastal bush: the first four in small reserves managed by the Esperance Shire and Thomas River in Cape Arid National Park. Other Crown land from Mungliginup Creek to Cape Arid National Park will be incorporated in a proposed nature reserve. East of Cape Arid the coastline changes direction to the north-east character and is devoid of water courses.

In Cape Arid National Park, creeks have carved valleys in the granite. Two of these, Jorndee Creek and Poison Creek, discharge to small estuaries on the eastern shore of the Cape Arid peninsula. The brackish or very salty water of these eastern estuaries is a particularly harsh environment for plants and animals. Nevertheless, like the riverine estuaries in the wetter western areas, they have a natural beauty and are a focus of wildlife and interest.

MAP 10.3 *The location of five estuarine rivers east of Esperance and the geology of their catchments.* Source Hodgkin & Clark, 1989 (ESS #5). Courtesy DoE.

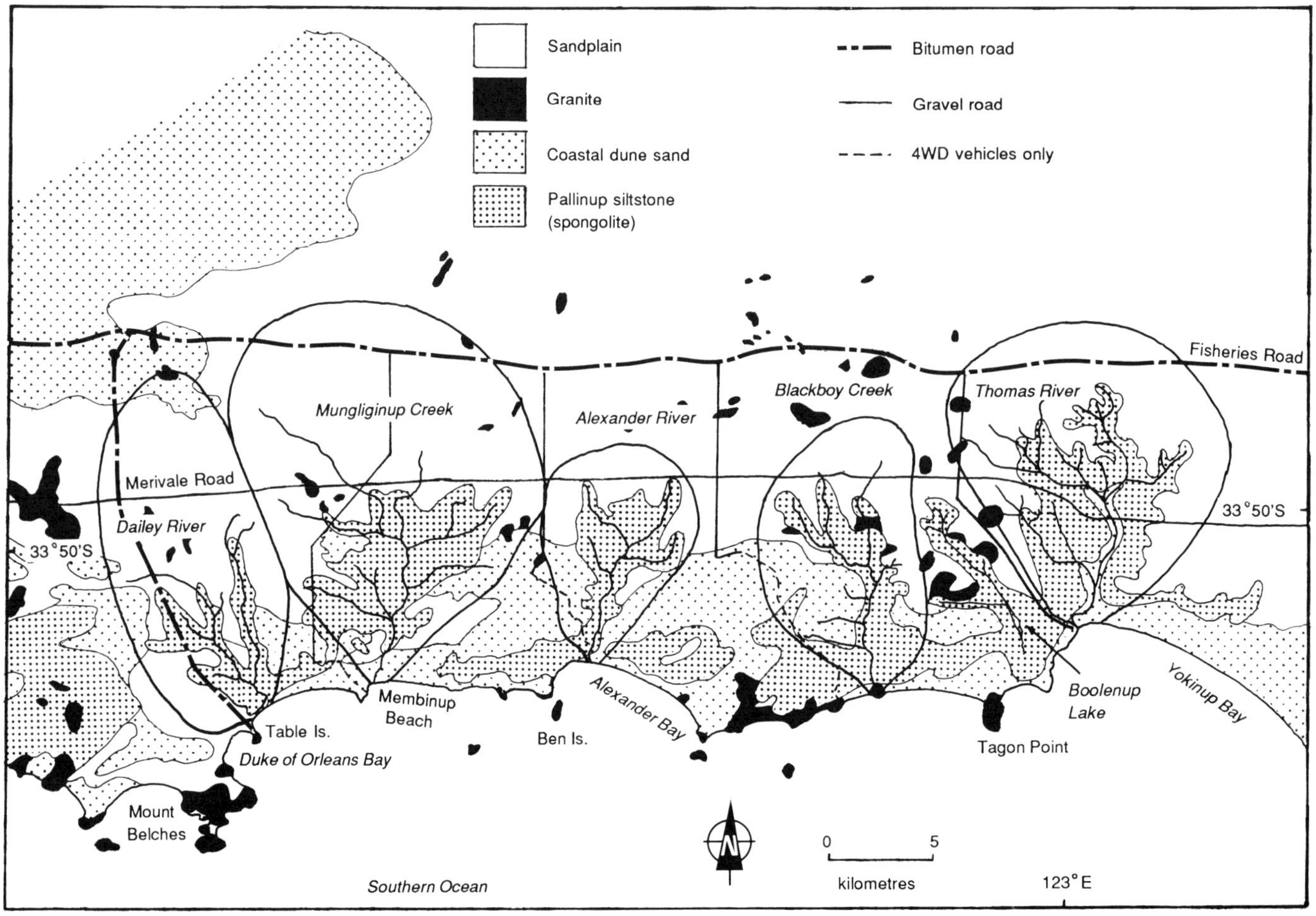

FIG. 10.56 *Estuaries to the east of Esperance: (a) Dailey River at Duke of Orleans Bay, May 1977; (b) Yorndee Creek, May 1987: (c) Poison Creek, aerial view, October 1978.*
E. P. Hodgkin Collection.

WHAT TYPE OF FUTURE?

As with other barred estuaries of the south coast, the issue of artificially breaching the bars to allow more frequent recruitment of fish from the coastal waters is a controversial issue. Stokes Inlet, which is open to commercial fishing, is of particular interest. If the opening is done before the water level has reached its maximum height and the bar is ready to break naturally, there may insufficient flow to scour a good channel. The result is that the natural rate of infilling of the estuary with marine sand washed in when the bar is open is accelerated.

The clearing in the catchments has led to increased river flow and, in the case of Stokes Inlet, the bar now appears to break more often; a condition favourable for fish recruitment. The other effect of increased run-off has been to increase the rate of erosion from the catchment and sedimentation within the estuary, so that the basin is shallowing rapidly. Currently the greater depth of water means that the estuary does not become extremely hypersaline and always retains some water, unlike the Oldfield Estuary and Barker Inlet.

Dune erosion and migration is currently being exaggerated by vehicle traffic and more frequent fires. The cascading sand on the west of Stokes Inlet is probably contributing to the infilling of the estuary, although the rate of this is unknown. Stabilisation and revegetation with dune plants may reduce sand movement.

In the most eastern estuaries there is little evidence of eutrophication from excessive nutrient input. Nevertheless experience in other estuaries of the south coast with cleared catchments, such as Wilson and Beaufort Inlets, indicates that there is potential for increased nutrient input and consequent greater algal growth when the catchments are cleared and used for agriculture. The importance of bushland in national parks and shire reserves as buffers helping in the maintenance of healthy estuarine systems as well as providing havens for a diverse flora and fauna cannot be overestimated.

Chapter 11 Enjoying and caring for Swanland

In the Prelude to this book about Swanland and its estuaries, a quotation from Ernest Hodgkin's final paper includes the words 'we need to understand' (Hodgkin & Hesp, 1998). I hope this book has done that by describing the estuaries of Swanland, the natural processes, things that have gone wrong and solutions undertaken.

Another quotation from the concluding remarks of the *Estuarine Study Series no. 3*, Wilson Inlet, Irwin Inlet and Parry Inlet in 1988, although referring in particular to those three much changed estuaries, is also appropriate at this the end of a large book and a journey around 2,000 kilometres of coastline.

> All the three estuaries, especially Wilson Inlet, are of great **social and economic value** and it is important that they should be maintained as **beautiful** and **healthy ecosystems** which give pleasure to people and support a varied plant and animal life, with an abundance of fish. They are **dynamic systems** subject to constant change, both as a result of natural processes and the increasing impact of human activities on the estuaries themselves and in their catchments. These activities make them vulnerable to dramatic and unwelcome events such as the recent *Nodularia* (blue green algal) blooms in the Peel Harvey.

Although the bold script is mine, and the statement was written by Hodgkin with reference to three specific estuaries, the message is clear and resounds throughout his other writings. He was emphasising the importance of the estuaries in Swanland and the themes that this book has attempted to convey—the value and beauty of the varied and abundant plant and animal life, and the dynamism of natural processes and the impacts of human activities. In many ways the estuaries of Swanland are unusual, but they are also typical of estuarine environments worldwide.

The estuaries of Swanland are unusual in that many can fluctuate between a freshwater to marine state with the seasons, not with daily tidal changes. Many have also reached a state of development where they are more akin to coastal lagoons or lakes, and that would be the terms used to describe them in other parts of the world. However in the northern and eastern estuaries of Swanland, we can see estuarine evolution taking place. They are now very shallow having filled with sediment from the catchment and beaches. Although infilling rates are not well documented, eroded sediments that fill the watercourses are brought downstream during major flood events accelerating the 'natural' infilling process.

The estuaries of Swanland are also typical because, like other areas around the world, population growth along the coast and modifications to the river catchments in the hinterland have affected how they function. It is widely agreed that these changes are manifest throughout Australia and in the Swanland of the South-West. Here the natural extremes of climate of winter rain and summer drought overlain with the human impacts of large scale clearing, additions of fertilisers and waste from cities and towns, drains we have dug and the efforts made to train and divert water flows, together with the structures we have built on and around them have markedly changed the hydrology, sediments and nutrient supply. As the ocean sandbars are rarely left undisturbed, many openings are deepened and those that close are breached prematurely to reduce flooding of low-lying margins with the result that natural scouring is reduced.

Perhaps a quotation from Griffith Taylor (1918), the originator of the term Swanland, puts the value of our estuaries and the surrounding environment into some perspective.

> With low summer humidity, sufficient rainfall, and temperate climate, the region of 'Swanland' is undoubtedly one of the healthiest in the world for people of our race.

Although written before World War II from a British Colonial Eurocentric viewpoint, before our appreciation of the natural variability of natural systems from 'droughts and flooding rains', now compounded by human induced Global Climate Change as well as the changes wrought on a more local scale around the estuaries and their catchments, this statement in 1918 encapsulates a great natural heritage and privilege for the multicultural society of Western Australia in the 21st Century.

For a sparsely populated state we have had some notable disasters in our estuaries: the unpleasant and toxic algal blooms in the Peel–Harvey, death of seagrasses in Princess Royal Harbour and Oyster Harbour overgrown with algae, closure of Princess Royal Harbour to fishing due to mercury contamination, and the toxic blue-green algal bloom in the Swan–Canning. Also worthy of note are the massive seagrass losses in Cockburn Sound near Perth, as the area lies in the same landscape as the estuaries of the Swan Coastal Plain and is sometimes classified as an estuary. These problems have been clearly linked to nutrient contamination due to artificial fertilisers applied to soils of cleared catchments, and to point sources around their margins. While some aspects of estuarine decline happened quite quickly and have taken us by surprise, the warning signs were there for those who knew how or cared to look. The sad thing is that we have known about some of the problems and not heeded the warnings.

We know that the soils are naturally infertile without the capacity to retain nutrients. We therefore know that a reduction in fertiliser use or the type of fertilisers used is critical. We know that nutrients leach from animal manures and that animals destroy riparian vegetation. We know that watertables are rising, and salinisation of the land has increased with clearing. We know that stream flows are now quite different, increased in some areas due to increased run-off and less in others due to dams or diversions. We know we need to reduce the amount of nutrients leaving the land and entering the watercourses by retaining or restoring the riparian vegetation. We know that we need to deal with waste from our communities more appropriately and effectively. We know that when nutrients have reached the estuaries we have to get rid of them, immobilise them or flush them away.

How to deal with some of the problems remains much debated. All the solutions are expensive: channels to the sea like that at Dawesville are particularly so and may bring other unpredicted outcomes. Chemical treatments to reduce nutrient availability are very expensive, and so is restoration of the riparian vegetation. However, maintenance of the natural attributes remaining and restoration of what is degraded are comparatively cheap in monetary terms, and they have added values of visual and structural landscape protection. Again in the words of Ernest Hodgkin (1978):

> In a State where a persistent pioneering urge is still pushing back the native bush for immediate economic gain with little thought for the welfare of future generations, these and other human values have somehow to be weighed against present gains to the community.

We could also add that we had disastrous floods in the catchments, and in this flat coastal landscape almost at current sea level our efforts to drain and alter river flows have to be viewed with caution. In the past the responses were obvious—move the water problem somewhere else. Now the options of somewhere else are limited—it's someone else's 'backyard'. We also know that flow to the sea can be debatable as the gradients are low and water levels are often high when flooding occurs. In addition, we have the likely scenario of increased flooding due to altered watertables, and predicted more intense periods of rain and higher sea levels. Perhaps we should view where we build by looking at the landscape and vegetation, and what it tells us about the likelihood of flooding events: added reasons for not building over or destroying the intact estuary margins.

We can also add the growing problem of mosquito-borne diseases with the capacity to affect health for a lifetime. A little distance from the problem is a most obvious remedy; leaving the estuary margins intact to filter the nutrients, reduce erosion, and buffer the floodwaters. In addition, reducing the need for spraying and use of chemicals would keep the food chain more complete for the fish, crabs and birds that we value.

Do we have to wait for disasters to know we have troubled waters?

How do we know the ecology has changed? Perhaps it is the increase in the magnitude and frequency of algal blooms, particularly the nuisance blooms of cyanobacteria, along with red tides and fish kills. More simply, it could be that some of the values we consider important can no longer be applied to an estuary. This may be classed as anecdotal evidence, but it serves as a benchmark for asking more detailed questions that need sophisticated equipment and analysis.

Some of these questions might include:

- Can we still catch fish? Are they the same types of fish? Can we still catch the same numbers? Are they as big as they once were (despite the jokes about fishers' stories)?
- Is the water still clear? Are there large accumulations of algae? Where do large accumulations of algae occur—is this pattern simply due to trapping, poor circulation or wind action?

Essentially, we are asking if the estuary is functioning in a healthy and sustainable manner, and what is estuarine function? The follow on from this being: Do we need it? If the answer is YES, then how do we maintain estuarine function in a changed and ever changing world?

Obviously we need to identify the core processes and set priorities for understanding them. The dynamism means that we do need information collected systematically over time so trends in declining health or improvements after remedial action can be assessed. Some of this data, such as nutrient concentrations and water flows, can be expensive to collect, analyse and interpret.

In Chapter 2 The Swan–Canning Estuary, I used the heading 'Keeping the jigsaw complete' to describe an ecosystem with many interlocking pieces that make a whole, so a piece missing is a process incomplete. Another analogy is a house of cards, one piece removed or an extra one added and the structure collapses.

So what do we know about the jigsaw or our house of cards?

We do know quite a lot or at least sufficient about which processes are important, but we do not know a lot about the size of the pieces or cards, or the rates of the processes that keep them together.

For example, we now know quite a lot about the types and biology of fish that inhabit the estuaries so changes to fish communities, one of the most obvious and widely valued aspects of estuaries, are very instructive. However, although we know what some of the fish and birds such as the Black Swans feed on, we do not really know how many of each fish there are or how many there should be, how much they eat over a particular time and how much this relates to what is available, or how one species of animal affects the abundance of others. This leads us back to the base of the food chain and the phytoplankton: the base of the food chain but also the organisms responsible for causing the nuisance and toxic algal blooms. We do know that nutrients have been the problem in fuelling algal blooms, but we need more detailed knowledge about the processes.

Some big questions remain. How do we reduce the nutrients coming into an estuary or other water body without compromising the land use or reducing production to feed the growing population? How do we deal with the nutrients stored in the system; those that fuelled the initial algal growths then died and decayed, releasing nutrients to fuel new generations of algae in a decay process that creates conditions favouring the release of sediment-stored nutrients in an ever increasing spiral. Basically we need to how the systems function over those highly variable natural cycles.

Why are estuaries important to us?

Why do we like to live around them or visit them? If we ask ourselves these questions some of the solutions to management become more obvious.

If it's the landscape, we have identified that we do not want the monotony of continuous housing, and that nature's gardens of marshes and trees are a framework for the landscape. Certainly there are some monetary costs of undoing problems, and change or disruption of what has been perceived as a 'right' of access to a place or activity (fishing, boating or skiing). However, the fencing of waterways, replanting of fringing vegetation, and limiting access points in rural areas or around densely populated urban areas overlooking estuary banks is relatively cheap with added benefits. It helps in keeping the ecological processes intact—the chemistry that dictates how the nutrients fuel the algal blooms, and keeps the organisms alive and active. Organisms range from the plants that oxygenate the sediment and fuel the food webs through to the tiny animals that eat algae, that are in turn food for larger animals including the fish we like to catch and the birds we like to see. In the late 20th century the term Integrated Catchment Management was coined to explain the management of catchment and watershed to maintain environmental and social values we desire, as well as providing us with adequate standards of living.

If it's healthy and abundant fish to catch, we want them untainted by oily taste or hidden poisons. We need to limit the pollutants that kill fish, but we also need to ensure that the tiniest and most insignificant plants and animals in the water column and in the sediment that fish eat are also unaffected. We need to understand about each organism's particular needs and requirements. We need some algae and the right types of algae for fish to feed on. This will require an acceptance of some level of algal drift on beaches, particularly in sheltered and shallow areas as these are feeding grounds for fish, and undisturbed areas for schooling and breeding. Healthy and abundant fish link back to management of the nutrient sources, and preservation and restoration of the landscape of fringing areas around the waterways leading to the estuaries.

If it's the birds as an indication that small animals are alive and well, or simply for the pleasure of our minds, we need again to ensure that the supporting plant and animal communities are healthy with the added requirement of enclaves free of noise and bustle. Again, this value links back to healthy landscapes in the catchments both far and near.

Around the highly developed estuaries, we should not regard the reserves we have as just 'museum exhibit pieces' of what was but as an intrinsic part of the currently relatively healthy estuarine system. Where development has not yet occurred or is as yet small in scale, we should be quite clear about what the whole community values—not just the short-term gains for a few that may later incur cost for a wider community. Indeed this is happening along the coast with Coastcare, and in the catchments with the Landcare movement. Although the values must come from the heart of the community, government commitment is vital to guide and foster community aspirations, and to undertake the scientific studies and archive the results and findings for the future.

There are no simple answers, no strict formulae of what can and cannot be done to protect the estuaries and their catchments, but an appreciation of their intrinsic value and dynamism will help with decision making so that, perhaps, we can have our fish and eat some of them too. The legacy of Ernest Hodgkin ('Hodge') is clear: from his endeavours the foundations of our scientific studies of estuaries are well founded, and his vision of bringing this knowledge to a wider audience, the whole community, the people of Western Australia, have been realised with the publication of this book *Swanland*.

DR ANNE BREARLEY
2005

Appendix

Estuaries and rivers of the south-west coast of Western Australia

Figures of catchment areas vary slightly depending on the boundary criteria and mapping techniques used. Sourced from Water and Rivers website and Pen, 1997 & 1999. Lagoon areas included only where these are a significant feature (riverine estuaries have only small basins or lagoons).

Estuary	Main river	River main channel (km)	Opening	Estuary area of lagoon (km²)	Catchment area (km²)	Cleared (%)	Median rainfall (mm)	Mean annual flow (ML)	Stream salinity	Oz estuaries classification
Murchison	Murchison	≈800	PO	3	82,300		≈300	162,800	Brackish	Modified
Hutt	Hutt	50			1,240	60	440	13,000	Brackish	Severely modified
	Bowes	40	NC/SO		640	80	440	8,000	Brackish	Modified
	Wollawar Gy	< 10	NC/SO						Brackish	
	Oakabella Ck	10	NC/SO		55	70	470	1,100	Brackish	
	Oakajee	10	NC/SO		50	40	470	510	Brackish	Modified
	Chapman	80	SO		1,600	65	470	19,700	Brackish	
	Greenough	250			11,850	50	350	21,000	Brackish	Severely modified
	Irwin	130			5,360	65	430	32,990	Brackish	Severely modified
	Arrowsmith	80			810	60	510	4,400	Marginal	
	Eneabba Ck	25			–	–	–	–	–	
	Bindoon Ck	25			–	–	–	–	–	
	Stockyard Gy	20			–	–	–	–	–	
	Cockleshell Gy	15			70	0	540	110	–	
	Hill	50–80			2,800	25	560	8,250	–	Modified
	Nambung	35			420	35	600	2,400	–	
	Mullering Bk	60			450	85	620	6,000	–	
	Minyulo Ck	60			610	85	610	8,000	–	
	Caren Caren Bk	35			425	82	640	5,700	–	
Moore	Moore	150	IO		14,400	82	500	98,000	–	Severely modified
	Gingin Bk									
Swan–Canning	Swan–Avon	280	PO	55	141,000	63	405	520,000	Brackish	Severely modified
	Canning	80			1,150	15	900	~80,000	Fresh	
	Coastal catchment				2,117	Largely urban				
Peel–Harvey	Serpentine	90	PO	75	1,700	11	1050	210,000	Fresh	Severely modified
	Murray	190			8,000	48	810	500,000	Brackish	
	Harvey	60	PO	61	450	40	1,080	100,000	Fresh	
Leschenault	Collie	110	PO	25	3,500	33	850	400,000	Marginal	Severely modified
	Preston	70			1,100	47	1,000	170,000	Fresh	
	Brunswick									
	Capel	45			620	30	950	62,000	Fresh	

Estuary	Main river	River main channel (km)	Opening	Estuary area of lagoon (km²)	Catchment area (km²)	Cleared (%)	Median rainfall (mm)	Mean annual flow (ML)	Stream salinity	Oz estuaries classification
Wonnerup	Ludlow	30			210	25	915	17,000	Fresh	Severely modified
Vasse–Wonnerup	Abba	20	IO	8	140	35	895	12,000	Fresh	
Vasse	Sabina	25			105	54	900	11,300	Fresh	
	Buayanup	25			160	70	1,010	30,000	Fresh	
Toby	Wetland system	≈6	SO	0.2	260	73	≈1,000	48,000	Fresh/brackish	Severely modified
	Carbanup (diverted)	30			170	55	1,050	30,000	Fresh	
	Mary (diverted)	30			90	60	1,050	15,000	Fresh	
	Gungyulup	10			50	70	975	8,000	Fresh	
	Wilyabrub	15			90	70	1,100	8,000	Fresh	
	Cowaramup	10			20	59	1,030	3,800	Fresh	
	Ellen	12			30	50	1,060	5,000	Fresh	
Margaret	Margaret	60	SO		470	23	1,075	65,000	Fresh	Modified
	Boodijup	12			60	40	1,105	11,000	Fresh	
	Calgarup	6			30	35	1,105	4,800	Fresh	
	Turner	6			50	65	1,175	14,000	Fresh	
Hardy	Blackwood	330	PO	9	21,400	85	400 inland 700–1,200 coastal	740,000	Brackish	Modified
	Scott	35			670	30	1,080	120,000	Fresh	
Donnelly	Donnelly Barlee	60	SO	< 1	1,690	11	1,200	310,000	Fresh	Largely unmodified
Warren	Tone & Perup	150	PO	< 1	4,350	35	600–1,400	380,000	Marginal	Largely unmodified
	Meerup	17	SO		120	5	1,405	24,000	Fresh	
	Doggerup	11	SO		80	0.5	1,390	14,000	Fresh	
	Gardner	35	PO	< 1	530	16	1,420	125,000	Fresh	Largely unmodified
Broke	Shannon	50	SO	48	610	4.5	1,320	145,000	Fresh	Near pristine
	Forth	10			20		1,420	5,000	Fresh	
	Inlet	14			60		1,410	12,000	Fresh	
Walpole–Nornalup	Deep	120	PO	15	1,060	3	1,120	140,000	Fresh	Largely unmodified
	Frankland	400			4,650	56	600 inland 1,200 coastal	200,000	Marginal brackish	
	Walpole	15		1	60	13	1,415	19,100	Fresh	
	Collier	6			15	20	1,320	4,000	Fresh	

Estuary	Main river	River main channel (km)	Opening	Estuary area of lagoon (km²)	Catchment area (km²)	Cleared (%)	Median rainfall (mm)	Mean annual flow (ML)	Stream salinity	Oz estuaries classification
Irwin	Kent (Owingup Swamp)	100	SO	11	2,040	42	780	123,000	Marginal	Largely unmodified
	Bow	20			250	15	1,200	41,000	Fresh	
Parry	Kordabup swamp lands	12	SO	1	120	50	1,150	28,000	Fresh	Largely unmodified
Wilson	Hay	80	SO	48	2,379	54	760	78,000	Marginal brackish	Modified
	Denmark	60			690	15	850	45,000	Fresh marginal	
	Sleeman	22			90	80	890	15,000	Fresh	
	Lake Saide Drain	8			30	20	1,050	3,800	Fresh	
	Cuppup Drain	10			60	80	1,005	13,00	Fresh	
	Little River	10			30	40	1,050	6,600	Fresh	
Torbay Inlet	Torbay Main Drain	20	SO	2	70	85	1,010	45,000	Fresh	Modified
	Marbelup Drain	20			210	50	920	30,000	Fresh	
Oyster Harbour	Kalgan	140	PO	16	256	70	500 inland 1,000 coastal	52,000	Brackish	Modified
	King	27			370	90	860	40,000	Fresh	
	Johnston Ck	18			16	38	800	1,500	Fresh	
	Yakamia Ck	8			20	95	900	3,700	Fresh	
Princess Royal Harbour	Robinson Drain Seepage	10	PO	29	15	80	940	3,600	Fresh	Modified
Taylor Inlet	Taylor	2	SO	< 1.0	10	70	800	1,400	Fresh	Modified
King Creek	King	12	SO		25	60	840	4,300	Fresh/Marg.	
Normans Inlet	Normans	6	SO	0.5	20	46	–	1,800	Fresh/Marg.	Modified
Waychinicup River	Waychinicup	17	PO		160	41	760	8,000	Fresh/Marg.	Modified
Cordinup River	Cordinup	21	SO		110	27	610	1,700	Fresh/Marg.	Modified
Cheyne Inlet	Eyre	13	SO	< 0.5	70	40	610	1,800	Fresh/Marg.	Modified
Beaufort Inlet	Pallinup	150	NC	7	4,970	80	410	36,000	Saline	Modified
Wellstead Estuary	Bremer	70	NC	> 2.5	720	75	465	14,000	Saline	Modified
Hunter	Hunter	5	NC		20	0	490	50	Fresh/Marg.	
Kellys Creek	Kellys Ck	8	NC		15	0	490	40	Fresh/Marg.	
Gordon Inlet	Gairdner	130	NC	3	1,770	40	430	10,700	Saline	Modified

Estuary	Main river	River main channel (km)	Opening	Estuary area of lagoon (km^2)	Catchment area (km^2)	Cleared (%)	Median rainfall (mm)	Mean annual flow (ML)	Stream salinity	Oz estuaries classification
Boondadup	Boondadup	10	NC		40	0	480	90	Fresh/Marg.	
Saint Mary Inlet	Saint Mary	15	NC	< 1	170	0	480	360	Fresh/Marg.	
Lake Nameless	Unnamed	8	PC		50	0	470	100	Fresh/Marg.	
Fitzgerald Inlet	Fitzgerald	80	NC	7	1,610	40	420	5,400	Brackish	Largely unmodified
Dempster Inlet	Dempster	15	NC	2.3	320	140	475	340	Fresh/Marg.	Modified
	Coppermine Ck	25			140	0	460	410		
Hamersley Inlet	Hamersley	50	NC	2.3	840	10	440	1,160	Brackish	Modified
Culham Inlet	Phillips	50	PC/NC	11	1,940	35	450	2,000	Saline	Severely modified
	Steere	30			360	30	450	1,400	Saline	
Jerdacuttup Lakes	Jerdacuttup	65	PC	3.8 & 1.5	2,320	30	415	8,800	Saline	Modified
Oldfield	Oldfield	95	NC	1	2,480	30	440	8,100	Brackish/Sal.	Modified
Torradup	Torradup	15	NC	< 1	90	50	550	1,200	Fresh/Marg.	Modified
Stokes	Young	120	NC	14	1,610	75	400	5,880	Brackish/Sal.	Severely modified
	Lort	100			2,800	60	375	6,000	Brackish/Sal.	
Barker	Coomalbidgup	13	BC	1.8	130	95	500	1,400	Brackish	Modified
Bandy	Mullet Lakes	30	NC		1,380	85	465	6,400	Brackish/Sal.	Severely modified
Blackboy		13	NC		90		595	1,100	Brackish/Sal.	
Thomas		10	NC		130	25	560	1,500	Brackish/Sal.	
Jenamullup		7	NC		20	30	580	150	Fresh/Marg.	
Cape Arid		5	NC		8	5	580	230	Fresh/Marg.	
Jorndee		7	NC		2	20	580	190	Fresh/Marg.	
Poison		15	NC		2	0	580	100	Fresh/Marg.	
Fern		8	NC		20	0	500	60	Fresh/Marg.	
Weanerjungup		5	NC		50	0	480	90	Fresh/Marg.	

Hodgkin & Clark, 1987; 1988a, b & c; 1989a & b; 1990a & b; 1999.

Salinity of river water: Fresh (0–500 mg/L), marginal 500–1,500 mg/L), saline (> 5,000 mg/L)

Permanently open PO, Seasonally open SO, Normally closed NC, Permanently closed PC, Intermittently open IO, Briefly closed BC.

Twenty-three dams on the rivers today collect about 32% of the total freshwater flow between Perth and Bunbury (only the brackish Avon and Murray have not been dammed) and another twenty scarp streams flow to the Swan Coastal Plain or into the Murray (Penn, 1999).

Glossary

AHD (Australian Height Datum) mean sea level (zero) based on thirty tide gauges around the country in 1966 and 1968, with minor modifications in 1971, and 1983 in Tasmania. 'The datum surface is that which passes through mean sea level at the thirty tide gauges and through points at zero AHD height vertically below the other basic junction points.' (Geocentric Datum of Australia Technical Manual)

Aerobic presence of free oxygen. *See also* Respiration.

Algae aquatic organisms containing chlorophyll and other photosynthetic pigments. Range in size from microscopic (generally single cells termed microalgae) to large seaweeds (macroalgae) metres in length. May live as single cells or aggregations.

Algal bloom rapid increase in the number of microalgae in the water column, usually in response to high nutrient concentrations and favourable conditions for growth, e.g. sunlight, calm water, temperature and salinity. Usually expressed as a concentration of chlorophyll *a*. The number of cells per mL used to define a bloom varies with cell size (biovolume), e.g. a large number of small cells or fewer large cells. Generally accumulations of cells visible to the naked eye are considered a bloom. Cell toxicity is also relevant to defining the importance of a bloom.

Alkaline *see* pH.

Alluvial deposits detrital material transported by a river and deposited, usually temporarily, along the flood plains of a river; commonly sands and gravel.

Anerobic absence of free oxygen. *See also* Respiration.

Anoxic deficiency of oxygen. *See also* Hypoxic (low oxygen).

Anticyclone region of relatively high atmospheric pressure and descending air. Marked on a map by lines or isobars. In the Southern Hemisphere, winds blow anti-clockwise and in the Northern Hemisphere, clockwise from an area of high pressure at the centre to areas of lower pressure. Compared to low pressure systems, highs cover a larger area and move more slowly. When located over land, the sky is usually clear and the weather dry. *See also* Cold or Warm front, Cyclone, Depression, Isobar, Tropical cyclone.

Apatite phosphorus; crystalline calcium phosphate containing varying amounts of fluoride, chloride and hydrogen in a crystal lattice.

Arbovirus a group of viruses with an RNA nucleic acid core, carried by arthropods (insects, spiders and crustacea).

Autotroph organism capable of manufacturing organic material from inorganic material, independent of outside sources of organic substance. May be phototrophic, possessing chlorophyll and using light energy (plants), or chemotrophic using energy from inorganic oxidation, e.g. hydrogen sulfide by sulfur bacteria or hydrogen by hydrogen bacteria. *See also* Heterotroph.

Bar an accumulation of sand or silt across the entrance of a river or estuary. *See also* Entrance bar.

Benthic associated with the bottom (associated organisms = benthos).

Berm narrow ledge or path between two slopes; at the edge of a slope to a low point or depression (swale).

Biomass weight of organic matter (plant or animal) in a given area.

Biological Oxygen Demand (BOD) indicates the dissolved oxygen used by bacteria that decompose organic material. Measured by the amount of oxygen used by one litre of sample stored in the dark for 5 days at 20°C. Can be used as an indicator of the polluting capacity of an effluent.

Bioturbation mixing action by burrowing invertebrates by which organic material is buried in the sediment.

Bivalve mollusc with a shell in two parts joined by a hinge.

Bloom rapid increase in the concentration of microalgae in the water column, often expressed in units of chlorophyll *a*; a bloom being > 20 μg/L or 20,000 cells/mL. *See also* Algal bloom.

Blue-green algae ancient group of photosynthetic bacteria without a nucleus, which produce energy from sunlight. Some can use dissolved nitrogen. Some species produce toxins or irritants.

Brackish water salinity between that of fresh and marine waters (1,500–5,000 mg/L). *See also* Salinity.

Catchment area of land that intercepts rainfall and contributes the collected water to surface water (streams, rivers, wetlands) or ground water.

Cenozoic Era most recent era in the earth's history, the last 65 million years that covers the Paleocene, Eocene, Oligocene, Miocene, Pliocene, Pleistocene and Holocene epochs, in the Tertiary and Quaternary periods. *See also* Mesozoic Era, Palaeozoic Era and Fig. 1.1.

Charophyta (Charophytes) commonly termed stoneworts due to the encrusting chalky material (calcium carbonate). Major phyla of the algae. Occurring in fresh and salty water; possess walls of cellulose; pigments of chlorophyll *a* and *b*; and store starch. Thallus (body) multicellular, anchored to the bottom by 'roots' rhizoids. Lateral branches in whorls around main axis. Possess multicellular sex organs antheridia and oogonia. *See also* Fig. 10.16.

Chlorophyta green algae—one of the main divisions of the algae. *See also* Table 1.10.

Chlorophyll *a* green pigment occurring with plant cells and the site for photosynthesis. Used as a measure of phytoplankton abundance. *See also* Algal bloom.

Cold front boundary between an advancing mass of cold air and a mass of warm air beneath which the cold air pushes like a wedge. Passage of a cold front is marked by a rise of atmospheric pressure, a fall in temperature, a change in wind direction, and may be accompanied by squalls, heavy rain, and thunder.

Colluvial deposits weathered material transported by gravity, e.g. scree (talus) slopes. *See also* Alluvial deposits, Eluvial deposits.

Compensation basin temporary storage of stormwater or run-off allowing water to soak into the soil profile, reducing the need for large drains.

Cretaceous period between 142 and 65.5 million years ago. *See also* Fig. 1.1.

Cyclone region of low pressure. In temperate regions referred to as a depression, but in the tropics as a tropical cyclone. In the Southern Hemisphere the rising air blows clockwise. Usually occurs where warm air meets colder air and ascends leading to the formation of a front. As the air rises it cools and clouds form; rain falls, often accompanied by strong winds.

Cyst resting stage or spore of some types of algae.

Depression a region where the atmospheric pressure is low relative to the surrounding areas. Also referred to as a low or cold front.

Destratification breakdown of stratification.

Detrital (detritus) term originally for weathered products of rocks. Now commonly used for decomposed organic material of plant or animal origin.

Dewatering removing water to lower the water table.

Diatoms microscopic algae with siliceous cell walls that belong to the algal phylum Bacillariophyta. *See also* Table 1.10.

Diffuse source pollution nutrients or other contaminants originating from a dispersed area and with sporadic occurrence, often from broad-scale fertiliser application or contaminated ground water. Individually sources may be quite minor but collectively significant. The main source of most water pollution and often too small to be licensed or controlled by regulation. *See also* Point source pollution.

Dinoflagellate Motile microscopic algae with flagellae that belong to the algal phylum Dinophyta. Some types have a shell (theca) of small plates. Some produce toxins. *See also* Table 1.10.

Diurnal daily. With reference to tides, only one high water and low water in 24 hours. Most common in south-west Australia. *See also* Semidiurnal.

Ebb tide falling tide.

Ecology study of the interrelationships between living organisms and their environment.

Eocene period between 55 and 33.7 million years ago. *See also* Fig. 1.1.

Epiphyte organism growing attached, but not parasitically, to other algae or seagrass. Can be photosynthetic (algal) or animal. *See also* Box 1.8.

Eluvial deposits weathered material ('float') that is still at or near the point of formation. The term is applied especially to deposits of economic substances.

Entrance bar an accumulation of sand or silt across the entrance of a river or estuary. *See also* Bar.

Estuary enclosed or semi-enclosed coastal water body with an opening (permanent or intermittent) to the sea that allows some penetration of marine water with tides, and receiving fresh water inflow from the land. Definition used in *Swanland*: 'an estuary is a place which was the mouth of a river where tidal effects were evident and where river water and sea water mixed when first flooded by the Holocene marine transgression'.

Euryhaline able to tolerate a wide variation of osmotic pressure of the environment. *See also* Stenohaline.

Eutrophic enriched with nutrients and therefore highly productive in terms of algal growth and organic matter production. *See also* Mesotrophic and Oligotrophic.

Flood tide rising tide.

Flushing time the time taken for the water in a semi-enclosed area to be removed. *See also* Residence time.

Forams (foraminfera) single-celled protists with shells (tests).

Gastropod mollusc with a single shell (includes snails). *See also* Bivalve.

Geological time periods see Fig. 1.1.

Geomorphology study of structure, origin and development of topographic (surface) or landform features of the earth's crust.

Gigalitre (GL) 1,000,000,000 litres or 1 million cubic metres (10^9) or 1 million kilolitres (KL).

Greywater water used for domestic purposes but not sewage.

Ground water water within the pores or crevices of rock or soil.

Ground water mound where the watertable slopes away from a higher central area and from which ground water flows outwards, eventually to the sea or river. The highest point of the Gnangara Mound on the Swan Coastal Plain is 70 m above sea level.

Halophyte plant that tolerates salty soils, a condition commonly found around the edges of estuaries and around salt lakes; likely to increase in extent as more land becomes saline.

Halocline also termed salocline, zone in the water column where the vertical change in salinity is rapid. May be due to freshwater inflow at the surface or to salt water intruding with tides (salt wedge) under fresh water. *See also* Stratification.

Herbivore plant eater.

Heterotroph organisms that are not autrophic and require food, living or dead, from the environment; may be herbivores, carnivores, parasites or saprophytes, therefore dependant on autotrophs. *See also* Autotroph.

Holocene Epoch since the Pleistocene, the last 10,000 years marking the end of the last ice age. *See also* Fig. 1.1.

Holoplankton Organisms such as diatoms and algae that spend all stages of their life cycle as plankton.

Hypersaline water salinity in excess of marine waters (35 ppt (‰)).

Hypoxic low-oxygen conditions.

Infauna animals living within the sediments.

Intertidal region between the low and high tide levels.

Invertebrate animals without backbones. The main groups are worms including nematodes and bristle worms or polychaetes, bivalve or gastropod molluscs or snails, crustaceans (prawns and crabs), and insects.

Isobar line drawn on a map to connect areas of equal atmospheric pressure.

Larvicide chemical used to kill larvae (e.g. mosquito larvae 'wrigglers').

Laterite red rock, rich in iron and aluminium oxides formed by the weathering of rocks.

Leachate product of leaching; may contain nutrients or other dissolved substances that can pollute ground water and water bodies.

Leaching the process by which organic matter and mineral salts are washed out of soil either in a dissolved or suspended state; the material being washed out is known as leachate.

Lithified loosely deposited sediment consolidated into a hard sedimentary rock. Many of the dunes along the coast have been lithified, the sand grains (predominantly quartz) cemented together by calcium carbonate (from shells and algae) that has been dissolved by acidic rain water and redeposited as the water evaporates. The process of lithification makes a 'rock'.

Macroalgae large algae that can be seen by the human eye.

Macrobenthos large organisms living on the bottom; usually those retained on an 0.5 mm mesh. *See also* Microbenthos (microfauna) and Meiofauna.

Macrophyte larger aquatic plants including seagrasses, macroalgae and charophytes.

Meiofauna small animals (0.1–1 mm in length) just large enough to be seen with the human eye that pass through a 0.5 mm sieve mesh but are retained on a 0.1 mm mesh; live in sediment, may form the base of a food chain. *See also* Macrobenthos and Microbenthos.

Meroplankton plankton consisting of organisms at a certain stage of their life cycles, especially larvae; the other stages not being spent as part of the plankton community.

Mesotrophic moderate concentrations of nutrients. *See also* Eutrophic and Oligotrophic.

Mesozoic Era era in the earth's history after the appearance of life: 'middle life' includes the Cretaceous, Jurassic and Triassic periods. *See also* Palaeozoic Era and Cenozoic Era, Fig. 1.1.

Metabolism chemical processes within an organism that break down organic compounds (catabolism) and release energy for the build up of organic compounds (anabolism) and use of energy. Energy may come from the breakdown processes or from the use of external energy such as light energy in photosynthesis. The rate of metabolism is usually referred to as the amount of oxygen consumed or carbon dioxide produced. *See also* Box 1.9 Processes for life.

Microbenthos (microfauna) small organisms visible only with a microscope (microscopic fauna). *See also* Macrobenthos and Meiofauna.

Microphytobenthos (MPB) benthic microscopic algae and bacteria capable of photosynthesis; therefore primary producers at the base of food webs. Also important mediators in nutrient cycling.

Microtidal small tidal range (< 1 m), typical of tides of south-western Australia.

Mollusc large group of animals including bivalves and gastropods (shellfish).

Nekton populations of free-swimming animals that inhabit the middle depths of an ocean or lake.

Nematode unsegmented worms.

Neogene period includes the Miocene and Pliocene epochs. *See also* Paleogene and Tertiary, Fig. 1.1.

Nitrification oxidation of ammonia to nitrite and nitrates by bacteria.

Nitrogen cycle description of the transformations and flow of different nitrogen-containing compounds in the atmosphere, water, soil and living matter. Inorganic nitrogen (ions or gas) is incorporated into soil or water in organic compounds (e.g. amino acids and proteins) by autotrophic organisms. In turn these die and decay, or are eaten by other organisms (heterotrophs) that die and decay, and are returned to the soil or water along with excretory products produced during life. In the water or soil they are converted to inorganic nitrogen compounds or gas by bacteria. *See also* Box 1.9 Processes for life and Box 3.3 Blue-greens—ancient life a modern day nuisance.

Nitrogen fixation conversion of atmospheric gaseous nitrogen into nitrogenous compounds. In nature this can occur during thunderstorms by electrical energy released by lightning, photochemically or by microorganisms such as bacteria. These may be blue-green algae, bacteria and fungi, free-living or living symbiotically within the tissues of higher terrestrial plants, e.g. bacteria in root nodules of legumes such as clover, native pea flowers or wattles (acacia), and roots of casuarina. Nitrogenous compounds (fertilisers) are also produced synthetically and account for much of the nutrient pollution of water bodies. *See also* Symbiosis.

Nodularia genus name of a toxic blue-green alga (cyanobacteria) that was common in the Peel–Harvey.

Non-vascular without a network of conducting vessels (vascular system), includes the algae, mosses and liverworts.

Nutrients the minerals dissolved in water; inorganic compounds of nitrogen (nitrate and ammonia) and phosphorus (phosphate) that provide nutrition for plant growth. Total nutrients include the organic and inorganic compounds.

Oligotrophic low concentrations of nutrients, therefore not very productive. In Australia, often a condition of undisturbed water bodies. *See also* Eutrophic and Mesotrophic.

Omnivore plant and animal eater.

Organic phosphorus phosphorus incorporated into organic matter.

Osmotic pressure depends on the number of dissolved particles in a given volume of solution; water for example will pass through a semi-permeable membrane by the process of osmosis from a weaker (less dense) solution to a stronger (more dense) solution until equilibrium occurs. In some cases fluids may contain different types of molecules but have the same total osmotic pressure because it is not the type of molecule that matters but the number of molecules.

Osmoregulation control of osmotic pressure or maintenance of suitable concentration of body fluids.

Osmoconformer an organism whose body fluids change with the osmotic concentration of the environment. They tend to be stenohaline species that only tolerate or live in environments with a narrow range of salinities, i.e. stable environments such as ocean or fresh water.

Osmoregulator organism whose body fluids remain relatively constant in spite of changes in the osmotic concentration of the environment. Euryhaline species tolerate a wide range of salinity; some may be hyperosmotic and maintain higher concentrations in low salinities or osmoconformers changing with the environment.

Oxidation loss of electrons accompanied by an increase in oxdidation number, e.g. oxidation of ammonium (NH_4^+) to nitrite (NO_2^-), or nitrite (NO_2^-) to nitrate (NO_3^-), or ferrous iron (Fe^{2+}) to ferric iron (Fe^{3+}). *See also* Redox reaction, Redox potential and Reduction.

Palaeozoic Era oldest era in the earth's history ('ancient life') from the evolution of life 570 million years ago to the Mesozoic era 225 million years ago; includes the Cambrian, Ordovician, Silurian, Devonian, Carboniferous (Mississippian and Pennsylvanian) and Permian periods. *See also* Cenozoic Era and Mesozoic Era.

Paleogene Period includes Paleocene, Eocene and Oligocene Epochs. *See also* Neogene and Tertiary, Fig. 1.1.

Parabolic dune hill or ridge of sand in the shape of a parabola (half cone lying on side), oriented and often moving in line with the main wind direction.

pH concentration of hydrogen (H^+) ions or H_3O^+ (indicating relationship with H_2O) ions in solution. A measure of acidity or alkalinity of aqueous solutions on a logarithmic scale of 1–14, in which pH 7 is neutral, values above 7 are alkaline and values below 7 are acid. Sea water is generally 7.5–8.4. Concentration of ions increases or decreases ten times for each unit of change. The pH of a water body depends on the amount of carbon dioxide dissolved in the water and may be influenced by the biological processes of photosynthesis (that uses carbon dioxide and releases oxygen), respiration and decay (that use oxygen and produce carbon dioxide) that in turn influence nutrient release from the sediment.

Photosynthesis conversion of light energy to sugar and oxygen in the presence of light.

Photic zone layer of water (generally near the surface) where there is sufficient light for photosynthesis.

Phytoplankton free-floating microscopic plants, e.g. diatoms, dinoflagellates and chlorophytes. *See also* Table 1.10.

Plankton organism inhabiting the surface layers of the sea or lake, consisting of small drifting plants (phytoplankton) (e.g. diatoms) and animals (zooplankton) such as nekton, holoplankton and meroplankton.

Pleistocene Epoch the recent ice ages when the ice expanded and contracted from the poles, and sea levels fell and rose: the last 1.8 million years. The last expansion is marked by the beginning of the Holocene Epoch. *See also* Fig. 1.1.

Point source pollution large amounts of nutrients or other pollutants from a single source, such as factories, abattoirs or livestock holding yards. Usually licences required for particular discharge by DEP, local government or other government agencies. *See also* Diffuse source pollution.

Polychaetes (Polychaeta) segmented worms with bristles.

Precambrian oldest era in the earth's history from formation of the earth's crust to the emergence of life in the Cambrian.

Quaternary most recent period, 0–2 million years from glacial time (Pleistocene) to post-glacial time (Holocene). Sometimes refers to the division of pre-human and post-human times and is therefore increasing in duration as knowledge of human ancestry becomes longer. In many ways, this is an artificial division. *See also* Tertiary, Paleogene and Neogene; Fig. 1.1.

Recruitment replenishment of the biological community by juveniles. In an estuary this can be from an existing population or from the ocean when the bar is open.

Redox reaction process in which electrons are transferred from one reactant (reducing agent) to another (oxidising agent).

Redox potential a scale representing the potential reduction (addition of electrons) and oxidation (removal of electrons) from a given material. Measured as an electric potential, and dependent on pH.

Reduction gain of electrons accompanied by a decrease in oxidation number, e.g. reduction of nitrate (NO_3^-) to nitrite (NO_2^-), nitrite (NO_2^-) to ammonium (NH_4^+), nitrate (NO_3^-) to nitrogen (N_2), or ferric iron (Fe^{3+}) to ferrous iron (Fe^{2+}). *See also* Oxidation and Redox.

Residence time time taken for a water body to be flushed to the ocean. *See also* Flushing time.

Respiration 1. The process of living organisms of taking in oxygen and releasing carbon dioxide: breathing. 2. The chemical breakdown of complex organic compounds such as carbohydrates and fats that takes place in cells of plants and animals during which carbon dioxide is produced and energy released. Aerobic respiration: consumption of oxygen and the release of carbon dioxide and energy. Anaerobic respiration: breakdown of molecules without oxygen, glycogen to lactic acid in vertebrate muscle or glucose to ethyl alcohol and carbon dioxide by yeasts. Anaerobic respiration does not release as much energy as aerobic respiration.

Ria a long narrow inlet of the sea coast, being a former valley that was submerged by a rise in sea level; common on the coasts of SW Ireland and NW Spain.

Riparian (vegetation or zone) growing in a fringe along the banks or margins of a river or estuary.

Salinisation the accumulation of salts in soil and water that causes degradation of vegetation and land.

Salinity commonly measured in units of mg/L or parts per thousand (‰), predominantly sodium chloride (NaCl); fresh (0–500 mg/L, 1‰), marginal (500–1,500 mg/L, 1.6‰) and saline water (> 5,000 mg/L, 6.2–13.4‰); brackish water has salinity between that of fresh and marine waters (1,500–5,000 mg/L, 2.7–5.5‰); marine (35,000 mg/L, 34–37‰ usually expressed as 35‰); hypersaline (> 35‰). *See also* Hypersaline and Tables 1.2, 1.3 and 1.4.

Salt wedge body of dense saline water at the bottom of an estuary or river moving upstream in a 'wedge' under the lighter fresh water downstream flow. *See also* Stratification.

Secchi depth measure of water clarity determined from the depth at which a Secchi disc can be seen from the surface.

Semidiurnal two high and two low tides in 24 hours; the most common tidal pattern. *See also* Diurnal.

Senescent cells dying or decaying cells.

Seston living and non-living particles in the water.

Species richness number of different types of species present in a given area.

Spring tide largest tide experienced during a lunar tidal cycle.

Stenohaline unable to tolerate wide variations of osmotic pressure of environment. *See also* Euryhaline.

Storm surge elevated water level occurring during storms far above predictions due to a high tide. Caused mainly by onshore winds.

Stratification division or layering of water column into discrete masses either of temperature (thermocline) or salinity (halocline and salocline). Leads to isolation of bottom water and anoxic (anerobic) conditions.

Superphosphate form of artificial fertiliser high in phosphate.

Swale moist depression. *See also* Berm.

Symbiosis living in close association for mutual benefit, e.g. bacterium in association with higher plant supplying ammonium and using carbohydrate produced in photosynthesis by the plant, or a dinoflagellate alga inside jellyfish tissue.

Taxa name given to scientific classifications (groups) of related organisms.

Tertiary period older term used to denote period of time after the Cretaceous to the present time (≈ 65 million years) now referred to as the Paleogene (Paleocene, Eocene and Oligocene) and the Neogene (Miocene and Pliocene). *See also* Fig. 1.1.

Thermocline zone of usually rapid temperature change, between the warmer surface and cooler deeper wasters. *See also* Stratification.

Tidal prism total volume of ocean water entering an estuary on a flood tide.

Tropical cyclone intense low or depression where the distance or gradient between pressures is very steep and winds are very strong. Winds of extreme speed circulate around the centre of the eye of the storm where the pressure is lowest and winds may be calm. The violent winds are often accompanied by black clouds and torrential rain. Tropical cyclones originate over oceans where humidity is high, and generally follow fairly well defined tracks travelling westwards and then polewards before tuning eastwards, dissipating over land. Referred to as a typhoon in the South China Sea, and hurricane in the West Indies.

Turbidity measure of water clarity.

Vascular plants plants containing vessels for conducting water or sap.

Warm front boundary between a mass of advancing warm air and cooler dense air over which it rises. Often accompanied by cloud and rain at the head of the front that diminishes or ceases as the front passes. As the warm front passes, temperatures rise, rain decreases, and winds change. Most common in high latitudes during winter when depressions are most common.

Wrack accumulation of algae or seagrass washed onto a beach or along the shore.

Zooplankton microscopic or small animals that live in the water column.

Bibliography

References for a specific estuary or area are listed under separate chapter bibliographies. Where the subject matter relates to a number of estuaries, is less specific in location or topic, or referred to in the prelude or chapter 1 the sources are listed under Chapter 1 Understanding Swanland. Sources for data in figures, graphs, maps, tables and boxes are referenced and indicate other publications by principal authors. As some authors have written on a number of topics and locations, their publications should be searched for under Chapter 1. If searching for additional references and sources, many secondary authors are the first or primary author on a closely related subject, e.g. on the subject of fish some papers by Professor Ian Potter have secondary authors who are primary authors of other papers on the topic. Similarly, authors may have other publications listed in a bibliography for a specific chapter or the general listing. Examination of more recent papers should provide additional sources for a particular topic.

1 Understanding Swanland

Abbott, I. (2001) 'Historic record of Australian Pelican, *Pelecanus conspicillatus*, breeding in southwest Western Australia', *The Western Australian Naturalist*, **23**, 1–7.

ANCA (2001) *A Directory of Important Wetlands in Australia*, 2nd edn, Australian Nature Conservation Agency, Canberra.

Angell, K. (1999) 'Fertilisers for the pastures on sandy soils of the Swan Coastal Plain', *Agriculture Western Australia Bulletin 4375*, 1–11.

ANZECC (1992) 'Australian Water Quality Guidelines for Fresh and Marine Waters', National Water Quality Management Strategy Paper no. 4, Australian and New Zealand Environment and Conservation Council, Canberra.

ANZECC (2000) 'Australian and New Zealand Guidelines for Fresh and Marine Water Quality', vol. 1, paper 4, ch. 1–7, Australia and New Zealand Environment and Conservation (ANZECC), & Agricultural and Resource Management Council of Australia and New Zealand (ARMCANZ).

Appleyard, R. T. & Manford, T. (1979) *The Beginning of European Discovery and Early Settlement of Swan River Western Australia*, UWA Press, WA.

Atkinson, M. J. & Smith, S. V. (1983) 'C:N:P ratios of benthic marine plants', *Limnology and Oceanography*, **28**, 568–74.

Ayvazian, S. & Nowara, G. (2000) 'West and South Coast Estuarine Fisheries Status Reports', in Penn, J. W. (ed.), *State of the Fisheries Report 1999–2000*, 18–21 & 87–91.

Ayvazian, S. & Nowara, G. (2001) 'West and South Coast Estuarine Fisheries Status Reports', in Penn, J. W. (ed.), *State of the Fisheries Report 2000–2001*, 22–5 & 100–3.

Ayvazian, S. & Nowara, G. (2002a) 'South Coast Estuarine Fisheries Status Report', in Penn, J. W. (ed.), *State of the Fisheries Report 2001–2002*, 106–10.

Ayvazian, S. & Nowara, G. (2002b) 'West Coast Estuarine Fisheries Status Report', in Penn, J. W. (ed.), *State of the Fisheries Report 2001–2002*, 25–8.

Ayvazian, S. & Nowara, G. (2003) 'West and South Coast Estuarine Fisheries Status Reports', in Penn, J. W. (ed.), *State of the Fisheries Report 2002–2003*, 28–31 & 119–22.

Ayvazian, S. G., Johnson, M. S. & McGlashan, D. J. (1994) 'High levels of genetic subdivision of marine and estuarine populations of the estuarine catfish *Cnidoglanis macrocephalus* (Plotosidae) in southwestern Australia', *Marine Biology*, **118**, 25–31.

Bastian, L. V. (1996) 'Residual soil mineralogy and dune subdivision, Swan Coastal Plain, Western Australia', *Australian Journal of Earth Sciences*, **43**, 31–44.

Baudin, N. (1974) *The Journal of Post Captain Nicolas Baudin, Commander-in-Chief of the Corvettes* Geographe *and* Naturaliste, *Assigned by Order of the Government to a Voyage of Discovery*, trans. C. Cornell (with foreword by J.-P. Faivre), Libraries Board of South Australia.

Beard, J. S. (1980) 'A new phytographic map for Western Australia', *Research Notes of the Western Australian Herbarium*, **3**, 37–58.

Beard, J. S. (1990) *Plant Life of Western Australia*, Kangaroo Press, Kenthurst, NSW.

Beard, J. S. (1999) 'Evolution of the river systems of the south-west drainage division', *Journal of the Royal Society of Western Australia*, **82**, 147–64.

Bennett, I. (1992) *Australian Seashores*, adapted from W. J. Dakin's 'Australian Seashores', Collins, & Angus & Robertson, Pymble, NSW.

Black, J. M. (1986*) Flora of South Australia, Part IV Alismataceae–Orchidaceae, Handbook of the Flora and Fauna of South Australia*, 4th edn, South Australian Government.

Bold, H. C. & Wynne, J. *Introduction to the Algae Structure and Reproduction*, 2nd edn, Prentice-Hall Inc., Englewood Cliffs, New Jersey.

Borch, C. C. von der (1968) 'Southern Australian submarine canyons: their distribution and ages', *Marine Geology*, **6**, 267–79.

Bureau of Meteorology (Australia) (1929) 'Results of rainfall observations made in Western Australia, including all available annual rainfall totals from 1374 stations for all years of record up to 1927', with maps, diagrams; record of notable meteorological events, presenting monthly and yearly meteorological elements of Perth, Broome and Kalgoorlie.

Chivas, A. R., Torgersen, T. & Plach, H. A. (1990) 'Growth rates and Holocene development of stromatolites from Shark Bay, Western Australia', *Australian Journal of Earth Sciences*, **37**, 113–21.

Chrystal, P. J., Potter, I. C., Loneragan, N. R. & Holt, C. P. (1985) 'Age structure, growth rates, movement patterns and feeding in an estuarine population of the cardinal fish *Apogon rueppelii*', *Marine Biology*, **85**, 185–97.

Chubb, C. F. & Potter, I. C. (1986) 'Age, growth and condition of the Perth herring, *Nematalosa vlaminghi* (Munro) (Dorosomatinae), in the Swan Estuary, south-western Australia', *Australian Journal of Marine and Freshwater Research*, **37**, 105–12.

Churchill, D. M. (1959) 'Late quaternary eustatic changes in the Swan River district', *Journal of the Royal Society of Western Australia*, **42**, 53–5.

Clarke, E. de C. (1926) 'Natural regions in Western Australia', *Journal of the Royal Society in Western Australia*, **12**, 117–32.

Collins, L. B. (1987) 'Geological evolution of the Swan–Canning estuarine system', in *The Swan River Estuary Ecology and Management*, ed. J. John, Curtin University of Technology, Environmental Studies Group, Report no. 1, 9–20.

Collins, L. B. (1988) 'Sediments and history of the Rottnest Shelf, southwest Australia: a swell dominated, non-tropical margin', *Sedimentary Geology*, **60**, 15–49.

Commander, D. P., Schoknecht, N., Verboom, W. & Casetta, P. (2001) 'The geology, physiography and soils of wheatbelt valleys', *Dealing with salinity in wheatbelt valleys Conference Papers on Processes, prospects and practical options, Merredin, WA*, 30 July – 1 August 2001, State Salinity Council & Avon Working Group, 1–39.

Congdon, R. A. & McComb, A. J. (1979) 'Productivity of *Ruppia*: seasonal changes and dependence on light in an Australian estuary', *Aquatic Botany*, **6**, 121–32.

Conolly, J. R. & von der Borch, C. C. (1967) 'Sedimentation and physiography of the sea floor of Australia', *Sedimentary Geology*, **1**, 181–220.

Dakin, W. J., Bennett, I. & Pope, E. (1952) *Australian Seashores: A Guide for the Beach-lover, the Naturalist, the Shore Fisherman, and the Student*, Angus & Robertson, Sydney.

Davidson, W. A. (1995) 'Hydrology and groundwater resources of the Perth Region, Western Australia', *Geological Survey Bulletin*, **142**.

Davies, P. M. & Lane, J. A. K. (1996) 'The impact of vegetated buffer zones on water and nutrient flow into Lake Clifton, Western Australia', *Journal of the Royal Society of Western Australia*, **79**, 155–60.

Deeley, D. (2001) Evaluating indicators of ecological health for estuaries in Southwestern Australia, PhD thesis, Murdoch University, WA.

Department of Conservation and Environment (1981) 'The Darling System, Western Australia', Proposals for Parks and Reserves, The System 6 Study Report to the EPA, Report no 8.

Department of Conservation and Land Management (1981) 'The Darling System, Western Australia: Proposals for Parks and Reserves', The System 6 Study Report to the Environmental Protection Authority.

Department of Conservation and Land Management (1994) 'A representative Marine Reserve System for Western Australia', Report of the Marine Parks and Reserves Selection Working Group.

Department of Environmental Protection (1974) 'Conservation Reserves in Western Australian', Report to the Environmental Protection Authority.

Department of Environmental Protection (1996) 'Southern Metropolitan Coastal Waters Study (1991–1994)', Report 17.

Department of Fisheries Western Australia (2002) 'State of the Fisheries Report 2001–2002', Ayvazian, S. & Nowara, G.

Diels, L. (1906) *Die Pflanzenwelt von West-Australien sudlich des Wendekreises, Vegetation der Erde* 7, Leipsig.

Dix, W. C. & Meagher, S. J. (1976) 'Fishtraps in the south-west of Western Australia', *Records of the Western Australian Museum*, **4**, 171–87.

Dortch, C. E. (1997a) 'New perceptions of the chronology and development of Aboriginal fishing in south-western Australia', *World Archaeology*, **29**, 15–35.

Dortch, C. E. (1997b) 'Prehistory down under: archaeological investigations of submerged Aboriginal sites at Lake Jasper, Western Australia', *Antiquity*, 116–23.

Dortch, C. E. (1999) 'Archaeological assessment of Aboriginal fishing on the Southern Ocean coast of Western Australia' in Hall, J. & McNiven, I. J. (eds) *Australian Coastal Archaeology*, ANH Publications, Research School of Pacific and Asian Studies, Australian National University, 25–35.

Dortch, C. E. & Hesp, P. A. (1994) 'Rottnest Island artifacts and palaeosols in the context of Greater Swan Region prehistory', *Journal of the Royal Society of Western Australia*, **77**, 23–32.

Dortch, C. E., Kendrick, G. W. & Morse, K. (1984) 'Aboriginal mollusc exploitation in southwestern Australia', *Archaeology Oceania*, **19**, 81–104.

Dring, M. J. (1992) *The Biology of Marine Plants*, Contemporary Biology, Edward Arnold, London.

Duyker, E. & Duyker, M. (eds & trans.) (2001) *Bruny d'Entrecasteaux: Voyage to Australia and the Pacific 1791–1793*, Melbourne University Press, Victoria.

Edgar, G. J. (1997) *Australian Marine Life: The Plants and Animals of Temperate Waters*, Reed Books, Victoria.

Environment Australia (2001) *A directory of important wetlands in Australia*, 3rd edn, Environment Australia, Canberra.

Environmental Protection Authority (1975) 'Conservation Reserves for Western Australia', Department of Conservation and Environment, WA Systems 4, 8, 9, 10, 11, 12.

Environmental Protection Authority (1976) 'Conservation Reserves for Western Australia', Department of Conservation and Environment, WA Systems 1, 2, 3, 5.

Environmental Protection Authority (1991) 'Reflections on 20 years', R. Hughes (ed.).

Fisheries Department of Western Australia (1990a) 'Prawns in WA Estuaries No. 1', Fisheries Department of Western Australia.
Fisheries Department of Western Australia (1990b) 'Tailor *Pomatomus saltratrix*', *Fishing WA no. 4*.
Fisheries Department of Western Australia (1993a) 'Black Bream *Acanthopagrus butcheri*', *Fishing WA no. 10*.
Fisheries Department of Western Australia (1993b) 'Mulloway *Argyrosomus hololepidotus*', *Fishing WA no. 2*. (Note the species name is now *A. japonicus*.)
Fisheries Department of Western Australia (1996) 'State of the Fisheries Report 1995–1996'.
Fisheries Department of Western Australia (1998) 'Recreational Net Fishing in Western Australia', Recreational Fishing Program.
Fisheries Western Australia (1998) 'State of the Fisheries Report 1997–1998'.
Fisheries Western Australia (1999) 'State of the Fisheries Report 1998–1999'.
Fisheries Western Australia (2000a) 'Introduced Freshwater Species in Western Australia', Fish and Fish Habitat Protection Program.
Fisheries Western Australia (2000b) 'State of the Fisheries Report 1999–2000'.
Fisheries Western Australia (2001) 'State of the Fisheries Report 2000–2001'.
Flannery, T. (2001) *Terra Australis: Matthew Flinder's Great Adventures in the Circumnavigation of Australia*, T. Flannery (ed.), Text, Melbourne.
Flinders, M. (1966) *A Voyage to Terra Australis: Undertaken for the Purpose of Completing the Discovery of that Vast Country, and Prosecuted in the Years 1801, 1802, 1803 in His Majesty's Ship* The Investigator, Facsimile edition, Libraries Board of South Australia, Adelaide.
Gardner, C. A. (1942) 'The vegetation of Western Australia with special reference to climate and soils', *Journal of the Royal Society of Western Australia*, **28**, 11–87.
Gaughan, D. J. & Potter, I. C. (1994) 'Relative abundance and seasonal changes in the macrozooplankton of the lower Swan estuary in South-Western Australia', *Records of the Western Australian Museum*, **16**, 461–74.
Gaughan, D. J. & Potter, I. C. (1995) 'Composition, distribution and seasonal abundance of zooplankton in a shallow, seasonally closed estuary in temperate Australia', *Estuarine and Coastal Shelf Science*, **41**, 117–35.
Geological Survey of Western Australia (1975) 'Geology of Western Australia', *WA Geol. Survey, Memoir 2*.
George, R., Weaver, D. & Terry, J. (1996) 'Environmental water quality: A guide to sampling and measurement', Agriculture Western Australia, 1–38.
Geoscience Australia, database of Australian estuaries <www.agso.gov.au>, 2003.
Government of Western Australia Fish Resources Management Act 1994, 'South Coast Estuarine Fishery (Interim) Management Plan (2001)', *Government Gazette No. 14*, (WA), 22 January 2002.
Green, J. (1968) *The biology of estuarine animals*, Sidgwick & Jackson, London, 401 pp.
Grey, K., Moore, L. S., Burne, R. V., Pierson, B. K. & Bauld, J. (1990) 'Lake Thetis, Western Australia: an example of saline lake sedimentation dominated by benthic microbial processes', *Australian Journal of Marine and Freshwater Research*, **41**, 275–300.
Haig, D. W. (2002) 'Guidebook for Post-Conference Field Excursion: Perth to Shark Bay', *International Symposium on Foraminifera, Perth, WA, Forams 2002, 10–17 February 2002*.
Hallam, S. J. (1987) 'Yams, alluvium and "villages" on the west coastal plain', paper delivered to 54th ANZASS Congress, Canberra, May 1984, Australian Institute of Aboriginal Studies, Canberra Archaeological Society, 116–32.
Hallam, S. J. (1989) 'Plant usage and management in Southwest Australian Aboriginal societies', in Harris, D. R. & Hillman, G. C. (eds) *Foraging and Farming: The Evolution of Plant Exploitation*, Unwin Hyman, London.
Hasluck, A. (1961) 'Yagan the patriot', *Journal Proceedings of the Western Australian Historical Society*, **5**, 33–48.
Hesp, P. A. (1984) 'Aspects of the Geomorphology of South Western Australia', in Hodgkin, E. P. (ed.) *Estuarine environments of the Southern Hemisphere*, Department of Conservation and Environment, Western Australia, Bulletin 161, 61–83.
Hill, A. L., Semeniuk, C. A., Semeniuk, V. & Del Marco, A. (1996) 'Wetlands of the Swan Coastal Plain Wetland Mapping', Classification and Evaluation, vols 2a (Main report) & 2b (Atlas).
Hill, N. A., Lukatelich, R. J. & McComb, A. J. (1991) 'A comparative study of some of the physical and chemical characteristics of the sediments from three estuarine systems in south western Australia', Report to the Waterways Commission, *Report no. 23*.
Hodgkin. E. P. (1987) 'The hydrology of the Swan River estuary: Salinity the ecological master factor', in *The Swan River Estuary Ecology and Management*, J. John (ed.), Curtin University, Environmental Studies Group, Report no. 1, 34–44.
Hodgkin E. P. (2005) Collected letters, diaries, notes and photographs pertaining to Estuaries of South-Western Australia, to be archived at The University of Western Australia Library.
Hodgkin, E. P., Birch, F. B., Black, R. E. & Humphries, R. B. (1980) 'The Peel–Harvey Estuarine System Study 1976–1980', A report to the Estuarine and Marine Advisory Committee, 1–72.
Hodgkin, E. P. & Clark, R. (1987) 'Wellstead Estuary, The estuary of the Bremer River, An inventory of information on the estuaries and coastal lagoons of south Western Australia', *Estuarine Studies Series no. 1*, Environmental Protection Authority, WA.
Hodgkin, E. P. & Clark, R. (1988a) 'Beaufort Inlet and Gordon Inlet, Estuaries of the Jerramungup Shire, An inventory of information on the estuaries and coastal lagoons of south Western Australia', *Estuarine Studies Series no. 4*, Environmental Protection Authority, WA.
Hodgkin, E. P. & Clark, R. (1988b) 'Nornalup and Walpole Inlets and the estuaries of the Deep and Frankland Rivers, An inventory of information on the estuaries and coastal lagoons of south Western Australia', *Estuarine Studies Series no. 2*, Environmental Protection Authority, WA, revised 1999.
Hodgkin, E. P. & Clark, R. (1988c) 'Wilson, Irwin and Parry Inlets, Estuaries of the Denmark Shire, An inventory of information on the estuaries and coastal lagoons of south Western Australia', *Estuarine Studies Series no. 3*, Environmental Protection Authority, WA.
Hodgkin, E. P. & Clark, R. (1989a) 'Broke Inlet and other estuaries of the Shire of Manjimup, An inventory of information on the estuaries and coastal lagoons of south Western Australia', *Estuarine Studies Series no. 6*, Environmental Protection Authority, WA.
Hodgkin, E. P. & Clark, R. (1989b) 'Stokes Inlet and other estuaries of the Shire of Esperance, An inventory of information on the estuaries and coastal lagoons of south Western Australia', *Estuarine Studies Series no. 5*, Environmental Protection Authority, WA.
Hodgkin, E. P. & Clark, R. (1990a) 'Estuaries of the Shire of Albany, An inventory of information on the estuaries and coastal lagoons of south Western Australia', *Estuarine Study Series no. 8*, Environmental Protection Authority, WA.
Hodgkin, E. P. & Clark, R. (1990b) 'Estuaries of the Shire of Ravensthorpe and the Fitzgerald River National Park, An inventory of information on the estuaries and coastal lagoons of south Western Australia', *Estuarine Studies Series no. 7*, Environmental Protection Authority, WA.
Hodgkin, E. P. & Di Lollo, V. (1958) 'The tides of south-western Australia', *Records of the Royal Society of Western Australia*, **41**, 42–54.

Hodgkin, E. P. & Hesp, P. (1998) 'Estuaries to salt lakes: Holocene transformation of the estuarine ecosystems of south-western Australia', *Marine and Freshwater Research*, **49**, 183–201.

Hodgkin, E. P. & Kendrick, G. W. (1984) 'The changing aquatic environment 7000 BP to 1983 in the estuaries of south western Australia', in *Estuarine Environments of the Southern Hemisphere*, E. P. Hodgkin (ed.), Bulletin 161, Department of Conservation and Environment, WA, 61–83; 85–95.

Hodgkin, E. P., Kendrick, G., Marsh, L. & Slack-Smith, S. (1966) 'The Shelled Gastropoda of South Western Australia', *Handbook no. 9*, Western Australian Naturalists' Club, Perth.

Hodgkin, E. P. & Majer, K. (1978) 'An index of ecological information on estuaries and marine embayments in Western Australia', CSIRO Division of Fisheries and Oceanography, Report no. 70.

Hodgkin, E. P. & Vicker, E. (1987) 'A history of algal pollution in the estuary of the Swan River', in *The Swan River Estuary Ecology and Management*, J. John (ed.), Curtin University of Technology, Environmental Studies Group, Report no. 1, 65–70.

Horwitz, P. & Wardell-Johnson, A. (1996) 'Historical associations of wetlands and rivers in the Busselton–Walpole Region', Water and Rivers Commission, *Report WRT 2*.

Hosja, W. & Deeley, D. (1994) 'Harmful phytoplankton surveillance in Western Australia' with appendix by Dr G. Hallegraeff, Waterways Commission, *Report No. 43*.

Huisman, J. M. (2000) *Marine Plants of Australia*, University of Western Australia Press, in association with the Australian Biological Resources Study.

Humphries, P., Hyndes, G. A. & Potter, I. C. (1992) 'Comparisons between the diets of distant taxa (Teleost and Cormorant) in an Australian estuary', *Estuaries*, **15**, 327–34.

Hutchins, B. & Swainston, R. (1996) *Sea Fishes of Southern Australia: Complete Field Guide for Anglers and Divers*, Swainston Publishing, & Gary Allen Pty Ltd, Victoria.

Hutchins, B. & Thompson, M. (1983) *The marine and estuarine fishes of South-western Australia: A field guide for anglers and divers*, Western Australian Museum, Perth.

Hyerdahl, T. (1958) *Aku-aku: The Secret of Easter Island*, Allen & Unwin, London.

Iasky, R. P. & Mory, A. J. (1999) 'Geology and petroleum potential of the Gascoyne Platform, southern Carnarvon Basin, Western Australia', Geological Survey of Western Australia, *Report 69*.

Jaensch, R. P. & Vervest, R. M. (1988) 'Ducks, Swans and Coots in South-western Australia: the 1986 and 1987 counts', Royal Australasian Ornithologists, *Union Report No. 31*.

Jones, D. S. & Morgan, G. J. (1994) *A Field Guide to Crustaceans of Australian Waters*, Reed, NSW.

Jutson, J. T. (1914) 'An outline of the physiographical geology of Western Australia, The physiography (geomorphology) of Western Australia', (2nd edn, WA Geological Survey, Bulletin No. 95 in 1934).

Kendrick, G. W. (1960) 'The fossil mollusca of the Peppermint Grove limestone, Swan River district of Western Australia', *Western Australian Naturalist*, **7**, 53–66.

Kendrick, G. W. (1976) 'The Avon: Faunal and other notes on a dying river in south-western Australia', *The Western Australian Naturalist*, **13**, 97–114.

Kendrick, G. W. (1977) 'Middle Holocene marine molluscs from near Guildford, Western Australia, and evidence for climatic change', *J. Roy. Soc. West. Aust.*, **59**, 97–104.

Kendrick, G. W., Wyrwoll, K. H. & Szabo, B. J. (1991) 'Pliocene–Pleistocene coastal events and history along the western margin of Australia', *Quaternary Science Reviews*, **10**, 419–39.

Knox, B., Ladiges, P., Evans, B. & Saint, R. E. (2001) *Biology*, 2nd edn, McGraw-Hill, Australia.

Lavery, P. S., Oldham, C. E. & Ghisalberti, M. (2001) 'The use of Fick's First Law for predicting pore water nutrient fluxes under diffuse conditions', *Hydrological Processes Special Issue*, **15**, 2435–51.

Lenanton, R. C. J. (1982) 'Alternative non-estuarine nursery habitats for some commercially and recreationally important fish species of south-western Australia', *Australian Journal of Marine and Freshwater Research*, **33**, 881–900.

Lenanton, R. C. J. (1984) 'The commercial fisheries of temperate Western Australia; early settlement to 1975', Department of Fisheries and Wildlife, *Report 62*.

Lenanton, R. C. J., Allison, R. & Ayvazian, S. G. (1996) 'Assessment of the effects of a trial period of unattended recreational netting in selected estuaries of temperate Western Australia, between 1 January 1994 and 30 June 1995', Fisheries Department of Western Australia, *Fisheries Research Report no. 104*.

Lenanton, R. C. J. & Hodgkin, E. P. (1985) 'Life history strategies of fish in some temperate Australian estuaries' in *Community Ecology in Estuaries and Coastal Lagoons: Towards an Ecosystem Integration*, Yanez-Aranciba, A. (ed.) UNAM Press, Mexico City, 267–84.

Lenanton, R. C. J., Loneragan, N. R., Potter, I. C. & Caputi, N. (1986) 'Spatial and seasonal differences in the fish fauna in the shallows of a large Australian estuary', *Marine Biology*, **92**, 575–86.

Lenanton, R. C. J. & Potter, I. C. (1987) 'Contribution of estuaries to commercial fisheries in temperate Western Australia and the concept of estuarine dependence', *Estuaries*, **10**, 28–35.

Loneragan, N. R. & Potter, I. C. (1990) Factors influencing community structure and distribution of different life cycle categories of fishes in shallow waters of a large Australian estuary, *Marine Biology*, **106**, 25–37.

Loneragan, N. R., Potter, I. C. & Lenanton, R. C. J. (1987) 'The fish and fisheries of the Swan Estuary', in *The Swan River Estuary Ecology and Management*, J. John (ed.), Curtin University, Environmental Studies Group, Report no. 1, 178–201.

Loneragan, N. R., Potter, I. C. & Lenanton, R. C. J. (1989) 'Influence of site, season and year on contributions made by marine, estuarine, diadromus and freshwater species to the fish fauna of a temperate Australian estuary', *Marine Biology*, **103**, 461–79.

Lowry, D. C. & Doepel, J. J. G. (1974) Malcolm–Cape Arid, Western Australia, Western Australia Geological Survey, 1:250,000 Geological Series Explanatory Notes.

McComb, A. J. (1984) 'Plant biomass and productivity in south-western Australian estuaries', in *Estuarine Environments of the Southern Hemisphere*, E. P. Hodgkin (ed.) Department of Conservation and Environment Western Australia, Bulletin 161, 97–111.

McNamara, K. J. (1992) 'Stromatolites', Western Australian Museum, Perth.

Malseed, B. E. & Sumner, N. R. (2001b) 'A 12-month survey of recreational fishing in the Swan–Canning Estuary Basin of Western Australia 1998–99', *Fisheries Research Report no.126*.

Marchant, L. R. (1982) *France Australie: A Study of French Explorations and Attempts to Found a Penal Colony and Strategic Base in South Western Australia 1503–1826*, Artlook Books, Perth.

Meagher, S. J. (1974) The food resources of the Aborigines of the south-west of Western Australia, *Records of the Western Australian Museum* **3**, 14-65.

Michaelsen, W. & Hartmeyer, R. (1907–1930) *Die Fauna Südwest-Australiens*, Ergebnisse der Hamburger sudwest-australischen Forschungsreise 1905, herausgegeben von Prof. Dr. W. Michaelsen und Prof. Dr. R. Hartmeyer, Jena (Germany), Gustav Fischer.

Mitrovic, S. (1995) 'What scum is that? Algal blooms and similar prolific plant growth', Department of Land and Water Conservation.

Moore, L., Knott, B. & Stanley, N. (1983) 'The stromatolites of Lake Clifton, Western Australia: Living structures representing the origins of life', *Search*, **14**, 309–14.

Moore, L. S. (1987) 'Water chemistry of the coastal saline lakes of the Clifton–Preston lakeland system and its influence on stromatolite formation', *Australian Journal of Marine and Freshwater Research*, **38**, 647–60.

Moore, W. G. (1988) *The Penguin Dictionary of Geography*, 7th edn, Penguin, London.

Morgan, D. L., Gill, H. S. & Potter, I. C. (1998) 'Distribution, identification and biology of freshwater fish in south-western Australia', *Records of the Western Australian Museum, Supplement No.56.*

Morrissy, N. M. (1978) 'The past and present distribution of the marron, *Cherax tenuimanus* (Smith) in Western Australia', *Fisheries Research Bulletin Western Australia*, **22**, 1–38.

Mueller, F. von (1867) 'Australian Vegetation', *Intercolonial Exhibition Essays*, 251–61.

National Land and Water Resources Audit (2002) 'Australian catchment, river and estuary assessment 2002: assessing the aggregate impact of resource use on key natural ecosystems', Natural Heritage Trust, <www.ozestuaries.org/>, 2005.

Neira, F. J. & Potter, I. C. (1992a) 'The ichthyoplankton of a seasonally closed estuary in temperate Australia. Does an extended period of opening influence species composition?', *Journal of Fish Biology*, **41**, 935–53.

Neira, F. J. & Potter, I. C. (1992b) 'Movement of larval fishes through the entrance channel of a seasonally open estuary in Western Australia', *Estuarine Coastal Shelf Sciences*, **335**, 213–24.

Nielsen, J. & Jernakoff, P. (1996) 'A review of the interaction of sediment and water quality with benthic communities', Report No. 25, CSIRO, Canberra.

O'Connor, R., Quartermaine, G. & Yates, A. (1995) 'An investigation into the Aboriginal significance of wetlands and rivers in the Busselton–Walpole Region', WP 226, Water Authority of Western Australia.

Olsen, G. & Skitmore, E. (1991) 'State of the rivers of the South West Drainage Division', Western Australian Water Resources Council.

Pate, J. S. & Beard, J. S. (1984) *Kwongan Plant Life of the Sandplain, Biology of a South-West Australian Shrubland Ecoystem*, UWA Press, WA.

Pearce, A. & Caputi, N. W. A. (1997) 'Not another El Nino?' Western Fisheries, Summer.

Pearce, A. F. & Walker, D. I. E. (1991) 'The Leeuwin Current: an influence on the coastal climate and marine life of Western Australia', *Journal of the Royal Society of Western Australia*, **74**, 1–140.

Pearn, R. & Cappelluti, T. (1999) 'The south-coast estuarine fishery: A discussion paper', Fisheries Management Paper no. 126, Fisheries Western Australia.

Pen, L. (1997) 'A systematic overview of environmental values of the wetlands, rivers and estuaries of the Busselton–Walpole Region', Water and Rivers Commission, Water Resource Allocation and Planning Report Series WRAP no. 7.

Pen, L. J. (1999) 'Managing our rivers: A guide to the nature and management of the streams of south-west Western Australia', Water and Rivers Commission, & Natural Heritage Trust.

Pizzey, G. (1985) *A Field Guide to the Birds of Australia*, illustrated by R. Doyle, Collins Australia.

Playford, P. (1998) *Voyage of Discovery to Terra Australis by Willem de Vlamingh in 1696–97*, Western Australian Museum, Perth.

Playford, P. E. (1988) *Guidebook to the Geology of Rottnest Island*, Geological Society of Australia, Western Australian Division, Excursion Guidebook 2.

Playford, P. E., Cockbain, A. E. & Low, G. H. (1976) 'Geology of the Perth Basin Western Australia', *Geological Survey of Western Australia Bulletin*, **124**.

Playford, P. E., Cope, R. N., Cockbain, A. E., Low, G. H. & Lowry, D. C. (1975) 'Phanerozoic', *The Geology of Western Australia, Western Australia Geological Survey Memoir 2*, 223–432.

Playford, P. E. & Leech, R. E. J. (1977) 'Geology and hydrology of Rottnest Island', Geological Survey Western Australia Report no. 6.

Potter, I. C., Cheal, A. J. & Loneragan, N. R. (1988) 'Protracted estuarine phase in the life cycle of the marine pufferfish *Torquigener pleurogramma*', *Marine Biology*, **98**, 317–29.

Potter, I. C., Chrystal, P. J. & Loneragan, N. R. (1983) 'The biology of the blue manna crab *Portunus pelagicus* in an Australian estuary', *Marine Biology*, **78**, 75–83.

Potter, I. C. & Hyndes, G. A. (1994) 'Composition of the fish fauna of a permanently open estuary on the south coast of Australia, and comparisons with a nearby seasonally closed estuary', *Marine Biology*, **121**, 199–209.

Potter, I. C. & Hyndes, G. A. (1999) 'Characteristics of the ichthyofaunas of southwestern Australian estuaries, including comparisons with holarctic estuaries and estuaries elsewhere in temperate Australia: a review', *Australian Journal of Ecology*, **24**, 395–421.

Potter, I. C. & Lestang, S. de (2000) 'Biology of the blue swimmer crab *Portunus pelagicus* in Leschenault Estuary and Koombana Bay in south-western Australia', *Journal of the Royal Society of Western Australia*, **83**, 443–58.

Potter, I. C., Penn, J. W. & Brooker, K. S. (1986) 'Life cycle of the western school prawn, *Metapenaeus dalli* Racek, in a Western Australian estuary', *Australian Journal of Marine and Freshwater Research*, **37**, 95–103.

Potter, I. C., Tiivel, D., Valesini, F. J. & Hyndes, G. A. (1997) 'Comparisons between the ichthyofaunas of a temperate lagoon-like estuary and the embayment into which that estuary discharges', *International Journal of Salt Lake Research*, **5**, 337–58.

Prince, J. D., Potter, I. C., Lenanton, R. C. J. & Loneragan, N. R. (1982) 'Segregation and feeding of Atherinid species (Teleostei) in South-western Australian estuaries', *Australian Journal of Marine Freshwater Research*, **33**, 865–80.

Rippey, E. & Rowland, B. (1995) *Plants of the Perth Coast and Islands*, 2004 as *Coastal Plants: Perth and the south-west region*, 2nd edn UWA Press, WA.

Robert, W. C. H. (ed.) (1972) *The explorations 1696–1697 of Australia by Willem de Vlamingh*, Philo Press, Amsterdam.

Robertson, E. L. (1986) *Flora of South Australia, Part IV, Alismataceae–Orchidaceae*, J. M. Black (ed.), South Australian Government Printing Division, Adelaide, 1711–41.

Rowe, M. (2002) 'Feathers, Flyways and Fast Food', Environment Australia, produced for the State of Victoria through the Teacher release to Industry Program.

Ruprecht, J. K., Bates, B. C. & Stokes, R. A. (1996) 'Climate variability and water resources workshop: summary of outcomes', Water and Rivers Commission, Water Resource Technical Report WRT 5.

Sarre, G. A., Platell, M. E. & Potter, I. C. (2000) 'Do the dietary compositions of *Acanthopagrus butcheri* in four estuaries and a coastal lake vary with body size and season and within and amongst these water bodies?', *Journal of Fish Biology*, **56**, 103–22.

Sarre, G. A. & Potter, I. C. (1999) 'Comparisons between the reproductive biology of black bream *Acanthopagrus butcheri* (Sparidae) in four estuaries with widely differing characteristics', *International Journal of Salt Lake Research*, **8**, 179–210.

Sarre, G. A. & Potter, I. C. (2000) 'Variations in the age comparisons and growth rates of *Acanthopagrus butcheri* (Sparidae) among estuaries: some possible contributing factors', *Fisheries Bulletin*, 785–99.

Saunders, D. & de Rebeira, P. (1985) *The Birdlife of Rottnest Island*, DAS & CpdeR, Western Australia.

Saunders, D. A. & Ingram, J. A. (1995) *Birds of Southwestern Australia: An atlas of changes in distribution and abundance of the wheatbelt fauna*, Surrey Beatty, & Western Australian Laboratory CSIRO Division of Wildlife and Ecology.

Schofield, N. J., Ruprecht, K. J. & Loh, L. C. (1988) 'The impact of agricultural development on the salinity of surface water resources of south-west Western Australia', Report WS 27 Western Australian Water Authority.

Searle, D. J. & Woods, P. J. (1986) 'Detailed documentation of a Holocene sea-level in the Perth region, southern Western Australia', *Quaternary Research*, **26**, 299–308.

Seddon, G. (1972) *Sense of Place: A Response to an Environment—the Swan Coastal Plain, Western Australia*, UWA Press, WA.

Semeniuk, V. (1985) 'The age of a Holocene barrier dune system and its implications for sea level history of reconstruction in southwestern Australia', *Marine Geology*, **67**, 197–212.

Semeniuk, V. & Searle, D. J. (1986) 'Variability of Holocene sea level history along the southwestern coast of Australia—evidence for the effect of significant local tectonism', *Marine Geology*, **72**, 47–58.

Semeniuk, V. & Semeniuk, C. A. (1991) 'Radiocarbon ages and some coastal landforms in the Peel–Harvey estuary, south-western Australia', *Journal of the Royal Society of Western Australia*, **73**, 61–71.

Semeniuk, V., Semeniuk, T. A. & Unno, J. (2000) 'The Leschenault Inlet estuary: an overview', *Journal of the Royal Society of Western Australia*, **83**, 207–28.

Serventy, D. L. (1938a) 'The feeding habits of Cormorants in south-western Australia', *The Emu*, **38**, 293–316.

Serventy, D. L. (1938b) 'Waders and other aquatic birds on the Swan River Estuary', *The Emu*, **38**, 18–29.

Serventy, D. L. (1939) 'Notes on Cormorants', *The Emu*, **38**, 357–71.

Serventy, D. L. (1948) 'The birds of the Swan River District, Western Australia', *The Emu*, **47**, 241–86.

Serventy, D. L. & Whittell, H. M. (1962) *Birds of Western Australia*, illustrations by O. Seymour, H. O. Webster, P. Slater & G. E. Binsted, Paterson Brokensha, Perth.

Shields, W. H. (1928) 'Sandbars on intermittent rivers—a new method of treatment', *Journal of the Royal Society of Western Australia*, **10**, 1–3.

Slack-Smith, S. M. & Brearley, A. (1987) '*Musculista stenhousia* (Benson, 1842); a mussel recently introduced to the Swan River estuary, Western Australia (Mollusca, Mytilidae)', *Records of the Western Australian Museum*, **13**, 225–30.

Smith, T. F., Sant, M. & Thom, B. (2001) 'Australian estuaries: A framework for management', Cooperative Research Centre for Coastal Zone, Estuary and Waterway Management, Indooroopilly, Qld.

Spencer, R. S. (1956) 'Studies in Australian Estuarine Hydrology', *Australian Journal of Marine and Freshwater Research*, **7**, 193–253.

Steckis, R. A., Potter, I. C. & Lenanton R. C. J. (1995) 'The commercial fisheries in three southwestern Australian estuaries exposed to different degrees of eutrophication', in *Eutrophic Shallow Estuaries and Lagoons*, A. J. McComb (ed.) CRC Press, Boca Raton, 189–203.

Stokes, R. A., Stone, K. A. & Loh, I. C. (1980) 'Summary of soil salt storage characteristics in the Northern Darling Range', Water Resources Branch, Public Works Department, WA Technical Report no. WRB 94.

Storey, A. W., Vervest, R. M., Pearson, G. B. & Halse, S. A. (1993) *Wetlands of the Swan Coastal Plain*, vol. 7, Waterbird usage of wetlands on the Swan Coastal Plain, Environmental Protection Authority, & Water Authority of Western Australia.

Tauss, C. (2002) Surveying Western Australia's Land Edge, Reference transects in coastal vegetation at Geraldton, Port Kennedy, Bunbury and Esperance, unpublished report for the Western Australian Herbarium (CALM), & Coastwest/Coastcare (DIP).

Taylor, G. (1911) *Australia in its Physiographic and Economic Aspects*, Clarendon Press, Oxford (5th edition 1928).

Taylor, G. (1915) 'Climatic control of Australian production (an attempt to gauge the potential wealth of the Commonwealth)', *Meteorological Bulletin 11*, Melbourne.

Taylor, G. (1918) 'The Australian Environment (especially as controlled by rainfall): A regional study of the topography, drainage, vegetation and settlement; and of the character and origins of the rains', Memoir no. 1, Advisory Council of Science and Industry, Melbourne. For the Commonwealth of Australia.

Tomas, C. R. (ed.) (1997) *Identifying Marine Phytoplankton*, Academic Press, San Diego.

Valesini, F. J., Potter, I. C., Platell, M. E. & Hyndes, G. A. (1997) 'Ichthyofauna of a temperate estuary and adjacent marine embayment. Implications regarding choice of nursery area and influence of environmental changes', *Marine Biology*, **128**, 37–328.

van de Graaf, W. J. E., Crowe, R. W. A. & Bunting, J. A. (1977) 'Relic early cainozoic drainage in arid Western Australia', *Zeitschrift fur Geomorhologie (Annals of Geomorphology)*, **21**, 379–400.

Vancouver, G. (1798) *A voyage of discovery to the North Pacific Ocean and round the World 1791–1795*, with introduction and appendices, vol. 1, Lamb, W. K. (ed.) The Hakluyt Society, London, 1–442.

Water and Rivers Commission (1997) 'East Gnangara Environmental Water Provisions Plan Public Environmental Review'.

Wells, F. E. & Bryce, C. W. (1984) *A guide to the common molluscs of south-western Australian estuaries*, Western Australian Museum, Perth.

Western Australian Water Resources Council (1992) 'The State of the Rivers of the South West', *WAWRC* **2**.

White, M. E. (1986) *The Greening of Gondwana: The 400 Million Year Story of Australia's Plants*, Reed Books, Frenchs Forest, NSW.

White, M. E. (1994) *After the Greening: The Browning of Australia*, Kangaroo Press, Kenthurst, NSW.

White, M. E. (1997) *Listen . . . Our Land is Crying, Australia's Environment: Problems and Solutions*, Kangaroo Press, Kenthurst, NSW.

Whitehurst, R. (1997) *Noongar Dictionary: Noongar to English and English to Noongar*, Compiled by Rose Whitehurst for the Noongar Language and Cultural Centre 1992, Published by the Noongar Language and Cultural Centre (Aboriginal Corporation) WA, 2nd Edition 1997. (http://www.noongar.org.au/noongar.php).

Womersley, H. B. S. (1984) *The Marine Benthic Flora of Southern Australia, Part 1 Handbook of the Flora and Fauna of South Australia*, issued by the Handbooks Committee on behalf of the South Australian Government.

Wood, A. (ed.) (2003) *If this should be farewell: a family separated by war: the journal and letters of Ernest and Mary Hodgkin, 1942–1945*, Fremantle Arts Centre Press, Fremantle, WA.

Wood, W. E. (1924) 'Increase in salt in soil and streams following destruction of the native vegetation', *Journal of the Royal Society of Western Australia*, **10**, 35–47.

Wyrwoll, K. H. (2003) 'The geomorphology of the wider Perth region' in Bayne, F. (ed) *The Engineering Geology of Perth, Part 1*, special publication of the Australian Geomechanics Society, vol. **38** (3) 17–32.

Yassini, I. & Kendrick, G. W. (1988) 'Middle Holocene ostracodes, foraminiferas and environments of beds at Point Waylen, Swan River Estuary, southwestern Australia', *Alcheringa*, **12**, 107–21.

Young, G. C. & Potter, I. C. (2002) 'Influence of exceptionally high salinities, marked variations in freshwater discharge and opening of estuary mouth on the characteristics of the ichthyofauna of a normally-closed estuary', *Estuarine, Coastal and Shelf Science*, **55**, 223–46.

Young, G. C. & Potter, I. C. (2003) 'Influence of an artificial channel on the ichthyofauna of a large estuary', *Marine Biology*, 1181–94.

Young, G. C., Potter, I. C., Hyndes, G. A. & de Lestang, S. (1997) 'The ichthyofauna of an intermittently open estuary: implications of bar breaching and low salinities on faunal composition', *Estuarine, Coastal and Shelf Science*, **45**, 53–68.

2 The Swan–Canning Estuary

Allender, B. M. (1981) 'The distribution of benthic macroflora in the Swan River Estuary', Western Australia, *Journal of the Royal Society of Western Australia*, **4**, 17–22.

Anderson, M. G., Peters, N. E. & Walling, D. (2001) 'Integrating research and management for an urban estuarine system: The Swan–Canning Estuary, Western Australia', *Hydrological Processes Special Issue:* **15**, 1–2741.

Astill, H. & Lavery, P. S. (2001) 'The dynamics of unattached benthic macroalgal accumulations in the Swan–Canning Estuary', *Hydrological Processes Special Issue*, **15**, 2387–99.

Backhouse, J. (1843) *A Narrative of a Visit to the Australian Colonies*, Illustrated by three maps, fifteen etching, and several wood-cuts, Hamilton Adams, and Co., London.

Balla, S. (1994) *Wetlands of the Swan Coastal Plain*, vol. 1, Their Nature and Management, Water Authority of Western Australia, & the Western Australian Department of Environmental Protection.

Bamford, M. (1999) 'Western Australian Bird Notes 89', Royal Australasian Ornithologist Union, WA Group.

Bamford, M. J., Bamford, A. R. & Bancroft, W. (2004) 'Swan Estuary Marine Park: Human Usage, Waterbirds and Disturbance Study 2002–2003', CALM, WA.

Beeton, M. (1936) *Mrs Beeton's Everyday Cookery*, Ward, Lock & Co. Ltd, London, Melbourne.

British Parliamentary Papers (1829) Copies of the correspondence of the Colonial Department with certain Gentlemen proposing to form a new Settlement in the Neighbourhood of the Swan River, in Western Australia, 1829. British Parliamentary Papers, Correspondence relating to a proposed Settlement at Swan River 1829 (23*) vol. XXIV.

Calder, M. E. (1977) *Early Swan River Colony*, Rigby, Melbourne, Perth.

Cameron, J. M. R. (1981) *Ambitions Fire: The Agricultural Colonization of Pre-convict Western Australia*, UWA Press, WA.

Cardon, F. G. (1968) *Along the Canning: A History of the Shire of Canning District, Western Australia*, Shire of Canning, Cannington, WA.

Chalmer, P. N., Hodgkin, E. P. & Kendrick, G. W. (1976) 'Benthic faunal changes in a seasonal estuary of south-western Australia', *Records of the Western Australian Museum*, **4**, 383–410.

Connell, E. L. & Walker, D. I. (2001) 'Nutrient cycling associated with the seagrass *Halophila ovalis* in the Swan–Canning Estuary based on seasonal variations in biomass and tissue nutrients', *Hydrological Processes Special Issue*, **15**, 2401–9.

Department of Conservation and Environment (1979) 'Report by the Environmental Protection Authority on a dredging proposal at Pelican Rocks, Swan River Estuary', *Bulletin* 77.

Department of Conservation and Land Management (1997) 'Canning River Regional Park, Management Plan 1997–2007', Management Plan 36.

Department of Conservation and Land Management (1999) 'Swan Estuary Marine Park and Adjacent Nature Reserves Management Plan 1999–2009', Management Plan 41.

Donohue, R., Davidson, A., Peters, N. E., Nelson, S. & Jakowyna, B. (2001) 'Trends in total phosphorus and total nitrogen concentrations of tributaries to the Swan–Canning Estuary, 1987–1998', *Hydrological Processes Special Issue*, **15**, 2411–34.

Finucane, K. J. & Forman, F. G. (1929) 'Observations on the load carried by the Swan River during the 1926 flood', *Journal of the Royal Society of Western Australia*, **15**, 57–61.

Frazer, C. (1827) Report inserted by James Stirling into the Admiralty document and appended to the report to Governor Darling, Proposals by Captain James Stirling for a Settlement at Swan River, *Historical Records of Australia*, series III, vol. VI, 18 April 1827, Despatches and papers relating to the settlement of the states, Frederick Watson (ed.) Sydney, Library Committee of the Commonwealth Parliament, 1921–1923, 581–2.

Gaughan, D. J., Neira, F. J., Beckley, L. E. & Potter, I. C. (1990) 'Composition, seasonality and distribution of the Icthyoplankton in the lower Swan Estuary, South-western Australia', *Australian Journal of Marine and Freshwater Research*, **41**, 529–43.

Gregory, J. (2003) *City of Light: A History of Perth since the 1950s*, City of Perth, WA.

Griffin, S. L. & Rippingdale, R. J. (2001) 'Zooplankton grazing dynamics: top-down control of phytoplankton and its relationship to an estuarine habitat', *Hydrological Processes Special Issue*, **15**, 2453–64.

Hamilton, D. P., Chan, T., Robb, M. S., Pattiaratchi, C. B. & Herzfeld, M. (2001) 'The hydrology of the upper Swan River Estuary with focus on an artificial destratification trial', *Hydrological Processes Special Issue*, **15**, 2465–80.

Hamilton, D. P. & Turner, J. V. (2001) 'Integrating research and management for an urban estuarine system: the Swan–Canning Estuary, Western Australia', *Hydrological Processes Special Issue*, **15**, 2383–5.

Hamilton, D. P. & Turner, J. V. (eds) (2001) *Hydrological Processes Special Issue*, **15**, 2383–714.

Hasluck, A. & Lukis, M. (1977) *Victorian and Edwardian Perth from Old Photographs*, John Ferguson, Sydney.

Hass, C. G. & Knott, B. (1998) 'Sphaeromatid isopods from the Swan River, Western Australia: Diversity, distribution and geographic sources', *Crustaceana*, **71**, 36–46.

Hay, J. G. (1906) 'The visit of Mr Charles Fraser, Colonial Botanist of New South Wales, with Captain Stirling, in HMS *Success*, to the Swan River in 1827, with his report on the Botany, Soil, and Capabilities of the Locality', *Journal of the Western Australian Historical Society*, **2**, 16–35. A booklet of this paper with photographs was privately published by Hay, with an account by A. H. Gilbert, Clerk of HMS *Success*, 'An account of the expedition of HMS *Success*, Captain James Stirling, RN, from Sydney, to the Swan River, in 1827'.

Hillman, K. (1987) 'The production ecology of the seagrass *Halophila ovalis* in the Swan River Estuary', in *The Swan River Estuary Ecology and Management*, ed. J. John, Curtin University of Technology, Environmental Studies Group, Report no. 1, 102–16.

Hillman, K., McComb, A. J. & Walker, D. I. (1995) 'The distribution, biomass and primary production of the seagrass *Halophila ovalis* in the Swan–Canning Estuary, Western Australia', *Aquatic Botany*, **51**, 1–54.

Hodgkin. E. P. (1987) 'The hydrology of the Swan River estuary: Salinity the ecological master factor', in *The Swan River Estuary Ecology and Management*, J. John (ed.), Curtin University, Environmental Studies Group, Report no. 1, 34–44.

Hodgkin, E. P. & Vicker, E. (1987) 'A history of algal pollution in the estuary of the Swan River', in *The Swan River Estuary Ecology and Management*, J. John (ed.), Curtin University of Technology, Environmental Studies Group, Report no. 1, 65–70.

Horner Rosser, S. M. J. & Thompson, P. A. (2001) 'Phytoplankton of the Swan–Canning Estuary: A comparison of nitrogen uptake by different bloom assemblages', *Hydrological Processes Special Issue*, **15**, 2579–94.

Imberger, J. (1976) 'Dynamics of a longitudinally stratified estuary', *Coastal Engineering Hawaii*, 3108–23.

Jaensch, R. (1987) 'Waterbirds in the Swan–Canning Estuary', in *The Swan River Estuary Ecology and Management*, J. John (ed.), Curtin University of Technology, Environmental Studies Group, Report no. 1, 202–12.

John, J. (1987a) 'Phytoplankton of the Swan River Estuary', in *The Swan River Estuary Ecology and Management*, J. John (ed.), Curtin University, Environmental Studies Group, Report no. 1, 71–90.

John, J. (ed.) (1987b) *The Swan River Estuary Ecology and Management*, Curtin University of Technology, Environmental Studies Group, Report no. 1.

Jones, D. S. (1987) 'Preliminary investigations on the Barnacles of the Swan–Canning Estuary', in *The Swan River Estuary Ecology and Management*, J. John (ed.), Curtin University, Environmental Studies Group, Report no. 1, 153–62.

Kanandjembo, A. N., Platell, M. E. & Potter, I. C. (2001) 'The benthic macroinvertebrate community of the upper reaches of an Australian estuary that undergoes marked seasonal changes in hydrology', *Hydrological Processes Special Issue*, **15**, 2481–501.

Kanandjembo, A. N., Potter, I. C. & Platell, M. E. (2001) 'Abrupt shifts in the fish community of the hydrologically variable upper estuary of the Swan River', *Hydrological Processes Special Issue*, **15**, 2503–17.

Lenanton, R. C. J. (1978) 'Fish and exploited crustaceans of the Swan–Canning Estuary', Department of Fisheries and Wildlife Western Australia, *Report No. 35*.

Linderfelt, W. R. & Turner, J. V. (2001) 'Interaction between shallow groundwater, saline surface water and nutrient discharge in a seasonal estuary: the Swan–Canning system', *Hydrological Processes Special Issue*, **15**, 2631–53.

Masini, R. J. (1987) 'Preliminary studies on the productivity of benthic microalgae in the Swan River Estuary', in *The Swan River Estuary Ecology and Management*, J. John (ed.), Curtin University, Environmental Studies Group, Report no. 1, 91–101.

Masini, R. J. & McComb, A. J. (2001) 'Production by microphytobenthos in the Swan–Canning Estuary', *Hydrological Processes Special Issue*, **15**, 2519–35.

Pen, L. (1987) 'Peripheral vegetation of the Swan–Canning Estuary', in *The Swan River Estuary Ecology and Management*, J. John (ed.), Curtin University, Environmental Studies Group, Report no. 1, 221–31.

Pen, L. J. (1983) 'Peripheral vegetation of the Swan and Canning Estuaries', Department of Conservation and Environment, *Bulletin no. 113*.

Pennifold, M. & Davis, J. (2001) 'Macrofauna and nutrient cycling in the Swan River Estuary, Western Australia: experimental results', *Hydrological Processes Special Issue*, **15**, 2537–53.

Peter, N. E. & Donohue, R. (2001) 'Nutrient transport to the Swan–Canning Estuary, Western Australia', *Hydrological Processes Special Issue*, **15**, 2555–77.

Riggert, T. L. E. (1978) 'The Swan River Estuary: development, management and preservation', Frank Daniels Pty Ltd, Perth, WA, Swan River Conservation Board 1958–1976.

Rippingale, R. (1987) 'Zooplankton of the Swan River Estuary', in *The Swan River Estuary Ecology and Management*, J. John (ed.), Curtin University, Environmental Studies Group, Report no. 1, 117–26.

Rippingale, R. J. & Kelly, S. H. (1995) 'Reproduction and survival of *Phyllorhiza punctata* (Cnidaria: Rhizostomae) in a seasonally fluctuating salinity regime in Western Australia', *Marine and Freshwater Research*, **46**, 1145–51.

Sassoon, J. (2003) 'The common cormorant or shag lays eggs inside a paper bag: A cultural ecology of fish-eating birds in Western Australia', *Environment and History*, **9**, 31–52.

Seddon, G. (1970) *Swan River Landscapes*, UWA Press, WA.

Seddon, G. (1972) *Sense of Place: A Response to an Environment—the Swan Coastal Plain, Western Australia*, UWA Press, WA.

Seddon, G. (1995) *Swan Song Reflections on Perth and Western Australia 1956–1995*, Centre for Studies in Australian Literature, The University of Western Australia.

Seddon, G. & Ravine, D. (1986) *A City and its Setting: Images of Perth, Western Australia*, Fremantle Arts Centre Press.

Smith, A. J. & Turner, J. V. (2001) 'Density-dependent surface water–groundwater interaction and nutrient discharge in the Swan–Canning Estuary', *Hydrological Processes Special Issue*, **15**, 2595–616.

Stannage, C. T. (1979) *The People of Perth: A Social History of Western Australia's Capital City*, Perth City Council, Perth, WA.

Statham-Drew, P. (2003) *James Stirling: Admiral and Founding Governor of Western Australia*, UWA Press, WA.

Stephens, R. & Imberger, J. (1996) 'Dynamics of the Swan River Estuary: The seasonal variability', *Marine and Freshwater Research*, **47**, 517–29.

Stirling, J. (1827) Narrative of Operations enclosed with a letter from Captain James Stirling to Governor Darling, Proposals by Captain James Stirling for a Settlement at Swan River, *Historical Records of Australia*, series III, vol. VI, 18 April 1827.

Swan River Trust (2002) Riverview September, Swan River Trust Newsletter no. 29.

Swan River Trust (1999) *Swan–Canning Cleanup Program*, Action Plan, Taskforce chair, D. Lord.

Thompson, P. A. & Hosja, W. (1996) 'Nutrient limitation of phytoplankton in the upper Swan River estuary, Western Australia', *Marine and Freshwater Research*, **47**, 659–67.

Thurlow, B. H., Chambers, J. & Klemm, V. (1986) 'Swan–Canning Estuarine System: Environment, use and the future', Waterways Commission Report no 9.

Twomey, L. & John, J. (2001) 'Effects of rainfall and salt-wedge movement on phytoplankton succession in the Swan–Canning Estuary, Western Australia', *Hydrological Processes Special Issue*, **15**, 2655–69.

Viney, N. R. & Sivapalan, M. (2001) 'Modelling catchment processes in the Swan–Avon river basin', *Hydrological Processes Special Issue*, **15**, 2671–85.

Wallace, J. (1977) 'The macrobenthic invertebrate fauna of Pelican Rocks, March–April 1977', Department of Conservation and Environment, Public Works Department, *Bulletin no. 35*.

Water and Rivers Commission (2000a) 1998–1999 'Canning River Oxygenation trial', *River Science*, **14**.

Water and Rivers Commission (2000b) 'Bringing back the Swans: The potential to encourage more Black Swans onto the Swan River', ATA Environmental, Bamford Consulting Ecologists, M. P. Rogers & Associates, Report no. 2000/63.

Water and Rivers Commission (2000c) 'Summer Surprise, The Swan River blue-green algal bloom', *River Science*, **2**.

Water and Rivers Commission (2001a) 'Aquatic plants in the Canning River', *River Science*, **19**.

Water and Rivers Commission (2001b) 'Phosphorus in the Canning 1999–2000 Phoslock™ trials', *River Science*, **17**.

Water and Rivers Commission (2002a) 'Seasonal nutrient dynamics in the Swan Estuary 1995–2000', *River Science* **8**.

Water and Rivers Commission (2002b) 'Yenyening Lakes Management Strategy 2002–2012', Prepared for Water and Rivers Commission, & Department of Conservation & Land Management in consultation with Yenyening Lakes Management Committee, *WRM Report no. 32*.

West Australian Newspapers Ltd (1979) *Swan River Colony Life in Western Australia since the Early Colonial Settlement*, illustrated from an exhibition mounted by West Australian Newspapers Ltd as a contribution to celebrations for the State's 150th year, Joll, T., Edmonds, J. & Tresham, A. (eds).

Wilson, B. R. (1967) 'A comparative account of the reproductive cycles of five species of marine mussels (Bivalvia: Mytilidae) in the vicinity of Fremantle, Western Australia', *Australian Journal of Marine and Freshwater Research*, **18**, 175–203.

Wilson, B. R. (1968) 'Survival and reproduction of the mussel *Xenostrobus securis* (Lamarck) (Mollusca: Bivalvia: Mytilidae) in a Western Australian estuary, Part I Salinity tolerance', *Journal of Natural History*, **2**, 307–28.

Wilson, B. R. (1969) 'Survival and reproduction of the mussel *Xenostrobus securis* (Lamarck) (Mollusca: Bivalvia: Mytilidae) in a Western Australian estuary, Part II Reproduction, growth and longevity', *Journal of Natural History*, **3**, 93–120.

Wilson, B. R. & Kendrick, G. W. (1968) 'The recent appearance of *Notospisula trigonella* (Lamarck) (Mollusca: Bivalvia: Mactridae) in the Swan estuary', *Journal of the Malacological Society of Australia*, **1**, 25–31.

3 The Peel–Harvey Estuarine System

Angell, K. (1999) 'Fertilisers for the pastures on sandy soils of the Swan Coastal Plain', *Agriculture Western Australia Bulletin 4375*, 1–11.

Bradby, K. (1997) 'Peel Harvey: The Decline and Restoration of an Ecosystem', Greening the Catchment Taskforce, 1–223.

Brown, R. G., Treloar, J. M. & Clifton, P. M. (1980) 'Sediments and organic detritus in the Peel–Harvey estuarine system', Report to Department of Conservation and Environment, WA, *Bulletin 90*.

Burgess, J. (1988) *Mandurah: Water under the Bridge*, South-West Printing & Publishing for the Town of Mandurah, WA.

Coate, K. H. & Jonstone, R. E. (2003) 'A failed breeding attempt by Banded Stilt in the Eastern Goldfields of Western Australia', *The Western Australian Naturalist*, 79–86.

D. A. Lord & Associates Pty Ltd (1998) 'Dawesville Channel Monitoring Programme Technical Review', Prepared for the Water and Rivers Commission, WRT 28.

Department of Agriculture (1993) 'Use of bauxite residue in the Peel–Harvey coastal plain catchment', Public Environmental Review.

Department of Conservation and Environment (1981) 'The Peel–Harvey Estuarine System Study', Report by the Estuarine and Marine Advisory Committee to the Environmental Protection Authority, March, *Bulletin 88*.

Department of Conservation and Environment (1983a) 'Peel–Harvey Estuarine System Study: Symposium, Prospects for Management', *Bulletin 136*.

Department of Conservation and Environment (1983b) 'Peel–Harvey Estuary Progress', *Bulletin 146*.

Department of Conservation and Environment (1984) 'Management of the Peel Inlet and Harvey Estuary: Report of research findings and options for management', *Bulletin 170*.

Department of Planning and Urban Development Western Australia (1994) 'Peel Regional Strategy: A strategic policy statement and land use plan'.

Donohue, R. B., Nelson, S., Wittenoom, P. & Bowyer, L. (1998) 'Phosphorus and nitrogen in tributary inflows to the Peel–Harvey estuary: a statistical analysis of trends in ambient concentration', Draft report of the Water and Rivers Commission.

Environmental Protection Authority (1988) 'Peel Inlet and Harvey Estuary Management Strategy: Environmental Review and Management Programme Stage 2', *EPA Bulletin 363 (parts 1 & 2)*, Perth.

Environmental Protection Authority (1994) 'Review of Peel Inlet–Harvey Estuary Management Strategy, Stage 2: Environmental Conditions', *EPA Bulletin 749*, Perth.

Environmental Protection Authority (2003) 'Peel Inlet and Harvey Estuary System Management Strategy: Progress and Compliance by the Proponents with the Environmental Conditions set by the Minister for the Environment in 1989, 1991 and 1993', Advice of the Environmental Protection Authority, *EPA Bulletin 1087*, Perth.

Gerritse, R. G., Wallbrink, P. J. & Murray, A. S. (1998) 'Accumulation of phosphorus and heavy metals on the Peel–Harvey Estuary in Western Australia: results of a preliminary study', *Estuarine, Coastal and Shelf Science*, **47**, 679–93.

Gordon, D. M., Birch, P. B. & McComb, A. J. (1980) 'The effect of light, temperature and salinity on photosynthetic rates of an estuarine *Cladophora*', *Botanica Marina*, **23**, 749–55.

Gordon, D. M. & McComb, A. J. (1989) 'Growth and production of the green alga *Cladophora montagneana* in a eutrophic Australian estuary and its interpretation using a computer model', *Water Research*, **23**, 633–45.

Hale, J. & Paling, E. I. (1999a) 'Sediment nutrient concentrations of the Peel–Harvey Estuary: Comparisons before and after the opening of the Dawesville Channel', Institute for Environmental Science, Murdoch Report No. MAFRA 99/6.

Hale, J. & Paling, E. I. (1999b) 'Water quality of the Peel–Harvey Estuary: Comparisons before and after the opening of the Dawesville Channel (July 1985 to June 1999)', Institute for Environmental Science, Murdoch Report No. MAFRA 99/4.

Hill, N. A., Lukatelich, R. J. & McComb, A. J. (1991) 'A comparative study of some of the physical and chemical characteristics of the sediments from three estuarine systems in south western Australia', Report to the Waterways Commission, *Report no. 23*.

Hodgkin, E. P., Birch, F. B., Black, R. E. & Hillman, K. (1985) The Peel–Harvey Estuarine System: proposals for management, Department of Conservation and Environment, WA, *Report no. 14*.

Hodgkin, E. P., Birch, F. B., Black, R. E. & Humphries, R. B. (1980) 'The Peel–Harvey Estuarine System Study 1976–1980', A report to the Estuarine and Marine Advisory Committee, 1–72.

Hodgkin E. P. & Yeates, J. S. (1993) 'Fertilizers and eutrophication in South-Western Australia', Hodgkin E. P. & Yeates, J. S. (guest eds) *Fertilizer Research Special Issue*, **36**, 95–192.

Humphries, R. B. & Croft, C. M. (1984) 'Management of the eutrophication of the Peel–Harvey Estuarine System', Report 3, Final report, Department of Conservation and Environment, *Bulletin 165*.

Jakowyna, B. N. (2000) 'Nutrients in tributary inflows to the Peel–Harvey Estuarine System', Western Australia status and trend, Water and Rivers Commission, Water Resource Technical Series, *Report No. WRT 23*.

Jasinska, E. M., Lindsay, M., Wright, A. & Smith, D. (1997) 'Changes in mosquito populations and arbovirus activity following construction of a channel to increase tidal flushing in an estuarine system in south-western Australia', *Arbovirus Research in Australia*, **7**, 116–22.

Kobryn, H., Glasson, R., Segal, R., Hale, J. & Paling, E. I. (2000) 'Assessment of the Current and Historical Extent of Shore Erosion in the Peel–Harvey Estuary', Institute for Environmental Science, *Report No. MAFRA 00/9*.

Lane, J. A. K., Clarke, A. G. & Pearson, G. B. (2002a) Waterbirds of the Peel–Harvey Estuary in 1996–1997, Department of Conservation and Land Management, unpublished report.

Lane, J. A. K., Clarke, A. G. & Pearson, G. B. (2002b) 'Waterbirds of Peel–Harvey Estuary in 1998–1999', Department of Conservation and Land Management.

Lane, J. A. K. & Pearson, G. B. (2002) 'Waterbirds of Peel–Harvey estuary in the mid 1970s', Department of Conservation and Land Management.

Lane, J. A. K., Pearson, G. B. & Clarke, A. G. (1997) 'Waterbird use of Peel–Harvey Estuary following opening of the Dawesville Channel in April 1994', Department of Conservation and Land Management.

Lavery, P., Bootle, S. & Vanderklift, M. (1999) 'Ecological effects of macroalgal harvesting on beaches in the Peel–Harvey Estuary, Western Australia', *Estuarine, Coastal and Shelf Science*, **49**, 295–309.

Lavery, P., Hillman, K., Bastyan, G., McComb, A. J. & Paling, E. I. (1995) Changes in macrophyte abundance and composition in the Peel–Harvey Estuarine System from 1978 to 1994, Report prepared for the Waterways Commission, Perth, by the Marine and

Freshwater Research Laboratory, Murdoch Report no. MAFRA 95/11.

Lenanton, R. C. J., Loneragan, N. R. & Potter, I. C. (1985) 'Blue-green algal blooms and the commercial fishery in a large Australian estuary', *Marine Pollution Bulletin*, **16**, 477–82.

Lindsay, M. D., Johansen, C. A., Broom, K., D'Ercole, M., Wallace, M., Wright, A. E., Codon, D. & Mackenzie, J. S. (1992) 'The epidemiology of outbreaks of Ross River virus infection in Western Australia in 1991–1992', *Arbovirus Research Australia*, **6**, 72–6.

Lukatelich, R. J. & McComb, A. J. (1986) 'Nutrient levels and the development of diatom and blue-green algal blooms in a shallow Australian estuary', *Journal of Plankton Research*, **8**, 597–618.

McComb, A. J. (1982) 'The effect of land use in catchments on aquatic systems: a case study from Western Australia', *Australian Society of Limnology Bulletin*, **9**, 1–19.

McComb, A. J. (ed.) (1995) *Eutrophic Shallow Estuaries and Lagoons*, CRC Press, Boca Raton.

McComb, A. J. & Lukatelich, R. J. (1995) 'The Peel–Harvey Estuarine System, Western Australia', in *Eutrophic Shallow Estuaries and Lagoons*, A. J. McComb, (ed.) CRC Press, Boca Raton, 5–17.

McComb, A. J., Atkins, R. P., Birch, P. B., Gordon, D. M. & Lukatelich, R. J. (1981) 'Eutrophication in the Peel–Harvey Estuarine System, Western Australia', *Estuaries and Nutrients*, B. J. Neilson & L. E. Cronin (eds) Humana Press, Clifton, New Jersey, 323–42.

McComb, A. J., Kobryn, H. T. & Latchford, J. A. (1995) 'Samphire marshes of the Peel–Harvey estuarine system, Western Australia', Peel Preservation Group, & Murdoch University.

McComb, A. J., Qiu, S., Lukatelich, R. J. & McAuliffe, T. F. (1998) 'Spatial and temporal heterogeneity of sediment phosphorus in the Peel–Harvey Estuarine System', *Estuarine, Coastal and Shelf Science*, **47**, 1–17.

Malseed, B. E. & Sumner, N. R. (2001a) 'A 12-month survey of recreational fishing in the Peel–Harvey Estuary of Western Australia 1998–99', *Fisheries Research Report no. 127*.

Wells, F. E. & Threlfall, T. J. (1982) 'Reproductive strategies of *Hydrococcus brazieri* (Tenison Woods, 1876) and *Arthritica semen* (Menke, 1843) in Peel Inlet, Western Australia', Journal of the Malacological Society of Australia, **5**, 157–166.

Wells, F. E., Threlfall, T. J. & Wilson, B. R. (1980) 'Aspects of the biology of molluscs in the Peel–Harvey estuarine system, Western Australia', Department of Conservation and Environment, *Bulletin 97*.

Whisson, C. S., Wells, F. E. & Rose, T. (2004) 'The benthic invertebrate fauna of the Peel–Harvey estuarine system in south-western Australia after completion of the Dawesville Channel', *Records of the Western Australian Museum*, **22**, 81–90.

Wilson, C., Hale, J. & Paling, E. I. (1999) 'Macrophyte abundance, distribution and composition of the Peel–Harvey Estuary: Comparisons before and after the opening of the Dawesville Channel (July 1985 to June 1999)', Institute for Environmental Science Report no. MAFRA 99/5.

4 Leschenault Inlet and Estuary and smaller estuaries of the Geographe Bay area

Barker, A. J. & Laurie, M. (1992) *Excellent Connections: A History of Bunbury, WA, 1836–1990*, City of Bunbury.

Bunbury, W. St Pierre and Morrell, W. P. (eds) (1930) *Early Days in Western Australia: Being the Letters and Journal of Lieut. H. W. Bunbury, 21st Fusiliers*, Oxford University Press, London, Humphrey Milford.

Chalmer, P. N. & Scott, J. K. (1984) 'Fish and benthic faunal survey of the Leschenault and Peel–Harvey estuarine systems of south-western Australia in December 1984', Department of Conservation and Environment, *Bulletin 149*.

Chester, E. T. & Klemm, V. V. (1990) 'Draft Integrated Mosquito Control Strategy for the Leschenault Estuary Region, Western Australia', Report to the Waterways Commission, no. 21.

Clark, G. E. (1946) *The Early History of Bunbury*, Colortype Press, Perth.

Clay, B. T. & Weston, A. S. (1997) 'Toby Inlet and associated wetlands draft management plan', Prepared for the Sussex Land Conservation District Committee, Funded by the Gordon Reid Foundation for Conservation.

Comer, S. & Clay, B. T. (1999) 'Toby Inlet and associated wetlands management plan', Toby Inlet Catchment Group Sub-group, Sussex Land Conservation District Committee.

Cresswell, I. D. (2000) 'Ecological significance of freshwater seeps along the western shore of the Leschenault Inlet estuary', *Journal of the Royal Society of Western Australia*, **83**, 285–92.

Cullity, T. B. (1992) 'Vasse: An account of the disappearance of Thomas Timothee Vasse', Private publication.

Davis, G. (1988) 'The biology of the Blue Manna crab (*Portunus pelagicus*) in the estuaries of South-west Western Australia, *Waterways Information No.1*.

Department of Conservation and Land Management (1995) 'Leschenault Peninsula Draft Management Plan 1995'.

Department of Conservation and Land Management (1998) 'Leschenault Peninsula Management Plan 1998–2008', Management Plan no. 38.

Department of Planning and Urban Development Western Australia in conjunction with the City of Bunbury (1993) 'Bunbury Coastal Plan'.

Hass, C. G. & Knott, B. (2000) 'Sphaeromatid isopods (Crustacea: Isopoda) from the Leschenault Estuary, Collie River and Bunbury Harbour', *Journal of the Royal Society of Western Australia*, **83**, 459–62.

Hillman, K., McComb, A. J., Bastyan, G. & Paling, E. (2000) 'Macrophyte abundance and distribution in Leschenault Inlet, an estuarine system in south-western Australia', *Journal of the Royal Society of Western Australia*, **83**, 349–56.

Hodgkin, E. P. & Smith, G. G. (1971) 'Leschenault Inlet: aspects of conservation', SWANS Dept, Fish and Fauna Western Australia, 54–6.

Hutchings, P. & Saenger, P. (1987) *Ecology of Mangroves*, University of Queensland Press, Brisbane.

Klemm, V. V. (1989) 'Organochlorine pesticide residues in the Preston River Western Australia, December 1985 to May 1986', Report to the Leschenault Inlet Management Authority, Waterways Commission Report no. 12.

Lane, J. A. K. (1990) 'Swamped with birds', *Landscope*, **5**, 17–22.

Lane, J. A. K., Hardcastle, K. A., Tregonning, R. J. & Holtfreter, G. J. (1997) 'Management of the Vasse–Wonnerup wetland system in relation to sudden, mass fish deaths', Report to the Vasse Estuary Technical Working Group.

Laporte Effluent Disposal Committee (1982) 'Summary report on Laporte Factory Effluent Disposal', Public Works Department, Western Australian Government.

Lear, R. & Turner, T. (1977) *Mangroves of Australia*, University of Queensland Press, Brisbane.

LeProvost, Semeniuk & Chalmer (1983) 'Effects of discharges of acid-iron effluent from production of titanium dioxide on the abundance of benthic biota of Leschenault Inlet', Western Australian Waterways Commission, Perth, no. 4.

Leschenault Inlet Management Authority (n.d.) 'Leschenault Inlet: Man and the environment', Leschenault Inlet Management Authority, Bunbury.

Leschenault Inlet Management Authority & Shire of Harvey (1995) 'Draft Management Plan', Waterways Commission Report No. 60 (for the eastern shore north of Australind).

Lukatelich, R. J. (1989) 'Leschenault Inlet: Macrophyte abundance and distribution', Waterways Commission, *Report No. 15*, Perth.

McAlpine, K. W., Spice, J. F. & Humphries, B. (1989) 'The environmental condition of the Vasse Wonnerup wetland system, and a discussion of management options', Environmental Protection Authority, Perth, *Technical Series no. 31*.

McComb, A. J., Qiu, S., Paling, E. I. & Hill, N. A. (2000) 'Sediments of Leschenault Inlet: a comparison with other estuaries in south-western Australia', *Journal of the Royal Society of Western Australia*, **83**, 275–84.

McMahon, K., Young, E., Montgomery, S., Cosgrove, J., Wilshaw, J. & Walker, D. I. (1997) 'Status of a shallow seagrass system, Geographe Bay, south-western Australia', *Journal of the Royal Society of Western Australia*, **80**, 255–62.

Malseed, B. E., Sumner, N. R. & Williamson, P. C. (2000) 'A 12-month survey of recreational fishing in the Leschenault Estuary of Western Australia 1998', *Fisheries Research Report no. 120*.

Meagher, T. D. (1970) 'Crab Research Program—Leschenault Inlet, Bunbury', *Fins*, **3**, 18–26.

Morrissy, N. M. (1970) 'Leschenault Inlet Development and Conservation', *Swans*, **2**, 12–18.

Morrissy, N. M. (1972) 'Conservation and biology of the white mangrove (*Avicennia marina*) on Leschenault Estuary near Bunbury', *Swans*, **3**, 66–9.

MP Rogers & Associates Pty Ltd, Coastal & Port Engineers (1999) 'Toby Inlet Ocean Entrance Management Study', produced for the Geographe Catchment Council, Shire of Busselton & Sussex Land Conservation District Committee.

Pen, L. (1992a) 'Fringing vegetation of the Leschenault Estuary 1941–1991', Report to the Leschenault Inlet Management Authority, Waterways Commission Report no. 31.

Pen, L. (1992b) 'Fringing vegetation of the lower Collie and Brunswick Rivers 1992', Waterways Commission Report no. 37.

Pen, L., Semeniuk, V. & Semeniuk, C. A. (2000) 'Peripheral wetland habitats and vegetation of the Leschenault Inlet estuary', *Journal of the Royal Society of Western Australia*, **83**, 293–316.

Potter, I. C., Chalmer, P. N., Tiivel, D. J., Steckis, R. A., Platell, M. E. & Lenanton, R. C. J. (2000) 'The fish fauna and finfish fishery of the Leschenault Estuary in south-western Australia', *Journal of the Royal Society of Western Australia*, **83**, 481–501.

Semeniuk, V., Kenneally, K. F. & Wilson, P. G. (1978) *Mangroves in Western Australia, Handbook No. 12*, Western Australian Naturalists' Club, Perth.

Semeniuk, V. & Meagher, T. D. (1981) 'The geomorphology and surface processes of the Australind–Leschenault Inlet coastal area', *Journal of the Royal Society of Western Australia*, **64**, 33–51.

Semeniuk, V., Milne, V. & Waterhouse, H. (2000) 'Contemporary and sub-recent pollen record for the Leschenault Inlet estuary', *Journal of the Royal Society of Western Australia*, **83**, 335–48.

Semeniuk, V., Semeniuk, T. A. & Unno, J. (2000) 'The Leschenault Inlet estuary: an overview', *Journal of the Royal Society of Western Australia*, **83**, 207–28.

Semeniuk, V., Tauss, C. & Unno, J. (2000) 'The white mangrove *Avicennia marina* in the Leschenaut Inlet area', *Journal of the Royal Society of Western Australia*, **83**, 317–33.

Semeniuk, V. & Withers, P. E. (eds) (2000) 'Leschenault Inlet Estuary Symposium, December 2001', *Journal of the Royal Society of Western Australia, Leschenault Inlet Special Issue*, **83**, 205–512.

Trudgen, M. E. (1984) 'The Leschenault Peninsula—A flora and vegetation survey with an analysis of its conservation value and appropriate uses', Department of Conservation and Environment, Perth, *Bulletin 157*.

Water and Rivers Commission (2001a) 'Water quality in the Leschenault Estuary and catchment 1995–1999', *South West Inflo*.

Water and Rivers Commission (2001b) 'Water quality in the Vasse–Wonnerup Estuarine System and Lower Vasse River 1996–2000', *South West Inflo*.

Western Australian Planning Commission, 'Busselton Wetlands Conservation Strategy for Public Comment', April 2003.

Wright, E., Derrington, C. & Barrett, L. (1995) Leschenault Inlet Management Authority & Shire of Harvey, Northern Leschenault Estuary Foreshore Draft Management Plan, Report no. 60, Waterways Commission.

5 Estuaries north of the Swan

Alan Tingay & Associates (ATA Environmental) (1996) 'Mid West Iron & Steel Project: Geraldton Steel Plant', public environmental review, Environmental Protection Authority, Western Australia, Report no. 94/65.

Alan Tingay and Associates (ATA Environmental) (1997) 'Oakajee Deepwater Port', public environmental review, Report to the Minister for Resources Development, Report no. 96/93.

Alan Tingay and Associates (ATA Environmental) (2000a) 'Consultative Environmental Review: new Kalbarri Airport, Shire of Northampton', Report to Western Australia, Environmental Protection Authority.

Alan Tingay & Associates (ATA Environmental) (2000b) 'Turquoise Coast Development, Jurien Bay', Ardross Estates Pty Ltd Environmental Report, final report to the Environmental Protection Authority, Western Australia, no. 99/07.

Arakel, A. V. (1979) Quaternary sedimentation in Hutt and Leeman Lagoons, Western Australia, PhD thesis, The University of Western Australia.

Bain, M. A. (1975) *Ancient Landmarks: A Social and Economic History of the Victoria District of Western Australia 1839–1894*, UWA Press, WA.

Battye, J. S. (1923) 'Development of the Pastoral Industry in Western Australia', Reprinted from *Elder's Weekly*, 1–41.

Bligh, K. J. (1987) 'Water erosion in the Geraldton area during June–July 1986', Technical Report No. 59, DAWA.

Borowitzka, L. J. & Borowitzka, M. A. (1990) 'Commercial production of β-carotene by *Dunaliella salina* in open ponds', *Bulletin of Marine Science*, **47**, 244–52.

Borowitzka, M. A. (1986) 'Micro-algae as sources of fine chemicals', *Microbiological Sciences*, **3**, 372–5.

Bott, G. & Maling, C. (1989) 'Moore River Estuary: Water Quality and Catchment Status, Soil and Land Uses in the Gingin Region', Australian Society of Soil Science WA, Branch Field Trip.

Brooker, J. (2001) 'Chapman Regional Wildlife Corridor Project Implementation Plan Draft'.

Burgh, W. J. de (1976) *Neergabby (where the brook and river meet): A History of the Moore River and Lower Gingin Brook 1830–1960*, Shire of Gingin.

Chalmers, C. E. (1983) 'Coastal Management Plan Kalbarri Townsite', Department of Conservation and Environment, *Bulletin 145*.

Clayton, D. M. & Elliott, J. C. (1985) 'Draft Coastal Management Plan Shire of Greenough', Report to Western Australian Department of Conservation and Environment, *Bulletin no. 189*.

Cousins, M. D. (2000) 'Hydrodynamics of the Moore River Estuary, Water Quality Sampling', *Annual Report (December 1998 – December 1999)*, Summary report to the Community, WRC.

Cousins, M. & Tracey, D. (2003a) 'How healthy is the Moore River Estuary?', Water quality study, pamphlet, Water and Rivers Commission.

Cousins, M. D. (2003) 'Water quality and hydrodynamics of the Moore River Estuary and surrounds: December 1998 – February 2002', Synthesis Report to the Community, WRC.

Cousins, M. D. & Tracey, D. P. (2003b) 'Water quality and hydrodynamics of the Moore River estuary and surrounds, March–November 2002', Final Report to the community, Department of Environment.

Cousins, M. D. & Tracey, D. P. (2003c) 'Water quality and hydrodynamics of the Moore River estuary and surrounds, March–November 2002', Synthesis Report to the community, Department of Environment.

Curry, P. J., Payne, A. L., Leighton, K. A., Hennig, P. & Blood, D. A. (1994) 'An inventory and condition survey of the Murchison river catchment, Western Australia', *Department of Agriculture Technical Bulletin No. 84.*

D. A. Lord & Associates Pty Ltd & Jesz Fleming & Associates Pty Ltd (2001) 'Dredging and dredge material management plan for maintenance dredging at mouth of Murchison River, Kalbarri', Report for the Department of Transport, Report no. 01/219/1.

Department of Conservation and Land Management (2000) 'Indicative Management Plan for the Proposed Jurien Bay Marine Park', Marine Parks and Reserves Authority.

Environmental Protection Authority (1991) 'Rural Subdivision for Horticulture Lot 2 Gingin Brook Road, Gingin', *EPA Bulletin 514*, Perth.

Environmental Protection Authority (1997) Oakajee Deepwater Port—Concept, Shire of Chapman Valley, Department of Resources Development, Environmental Protection Authority, *Bulletin 849*, Perth.

Feldtmann, F. R. (1963) 'Some Pelecypods from the Cretaceous Gingin Chalk, Western Australia, together with descriptions of the Principal Chalk Exposures', *Journal of the Royal Society of Western Australia*, **46**, 101–23.

Geddes, M. C. (1981) 'The brine shrimps *Artemia* and *Parartemia*: comparative physiology and distribution in Australia', *Hydrobiologia*, **81**, 169–79.

Gibbs, M. (1988) *The Greenough Flats: A Bibliography of Historical Sources*, Centre for Prehistory, The University of Western Australia.

Gopurenko, D., Hughes, J. M. & Bellchambers, L. (2003) 'Colonisation of the south-west Australian coastline by mud crabs: evidence for a recent range expansion or human-induced translocation?', *Marine and Freshwater Research*, **54**, 833–40.

Gorter, J. D., Nicoll, R. S. & Foster, C. B. (1994) 'Lower palaeozoic facies in the Carnarvon Basin, Western Australia: Stratigraphy and hydrocarbon prospectivity', in Purcell, P. G. & Purcell, R. R. (eds) *The Sedimentary Basins of Western Australia: Proceedings of Petroleum Exploration Society of Australia Symposium, Perth, Western Australia*, Petroleum Exploration Society of Australia Limited, Perth, pp. 373–96.

Grant, B. (2000) Zooplankton assemblage of the Moore River Estuary, Honours thesis, Curtin University of Technology, WA.

Gregory, A. C. & Gregory, F. T. (1884) *Journals of Australian Explorations*, Government Printer, James Beal, Brisbane, Australiana Facsimile (1969) no. 14, Libraries Board of South Australia.

Grey, S. G. (1841) *Expeditions in Western Australia 1837–1839*, Journals of two expedition of discovery in the north-west and Western Australia, during the years 1837, 38 and 39, under the authority of Her Majesty's Government, vols 1 & 2, T. & W. Bonne, London.

Griffin, E. A. & Keighery, B. J. (1989) 'Moore River to Jurien Sandplain Survey', Western Australian Wildflower Society Inc., 1–90.

Grimes, S. L. (1997) The influence of European settlement and activities on the sedimentology, hydrology and floodplain characteristics of the Lower Irwin River, Western Australia, Honours Thesis, Department of Geography, The University of Western Australia.

Grimes, S. L. & Eliot, I. G. (1998) *A History of Change: The mouth of the Moore River, WA*, Geography Department, UWA, for the Mid West Avon Region of the Water and Rivers Commission Western Australia, 1–56 & 97 plates.

Gutteridge, Haskins & Davey Pty Ltd (2001) 'Chapman Regional Wildlife Corridor Vegetation Quality Assessment and Analysis'.

Hocking, R. M. (1991) 'The Silurian Tumblagooda Sandstone, Western Australia', Geological Survey Western Australia, *Report 27.*

McCotter, M. (1993) Chapman Regional Park Management Plan, unpublished report for the City of Geraldton, Western Australia.

McMillan, R. P. & Foulds, W. (1982) 'Greenough Landscape Protection Area, Western Australia', Report on a biological survey of Greenough Front Flats, Claremont Teachers College, Report for the National Trust of Australia (Western Australia).

McNamara, K. & Trewin, N. H. (1993) 'Footprints on the sands of time', *Landscope*, **8**, 45–8.

Mann, A. W. & Lintern, M. (1983) 'Heavy metal dispersion patterns from tailing dumps, Northampton District, Western Australia', *Environmental Pollution Series B*, **6**, 33–49.

Rippingale, R. J. (ed.) (1988) 'The Moore River Estuary', Report prepared for Mappin Butcher Consultants, School of Biology, Curtin University, WA.

Schmidt, P. W. & Hamilton, P. J. (1990) 'Palaeomagnetism and the age of the Tumblagooda Sandstone, Western Australia', *Australian Journal of Earth Sciences*, **37**, 381–5.

Stafford R. A. (1989) *Scientist of the Empire: Sir Roderick Murchison, Scientific Exploration and Victorian Imperialism*, Cambridge University Press.

Stelfox, L. (2001) 'Assessment of potential groundwater contamination from the Moore River', Water and Rivers Commission, Hydrology Report no. 189.

Treadgold, C. (1941) 'The Greenough Flood—A disaster of the eighties', *Early Days Journal and Proceedings of the Western Australian Historical Society*, **3**, 19–20.

Trewin, N. H. & McNamara, K. J. (1994) 'Arthropods invade the land: trace fossils and palaeoenvironments of the Tumblagooda Sandstone (?late Silurian) of Kalbarri, Western Australia', *Transactions of the Royal Society of Edinburgh: Earth Sciences*, **85**, 177–210.

Water and Rivers Commission (1997a) 'Moore River Flood Management Plan, Bidaminna', *Water Resources Management Series WRM 7.*

Water and Rivers Commission (1997b) 'The state of the northern rivers: A report designed to inform the community of the state of Western Australia's rivers in the Indian Ocean, Timor Sea and Western Plateau drainage divisions', *Water Resource Allocation and Planning Series WRAP* **10**.

Water and Rivers Commission (2000) 'Flooding in Western Australia', *Water Facts*, **13**.

Water and Rivers Commission (2001a) 'Chapman River Foreshore Assessment', *Water Resource Management Report, WRM* **23**.

Water and Rivers Commission (2001b) 'Greenough River Foreshore Assessment', *Water Resource Management Report, WRM* **24**.

Wilson, B. & Eliot, I. (2000) Sediments of the Moore River Estuary, Project for the Water and Rivers Commission of Western Australia.

Wilson, C. C. (1991) Dissolved organic matter and phytoplankton growth in aquatic systems, Honours thesis, Murdoch University.

Wyrwoll, K. H. (1984) 'The sedimentology, stratigraphy and palaeoenvironmental significance of a late Pleistocene alluvial fill: central coastal areas of Western Australia', *Catena*, **11**, 201–18.

Wyrwoll, K. H. (1988) 'Determining the causes of Pleistocene stream-aggradation in the central coastal areas of Western Australia', *Catena*, **15**, 329–51.

6 Hardy Inlet and estuaries of the Naturaliste–Leeuwin Ridge

Congdon, R. A. (1981) 'Zonation in the marsh vegetation of the Blackwood River Estuary in south-western Australia', *Australian Journal of Ecology*, **6**, 267–78.

Congdon, R. A. & McComb, A. J. (1980) 'Nutrient pools of an estuarine ecosystem—the Blackwood River Estuary in south-western Australia', *Journal of Ecology*, **68**, 287–313.

Congdon, R. A. & McComb, A. J. (1981) 'The vegetation of the Blackwood River Estuary, south-west Australia', *Journal of Ecology*, **69**, 1–16.

Cresswell, G. J. (1990) *The Light of the Leeuwin: The Augusta–Margaret River Shore History*, The Augusta–Margaret River Shire History Group.

Department of Fisheries Western Australia (2004) 'The Hardy Inlet Estuarine Fishery Management Issues and Options', Fisheries Management Paper no. 169.

Hailes, G. (1844) Letter from George Hailes to Captain James Mangles, 5 June 1844, Battye Library 479A II Letter–Books, 1837–1844 Entitled 'Letters &c., Botany, Gardening, Seeds, Horticulture, the Parks', p. 222.

Hardcastle, K. & Cousins, M. (2000) 'The Hardy Inlet: Water quality sampling', *Annual Report (January 1999 – January 2000)*, Water and Rivers Commission.

Hasluck, A. (1955) *Portrait with Background: A Life of Georgiana Molloy*, Oxford University Press, Melbourne, Victoria.

Hodgkin, E. P. (1976) 'The history of two coastal lagoons at Augusta, Western Australia', *Journal of the Royal Society of Western Australia*, **59**, 39–45.

Hodgkin, E. P. (1978) 'An environmental study of the Blackwood River estuary, Western Australia 1974–1975', *Report No. 1*, Department of Conservation and Environment.

Imberger, J., Agnew, H. J., Billings, J. & Wallace, D. F. (1976) 'The dynamics of the Blackwood River System', part of the environmental study of the Blackwood River Estuary, report to the Estuarine and Marine Advisory Committee of the Environmental Protection Authority.

Kuo, J. & Kirkman, H. (1995) '*Halophila decipiens* Ostenfeld in estuaries of southwestern Australia', *Aquatic Botany*, **51**, 335–40.

Lenanton, R. C. J. (1977) 'Aspects of the ecology of fish and commercial crustaceans of the Blackwood River estuary Western Australia', *Fisheries Research Bulletin No. 19*.

Lines, W. J. (1994) *An All-consuming Passion: Origins, Modernity, and the Australian Life of Georgiana Molloy*, Allen & Unwin, St. Leonards, NSW.

Molloy, G. (1832) Letter from Georgiana Molloy to her sister Elizabeth, 7 November 1832, Battye Library 501A/1, Manuscript 678.

Molloy, G. (1838) Letter from Georgiana Molloy to Captain James Mangles, January 1838, Battye Library 479A 879. Letter–Books, 1837–1844 Entitled 'Letters &c., Botany, Gardening, Seeds, Horticulture, the Parks', p. 305.

Turner, T. (1956) *Turners of Augusta*, Paterson Brokensha, Perth, WA.

7 Estuaries of the wet Western South Coast

Bannister, C. T. (1831) A Report of Captain Bannister's Journey to King George's Sound Overland, February 1831, *Exploration Diaries*, Battye Library, **1**, 96–106.

Bellanger, B. J. A. (1980) *Champagne and Tingle Trees: A Memoir of the Bellanger Family Who First Settled the Nornalup District, Western Australia*, Apollo Press, Nedlands, WA, 1–88.

Clark, W. N. (1841) Journal of an Expedition to Nornalup, or the Deep River of the Sealers, in the months of March and April, 1841, By W. M. Nairne Clark, *The Inquirer: A Western Australian Journal of Politics and Literature*, August 25, cont. September 1, 1841, vols 56, 57.

Crawford, P. and Crawford, I. (2003) *Contested Country: History of the Northcliffe Area, Western Australia*, UWA Press, WA.

Department of Conservation and Land Management (1991b) 'South Coast Regional Management Plan 1999–2002', Management Plan 24.

Department of Conservation and Land Management (1992) 'Walpole–Nornalup National Park Management Plan 1992–2002', Management Plan 22.

Fernie, L. & Fernie, G. (1989) *In Praise of a National Park: The Origins and History of the Walpole–Nornalup National Park*, L. & G. Fernie, Walpole, WA.

Hassell, G. W. (1962) Estuarine sedimentation on the south coast of Western Australia with particular reference to Nornalup Inlet, PhD thesis, The University of Western Australia.

Hillman, A. (1833) 'Transcript of the Journal of an Expedition to Nornalup undertaken by Mr. Alfred Hillman, 3rd July 1833, from the settlement at Albany–K. G. Sound', *Exploration Diaries*, Battye Library, 1834–1835, vol. 2, 269–77.

Hodgkin, E. P. & Clark, R. (1988b revised 1999) 'Nornalup and Walpole Inlets and the estuaries of the Deep and Frankland Rivers, An inventory of information on the estuaries and coastal lagoons of south Western Australia', *Estuarine Studies Series no. 2*, Environmental Protection Authority, WA.

Hodgkin, E. P. & Clark, R. (1988c) 'Wilson, Irwin and Parry Inlets, Estuaries of the Denmark Shire, An inventory of information on the estuaries and coastal lagoons of south Western Australia', *Estuarine Studies Series no. 3*, Environmental Protection Authority, WA.

Hodgkin, E. P. & Clark, R. (1989a) 'Broke Inlet and other estuaries of the Shire of Manjimup, An inventory of information on the estuaries and coastal lagoons of south Western Australia', *Estuarine Studies Series no. 6*, Environmental Protection Authority, WA.

Moran, C. J. (1930) 'Dairying at Nornalup', The *West Australian*, 17 October 1930.

Mueller, F. von (1882) 'General information respecting the present condition of the forests and timber trade of the southern part of the colony: together with a Report on the forest resources on the colony' compiled by direction William C. F. Robinson, Government Printer, Perth, WA.

Neira, F. J. & Potter, I. C. (1994) 'The larval fish assemblage of the Nornalup–Walpole estuary, a permanently open estuary of the southern coast of Western Australia', *Australian Journal of Marine and Freshwater Research*, **45**, 1193–207.

Potter, I. C., Hilliard, R. W. & Neira, F. J. (1986) 'The biology of the Australian lamprey', in *Limnology in Australia*, P. De Dekker & W. D. Williams (eds), CSIRO & Dr W. Junk, Dordrecht, Netherlands.

Potter, I. C. & Hyndes, G. A. (1994) 'Composition of the fish fauna of a permanently open estuary on the south coast of Australia, and comparisons with a nearby seasonally closed estuary', *Marine Biology*, **121**, 199–209.

Wilson, E. H. (1920) 'Our forest wealth', *The West Australian*, 16 November 1920.

8 Wilson Inlet

APACE Green Skills & Pen, L. (1995) 'The condition of the Denmark and Hay River foreshores 1995', Report prepared for the Wilson Inlet Management Authority, Waterways Commission **60**.

Bastyan, G., Latchford, J. & Paling, E. I. (1995) 'Macrophyte distribution and sediment chemistry in Wilson Inlet', Report to the Waterways Commission, Institute for Environmental Science, Murdoch Report no. MAFRA 95/8.

Boughton, N. (1999) 'Wilson Inlet Nutrient Export to the Ocean in 1995, 1996 & 1997', Report prepared for the Water and Rivers Commission, Western Australia.

Boughton, N. (2000) 'Wilson Inlet Nutrient Export to the Ocean in 1998 & 1999', Report prepared for the Water and Rivers Commission, Western Australia.

Carruthers, T. J. B. (1998) Ecology of *Ruppia megacarpa* Mason in Wilson Inlet: a perspective on survival in a highly variable aquatic environment, PhD thesis, The University of Western Australia.

Carruthers, T. J. B., Walker, D. I. & Kendrick, G. A. (1999) 'Abundance of *Ruppia megacarpa* Mason in a seasonally variable estuary', *Estuarine, Coastal and Shelf Science*, **48**, 497–509.

Carruthers, T. J. B., Wilshaw, J. & Walker, D. I. (1997) 'Ecology of *Ruppia megacarpa* and its Epiphytes in Wilson Inlet: The Influence of Physical Factors', Report to the WRC, Western Australia.

Cliff, M. & Lenanton, R. (1993) 'The case of the burrowing cobbler', *Western Australia's Journal of Fishing and Aquatic Environment*, Western Fisheries, Spring 1993.

Congdon, R. A. & McComb, A. J. (1979) 'Productivity of *Ruppia*: seasonal changes and dependence on light in an Australian estuary', *Aquatic Botany*, **6**, 121–32.

Donohue, R. B., Jakowyna, B. N. & Nelson, S. W. (1999) 'Nitrogen and phosphorus in tributary inflows to the Wilson Inlet, Western Australia', Water and Rivers Commission Report WRT 21.

Dudley, B. J. (2002) The role of benthic vegetation as a nutrient buffer in Wilson Inlet, PhD thesis, The University of Western Australia.

Dudley, B. J., Gahnstrom, A. M. E. & Walker, D. I. (2001) 'The role of benthic vegetation as a sink for elevated inputs of ammonium and nitrate in a mesotrophic estuary', *Marine Ecology Progress Series*, **219**, 99–107.

Forbes, V. (2004) The role of benthic vegetation in the ecology of south coast estuaries in Western Australia, PhD thesis, The University of Western Australia.

Hodgkin, E. P. & Clark, R. (1988c) 'Wilson, Irwin and Parry Inlets, Estuaries of the Denmark Shire, An inventory of information on the estuaries and coastal lagoons of south Western Australia', *Estuarine Studies Series no. 3*, Environmental Protection Authority, WA.

Humphries, P., Potter, I. C. & Loneragan, N. R. (1992) 'The fish community in the shallows of a temperate Australian estuary: relationships with the aquatic macrophyte *Ruppia megacarpa* and environmental variables', *Estuarine, Coastal and Shelf Science*, **34**, 325–46.

Humphries, R. B., Robertson, J. G. M. & Roberston, F. E. (1982) 'A resource inventory and management information system for Wilson Inlet, Western Australia', Department of Conservation and Environment, *Bulletin 132*.

Hyndes, G. A., Neira, F. J. & Potter, I. C. (1992) 'Reproductive biology and early life history of the marine teleost *Platycephalus speculator* (Platycephalidae) in a temperate Australian estuary', *Journal of Fish Biology*, **40**, 859–74.

Kalnejais, L. & Robb, M. (1999) 'Water quality in Wilson Inlet 1995–1997, A report on the monitoring data collected between 1995 and 1997', Water and Rivers Commission, Water Resources Technical Series Report WRT 14.

Laurenson, L. J. B., Neira, F. J. & Potter, I. C. (1993) 'Reproductive biology and larval morphology of the marine plotosid *Cnidoglanis macrocephalus* (Teleostei) in a seasonally closed Australian estuary', *Hydrobiologia*, **268**, 179–92.

Lenanton, R. C. J. (1974) 'Wilson Inlet: A seasonally closed Western Australian South Coast Estuary', Department of Fisheries and Fauna Western Australia, *Report No. 14*.

Lukatelich, R. J. & Schofield, M. A. J. (1984) 'The nutrient status of Wilson Inlet 1982–1983', Department of Conservation and Environment Western Australia, *Bulletin 159*.

Lukatelich, R. J., Schofield, N. J. & McComb, A. J. (1987) 'Nutrient loading and macrophyte growth in Wilson Inlet, a bar-built Southwestern Australian estuary', *Estuarine, Coastal and Shelf Science*, **24**, 141–66.

Mumford, R. W. (1996) 'Denmark Western Australia: A history to 1905, Exploration and early settlement', Denmark Historical Society.

Platell, M. E. & Potter, I. C. (1996) 'Influence of water depth, season, habitat and estuary location on the macrobenthic fauna of a seasonally closed estuary', *Journal of Marine Biological Association of the United Kingdom*, **76**, 1–21.

Potter, I. C. & Hyndes, G. A. (1999) 'Characteristics of the ichthyofaunas of southwestern Australian estuaries, including comparisons with holarctic estuaries and estuaries elsewhere in temperate Australia: a review', *Australian Journal of Ecology*, **24**, 395–421.

Twomey, L. & Thompson, P. (2001) 'Nutrient limitation of phytoplankton in a seasonally open bar-built estuary: Wilson Inlet, Western Australia', *Journal of Phycology*, **37**, 16–29.

Volkman, J. K., Holdsworth, D. G. & Revill, A. T. (1998) 'Lipid biomarker and isotope studies of sediments and seagrass from Wilson Inlet, Western Australia', CSIRO Division of Marine Research, Hobart.

Walker, D. I., Dudley, B., Carruthers, T. B. J. & Gahnstrom, A. (2001) 'Nutrient cycling by *Ruppia megacarpa* and epiphytes in Wilson Inlet', Report to the National Eutrophication Management Program, Department of Botany, The University of Western Australia.

Water and Rivers Commission (1998) 'Summary of Wilson Inlet Studies from 1994–1997', Report to the Community no. 1.

Water and Rivers Commission (1999) 'Summary of the Estuarine Monitoring Programme conducted in Wilson Inlet 1995 to 1998', Report to the community no. 2.

Water and Rivers Commission (2000) 'Summary of Wilson Inlet Catchment Monitoring Program', Report to the community no. 3.

Water and Rivers Commission (2002a) 'Water quality in Wilson Inlet from 1995 to 2002', Report to the Community no. 5.

Water and Rivers Commission (2002b) 'Wilson Inlet: Managing the bar and the Inlet', Report to the Community no. 6.

Water and Rivers Commission (2003) '*Ruppia* in Wilson Inlet, Wilson Inlet', Report to the Community no. 7.

Wilshaw, J. (1996) Epiphytic algal assemblages on *Ruppia megacarpa* in Wilson Inlet, Honours thesis, The University of Western Australia.

9 Albany Harbours and the smaller estuaries of the Albany region

Adeney, J. A. (2001) 'Cyanobacterial issues in the Lake Powell / Torbay Inlet Drainage System', CSIRO Australia Land and Water, Floreat, W.A., prepared for the Water and Rivers Commission, CSIRO Technical Report 44/01.

Albany Harbours Planning Committee (1999) Albany Harbours Planning Strategy Guidelines, 1–92.

Bastyan, G. M., Deeley, D. M., White, K. S. & Paling, E. I. (1996) 'Seagrass and Macroalgal Distribution in Princess Royal and Oyster Harbours, Albany, 1996 Distribution and Comparisons with Previous Surveys', Report to the WRC, Perth, WA, Institute for Environmental Sciences (Report no. MAFRA 96/4) Murdoch University, 1–38.

Bastyan G. R. (1986) 'Distribution of seagrasses in Princess Royal Harbour and Oyster Harbour, on the southern coast of Western Australia', Department of Conservation and Environment, Technical Series 1.

Cambridge, M. L., Bastyan, G. R. & Walker, D. I. (2002) 'Recovery of *Posidonia* meadows in Oyster Harbour, Southwestern Australia', *Bulletin of Marine Science*, **71**, 1279–89.

Coupland, G. & Crawley, K. (1997) 'Seagrass Regeneration in Oyster Harbour, Albany: Mechanisms for Regrowth', Report to the Water and Rivers Commission; Botany Department, UWA.

D. A. Lord & Associates Pty Ltd (1997) 'Review of algal harvesting operation in Princess Royal Harbour', Albany Technical Report 97/041/1, Water and Rivers Commission.

Environmental Protection Authority (1988) 'What's happening in our harbours? Princess Royal and Oyster Harbours–Albany', *EPA Bulletin 341*, Perth.

Environmental Protection Authority (1990a) 'Albany Harbours Environmental Study 1988–1989', A Report to the Environmental Protection Authority from the Technical Advisory Group, *EPA Bulletin 412*, Perth.

Environmental Protection Authority (1990b) 'Recommendations of the Environmental Protection Authority in relation to the environmental problems of the Albany Harbours', *Bulletin 442*, Perth.

Francesconi, K. A., Lenanton, R. C. J., Caputi, N. & Jones, J. (1997) 'Long-term study of mercury concentrations in fish following cessation of a mercury-containing discharge', *Marine Environmental Research*, **43**, 27–40.

Hillman, K., Lukatelich, R. J., Bastyan, G. & McComb, A. J. (1991) 'Distribution and biomass of seagrass and algae, and nutrient pools in water, sediments and plants in Princess Royal Harbour and Oyster Harbour', Report to the Environmental Protection Authority, Technical series no. 40.

Hodgkin, E. P. & Clark, R. (1990a) 'Estuaries of the Shire of Albany, An inventory of information on the estuaries and coastal lagoons of south Western Australia', *Estuarine Study Series no. 8*, Environmental Protection Authority, WA.

Hutchings, P. A., Wells, F. E., Walker, D. I. & Kendrick, G. A. (1991) 'Seagrass, sediment and infauna—a comparison of *Posidonia australis*, *Posidonia sinuosa* and *Amphibolis antarctica* in Princess Royal Harbour, South-Western Australia, II Distribution, abundance and composition of macrofauna', in Wells, F. E., Walker, D. I., Kirkman, H. & Lethbridge, R. (eds) *The Marine Flora and Fauna of Albany, Western Australia*, vol. 2, Western Australian Museum, Perth, 611–33.

LeProvost, Dames and Moore (1997) 'Albany Shoreline Survey April 1997', Report for the Fisheries Department of Western Australia.

McKenzie, K. G. (1962) 'A record of foraminfera from Oyster Harbour, near Albany', *Journal of the Royal Society of Western Australia*, **45**, 117–32.

McKenzie, K. G. (1964) 'The ecological associations of an ostracod fauna from Oyster Harbour, a marginal marine environment near Albany, Western Australia', *Publication Stazione Zoologica, Napoli Supplement*, **33**, 1–41.

Masini, R. J., Cary, J. L., Simpson, C. J. & McComb, A. J. (1990) 'Effects of light and temperature on the photosynthesis of seagrasses, epiphytes and macroalgae and implications for management of the Albany Harbours', Environmental Protection Authority, Technical Series no. 32.

Peterson, C. H., Irlandi, E. A. & Black, R. (1994) 'The crash in suspension-feeding bivalve populations (*Katelysia* spp) in Princess Royal Harbour—An unexpected consequence of eutrophication', *Journal of Experimental Marine Biology and Ecology*, **176**, 39–52.

Phillips, J. & Lavery, P. (1997) 'Waychinicup Estuary, Western Australia: the influence of freshwater inputs on the benthic flora and fauna', *Journal of the Royal Society of Western Australia*, **80**, 63–72.

Roberts, D. & Wells, F. E. (1980) 'The marine and estuarine molluscs of the Albany area of Western Australia', *Records of the Western Australian Museum*, **8**, 335–57.

Sellick, D. R. G. (1997) *First Impressions: Albany Traveller Tales 1791–1901*, Western Australian Museum, Perth.

Simpson, C. J. & Masini, R. J. (1990) 'Albany Harbours Environmental Study (1988–1989)', Environmental Protection Authority, *Bulletin 412*.

Smit, A. J. & Walker, D. I. (1999) 'Seagrass and macroalgal distribution in Princess Royal Harbour, Albany', Report to the Water and Rivers Commission, Western Australia.

Walker, D. I., Hutchings, P. A. & Wells, F. E. (1991) 'Seagrass, sediment and infauna—a comparison of *Posidonia australis*, *Posidonia sinuosa* and *Amphibolis antarctica* in Princess Royal Harbour, South-Western Australia, I Seagrass biomass, productivity and contribution to sediments', in Wells, F. E., Walker, D. I., Kirkman, H., & Lethbridge, R. (eds) *The Marine Flora and Fauna of Albany, Western Australia*, vol. 2, Western Australian Museum, Perth, 597–610.

Water and Rivers Commission (Albany Waterways Management Authority) (2000) 'Woolstores to Frenchman Bay Foreshore Management Plan', *Water Resource Management Series WRM* **9**.

Weaver, D. M., Pen, L. J. & Reed, A. E. G. (1994) 'Modifying the phosphorus cycle to achieve management objectives in the Oyster Harbour Catchment', *Water*, 28–32.

Weaver, D. M. & Prout, A. L. (1993) 'Changing farm practice to meet environmental objectives of nutrient loss to Oyster Harbour', *Fertiliser Research*, **36**, 177–84.

Wells, F. E., Walker, D. I. & Hutchings, P. A. (1991) 'Seagrass, sediment and infauna—a comparison of *Posidonia australis*, *Posidonia sinuosa* and *Amphibolis antarctica* in Princess Royal Harbour, South-Western Australia, III Consequences of seagrass loss', in Wells, F. E., Walker, D. I., Kirkman, H., & Lethbridge, R. (eds) *The Marine Flora and Fauna of Albany, Western Australia*, vol. 2, Western Australian Museum, Perth, 635–40.

Western Australia Government Chemical Laboratories (1978) 'Water quality monitoring investigation Princess Royal Harbour, Albany, April 1977 – January 1978', report by the Government Chemical Laboratories for the Albany Waterways Management Advisory Committee.

Woodward, H. P. (1917) 'Investigations into the cause of the mineralisation of the "Seven-Mile" Swamp at Grassmere, near Albany, South-West Division', Geological Survey, WA, Miscellaneous Report 64, Bull. 74, 49–57.

10 Estuaries to the east of Albany

Abbott, I. (1979) 'The past and present distribution of Sea Lions and Fur Seals in Western Australia', *Records of the Western Australian Museum*, **7**, 375–90.

Bignell, M. (1977) *The Fruit of the Country: A History of Gnowangerup, Western Australia*, UWA Press, WA.

Bradby, K. (1989) 'A Park in Perspective: A Report on the Past, Present and Future of Fitzgerald River National Park', Report commissioned by the Fitzgerald River National Park Association & paid for by a grant from the Western Australian Heritage Committee, 1–136.

Chapman, A. & Newbey, K. R. (1995) 'A biological survey of the Fitzgerald area, Western Australia', *CALM Science Supplement* **3**, 1–258.

Craig, G., Burkin, B. & Van Steveninck, A. (1984) 'Ravensthorpe District Draft Coastal Management Plan', Department of Conservation and Environment, WA, *Bulletin 152*.

Craig, G. F. (1999) 'Oldfield Estuary Vegetation and Flora Survey', Report prepared for WRC South Coast Region.

Craig, G. F. & Oma, V. (1984) 'Esperance East Coast Planning and Management Report', Department of Conservation and Environment, WA, *Bulletin 173*.

Department of Conservation and Land Management (1991a) 'Fitzgerald River National Park Management Plan 1991–2001', Management Plan no. 15.

Department of Conservation and Land Management (1999) 'Esperance Lakes Nature Reserves Management Plan 1999–2009'.

Eyre, E. J. (1845) *Journal of Discovery into Central Australia and Overland from Adelaide to King George Sound in the years 1840–41. Sent by the colonists of South Australia, with the Sanction and Support of the Government: Including an Account of the Manners and Customs of the Aborigines and the State of their Relations with Europeans*, Vol. II, T. W. Boone, 29 New Bond St, London.

Fitzgerald Biosphere Project (1989) The bush comes to the city, *Papers from a seminar at Murdoch University, 25 Sept. 1987*, Environmental Protection Authority, WA.

Forest, R. & Crow, S. (1996) *Yarra-mo-up: place of the tall yate trees*, Commonwealth of Australia, AGPS, Canberra.

Gregory, A. C. (1849) Assistant Surveyor A. C. Gregory during an expedition in April 1849 to the Fitzgerald River to trace the Coal discovered by the Surveyor General towards the coast, to ascertain the most eligible place for shipment, the facilities for conveying Coal to the coast, the Survey of the Doubtful Island Bay and to procure half a ton of the best specimens of Coal, *Exploration Diaries*, vol. 4, Battye Library.

Hassell, E. (1975) *My Dusky Friends*, C. W. Hassell, General Printing, Fremantle, WA.

Heald, D. I. (1984) 'Amateur net fishing survey of two Western Australian south coast estuaries in January 1981', Department of Fisheries and Wildlife Western Australia, Report no. 60.

Hodgkin, E. P. (1997) 'History and management of Culham Inlet, a coastal salt lake in south-western Australia', *Journal of the Royal Society of Western Australia*, **80**, 239–47.

Hodgkin, E. P. & Clark, R. (1987) 'Wellstead Estuary, The estuary of the Bremer River, An inventory of information on the estuaries and coastal lagoons of south Western Australia', *Estuarine Studies Series no. 1*, Environmental Protection Authority, WA.

Hodgkin, E. P. & Clark, R. (1988a) 'Beaufort Inlet and Gordon Inlet, Estuaries of the Jerramungup Shire, An inventory of information on the estuaries and coastal lagoons of south Western Australia', *Estuarine Studies Series no. 4*, Environmental Protection Authority, WA.

Hodgkin, E. P. & Clark, R. (1989b) 'Stokes Inlet and other estuaries of the Shire of Esperance, An inventory of information on the estuaries and coastal lagoons of south Western Australia', *Estuarine Studies Series no. 5*, Environmental Protection Authority, WA.

Hodgkin, E. P. & Clark, R. (1990b) 'Estuaries of the Shire of Ravensthorpe and the Fitzgerald River National Park, An inventory of information on the estuaries and coastal lagoons of south Western Australia', *Estuarine Studies Series no. 7*, Environmental Protection Authority, WA.

Jerramungup Landcare Centre (2000) 'Stories of the Bremer River and Wellstead Estuary', Compiled by J. Cappelluti.

Kalnejais, L. & Hosja, W. (1998) 'Water quality update for four south coast inlets', Summary of current monitoring data collected by the Water and Rivers Commission, October 1997 – March 1998.

Kohn, A. J. & Blahm, A. M. (in press) 'Anthropogenic enhancement of marine invertebrate diversity and abundance: Intertidal infaunal invertebrates along an exposure gradient at Esperance, Western Australia', in Kendrick, G. E., Wells, F. E. & Walker, D. I. (eds) *The Marine Flora and Fauna of Esperance, Western Australia*, Western Australian Museum, Perth.

McNamara, K. J. (1992) 'A guide to the fossils of the Albany region', Western Australian Museum, Perth.

Murray-Wallace, C. V., Beu, A. G., Kendrick, G. W., Brown, L. J., Belperio, A. P. & Sherwood, J. E. (2000) 'Palaeoclimatic implications of the occurrence of the arcoid bivalve *Anadara trapezia* (Deshayes) in the Quaternary of Australasia', *Quaternary Science Reviews*, **19**, 559–90.

Parker, K. (2002) 'Stories of the Oldfield', Oldfield Catchment Group.

Shire of Jerramungup (1995) 'Jerramungup Coastal Management Plan', Amalgamation of the draft Doubtful Islands Area Coastal Plan, and the draft Beaufort Inlet to Wellstead Estuary Coastal Plan.

Tomlinson, K. (2000a) '1999 Draft, Bremer River Foreshore Condition and Action Plan 2000', prepared for the Jerramungup Land Conservation District Committee, funded by Water and Rivers Commission, & the Natural Heritage Trust.

Tomlinson, K. (2000b) 'Fitzgerald River Foreshore Condition and Action Plan 2000', prepared for the Jerramungup Land Conservation District Committee, funded by Water and Rivers Commission, & the Natural Heritage Trust.

Tomlinson K. & Parker, K. (2001) 'Bremer River and Devils Creek River Action Plan', Water and Rivers Commission, *Water Resources Management Series WRM 25*.

List of Figures

Chapter 3

Chapter 4

Chapter 5

Chapter 6

Chapter 7

Chapter 8

Chapter 9

Chapter 10

List of Graphs

Chapter 1

Chapter 2

Chapter 3

Chapter 4

Chapter 5

Chapter 6

List of Maps

Chapter 2

Chapter 3

Chapter 4

Chapter 5

List of Tables

Chapter 2

Chapter 3

Chapter 4

Chapter 5

Chapter 6

Chapter 7

Chapter 8

Chapter 9

Index

Page numbers in **bold** refer to illustrations.
* denotes alien or introduced species.

COMMON ESTUARINE PLANTS AND ANIMALS

Approximate sizes in mm.

Plants
- A Rush - *Juncus kraussii*
- B Samphire - *Sarcocornia*spp.
- C Paperbark tree - *Melaleuca cuticularis*
- D Seagrass - *Ruppia megacarpa*
- E Diatoms 0.01

F Tubeworms - *Ficopomatos enigmaticus* 20

Bivalve molluscs
- G Estuarine mussel - *Xenostrobus securis* 30
- H Edible mussel - *Mytilus edulis* 100
- I *Arthritica semen* 3
- J *Sanginolaria biradiata* 50
- K Cockle - *Katelysia* 3 spp. 40
- L *Spisula trigonella* 20

Gastropod molluscs
- M Snail - *Hydrococcus brazieri* 4

Crustacea
- N Amphipod - *Corophium minor* 15
- O Shrimp - *Palaemonetes australis* 40
- P Copepod - *Gladioferens imparipes* 2
- Q King Prawn - *Penaeus latisulcatus* 100

R Foraminifera 0.02

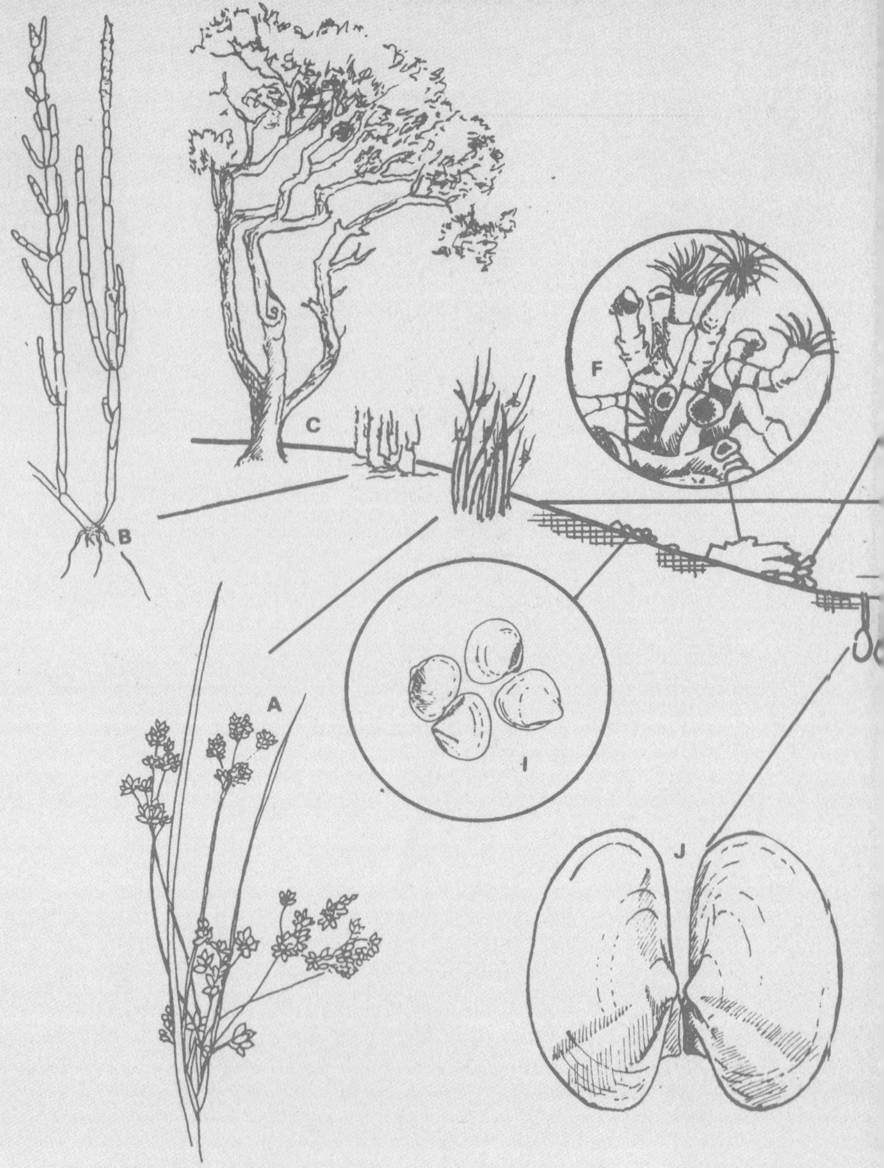

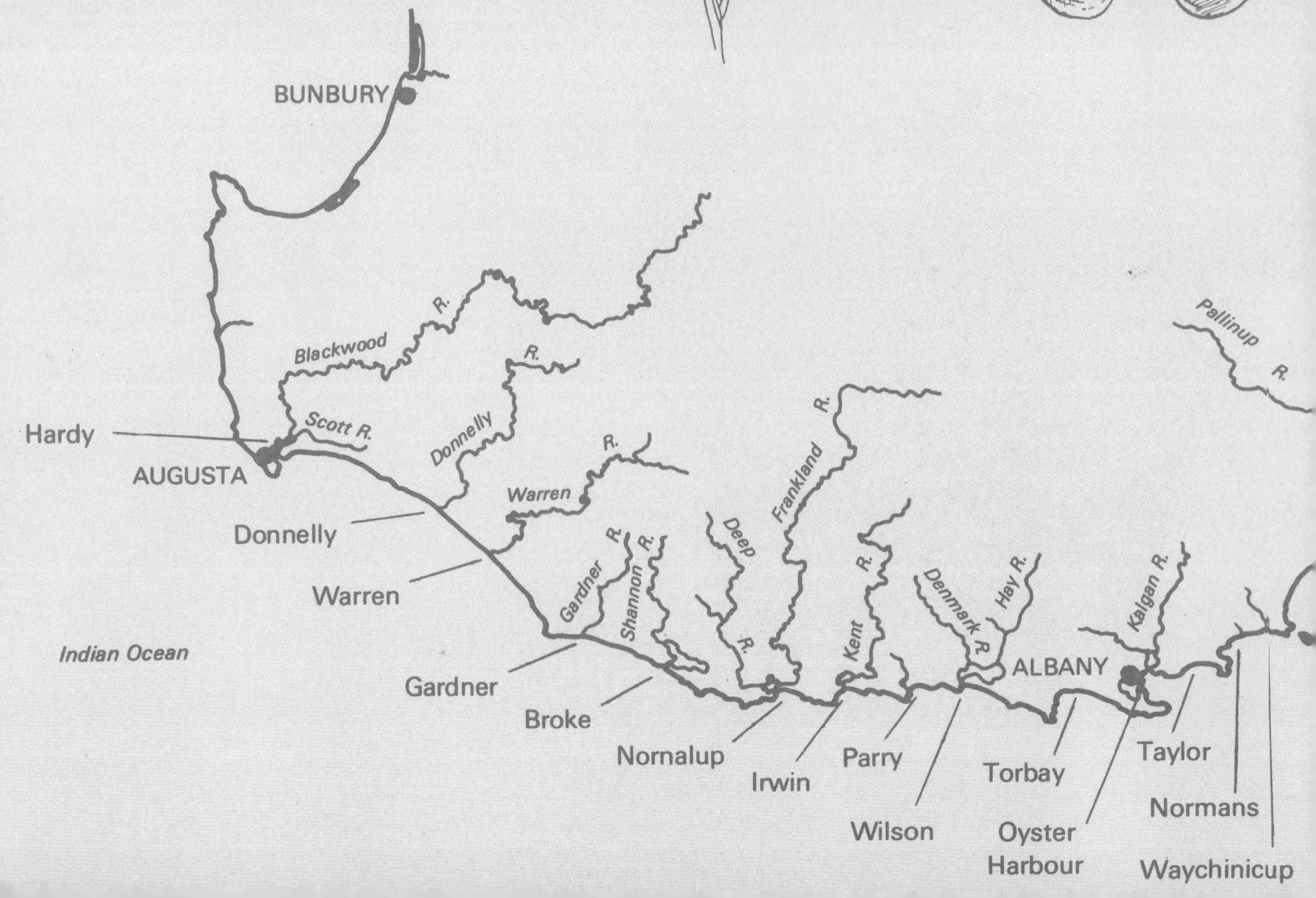